PERIODIC ORBITS, STABILITY AND RESONANCES

PERIODIC ORBITS, STABILITY AND RESONANCES

PROCEEDINGS OF A SYMPOSIUM CONDUCTED BY THE UNIVERSITY OF SÃO PAULO, THE TECHNICAL INSTITUTE OF AERONAUTICS OF SÃO JOSÉ DOS CAMPOS, AND THE NATIONAL OBSERVATORY OF RIO DE JANEIRO, AT THE UNIVERSITY OF SÃO PAULO, SÃO PAULO, BRASIL, 4–12 SEPTEMBER, 1969

Edited by

G. E. O. GIACAGLIA

Escola Politécnica, University of São Paulo and
Instituto de Matemática, University of Campinas,
São Paulo, Brasil

D. REIDEL PUBLISHING COMPANY

DORDRECHT-HOLLAND

Library of Congress Catalog Card Number 74–124848

SBN Number 90 277 0170 9

ISBN-13: 978-94-010-3325-1 e-ISBN-13: 978-94-010-3323-7
DOI: 10.1007/ 978-94-010-3323-7

Softcover reprint of the hardcover 1st edition 1970

"There once was a comet quite old
Whose core grew gradually cold
It began to librate
With amplitude great
And now has a shell of green mould."

W. T. KYNER

PREFACE

The subjects of resonance and stability are closely related to the problem of evolution of the solar system. It is a physically involving problem and the methods available to mathematics today seem unsatisfactory to produce pure non linear ways of attack. The linearization process in both subjects is clearly of doubtful significance, so that, even if very restrictive, numerical solutions are still the best and more valuable sources of informations. It is quite possible that we know now very little more of the entire problem that was known to Poincaré, with the advantage that we can now compute much faster and with much more precision.

We feel that the papers collected in this Symposium have contributed a step forward to the comprehension of Resonance, Periodic Orbits and Stability. In a field like this, it would be a surprise if one had gone a long way toward that comprehension, during the short time of two weeks. But we are sure that the joint efforts of all the scientists involved has produced and will produce a measurable acceleration in the process.

If this is true it will be a great satisfaction to us that this has happened in Brasil. The Southern Hemisphere in America has now begun to participate actively in the Astronomical Society and for this, we are grateful to everyone who has helped.

G. E. O. GIACAGLIA

December, 1969

FOREWORD

This volume presents the invited and contributed papers of the International Symposium on 'Periodic Orbits, Stability and Resonances' held in São Paulo from the 4th to the 12th of September, 1969. The papers are arranged in the same order as they were presented.

The purpose of the Symposium was to congregate scientists from all over the world to discuss one of the most exciting topics in Celestial Mechanics. It served to provide stimulation for researches in Astronomy in the Southern Hemisphere. In the overall, it resulted also as a catalyzer for development of new methods in Applied Mathematics.

We extend our thanks to the following Institutions who have financially supported the expenses for the participants of the Symposium:

Fundação de Amparo à Pesquisa do Estado de São Paulo
Conselho Nacional de Pesquisas
Prefeitura do Município de São Paulo
Horsa – Hotéis Reunidos S.A.
Varig – Viação Aérea Riograndense
Construções e Comércio Camargo Corrêa S.A.
Cerâmica Sanitária Porcelite S.A.
Ford Motors do Brasil S.A.
Chrysler do Brasil S.A.

We have deeply appreciated the prompt collaboration of the invited lecturers and the scientific support received by the International Astronomical Union. It is our pleasure to include in this volume a letter by Prof. C. de Jager, Assistant Secretary of the IAU, which was read in the opening session of the Symposium.

The Executive Committee of the Symposium was composed by

Prof. Abrahão de Moraes, President
Prof. Luiz Muniz Barreto
Prof. Sylvio Ferraz Mello
Prof. Victor Szebehely, Coordinator in U.S.A.
Prof. G. E. O. Giacaglia, General Coordinator
Dr. Iussef Hana Abduch, General Secretary

It is our duty to acknowledge the assistance of Mrs. M. Olympia A. F. França who arranged the social activities, Mrs. Mary Yohanna who arranged the reception scheme and Mr. G. Schmidt, Assistant Secretary of the Symposium.

LETTER OF PROFESSOR DE JAGER
TO THE ORGANIZING COMMITTEE

At the occasion of the Symposium on 'Periodic Orbits, Stability and Resonances', to be held in São Paulo, from 4 till 12 September 1969, the International Astronomical Union wishes to congratulate you and the other Members of the Executive Committee of this Symposium for the interesting meeting that you have organized. We send you our best wishes for a successful and highly stimulating scientific meeting.

The International Astronomical Union regrets that too few large astronomical meetings take place in Latin America and very much welcomes the present Symposium. The Union expresses the hope that this may stimulate young scientists in Brasil and other Latin-American countries in their study of astronomy, and hopes that the Symposium may contribute to the further development of astronomy in your continent.

For the Executive Committee of the
International Astronomical Union,

C. DE JAGER,
Assistant General Secretary

TABLE OF CONTENTS

THE TROJAN MANIFOLD - SURVEY AND CONJECTURES

ANDRÉ DEPRIT and JACQUES HENRARD

Boeing Scientific Research Laboratories, Seattle, Wash., U.S.A.

Abstract. Recent results concerning the families of periodic orbits emanating from the triangular equilibrium L_4 are interpreted in an attempt to establish the evolution of these manifolds as the mass ratio varies from Routh's critical value down to its value in the system sun-jupiter.

We show in this report how recent results concerning the equilateral equilibrium L_4 in the planar restricted problem of three bodies point to some conjectures about the genealogy of families of periodic orbits, and especially about their evolution through a resonance.

The new results discussed here are related, in the local, to Liapunov's theorem about the emergence of a family of periodic orbits from an equilibrium, and, in the global, to conjectures about the natural termination of the family of Trojan orbits. These results clarify in part the role of resonances.

The state of the problem prior to the year 1966 has been covered by Professor Szebehely in his *Theory of Periodic Orbits*. The present report restricts itself to contributions made in the last three years.

1. Introduction

μ is the mass ratio so dimensioned that the interval $0<\mu\leqslant\frac{1}{2}$ covers the range of values considered in the restricted problem from the degenerate case $\mu=0$ which is the planar problem of two bodies to the symmetric case $\mu=\frac{1}{2}$ which is the Copenhagen case. C is the Jacobian constant defined so that, whatever the mass ratio μ may be, it is equal to 3 at the equilibrium L_4.

For mass ratios in the interval

$$I: 0 < \mu < \mu_1 = 0.038\,520\,897\ldots$$

the flow of trajectories in the vicinity of L_4 is described by a conservative irreversible Hamiltonian function with two degrees of freedom

$$\mathscr{H} \equiv \mathscr{H}(l, s, L, S; \mu) = \mathscr{H}_2 + \sum_{n\geqslant 3} \mathscr{H}_n \tag{1}$$

where

$$\mathscr{H}_2 \equiv \mathscr{H}_2(-, -, L, S; \mu) = \sigma(\mu)\cdot S - \lambda(\mu)\cdot L$$

and, for any $n\geqslant 3$, $\mathscr{H}_n$ is a homogeneous polynomial of degree n in $L^{1/2}$ and $S^{1/2}$, its coefficients being finite trigonometric sums in the arguments l and s with the d'Alembert characteristic. The frequencies being such that

$$0 < \lambda < 1/\sqrt{2} < \sigma < 1\,,$$

there exists a sequence of critical mass ratios

$$\mu_k = \tfrac{1}{2}\left[1 - \sqrt{1 - \frac{16k^2}{27(k^2+1)^2}}\right] \quad (k \geqslant 1),$$

such that

$$\sigma(\mu_k) = k\lambda(\mu_k).$$

Let K be the subset of I cleared of the critical mass ratios.

G. E. O. Giacaglia (ed.), Periodic Orbits, Stability and Resonances, 1–18.

Liapunov's theorem establishes that

(i) for any μ in I, there emanates from L_4 a family $\mathscr{L}_4^s$ of *short period* orbits parametrized analytically by the action S; when S goes to zero, the period T_s tends to $2\pi/\sigma$.

(ii) for any μ in K, there emanates from L_4 a family $\mathscr{L}_4^l$ of *long period* orbits parametrized locally by the action L; when L goes to zero, the period T_l tends to $2\pi/\lambda$.

As a *local* proposition, Liapunov's theorem is not exhaustive: beside $\mathscr{L}_4^s$ and $\mathscr{L}_4^l$, there may emanate from the equilibrium other families of periodic orbits. The classical proofs of the theorem fail in case the mass ratio is a critical one; this does not mean that, in the critical cases, no family emanates from L_4.

Being purely local, Liapunov's theorem contains no global information as to the evolution of the families $\mathscr{L}_4^s$ and $\mathscr{L}_4^l$ with (i) the Jacobi constant for a fixed mass ratio, or (ii) the mass ratio running through the interval I.

2. The Non-Critical Case

One non-critical mass ratio has been analyzed in great detail, namely that of the system sun-jupiter, in which case

$$\mu_{13} < \mu < \mu_{12}.$$

The value $\mu = 0.000953875$ having been inserted in the Hamiltonian (1), a canonical transformation $(s, l, L, S,) \to (s^*, l^*, L^*, S^*)$ has been constructed to transform (1) into the function

$$\mathscr{H}(s^*, l^*, L^*, S^*) = \sigma S^* - \lambda L^* + \sum_{2 \leqslant n \leqslant 7} \mathscr{H}_{2n} + \mathscr{R},$$

where, for $2 \leqslant n \leqslant 7$, $\mathscr{H}_{2n}$ is a homogeneous polynomial of degree n in L^* and S^* with real coefficients, and $\mathscr{R}$ is a series in $L^{*1/2}$ and $S^{*1/2}$ beginning with terms of degree 15, its coefficients being finite trigonometric sums in l^* and s^* with the d'Alembert characteristic. The residual function $\mathscr{R}$ being dropped, the normalized system

$$\mathscr{N} = \mathscr{N}(-, -, L^*, S^*) = \sigma S^* - \lambda L^* + \sum_{2 \leqslant n \leqslant 7} \mathscr{H}_{2n}$$

has trivial solutions:

$$L^* = \text{constant}, \quad s^* = \sigma^*(L^*, S^*)\, t + s_0^*,$$
$$S^* = \text{constant}, \quad l^* = \lambda^*(L^*, S^*)\, t + l_0^*.$$

The empirical conviction being that the normalized Hamiltonian $\mathscr{N}$ would reflect some properties of the $\mathscr{H}$-phase space around L_4, the trivial solutions of $\mathscr{N}$ have been analyzed in great detail. Then the remainder $\mathscr{R}$ has been reintroduced into the problem to follow its dissolving effect on the separable phase space generated by $\mathscr{N}$. The overall effect has been to break down the ordinary families of periodic orbits in $\mathscr{N}$ into pairs of natural families, one branch consisting of orbits with stable characteristic exponents, the other of orbits with unstable characteristic exponents. Such families have been traced to their natural terminations. Here are some of the findings.

(a) The orbits of short period emanating from L_4 have characteristic exponents of the stable type all the way (Deprit and Palmore, 1966); they keep the shape of simple loops. As the Jacobi constant C decreases, they cross the line of syzygies (Goodrich, 1966), and the crossing angles tend to $\pi/2$. Eventually the family terminates on a symmetric orbit $B_{4,5}$. This bifurcation orbit belongs to the family $\mathscr{L}_3$ of periodic orbits emanating from the collinear equilibrium L_3. By virtue of the symmetry inherent to the restricted problem, the termination orbit $B_{4,5}$ also belongs to the family $\mathscr{L}_5^s$ of short period orbits emanating from L_5.

This evolution invalidates that part of Brown's conjecture suggesting that $\mathscr{L}_4^s$ would eventually contain a periodic orbit of ejection from the sun, which Brown looked at as the termination of the family. Instead, we submit the following conjecture:

For any mass ratio μ in the interval $0<\mu\leqslant\mu_1$, the families $\mathscr{L}_4^s(\mu)$ and $\mathscr{L}_5^s(\mu)$ of asymmetric periodic orbits emanating from the triangular equilibrium points L_4 and L_5 terminate on a symmetric periodic orbit $B_{4,5}(\mu)$ that belongs to the family $\mathscr{L}_3(\mu)$ of periodic orbits emanating from the collinear equilibrium point L_3.

This statement of a global nature has been checked numerically, not only in the system sun-jupiter, but also in the system earth-moon, and for the critical mass ratios μ_1, μ_3, and μ_4.

(b) The family $\mathscr{L}_3$ itself terminates by duplication on an orbit $R_{2/1}$ in the family $\mathscr{R}$ of periodic retrograde orbits of the first kind for inferior planets. Actually $R_{2/1}$ is one end of the interval of instability along $\mathscr{R}$ that had been anticipated by Wintner (1936). The other end being an orbit $R'_{2/1}$ out of which bifurcates, also by duplication, the family $\mathscr{L}_1$ of periodic orbits emanating from L_1.

From this particular instance can we extrapolate a general statement?

For any mass ratio μ in the interval $0<\mu\leqslant\frac{1}{2}$, the families $\mathscr{L}_1(\mu)$ and $\mathscr{L}_3(\mu)$ of periodic orbits emanating respectively from the collinear equilibrium points L_1 and L_3 terminate by duplication on the orbits of indifferent stability at both ends of the interval of instability conjectured by Wintner along the family $\mathscr{R}$ of retrograde periodic orbits of the first kind for inferior planets.

The proposition has been checked numerically in the system sun-jupiter, also in the system earth-moon (Broucke, 1968); it can be shown that this is also the case in the Copenhagen problem. Indeed, in the notations of Hénon (1965), the family $\mathscr{L}_1$ – which is Stromgren's class (*c*) – consists of orbits symmetric with respect to both coordinate axes Ox and Oy. But, at the orbit '*c*6', there occurs a triple bifurcation analogous to the ramification of $\mathscr{L}_3$, $\mathscr{L}_4^s$, and $\mathscr{L}_5^s$ we mentioned earlier. The family $\mathscr{L}_1$ continues, along what Stromgren has called group '*n*', by orbits which are no longer symmetric with respect to the axis Oy; Hénon checked that, indeed, Stromgren's group (*n*) terminates by duplication on orbit *f*1 in Stromgren's class (*f*), i.e., in the family $\mathscr{R}$ of retrograde orbits of the first kind for inferior planets. The symmetric of group (*n*) which Hénon indicates by n', terminates similarly on orbit *h*1 in the symmetric (*h*) of the family $\mathscr{R}$, which is the family of retrograde orbits of the first kind for inferior planets around the other primary. The third branch, to be denoted (n'') still consisting of orbits symmetric with respect to both axes, is identical to what comes after orbit *c*6 in Stromgren's group (*c*). Figure 1 may help in clarifying this interpretation of Stromgren's results as they have been improved by Hénon.

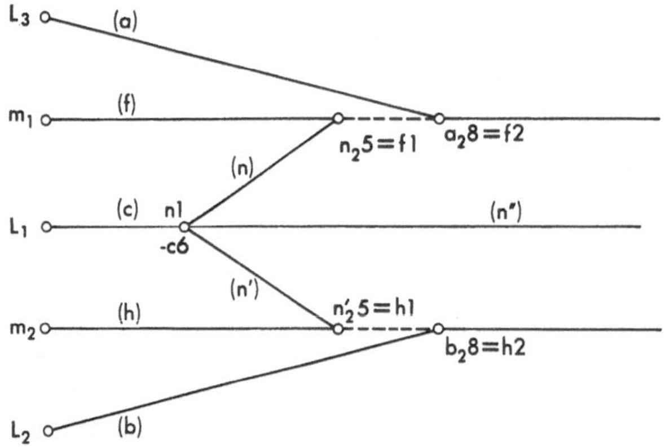

Fig. 1. Genealogy of Stromgren's groups (*a*), (*b*), (*c*), (*f*), (*g*), and (*n*).

As the symmetry with respect to the ordinate axis Oy fades away as soon as μ departs from the exceptional value $\frac{1}{2}$, the genealogy of the families (*a*), (*b*), (*c*), (*f*), (*g*), and (*n*) is not a *generic* property of the restricted problem. Bartlett (1965) provides profiles that suggest how this non-generic pattern evolves away from $\mu=\frac{1}{2}$ into the classification stated in the conjecture. Here is how his results could be interpreted. The orbit *c*6 ceasing to be a bifurcation, the families (*c*) and (*n*) merge to become the usual family $\mathscr{L}_3$ or, in Bartlett's terminology, the Upper (*n*) class; the branches (n') and (n'') merge into one family while breaking away from the orbit (*c*6).

(c) For the sake of completeness, we should add that, in the system Sun-Jupiter, the family $\mathscr{D}$ of direct periodic orbits of the first kind for inferior planets terminates by triplication on an orbit $R_{3/1}$ (index of stability $\mathrm{Tr}=2\cos 2\pi/3$) in the family $\mathscr{R}$. Thus the remarkable assemblage: the classical families $\mathscr{R}$, $\mathscr{D}$, $\mathscr{L}_1$, $\mathscr{L}_3$, $\mathscr{L}_4^s$, $\mathscr{L}_5^s$, and, as we shall see in a moment, $\mathscr{L}_4^l$ and $\mathscr{L}_5^l$ constitute but branches of a single network of periodic orbits in the restricted problem.

The termination of $\mathscr{D}$ onto an orbit of $\mathscr{R}$ has also been established by Broucke in the system earth-moon. The following conjecture is pleasing to entertain

?

For any mass ratio μ in the interval $0<\mu\leqslant\frac{1}{2}$, the family $\mathscr{D}$ of direct periodic orbits of the first kind for inferior planets terminates by triplication on an orbit in the family $\mathscr{R}$ of periodic orbits of the first kind for inferior planets.

(d) In the system Sun-Jupiter, the family $\mathscr{L}^l_4$ begins with simple ovals having exponents of the stable type. After a maximum of C is reached, the exponents become unstable and the orbits develop loops by contact with the curves of zero velocity. The Jacobian constant having fallen below the value $C=3$, the orbits cannot acquire new loops, but the 12 loops already acquired inflate until they come to coincidence. At the terminal stage, the long period orbit has become a simple oval traveled 13 times, and it is identical to the orbit B^s_{13} of $\mathscr{L}^s_4$ whose stability index is $2\cos 2\pi/13$.

(e) Out of B^s_{13} bifurcates a second family of orbits with long periods, consisting of 13 loops almost identical, the characteristic exponents being of the stable type. After C has become greater than 3, the loops disappear one after the other. Around a maximum of C, all loops have gone, and the orbits are now simple ovals – several Trojan orbits computed by Rabe fall into this family. Beyond the maximum, the orbits reacquire loops, 13 in total, while the characteristic exponents have become unstable. The family terminates like $\mathscr{L}^l_4$: the loops increase in size and tend to coincide. The end is an element B^s_{14} of $\mathscr{L}^s_4$ traveled 14 times. For this reason, the family is denoted $\mathscr{B}(13S, 14S)$, and it is called a bridge from 13 times the short period to 14 times the short period.

The branchings found at B^s_{13} are repeated at B^s_{14}: another family $\mathscr{B}(14S, 15S)$ has been followed throughout, and then a bridge $\mathscr{B}(15S, 16S)$ as well as a bridge $\mathscr{B}(18S, 19S)$.

On the face of these results we venture the following conjecture.

Let p be any integer >12. Indicate by B^s_p the orbit in $\mathscr{L}^s_4$ whose index of stability is $2\cos 2\pi/p$. We know that, as $p\to\infty$, the sequence (B^s_p) tends to the termination orbit $B_{4,5}$ at the junction of $\mathscr{L}^s_4$, $\mathscr{L}^s_5$ and $\mathscr{L}_3$.

?

In the system Sun-Jupiter ($\mu_{13}<\mu<\mu_{12}$), the chain of bridges and bifurcations $\cdots\to B^s_p \xrightarrow{\mathscr{B}(pS,(p+1)S)} B^s_{p+1} \to\cdots$ goes on indefinitely as p tends to ∞. Moreover, for any $p>12$, the Jacobi constant reaches a maximum value C_p along the bridge $\mathscr{B}(pS,(p+1)S)$. The sequence C_p is strictly increasing, and, as $p\to\infty$, C_p tends to the value of the Jacobi constant at L_3.

Should this conjecture turn out to be true, it would singularly clarify the structure of the phase space around the collinear equilibrium L_3.

3. The Critical Resonance $\sigma=\lambda$

For the mass ratio μ_1 at which $\sigma=\lambda$, Poincaré's method of continuation establishes that two families of periodic orbit emanate from L_4 (Buchanan). Contrary to the families established by Liapunov's theorem in the non-critical cases, these families are in the analytical continuation of one another. One of them which begins below $C=3$ has periods strictly decreasing from $2\pi/\sigma$; thus we call it $\mathscr{L}_4^s$; the other, which begins above $C=3$ has periods strictly increasing from $2\pi/\sigma$; thus we call it $\mathscr{L}_4^l$ (Deprit and Henrard, 1968).

As we said earlier, $\mathscr{L}_4^s$ terminates by meeting $\mathscr{L}_3$ and $\mathscr{L}_5^s$ on a symmetric periodic orbit. Also $\mathscr{L}_4^l$ terminates on an orbit B_2^s of $\mathscr{L}_4^s$ circuited twice.

There are, close to L_4, two long period orbits B_2^l and $\bar{B}_2^l$ of indifferent stability which make the endpoints of an interval of instability along $\mathscr{L}_4^l$. Between L_4 and B_2^l, for any $p>2$, designate by B_p^l the long period orbit whose index of stability is $2\cos 2\pi/p$. As $p\to\infty$, the sequence (B_p^l) tends to collapse into L_4. Similarly, for any $p>2$, designate by B_p^s the short period orbit between B_2^s and L_4 whose index of stability is $2\cos 2\pi/p$; as $p\to\infty$, the sequence (B_p^s) tends to collapse into L_4.

We found numerically that out of B_3^l, branch off by triplication a family of periodic orbits with stable characteristic exponents, and a second one, with unstable characteristic exponents. Both families terminate by quadruplication on the short period orbit B_4^s, thus constructing a two-lane bridge $\mathscr{B}(3L, 4S)$.

Similarly, by numerical analysis, there has been established a two-lane bridge $\mathscr{B}(4L, 5S)$ branching off of B_3^l and terminating at B_4^s.

We also found a two-prong bridge originating from B_3^s by triplication. Both prongs come out of B_3^s at first with unstable characteristic exponents. One of them terminates at B_2^l with unstable characteristic exponents; the other at $\bar{B}_2^l$ with stable characteristic exponents.

This remarkable nesting of two-lane bridges connecting the families $\mathscr{L}_4^l$ and $\mathscr{L}_4^s$ has been justified mathematically at least in the local. To this effect, Meyer and Palmore (1969) request that the Hamiltonian (1) of the restricted problem around L_4 be normalized through the fourth order. Thus a completely canonical transformation $(l, s, L, S)\to(l^*, s^*, L^*, S^*)$ is constructed to change (1) into a function of the type

$$\mathscr{H} \equiv \mathscr{H}(l^*, s^*, L^*, S^*) = \mathscr{H}_2 + \mathscr{H}_4 + \sum_{n\geqslant 5} \mathscr{H}_n,$$

where

$$\mathscr{H}_2 \equiv \mathscr{H}_2(-, -, L^*, S^*) = \sigma(\mu)\, S^* - \lambda(\mu)\, L^*,$$
$$\mathscr{H}_4 \equiv \mathscr{H}_4(-, -, L^*, S^*) = \tfrac{1}{2}\left[A(\mu)\, S^{*2} + 2B(\mu)\, S^*L^* + C(\mu)\, L^{*2}\right]$$

and, for any $n\geqslant 5$, $\mathscr{H}_n$ is a homogeneous polynomial of degree n in $L^{*1/2}$ and $S^{*1/2}$, its coefficients being finite trigonometric sums in the angles s^* and l^* with the d'Alembert characteristic. This Birkhoff's normalization through the fourth order is valid at all mass ratios in the open interval $0<\mu<\mu_1$ except at μ_2 and μ_3. The coefficients are as follows (Deprit and Deprit-Bartholomé, 1967):

$$C = \frac{\sigma^2(81 - 696\lambda^2 + 124\lambda^4)}{72(1 - 2\lambda^2)^2 (1 - 5\lambda^2)},$$

$$B = -\frac{\lambda\sigma(43 + 64\sigma^2\lambda^2)}{6(4\sigma^2\lambda^2 - 1)(25\sigma^2\lambda^2 - 4)},$$

$$A = \frac{\lambda^2(81 - 696\sigma^2 + 124\sigma^4)}{72(1 - 2\sigma^2)^2 (1 - 5\sigma^2)}.$$

In the set $I-\{\mu_1, \mu_2\}$, the function $\mu \rightarrow B(\mu)\,\lambda(\mu)+C(\mu)\,\sigma(\mu)$ has a unique zero, namely $\nu=0.0127\ldots$; in the same set, the function $\mu \rightarrow A(\mu)\,\sigma^2(\mu)+2B(\mu)\,\sigma(\mu)\,\lambda(\mu)+C(\mu)\,\lambda^2(\mu)$ has a unique zero, namely $\nu'=0.0109\ldots$. Notice that

$$\mu_4 < \nu' < \nu < \mu_3 < \mu_2 < \mu_1.$$

Now let μ_r designate the mass ratio μ for which

$$\sigma(\mu) = r\cdot\lambda(\mu).$$

Meyer's 'theorem of the bridges' deals with any mass ratio μ_r in the interval $\nu<\mu<\mu_1$ for which $r=p/q$ with p and q relatively prime integers restricted by the inequalities $p>q>3$. Roughly speaking the theorem states that:

(a) for any such $\mu_r>\mu_2$, there exists at any $\mu>\mu_r$ and sufficiently close to μ_r a two-lane bridge $\mathscr{B}(qL, pS)$ emanating from a long period orbit circuited q times and terminating on a short period orbit circuited p times. As μ goes decreasing toward μ_r, the bridge tends to collapse onto the equilibrium L_4;

(b) for any such μ_r in the interval $\nu<\mu_r<\mu_2$, there exists, at any $\mu<\mu_r$ and sufficiently close to μ_r, a two-lane bridge $\mathscr{B}(qL, pS)$ connecting the Liapunov families $\mathscr{L}^l_4$ and $\mathscr{L}^s_4$. As μ goes increasing toward μ_r, the bridge tends to collapse onto the equilibrium L_4.

The existential conclusions by Meyer and Palmore have been deepened – but only formally – by Henrard (1969a) even in the case when $q=2$.

Let's see how Part (a) of Meyer's theorem coupled with our partial findings at μ_1 could be expanded in order to give a hint as to the evolution of the phase space around L_4 when μ varies in the interval $I_1: \mu_2 \leqslant \mu \leqslant \mu_1$.

On the branch of $\mathscr{L}^s_4(\mu_1)$ extending from L_4 to B^s_2, we designate by $B^s_{p/q}$ the orbit whose stability index is $\mathrm{Tr}=2\cos 2\pi q/p$; similarly, on the branch of $\mathscr{L}^l_4(\mu_1)$ extending from L_4 to where the family acquires unstable characteristic exponents, we designate by $B^l_{q/p}$ the orbit whose stability index $\mathrm{Tr}=2\cos 2\pi p/q$. (If p/q is $<\frac{3}{2}$, the orbit is located between L_4 and B^l_2, but if p/q is $>\frac{3}{2}$, it is located beyond $\bar{B}^l_2$.)

At the critical mass ratio μ_1 itself, we conjecture the following situation:

?

> *For any pair (p, q) of relatively prime integers such that $1<q<p<2q$,*
> (a) *when $q \neq 2$, there exists a two-lane bridge $\mathscr{B}(qL, pS)$ connecting the short period orbit $B^s_{p/q}$ traveled p times to the long period orbit $B^l_{q/p}$ traveled q times. One family has characteristic exponents of the stable type, the other of the unstable type.*
> (b) *when $q=2$ (hence $p=3$), as Henrard (1969) discussed it, the connection should be made of the two-prong bridge $\mathscr{B}(2L, 3S)$ as indeed it was found numerically.*

We summarize in Figure 2 what the conjecture implies in the case when $q=5$. Then p could take the values 6, 7, 8, 9. Notice that

$$1 < \tfrac{6}{5} < \tfrac{7}{5} < \tfrac{3}{2} < \tfrac{8}{5} < \tfrac{9}{5} < 2.$$

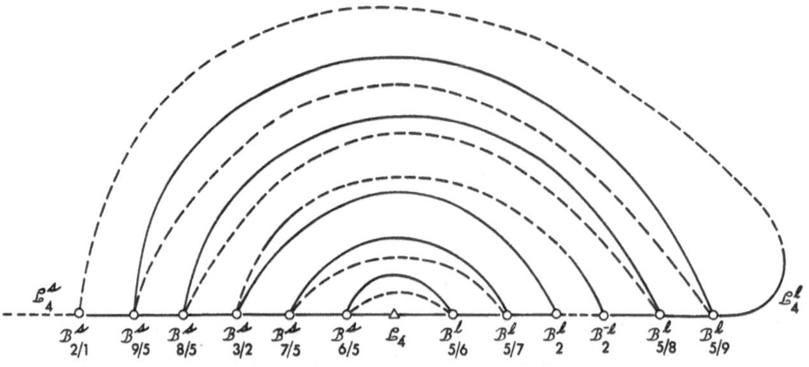

Fig. 2. Bridges from $\mathscr{L}_4{}^l$ to $\mathscr{L}_4{}^s$ for the critical mass μ_1.

Thus the two-lane bridges relative to $\frac{8}{5}$ and $\frac{9}{5}$ would envelop the two-prong bridge $\mathscr{B}(2L, 3S)$, whereas the two-lane bridges relative to $\frac{7}{5}$ and $\frac{6}{5}$ would be enveloped by it. For $p/q=\frac{6}{5}$, the bridge begins at $B^l_{5/6}$ with the stability index $\mathrm{Tr}=2\cos 12\pi/5$ $2\cos 2\pi/5$, and ends at $B^s_{6/5}$ with the stability index $\mathrm{Tr}=2\cos 10\pi/6=2\cos 2\pi/6$. For $p/q=\frac{9}{5}$, it begins at $B_{9/5}$ with the stability index $\mathrm{Tr}=2\cos 18\pi/5=2\cos 8\pi/5=$ $2\cos 2\pi/5$, and ends at the orbit $B^s_{9/5}$ with the index $\mathrm{Tr}=2\cos 10\pi/9=2\cos 8\pi/9$.

Consider now this maze of bridges connecting at μ_1 the families $\mathscr{L}^s_4$ and $\mathscr{L}^l_4$, and let us make a guess as to what happens to it when μ decreases through the interval I_1.

At any mass ratio μ in the interval $\mu_2<\mu<\mu_1$, the family $\mathscr{L}^l_4$ maintains its termination at the short period orbit $B^s_{2/1}$, but as μ approaches μ_2 from above, this bifurcating orbit moves closer to the equilibrium L_4 from below, and shrinks in size. As a matter of fact, at μ_2, it collapses into L_4, together with the family $\mathscr{L}^l_4$.

As μ decreases through the interval I_1, the network of bridges from $\mathscr{L}^l_4$ to $\mathscr{L}^s_4$ vanishes by gradual depletion. More precisely, as μ approaches a mass ratio $\mu_{p/q}$, the long period orbit $B^l_{q/p}$ moves closer to the equilibrium L_4 from above, whereas the short period $B^s_{p/q}$ comes closer to L_4 from below. At $\mu_{p/q}$ exactly, both orbits collapse onto L_4, which results in the vanishing of the bridge $\mathscr{B}(B^l_{q/p}, B^s_{p/q})$.

For instance, in reference to Table I, we would expect that, as μ decreases from μ_1 to μ_2, the bridge relative to $\frac{6}{5}$ has disappeared after $\mu=0.037$, the one relative to $\frac{7}{5}$ after $\mu=0.034$, the one for $p/q=\frac{8}{5}$ below $\mu=0.03$, and finally, the bridge for $p/q=\frac{9}{5}$ vanishes $\mu=0.027$.

TABLE I

Some singular mass ratios between μ_1 and μ_2

p	q	$\mu_{p/q}$		
1	1	0.038	520	897
6	5	0.037	217	828
7	5	0.034	319	046
8	5	0.030	878	630
9	5	0.027	453	593
2	1	0.024	293	897

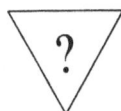

Should these conjectures prove true, one can anticipate how intricate the phase space can be around L_4 for the mass ratio between μ_1 and μ_2. For, in the invariant sections of the energy manifolds close to L_4, each lane of periodic orbits that penetrates the section leaves on it a finite set of fixed points; of the elliptic type if the intersecting orbit has characteristic exponents of the stable type, of the hyperbolic type if the characteristic exponents are of the unstable type. The closer we approach the equilibrium, the greater the number of islands and of zones of instability generated respectively by the elliptic and hyperbolic fixed points, yet the smaller the measure of the phase subsets 'influenced' by these singularities. As a matter of fact, the condensation may become so great that it turns out to be numerically impossible to separate the actual phase space from its normalizations. Hence the paradox of the 'third integral by proxy': increasing clustering of fixed points at the equilibrium rules out the mathematical existence of another integral beside the energy, yet the best numerical integrations are unable to unravel this complexity, to which their built-in insensitiveness below a certain level of accuracy substitutes quasi-imperceptibly a separable model, thus willy-nilly pretending that a third integral is attending the problem 'by proxy'.

4. The Critical Resonance $\sigma = 2\lambda$

At the mass ratio μ_2, according to Liapunov's theorem, there emanates from L_4 a family of short period orbits; it begins from L_4 with orbits having characteristic exponents of the unstable type (Henrard, 1969a). This unstable initial portion ends with an orbit B_2^s of indifferent stability after which $\mathscr{L}_4^s$ goes through an interval of stability. Out of B_2^s bifurcates by duplication a bridge $\mathscr{B}(2S, 3S)$, initially stable; after passing through a maximum $C=3.01597$ ..., it decreases in Jacobian constants and terminates by triplication on a short period orbit B_3^s (index of stability $\mathrm{Tr}=2 \cos 2\pi/3$) (Henrard, 1969b). These are the only facts established numerically at exactly the critical mass ratio of the second order μ_2.

What about a family of long period orbits possibly emanating from L_4? A chain of formal treatments and numerical confirmations points to a reasonable answer.

(a) For mass ratios on both sides of μ_2, not equal to μ_2 but close to it, there exists on $\mathscr{L}_4^l$ an orbit $\bar{B}_2^s$ with indifferent stability. Pedersen (1939) makes a formal expansion for an element of periodic orbits branching from $\bar{B}_2^s$ by duplication. As μ tends μ_2, the orbit $\bar{B}_2^s$ shrinks in size and goes to collapse onto the equilibrium L_4; simultaneously the orbits in the incipient branch bifurcating off of $\bar{B}_2^s$ collapse onto L_4.

(b) As a matter of fact, for the mass ratios $\mu=0.023\,813\,674$ and $\mu=0.024\,826\,698$ on both sides of $\mu_2=0.024\,293\,897$, the families $\mathscr{L}_4^l(\mu)$ have been continued numerically; they were both found to terminate by duplication on a short period orbit (Roels, 1966), which is none other than the orbit $\bar{B}_2^s$ of indifferent stability.

(c) The numerical investigations of Palmore (1967) indicate that, if it exists, the family $\mathscr{L}_4^l$ for μ_2 could not be in the analytical continuation of the families $\mathscr{L}_4^l$ for μ on both sides of μ_2; nonetheless they do not exclude the existence at μ_2 of a family $\mathscr{L}_4^l$ emanating from L_4, but totally isolated.

(d) In this respect, the latest analysis by Henrard (1969a) and Alfriend (1969) takes the definitive step. Whereas Alfriend examines μ_2 and its neighborhood by the method of the two variables, Henrard proceeds by normalization of the Hamiltonian in a manner somewhat analogous to the way Breakwell and Pringle (1966) have treated the resonance $\sigma=3\lambda$. Both authors, independently, conclude that their formalism does not warrant the existence of a family of long period orbits emanating from L_4. But Henrard is able to assemble the mechanism which could explain how, as $\mu \to \mu_2$, the family $\mathscr{L}_4^l$ vanishes by collapsing at L_4.

(e) Numerical verification by Palmore, started at Yale (Palmore, 1967) and pursued at the Control Sciences Center of the University of Minnesota confirm the formal findings of the previous authors.

As a result we feel confident that the following conjecture should not wait the sanction of a mathematical proof to be taken as true.

> *There exists an open interval K of mass ratios containing μ_2 such that, at each μ in $K-\{\mu_2\}$, the family $\mathscr{L}^s_4$ comes out of L_4 with characteristic exponents of the stable kind until at the orbit $\bar{B}^s_2$ of indifferent stability it enters an interval of instability. For such mass ratios, the family $\mathscr{L}^l_4$ terminates by duplication onto $\bar{B}^s_2$.*
>
> *As $\mu \to \mu_2$, the orbit $\bar{B}^s_2$ moves toward L_4 where it collapses for $\mu = \mu_2$; simultaneously the incipient interval of stability along L^s_4 vanishes, as does also the family $\mathscr{L}^l_4$.*
>
> *At μ_2 itself, there emanates from L_4 NO family of long period orbits, only a family $\mathscr{L}^s_4$ of short period orbits.*

Henrard (1969b) has followed the evolution of the family $\mathscr{L}^l_4(\mu)$ as μ departs from μ_2 and decreases in the interval I_2. This numerical study has brought out an extraordinary event. In the vicinity of μ_2, on both sides of this critical mass ratio, $\mathscr{L}^l_4$ terminates on a short period orbit B^s_2 traveled *twice*; on the other hand, at μ_3 and around the critical mass ratio of order three, it ends by *triplication* on a short period orbit. At what mass ratio does the terminal bifurcation of $\mathscr{L}^l_4$ change from a duplication to a triplication, and in virtue of what mechanism?

For μ less than, but close to, μ_2, the family $\mathscr{L}^s_4$ emanates from L_4 with stable characteristic exponents; then comes along $\mathscr{L}^s_4$ an interval of instability whose end orbits of indifferent stability we denote by $\bar{B}^s_2$ and B^s_2, with the understanding that $\bar{B}^s_2$ appears on $\mathscr{L}^s_4$ before B^s_2. The family $\mathscr{L}^l_4$ terminates by duplication at $\bar{B}^s_2$; out of B^s_2 by duplication there emanates a bridge $\mathscr{B}(2S, 3S)$ which terminates on $\mathscr{L}^s_4$ by triplication from an orbit B^s_3 (index of stability $\mathrm{Tr} = 2\cos 2\pi/3$). Both families undergo a maximum Jacobi constant; let Γ designate the orbit of maximum C on $\mathscr{L}^l_4$.

At a mass ratio μ^* such that

$$0.020\,70 < \mu^* < 0.020\,75\,,$$

the orbit Γ comes to coincidence with an orbit Γ' of the bridge $\mathscr{B}(2S, 3S)$. At which mass ratio, Γ constitutes also along $\mathscr{L}^l_4$ an orbit at which the period passes through a minimum, whereas Γ' turns out to be along $\mathscr{B}(2S, 3S)$ an orbit at which the period undergoes a maximum. Actually this common orbit can be looked at as the bifurcation of four branches of periodic orbits: the branch (I) made of long period orbits and collapsing at L_4, the branch (II) of long period orbits ending at $\bar{B}^s_2$, the branch (III) of the bridge $\mathscr{B}(2S, 3S)$ terminating at B^s_2, and finally the branch (IV) of the bridge $\mathscr{B}(2S, 3S)$ terminating at B^s_3.

As soon as μ decreases away from μ^*, this extraordinary junction breaks apart, and the branches regroup themselves differently: the branches (I) and (IV) merge to form the family $\mathscr{L}^l_4$, thus terminating by triplication on the short period orbit B^s_3, whereas the branches (II) and (III) merge to produce a one-lane bridge from the short period orbit $\bar{B}^s_2$ to the short period orbit $\bar{B}^s_2$ (see Figure 3).

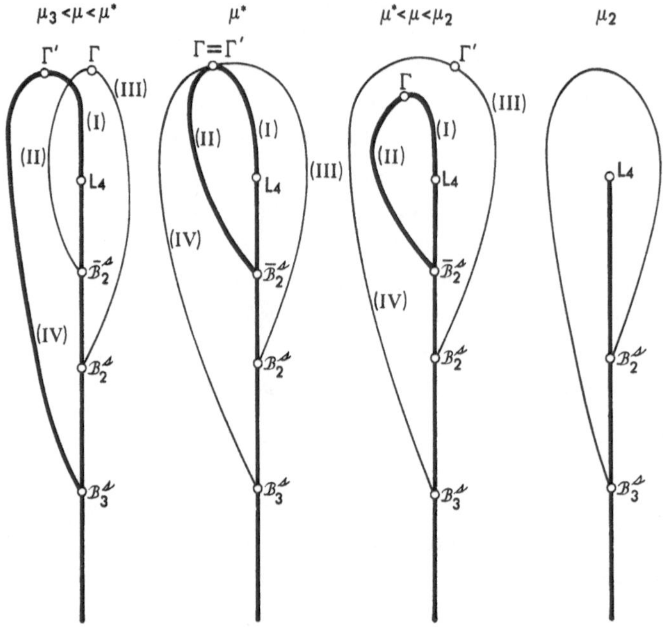

Fig. 3. Schematic representation of the evolution of the family of long period Trojan orbits between μ_2 and μ_3.

5. The Critical Resonance $\sigma = 3\lambda$

At μ_3, like at μ_2, Liapunov's theorem ensures a family $\mathscr{L}_4^s$ of short period orbits emanating from L_4. It begins with stable characteristic exponents, goes through an interval of stability, comes back to stability and terminates on a symmetric orbit $B_{4,5}$ in the family $\mathscr{L}_3$ (Deprit and Price, 1969).

The existence of a long period family out of L_4 has long been in doubt.

(a) At the critical resonance $\frac{1}{3}$, the matrizant of the variational equations around the equilibrium is periodic in the time. As a matter of fact, even the equations for the second order variations have a general solution that is purely periodic, and the resonance generates possible secular terms only from order 3 onward. At this order, Roels (1969a) indicates how formally the bifurcation relation can be treated by a proper adjustment of the orbital parameters to bring out of the general solution a family of periodic orbits still emanating from L_4. Therefrom, a consistent formalism can be set up to expand recursively $\mathscr{L}_4^l$ in power series of a unique orbital parameter. The convergence of these series is an open question.

(b) But the series produce in a narrow neighborhood of L_4 initial conditions which, after isoenergetic corrections somewhat significant, correspond to periodic orbits. Remarkably enough, $\mathscr{L}_4^l$ begins with Jacobian constants smaller than 3 and with unstable characteristic exponents. In fact, the family keeps these characters until its termination by triplication on a short period orbit B_3^s (Deprit and Price, 1969).

(c) Out of B_3^s, also by triplication, branches off another family $\mathscr{B}(3S, 4S)$ that terminates by quadruplication on a short period orbit B_4^s.

(d) A sort of normalization at L_4 for μ_3 elucidates the mechanisms of the resonance; among other formal results, it gives ground to exclude the possibility of other long period families emanating from L_4.

In sum, consistent formalisms and numerical investigations point jointly to the conjecture.

> *At the critical mass ratio μ_3, there emanates from L_4, beside the family $\mathscr{L}_4^s$ of short period orbits, only one other family of periodic orbits. These are of long period; they come out initially with unstable characteristic exponents.*
> *The family terminates by triplication on an orbit of $\mathscr{L}_4^s$.*

The evolution of the families $\mathscr{L}_4^l$ with the mass ratio μ decreasing in the interval $\mu_4 < \mu \leqslant \mu_3$ has been followed numerically (Henrard, 1969b). The story elucidates the status of the family $\mathscr{L}_4^l$ at μ_3 itself; in particular it indicates the nature of the singularity caused by the resonance $\frac{1}{3}$.

While μ decreases toward μ_3, the family $\mathscr{L}_4^l(\mu)$ tends to the family $\mathscr{L}_4^l(\mu_3)$ that we found there. But, on the other side of μ_3, beside B_3^s, there appears on $\mathscr{L}_4^s$ between L_4 and B_3^s another orbit $\bar{B}_3^s$ whose index of stability is $2\cos 2\pi/3$. Actually, as μ decreases away from μ_3, the family $\mathscr{L}_4^l(\mu_3)$ evolves into a single-lane bridge $\mathscr{B}(3S, 3S)$ originating from $\bar{B}_3^s$ and terminating at B_3^s; naturally the bridge $\mathscr{B}(3S, 4S)$ maintains itself. At the same time, there now appears a family $\mathscr{L}_4^l$ terminating by triplication at $\bar{B}_3^s$. Let us contrast this transition with what we found at μ_2. There, from both *sides* of μ_2, as μ moves toward μ_2, the family $\mathscr{L}_4^l$ vanishes by collapsing at L_4. But at μ_3, this occurs only on one side of the critical mass ratio, namely as μ increases toward μ_3.

For $\mu < \mu_3$, there exists on $\mathscr{L}_4^l$ an orbit of maximum Jacobi constant that we designate by Γ. At a mass ratio μ^{**} such that

$$0.012\,266 < \mu^{**} < 0.012\,316\,,$$

the orbits Γ come into coincidence with an orbit of $\mathscr{B}(3S, 4S)$. Thereafter, like in the case of μ^* in the interval $\mu_3 < \mu < \mu_2$, an exchange of branches takes place (see Figure 4): the family $\mathscr{L}_4^l$ terminates on the short period B_4^s, and, at the same time, a two lane bridge $\mathscr{B}(3S, 3S)$ from $\bar{B}_3^s$ to B_3^s has been established.

This analysis reconciles conflicting views on how the family $\mathscr{L}_4^l$ develops in the system earth-moon: Breakwell and Pringle (1966) assume that the mass ratio for the system earth-moon would lie between μ_3 and μ^{**} (hence $\mathscr{L}_4^l$ would terminate by triplication on $\bar{B}_3^s$), whereas others (Deprit *et al.*, 1967) ascribe to this mass ratio the value $0.01215 < \mu^{**}$ (There $\mathscr{L}_4^l$ terminates by quadruplication on B_4^s.)

6. The Critical Resonance $\sigma = 4\lambda$

The situation here is radically different from the ones we encountered at μ_2 and μ_3.

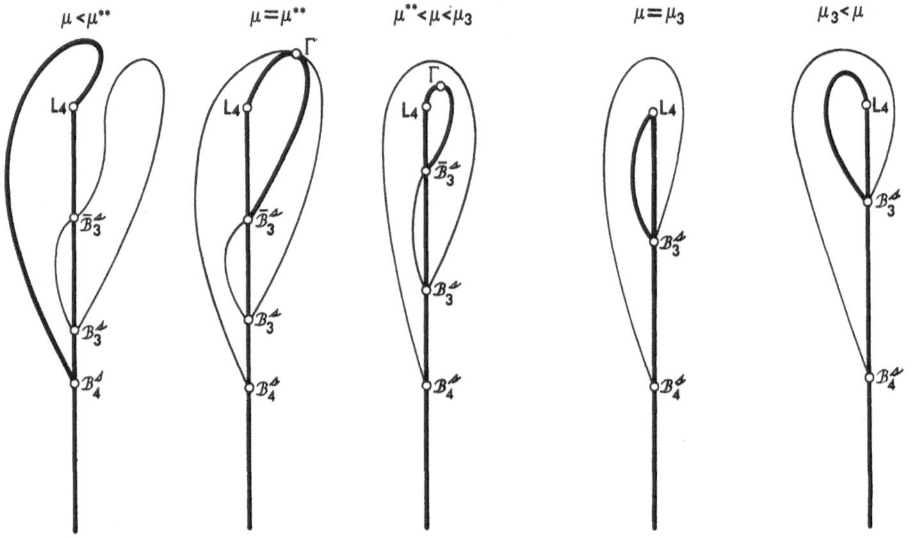

Fig. 4. Schematic representation of the evolution of the family of long period Trojan orbits between μ_3 and μ_4.

(a) By formal series presenting the d'Alembert characteristic, Roels (1969b) indicates a branch of long period orbits emanating from the equilibrium.

(b) The initial conditions produced by these series have been found to correspond to periodic orbits around L_4 with stable characteristic exponents. This family which we shall indicate by $\mathscr{L}^l_4$, goes back to the equilibrium where it terminates with stable characteristic exponents (Deprit and Henrard, 1969). Moreover a third branch has been found to leave the equilibrium: all the orbits lie below $C=3$, have long periods and unstable characteristic exponents. We indicate this branch by $\mathscr{B}(L_4, 4S)$ because it terminates by quadruplication on a short period orbit B^s_4.

(c) At the time this configuration was elucidated by numerical continuation, Roels (1969c) reexamined the bifurcation condition imposed by the resonance, and produced formal series corresponding to two branches of long period orbits emanating from L_4, but different from $\mathscr{L}^l_4$.

These independent results supporting one another lead to the conjecture:

?

At the critical mass ratio μ_4, there emanate from L_4, beside the family $\mathscr{L}^s_4$ of short period orbits, three branches of long period orbits. The two branches that leave the equilibrium with stable characteristic exponents constitute but one family, $\mathscr{L}^l_4$, of periodic orbits arising from and terminating at L_4. The third branch, which leaves the equilibrium with unstable characteristic exponents terminates by quadruplication on a short period orbit.

The evolution of the family $\mathscr{L}^l_4$ by transition through μ_4 has not yet been studied numerically. Nonetheless, reasoning by analogy with the transitions through μ_2 and μ_3, we can make use of an information released by Palmore to propose a working hypothesis.

Between μ^{**} and μ_4, Palmore found a family of orbits with a period close to four times the short period; the orbits lie in the vicinity of L_4; the family is closed upon itself. This being so, we submit that, as μ decreases toward μ_4, Palmore's closed family moves toward the equilibrium. At μ_4 itself, it will attach to the equilibrium where it constitutes the family $\mathscr{L}^l_4(\mu_4)$ that we just mentioned. On the other hand, with μ decreasing toward μ_4, the families $\mathscr{L}^l_4(\mu)$ continue one another, until at μ_4 itself, they became the bridge $\mathscr{B}(L_4, 4S)$ that we discovered there (see Figure 5).

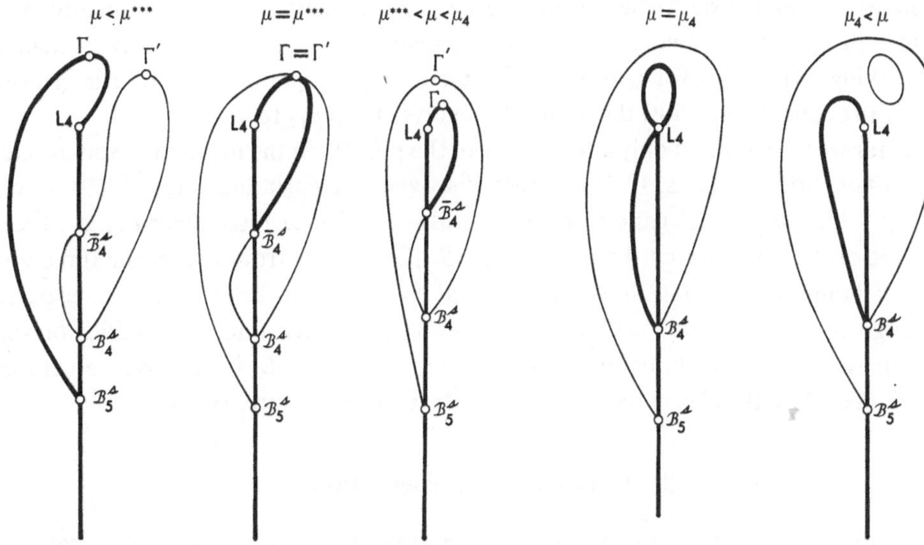

Fig. 5. Suggested evolution of the family of long period Trojan orbits between μ_4 and μ_5.

We feel confident that, out of B^s_4 where $\mathscr{L}^l_4$ terminates, there branches a bridge $\mathscr{B}(4S, 5S)$ ending on a short period orbit B^s_5 traveled five times.

With μ decreasing away from μ_4, there appears on $\mathscr{L}^s_4$ between L_4 and B^s_4 another orbit $\bar{B}^s_4$ whose stability index is $2\cos 2\pi/4$. It will be the end orbit of the long period Trojan family $\mathscr{L}^l_4(\mu)$. Accordingly, immediately to the left of μ_4, we expect to find the following chain

$$L_4 \xrightarrow{\mathscr{L}^l_4} \bar{B}^s_4 \xrightarrow{\mathscr{B}(4S,\,4S)} B^s_4 \xrightarrow{\mathscr{B}(4S,\,5S)} B^s_5$$

Then will come a mass ratio μ^{***} at which the orbits of maximum Jacobian constant on $\mathscr{L}^l_4$ will come to coincide with an orbit of $\mathscr{B}(4S, 5S)$. As μ departs from μ^{***}, the families $\mathscr{L}^l_4$ and $\mathscr{B}(4S, 5S)$ exchange branches in such a way that $\mathscr{L}^l_4$ now terminates at B^s_5, and the single bridge $\mathscr{B}(4S, 5S)$ evolves into a second bridge $\mathscr{B}(4S, 4S)$.

7. The Critical Resonances $\sigma = k\lambda$ for $k > 4$

Not much is known about the families of long period orbits possibly emanating from L_4 at critical resonances below the cases we have so far examined.

The formalism proposed by Roels is consistent at any critical mass ratio $\mu_k (k \geqslant 4)$, and it suggests the existence at μ_k of a family emanating from L_4 with long period orbits, the representative in the vicinity of the equilibrium, having the d'Alembert characteristic in relation to the orbital parameter. The situation looks even more involved. For Hamiltonian systems with two degrees of freedom, to an equilibrium whose normal modes of vibration are in resonance, the generic case seems to be that three families of long period orbits emanate from it (Henrard, 1969a).

In that respect, for instance, at μ_{12} and μ_{13}, there would come out of L_4 three families of long period Trojan orbits, beside, of course, the family $\mathscr{L}_4^s$ already guaranteed by Liapunov's theory. It would be interesting to make of this tentative statement a working hypothesis for a thorough numerical investigation. It would lead eventually to conjectures as how the family $\mathscr{L}_4^l$ evolves from μ_{12} to μ_{13}.

So far we have located only a few pieces of the puzzle. At the mass ratio Sun-Jupiter, a two-lane bridge $\mathscr{B}(12S, 12S)$ has been discovered (Deprit and Rabe, 1969), a two-prong bridge $\mathscr{B}(25S, 2L)$ and a two-lane bridge $\mathscr{B}(35S, 3L)$ (see Deprit *et al.*, 1969). We are of the opinion that the bridge $\mathscr{B}(12S, 12S)$ results from two of the three long period families that could have appeared at μ_{12}; also we believe that, as μ passes through $\mu_{25/2}$, the bridge $\mathscr{B}(25S, 2L)$ possibly transforms into the families of very long period orbits which, beside $\mathscr{L}_4^l(\mu_{25/2})$, emanate from L_4 in accordance with the theorem of the natural centers of libration (Meyer and Palmore, 1969).

8. Toward the Copenhagen Problem

Soon after it had been recognized that, in the range $0 < \mu \leqslant \mu_1$, L_4 was generically the source of two families of periodic orbits, inquiries began to determine how these families survived for $\mu > \mu_1$.

Formal series representing periodic orbits around L_4 for mass ratios greater than, but close to, μ have been produced, (Deprit, 1966). The formalism consists in keeping the period at $T = 2\pi\sqrt{2}$, and in expanding the mass ratio around μ_1 in powers of an orbital parameter so as to extract a periodic solution from the differential equations. Actually the series obtained in this way converge too slowly to yield extrapolations valid all the way to the Copenhagen problem in which $\mu = \frac{1}{2}$.

Where analytical methods lose their capabilities, numerical continuation takes the relay. Starting from several orbits of $\mathscr{L}_4^s$ and $\mathscr{L}_4^l$ at μ_1, Palmore (1967) pursued by numerical continuation, for fixed Jacobian constants, to the mass ratio $\mu = 0.04$. At this point, he continued the natural families to which these orbits belong. He found that the family – to be indicated by $\mathscr{L}_4$ – goes through a maximum Jacobian constant; one branch terminates on an orbit of $\mathscr{L}_4$ traveled twice, whereas the other branch meets its symmetric $\mathscr{L}_5$ at a symmetric orbit B_5^4 in the family $\mathscr{L}_3$.

Then, again fixing the Jacobian constant, Palmore continued an orbit of $\mathscr{L}_4$ through increasing mass ratios until $\mu=0.5$. At that mass ratio, he resumed analytical continuation with respect to the Jacobian constant. He found again that the family $\mathscr{L}_4$ meets its counterpart $\mathscr{L}_5$ on a symmetric orbit in the family $\mathscr{L}_3$, actually at the orbit of indifferent stability where Hénon (1965) anticipated branching off to a family of non-symmetric orbits. Palmore has not determined whether, at the other end, the family $\mathscr{L}_4$ terminates upon itself by duplication.

There is no reason why a numerical continuation with respect to the mass ratio should stop at $\mu=\frac{1}{2}$. As a matter of fact, Palmore discovered that the extension through $\mu=\frac{1}{2}$ does not fall back upon itself: an orbit of $\mathscr{L}_4$ at $\mu'<\frac{1}{2}$ generates at the mass ratio $1-\mu'$ an orbit that is not its mirror-like image.

Palmore's investigations are too incomplete to lead to firm conjectures. But they raise interesting questions concerning the permanence of the Trojan manifold beyond μ_1.

Acknowledgements

The idea of collecting *bona fide* conjectures about families of periodic orbits in the restricted problem of three bodies was suggested to us by Professor Allen Goldstein, Department of Mathematics, The University of Washington, Seattle.

We are grateful for helpful discussions and comments to Professors Breakwell, Danby, Hale, LaSalle, Moser, Meyer and Rabe, and to Drs Palmore and Roels.

References

Alfriend, K. T.: 1969, 'The Stability of the Triangular Lagrangian Points for Commensurability of Order Two', *Celes. Mech.* **1**, 351.

Bartlett, J. H. and Wagner, C. A.: 1965, 'The Restricted Problem of Three Bodies (II)', *Skrifter. Dan. Videnskap. Selsk. Mat.-Fys. Kl.* **3**, No. 1, 53 pp.

Breakwell, J. V. and Pringle, R.: 1966, 'Resonances Affecting Motion near the Earth-Moon Equilateral Libration Points' in *Methods in Astrodynamics and Celestial Mechanics* (ed. by R. L. Duncombe and V. G. Szebehely), Academic Press, New York, pp. 55–74.

Broucke, R.: 1968, 'Periodic Orbits in the Restricted Three-Body Problem With Earth-Moon Masses'. NASA TN 32-1168, Jet Propulsion Laboratory, 92 pp.

Deprit, A.: 1966, 'Limiting Orbits Around the Equilateral Centers of Libration', *Astron. J.* **71**, 77–87.

Deprit, A. and Deprit-Bartholomé, A.: 1967, 'Stability of the Triangular Lagrangian Points', *Astron. J.* **72**, 173–179.

Deprit, A. and Henrard, J.: 1968, 'A Manifold of Periodic Orbits', *Advan. Astron. Astrophys.* **6**, 2–124.

Deprit, A. and Henrard, J.: 1969, 'Sur les orbites périodiques issues de L_4 à la résonance interne $^1/_4$', *Astron. Astrophys.*, to appear.

Deprit, A. and Palmore, J.: 1966, 'Analytical Continuation and First Order Stability of the Short Period Orbits at L_4 in the Sun-Jupiter System', *Astron. J.* **71**, 94–98.

Deprit, A. et Price, J. F.: 1969, 'L'espace de phase autour de L_4 pour la résonance interne $^1/_3$', *Astron. Astrophys.* **1**, 427–430.

Deprit, A. and Rabe, E.: 1969, 'Periodic Trojan Orbits for the Resonance $^1/_{12}$', *Astron. J.* **74**, 317–320.

Deprit, A., Henrard, J. and Price, J. F.: 1969, 'The Periodic Trojan Orbits in the System Sun-Jupiter' (in preparation).

Deprit, A., Henrard, J., Palmore, J., and Price, J. F.: 1967, 'The Trojan Manifold in the System Earth-Moon', *Monthly Notices Roy. Astron. Soc.* **137**, 311–335.

Deprit, A., Henrard, J., Price, J. F., and Rom, A.: 1969, 'Birkhoff's Normalization', *Celes. Mech.* **1**, 222.

Goodrich, E.: 1966, 'Numerical Determination of Short-Period Trojan Orbits in the Restricted Three-Body Problem', *Astron. J.* **71**, 88–93.

Hénon, M.: 1965, 'Exploration numérique du problème restreint. II: Masses égales, stabilité des orbites périodiques', *Ann. Astrophys.* **28**, 992–1007.

Henrard, J.: 1969a, 'Periodic Orbits Emanating from a Resonant Equilibrium', *Celes. Mech.* **1**, 437.

Henrard, J.: 1969b, 'Concerning the Genealogy of Long Period Families at L_4', *Astron. Astrophys.* **5**, 45–52.

Meyer, K. R. and Palmore, J.: 1969, 'A New Class of Periodic Solutions in the Restricted Three-Body Problem', Proceedings of the Symposia in Pure Mathematics, **14**, *Global Analysis.*

Palmore, J.: 1967, 'Numerical Experimentation into Effects of Varying the Mass Ratio on Periodic Solutions of the Restricted Problem of Three Bodies', Ph.D. Dissertation, Graduate School of Yale University, New Haven, Conn., 122 pp.

Roels, J.: 1966, 'Le phénomène de résonance (2-1) au voisinage des points d'équilibre équilatéraux du problème restreint', *Bull. Astron.* **1**, 241–255.

Roels, J.: 1969a, 'Orbites de longues périodes résonantes autour des points équilatéraux de Lagrange. I: Rapport de masses critiques μ_3', *Astron. Astrophys.* **1**, 77–90.

Roels, J.: 1966b, 'Orbites de longues périodes résonantes autour des points équilatéraux de Lagrange. II: Rapports de masses critiques $\mu_k (k \geq 4)$. *Astron. Astrophys.* **1**, 380–387.

Roels, J.: 1969c, 'Nouvelle famille d'orbites périodiques autour des points équilatéraux pour la résonance $^1/_4$, III', *Astron. Astrophys.* **2**, 52–62.

Wintner, R. A.: 1936, 'On the Periodic Analytic Continuation of the Circular Orbits in the Restricted Problem of Three Bodies', *Proc. Nat. Acad. Sci.* **22**, 435–439.

ON ASYMMETRIC PERIODIC SOLUTIONS OF THE PLANE RESTRICTED PROBLEM OF THREE BODIES

P. J. MESSAGE

Dept. of Applied Mathematics, Mathematical Institute, The University, Liverpool, England

Abstract. For a Hamiltonian system of two degrees of freedom, of such a type that one co-ordinate may be made ignorable by a transformation of Von Zeipel's type, the linear equations of variation from a solution of a certain type are set up to examine changes of stability along a series of such solutions, and some results are derived. In the application to the plane restricted problem of three bodies, in a near-commensurability case, the solutions in question are the symmetric periodic solutions, and one of the results supplies a criterion for the bifurcation of a series of asymmetric periodic solutions from a symmetric series. This is used to examine some small integer commensurability cases, and some indications of such bifurcations are found, in addition to those already known.

Numerical integration is used to trace the complete series of asymmetric periodic solutions associated with the 2:1 exterior commensurability, from its bifurcation from a series of symmetric solutions, to its termination in a second bifurcation with the same series.

The extension is indicated of the general discussion to a type of dynamical system of which the general plane problem of three bodies gives an example.

1. Some Results on Linear Equations of Variation and Changes of Stability

The remarks of this section refer to a conservative dynamical system of two degrees of freedom, described by co-ordinates y_1 and y_2, whose conjugate momenta are x_1 and x_2 respectively, and by the Hamiltonian function

$$H(y_1, y_2, x_1, x_2) = H_0(x_1, x_2) + \varepsilon \sum_{j \in J} K_j(x_1, x_2) \cos(j_1 y_1 + j_2 y_2), \quad (1.1)$$

where ε is a small parameter, J is the set of integer pairs $j=(j_1, j_2)$ with $j_1 \geqslant 0$, and $\partial H_0/\partial x_2$ is not large compared with ε. The system may be related to one with one co-ordinate ignorable by the Von Zeipel type of formal transformation $(y_i, x_i) \rightarrow (y_i^*, x_i^*)$, given by the determining function

$$S(y_1, y_2, x_1^*, x_2^*) = y_1 x_1^* + y_2 x_2^* + \varepsilon \sum_{j \in J'} L_j(x_1^*, x_2^*) \sin(j_1 y_1 + j_2 y_2), \quad (1.2)$$

where J' is the set of integer pairs $j=(j_1, j_2)$ with $j_1 > 0$, and the coefficients L_j are chosen so that the Hamiltonian function of the new system has the form

$$H^*(y_2^*, x_1^*, x_2^*) = H_0(x_1^*, x_2^*) + \varepsilon \sum_{j \in J^*} K_j^*(x_1^*, x_2^*) \cos(j_2 y_2^*), \quad (1.3)$$

J^* being the set of integer pairs $j=(0, j_2)$. This is found to entail the expression of the L_j and K_j^* as power series in ε, the leading terms being

$$L_j = K_j/(j_1 n_1 + j_2 n_2) + \mathrm{O}(\varepsilon), \quad \text{for} \quad j \in J', \quad (1.4)$$

where

$$n_i = \frac{\partial H_0}{\partial x_i} \quad \text{for} \quad i = 1 \text{ or } 2,$$

G. E. O. Giacaglia (ed.), Periodic Orbits, Stability and Resonances, 19–32.

and

$$K_j^* = K_j + O(\varepsilon) \quad \text{for} \quad j \in J^*. \tag{1.5}$$

The coordinate y_1^* is ignorable, so x_1^* is constant, equal to D, say. We can regard the new system as of one degree of freedom, with y_2^* and x_2^* as the variables, and D as a parameter of the Hamiltonian.

There are solutions

$$y_2^* \equiv 0,\ x_2^* \equiv X(D), \quad \text{for} \quad D_1 < D < D_2, \tag{1.6}$$

where

$$\frac{\partial H^*}{\partial x_2}(0, D, X(D)) = 0 \quad \text{for this range of } D. \tag{1.7}$$

The linear equations of variation from these solutions are

$$\frac{\mathrm{d}\eta_2}{\mathrm{d}t} = \frac{\partial^2 H^*}{\partial x_2^{*2}}(0, D, X(D))\cdot\xi_2 \quad \text{and} \quad \frac{\mathrm{d}\xi_2}{\mathrm{d}t} = -\frac{\partial^2 H^*}{\partial y_2^{*2}}(0, D, X(D))\cdot\eta_2, \tag{1.8}$$

since (1.3) shows that

$$\partial^2 H^*/(\partial x_2^* \partial y_2^*)(0, D, X(D)) = 0. \tag{1.9}$$

The auxiliary equation is

$$\lambda^2 + \frac{\partial^2 H^*}{\partial x_2^{*2}}\cdot\frac{\partial^2 H^*}{\partial y_2^{*2}} = 0, \tag{1.10}$$

whose roots are the characteristic exponents of the solution (1.6), and so the non-zero exponents of the corresponding solution of the original system (1.1). Thus a simple change of stability is associate with a simple zero of either

$$\frac{\partial^2 H^*}{\partial x_2^{*2}} \quad \text{or} \quad \frac{\partial^2 H^*}{\partial y_2^{*2}}.$$

Suppose now that a series of solutions

$$y_2^* \equiv \phi(D), \quad x_2^* \equiv \psi(D), \quad \text{for} \quad D_0 \leqslant D < D_3, \tag{1.11}$$

bifurcates with the series (1.6) at $D = D_0$, where $D_1 < D_0 < D_2$, so that $\psi(D_0) = X(D_0)$ and $\phi(D_0) = 0$. We suppose first that $\phi'(D_0) \neq 0$. We must have

$$\frac{\partial H^*}{\partial y_2^*}(\phi(D), D, \psi(D)) = \frac{\partial H^*}{\partial x_2^*}(\phi(D), D, \psi(D)) = 0 \quad \text{for} \quad D_0 \leqslant D < D_3,$$

and differentiating with respect to D, putting $D = D_0$, and using (1.9), we obtain from these

$$\frac{\partial^2 H^*}{\partial y_2^{*2}}(0, D_0, X(D_0))\cdot\phi'(D_0) = 0$$

and

$$\frac{\partial^2 H^*}{\partial x_1^* \partial x_2^*}(0, D_0, X(D_0)) + \frac{\partial^2 H^*}{\partial x_2^{*2}}(0, D_0, X(D_0))\cdot\psi'(D_0) = 0. \tag{1.12}$$

Likewise from (1.7),

$$\frac{\partial^2 H^*}{\partial x_1^* \partial x_2^*}(0, D_0, X(D_0)) + \frac{\partial^2 H^*}{\partial x_2^{*2}}(0, D_0, X(D_0)) \cdot X'(D_0) = 0 . \tag{1.13}$$

The first of (1.12) shows that

$$\frac{\partial^2 H^*}{\partial y_2^{*2}}(0, D_0, X(D_0)) = 0 . \tag{1.14}$$

Thus there is in general a change of stability at the bifurcation. Also the second of (1.12), and (1.13), give

$$\frac{\partial^2 H^*}{\partial x_2^{*2}}(0, D_0, X(D_0))\,(\psi'(D_0) - X'(D_0)) = 0 . \tag{1.15}$$

so that, unless $\partial^2 H^*/\partial x_2^{*2}$ also has a zero at the bifurcation,

$$\psi'(D_0) = X'(D_0) . \tag{1.16}$$

Suppose now that the series meeting (1.6) has $y_2^* \equiv 0$, $x_2^* \equiv \psi(D)$, for $D_0 \leqslant D \leqslant D_3$, with $\psi(D_0) = X(D_0)$, but $\psi'(D_0) \neq X'(D_0)$. Now (1.15) shows that $\partial^2 H^*/\partial x_2^{*2}$ is zero, and so there is once again a change of stability, but since now $\phi(D) \equiv 0$, the first of (1.12) no longer requires $\partial^2 H^*/\partial y_2^{*2}$ to vanish. The vanishing of the latter quantity is therefore characteristic of bifurcation with a series along which y_2^* changes value.

Suppose that, along a series of solutions with $y_2^* = 0$, D has an extreme value, that is, the solutions of the series may be represented by $y_2^* \equiv 0$, $x_2^* \equiv X(\alpha)$, $D = x_1 \equiv \theta(\alpha)$, for $\alpha_1 \leqslant \alpha \leqslant \alpha_2$, with

$$\theta'(\alpha_0) = 0 \quad \text{and} \quad X'(\alpha_0) \neq 0 \tag{1.17}$$

for some α_0 in $\alpha_1 < \alpha_0 < \alpha_2$. We must have then

$$\partial H^*/\partial x_2^*(0, \theta(\alpha), X(\alpha)) = 0$$

for this range of α, and so

$$\frac{\partial^2 H^*}{\partial x_1^* \partial x_2^*}(0, \theta(\alpha), X(\alpha)) \cdot \theta'(\alpha) + \frac{\partial^2 H^*}{\partial x_2^{*2}}(0, \theta(\alpha), X(\alpha)) \cdot X'(\alpha) = 0 .$$

Then from (1.17),

$$\frac{\partial^2 H^*}{\partial x_2^{*2}}(0, \theta(\alpha_0), X(\alpha_0)) = 0 . \tag{1.18}$$

so that an extreme value of D is associated with a change of stability in general.

The results of this section also follow if a series of solutions with $y_2^* = \pi$ is considered in place of one with $y_2^* = 0$.

2. Application to Near-Commensurability of Period in the Plane Restricted Problem of Three Bodies

Consider, in the plane restricted problem, the case where n', the mean motion of the second body (moving relative to the primary in a circle of radius a'), and n, the mean motion of the third (and massless) body, are related by

$$(p + q)\, n \doteqdot pn', \tag{2.1}$$

where p and q are integers of small absolute value.

Put $y_1 = \lambda - \lambda'$, the difference between the mean longitudes of the third and second bodies, and $y_2 = py_1 + ql$, l being the mean anomaly of the third body. Then the momenta are (as shown in Message (1966) p. 200)

$$x_1 = \sqrt{(\mu a)}\,\{(p + q)\sqrt{(1 - e^2)} - p\}/q,$$

and

$$x_2 = \sqrt{(\mu a)}\,\{1 - \sqrt{(1 - e^2)}\}/q. \tag{2.2}$$

(here μ is the product of the mass of the primary with the constant of gravitation, and a and e are the major semi-axis and eccentricity, respectively, of the orbit of the third body about the primary), and the Hamiltonian function is

$$H = -\tfrac{1}{2}\mu^2/\{x_1 + (p + q)\, x_2\}^2 - n'(px_2 + x_1) - \mu m'(1/\Delta - r\cos\chi/r'^2), \tag{2.3}$$

where r' and r are the distances of the second and third bodies, respectively, from the primary, χ is the angle they subtend at the primary, m' is the mass of the second body in terms of that of the primary, and

$$\Delta = \sqrt{(r^2 + r'^2 - 2rr'\cos\chi)} \tag{2.4}$$

the distance between the second and third bodies. Also

$$\frac{\partial H}{\partial x_2} = (p + q)\, n - pn' + \mathrm{O}(m'),$$

which is small by (2.1), so that the transformation of Section 1 may be applied, with the mass-ratio m' for ε.

Periodic solutions of Poincaré's 'premier genre' have constant values of y_2^* and x_2^*. Such solutions exist both for $y_2^* = 0$ and $y_2^* = \pi$, and these are the symmetric solutions (Message, 1966, p. 202). The result (1.14) shows that a bifurcation of such a series with a series of asymmetric solutions will be marked by a change of sign of $\partial^2 H^*/\partial y_2^{*2}$. Some series of symmetric solutions, corresponding to some small integer commensurabilities, were examined for such changes of sign, in the following manner. Now if r, the radius vector of the third body, and ψ, its true longitude, are expressed in terms of the y_i and x_i, we find that

$$\frac{\partial r}{\partial y_1} = -\, pae \sin f / \{q \sqrt{(1-e^2)}\} = -\, p \frac{\partial r}{\partial y_2}, \tag{2.5}$$

and

$$\frac{\partial \psi}{\partial y_1} = (p+q)/q - p^2 a^2 \sqrt{(1-e^2)}/(qr^2),$$
$$\frac{\partial \psi}{\partial y_2} = -\, 1/q + a^2 \sqrt{(1-e^2)}/(qr^2), \tag{2.6}$$

where f is the true anomaly. From these

$$\begin{aligned}\frac{\partial^2 H}{\partial y_2^2} &+ \left(\frac{1}{p}\right)\frac{\partial^2 H}{\partial y_1\, \partial y_2} + \left(\frac{1}{p^2}\right)\frac{\partial}{\partial y_1}\left\{\mu m' r \sin\chi \left(\frac{a'}{\Delta^3} + \frac{1}{a'^2}\right)\right\} \\ &= (\mu m'/p^2)\,\{r \cos\chi (a'/\Delta^3 - 1/(a')^2 - 3r^2 (a')^2 \sin^2\chi/\Delta^5\} \\ &= (\mu m'/p^2)\, F(y_1, y_2, x_1, x_2), \quad \text{say.}\end{aligned} \tag{2.7}$$

Now (1.1) and (1.5) show that H^* is, to first order in m', the mean of H over y_1, and we see that, since q revolutions of y_1 for a fixed value of y_2 are required to complete an integral number of revolutions of l,

$$\frac{\partial^2 H^*}{\partial y_2^{*2}}(y_2^*, x_1^*, x_2^*) = \mu m'/(2\pi q p^2) \int_0^{2\pi q} F(y_1, y_2^*, x_1^*, x_2^*)\, \mathrm{d}y_1 + \mathrm{O}((m')^2). \tag{2.8}$$

The quantity computed was

$$F^*(y_2, x_1, x_2) = 1/(2\pi q) \int_0^{2\pi q} F(y_1, y_2, x_1, x_2)\, \mathrm{d}y_1, \tag{2.9}$$

and the value of a' used was that making (2.1) an exact equality. For $p=q=1$, $y_2=\pi$, the values obtained for a series of values of e are given in Table I.

By computing F^* for further values of e near the changes of sign, and interpolating, the zeros of this quantity were found to correspond to $e=0.03652$ and $e=0.95927$. The first value corresponds to the bifurcation described in an earlier paper (Message, 1958), and the second value corresponds to the bifurcation found in the numerical integrations described in the next section, and is in good agreement with the values of the mean eccentricity calculated there. For $p=1$, $q=2$, $y_2=\pi$, values of e up to 0.999 were taken, and a single zero of F^* was found, at $e=0.03963$. Asymmetric periodic solutions corresponding to this case were found in numerical investigations by Fragakis (1968). For $p=1$, $q=3$, $y_2=\pi$, zeros for F^* were found at $e=0.29860$ and $e=0.97415$. No zeros were found for $p=1$, $q=4$, $y_2=\pi$, values of e being taken up to 0.999. For $p=1$, $q=5$, $y_2=\pi$, values of e up to 0.97 were taken, and a single zero found, at $e=0.24329$. (The values computed in this way are of course the limiting values as $m'\to 0$ of the values of e corresponding to the zeros of $\partial^2 H^*/\partial y_2^{*2}$.) A few cases with $y_2=0$ and $p>0$ were studied, but no zeros were found. In these cases F^* was found to diverge at values of e for which a collision between the second and third

TABLE I

Values of F^* for the series $p=q=1$, $y^*{}_2=\pi$

e	F^*	e	F^*
0.01	−0.01159	0.8	1.61651
0.02	−0.01362	0.9	0.99383
0.03	−0.00762	0.95	0.23600
0.04	0.00513	0.955	0.11509
0.05	0.02356	0.956	0.07601
0.06	0.04676	0.957	0.06266
0.1	0.17389	0.958	0.03547
0.2	0.59692	0.959	0.00760
0.3	1.02920	0.96	−0.0210
0.4	1.40027	0.961	−0.0503
0.5	1.67612	0.965	−0.1760
0.6	1.83005	0.97	−0.3554
0.7	1.83010	0.975	−0.5672

(Values are given to as many significant figures as agreed with the results obtained with twice the interval of integration.)

bodies occurs for some value of y_1. No zero was found for $p=2$, $q=1$, $y_2=\pi$, e being taken up to 0.99. (The results given in 1958 for $e\geqslant 0.1$ for this pair of values of p and q were in error, due to the neglect of terms in the disturbing function expansion which are not in fact negligible: there is no bifurcation here. This has been made clear by work by Fragakis (1968).) For $p=2$, $q=3$, $y_2=\pi$, zeros of F^* were found for $e=0.27807$, $e=0.69524$, and $e=0.95754$. No zeros were found for $p=3$, $q=1$ or 2 with $y_2=\pi$ (for $q=2$, F^* diverges near $e=0.67$). For $p=3$, $q=4$, $y_2=\pi$, a zero was found at $e=0.878381$.

These results suggest that there are series of asymmetric periodic solutions associated with many exterior commensurabilities. In the case of interior commensurabilities, on the other hand, none of the cases with $-4\leqslant p\leqslant -2$, and $y_2=0$, provided a zero of F^*.

3. Numerical Integrations in the Case $p=q=1$

In an earlier paper (Message, 1959), the series of asymmetric periodic solutions of the plane restricted problem, associated with the commensurability $p=q=1$, was described, from its bifurcation near $e=0.037$, with the series of symmetric solutions with $y_2=\pi$ as far as the orbit for which the osculating eccentricity at conjunction is 0.4. This section now describes the continuations of this series of solutions, as explored by more recent numerical integrations of the equations of motion of the third body, in the usual rotating coordinate system, which are

$$\frac{d^2x}{dt^2} - 2n'\frac{dy}{dt} - n'^2x = -\mu(x + a_0m')/r^3 - \mu m'(x - a_0)/\Delta^3\,,$$
$$\frac{d^2y}{dt^2} + 2n'\frac{dx}{dt} - n'^2y + -\mu y/r^3 - \mu m'y/\Delta^3\,, \tag{3.1}$$

where

$$r = \sqrt{\{(x + a_0 m')^2 + y^2\}},$$

$$\Delta = \sqrt{\{(x - a_0)^2 + y^2\}},$$

and

$$a_0 = a'/(1 + m').$$

Simultaneously with the integration of these equations, integrations were also carried out of Hill's equation for the differential normal displacement, q,

$$\frac{d^2 q}{dt^2} + \Theta q = -\Delta C\left(n' + \frac{d\psi}{dt}\right)/V, \tag{3.2}$$

where

$$\begin{aligned}\Theta = {} & 3n'^2 + 6n'\frac{d\psi}{dt} + 3\left(\frac{d\psi}{dt}\right)^2 + \mu/r^3 + \mu m'/\Delta^3 \\ & + 3\mu\Bigg[\{(x + a_0 m')/r^5 + (x - a_0)\, m'/\Delta^5\}\, y\,\frac{dx}{dt}\frac{dy}{dt} \\ & - \{(x + a_0 m')^2/r^5 + (x - a_0)^2\, m'/\Delta^5\}\left(\frac{dy}{dt}\right)^2 \\ & - (1/r^5 + m'/\Delta^5)\, y^2\left(\frac{dx}{dt}\right)^2\Bigg]/V^2,\end{aligned} \tag{3.3}$$

$$V = \sqrt{\left\{\left(\frac{dx}{dt}\right)^2 + \left(\frac{dy}{dt}\right)^2\right\}}, \tag{3.4}$$

and

$$\begin{aligned}\frac{d\psi}{dt} = {} & \mu\Bigg[(n'^2 - 1/r^3 - m'/\Delta^3)\, y\,\frac{dx}{dt} \\ & - \{n'^2 x - (x + a_0 m')/r^3 - m'(x - a_0)/\Delta^3\}\frac{dy}{dt}\Bigg]/V^2 - 2n'\end{aligned} \tag{3.5}$$

After integrating an approximately closed orbit, linear corrections to the initial conditions, to approximate more closely to those of a periodic orbit, were made, using formulae equivalent to those given by Deprit *et al.* (1967, Equations (34) and (35), pp. 319 and 320) with $\Delta C = 0$. When an orbit, which was closed to the assigned accuracy, had been found, the same formulae were used to begin the search for a further closed orbit corresponding to an increase of ΔC in the constant of Jacobi's integral

$$\tfrac{1}{2}\left\{\left(\frac{dx}{dt}\right)^2 + \left(\frac{dy}{dt}\right)^2\right\} - \mu/r - \mu m'/\Delta - \tfrac{1}{2}n'^2(x^2 + y^2) = C \tag{3.6}$$

Osculating elliptic elements were computed from the coordinates and velocity components during the final integration of each closed orbit, and the mean values of these, and the coefficients of the leading terms in their Fourier series.

The formulae used for the integration, based on the Gauss-Jackson formula (given by Cowell and Crommelin (1909, p. 84)), and the Adams formula,

$$x_i = h^2 \{\delta^{-2} f_i + \tfrac{1}{12} f_i - \tfrac{1}{120}\delta^2 f_{i-1} + \tfrac{1}{240}\delta^2 f_{i-2} - \tfrac{137}{20160}\delta^4 f_{i-2} + \tfrac{19}{6048}\delta^4 f_{i-3} - \tfrac{9829}{3628800}\delta^6 f_{i-3} - \tfrac{387}{44800}\delta^7 f_{i-3\frac{1}{2}}\}$$

$$\dot{x}_i = h \{\delta^{-1} f_i + \tfrac{5}{12} f_i + \tfrac{1}{12} f_{i-1} - \tfrac{49}{720}\delta^2 f_{i-1} + \tfrac{19}{720}\delta^2 f_{i-2} - \tfrac{1997}{60480}\delta^4 f_{i-2} + \tfrac{863}{60480}\delta^4 f_{i-3} - \tfrac{275}{24192}\delta^6 f_{i-3} - \tfrac{33953}{3628800}\delta^7 f_{i-3\frac{1}{2}}\} \tag{3.7}$$

where f is $\ddot{x}$ as given by the differential equation, and h is the length of the integration step. This pair of formulae was used to provide a predictor, taking $\delta^7 f_{i-4\frac{1}{2}}$ as an estimate of $\delta^7 f_{i-3\frac{1}{2}}$, and was used again as a corrector if the calculation of f changed $h\delta^7 f_{i-3\frac{1}{2}}$ by more than twice the accuracy desired. After trials it was arranged for the step of integration to be halved after any step, if any $h|\delta^7 f_{i-3\frac{1}{2}}|$ there exceeded the assigned accuracy by more than five or ten times (this factor could probably have safely been taken larger), or if the correcting process (which involves a second calculations of the second derivatives from the differential equation) was called upon in more than seven steps out of the previous twelve. To examine the stability of the method, consider the equations

$$\frac{d^2x}{dt^2} - 2\omega \frac{dy}{dt} + (\nu^2 - \omega^2)x = 0$$

and

$$\frac{d^2y}{dt^2} + 2\omega \frac{dx}{dt} + (\nu^2 - \omega^2)y = 0. \tag{3.8}$$

The difference equations effectively being used by using the formulae (3.7), and keeping differences up to the second, will have solutions of the form

$$x_i = A\varrho^i, \quad \frac{dx_i}{dt} = B\varrho^i, \quad y_i = C\varrho^i, \quad \frac{dy_i}{dt} = E\varrho^i,$$

where

$$(\varrho - 1)^2 [\varrho^2 \{(\varrho - 1)^2 + h^2(\nu^2 - \omega^2)(\varrho + \tfrac{1}{12}(\varrho - 1)^2)\}^2 + h^2\omega^2(\varrho - 1)^2 (\tfrac{5}{6}\varrho^2 + \tfrac{4}{3}\varrho - \tfrac{1}{6})^2] = 0.$$

This has two roots $\varrho = 1$, two roots $\varrho = \pm(\frac{1}{6})\,h\omega + O(h^2)$, and four which satisfy

$$\{(\varrho - 1)^2 + h^2(\nu^2 - \omega^2)\}^2 + \{2h\omega(\varrho - 1)\}^2 = O(h^4).$$

Thus these four are

$$\varrho = 1 \pm \sqrt{(-1)}\,h\omega \mp \sqrt{(-1)}\,h\nu + O(h^2),$$

with all four combinations of signs. These four roots correspond to the correct solution of the Equations (3.8) in the limit $h \to 0$. The roots $\varrho = 1$ are found to correspond to constant solutions of the difference equations. Therefore the method is stable.

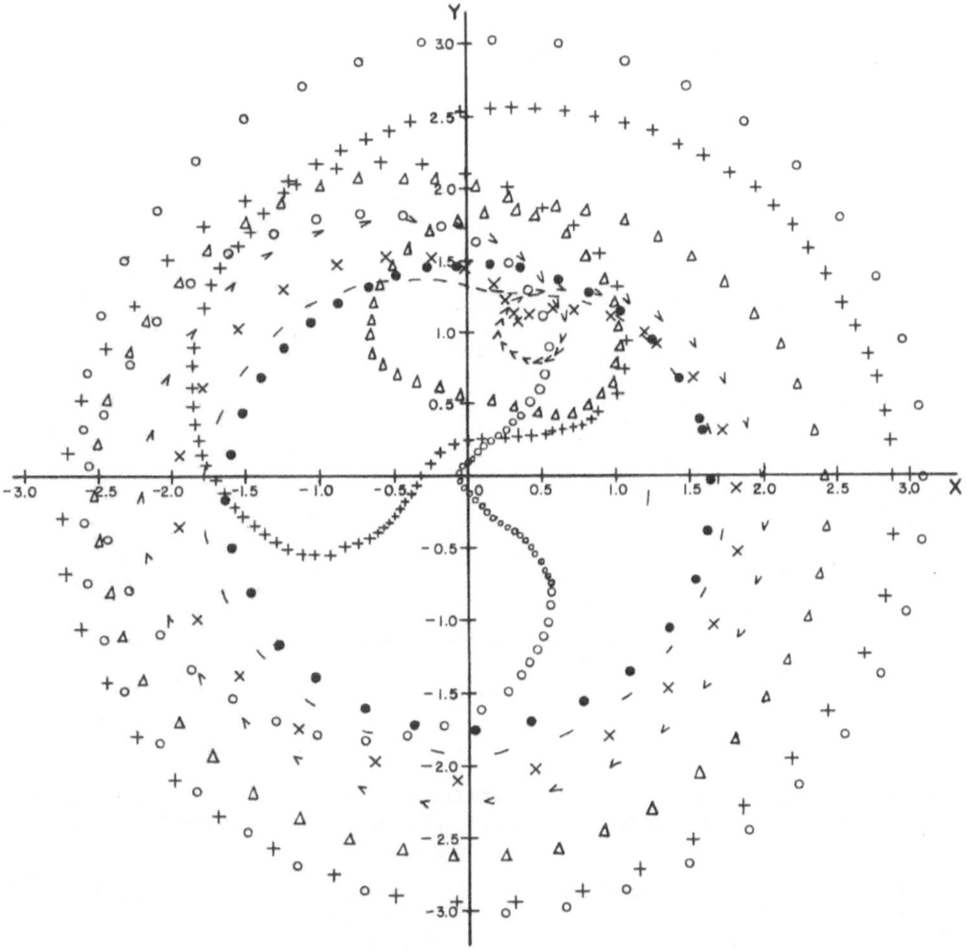

Fig. 1.

Figure 1 shows some asymmetric periodic orbits belonging to the series with $p=q=1$, with $m'=0.000\,954\,927$, illustrating the development along the series. It is seen that with increasing eccentricity and size of orbit, the pericentre, at first coinciding with opposition, moves at first towards conjunction, until after an extreme it returns, until the series terminates in a bifurcation with the same series of symmetric orbits, with pericentre at opposition, from which it had arisen. Figure 2 shows the run of y_2^* with the mean eccentricity, and Figure 3 that of cosh $(2\pi c)$, where $\pm c$ are the non-zero characteristic exponents. Table II gives, for a few of the orbits, the mean values and leading Fourier coefficients of the elliptic elements.

4. An Extension of the Results of Section 1

The results of Section 1 may be extended, with a view to application to the gravitat-

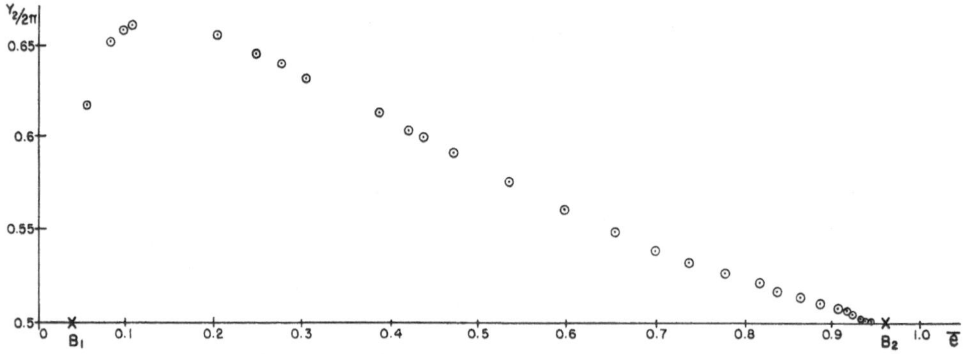

Fig. 2.

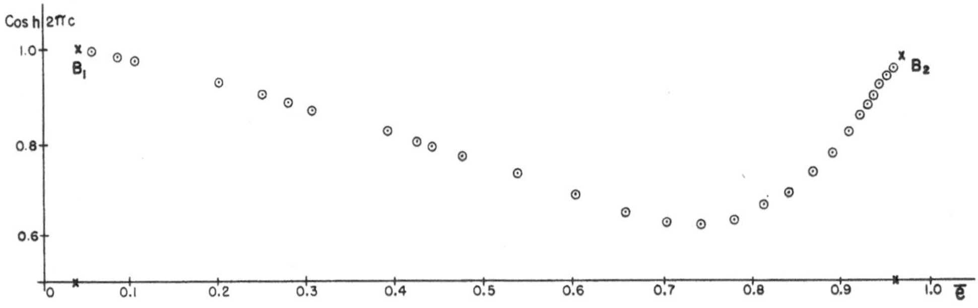

Fig. 3.

TABLE II

Underneath the mean values of each of the three elements given, are 10^4 times the coefficients of $\cos u$, $\sin u$, $\cos 2u$, and $\sin 2u$, where u is the phase in the orbit, measured from the crossing of the x-axis near conjunction

Eccentricity	Major semi-axis	Apse longitude (in units of one revolution)	Period	$\cosh(2\pi c)$
0.0519	1.5919	0.6170	1.99553	0.9969
−1, −8, −6, −7	27, 0, −9, 8	0, −15, 12, −19		
0.1033	1.5917	0.6611	1.99573	0.9808
−4. −17, −12, −15	39, 18, −60, 57	−5, 17, −4, 12		
0.2485	1.5910	0.6469	1.99680	0.9017
−3, −6, −6, −6	−19, −20, −6, 11	−1, 5, 2, 3		
0.3011	1.5907	0.6347	1.99736	0.8704
−5, −4, −6, −3	13, −21, −5, 17	−2, 4, 2, 5		
0.4705	1.5891	0.5914	1.99937	0.7736
−4, 0, −6, −3	11, −20, 2, 2	2, 0, 0, −2		
0.5972	1.5887	0.5606	2.00057	0.6879
−10, −1, −9, 2	−9, −17, −1, −10	−4, −4, −3, 2		
0.6985	1.5882	0.5397	2.00102	0.6326
−11, −10, −2, 3	−8, −11, −2, 7	−4, −5, −4, 2		
0.7753	1.5879	0.5264	2.00099	0.6396
−16, −4, −10, 7	−18, −11, −3, 10	−7, −6, −4, 3		
0.8613	1.5875	0.5139	2.00061	0.7422
−19, −2, −5, 7	−23, −5, 5, 8	−9, −3, −1, 3		
0.9452	1.5871	0.5030	1.99998	0.9550
−14, 1, 0, −2	−15, 2, 9, −5	−6, 1, 1, −1		

ional problem of three bodies in the plane, to a system of four degrees of freedom, with coordinates y_1, y_2, y_3, and y_4 (y_4 being ignorable), their conjugate momenta being x_1, x_2, x_3, and x_4 respectively, the Hamiltonian function having the form

$$H(y_1, y_2, y_3, x_1, x_2, x_3, x_4) = H_0(x_1, x_2, x_3, x_4) + \varepsilon \sum_{j \in J} K_j(x_1, x_2, x_3, x_4) \cos(j_1 y_1 + j_2 y_2 + j_3 y_3), \tag{4.1}$$

where J is the set of integer triples $j = (j_1, j_2, j_3)$ with $j_1 \geqslant 0$. If $\partial H_0 / \partial x_1$ is not large compared with ε, then a transformation $(y_i, x_i) \rightarrow (y_i^*, x_i^*)$ may be defined, having determining function

$$S(y_1, y_2, y_3, x_1^*, x_2^*, x_3^*, x_4^*) = \sum_{i=1}^{4} y_1 x_i^* + \varepsilon \sum_{j \in J'} L_j \sin(j_1 y_1 + j_2 y_2 + j_3 y_3), \tag{4.2}$$

where J' is the set of integer triples $j = (j_1, j_2, j_3)$ with $j_1 > 0$, and

$$L_j = K_j(x_1^*, x_2^*, x_3^*, x_4^*) \Big/ \left(\sum_{i=1}^{3} j_i \frac{\partial H_0}{\partial x_i}(x_i^*) \right) + O(\varepsilon). \tag{4.3}$$

The Hamiltonian function for the (y_i^*, x_i^*) has the form

$$H^*(y_2^*, y_3^*, x_1^*, x_2^*, x_3^*, x_4^*) = H_0(x_1^*, x_2^*, x_3^*, x_4^*) + \varepsilon \sum_{j J^*} K_j^*(x_1^*, x_2^*, x_3^*, x_4^*) \cos(j_2 y_2^* + j_3 y_3^*), \tag{4.4}$$

where J^* is the set of integer pairs $j = (j_2, j_3)$ with $j_2 \geqslant 0$, and

$$K_j^* = K_{(j_2, j_3)} + K_{(-j_2, j_3)} + O(\varepsilon). \tag{4.5}$$

Then $x_i^* = D$ and $x_4^* = E$ are integrals of the motion, and we may consider these as parameters of the system, now thought of as having two degrees of freedom. There will be in general series of solutions of the form

$$y_2^* = y_3^* = 0, \quad x_2^* = X_2(D, E), \quad x_3^* = X_3(D, E), \tag{4.6}$$

where X_2 and X_3 are functions such that

$$\frac{\partial H^*}{\partial x_i^*}(0, 0, D, X_2(D, E), X_3(D, E), E) = 0 \tag{4.7}$$

The linear equations of variation from these solutions are (noting that derivatives of the form $\partial^2 H^* / \partial y_i^* \partial x_k^*$ vanish on the solution)

$$\frac{d\eta_i}{dt} = \sum_{k=2}^{3} \frac{\partial^2 H^*}{\partial x_i^* \, \partial x_k^*} \cdot \xi_k, \qquad \frac{d\xi_i}{dt} = \sum_{k=2}^{3} \frac{\partial^2 H^*}{\partial y_i^* \, \partial y_k^*} \cdot \eta_k \qquad (i = 2, 3) \tag{4.8}$$

The auxiliary equation is

$$\lambda^4 + \left(\frac{\partial^2 H^*}{\partial x_2^{*2}} \frac{\partial^2 H^*}{\partial y_2^{*2}} + 2 \frac{\partial^2 H^*}{\partial x_2^* \partial x_3^*} \cdot \frac{\partial^2 H^*}{\partial y_2^* \partial y_3^*} + \frac{\partial^2 H^*}{\partial x_3^{*2}} \frac{\partial^2 H^*}{\partial y_3^{*2}}\right) \lambda^2 + \frac{\partial^2 H^*}{\partial (x_2^*, x_3^*)} \cdot \frac{\partial^2 H^*}{\partial (y_2^*, y_3^*)} = 0, \tag{4.9}$$

where $\partial^2 f/\partial (x, y)$ denotes the Hessian determinant

$$\begin{vmatrix} \frac{\partial^2 f}{\partial x^2} & \frac{\partial^2 f}{\partial x \, \partial y} \\ \frac{\partial^2 f}{\partial x \, \partial y} & \frac{\partial^2 f}{\partial y^2} \end{vmatrix}.$$

Then if the two pairs of non-zero characteristic exponents are $\pm c_1$ and $\pm c_2$, we have

$$c_1^2 c_2^2 = \frac{\partial^2 H^*}{\partial (x_2^*, x_3^*)} \cdot \frac{\partial^2 H^*}{\partial (y_2^*, y_3^*)}. \tag{4.10}$$

Suppose we have a series of solutions

$$\begin{aligned} &y_i^* = \phi_i(\alpha), \quad x_i^* = \psi_i(\alpha), \quad (i = 2, 3) \\ &x_1^* = D = \theta(\alpha), \quad x_4^* = E = \chi(\alpha), \quad \text{for} \quad \alpha_0 \leqslant \alpha < \alpha_1 \end{aligned} \tag{4.11}$$

which meets the solutions of type (4.6) at $\alpha = \alpha_0$, so that

$$\begin{aligned} &\phi_i(\alpha_0) = 0 \\ &\psi_i(\alpha_0) = X_i(\theta(\alpha_0), \chi(\alpha_0)). \quad (i = 2, 3) \end{aligned} \tag{4.12}$$

That (4.11) should be solutions requires that

$$\frac{\partial H^*}{\partial y_i} (\phi_2(\alpha), \phi_3(\alpha), \theta(\alpha), \psi_2(\alpha), \psi_3(\alpha), \chi(\alpha)) = 0$$

and

$$\frac{\partial H^*}{\partial x_i} (\phi_2(\alpha), \phi_3(\alpha), \theta(\alpha), \psi_2(\alpha), \psi_3(\alpha), \chi(\alpha)) = 0, \quad (i = 2, 3) \tag{4.13}$$

for $\alpha_0 \leqslant \alpha < \alpha_1$. Differentiating these with respect to α, and putting $\alpha = \alpha_0$ gives

$$\sum_{k=2}^{3} \left(\frac{\partial^2 H^*}{\partial y_i^* \partial y_k^*}\right)_{\alpha=\alpha_0} \cdot \psi_k'(\alpha_0) = 0, \quad (i = 2, 3) \tag{4.14}$$

and

$$\sum_{k=2}^{3} \left(\frac{\partial^2 H^*}{\partial x_i^* \partial x_k^*}\right)_{\alpha=\alpha_0} \cdot \phi_k'(\alpha_0) + \left(\frac{\partial^2 H^*}{\partial x_i^* \partial x_1^*}\right)_{\alpha=\alpha_0} \cdot \theta'(\alpha_0) + \left(\frac{\partial^2 H^*}{\partial x_i^* \partial x_4^*}\right)_{\alpha=\alpha_0} \cdot \chi'(\alpha_0) = 0 \quad (i = 2, 3) \tag{4.15}$$

Differentiating (4.7) with respect to D,

$$\left(\frac{\partial^2 H^*}{\partial x_i^* \, \partial x_1^*}\right)_{\alpha=\alpha_0} + \sum_{k=2}^{3} \left(\frac{\partial^2 H^*}{\partial x_i^* \, \partial x_k^*}\right)_{\alpha=\alpha_0} \frac{\partial X_k}{\partial D} = 0, \quad (i = 2, 3) \tag{4.16}$$

and with respect to E,

$$\left(\frac{\partial^2 H^*}{\partial x_i^* \, \partial x_4^*}\right)_{\alpha=\alpha_0} + \sum_{k=2}^{3} \left(\frac{\partial^2 H^*}{\partial x_i^* \, \partial x_k^*}\right)_{\alpha=\alpha_0} \frac{\partial X_k}{\partial E} = 0, \quad (i = 2, 3). \tag{4.17}$$

From (4.14) we see that if the $\psi_k'(\alpha_0)$ are not all zero, then we must have

$$\left(\frac{\partial^2 H^*}{\partial (y_2^*, y_3^*)}\right)_{\alpha=\alpha_0} = 0, \tag{4.18}$$

and so one of the c_i will in general change from stable to unstable type, or vice versa. This is the result corresponding to (1.14). From the other three equations,

$$\sum_{k=2}^{3} \left(\frac{\partial^2 H^*}{\partial x_i^* \, \partial x_k^*}\right)_{\alpha=\alpha_0} \left\{\phi_k'(\alpha_0) - \frac{\partial X_k}{\partial D}\theta'(\alpha_0) - \frac{\partial X_k}{\partial E}\chi'(\alpha_0)\right\} = 0, \quad (i = 2, 3)$$

so that unless $(\partial^2 H^*/\partial (x_2^*, x_3^*))\, \alpha = \alpha_0$ is also zero,

$$\phi_k'(\alpha_0) = \frac{\partial X_k}{\partial D}\theta'(\alpha_0) + \frac{\partial X_k}{\partial E}\chi'(\alpha_0), \quad (k = 2, 3). \tag{4.19}$$

Corresponding to (1.18), suppose that

$$D = \lambda(x_2^*, x_3^*), \quad E = \mu(x_2^*, x_3^*), \tag{4.20}$$

are the functions inverse to those of (4.6), so that (4.7) becomes

$$\frac{\partial H^*}{\partial x_i^*}(0, 0, \lambda(x_2^*, x_3^*), x_2^*, x_3^*, \mu(x_2^*, x_3^*)) = 0, \quad (i = 2, 3) \tag{4.21}$$

and that the mapping defined by (4.20) becomes singular on the curve

$$x_2^* = g(\beta), \quad x_3^* = h(\beta), \quad \beta_1 \leqslant \beta \leqslant \beta_2, \tag{4.22}$$

so that

$$\frac{\partial(\lambda, \mu)}{\partial(x_2^*, x_3^*)} = 0$$

on this curve. Differentiating (4.21) with respect to x_k^*,

$$\frac{\partial^2 H^*}{\partial x_k^* \, \partial x_i^*} + \frac{\partial^2 H^*}{\partial x_1^* \, \partial x_i^*}\frac{\partial \lambda}{\partial x_k^*} + \frac{\partial^2 H^*}{\partial x_4^* \, \partial x_i^*}\frac{\partial \mu}{\partial x_k^*} = 0 \quad (k = 2, 3, i = 2, 3)$$

and so

$$\frac{\partial^2 H^*}{\partial(x_2^*, x_3^*)} + M\frac{\partial(\lambda, \mu)}{\partial(x_2^*, x_3^*)} = 0,$$

where

$$M = \begin{vmatrix} \dfrac{\partial^2 H^*}{\partial x_1^* \, \partial x_2^*} & \dfrac{\partial^2 H^*}{\partial x_4^* \, \partial x_2^*} \\ \dfrac{\partial^2 H^*}{\partial x_1^* \, \partial x_3^*} & \dfrac{\partial^2 H^*}{\partial x_4^* \, \partial x_3^*} \end{vmatrix} .$$

Thus on the curve (4.22),

$$\frac{\partial^2 H^*}{\partial (x_2^*, x_3^*)} = 0 ,$$

and so there is a change of stability type of one of the characteristic exponent pairs across the curve.

References

Cowell, P. H. and Crommelin, A. C. D.: 1909, *Appendix to Greenwich Observations.*

Deprit, A., Henrard, J., Palmore, J., and Price, J. F.: 1967, *Monthly Notices Roy. Astron. Soc.* **137**, 311–335.

Fragakis, C. N.: 1968, Ph.D. Thesis, University of Liverpool.

Message, P. J.: 1958, *Astron. J.* **63**, 443–448.

Message, P. J.: 1959, *Astron. J.* **66**, 226–236.

Message, P. J.: 1966, IAU Symposium No. 25, pp. 197–222.

TWO NEW CLASSES OF PERIODIC TROJAN LIBRATIONS IN THE ELLIPTIC RESTRICTED PROBLEM AND THEIR STABILITIES

EUGENE RABE

University of Cincinnati Observatory, Cincinnati, O., U.S.A.

Abstract. In the ordinary restricted problem, the well-known long- and short-period solutions of the frequencies n_l and n_s exist for small oscillations about the equilateral points L_4 and L_5, for sufficiently small values of the mass parameter μ. In the elliptic restricted problem, two new classes of periodic librations about L_4 and L_5 exist (at least for sufficiently small eccentricities e of the basic elliptic motion of the two primaries), for two associated infinite series of specific, discrete values of μ. The orbital frequencies (in the nonuniformly rotating reference frame of the elliptic problem) are n_l and $N-n_s$, respectively, where N is the basic mean angular motion of the primaries. In both classes, a twofold infinite number of periodic librations exists for each admissible value of μ, because e as well as the initial mean anomaly M_0 of 'Jupiter' remain arbitrary. The stability analysis reveals the existence of *four* essentially different characteristic exponents for each class of periodic solutions, in contrast to the *two* for periodic librations in the ordinary restricted problem. For $e \to 0$, the associated four oscillation frequencies for motions deviating from the new periodic solutions tend to $N-n_s$, n_l, $N-n_l$, and n_s, respectively. All exponents are purely imaginary, indicating first-order stability for both classes of periodic librations. The only exception occurs for the critical $\mu_c = 0.029437$, where, with $n_l = N-n_l = N/2$, no periodic solution exists for $e \neq 0$. While the results of this first-order analysis are valid only for librations of infinitesimal size, additional numerico-analytical findings indicate the existence of both classes of orbits also at finite amplitudes, but for μ-values differing from those of the infinitesimally small solutions, in accordance with the additional dependence of $n_l(\mu)$ and $n_s(\mu)$ on *amplitude*.

1. Introduction

The elements of a theory of librational motions in the plane elliptic restricted problem have been developed in an earlier paper (Rabe, 1966). It was noted there that isolated long-periodic librations about L_4 and L_5 of substantial amplitudes exist in the elliptic problem, whenever the period T_l is an integral multiple of the basic period P of the relative elliptic motion of the two primaries. In the present study it will be seen that such solutions, of period T_l or frequency n_l, exist also for infinitesimally small oscillations about L_4 or L_5, for an infinite sequence of discrete values of the mass parameter μ, and that a second class of periodic oscillations of even longer periods $T_l^* = 2\pi/(N-n_s)$ is generated on the basis of the restricted problem solutions of short period T_s. It is well known that the difference between N and $n_s = 2\pi/T_s$ is only of the order of μ. Elliptic problem solutions of period T_l^* become possible whenever $n_s(\mu)$ is an integral multiple of $N-n_s$. Even though the n_l and n_s of the ordinary restricted problem of three bodies are functions not only of μ, but also of the amplitude parameter λ, the commensurability condition which is satisfied for an infinitesimally small value of λ will in general still be satisfied for finite λ-values, in conjunction with correspondingly adjusted values of μ. For the actual Trojan case with $\mu = 1/1\,047.355$, the existence of elliptic problem solutions of the very long periods T_l^* but of substantial amplitudes λ has already been shown by means of numerico-analytical studies based on orbital

G. E. O. Giacaglia (ed.), Periodic Orbits, Stability and Resonances, 33–44.

data for the restricted problem family of short period T_s (Rabe, 1969). An infinite number of discrete periods T_l^* was found to exist even for this fixed value of μ, because with an increasing amplitude λ of the basic short period component, n_s increases from $N-\varepsilon_1$ to $N+\varepsilon_2$ (ε_1 and ε_2 positive quantities of order μ), so that the required integral values of $n_s/(N-n_s)$ can be achieved by an infinite number of n_s-values converging towards N.

While the existence of periodic elliptic problem librations has thus been recognized first for large amplitudes and for motions based on arbitrarily selected specific values of μ, it is obvious that a study of the fundamental orbital and stability characteristics of these elliptic problem solutions will be greatly simplified when the orbits considered are of infinitesimal dimensions, so that a first-order analysis is sufficient. Retaining the functional dependence of n_l and n_s on μ, such an analysis will in effect trace the previously known results back to their origin at L_4 and L_5, and thus facilitate a systematic exploration of these periodic librations of the elliptic problem.

2. Basic Restricted Problem Solutions

In the ordinary restricted problem, any periodic libration of long or short period is representable in the form

$$\begin{aligned} x &= x_{c,0} + \sum_{j=1}^{\infty} x_{c,j} \cos(j\sigma) + \sum_{j=1}^{\infty} x_{s,j} \sin(j\sigma) \\ y &= y_{c,0} + \sum_{j=1}^{\infty} y_{c,j} \cos(j\sigma) + \sum_{j=1}^{\infty} y_{s,j} \sin(j\sigma), \end{aligned} \tag{1}$$

with

$$\sigma = n(t - t_0), \quad n = n_l \quad \text{or} \quad n = n_s, \tag{2}$$

where the notations, units, and reference frame are those of the earlier paper (Rabe, 1966). Librations about L_4 are symmetric to those about L_5 with respect to the x-axis. For infinitesimally small orbits about L_5, the point $(x_{c,0}, y_{c,0})$ can be identified with the position of L_5 itself, while the periodic terms with $j=1$ of Equations (1) describe an elliptic oscillation about L_5 which satisfies the linearized differential equations with constant coefficients (evaluated at L_5). The relevant frequency n is given by

$$n^2 = \tfrac{1}{2} N^2 \left[1 \mp \sqrt{1 - 27\mu/N^4}\right], \quad N^2 = 1 + \mu, \tag{3}$$

leading to

$$\begin{aligned} n_l^2 &= \tfrac{27}{4}\mu + \tfrac{621}{16}\mu^2 + \cdots \\ n_s^2 &= 1 - \tfrac{23}{4}\mu - \tfrac{621}{16}\mu^2 + \cdots. \end{aligned} \tag{4}$$

The resulting n_l, n_s are real as long as $\mu \leqslant 0.040\,1$. The limiting or critical value 0.040 1.. reduces the radicand of Equation (3) to zero, producing the double root $n_l = n_s = N/\sqrt{2}$.

For convenience, let the amplitude parameter λ denote the (very small) distance of the 'Trojan' from L_5 at the moment t_0 when the straight line connecting the 'sun' with L_5 is intersected outside of L_5, at the solar distance $1+\lambda$. According to Equation (2), this moment $t=t_0$ is associated with $\sigma=0$. With these provisions, the principal periodic terms of Equations (1) for a first-order representation of the well-known two families of very small periodic librations about L_5 have the following sets of coefficients:

(1) For $n=n_l$:

$$\begin{aligned} x_{c,1} &= -\tfrac{1}{2}\lambda, \quad x_{s,1} = \left(\tfrac{1}{2}\mu^{-\frac{1}{2}} - \tfrac{5}{16}\mu^{\frac{1}{2}} + \cdots\right)\lambda \\ y_{c,1} &= \tfrac{1}{2}\sqrt{3}\lambda, \quad y_{s,1} = \left(\tfrac{1}{6}\mu^{-\frac{1}{2}} - \tfrac{13}{48}\mu^{\frac{1}{2}} + \cdots\right)\sqrt{3}\lambda, \end{aligned} \tag{5}$$

(2) For $n=n_s$:

$$\begin{aligned} x_{c,1} &= -\tfrac{1}{2}\lambda, \quad x_{s,1} = \left(1 - \tfrac{17}{16}\mu + \cdots\right)\sqrt{3}\lambda. \\ y_{c,1} &= \tfrac{1}{2}\sqrt{3}\lambda, \quad y_{s,1} = \left(1 - \tfrac{73}{208}\mu + \cdots\right)\lambda. \end{aligned} \tag{6}$$

The ratio q of the major to the minor axis of the ellipse represented by these coefficients has the respective values:

$$q_l = \frac{1}{\sqrt{3\mu}}\left(1 - \tfrac{7}{8}\mu + \cdots\right), \quad q_s = 2\left(1 - \tfrac{23}{26}\mu + \cdots\right). \tag{7}$$

In each family, the value of the semi*minor* axis differs from λ only by an amount of the order $\mu^2\lambda$.

The preceding data for the restricted problem librations of small amplitude are needed as the basis for the derivation of the forced oscillations which have to be superposed in order to satisfy the differential equations of the elliptic restricted problem.

3. Forced Oscillations u_0, v_0 in Elliptic Problem

The method proposed in 1963 (Summer Seminar in Space Mathematics, Cornell University) for the elliptic restricted problem represents the motion of any librating Trojan in terms of a 'pulsating' restricted problem solution x, y, incorporating unknown increments u, v which have to be determined from the differential equations for the coordinates

$$\xi = r(x+u), \quad \eta = r(y+v), \tag{8}$$

with

$$r = 1 + \tfrac{1}{2}e^2 - \left(e - \tfrac{3}{8}e^3\right)\cos M - \tfrac{1}{2}e^2\cos 2M - \tfrac{3}{8}e^3\cos 3M + \cdots \tag{9}$$

and

$$M = M_0 + N(t - t_0). \tag{10}$$

The ξ, η system rotates nonuniformly with 'Jupiter', so that the two primaries stay on the ξ-axis, on which they oscillate in accordance with Equation (9) for their mutual distance. Since the differential equations for μ, v have been derived in the earlier publication (Rabe, 1966), it suffices to say that the general solution consists of a *forced* oscillation u_0, v_0, which is a function of M_0 and e and vanishes only with $e=0$, and of an arbitrary or *free* oscillation u_f, v_f. The exponents or arguments of the series for u_0, v_0 are of the general form $(jM+l\sigma)$, and the coefficients can be determined from nonhomogeneous linear equations resulting from the substitution of the assumed solution into the differential equations. For u, v arising on the basis of the *long* period x, y, the principal terms are

$$\begin{aligned} u_0 &= a_{-1}\cos(M-\sigma)+b_{-1}\sin(M-\sigma) \\ &\qquad + a_1\cos(M+\sigma)+b_1\sin(M+\sigma) \\ v_0 &= c_{-1}\cos(M-\sigma)+d_{-1}\sin(M-\sigma) \\ &\qquad + c_1\cos(M+\sigma)+d_1\sin(M+\sigma), \end{aligned} \tag{11}$$

and the linear equations for a first-order determination of the coefficients $a_{-1}, \ldots, d_1$ have been given in the earlier discourse.

While librations representable as oscillations about a *short* period basic solution x, y have not been considered in the earlier discussion of the elliptic restricted problem, the same method is applicable at once, because the only assumption was that x, y is a periodic solution of the ordinary restricted problem. However, certain terms multiplied by n^2 (arising from the involvement of $\ddot{x}$ and $\ddot{y}$ in the differential equations), which were of negligible size in the case of the n_l^2 of order μ, have to be retained for the n_s^2 of order N^2 or unity. On the other hand, the $N-n_s$ of order μ becomes a small divisor in the determination of the coefficients $a_{-1}, b_{-1}, c_{-1}, d_{-1}$ of Equations (11), with the result that the terms depending on $(M-\sigma)$, of very long period, dominate the much smaller terms with $(M+\sigma)$, and all other terms of u_0, v_0. Neglecting the terms with $(M+\sigma)$, the remaining coefficients $a_{-1}, \ldots, d_{-1}$ may be determined from the four Equations (43) of the earlier paper (Rabe, 1966), in which the subscript 2 has to be replaced by -1, but with the augmented right-hand sides:

$$\begin{aligned} &en_sN(-x_{c,1}+y_{s,1})+\tfrac{3}{2}en_s^2x_{c,1} \\ &en_sN(\ \ x_{s,1}+y_{c,1})-\tfrac{3}{2}en_s^2x_{s,1} \\ -&en_sN(\ \ x_{s,1}+y_{c,1})+\tfrac{3}{2}en_s^2y_{c,1} \\ &en_sN(-x_{c,1}+y_{s,1})-\tfrac{3}{2}en_s^2y_{s,1}\,. \end{aligned} \tag{12}$$

For infinitesimally small x, y of both families of restricted problem librations, the $x_{c,1}, x_{s,1}, y_{c,1}, y_{s,1}$ of Equations (5) and (6) may now be substituted into the right-hand sides of the relevant linear equations for the coefficients a_j, b_j, c_j, d_j. For the principal terms of u_0, v_0 according to Equations (11), the resulting coefficients are as follows.

(1) For $n=n_l$:

$$\begin{aligned} a_{\mp1} &= \left(\pm\frac{\sqrt{3}}{12}\mu^{-\frac{1}{2}}+\frac{11}{8}+\cdots\right)e\lambda \\ b_{\mp1} &= \left(\mp\frac{1}{2}\mu^{-\frac{1}{2}}-\frac{11}{4}\sqrt{3}+\cdots\right)e\lambda \\ c_{\mp1} &= \left(\mp\frac{1}{4}\mu^{-\frac{1}{2}}-\frac{11}{8}\sqrt{3}+\cdots\right)e\lambda \\ d_{\mp1} &= \left(\mp\frac{\sqrt{3}}{6}\mu^{-\frac{1}{2}}-\frac{11}{4}+\cdots\right)e\lambda, \end{aligned} \tag{13}$$

(2) For $n=n_s$:

$$a_{-1} = +\tfrac{1625}{1992}e\lambda + \cdots, \qquad b_{-1} = -\tfrac{2777}{2158}\sqrt{3}e\lambda + \cdots$$
$$c_{-1} = -\tfrac{367}{5976}\sqrt{3}e\lambda + \cdots, \quad d_{-1} = -\tfrac{2777}{2158}e\lambda + \cdots. \tag{14}$$

The general order of the coefficients is e-times the amplitude of the corresponding basic x, y, so that the usefulness of the method is limited to relatively small values of e, such as in the actual Trojan case. In the solution for $n=n_s$, the small divisor $N-n_s$ has not produced any drastic magnification of coefficients, because the relevant nominators are very small, too, in consequence of larger quantities nearly offsetting each other in the balance of terms. Since each set of a_j, b_j, c_j, d_j represents an elliptic oscillation of the period of the related argument, the axes ratios q_j are of interest:

(1) For $n=n_l$:

$$q_{\mp 1} = 2(1 + \tfrac{45}{4}\mu \mp \cdots) \sim 2. \tag{15}$$

(2) For $n=n_s$:

$$q_{-1} = (2.658\, e\lambda): \quad (0.484\, e\lambda) = 5.49. \tag{16}$$

The semimajor axis of each of the two oscillations superposed on the long period x, y, is $e\lambda/\sqrt{3\mu}$, or e-times the corresponding amplitude of x, y itself.

4. Periodic Librations in the Elliptic Problem

In general, the motions represented by

$$\xi = r(x + u_0), \quad \eta = r(y + v_0) \tag{17}$$

will be nonperiodic, because the frequency N of M will be noncommensurable with the frequency n_l or n_s of the relevant $\sigma = n(t-t_0)$. However, for

$$N = kn_l \quad \text{or} \quad N = K(N - n_s), \tag{18}$$

where k and K are integers, an integral number of periods of M will coincide with one completed period of σ (for $n=n_l$) *or* of the very long period terms involving $(M-\sigma)$, and thus all the expansion terms of ξ and η (including those of any higher approximations considering the second and higher powers of u_0, v_0) with arguments $(jM+l\sigma)$ will become periodic. The period will be that of σ_l in the first class of periodic solutions, but equal to the much longer period of $(M-\sigma_s)$ in the second class. The first class is the rather direct equivalent, for appropriate values of n_l or μ, of the corresponding solutions of the ordinary restricted problem, while the second class is a consequence of the almost identical periods of M and σ_s, leading to a 'beat period' of order $1/\mu$. It should be noted that the second Equation (18) is equivalent with

$$n_s = (K-1)(N - n_s), \tag{19}$$

so that $(K-1)$ periods of the basic x, y are completed during K periods of M. Consequently, the period of $(M-\sigma_s)$ is identical with the period of the $x, y-$ 'perihel', which has a forward motion of the rate $(N-n_s)$.

On the basis of Equation (3), it is easily verified that the periodicity requirement $N=kn_l$ imposes on μ the condition

$$\frac{27}{4}\frac{\mu_k}{(1+\mu_k)^2}=\frac{1}{k^2}-\frac{1}{k^4}, \tag{20}$$

while $N=K(N-n_s)$ demands that

$$\frac{27}{4}\frac{\mu_K}{(1+\mu_K)^2}=\frac{(K-1)^2}{K^2}-\frac{(K-1)^4}{K^4}. \tag{21}$$

Table I indicates the infinite set of μ-values which satisfies

TABLE I

μ-values required for periodic solutions

k	μ_k	K	μ_K
1	-------	2	-------
2	0.029437*	3	-------
3	0.015077	4	0.039387
4	0.008835	5	0.036684
.	....	.	....
.	....	.	....
12	0.001024	308	0.000956
13	0.000873	309	0.000953
14	0.000753	310	0.000950
..	0.......	...	0.......
..	0.......	...	0.......

* $n_l=N-n_l=N/2$ for $\mu_k=0.029\,437$

each condition. It is seen that the mass of Jupiter lies between those values required for $k=12$ and $k=13$, and also between those necessary for $K=308$ and $K=309$. No admissible values of μ exist for $k=1$ and for the K-values 1, 2, and 3. For each admissible value of k and K, however, a twofold infinite number of periodic solutions exists, because e and M_0 are arbitrary.

Of particular interest is the case $k=2$, because it is the only one in which the normally nonvanishing determinant

$$D_{-1}=[(N-n_l)^4-N^2(N-n_l)^2+\tfrac{27}{4}\mu]^2, \tag{22}$$

of the four equations for $a_{-1}, b_{-1}, c_{-1}, d_{-1}$, becomes equal to zero, so that the solution for u_0, v_0 blows up. Since the Equation (3) for $n=n_l$ itself can be written in the form

$$n_l^4-N^2n_l^2+\tfrac{27}{4}\mu=0, \tag{23}$$

it is evident that for $n_l=N-n_l$, or for $\mu=0.029\,437$, the D_{-1} given by Equation (22) vanishes, because the expression in the square bracket becomes identical with the left-hand side of Equation (23). The resulting instability of this particular elliptic problem solution of very small amplitude, for any $e \neq 0$, is in agreement with the findings of Danby (1964) for the stability of the triangular points themselves. Allowing for the different units used, the present critical mass value 0.029 437 is identical with the one found by Danby in a quite different way. Additional light will be shed on this in the subsequent discussion of the free oscillation u_f, v_f, where it will be seen that for this value of μ two of the altogether *four* characteristic exponents become equal, so that a double root of the determinant equation $\Delta = 0$ exists for $\mu = 0.029\,437$ and $e = 0$.

5. Free Oscillations and Stability

The differential equations governing the free oscillations u_f, v_f superposed on the forced $(x+u_0)$, $(y+v_0)$ have been given as Equations (51) in the basic investigation (Rabe, 1966). For the infinitesimally small x, y considered here, all higher-order terms can be neglected, and the partials Ω_{xx}, Ω_{yy}, Ω_{xy} of the potential function Ω can be approximated by their constant terms α_0, β_0, γ_0, respectively, evaluated at L_5. Omitting the subscript 'f', the resulting differential equations for the free part of u, v can be written in the form

$$
\begin{aligned}
P(M)\ddot{u} - Q(M)\dot{v} - T(M)\dot{u} &= \alpha_0 u + \gamma_0 v \\
P(M)\ddot{v} + Q(M)\dot{u} - T(M)\dot{v} &= \gamma_0 u + \beta_0 v\,,
\end{aligned}
\tag{24}
$$

where the $P(M)$, $Q(M)$, $T(M)$ contain the terms due to the ellipticity of 'Jupiter's' orbit. For periodic $(x+u_0)$, $(y+v_0)$, these functions take the form

$$
\begin{aligned}
P(M) &= 1 + e \sum_{-\infty}^{\infty} p_r \exp(ir\tau) \\
Q(M) &= 2N + e \sum_{-\infty}^{\infty} q_r \exp(ir\tau) \\
T(M) &= e \sum_{-\infty}^{\infty} t_r \exp(ir\tau),
\end{aligned}
\tag{25}
$$

where the independent variable τ has been introduced as representing either σ_l or $M - \sigma_s$. The values of the p_r, q_r, t_r depend on the adopted value of M_0. For $M_0 = 0$, the p_r and q_r are real, and the t_r imaginary. For $M_0 \neq 0$, these coefficients of Equations (25) become complex. A detailed analysis reveals that the roots of the determinant equation $\Delta = 0$ for the characteristic exponents remain unaffected by the choice of M_0, thanks to the conjugate symmetry of the complex coefficients outside of the principal diagonal of Δ with respect to that diagonal. M_0 affects the coefficients u_r, v_r (except for the arbitrary one, say u_0) of the solution

$$
u = \sum_{-\infty}^{\infty} u_r \exp[i(r+c)\tau]\,, \quad v = \sum_{-\infty}^{\infty} v_r \exp[i(r+c)\tau]\,,
\tag{26}
$$

but not its characteristic exponent ic. Therefore, the stability investigation may be simplified by assuming $M_0 = 0$.

With the introduction of τ, the further discussion covers both classes of periodic elliptic problem solutions, and the respective commensurability ratios k and K may both be represented from now on by the symbol k. In either case, each periodic solution $(x+u_0)$, $(y+v_0)$ under consideration will be associated with one specific integer k. It is easily seen that in Equations (25) only those p_r, q_r, t_r with $r=0, \pm k, \pm 2k, \ldots, \pm jk, \cdots$ are different from zero, because they originate in the terms involving M, $2M$, $3M$, etc. in the expansions for the motion of 'Jupiter'. For $k=12$, for instance, the only p_r etc. involved would be those with $r=0, \pm 12, \pm 24$, etc. Furthermore, the p_{jk}, q_{jk}, t_{jk} are at least of the order e^{j-1}, in consequence of the progressive powers of e involved in Equation (9) and similar elliptic expansions. If the solution (26) is substituted into the linearized and abbreviated Equations (24), a system of homogeneous linear equations for the coefficients u_r, v_r results from the identities which have to be satisfied for each of the exponents involved, or for each actually occurring integer r.

If the right-hand sides of Equations (24) would consider the coefficients α_{-1}, α_1, β_{-1}, β_1, γ_{-1}, γ_1 of the principal periodic terms of Ω_{xx} etc., in addition to the constant parts α_0 etc., then the solution would contain nonvanishing terms with

$$\pm k = 0, 1, \quad k-1, k, k+1, \quad 2k-1, 2k, 2k+1.$$

Terms with $\pm k=2, k-2, k+2, 2k-2, 2k+2$, etc., would have to be included if the α_{-2}, α_2, etc., would also need consideration, as perhaps for basic librations x, y of substantial amplitude. Even then, however, numerous terms with r-values between these 'clusters' around integral multiples of k would remain negligible, if k is of the order 12 or larger. For the present study of infinitesimally small librations about L_5, the abbreviated Equations (24) should be fully adequate. While they neglect the minute effects of infinitesimally small α_{-1}, α_1, etc., they retain the basic powers of e, which may well be of the order 0.1.

The substitution of Equations (26) into Equations (24) produces two homogeneous linear equations for each value of r, so that the consideration of the r-values $-k$, 0, $+k$ leads to six equations for the six unknowns u_{-k}, v_{-k}, u_0, v_0, u_k, v_k. The coefficients are functions of c, or, if z denotes either cn_l or $c(N-n_s)$, of z, which has to be determined from the condition

$$\Delta = 0 \tag{27}$$

for the determinant Δ of these coefficients. The coefficients of the six principal u_r, v_r form the matrix indicated below:

$$\begin{array}{llllll}
u_{-k} & v_{-k} & u_0 & v_0 & u_k & v_k \\
A(z-N) & C(z-N) & a_{-k}(z) & c_{-k}(z) & a_{-2k}(z+N) & c_{-2k}(z+N) \\
\bar{C}(z-N) & B(z-N) & c^*_{-k}(z) & b_{-k}(z) & c^*_{-2k}(z+N) & b_{-2k}(z+N) \\
a_k(z-N) & c_k(z-N) & A(z) & C(z) & a_{-k}(z+N) & c_{-k}(z+N) \\
c^*_k(z-N) & b_k(z-N) & \bar{C}(z) & B(z) & c^*_{-k}(z+N) & b_{-k}(z+N) \\
a_{2k}(z-N) & c_{2k}(z-N) & a_k(z) & c_k(z) & A(z+N) & C(z+N) \\
c^*_{2k}(z-N) & b_{2k}(z-N) & c^*_k(z) & b_k(z) & \bar{C}(z+N) & B(z+N).
\end{array} \tag{28}$$

As indicated, the coefficients of u_0 and v_0 are functions of z, while those of u_{-k}, v_{-k} are functions of $z-kn_l$ or of $z-K(N-n_s)$ and thus, according to Equations (18), of $z-N$. Similarly, the coefficients of u_k, v_k are functions of $z+N$. The expressions involved are the following ones, given here as functions of z.

$$\begin{aligned}
A(z) &= \alpha_0 + (1+3e^2)z^2 & B(z) &= \beta_0 + (1+3e^2)z^2 \\
C(z) &= \gamma_0 + 2iNz & \bar{C}(z) &= \gamma_0 - 2iNz \\
a_k(z) &= ep_k z^2 + iet_k z & b_k(z) &= ep_k z^2 + iet_k z \\
c_k(z) &= ieq_k z & c_k^*(z) &= -ieq_k z \\
a_{2k}(z) &= e^2 f_{2k}(z^2, z) & b_{2k}(z) &= e^2 f_{2k}(z^2, z) \\
c_{2k}(z) &= e^2 i g_{2k}(z) & c_{2k}^*(z) &= -e^2 i g_{2k}(z).
\end{aligned} \tag{29}$$

Since the coefficients factored by e^2 will be neglected, their actual expressions in terms of z are not needed. The corresponding functions of $z-N$ and $z+N$ will be obtained by simply replacing z by $z-N$ or $z+N$, respectively. It will be noted that the above expressions for $a_{jk}(z)$ are identical with those for $b_{jk}(z)$. This would be no longer true if the higher-order α_j, β_j, γ_j would be considered in Equations (24), in which case the $a_k(z)$, $b_k(z)$, for instance, would be given by

$$a_k(z) = ep_k z^2 + iet_k z + \alpha_k^*, \quad b_k(z) = ep_k z^2 + iet_k z + \beta_k^*,$$

where in α_k^* and β_k^*, in addition to α_k and β_k, even the relevant contributions of the functions $(\Omega_{xxx}u_0 + \Omega_{xxy}v_0)$ etc. may have to be incorporated, in a high-accuracy study of more substantial librations. Similarly, $c_k(z)$ and $c_k^*(z)$ would no longer be conjugates, if the (complex) increment γ_k^* is added to *both* of them. The matrix (28) would retain conjugate symmetry with respect to its principal diagonal (except for the replacement of z by $z-N, z+N, z-2N, z+2N, \ldots$, in the relevant columns), however, even when extended to an infinite number of coefficients, because the relevant pairs would involve α_{-k}^* and α_k^*, β_{-k}^* and β_k^*, etc. For the present purpose, the consideration of the approximated coefficients represented by Equations (29) will be sufficient, neglecting those with the subscripts $2k$ and $-2k$, as of order e^2. Their retainment would produce in Δ terms multiplied by at least e^4. These would have to be considered in the determination of the solution (26) up to $r = \pm 2k$, or to the order of e^2, but for this purpose the matrix (28) would have to be extended to that of a 10×10 system of equations.

For convenience, put

$$Z = z/N \quad \text{and} \quad \nu = \mu/N^4. \tag{30}$$

Neglecting terms of order e^4 and higher, the determinant Δ of the linear Equations (28) for the six principal u_j, v_j can be written

$$\begin{aligned}
\Delta/N^{12} = {} & (1+9e^2)Z^{12} - (7+57e^2)Z^{10} + (15 + 99e^2 + \tfrac{81}{4}\nu \\
& + 243e^2\nu)Z^8 - (13 + 63e^2 - \tfrac{27}{2}\nu - \tfrac{513}{4}e^2\nu)Z^6 \\
& + (4 + 12e^2 - \tfrac{27}{4}\nu + \tfrac{27}{4}e^2\nu + \tfrac{2187}{16}\nu^2 + \tfrac{19683}{16}e^2\nu^2)Z^4 \\
& - 27\nu(1 + 14e^2 - \tfrac{243}{16}\nu - \tfrac{1431}{8}e^2\nu)Z^2 + \tfrac{729}{8}\nu^2(15e^2 + \tfrac{27}{8}\nu).
\end{aligned} \tag{31}$$

This expression has the property that not only the numerical coefficients of the various Z^{2j} add up to zero, but that the same is true also for those coefficients multiplied by e^2, for those factored by ν, and even those multiplied by $e^2\nu$. Consequently, Δ is very small not only for $Z=0$, but also for $Z=\pm 1$. While the solution of the equation $\Delta=0$ with the Δ of Equation (31) presents some algebraic difficulties, it is relatively easy to obtain sufficient approximations to the roots near zero, in the form

$$\begin{aligned} Z_1^2 &= \frac{27}{4}\nu\left(1+\frac{27}{4}\nu+\frac{7}{2}e^2\right) \\ Z_2^2 &= \frac{27}{8}\nu\left(\frac{27}{8}\nu+15\,e^2\right)=\frac{729}{64}\nu^2\left(1+\frac{40}{9}\frac{e^2}{\nu}\right). \end{aligned} \tag{32}$$

Here terms of the order $e^2\nu^2$ have been neglected, in line with the neglection of e^4. For large e, the order of $e^2\nu^2$ may approach that of ν^2, but in such cases the omitted $e^4\nu$-terms should also be established from an extended 10×10 system of linear equations, in order to find the exponents represented by Equations (32) to higher orders of e and ν. To the present degree of approximation, and in terms of the original z and μ, the two roots z_l and z_l^* corresponding to the above Z_1 and Z_2 are given by

$$\begin{aligned} |z_l| &= \frac{3}{2}\sqrt{3\mu}\left(1+\frac{23}{8}\mu+\frac{7}{4}e^2\right)=n_l+\frac{21}{8}e^2\sqrt{3\mu} \\ |z_l^*| &= \frac{27}{8}\mu\sqrt{1+\frac{40}{9}\frac{e^2}{\mu}}=(N-n_s)+\frac{27}{8}\mu\left(\sqrt{1+\frac{40}{9}\frac{e^2}{\mu}}-1\right). \end{aligned} \tag{33}$$

From the structure of the matrix (28), even when extended to an infinite number of members and higher powers of e, it is evident that any solution z of $\Delta=0$ is accompanied by identical solutions for $(z\pm N)$, $(z\pm 2N)$, etc. Consequently, Equation (27) has *between* 0 *and* $+1$ the additional roots

$$\begin{aligned} z_s &= N-|z_l^*|=n_s-\frac{27}{8}\mu\left(\sqrt{1+\frac{40}{9}\frac{e^2}{\mu}}-1\right) \\ z_s^* &= N-|z_l| = (N-n_l)-\frac{21}{8}e^2\sqrt{3\mu}\,. \end{aligned} \tag{34}$$

Thus in the elliptic restricted problem the periodic librations under consideration have *four* essentially different characteristic exponents, in contrast to only *two* for periodic solutions of the ordinary restricted problem. For small e, two of these roots correspond to free oscillations of long, respectively of very long period, while the other two roots are associated with short-period oscillations about the periodic orbit. Since the amplitude parameter λ of this orbit may be as small as desired, the results are valid, in the limit, also for free oscillations about the equilateral points themselves, and for the stability of these points in the elliptic restricted problem. No instability is indicated, with one exception to be discussed below, for small values of e. It may be noted, though, that according to Equations (33) the effect of e on the roots becomes substantial when e approaches the order of $\sqrt{\mu}$. However, the precise effects of more substantial values of e can be established only when higher orders of e are retained in the differential equations and their solution, and in the resulting expression for Δ.

For $e \to 0$, the four frequencies $z = cn_l$ *or* $z = c(N - n_s)$ are seen to approach, according to Equations (33) and (34), the limiting values

$$N - n_s, \quad n_l, \quad N - n_l, \quad n_s, \tag{35}$$

in order of their increasing values between 0 and +1. The related limiting c-values, of the relevant characteristic exponents ic, may be obtained from the values (35) by division, by either n_l *or* $N - n_s$. For $e = 0$ *and* $\mu = 0.029\,437$, the two frequencies n_l and $N - n_l$ in the sequence (35) for the free oscillations become identical and equal to $N/2$. Such a double root indicates instability. It has already been noted, in connection with Table I, that even the basic or forced oscillation u_0, v_0 becomes unstable, for any $e \neq 0$, in the case of this singular μ-value, which is the largest one compatible with the periodicity requirement of Equation (20). For slightly different values of μ, the numerical identity of the two roots z_l and z_s^* may be formally obtainable by means of some appropriate small value of e in the relevant Equations (33) and (34), but it has to be remembered that only those discrete μ-values indicated in Table I are associated with the existence of periodic solutions of the elliptic problem. Therefore, no other double root of $\Delta = 0$ exists for admissible μ, at least for small values of e.

With $e \to 0$, all those coefficients multiplied by e vanish in the basic matrix (28), and the determinant Δ of Equation (31) reduces to the product of its three principal 2×2 subdeterminants. It is easily verified that the four limiting roots (35) are indeed obtainable, for instance, from the two separate conditions

$$\begin{aligned} \Delta_0 &= z^4 - N^2 z^2 + \tfrac{27}{4}\mu = 0 \\ \Delta_{-k} &= (z - N)^4 - N^2 (z - N)^2 + \tfrac{27}{4}\mu = 0 \,. \end{aligned} \tag{36}$$

In the subsequent solution of Equations (28), for five of the various u_j, v_j in terms of the arbitrary u_0 (not to be confused with the part u_0 of the forced solution u_0, v_0), the assumption $e = 0$ leads to the results $u_{-k} = v_{-k} = u_k = v_k = 0$, so that in Equations (28) actually the two pairs of equations for $r = -k$ and $r = k$ have become meaningless, and only the *two* roots from the first Equation (36) enter the remaining solution, which of course is that of the ordinary restricted problem.

Note added in proof. An extension of the stability analysis, to the consideration of a 10×10 determinant Δ and of terms multiplied by e^4, confirmed to the order $e^2 v$ the preceding result for Z_1^2, as given by the first Equation (32). For Z_2^2, however, the term of order $e^2 v$ vanishes when the solution is based on the extended Δ from the 10×10 matrix, and the preceding second Equation (32) has to be replaced by

$$Z_2^2 = \tfrac{729}{64} v^2 \left[1 + \tfrac{135}{8} v - \tfrac{1}{3} e^2\right], \tag{32b}$$

valid to the order of $e^2 v^2$. Consequently, the second Equation (33) and the first Equation (34) have to be replaced, respectively, by the following two expressions:

$$|z_l^*| = (N - n_s) - \tfrac{9}{16}e^2\mu, \tag{33b}$$

$$z_s = n_s + \tfrac{9}{16}e^2\mu. \tag{34a}$$

It is seen, thus, while as Δ based on a 6×6 matrix is sufficient to find Z_1^2 to the order $e^2\nu$, a higher approximation based on the 10×10 matrix is necessary to determine the principal e^2-term of Z_2^2, which is of the order $e^2\nu^2$.

References

Danby, J. M. A.: 1964, 'Stability of the Triangular Points in the Elliptic Restricted Problem of Three Bodies', *Astron. J.* **69**, 165.

Rabe, E.: 1966, 'Elements of a Theory of Librational Motions in the Elliptical Restricted Problem', *Lectures in Applied Mathematics* (American Mathematical Society), Vol. 6, p. 102.

Rabe, E.: 1969, Communication at the Seventh Annual Seminar in Celestial Mechanics held at the University of Texas, Austin, January 1969.

MINOR PLANETS ON COMMENSURABLE ORBITS WITH APPROACHES TO JUPITER

J. SCHUBART

Astronomisches Rechen-Institut, Heidelberg, Germany

1. It is well known, that the Trojan planets avoid close approaches to Jupiter by a mechanism of libration, although parts of their orbits are very close to the orbit of Jupiter. The difference of the longitudes of a Trojan and Jupiter librates in such a way, that it does not come close to 0°. Belyaev and Tchebotarev [1] have recently investigated the orbital motion of 14 Trojans and of 40 other interesting minor planets by numerical integrations extended over an interval of 400 years. They find the minimum distance of a Trojan to Jupiter as 2.6 AU, while in the contrary (944) Hidalgo can approach Jupiter to a distance of 0.4 (AU). Among 20 planets of the Hilda group and (279) Thule, the closest encounter with Jupiter occurs in case of (334) Chicago (1.1 AU). The planet Chicago has a small orbital eccentricity, while other planets in the Hilda group have much larger eccentricities. Since these planets have also a large mean distance from the sun, some of them come close to the orbit of Jupiter, if they pass the aphelion of their orbit. In spite of this, there are no very close encounters with Jupiter in such a case. The planets in the Hilda group are characterized by a mean motion which is nearly commensurable to that of Jupiter according to the commensurability ratio $\frac{3}{2}$. The near commensurability allows a mechanism of libration, which acts in the more eccentric cases among the orbits. As a consequence of this mechanism, a conjunction of the minor planet with Jupiter takes only place, if the minor planet is not too far from the perihelion of its orbit. In this way the minor planet avoids very close encounters with Jupiter.

My former paper on the motion of Hilda-type planets [2] contains detailed results on this mechanism of libration. 21 out of 23 members of the Hilda group show libration. Among the 21 librating members (1269) Rollandia has the closest encounters with Jupiter at a distance of 1.4 AU, while the corresponding minimum distance is as large as 2.1 AU in case of (1212) Francette. Although this result is based on a method of approximation, I expect it to be valid for at least very long intervals of time. I believe that the 21 librating planets are permanent members of the Hilda group. I was not so sure about this in case of the two remaining planets, (334) Chicago and (1256) Normannia, when I published my former paper. These two planets do not show libration. Their comparatively small orbital eccentricity and some effect of the near commensurability prevent too close an encounter with Jupiter, but especially Chicago can come as close as 1.1 AU to Jupiter, as mentioned above. Doubt arises about the validity of the approximate method used in the former paper [2], if the minimum distance to Jupiter is so small.

Another planet with a near commensurability to the mean motion of Jupiter and

G. E. O. Giacaglia (ed.), Periodic Orbits, Stability and Resonances, 45–52.

with moderately close approaches to Jupiter, is (279) Thule. The mean motion of Thule closely corresponds to the $\frac{4}{3}$ commensurability ratio. The minimum distance to Jupiter is 1.13 AU, according to an extended numerical integration by Marsden [3]. Takenouchi [4] found Thule to show libration in a way, which is analogous to the libration shown by most of the Hilda-type planets. The minimum distance is small, although the process of libration tends to keep it large. The reason for this is mainly given by the mean heliocentric distance of Thule, which is larger than that of a Hilda-type planet.

In the present paper I shall describe investigations about the long-period effects in the motion of planets like Chicago and Thule, which correspond to commensurability cases with respect to Jupiter, but where the application of the method of the former paper [2] seems to be doubtful according to the close approaches to Jupiter. At first I shall give a theoretical study on the effects caused by a sequence of such approaches. Then I shall compare the theory with the result of extended numerical integrations in case of Chicago and Thule.

2. I am mainly concerned here in planets like Chicago, Normannia or Thule, but the following considerations may also be applied as an approximation to planets like Rollandia, where the minimum distance to Jupiter is not quite as small. All these planets have closely a rational ratio of the mean motion to that of Jupiter. This ratio is given by $(p+1)/p$, where $p=2$ in case of the Hilda group, while $p=3$ refers to Thule. Since the inclination of the orbits of these planets to the orbit of Jupiter is small, the following theory is based on the planar three-body problem with Jupiter moving on a fixed ellipse around the sun. Since further the orbits of the planets under consideration are not very eccentric, there will be a more or less close approach to Jupiter at every heliocentric conjunction. The largest perturbations in such an orbit will occur during the encounters with Jupiter. To get an insight into the action of these strong perturbations, they will be studied alone, neglecting the smaller perturbations, which act on the planet during the remaining part of its synodic revolution with respect to Jupiter. Therefore, the following approximate theory will give the isolated influence of a sequence of close encounters with Jupiter on the orbit of a planet.

The same designations, constants and units will be used as in [2]. l and l_J are the mean longitudes of a small planet and Jupiter. $\mu=l-l_J\,(p+1)/p$ is a combination of these longitudes which changes slowly according to the near commensurability. Let l_C be the common mean longitude of the asteroid and Jupiter at the moment of a conjunction $l=l_J$. Then it follows $\mu=-l_C/p$; $l_C=-p\mu$. The perturbing force of Jupiter during an encounter may be represented by the attraction in the moment of the conjunction, if the minimum distance is not too small. This attraction depends on the corresponding value of l_C and on the osculating values of a, e and $\tilde{\omega}$. Since μ changes slowly, l_C will not change too much from one conjunction to the next. According to this, it is permissible in an approximate theory to study the action of the perturbing forces by a continuous variation of l_C, although in nature there is a sequence of encounters and, therefore, a discontinuous sequence of values l_C. If the forces are taken to act continuously, they must be multiplied by an appropriate small factor, since in nature they act only during short intervals. Let this factor be κ, a constant.

In an encounter with Jupiter, the minimum distance depends mainly on the difference of the heliocentric distances of the small planet and Jupiter. Although the orbital eccentricities, e and e_J, are small, they are important in determining the minimum distance, while in all other instances they may be neglected in this approximate theory. According to this, the perturbing forces in the moment of conjunction are computed under the assumption, that the true longitudes coincide. Then the differential equations of the problem take a simple form. Let G, μ, ψ_1, ψ_2, replace the osculating orbital elements according to the formulas

$$G^2 = a(1 - e^2), \quad \psi_1 = e \cos \tilde{\omega}, \quad \psi_2 = e \sin \tilde{\omega}.$$

The former paper [2] contains the differential equations for these quantities. I shall not repeat these equations here. They contain auxiliary variables S_J and T_J, for which I get here

$$S_j = m_J \kappa G(\Delta^{-2} - 1); \; T_j = 0,$$

where m_J is the mass of Jupiter and Δ equals the minimum distance mentioned before. T_J changes the sign during an encounter. Since $\dot{G}$ is proportional to T_J, G will attain an extremal value at the conjunction. The element a closely follows G^2 according to the smallness of e. Therefore, a will also have an extreme during the encounter, and the perturbations before and after the moment of this extreme will partly cancel each other. The sum of the changes in a during the encounter is expected to be small. Practical examples like Takenouchi's treatment of Thule [4] confirm this. The mean distance a is taken as a constant in this theory.

The remaining differential equations in [2] change to:

$$\dot{\psi}_1 = S_J \sin l_C, \quad \dot{\psi}_2 = - S_J \cos l_C, \quad \dot{\mu} = a^{-3/2} - n_J(p+1)/p - 2S_J,$$

where $n_J \approx 1$ is the mean motion of Jupiter. Therefore,

$$\dot{l}_C = n_J(p+1) - pa^{-3/2} + 2pS_J.$$

Δ equals the difference of the heliocentric distance of Jupiter minus that of the asteroid in the cases treated here.

Since $\tilde{\omega}_J = 0°$ and $a_J = 1$, it results $\Delta \approx 1 - e_J \cos l_C - a[1 - e \cos(l_C - \tilde{\omega})]$, or

$$\Delta = 1 - a + a(\psi_1 - \bar{\psi}) \cos l_C + a\psi_2 \sin l_C, \quad \text{if} \quad \bar{\psi} = e_J/a.$$

It is interesting to note, that e_J enters the equations only by $\bar{\psi}$. If $\psi_1 - \bar{\psi}$ is introduced as a new variable instead of ψ_1, e_J formally disappears from the equations.

If $(1-a)$ is not too small, Δ^{-2} entering in the formula for S_J may be developed according to powers of $\psi_1 - \bar{\psi}$ and ψ_2. Let

$$S_J = S_0 - S_1(\psi_1 - \bar{\psi}) \cos l_C - S_1 \psi_2 \sin l_C$$

be a first approximation with constants S_0, S_1, that are proportional to κm_J. Substitute

$$y_1 = (\psi_1 - \bar{\psi}) \cos l_C + \psi_2 \sin l_C$$

and

$$y_2 = -(\psi_1 - \bar{\psi}) \sin l_C + \psi_2 \cos l_C$$

instead of the eccentricities. Then it follows

$$\dot{y}_1 = y_2 \dot{l}_C \quad \text{and} \quad \dot{y}_2 = -S_J - y_1 \dot{l}_C, \quad \text{while} \quad S_J = S_0 - S_1 y_1 .$$

From practical examples μ and l_C are known to perform whole revolutions, even in the cases of libration. $\dot{l}_C$ will vary according to the changes of y_1, but in a treatment of the equations for $\dot{y}_1$ and $\dot{y}_2$, $\dot{l}_C$ can be approximated by a constant, since the variable part will only cause small quadratic terms. In this way the linear equations result:

$$\dot{y}_1 = y_2 \dot{l}_C ; \quad \dot{y}_2 = -S_0 + y_1 (S_1 - \dot{l}_C), \quad \text{or} \quad \ddot{y}_1 = -S_0 \dot{l}_C + y_1 (S_1 - \dot{l}_C) \dot{l}_C .$$

Evidently y_1 and y_2 are periodic functions of the time, if $(S_1 - \dot{l}_C)\, \dot{l}_C < 0$. Since $S_1 > 0$, this is true for $\dot{l}_C < 0$ or for $\dot{l}_C > S_1$. Then, the solution is $y_1 = A \cos \alpha t + \bar{y}$; $y_2 = -B \sin \alpha t$, where $\bar{y} = S_0 / (S_1 - \dot{l}_C)$, $\alpha^2 = -(S_1 - \dot{l}_C)\, \dot{l}_C$ and $\alpha A = \dot{l}_C B$, $A \geqq 0$. Inverting the substitution, it follows:

$$\begin{aligned} \psi_1 - \bar{\psi} &= \bar{y} \cos l_C + C_1 \cos(l_C + \alpha t) + C_2 \cos(l_C - \alpha t), \\ \psi_2 \quad &= \bar{y} \sin l_C + C_1 \sin(l_C + \alpha t) + C_2 \sin(l_C - \alpha t), \end{aligned}$$

where $C_1 + C_2 = A$ and $C_1 - C_2 = -B$. According to these formulas, the osculating elements of the minor planet change periodically in spite of the sequence of encounters with Jupiter.

3. Looking at the minor planets to which the theory will be applied, it turns out that the mean values of $\dot{l}_C$ are all negative. Therefore, the condition of periodicity for y_1 and y_2 is fulfilled. A qualitative comparison of the theory obtained here with the results of the former paper [2] is possible. Since A is optional, it can be equal zero. Then, $y_1 = \bar{y}$ and $y_2 = 0$ give a periodic solution:

$$\psi_1 = \bar{\psi} + \bar{y} \cos l_C, \quad \psi_2 = \bar{y} \sin l_C .$$

In the paper [2] I listed periodic solutions, which very closely obey to these formulas. $\bar{\psi}$ was determined empirically in [2], but it fits nicely to the theoretical formula found here. The more general solution with small $A \neq 0$ will correspond to the cases of libration. They are described in [2] by means of the librating angle $\bar{\sigma}$. If l_C is introduced instead of μ, the empirical formulas for ψ_1 and ψ_2 become:

$$\psi_1 - \bar{\psi} = e_p \cos(l_C - \bar{\sigma}), \quad \psi_2 = e_p \sin(l_C - \bar{\sigma}),$$

where e_p is a positive variable. This leads to $y_1 = e_p \cos \bar{\sigma}$, $y_2 = -e_p \sin \bar{\sigma}$. In the formulas found here $\dot{l}_C < 0$ means $\bar{y} > 0$. If A is small enough, y_1 cannot become negative.

This means libration of $\bar{\sigma}$ around 0°, since $\cos\bar{\sigma} > 0$. The formulas

$$e_p \cos\bar{\sigma} = \bar{y} + A \cos\alpha t\,, \quad e_p \sin\bar{\sigma} = B \sin\alpha t$$

describe qualitatively the libration and an oscillation of e_p around $\bar{y}$. α determines the length of the period of libration.

The two Hilda-type planets without libration have a mean motion, which is larger in the average than that of the librating planets in the Hilda group. Therefore, l_C decreases comparatively fast, and $|\dot{l}_C|$ does not differ very much from α, which may be positive. $\beta = (l_C + \alpha t)$ is then a slowly increasing argument. The formulas for ψ_1 and ψ_2 change to:

$$\begin{aligned} \psi_1 - \bar{\psi} &= C_1 \cos\beta + \bar{y} \cos l_C + C_2 \cos(2\,l_C - \beta),\\ \psi_2 &= C_1 \sin\beta + \bar{y} \sin l_C + C_2 \sin(2\,l_C - \beta). \end{aligned}$$

Since $C_2/C_1 = (A+B)\,/\,(A-B) = (\dot{l}_C + \alpha)\,/\,(\dot{l}_C - \alpha)$, C_2 is small in comparison with C_1. These formulas give an excellent description of the former results found for Chicago and Normannia in [2]. In both cases the changes of ψ_1 and ψ_2 result from a slowly increasing argument β and from a second argument decreasing much faster in accordance with the above formulas. Before this theory was compiled, I discovered empirically the small third terms at the end of the formulas, in analysing the results for Chicago mentioned in [2]. Therefore, the theory developed here predicts qualitatively the same effects as they were found in [2]. The method of the former paper is confirmed qualitatively in the cases, where its application seems to be doubtful.

I treated empirically Chicago and Normannia as planets without libration in [2]. This was due to the uninterrupted decrease of a quantity σ, which showed libration in the other cases, and to the special type of variation found for ψ_1 and ψ_2. The definition of $\bar{\sigma}$, which coincides with the one for σ only in case of $\bar{\psi} = 0$, was not applied to the two exceptional planets. Since here I describe the changes of ψ_1 and ψ_2 in all cases by only one set of formulas, I shall also use the same formal definition of $\bar{\sigma}$ for all the planets. The above formulas in comparison with the numerical results obtained in [2] show, that A is much larger than $\bar{y}$ in case of Chicago, while A is a little bit smaller than $\bar{y}$ for Normannia. According to this, Normannia shows libration in $\bar{\sigma}$. Chicago remains as the only non-librating planet, if the variations of $\bar{\sigma}$ are considered.

It is better to compare the qualitative theory obtained here with the motion of real planets, since the former results [2] will not give more than a qualitative approach to the real motion, if close encounters with Jupiter occur. Takenouchi [4] integrated the motion of Thule on the basis of the planar three-body problem with Jupiter as the perturbing body. His computations cover an interval of about 400 yr. I extended these computations on the same theoretical basis, so that the interval became as long as 600 yr. Further, I integrated the motion of Chicago over an interval of almost 2500 yr, again using the planar three-body problem as the basis. I took the starting values for Chicago from a list of osculating elements.

Among the results of these two computations, the sequence of minimum distances to Jupiter is interesting. Thule approaches Jupiter to a distance of less than 1.25 AU near every conjunction. The smallest distance of 1.13 AU was mentioned before. The sequence of minimum distances from Chicago to Jupiter shows larger changes from 1.1 to 1.7 AU about. The variations of a are generally small in both cases, as it was predicted for Chicago in [2]. The curves corresponding to these variations show steep and narrow peaks near the times of conjunction. This agrees to the remarks made above.

The functional relation between ψ_1 and ψ_2 found from the computations allows a comparison with the theoretical formulas, since the synodic period causes only small effects. A graphical representation of the relation is possible in rectangular coordinates ψ_1, ψ_2. In this way Thule gives a curve, which does not differ very much from a circle. Since a circle corresponds to a periodic solution with $A=0$, Thule evidently represents the type of orbit, which deviates from a periodic solution according to a small value of $A>0$. The curve of Chicago agrees qualitatively to the former results [2], which were discussed before and correspond to a comparatively large value of A.

Since $\Delta=(1-a)+ay_1$ and $y_1=\bar{y}+A\cos\alpha t$, A determines the amplitude of the changes in Δ. This explains the larger changes of the minimum distance in case of Chicago. The positive values of $\bar{y}$ tend to keep Δ large. Therefore, an effect of the near commensurability prevents too small values of Δ even in the case of Chicago. A reverse effect would act on planets with a sufficiently large positive value of $\dot{l}_C$.

Δ^{-2} was developed above according to powers of y_1, retaining only the first power. It is interesting to note, that one can also develop Δ^{-2} with respect to powers of (y_1-c), where c is a positive constant. This leads to the same formal approximation for S_J, but since one can put $c=\bar{y}$ in a practical case, the amount of the neglected terms depends only on A, not on $\bar{y}$. y_1 causes changes of $\dot{l}_C$, which were neglected here. These changes depend also only on the amount of A.

Numerical values for some of the constants follow from a more detailed comparison between the curves $\psi_2=\psi_2(\psi_1)$ and the formulas. I list some approximate values here. If I take $\bar{\psi}=0.057$ in case of Thule, e_p oscillates between 0.067 and 0.078. This gives the values $\bar{y}=0.073$ and $A=0.006$. According to [4], the long period corresponding to the revolution of l_C equals 460 yr, while the period of libration is close to 170 yr. The curve of Chicago is best described with $\bar{\psi}=0.055$. The extremes of e_p are 0.035 and 0.081 about and lead to $\bar{y}=0.023$ and $A=0.058$, since $\bar{\sigma}$ performs revolutions. The period following from the oscillations of e_p equals 123 yr. It agrees with the value found for an analogous period in [2]. The slowly moving argument β increases according to a period of 1540 yr. In [2] this period was still larger, so that the picture of the former curve differs from the one obtained here. Both the theory and the numerical integrations indicate, that the long-period effects in the motion of Chicago and Thule depend on two very long periods. There are no indications for an instability of orbits of this type.

Acknowledgements

I am grateful to the Deutsche Forschungsgemeinschaft and to the Deutsches Rechenzentrum in Darmstadt for financial and other aid with regard to the numerical computations.

References

[1] Belyaev, N. A. and Tchebotarev, G. A.: 1968, 'The Orbit Evolution of 54 Minor Planets for 400 Years (1660–2060)', Astron. Circ. Moscow No. 480, Aug. 22, 1968.
[2] Schubart, J.: 1968, 'Long-period Effects in the Motion of Hilda-type Planets', *Astron. J.* **73**, 99.
[3] Marsden, B. G.: 1969, Personal communication.
[4] Takenouchi, T.: 1962, 'On the Characteristic Motion and the Critical Argument of Asteroid (279) Thule', *Ann. Tokyo Astron. Obs.* 7, 191.

Discussion

J. V. Breakwell: Are there cases in which it may be worthwhile to include the perturbation by Jupiter throughout the synodic period? And if so, could this add certain constants into your differential equations?

J. Schubart: The mean distance a is taken as a constant in this theory, but it would be important to study the effects, which Jupiter causes in a during the whole synodic period. An inclusion of these effects could add new constants into the equations. Another important extension of the theory consists in replacing the continuous variation of l_C by a sequence of l_C-values.

E. Rabe: Could the assumption $a = const.$ in your equations suppress any important effects, perhaps of long period? (For the Trojans the periodic variability of a is one of the most important features.)

J. Schubart: I do not believe that this assumption suppresses important qualitative effects, according to my experience. The Trojan planets have large minimum distances. The Trojan-case is quite different from the cases studied here, and my theory is not applicable to the Trojan-case.

E. Rabe: Does the eccentricity e_p vary considerably?

J. Schubart: In a graphical representation of a curve $\psi_2 = \psi_2(\psi_1)$, e_p is the distance from a point of the curve to the point $(\psi_1 = \bar{\psi}, \psi_2 = 0)$. The changes of e_p are comparatively large in case of Chicago. The changes of e are much larger. The e-values of Chicago and Thule vary between very small amounts and a maximum of almost 0.14.

A. Deprit: Would your conclusions substantially change if, from the start, Jupiter's eccentricity is set equal to zero?

J. Schubart: No; the only change comes from $\bar{\psi} = 0$, if $e_J = 0$. My numerical integrations for Thule and Chicago are based on the value $e_J = 0.048$. In the comparison of these computations with the theory, I used empirical values of $\bar{\psi}$, which differ somewhat from the theoretical values.

A. Deprit: Since you restrict yourself to minor planets that stay close to the Sun within 80% of the distance Sun-Jupiter, why not simplify further the model by simply treating it as the problem of two bodies in a rotating coordinate system?

J. Schubart: I believe, that this is too strong a simplification, and that it becomes impossible then to describe the libration of $\bar{\sigma}$ by formulas. (This is indeed so: In case of $m_J = 0$, ψ_1, ψ_2, e_p and $(l_C - \bar{\sigma})$ are constants. Therefore, $\dot{l}_C$ is the constant angular velocity of $\bar{\sigma}$.)

J. Kevorkian: What is your definition of close approach? If the eccentricities of

Jupiter and the minor planet are small and one excludes the 1:1 commensurability case, the minor planet cannot have a close approach in the conventional sense where the gravitational attraction of Jupiter dominates over the Solar attraction.

J. Schubart: I treated only moderately close approaches, which are something intermediate between the very close approaches mentioned in your question and the usual case of minor-planet motion, where the disturbing force does not change too much in the amount during a synodic period.

J. Kevorkian: Can you use your definition of 'close approach' to derive approximate differential equations valid when the minor planet is close to Jupiter by means of a formal limit process?

J. Schubart: No; I cannot even derive an analytical lower limit to the minimum distances appearing in moderately close approaches. From numerical experiences I come to the assumption, that my theory is applicable to minor planets with a minimum distance to Jupiter of not less than 1 AU about.

B. G. Marsden: I was very interested to hear that you have discovered that (1256) Normannia librates (as far as $\bar{\sigma}$ is concerned) and that (334) Chicago does not. Have you considered the case of (1144) Oda? This is rather close to the $\frac{3}{2}$ resonance, but maybe it is also influenced by the $\frac{5}{3}$ resonance.

J. Schubart: I do not think that Oda is close enough to the $\frac{3}{2}$ resonance for an application of my theory. I cannot treat the $\frac{5}{3}$ resonance, since my theory is restricted to ratios of the form $(p+1)/p$.

B. Garfinkel: How is the librating argument $\bar{\sigma}$ related to the slowly changing quantity μ?

J. Schubart: If I introduce the definition of $\tilde{\omega}_p$, which was given in my former paper [2], I can write

$$\tilde{\omega}_p = l_C - \bar{\sigma} = -(\bar{\sigma} + p \cdot \mu).$$

B. Garfinkel: Is $\bar{\sigma}$ a linear combination, with integral coefficients, of the mean anomalies and the perihelion?

J. Schubart: $\bar{\sigma}$ is a linear combination of the mean longitudes and of $\tilde{\omega}_p$. The coefficients are integers in the cases treated here. The relation between $\tilde{\omega}_p$ and $\tilde{\omega}$ is given by

$$e \cos\tilde{\omega} = \bar{\psi} + e_p \cos\tilde{\omega}_p,$$
$$e \sin\tilde{\omega} = e_p \sin\tilde{\omega}_p.$$

($\bar{\psi}$ vanishes with e_J. Then one gets simply $\tilde{\omega}_p = \tilde{\omega}$.)

DISINTEGRATION AND ESCAPE

HARRY POLLARD

Purdue University, Lafayette, Ind., U.S.A.

For the purpose of this paper a gravitational system is a set of a finite number n of mass-particles governed by Newton's Law, and moving through all future time without singularity.

A particle is said to *escape* if its distance from the center of mass becomes infinite with the time. Because of a general confusion in the literature it is important to emphasize the distinction between a function which becomes infinite, and one which is merely unbounded. Compare, for example, the function t, which becomes infinite, with the unbounded function $t(1-\sin t)$, which does not.

I shall say that system *disintegrates* if all the mutual distances between pairs of particles become infinite. *It has apparently never been proved that a particle must escape from a system that disintegrates.* Nor is it true that if all the particles escape the system disintegrates; an example is the hyperbolic-elliptic case of the three-body problem [1] in which a bound pair escape in one direction and the third one in the opposite direction.

It is my purpose to present some criteria for the occurrence of these phenomena. Conditions which are both necessary and sufficient are unknown and we shall settle for considerably less.

Notation and Formulas

$$0 = \text{center of mass},$$
$$\mathbf{r}_k = \text{position of } k\text{th particle relative to } 0,$$
$$r_k = |\mathbf{r}_k|, \quad r_{jk} = |\mathbf{r}_j - \mathbf{r}_k|, \quad \mathbf{v}_k = \dot{\mathbf{r}}_k, \quad v_k = |\mathbf{v}_k|,$$
$$a_{jk} = r_{jk}(0), \quad b_{jk} = \dot{r}_{jk}(0), \quad m_{jk} = m_j m_k,$$
$$g_{jk}^{(t)} = a_{jk} + b_{jk}t,$$
$$v_{jk} = \frac{m_j + m_k}{m_{jk}}, \quad \sum_{jk}' = \sum_{\substack{jk \\ j \neq k}},$$
$$T = \frac{1}{2}\sum m_k v_k^2, \quad U = \frac{1}{2}\sum_{jk}' \frac{m_{jk}}{r_{jk}}, \quad I = \frac{1}{2}\sum m_k r_k^2,$$

h = total energy with 0 fixed at center of mass, gravitational constant $= 1$,

$$T = U + h \qquad \text{(conservation of energy)},$$
$$\ddot{I} = U + 2h \qquad \text{(Lagrange-Jacobi identity)},$$
$$m_k \ddot{\mathbf{r}}_k = \sum_{j \neq k} \frac{m_{jk}}{r_{jk}^3}(\mathbf{r}_j - \mathbf{r}_k), \quad k = 1, \cdots, n.$$

(Equations of motion)

G. E. O. Giacaglia (ed.), Periodic Orbits, Stability and Resonances, 53–55.

1. Khilmi's Criterion

In his book on qualitative methods Khilmi has shown [2, pp. 34–40] that if the initial values of r_{jk} and $\dot{r}_{jk}$ satisfy the inequalities $b_{jk}>0$ and

$$b_{jk} > 8v_{jk} \sum{}' \frac{m_{jk}}{a_{jk}b_{jk}} \tag{1.1}$$

then the system disintegrates. He says nothing about escape of particles. We offer an improved version as follows.

THEOREM 1: *If* $b_{jk}>0$ *and*

$$b_{jk} > \frac{27}{4} v_{jk} \left(\sum{}' \frac{m_{jk}}{a_{jk}b_{jk}} \right), \tag{1.2}$$

then the system disintegrates. Moreover, all except possibly one of the particles escape.

The proof of the first part requires only a slight modification of Khilmi's [2, p. 37]. Instead of his choice of f_{jk}, let $f_{jk}=\varepsilon g_{jk}$, where $0<\varepsilon<1$. Mimicking his argument, it is readily seen that the condition for disintegration is

$$b_{jk} > \frac{1}{(1-\varepsilon)\,\varepsilon^2} \sum{}' \frac{m_{jk}}{a_{jk}b_{jk}}.$$

His criterion (1.1) corresponds to the choice $\varepsilon=\frac{1}{2}$, but clearly it is best to choose ε so that the constant preceding the summation sign is as small as possible. The choice $\varepsilon=\frac{2}{3}$ accomplishes this and yields (1.2). The reduction of the constant from 8 to $\frac{27}{4}$ may have numerical advantages.

More important, however, is the fact that his reasoning now shows that

$$r_{jk}(t) > \tfrac{2}{3}(a_{jk} + b_{jk}t),$$

so that $U=O(1/t)$. Consequently [3, p. 609] it follows that $U\sim A/t$, $t\to\infty$ for some positive value of A, and hence [3, p. 610] that at least $(n-1)$ particles escape.

2. A New Criterion for Escape

We begin with the identity

$$2\sum m_k \left[\left(\frac{\mathbf{r}_k}{\sqrt{I}} \right)' \right]^2 = \frac{4IT - \dot{I}^2}{I^2},$$

which follows from a computation of the left-hand side. Hence for each value of k

$$\left| \left(\frac{\mathbf{r}_k}{\sqrt{I}} \right)' \right| \leqq A_k \frac{\sqrt{4IT - \dot{I}^2}}{I}. \tag{2.1}$$

Now suppose that

$$\int_0^\infty \frac{\sqrt{4IT - \dot{I}^2}}{I} \, d\tau < \infty . \tag{2.2}$$

It follows from (2.1) that for each k

$$\lim_{\tau \to \infty} \frac{\mathbf{r}_k}{\sqrt{I}} = \mathbf{L}_k \tag{2.3}$$

exists. An immediate consequence is $\frac{1}{2}\sum m_k L_k^2 = 1$. Therefore at least one vector $\mathbf{L}_k$ is not zero. Moreover, because $\sum m_k \mathbf{r}_k = 0$ it is true that $\sum m_k \mathbf{L}_k = 0$. Consequently, at least two $\mathbf{L}_k$ are not zero. Then if $I \to \infty$ as $t \to \infty$, it follows from (2.3) that at least two particles escape.

THEOREM 2.2: *If $I \to \infty$ and the condition* (2.2) *is satisfied at least two particles escape from the system.*

This theorem contains the criteria previously given in [3] and [4]. Observe that there is no *à priori* condition on the sign of the energy. The condition $I \to \infty$ is automatically satisfied if $h \geqq 0$, [3].

References

[1] Chazy, J.: 1922, *Ann. Sci. Ecole Norm.* **39**, 124.
[2] Khilmi, G. F.: 1956, *Qualitative Methods in the Many-Body Problem*, Gordon and Breach, New York.
[3] Pollard, H.: 1967, *J. Math. Mech.* **17**, 601–612.
[4] Pollard, H. and Saari, D. G.: 1970, 'Escape From a Gravitational System of Positive Energy', *Celes. Mech.* **1**, 347.

SECULAR VARIATIONS DETERMINED BY A SURFACE OF SECTION

O. GODART

Université de Louvain, Institut d'Astronomie et de Géophysique, Georges Lemaître, Heverle-Louvain, Belgium

1. Introduction

In many astrophysical and geophysical problems, the invariance of magnetic mirrors is only correct to the first order. Their slow variations are important for the maintenance of radiation belts. They could be determined by a method inspired by the classical surface of section. This paper will consist of three parts. In the first, we define the semi-surface of section, determine the invariant curves of their mapping.

In the second, the special dynamical problems of two dimensions having such sections are studied in their integrable approximation and the modifications introduced when one passes to the general case.

In the third, we study the case of motion of charged particles in an axi-symmetric magnetic field and in particular, when it is reduced to a dipole.

It is felt that this method could be used in other dynamical problems that are encountered in celestial mechanics.

2. Semi- and Quasi-Surfaces of Section

Let $H(\mathbf{x}, \mathbf{p})$ the Hamiltonian of a regular dynamical system [1] of n degrees of freedom; $\mathbf{x}$ the coordinates, $\mathbf{p}$ the momenta. The kinetic energy K will be a function positive definite of $\mathbf{x}$ and $\mathbf{p}$. A choice of variables is always possible such that K is homogeneous of order 2 in $\mathbf{p}$ and that the potential function does not depend on $\mathbf{p}$.

Let

$$H \equiv K(\mathbf{p}, \mathbf{x}) + V(\mathbf{x}) = a/2 \tag{1}$$

define a constant energy surface of $2n-1$ dimensions.

In the n dimensional space of the $\mathbf{x}$, the motions will be restricted to the space configuration $\mathcal{O}$ where $V(\mathbf{x}) \leqslant a/2$. Let us isolate one of the components of $\mathbf{x}$ and call it y as well its momentum p_y; $\mathbf{x}$, $\mathbf{p}$ will then be written for the $n-1$ other components. The $n-1$ dimension surface $y = y_0$ (constant) will partition the configuration space in two subspaces $\mathcal{O}^+ (y > y_0)$, $\mathcal{O}^- (y < y_0)$ and the boundary $\Sigma (y = y_0)$. Let us suppose that it is possible to choose y such that in the upper part $(\mathcal{O}^+)$ when y reaches an extremum along any trajectory, it is necessarily a maximum in y.

Let us define as semi-orbit, the segment of an orbit lying in $\mathcal{O}^+$ and having two boundary points on $y = y_0$. Different semi-orbits may belong to the same orbit (see Figure 1).

G. E. O. Giacaglia (ed.), Periodic Orbits, Stability and Resonances, 56–75.

From each regular point of the surface at $2n-2$ dimensions defined by $2H=a$ and $y=y_0$, may start one and only one trajectory. Two trajectories going in the $\mathcal{O}^+$ subspace may lie on the same semi-orbit. We can establish between their respective starting points A and $B\in\Sigma$ an automorphism.

The set of all these couples of points will be called the semi-surface of section. The

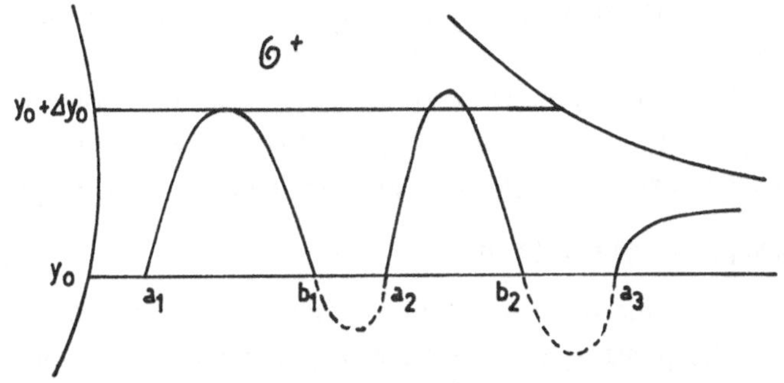

Fig. 1. Semi-orbits a_1b_1, a_2b_2, a_3 are on the same orbit; a_1b_1 is on the boundary of the surface of section $y_0+\Delta y_0$; a_3 is not on the y_0 surface of section ($b_3\to\infty$).

operation of passing from one trajectory of the semi-orbit to the other, will be called the mapping M defining the law of automorphism of the semi-surface of section designated by $\mathcal{S}^+$. In particular, we have $M(A)=B$ and $M(B)=A$. The operation M is symmetric: $M^{2n}=1$.

Let us call $\mathcal{O}_0\subset\mathcal{O}^+$ the subspace filled by the semi-orbits. If the Hamiltonian has continuous bounded first and second partial derivatives in $\mathbf{x}$ and $\mathbf{y}$ in $\mathcal{O}_0$, the mapping is continuous and thus topological. The proof can be extended fairly easily utilizing similar arguments than those for the surface of section [1] [2]. $\mathcal{S}^+$ will not necessarily include the whole surface Σ as well as the whole set of the semi-orbits will not fill completely $\mathcal{O}^+$. As example, all orbits going to infinity in $\mathcal{O}^+$ will be excluded; also, those lying completely in Σ. For continuity reason, if a set of these points forms a closed curve, it will be the boundary of one domain of $\mathcal{S}^+$ which may possess other unconnected bounded domains. Taking account of the Hamiltonian characteristics, the mapping of $\mathcal{S}^+$ in Σ will be area-preserving*.

The semi-surfaces of section will exist by families. In fact, the plane $y_1=y_0+\Delta y_0$ will also have the same property but only part of the subspace $\mathcal{O}_0$ will cut the plane $y=y_1$. All the semi-orbits having their maxima y_m between both planes $y_0<y_m\leqslant y_0+\Delta y_0$, will not cut that plane. On the other side, it is also possible by a choice of variable to limit the extent of y, let us say to $\pi/2$. Then if we call $\mathcal{G}_m$, the subset of all semi-orbits having the same maximum y_m, we can write: $\mathcal{O}_0=\cup\mathcal{G}_m$ for y_m varying continuously from y_0 to $\pi/2$.

Let us consider a closed connected part of Σ_0, the boundary will be formed by the intersection of y_0 by $\mathcal{G}_0$ in the same domain. Inside that boundary, will lie the inter-

* A more detailed study of these general properties will be published elsewhere.

section $y_0 \cap \mathscr{G}_1 = \mathscr{J}_{01}$ and it can be seen that each intersection with higher maximum will fit inside, with at the centre, the intersection $y_0 \cap \mathscr{G}_{\pi/2}$. In fact, $\mathscr{S}_0^+$ will be the union of all the intersections $\mathscr{S}_0^+ = \cup \mathscr{G}_m$, for $y_0 < y_m \leqslant \pi/2$. The intersections $\mathscr{J}_{0m}$ are invariant under M. Indeed, if $A \in \mathscr{J}_{0m}$, $M(A) = B \in \mathscr{J}_{0m}$ both being on the same semi-orbit.

Let us now suppose that y_0 is the lowest value of y for an upper semi-surface of section. Then between the plane y_0 and $y_0 - \Delta y_0$, some trajectories may have also their minima. If moreover, maxima are there impossible and if it is the case for all values of $y < y_0$, we can define in the same way, lower semi-surface of section and in that case, the above concepts will have less limited applications. It will happen generally when y_0 is a plane of symmetry. We can always take $y_0 = 0$, in that case, we can write:

$$H(\mathbf{x}, \mathbf{p}_x; y, p_y) = H(\mathbf{x}, \mathbf{p}_x; -y, p_y) \tag{2}$$

We need now to examine the passage of a trajectory from the $\mathcal{O}^+$ subspace to the $\mathcal{O}^-$ subspace. In the phase space, the trajectory starting from Σ on A will cross it again, not on B starting point of a trajectory in the $\mathcal{O}^+$ subspace, but on its prolongation in the $\mathcal{O}^-$ subspace, which means that we have to reverse the sign of time. In general, this will mean a reversal of sign of the momenta $\mathbf{p}$. If this operation is called R; $R(B) = B'$ will be the continuation in the $\mathcal{O}^-$ subspace of the trajectory starting in A in the $\mathcal{O}^+$ subspace. The reflection R will also be a symmetric operation and its application will have no restriction in the whole allowed region. From now on, we shall limit our considerations to the case where $y = 0$ is a surface of symmetry. In that case the mapping M for the lower sections will be identical to that for the upper section. The surface Σ_0 will be the support of both semi-surfaces of section. If $R(\mathscr{S}^+) = \mathscr{S}^-$ and $R(\mathscr{S}^-) = \mathscr{S}^+$, any trajectory crossing the equator twice will cross an infinite number of time $\mathscr{S}^+ \cup \mathscr{S}^-$, that is a real surface of section. The dynamical transformation of the surface of section in itself will be $T = RM$.

If it is not the case: $R(\mathscr{S}^+) \neq \mathscr{S}^-$ some trajectories after a finite number of crossings will not cross the surface $y = 0$. The union $\mathscr{S}^+ \cup \mathscr{S}^-$ will then be called a quasi-surface of section and the operation T cannot be iterated indefinitely for all trajectories but it will still be the case for periodic or quasi-periodic orbits. It will be interesting to determine the 'leakage' phenomenon.

Let us now examine the law of mapping along the invariant intersections $\mathscr{J}_{0m}$ for the particular simpler case of two dimensions dynamical problems.

3. Two-Dimensional Surfaces of Section

Let the Hamiltonian be

$$2H = p_R^2 + p_z^2 + Q^2(R, z) = a\,, \tag{3}$$

where $Q(R, z) = Q(R, -z)$. Let us suppose that the line $Q = 0$ can be expressed in a parametric form: $R = e^{g(y)} \cos y$, $z = e^{g(y)} \sin y$. For symmetry reason, $g(y)$ is an even periodic function of y or an holomorphic function of $\mu = \cos y$, in the interval $0 < \mu \leqslant 1$.

If we introduce with y a new coordinate u distance of the point (R, z) to the line $Q=0$ (see point T on Figure 2), we can express:

$$R = e^{g(y)} \cos y + u \cos(y+v) \quad z = e^{g(y)} \sin y + u \sin(y+v), \tag{4}$$

where the auxiliary angle v is defined by the odd function of y

$$\tan v = -\,\mathrm{d}g/\mathrm{d}y \tag{5}$$

(see Figure 2).

u will be unique at the condition that the point P and the intersection T of the normal with the line $Q=0$ are in the same hemisphere, giving:

$$\frac{z}{\sin y \cos v} = \frac{e^{g(y)}}{\cos v} + u(1 + \tan v \tan y) \geqslant 0. \tag{6}$$

The canonical transformation $p_R \mathrm{d}R + p_z \mathrm{d}z = p_u \mathrm{d}u + p_y \mathrm{d}y$ applied to H gives:

$$2H = p_u^2 + \frac{p_y^2}{J^2} + Q^2(u, y) = a, \tag{7}$$

where

$$J = \frac{e^{g(y)}}{\cos v} + u\left(1 + \frac{\mathrm{d}v}{\mathrm{d}y}\right) \tag{8}$$

is the Jacobian of the transformation and must be >0. The expansion in u of Q starts

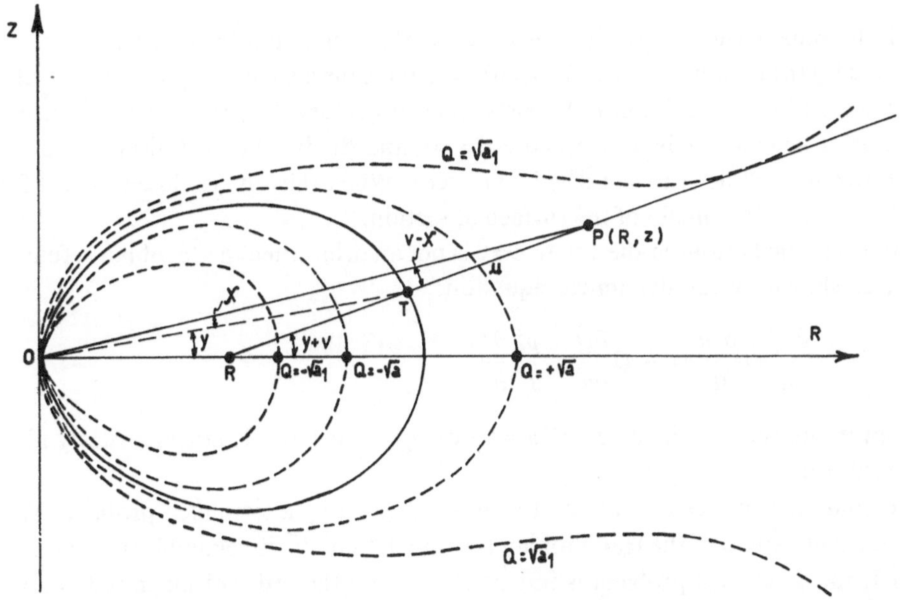

Fig. 2. Consists in the region of allowed motions, limited by both curves $+\sqrt{a}$, $-\sqrt{a}$ $(a<1/16)$ giving a finite area; for the value $a_1>1/16$, this area is infinite. The point P could have been chosen anywhere in the allowed area. It has been chosen on the latitude y_v of its guiding centre corresponding to the beginning of the valley.

with the first power

$$Q = uf(y) + O(u^2) \tag{9}$$

For 'a' small, u must remain small and it will always be possible to satisfy both inequalities (6) and (8).

What is the condition such that on each trajectory, the maxima in y will be reached in the northern hemisphere only and the minima in the southern? Noticing that Q is an even periodic function of y, it could be considered as continuous and twice differentiable in $(0<\mu\leqslant 1)$ the condition will be that

$$\partial Q^2/\partial\mu \leqslant 0 \tag{10}$$

in the whole allowed region. It is evident for all points where $Q\neq 0$, but it could also be verified that the extrema along $Q=0$ do not make exception. By successive derivations, we find $\mathrm{d}^2y/\mathrm{d}t^2=\mathrm{d}^3y/\mathrm{d}t^3=0$ but $\mathrm{d}^4y/\mathrm{d}t^4$ has a non-zero term:

$$-\frac{\partial^2}{\partial u^2}Q\left(\frac{\partial Q}{\partial y}\right)\frac{a}{J^2} = -\frac{a}{J^2}\frac{\mathrm{d}f^2(y)}{\mathrm{d}y} \quad \text{for} \quad u=0.$$

As a consequence of (10), $f(y)$ which is also expressible in analytic function of μ, will have to obey the same condition.

When a is very small $2\sqrt{a}/f(\mu)$ will be the approximate extent of the semi-surface of section in space coordinates.

From

$$\frac{\mathrm{d}f^2}{\mathrm{d}\mu} < 0 \tag{10}$$

it will become narrower with increasing y. As the valley ends for $y=\pi/2$, this may be satisfied if $f(\mu)$ becomes infinite for $\mu=0$, which means a singularity somewhere along the z axis. If such condition (10) exists, each trajectory will cross at least once the equator or lie completely in it because if $y=0$ and $\mathrm{d}y/\mathrm{d}t=0$ on a trajectory, it could be shown that all derivatives: $\mathrm{d}^ny/\mathrm{d}t^n$ are zero. When this one is closed, it will form the boundary of a domain of the surface of section.

However the bottom of the valley $u=0$ is not an orbit when the Jacobian is function of u, as shown by the dynamical equation:

$$\frac{\mathrm{d}p_u}{\mathrm{d}t} = \frac{\mathrm{d}^2u}{\mathrm{d}t^2} = -Q\frac{\partial Q}{\partial u} + \frac{p_y^2}{J^3}\frac{\partial J}{\partial u} \tag{11}$$

Except in the very special case $\partial J/\partial u = 1 + \mathrm{d}v/\mathrm{d}y = 0$, $\mathrm{d}^2u/\mathrm{d}t^2$ is not necessarily nil for $u=\mathrm{d}u/\mathrm{d}t=0$.

To study the surface of section, it is useful to start by a simplified problem. When the constant value a of the Hamiltonian is so small that $a^{1/2}$ is negligible in comparison with 1, the dynamical problem is integrable. To put the order of magnitude more in evidence; let us change the scale of the variables

$$u=\sqrt{a}U \quad p_u=\sqrt{a}P_u \quad y=\sqrt{a}Y \quad p_y=\sqrt{a}P_y\,;$$

$H'=H/a$ will be the new Hamiltonian but with energy constant equal to 1.

We could then expand $H' = H_0 + \sqrt{a}H_1 + aH_2 \ldots$ where the H_n depend only of the canonical variables U, P_u, P_y and y. The restricted problem will be then defined by the Hamiltonian

$$2H_0 = P_u^2 + P_y^2 \cos^2 v e^{-2g} + U^2 f^2 = 1 \tag{12}$$

Let us apply to this Hamiltonian, a canonical transformation keeping Y, as one of the canonical variables but changing U, P_u, P_y in q_1, p_1, P'_y defined by:

$$U = \sqrt{\frac{2p_1}{f(y)}} \cos q_1 \quad P_u = \sqrt{2p_1 f(y)} \sin q_1$$
$$P_y = P'_y - p_1 \sin 2q_1 \times \frac{\sqrt{a}}{f} \frac{\mathrm{d}f}{\mathrm{d}y} \tag{13}$$

The difference between P_y and P'_y depending on $\sqrt{a}$ will then be of a superior order and included in the new H_1

$$2H_0 = 2p_1 f(y) + P_y^2 e^{-2g} \cos^2 v = 1 \tag{14}$$

U is then oscillating with a phase q_1 changing with y: $\mathrm{d}q_1/\mathrm{d}t = f(y)$. In accordance with the main application of this theory, we shall call y the 'latitude' and the extrema in y the 'mirror' of the trajectory. The motion in latitude can be determined by elimination of P_y, $\mathrm{d}t$ from the Hamiltonian and the equation of motion in Y: $\mathrm{d}Y/\mathrm{d}t = P_y e^{-g} \cos^2 v$.

$$\frac{\mathrm{d}Y}{\mathrm{d}q_1} e^{2g} \left(1 + \left(\frac{\mathrm{d}g}{\mathrm{d}y}\right)^2\right) f = \sqrt{a(1 - 2p_1 f(y))} \tag{15}$$

The maximum value reached by $f(y)$ is $f(y_m) = 1/2p_1$. As $2p_1$ is a constant in this approximation, the mirror latitude y_m will also remain constant.

The phase angle of the oscillation in latitude could be introduced by

$$f(y) = f(y_m) + [f(0) - f(y_m)] \cos^2 q_2 \tag{16}$$

where $f(0)$ is the value of f for $y=0$. Then $q_2 = 0$ (mod. π) at the equator and $\pi/2$ (mod. π) at the mirror point. After expansion in trigonometric series, integration of (15) with a convenient constant q_m, we get:

$$q_1 - q_m = k_0 \left[\left(q_2 - \frac{\pi}{2}\right) + \sum_{n=1}^{\infty} \beta_{2n} \sin 2n\, q_2\right], \tag{17}$$

where q_m is the phase of the oscillation at the first mirror $q_2 = \pi/2$, k_0 and β_{2n} are parameters, depending only of y_m and a, becoming infinite with $a=0$ or for $y_m = \pi/2$; k_0 is the ratio of the number of oscillation to the number of mirroring. The equator is a real surface of section having for coordinates U, P (from now on we shall neglect the subscript u). The nth crossing given by $q_2 = n\pi$ will be represented by the point A_n

of coordinates:

$$U_n = \frac{1}{\sqrt{f(y_m)\,f(0)}} \cos\left(q_m + \left(n - \tfrac{1}{2}\right) k_0\pi\right)$$
$$P_n = \sqrt{\frac{f(0)}{f(y_m)}} \sin\left(q_m + \left(n - \tfrac{1}{2}\right) k_0\pi\right) \tag{18}$$

The intersection $\mathscr{I}_{0m}$ that we shall call: mirror curves will then be ellipses symmetric around the U axis. They are then also invariant for R and T. At each crossing the position angle will advance on the ellipse of a constant quantity $k_0\pi$, but varying from one ellipse to the other with the value of y_m. The boundary curve of the surface of section is the equatorial orbit. If $k_0(y_m)$ is the ratio of two integers prime to each other such that 1) $k_0(y_m)=2l/2r+1$ or 2) $k_0(y_m)=2l+1)/r$, the mirror curves are formed of all sets of periodic points belonging to orbits closing after the number of crossings $n=2r+1$ in the first case; $n=2r$ in the second. We shall classify periodic orbits following those values of n. It is well known that those periodic orbits are ordinary [3]. In this particular case, it is to be noticed that those denumerable mirror curves accumulate near the origin where the periodic points become everywhere dense. To parametrize completely the surface of section, we shall have to define another parameter. Noting that the coordinates of the first mirror are

$$U = \frac{\cos q_m}{f(y_m)} \qquad P = \sin q_m \tag{19}$$

We shall take the mirror phase q_m (mod. 2π) as second parameter. Writing in (18) $n=0$, we get the parametric expression of the point A_0 such that $T^n(A_0)=A_n$. The operation $RT^n(A_0)=R(A_n)=B_n$ can be obtained simply in changing the sign of P. If we put $n=1$, we obtain the law of mapping;

$$U_A = \frac{1}{\sqrt{f(0)\,f(y_m)}} \cos\left(q_m - \frac{k_0\pi}{2}\right) \qquad U_B = \frac{1}{\sqrt{f(0)\,f(y_m)}} \cos\left(q_m + \frac{k_0\pi}{2}\right)$$
$$P_A = \sqrt{\frac{f(0)}{f(y_m)}} \sin\left(q_m - \frac{k_0\pi}{2}\right) \qquad P_B = -\sqrt{\frac{f(0)}{f(y_m)}} \sin\left(q_m + \frac{k_0\pi}{2}\right) \tag{19}$$

We see that for $q_m=0$ or π, M is an identity. These semi-orbits will be called self-reversing because at the mirror point their speed becomes zero, and after they come back on themselves. All those points will form an invariant curve $\mathscr{M}_1$. It will be composed of two spirals winding indefinitely from a point on the equatorial orbit to the limiting point for $y_m=\pi/2$, one for the phase $q_m=0$, the other for the other phase $q_m=\pi$. For the simplified problem, where the singular point is at the origin, each branch will be the symmetric of the other with respect to the origin. The other curves $q_m=$constant are also spirals; they will appear like some kind of twisted polar radii. When deformed in such a way that they are straightened out; the mapping M will appear also as some reflection around the radius $\mathscr{M}_1$ then axis of symmetry (Figure 3).

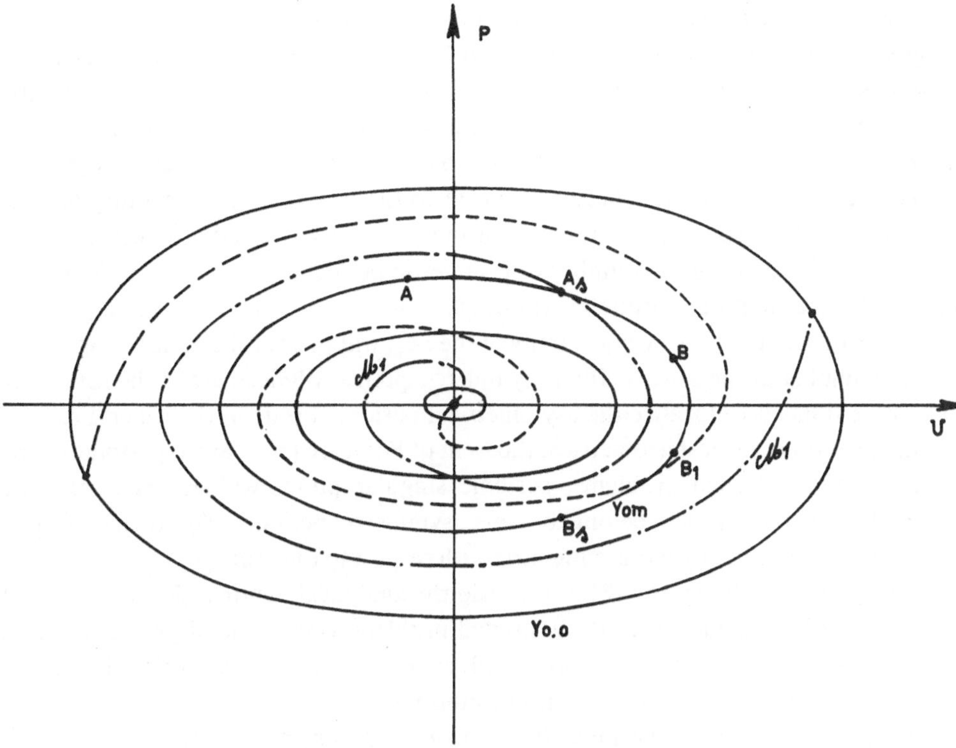

Fig. 3. M applied to A gives B, then $T(A)=B_1$ $T(A_s)=B_s$.
The angular distance $AB_1=A_sB_s=k_0(y_m)\,\pi$.

If we write in general $M_n=RT^n$, we may consider generally the invariants $\mathcal{M}_n$; $\mathcal{M}_0$ will be the segment of the U axis of the surface of section; in this case, the phase

$$q_m = k_0\pi/2\,. \tag{20}$$

It is quite easy to see that $T^n(\mathcal{M}_1)=\mathcal{M}_{2n-1}$ and $T^n(\mathcal{M}_0)=\mathcal{M}_{2n}$ which can be easily determined when the law of mapping is known for all mirror curves. They could also be determined in continuing the integration up to n crossings of all self-reversing orbits for $\mathcal{M}_{2n+1}$ or of all symmetric orbits for $\mathcal{M}_{2n}$.

In the simplified case, their analytic expressions could be determined by (18) in introducing the correct values of q_m and n. Those curves are interesting because their intersections with the U axis (also $\mathcal{M}_0$) give the periodic symmetric points having the phase values $k_0(y_m)=N/(r-\frac{1}{2})$ for $2n+1$ and $k_0(y_m)=N/r$ for $2n$ where N is an integer taking all values from the minimum of $k_0(y_m)$ to infinity. Those points will be given by

$$U_N = \frac{(-1)^N}{\sqrt{f(0)\,f(y_m)}}\,. \tag{21}$$

When we pass to the general form of the Hamiltonian, it is possible to continue analytically the mirror curve still invariant for M but not for R because, in particular, u appears in odd power specially in the Jacobian, the mirror curves are not symmetric with respect to the U axis. By arguments similar to those used by Malmquist [4], the singular orbit ($y_m = \pi/2$) will still exist but differ from $u=0$ and will not in general be a symmetric periodic orbit. The spiral $\mathscr{M}_1$ will not cut any more an infinite number of times the axis U. When the singular point $y_m = \pi/2$ gets further away, with increasing a, branches of spiral cutting at first $\mathscr{M}_0$ at two periodic singular symmetric points, one stable and the other unstable become tangent to the U axis giving a limiting periodic orbit of characteristic exponent zero. Further, these two periodic points will have disappeared, decreasing the number of symmetric periodic orbits. As a consequence, there will be a limiting latitude mirror y_l such that the following mirroring latitude is always inferior to y_l. The whole family of the $\mathscr{M}_n$, will present the same configuration and the number of the other periodic symmetric points will decrease in the same way. It is more difficult to localize the asymmetrical periodic orbits in the general case. In the application of the last geometrical theorem of Poincaré [5], near the periodic mirror curves of the restricted problems, periodic singular points will still exist, and it is probable that some of them belong to some asymmetric periodic trajectories although those curves are not any more symmetric with respect to the U axis.

A theorem of Kolmogorov [6] proves also the analytical continuation of almost all the quasi-periodic solutions of the restricted problem to the general problem. As the frequencies of the quasi-periodicities will change continuously, isolated cases of commensurability will give rise to periodic motions.

As for (18), the mirror curves may be expressed in parametric form:

$$
\begin{aligned}
U &= U_0(a, y_m) + \cos y_m \sum_{r=1}^{\infty} \left[U_r(a, y_m) \cos r q_m + U_r'(a, y_m) \sin r q_m\right] \\
P &= P_0(a, y_m) + \cos y_m \sum_{r=1}^{\infty} \left[P_r(a, y_m) \cos r q_m + P_r'(a, y_m) \sin r q_m\right],
\end{aligned}
\tag{21}
$$

where U_m, U_m', P_m, P_m' could be expanded in Fourier series in y_m. The knowledge of a finite number of semi-orbits of well chosen y_m and q_m will give by harmonic analysis in the two angular parameters q_m and y_m the expressions (21). The operation M will consist in changing q_m in $-q_m$; R in changing P in $-P$. The determination of the next mirror point (y_m', q_m') will not present great difficulties numerically or graphically (see Figure 4), but is analytically difficult due to the presence of periodic points. It is given by the relations:

$$
\begin{aligned}
& U_0(y_m') + \cos y_m' \left[\sum_{r=1}^{\infty} U_r(y_m') \cos r q_m' + U_r'(y_m') \sin r q_m'\right] \\
&= U_0(y_m) + \cos y_m \left[\sum_{r=1}^{\infty} U_r(y_m) \cos r q_m - U_r'(y_m) \sin r q_m\right] \\
& P_0(y_m') + \cos y_m' \left[\sum_{r=1}^{\infty} P_r(y_m') \cos r q_m' + P_r'(y_m') \sin r q_m'\right] \\
&= - P_0(y_m) + \cos y_m \left[\sum_{r=1}^{\infty} P_r'(y_m) \sin r q_m - P_r(y_m) \cos r q_m\right].
\end{aligned}
\tag{23}
$$

Periodic orbits of class one will maintain their mirror latitudes and if they are stable,

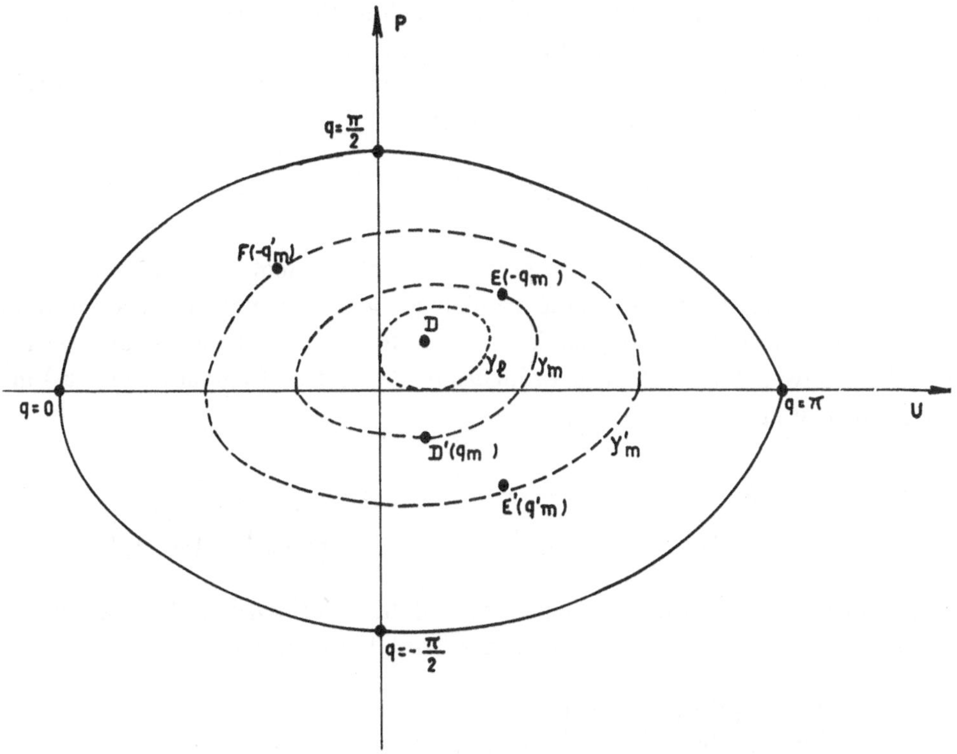

Fig. 4. On the diagram of the surface of section, are drawn different curves of constant latitude mirrors. All the particles having their first mirror at a latitude superior to y_l will have their next mirror latitude under y_l.

the orbits nearby will have very small variations of their mirror latitudes. To study the problem more in detail, it seems necessary to consider more specific cases.

4. Axisymmetric Magnetic Field, the Dipole Case

The motion of charged particles in a magnetic field has been mainly studied by the approximate theory of Alfvén. Neglecting the deviations due to collisions, it could be decomposed to:

(1) a gyration about a guiding centre;
(2) a translation of this centre along a magnetic line of force;
(3) a slow drift of this centre from one magnetic line to another.

The same problem could be studied by the classical Hamiltonian mechanics. In the axisymmetric field, the Hamiltonian will be

$$2H = p_R^2 + p_z^2 + (p_\phi/R - A_\phi)^2 = 1, \tag{24}$$

where p_R, p_z, p_ϕ are conjugate momenta of the cylindrical coordinates R, z, ϕ. The mass of the charged particles is taken as unity and as the speed is constant, the length of arc along the trajectories is the independent variable. A_ϕ is the only component of the potential vector different from zero multiplied by the charge of the particles and will depend only on the coordinates R, z.

The ignorable variable ϕ can be determined when the motion in the meridian plane R, z is known by the quadrature of the equation

$$R^2 \frac{\mathrm{d}\phi}{\mathrm{d}s} = p_\phi - RA_\phi = U(R, z). \tag{25}$$

The Hamiltonian (24) could be considered as one of a system of two degrees of freedom. The coordinates R, z are in the meridian plane following the particle. From the point of view of the magnetodynamics, we must first determine the curvature centre. In [7], we have shown that the coordinates of the centre of curvature (R_0, z_0) in a meridian plane are:

$$R_0 = R - \frac{U \dfrac{\partial U}{\partial R}}{\left(\left(\dfrac{\partial U}{\partial R}\right)^2 + \left(\dfrac{\partial U}{\partial z}\right)^2\right)} \qquad z_0 = z - \frac{U \dfrac{\partial U}{\partial z}}{\left(\left(\dfrac{\partial U}{\partial R}\right)^2 + \left(\dfrac{\partial U}{\partial z}\right)^2\right)}. \tag{26}$$

Instead of R, z; it is sometimes convenient to use the coordinate lines U=constant and the set of their orthogonal lines V=constant. Then, by definition

$$\left(\frac{\partial R}{\partial U}\right)_V \left(\frac{\partial R}{\partial V}\right)_U + \left(\frac{\partial z}{\partial U}\right)_V \left(\frac{\partial z}{\partial V}\right)_U = 0.$$

From the differential relations

$$\mathrm{d}R = \left(\frac{\partial R}{\partial V}\right)_U \mathrm{d}V + \left(\frac{\partial R}{\partial U}\right)_V \mathrm{d}U \qquad \mathrm{d}z = \left(\frac{\partial z}{\partial V}\right)_U \mathrm{d}V + \left(\frac{\partial z}{\partial U}\right)_V \mathrm{d}U$$

we can transform (26) to

$$R_0 = R - U\left(\frac{\partial R}{\partial U}\right)_V \qquad z_0 = z - U\left(\frac{\partial z}{\partial U}\right)_V \tag{27}$$

If we expand $R(U, V)$; $z(U, V)$ in power series of U and write $R(0)$, $z(0)$ for the coordinates of the line U=0 (bottom of the valley as we called it previously), we get up to the second order term in U

$$R_0 = R(0) - \frac{U^2}{2}\left(\frac{\partial^2 R}{\partial U^2}\right)_{V,0} \cdots \qquad z_0 = z(0) - \frac{U^2}{2}\left(\frac{\partial^2 z}{\partial U^2}\right)_{V,0} \cdots. \tag{28}$$

Then along the special line of force U=0, particles and meridian projection of their centre of curvature coincide; for gyrations where U remains small (U^2 negligible), the bottom of the valley could be considered as the guiding centre. As it will be in another meridian plane than the particle, if we want to take the maximum advantage of the rigorous integral of kinetic momentum, it is advantageous to abandon the strict magnetodynamical point of view and instead of the circulatory motion of Larmor, envisage an oscillatory motion in the meridian plane following the particles. The oscillation will be around the points of coordinates R_0, z_0. In the case where Section 3 and Section 4 can be applied, U will be sufficiently small to approximate R_0, z_0 by $R(0)$ and $z(0)$ and then, the bottom of the valley could be used as one of the axes of reference.

An interesting case of such magnetic field is the dipole field

$$A_\phi = \frac{eM}{c} \frac{R}{(R^2 + z^2)^{3/2}}, \tag{29}$$

where e is the electrical charge, M the magnetic moment, c the speed of light. A change of length scale and independent variable will combine the physical constants in the only constant of energy [8]. It will consist in using for scale length

$$L = eM/cp_\phi \quad \text{and writing} \quad t = (p_\phi)^3 (c/Me)^2$$

and defining

$$a = p_\phi^{-4} \left(\frac{eM}{c}\right)^2 \qquad Q = \frac{1}{R} - \frac{R}{(R^2 + z^2)^{3/2}} \tag{30}$$

we obtain a Hamiltonian of the type envisaged in Section 3 (3). This type of transformation could be used each time where A_ϕ is homogeneous in the coordinates R and z.

In the case of the dipole

$$e^{g(y)} = \cos^2 y, \qquad \tan v = 2 \tan y^*, \qquad f(y) = 1/\cos^5 y \cos v$$

satisfies the condition for the existence of the surface of section $\mathrm{d}f/\mathrm{d}\mu < 0$.

The Jacobian J can be written

$$J = \frac{\cos^2 y}{\cos v} + 3u \times \frac{1 + \sin^2 y}{1 + 3 \sin^2 y} \tag{31}$$

J will always be positive in the case of the unicity determination of u (6): $\cos^2 y/\cos v + 3u > 0$ in the dipole case.

To verify the general condition $\partial Q^2/\partial\mu \leqslant 0$ let us compute

$$\frac{\partial Q}{\partial \mu} = - \frac{\partial Q}{\partial R} \frac{\partial R}{\sin y\, \partial y} - \frac{\partial Q}{\partial z^2} \frac{\partial z}{\partial y} \frac{2z}{\sin y}.$$

From $\quad \partial R/\partial y = -J \sin(y + v) \quad \partial z/\partial y = J \cos(y + v)$

$$\frac{z \cos(v + y)}{\sin y} = \cos v \left[3R - 2 \cos y\right]$$

and performing the partial derivatives, one gets

$$\frac{\partial Q}{\partial \mu} = 3J \cos v \left(\frac{2R \cos y}{(R^2 + z^2)^{5/2}} - \frac{1}{R^2} - \frac{1}{(R^2 + z^2)^{3/2}}\right). \tag{32}$$

* The same angle y has been used by Lemaître and Bossy [9] for the study of orbits in the valley. They used also the variable u with a different meaning as the one of this paper.

Going back to Figure 2 where lines have been drawn for two typical constant values of Q taking for example the dipole problem; it is quite clear from their definition that u/Q is always positive.

Instead of u, a parametric angle χ can be used, being defined as the difference of position angle of the point P and its guiding centre T. From the properties of the triangle OTP, one deduces:

$$u = \frac{\cos^2 y \sin \chi}{\sin(v - \chi)} \qquad (R^2 + z^2)^{1/2} = \frac{\cos^2 y \sin v}{\sin(v - \chi)}$$
$$R = \frac{\cos^2 y \sin v \cos(y + \chi)}{\sin(v - \chi)}. \tag{33}$$

From (32), one deduces:

$$\frac{\partial Q}{\partial \mu} = -3J \cos v \left(\frac{Q}{R} + \frac{2}{(R^2 + z^2)^2}\left(\sqrt{R^2 + z^2} - \frac{R \cos y}{\sqrt{R^2 + z^2}}\right)\right)$$
$$= -3JQ \cos v \left(\frac{1}{R} + \frac{2u}{Q(R^2 + z^2)^2}\frac{\cos(y + \chi - v)}{\cos y}\right). \tag{34}$$

As the angle $|y+\chi-v|<\pi/2$, $\partial Q/\partial\mu$ has the contrary sign of Q and the existence of semi-surfaces of sections is proved following the theorem of previous paragraph (10).

To determine the maximum value of a where such theory is applicable it is sufficient from (6) that the point where $u=-\frac{1}{3}(\cos^2 y/\cos v)$ is on the boundary line of the allowed region $Q^2=a_1$.

One obtains $z_1=0$, $R_1=\frac{2}{3}\cos y$ giving $\frac{9}{16}$ for the maximum value of a including all values of physical interest.

When $a<\frac{1}{16}$, the allowed region is formed of two unconnected domains, one finite in our coordinates where the equator is a surface of section in the sense of Birkhoff, the other extending to infinity giving rise to a quasi-surface of section but no periodic orbits.

When $a>\frac{1}{16}$, the two domains are connected and there exists a quasi-surface of section. It can be shown that all the periodic orbits disappear for $a=0.161\,653$.

De Vogelaere [2] has localized the symmetric periodic orbits mainly for $a>\frac{1}{16}$ using the Lemaître and Bossy coordinates. Although introduced in another way, the same curves $\mathcal{M}_n$ of the last paragraph were studied in great detail with respect to the determination of all periodic points.

The increase of the number of periodic orbits with the decrease of a is well illustrated in this case; we do not consider periodic orbits of the second kind usually of higher class but only those of class one or two.

For $a<0.161\,653$, only two orbits exist of class 1, at $a=0.124\,9$ appear two others of class 2 down to $a=0.0794$ where appear two supplementary ones of class 1; for $a=0.071\,2$ appear two new periodic orbits of class 2, and so on. Those families may disappear later [2] but on the whole there will be an increase of the number of periodic orbits.

The geophysical problem origin of this work imposes on us the study of the Störmer problem for another range of values of a; a smaller than $\frac{1}{16}$. The variations of the mirror latitude require the study of the mirror curves. For that purpose, two preliminary problems have to be solved: 1) the restricted problem; 2) the determination of the centre of the surface of section.

4.1. The Restricted Problem

In this case, (13) can be written

$$\frac{u}{\sqrt{a}} = \sqrt{\cos^5 y_m \cos v_m \cos^5 y \cos v \cos q_1}$$
$$\frac{p_u}{\sqrt{a}} = \sqrt{\frac{\cos^5 y_m \cos v_m \sin q_1}{\cos^5 y \cos v}} \tag{35}$$

We can express (15) in the form:

$$\left(\frac{\mathrm{d}}{\mathrm{d}q_1}\left(\frac{\sin y}{\cos^4 y}\right)\right)^2 = a\left(1 - \frac{\cos^5 y_m \cos v_m}{\cos^5 y \cos v}\right). \tag{36}$$

Instead of the definition (16), we have used [11]

$$\frac{\sin y}{\cos^4 y} = \frac{\sin y_m}{\cos^4 y_m} \sin q_2$$

the actual q_2 will only differ from the one defined in the previous paragraph by periodic terms of $2m\,q_2$ arguments. $k_0(y_m, a)$ has been expressed in the form $k_0 = \omega/(\sqrt{a}\cos^4 y_m)$ where ω can be approximated in truncated Fourier series in y_m.

In the same manner, we put $\beta_{2m} = \sin^{2m} y_m \times b_{2m}$, where b_{2m} are also Fourier series of the same type. Their values are given in Table I. It is now quite easy to pass from the ordinary orbits $y_m \neq \pi/2$ to the singular one $y_m = \pi/2$.

Writing for $q_m + k_0\pi/2 = q_e$ the phase of oscillation at the equator and making the inversion of (17) we get an expression of the type

$$q_2 = \left(\frac{q_1 - q_e}{k_0}\right) + \sum_{n=1}^{\infty} a_{2n} \sin 2n\left(\frac{q_1 - q_e}{k_0}\right),$$

TABLE I

	ω	b_2	b_4	b_6	b_8	b_{10}	
	0.802639	0.102256	0.047971	0.046204	0.050546	0.055574	
$\cos 2y_m$	−0.312062	0.096465	0.065452	0.072378	0.082604	0.092179	$\cos 2y_m$
$\cos 4y_m$	−0.017952	0.031375	0.029583	0.039656	0.047508	0.050605	$\cos 4y_m$
$\cos 6y_m$	−0.000970	0.007438	0.009717	0.016239	0.019519	0.017328	$\cos 6y_m$
$\cos 8y_m$	−0.000154	0.001463	0.002472	0.005090	0.005574	0.002319	$\cos 8y_m$
$\cos 10y_m$	−0.000103	0.000307	0.000464	0.001028	0.000947	−0.001360	$\cos 10y_m$

we see that the limit $y_m \to \pi/2$ of $\sin y/\cos^4 y$ is given by

$$\frac{\sqrt{a}(q_1 - q_e)}{\omega}(1 + \Sigma 2na_{2n}), \tag{37}$$

where the phase q_e does not matter because the amplitude of oscillations becomes 0.

This shows how the infinite number of trajectories with mirror point y_m as one limiting trajectory: the line $u=0$ in the restricted problem. It is most important to compute it down to the equator because it forms the 'spinal column' of the surface of section. Unhappily, near the dipole, numerical integration is not suitable because the neighbouring orbits are strongly oscillating and a slight inevitable error of approximation will throw it out of the non-oscillatory state. Lower down near the equator, such a difficulty will not exist any more.

4.2. The non-oscillating curve

When u is not small, the guiding centre cannot be approximated by the bottom of the valley. In fact, it might vary for different oscillations on the same latitude y. Its average however will form a curve near the line $u=0$. Because of the symmetry of the problem, this curve could be expressed in even function of y and will be with a very high approximation a solution of the dynamical problem and as a consequence asymptotic to the singular orbit. In reality, this non-oscillating curve is not a rigorous solution and trial solution expressing R, z in series of even periodic function of y starting from the origin is diverging [8]. The divergence of the process could be illustrated by the fact that $\partial H/\partial p_1$ contains a term in $1/\sqrt{(2p_1)}$ appearing in the H_3 term and diverging for the non-oscillating curve. It is clear that the dipole orbit will also finally oscillate slightly and it will not be symmetric with respect of the equator, in general, two of them will exist, one coming from the north pole, the other from the south pole crossing the equator at the same point but with an angle different from $\pi/2$ giving a value of P different from 0 for the centre of the surface of section.

To evaluate the non-oscillating curve first step for an accurate determination of the dipole orbit, let us take as variables $r=\sqrt{(R^2+z^2)}$; $\lambda=y+\chi$ polar coordinates of this curve in the meridian plane, but take rather that the polar angle, the variable $c=\cos^2\lambda$. The Hamiltonian (24) is transformed:

$$2H = p_r^2 + \frac{4c(1-c)}{r^2}p_c^2 + \frac{1}{r^2 c}\left(1 - \frac{c}{r}\right)^2 = a\,. \tag{38}$$

If we use c as independent variable, we can reduce the dynamical system to one degree of freedom, the new Hamiltonian $-p_c(r, p_r, c)$ can be evaluated in solving the Equation (38).

The equations of movement will be

$$\frac{\mathrm{d}r}{\mathrm{d}c} = \frac{\partial p_c}{\partial p_r} \quad \text{and} \quad \frac{\mathrm{d}p_r}{\mathrm{d}c} = \frac{\partial p_c}{\partial r} \tag{39}$$

As the total derivative of the Hamiltonian is equal to its partial derivative $\mathrm{d}p_c/\mathrm{d}c = \partial p_c/\partial c$, we deduce from the energy integral

$$\frac{\mathrm{d}}{\mathrm{d}c}\left(4p_c^2 c(1-c)\right) = \frac{1}{c^2}\left(1 - \frac{c^2}{r^2}\right) \tag{40}$$

From the first equation of motion, one gets

$$\frac{\mathrm{d}r}{\mathrm{d}c} = \frac{p_r}{p_c}\frac{r^2}{4c(1-c)}. \tag{41}$$

Evaluating p_r, p_c from (41) and (38) and introducing in (40) we obtain the trajectory equation:

$$4(1-c)\frac{\mathrm{d}^2 r}{\mathrm{d}c} - 2\frac{\mathrm{d}r}{\mathrm{d}c} - \frac{4(1-c)}{r}\left(\frac{\mathrm{d}r}{\mathrm{d}c}\right)^2 + \frac{8(1-c)^2}{r^2}\left(\frac{\mathrm{d}r}{\mathrm{d}c}\right)^3$$

$$= \frac{\left(r^2 + 4c(1-c)\left(\frac{\mathrm{d}r}{\mathrm{d}c}\right)^2\right)\left(ar\left(1 - \frac{2(1-c)}{r}\left(\frac{\mathrm{d}r}{\mathrm{d}c}\right)\right) - \left(\frac{r-c}{r}\right)\left(1 + \frac{4(1-c)}{r}\left(\frac{\mathrm{d}r}{\mathrm{d}c}\right)\right)\right)}{ar^2 c - \left(\frac{r-c}{r}\right)^2} \tag{42}$$

When a tends to 0, the singular solution tends to $r = c$ valid for the restricted problem. When we take a into account but neglect a^2, one gets for solution

$$r = c\left(1 + 6\frac{ac^4}{4(4-3c)}\left(1 + \frac{c}{4-3c}\right)\right) \tag{43}$$

This is equivalent to the solution obtained by Störmer [8] (p. 248) but can be improved greatly in noticing that the corrections introduced by the consideration of the terms in $a^2, a^3 \cdots$ are rational functions of c having as denominators powers of $(4-3c)$.

It is then interesting to introduce two new variables

$$\frac{c}{4-3c} = x. \tag{44}$$

As c is nearly $\mu^2 = \cos^2 y$; x is very near $\cos^2 v$.

The second is some kind of parameter

$$\alpha = \frac{ac^4}{4(4-3c)}. \tag{45}$$

It can be shown that r/c may be expressed in a potential series of both x and α with integer coefficients. As both auxiliary variables are functions of c, we shall utilize the two relations

$$c\frac{\mathrm{d}x}{\mathrm{d}c} = x(1+3x) \qquad \frac{c}{\alpha}\frac{\mathrm{d}\alpha}{\mathrm{d}c} = 4 + 3x. \tag{46}$$

Let us write

$$\frac{r}{c} - 1 = 2R = 2 \sum_{m=1}^{\alpha} \alpha^m R_m(x) \tag{47}$$

where $R_m(x)$ are polynomials in x.

To simplify the writings, let us pose:

$$\frac{\mathrm{d}}{\mathrm{d}c}(r - c) = 2P \qquad c\,\frac{\mathrm{d}^2 r}{\mathrm{d}c^2} = 2S\,. \tag{48}$$

P and S can also be expanded in powers of α and x as the expression of R.

Taking into account the definitions (48) and the relations (46) one gets the recurrent formulae:

$$P_m = \{m(4 + 3x) + 1\}\, R_m + x(1 + 3x)\,\frac{\mathrm{d}R_m}{\mathrm{d}x}$$

$$S_m = m(4 + 3x)\, P_m + x(1 + 3x)\,\frac{\mathrm{d}P_m}{\mathrm{d}x}\,.$$

Introducing (47) in (42) after replacing c by x (44), a by α (45) and the derivatives by means of (48), one obtains finally:

$$\begin{aligned}
&\alpha(1 + 2R^5) \\
&\times \left[\left\{\begin{array}{ll} 3(1 + x) & + 2(7 - 3x) \\ 2(-1 + 9x)\, R & + 4(-1 + 5x)\, R \\ 8xR^2 & \end{array}\right\} P + 16(1 - x)\, P^2 - 4(1 - x)(1 + 2R)S\right] = \\
&\left|\begin{array}{r} R \\ + (1 + 6x + 3x^2)\, R^2 \\ + 4x(2 + 5x)\, R^3 \\ + 24x^2 R^4 \end{array}\right|
\left.\begin{array}{r} + 6(1 - x)\, R \\ + 2(3 + 6x - 5x^2)\, R^2 \\ + 8x(3 + x)\, R^3 \\ + 32x^2 R^4 \end{array}\right| P
\quad \left.\begin{array}{r} 4(1 - x)(3 - 2x)\, R \\ + 4(3 - 2x - 9x^2)\, R^2 \\ - 16xR^3 \end{array}\right| P^2 \\
&\qquad\qquad \left.\begin{array}{r} + 8(1 - x)^2\, R \\ + 8(1 - x)^2\, R^2 \end{array}\right| P^3 - 4x(1 - x)\, R^2 (1 - 2R)^2\, S
\end{aligned} \tag{49}$$

Introducing the expressions of R, P, S one can compute the polynomials R_m, P_m, S_m by identification of the different terms of α^m. The calculation has been performed up to R_6 and the coefficients of the polynomials are given in Table II in exact integers.

The increase of the coefficients shows a formal divergence of the series near the equator but the polynomial sum can be effectively used down to near the equator for small values of a, in particular for those where a finite surface of section exists.

The exact dipole trajectories will be published elsewhere with other quantitative features of the surface of section.

As preliminary results, we give in Figure 5, the position of the centre of the surface of section for different values of $\sqrt{a}$. When $\sqrt{a}>0.25$, we are in the case of quasi-surface of section. For $a<0.25$, it is the geophysical case of the Van Allen Belt with

TABLE II

	R_1
	3
x	3
x^2	
x^3	

	R_2
	-225
x	-522
x^2	-324
x^3	702
x^4	945
x^5	
x^6	

	R_3
	81270
x	281610
x^2	333774
x^3	-505638
x^4	-1933470
x^5	-1330722
x^6	1475658
x^7	1818126
x^8	
x^9	

	R_4
	-59687685
x	-275252580
x^2	-483969195
x^3	458013204
x^4	3745676682
x^5	5652939564
x^6	-1933429338
x^7	-13767970788
x^8	-9127789317
x^9	7604933832
x^{10}	8438353605
x^{11}	
x^{12}	

	R_5
	72810408330
x	418713709050
x^2	976837762260
x^3	-420278955156
x^4	-8924656265226
x^5	-21316241372130
x^6	-6280145858520
x^7	61785741068856
x^8	103025486714022
x^9	-8432824263930
x^{10}	-162263802551868
x^{11}	-100791968025060
x^{12}	71261633813274
x^{13}	71434447876386
x^{14}	
x^{15}	

	R_6
	-132945389709282
x	-914885035445772
x^2	-2658151904487660
x^3	-135931972240932
x^4	26894546719125680
x^5	89811085550029152
x^6	79630133481525636
x^7	-257753371270844772
x^8	-801361271805228936
x^9	-469050654083229540
x^{10}	1376564053055749308
x^{11}	2430194927428108692
x^{12}	36819128132213940
x^{13}	-2863994038587331236
x^{14}	-1653670012653320148
x^{15}	1052614566698742324
x^{16}	959040586646914986

a classical surface of section. As the dipole orbit does not cut the equator on the axis U, the mirror points will not be invariant but their motions will be known when the parametric representation (y_m, q_m) of the surface of section is determined after numerical integrations for a short interval of time from mirror to equator of a certain number of well chosen trajectories.

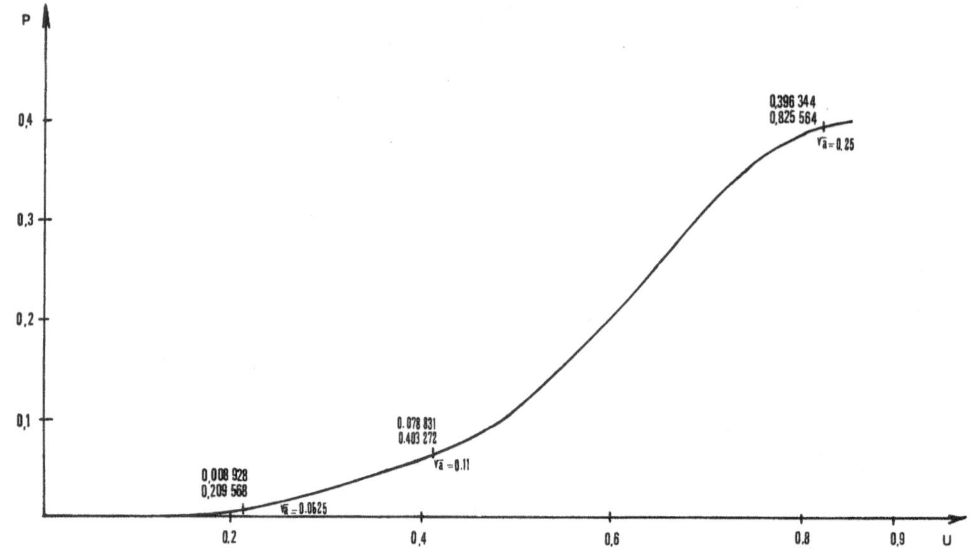

Fig. 5. Singular point of the surface of section for different values of *a*.

5. Conclusion

We have shown that in a class of dynamical problems, it is possible to construct a surface of section or in special cases (infinite extent) a quasi-surface of section.

The transformation of the surface of section can be decomposed in two operations M and R both of them being in fact a reflection consisting in the change of sign of two parameters (q_m and P). The curves invariant following M are the mirror curves intersections of all the trajectories having the same mirror latitude with the surface of section. The set of points invariant following $M:\mathscr{M}_1$ forms a twisted radius from where is measured the angular variable q_m. The set of points invariant following R forms the axis U of the representation of the surface of section. In an integrable case of this type of problem, the mirror curves are also invariant following R because they are symmetric with respect to the axis U. As a consequence mirror latitudes are invariant. This symmetry disappears on the general case. The periodic orbits become singular and the mirror latitudes will remain only invariant at and near symmetric stable periodic orbits of the first class.

References

[1] Birkhoff, G.: 1927, *Dynamical Systems*, American Math. Soc. Colloquium Publ., Vol. IX.

[2] De Vogelaere: 1957, *Contributions to the Theory of Non Linear Oscillations*, Vol. IV (ed. by Lefschetz).

[3] Whittaker, E. T.: 1944, *Analytical Dynamics*, Dover Publications, New York, p. 395.

[4] Malmquist, J.: 'Sur un système d'équations différentielles étudiées par Mr. Störmer', *Archiv Math. Astron. Fysik* **30A**, No. 5.

[5] Poincaré, H.: 1912, 'Sur un théorème de géométrie', *Rediconti circolo math. di Palerma* **33**, 375.

[6] Kolmogorov, A.N.: 1954, 'On the Conservation of Quasi Periodic Motions for a Small Change in the Hamiltonian', *Dokl. Akad. Nauk.* **98**, No. 4.

[7] Godart, O.: 1964, 'Modification de la méthode de l'Intégrale adelphique et son application au problème de Störmer', *Ann. Soc. Sci. Bruxelles* **78**, 1.
[8] Störmer, C.: 1965, *The Polar Aurora*, Clarendon Press, Oxford.
[9] Lemaître, G. and Bossy, L.: 1945, 'Sur un cas limite du problème de Störmer', *Bull. Classe Sci. Acad. Roy. Belg.* **31**, 357–364.
Bossy, L.: 1962, 'Le problème de Störmer et le mouvement des particules dans les ceintures de radiation', *Ann. Geophys.* **18**, No. 2.
[10] Godart, O.: 1941, 'Mouvement de particules chargées dans le champ d'un dipole magnétique: sur une famille d'orbites périodiques', *Revista* – Serie A, **2**, No. 1 and 2, Tucuman – Rep. Argentina.
[11] Godart, O. and Dejaiffe, R.: 1967, 'Solution du problème de Störmer pour un très grand moment de quantité de mouvement', *Ann. Soc. Sci. Bruxelles*, **81**, no. 3.

ON BOUNDED SOLUTIONS OF THE n-BODY PROBLEM

DONALD G. SAARI

Dept. of Mathematics, Northwestern University, Evanston, Ill., U.S.A.

1. Introduction

Very little is known about the solutions of a Newtonian gravitational system of n mass particles where the mutual distances between the particles are bounded as time, t, approaches infinity. In fact it is not even known whether or not the velocities must be bounded as $t \to \infty$.

We discuss here some contributions that are obtained by employing two techniques. The first is the application of analysis to the deterministic approach to the problem, i.e., the study of the behavior of motion corresponding to given initial conditions. Here some direct results and observations that seem to have escaped attention are presented. The second is the measure theoretical approach discovered by Poincaré. In both sections, some of the results are verified while others are announced with the proofs to appear elsewhere.

We assume that the number of particles, n, is finite and the center of mass O is fixed. Secondly, in Sections 3 and 4 we consider only those initial conditions where the solution exists for all time.

The following notation will be used. The symbols m_k, $\mathbf{r}_k$, $\mathbf{v}_k$ denote respectively the mass, position and velocity of the kth particle. We employ the same letter to denote the magnitude of a vector, e.g., $r_k = |\mathbf{r}_k|$, $r_{ki}^2 = (\mathbf{r}_k - \mathbf{r}_i)^2$.

We define further

$$T = \tfrac{1}{2} \sum m_k v_k^2 , \quad U = \sum_{1 \leqslant j < k \leqslant n} \frac{m_j m_k}{r_{jk}}$$

and $I = \frac{1}{2} \sum m_k r_k^2$.

If we assume the gravitational constant to be unity, the law of conservation of energy becomes

$$T = U + h \tag{1.1}$$

where h is the constant total energy. The Lagrange-Jacobi formula is simply [6].

$$\ddot{I} = U + 2h . \tag{1.2}$$

Combining (1.1) and (1.2) we get

$$\ddot{I} = 2T - U .$$

2. Bounded Solutions

Clearly $I^{1/2}$ can be considered as a measure of the expansion of the system. To see that this is so is almost a tautology. If lim sup $r_i = \infty$ for some i, then lim sup $I = \infty$.

G. E. O. Giacaglia (ed.), Periodic Orbits, Stability and Resonances, 76–81.

Likewise, if lim sup $I=\infty$, then for some i, lim sup $r_i=\infty$. (In the first, but not the second, statement 'lim sup' can be replaced with the word limit.)

So the study of boundedness of a system can be shifted to the study of the boundedness of I. The Lagrange-Jacobi relationship, $\ddot{I}=2T-U$, lends the suggestion that possibly sufficient conditions to assure the boundedness of I can be expressed in terms of $2T-U$.

If this term is positive for all time, then I is concave up, and, at the very best, $I\to\infty$ as $t\to\infty$ or $t\to-\infty$.

This term cannot be negative for all time, for if $2T-U<0$, $U<2|h|$ (by (1.1)) and I is concave down. This implies we have a complete collapse of the system at some finite time. To have a complete collapse of the system, $r_{ij}\to 0$ which implies $U\to\infty$. This contradicts the boundedness of U.

What we occasionally do see in the literature is the statement that if $2T(t)=U(t)$ for all t in some open interval of time, then *hopefully* the solution will be bounded.

Let us assume $2T(t)=U(t)$ on any nondegenerate interval of time. (Or even weaker, that there exist a sequence $\{t_i\}$ such that t_i converges to finite t_0 and $2T(t_i)=U(t_i)$.)

From the theory of differential equations any solution of the n-body problem is an analytic function of time, t, in the domain of definition. Hence T and U are analytic functions of t. As $2T$ and U agree on a nondegenerate interval, it follows that $2T(t)=U(t)$ for all t in the domain of definition. By comparing this result with (1.1) it follows that for t in the domain of definition $U=-2h$ and $T=-h$. (Note, h must be negative.)

Painlevé [4] has shown that for a real singularity to occur at real time t_0, $U\to\infty$ as $t\to t_0$. Hence the solution $U=-2h$ can be continued to $(-\infty, \infty)$.

This one assumption increases the number of "integrals" from 10 to at least 11. It also guarantees the existence of the solution for all time. This is the first half of:

THEOREM: If $2T=U$ on any nondegenerate interval of time, then $T=-h$, $U=-2h$, $\dot{U}=0$, $\ddot{I}=0$ and $\dot{I}=\dot{I}(0)$ for all time.

PROOF: From the above assumptions and the preceding paragraph, $T=-h$, $U=-2h$ and $\dot{U}=0$ for all time. This implies from (1.2), that $\ddot{I}=0$, $\dot{I}(t)=\dot{I}(0)$ and $I(t)=\dot{I}(0)t+I(0)$ for all time. If $\dot{I}(0)\neq 0$, then either as $t\to-\infty$ or $t\to\infty$ (depending on the sign of $\dot{I}(0)$) $I\to-\infty$. This is a contradiction to the definition of I which implies $I\geqq 0$. Hence $\dot{I}(0)=0$ and the theorem is proved.

COROLLARY: If $2T=U$ on any nondegenerate interval, then for all time

$$\frac{m_i m_j}{2|h|}\leqslant r_{ij}(t)\leqq\sqrt{\frac{I(0)}{m_i}}+\sqrt{\frac{I(0)}{m_j}}.$$

This simply states that close approach or escape is impossible.

PROOF:

$$2|h|=U\geqq\frac{m_i m_j}{r_{ij}(t)}.$$

This implies

$$r_{ij}(t) \geqq \frac{m_i m_j}{2|h|}.$$

Likewise

$$I(0)=I(t)>m_i r_i^2, \quad \text{so} \quad r_i(t) \leqq \sqrt{\frac{I(0)}{m_i}}.$$

By the triangle inequality,

$$r_{ij} \leqslant \sqrt{\frac{I(0)}{m_i}} + \sqrt{\frac{I(0)}{m_j}}.$$

Any relative equilibrium solution [9] would be an example of the above type motion. The author further conjectures that any solution of the above form is an equilibrium solution. (If this conjecture is true, it may lend some insight as to the reasons the 'Virial Theorem' approach to determine the mass of a galaxy is sometimes at such odds with other techniques [1].)

An example, due to Maxwell [9], would be any n masses with $n-1$ of them equal and situated at the vertices of a regular $(n-1)$ gon. The last particle is located at the center. The system is given initial conditions to place the motion into a circular orbit.

The above can be partially generalized to allow $U(t)$ to agree with an analytic periodic function f of period w on some nondegenerate interval. Again using the analyticity of U it follows that $U(t)=f(t)$, i.e., $U(t)$ is periodic and the solution exists for all time. Here again we might expect that $I=0(1)$ for all t, but the best I have been able to do is the following [8]:

THEOREM: If $U(t)$ is periodic with period w, then either $I(t)$ is periodic with period w or $I \sim ct^2$ as $t \to \infty$ and there exist $\mathbf{r}_i$, $\mathbf{r}_k$ and $\mathbf{r}_j$ such that $\liminf r_{ki}/r_{ij}=0$, $\limsup (r_{ki}/r_{ij})>0$ and $\limsup r_{ij}=\infty$ as $t \to \infty$. c is some positive constant. If the second case occurs, a similar condition holds for the same choice of c as $t \to -\infty$.

An example where $U(t)$ is periodic would be the $(n-1)$-gon example where the initial conditions are such that we have elliptic motion. In this case $I(t)$ is periodic.

3. Bounded Solutions and Poisson Stability

Poincaré [5] was able to obtain results about bounded solutions of the n-body problem by studying the 'flow' of a set of initial conditions. Namely, if p is an initial condition and $f(p, t)$ is the position on the trajectory at time t in phase space, he studied the relationship and behavior of set A and $f(A, t)=\cup_{p\in A} f(p, t)$.

In particular, if we restrict our attention to incompressible systems ($\dot{\mathbf{X}}=\mathbf{g}(\mathbf{X})$ is incompressible if Div $\mathbf{g}(\mathbf{X})=0$) then for (Lebesgue) measurable A, meas $A=$ meas $f(A, t)$. This applies directly to the n-body problem as conservative dynamical systems are clearly incompressible.

Poincaré has shown the following: Let A be any bounded measurable set, meas $A>0$, with the property that $p\in A$ implies $I \leqslant D_1$ and $U \leqslant D_2$ as $t\to\infty$, where D_1 and D_2 are some positive constants. With the possible exception of a subset of measure zero, all motion with initial conditions in A is Poisson stable, i.e., the motion returns infinitely often arbitrarily close to any given point on the trajectory.

This result can be strengthened, as is undoubtedly known.

THEOREM: Let A be any measurable set of initial conditions with the properties meas $A>0$ and $p\in A$ implies $I=0(1)$ as $t\to\infty$. Then for almost all initial conditions of A, $I=0(1)$ as $t\to-\infty$ and the motion is Poisson stable.

PROOF: By Hopf [3], almost all initial conditions are either future Poisson stable or have the property that for any sequence $\{t_i\}$, $t_i\to\infty, f(p, t_i)$ does not have a limit point. We assume the latter holds and show it leads to a contradiction. Consider $J=\sum(\mathbf{r}_i^2+\dot{\mathbf{r}}_i^2)$. If $\lim\inf J<\infty$ then we can find positive $D<\infty$ and $\{t_i\}$ such that $t_i\to\infty$ and $J(t_i)\leqslant D$. As D defines a compact set in E^{6n} a subsequence can be found such that $\{f(p, t_i)\}$ has a limit point. As this is a contradiction, limit $J=\infty$. The condition $I=0(1)$, implies $\sum\dot{\mathbf{r}}_i^2\to\infty$ as $t\to\infty$, i.e. $T\to\infty$ as $t\to\infty$. From (1.1) and (1.2) this implies that after some time t, $\ddot{I}\geqslant 2$, or $I\geqslant(t-t_1)^2+o(t^2)$. As this is a contradiction to the boundedness of I, we have that almost all motion is future Poisson stable. By being future Poisson stable, this implies [7] $f(p, I)\subset\overline{f(p, I^+)}$ where $f(p, I^+)=\cup_{0<t<\infty}f(p, t)$ and $f(p, I)=\cup_{-\infty<t<\infty}f(p, t)$. The bar denotes the closure of the set. Although $\overline{f(p, I^+)}$ may be unbounded, it is bounded in the coordinates $\mathbf{r}_i$. Hence $I=0(1)$ as $t\to-\infty$. A similar argument shows that almost all motion is past Poisson stable $\Rightarrow$ almost all motion is Poisson stable. The proof is completed.

The next problem is to develop a criterion which will ensure that for a given initial condition, the motion is Poisson stable, i.e., that this initial condition does not belong to the set of measure zero. We first define orbital stability.

DEFINITION: $X(t)$ is future orbital stable if for any $\varepsilon>0$, there exists a $\delta>0$ such that if at time $t_0=0$, $|y(0)-X(0)|<\delta$ then $|y(t)-X(I^+)|<\varepsilon$. (We assume that the solutions exist on $(0, \infty)$.)

That is, slight perturbations of the initial condition result in only small deviations from the *trajectory* of the given solution. In what follows let g be of class C^1.

THEOREM: If $\dot{\mathbf{X}}=\mathbf{g}(\mathbf{X})$ is an incompressible system and if $\boldsymbol{\varphi}(t)$ is a bounded and future orbitally stable solution, then $\boldsymbol{\varphi}(t)$ is future Poisson stable.

Of course, the corresponding results about past Poisson stability and Poisson stability hold also. In the n-body case this reads:

THEOREM: If for a given initial condition, the resulting solution has $I=0(1)$ as $t\to\infty$ and is future orbital stable, then it is future Poisson stable and $I=0(1)$ as $t\to-\infty$.

4. Boundedness and Almost Periodicity

Motion that is Poisson stable seems to have a certain degree of predictability. It seems reasonable to conjecture that under more stringent conditions such motion may be almost periodic (in the sense of Bohr). A result, due to Helms and Putnam exists to this effect [2].

THEOREM: Let system $\dot{\mathbf{X}}=\mathbf{g}(\mathbf{X})$ be incompressible and suppose that $\mathbf{X}=\mathbf{X}(t)$ is a bounded, Liapounoff stable solution. Then $\mathbf{X}(t)$ must be almost periodic.

The condition that the motion be Liapounoff stable is too strong for this result to be applied even to the two body problem. To see this, assume the two body problem is in elliptic motion. Slight changes in initial conditions can be made which lead to a change in the frequency of the motion. Hence, at some time the distance between the two solutions will exceed any small $\varepsilon > 0$.

We generalize the above result to include motion such as the two body problem by introducing the following:

DEFINITION: $X(t)$ is said to be *almost Liapounoff stable if* it is orbital stable and for any $\varepsilon > 0$ there exists a positive T and a measurable set A, meas $A > 0$, such that if $y(0) \in A$, then for all t

$$|y(t_1) - X(t)| < \varepsilon,$$

where t_1 satisfies

$$|t_1 - t| < T.$$

THEOREM: If $\mathbf{X}(t)$ is bounded, almost Liapounoff stable solution of an incompressible system, with the additional property that $\mathbf{X}(t)$ is Liapounoff stable with respect to its trajectory, then $\mathbf{X}(t)$ must be almost periodic.

References

[1] Contopoulos, G.: 1966, 'Problems of Stellar Dynamics', *Space Math.*, Vol. 5, Appl. Math. A.M.S., pp. 178–181.

[2] Helms, L. L. and Putnam, G. R.: 1958, 'Stability in Incompressible Systems', *J. Math. Mech.* **7**, pp. 901–904.

[3] Hopf, E.: 1930, 'Zwei Sätze über den wahrscheinlichen Verlauf der Bewegungen dynamischer Systeme', *Math. Ann.* **103**, 710–719.

[4] Painlevé, M. P.: 1907, *Leçons sur la Théorie Analytique des Équations Différentielles* (Libraire Scientifique), Hermann, Paris, 1907.

[5] Poincaré, H.: 1957, *Méthodes nouvelles de la mécanique céleste*, Vol. III, Dover.

[6] Pollard, H.: 1966, *Mathematical Introduction to Celestial Mechanics*, Prentice-Hall, New York, chapter 2.

[7] Nemytskii, V. V. and Stepanov, L. L.: 1960, *Qualitative Theory of Differential Equations*, Princeton University Press.

[8] Saari, D. G.: 'Expanding Gravitational Systems', to appear.

[9] Wintner, A.: *The Analytical Foundations of Celestial Mechanics*, Princeton University.

Discussion

W. H. Jefferys: In the actual determination of masses of galaxies by the Virial Theorem, the rather vague assumption is made that the time average of $2T$ approximately equals that of U over some fixed, large time interval (large because of the assumption of long-term stability of the cluster). Thus this is not the same assumption that you have of $2T = U$ over some interval. Certainly they do not think that the motion is anything like equilibrium orbits.

D. Saari: Yes, but I believe that after this assumption of time averages is made, embedded in their analysis is the further unintentional assumption that $2T = U$. What leads me to this conclusion is that those authors who assume that $2T=U$, or $I=D$, over some interval of time end up with similar final equations for mass. In some cases the error comes in by equating the time averages, and hence assuming

$$2\int_0^t T(s)\,\mathrm{d}s=\int_0^t U(s)\,\mathrm{d}s$$

over some interval of time. As the integrals are also analytic functions of time, this implies that this relationship must hold for all time. Hence their derivatives agree and $2T=U$ for all time.

Notice how a very innocent sounding assumption may carry with it undesired conclusions. This is due to the analyticity of the solutions. Notice also the possible implications resulting from the assumption of periodicity of U.

M. Hénon: My comment is similar to Jefferys's. I do not think that your proof can be applied to real galaxies or clusters. Numerical experiments which have been made to simulate these systems have shown that the quantity $2T=U$ never stays equal to zero, but rather fluctuates around this value.

The relative amplitude of these fluctuations seems to be of the order of $n^{-1/2}$, where n is the number of bodies. Thus if n is large, it can be assumed for practical purposes that $2T-U=0$. But this is only an approximation.

D. Saari: This is exactly my point. Of course real galaxies or clusters do not behave like equilibrium solutions. Hence if we try to approximate a galaxy by an equilibrium solution we would expect the fit to be quite poor and any resulting conclusions, such as the value of the mass of the galaxy, to be subject to question. This is what I believe happens in the Virial Theorem approach of mass determination. For example, in the three body problem my conjecture is true. The only equilibrium solutions are the Euler and Lagrange solutions. Obviously these collinear and equilateral triangle cases are a poor fit for most solutions. Secondly, you state that if n is large, for practical purposes it can be assumed that $2T=U$. But if this assumption of equality is ever used, even that it holds for only one second of time, then you are tacitly approximating your solution with a second one in which escape or very close approaches are automatically ruled out. If my conjecture is true, you run again into the problem of approximating your solution by an equilibrium solution.

E. M. Standish: Do you have any estimate for the maximum length of time during which the inequality, $2T<U$, can hold?

D. Saari: No.

STABILITY OF MOTION NEAR SUN-PERTURBED EARTH-MOON TRIANGULAR LIBRATION POINTS*

AHMED A. KAMEL and JOHN V. BREAKWELL
Stanford University, Stanford, Calif., U.S.A.

Abstract. An extension is presented of an analysis by Schechter [1] of sun-perturbed motion near the earth-moon L_4 point. In the present paper motion is confined to two dimensions. With the help of computerized symbolic manipulation and some new recurrence formulae [6] inspired by Deprit [4], a canonical perturbation treatment is carried to 4th order in the small quantities: lunar eccentricity, distance from L_4/earth-moon distance, moon mass/earth mass, solar mean motion/lunar mean motion; (earth-moon distance/earth-sun distance) is included as a second order small quantity. Ignoring lunar eccentricity, stable one-month periodic orbits, synchronized with the sun, are found, from the 3rd-order Hamiltonian, somewhat larger than in Schechter's second order treatment, and in very good agreement (3%) with Kolenkiewicz and Carpenter. Unstable three-month periodic orbits are also found in fairly close proximity to the stable one-month orbits, and the limited extent of the region of the stability is estimated in suitable phase coordinates. The inclusion of the 4th-order terms gave a slight deterioration rather than a further improvement in the agreement on the stable orbits. The inclusion of the lunar eccentricity up to 3rd-order yielded stable quasi-periodic orbits slightly larger than the previous stable orbits.

1. Linear Theory

It is very well-known in the restricted three-body problem that the equilateral libration point L_4 (and also L_5) is 'linearly stable' if $\mu' = \mu(1-\mu)^{**} < \frac{1}{27}$. This follows easily from local linearization of the equations of motion in the rotating frame of the two masses:

$$\ddot{x} - 2n\dot{y} = \partial U^*/\partial x$$
$$\ddot{y} + 2n\dot{x} = \partial U^*/\partial y$$
$$\ddot{z} + n^2 z = \partial U^*/\partial x,$$

where $U^* = (\mu_1/d_1 + \mu_2/d_2 + (n^2/2)\, d_B^2$, $n^2 = (\mu_1 + \mu_2)/d_{12}^3$, and d_1, d_2, d_B denote distances from m_1, m_2 barycenter B, respectively. Stability of the linearized equations for $\mu' < \frac{1}{27}$ occurs in spite of the fact that the effective potential energy $[-U^*(x, y, 0)]$ in the plane of the two masses has a maximum rather than a minimum at L_4, the stability depending on the influence of the Coriolis accelerations.

The 'normal modes' of the linearized problem are ellipses, centered at L_4 with minor axes almost along L_4B. The motions on the corresponding auxiliary circles are at uniform retrograde rates ω_1, ω_2. In the earth-moon case, of course, $\mu' < \frac{1}{27}$, and ω_1, ω_2 are found to be approximately $0.95n_☽$, $0.30n_☽$, respectively. The shapes of these ellipses are $b_1/a_1 \cong \frac{1}{2}$, $b_2/a_2 \cong \frac{1}{5}$, respectively.

The inclusion of the solar perturbation and the mean lunar eccentricity $e_☽$ in this linearized account may be accomplished 'to first order' as follows: the 'direct' solar

* This research was sponsored by the National Aeronautics and Space Administration under Research Grant NsG 133-61.

** μ denotes $m_2/(m_1+m_2)$.

G. E. O. Giacaglia (ed.), Periodic Orbits, Stability and Resonances, 82–90.

effect is the linearized difference ('gravity gradient' effect) between the solar attractions at L_4 and B. The effect of $e_☽$ and the 'indirect' solar effect arise since both $e_☽$ and the sun contribute fluctuations to the earth-moon distance and direction, the sun also contributing a constant correction term to the 'mean distance' $\bar{d}_{\oplus☽}$ related to the mean angular rate $n_☽$ by: $\bar{d}^3_{\oplus☽}=(\mu_\oplus+\mu_☽)/n^2_☽$. The solar and eccentricity effects thus introduce small forcing terms to the linear equations of motion, these to be evaluated as known time-functions at L_4. Here we may define L_4, to be precise, as forming an equilateral triangle with a 'mean earth' and 'mean moon' which revolve about their barycenter B at uniform rate $n_☽$ and mutual distance $\bar{d}_{\oplus☽}$.

The response of the linear system to a small eccentricity forcing term is found to be an exact imitation of the moon's fluctuation about the mean 'moon'; i.e., a possible motion is such as to form an equilateral triangle with the instantaneous earth and moon in their plane of relative motion, this well-known result being indeed not confined to small eccentricities. The response to the solar forcing terms includes constant and twice-monthly fluctuations of order 1 000 km with frequency $2(1-m)\,n_☽$, where $m=0.074\,801$ is the ratio of the sidereal month to the sidereal year. Note that the sun's angular motion in our rotating frame is retrograde with angular rate $\omega_\odot=(1-m)\,n_☽$ but that the solar gravity-gradient matrix has twice this frequency. The sun also causes a secular motion of the moon's perigee and this distorts somewhat the eccentricity imitation.

2. Non-Linear Features and Schechter's Results [1]

So far, nothing has been said to indicate the relatively unstable nature of L_4 which will appear later. If, however, terms are included non-linear in m and the displacements from L_4, certain of the effective forcing functions will have frequencies $2\omega_\odot-\omega_1$ or $\omega_1-2\omega_2$, which are found to lie rather close to ω_1 and ω_2, respectively. These non-linear near-resonances $2\omega_\odot\cong 2\omega_1$ ('external') and $\omega_1\cong 3\omega_2$ ('internal') may have a profound influence on stability near L_4, as was shown by Schechter [1] who included Hamiltonian perturbation terms up to 4th degree in m and the displacements from L_4, equivalent to forces up to the 3rd degree.

Without including $e_☽$, Schechter found that the solar perturbation was just strong enough to give rise to parametric excitation of mode 1, because of the external near-resonance, so that the relatively small forced solution of the linear theory is necessarily unstable. He also found a stable size for mode 1 with $a_1\cong 95\,000$ km, at which size the non-linear effect on frequency changes 'ω_1' to coincide with $\omega_\odot$, the uniform monthly retrograde motion around the auxiliary circle being, moreover, aligned directly towards or away from the sun. He also found a slightly larger unstable three-month periodic motion with mode '2' included but to a much smaller extent.

Incidentally, Schechter's application of a von Zeipel procedure to derive a long-period Hamiltonian was more successful than an earlier paper by Breakwell and Pringle [2], in which the 2nd-order (4th degree) Hamiltonian was subsequently found to be misleading, after comparison with a formal expansion by Deprit, Henrard and

Rom [3]. Schechter's formula for the 2nd-order long-period Hamiltonian was subsequently justified by Deprit [4].

Schechter also found that non-linear coupling with any small out-of-plane motion was insignificant, and that the motion will remain in the vicinity of the gradually changing earth-moon plane.

3. The Periodic Orbits of Kolenkiewicz and Carpenter [5]

Kolenkiewicz and Carpenter used a periodic coplanar earth-sun-moon model, without any mean orbital eccentricities, and investigated by trigonometric series the possibility of a coplanar monthly periodic motion in their presence in the general vicinity of L_4. They found, in addition to a small unstable orbit, two similar but not identical stable orbits about 50% larger than Schechter's stable orbits, one synchronized with the sun in its motion around L_4; and the other 180° out-of phase.

The slight difference between these two stable orbits is explainable, as we shall see, by the fact that Kolenkiewicz and Carpenter, unlike Schechter, included more than the 2nd harmonic (gravity-gradient term) in the expression for the solar disturbing potential. The next term, in fact, introduces a force field rotating in our coordinate system at rate $\omega_\odot$ rather than $2\omega_\odot$.

The substantial difference in size between the stable orbits in [1] and [5] prompted a higher-order study in two dimensions, reported in the remaining sections.

4. Outline of a Higher-Order Two-Dimensional Treatment

Adopting units of distance and time so that $n_☽ = \bar{d}_{\oplus ☽} = 1$, the earth-moon vector is expressed in our uniformly rotating frame $(\hat{\imath}_x, \hat{\imath}_y)$ by:

$$\mathbf{r}_☽ = (1 + x_☽)\,\hat{\imath}_x + y_☽ \hat{\imath}_y,$$

$\hat{\imath}_x$ being along the 'mean' rotating earth-moon line. $x_☽$ and $y_☽$ are known from lunar theory and need to be carried only to the 4th power in m and $e_☽$. Motion near L_4 is described by the position vector relative to earth:

$$\mathbf{r} = (\tfrac{1}{2} + \mu x_☽ + x)\,\hat{\imath}_x + (\sqrt{\tfrac{3}{2}} + \mu y_☽ + y)\,\hat{\imath}_y,$$

where μ, as usual, denotes $\mu_☽/(\mu_\oplus + \mu_☽)$, and (x, y) indicates position relative to L_4.

The Lagrangian for the motion is

$$\mathscr{L} = \tfrac{1}{2}|\dot{\mathbf{r}}|^2 - V,$$

in which

$$\dot{\mathbf{r}} = (\mu \dot{x}_☽ + \dot{x} - \sqrt{\tfrac{3}{2}} - \mu y_☽ - y)\,\hat{\imath}_x + (\mu \dot{y}_☽ + \dot{y} + \tfrac{1}{2} + \mu x_☽ + x)\,\hat{\imath}_y,$$

and

$$-V = \frac{1-\mu}{d_\oplus} + \left(\frac{\mu}{d_☽} - \frac{\mu \mathbf{r}_☽ \cdot \mathbf{r}}{r_☽^3}\right) + \left(\frac{\mu_\odot}{d_\odot} - \frac{\mu_\odot \mathbf{r}_\odot \cdot \mathbf{r}}{r_\odot^3}\right).$$

The direct solar terms in $-V$, i.e., the last bracket, may be approximated by

$$-V_\odot = m^2 \left\{ [\tfrac{3}{2}(\hat{r}_\odot \cdot \mathbf{r})^2 - \tfrac{1}{2} r^2] + \frac{\hat{r}_\odot \cdot \mathbf{r}}{\bar{r}_\odot} [\tfrac{5}{2}(\hat{r}_\odot \cdot \mathbf{r})^2 - \tfrac{3}{2} r^2] \right\},$$

where $\bar{r}_\odot$ is the mean earth-sun distance and $\hat{r}_\odot$ a unit-vector rotating in retrograde direction at rate $\omega_\odot = 1 - m$. The first square bracket in $V_\odot$ comprises the gravity-gradient terms.

The Hamiltonian is

$$\begin{aligned} H(x, y, p_x, p_y) &= (p_x - \tfrac{1}{2}\sqrt{3})\dot{x} + (p_y + \tfrac{1}{2})\dot{y} - \mathscr{L} \\ &= \tfrac{1}{2}(p_x^2 + p_y^2) + (yp_x - xp_y) - \tfrac{1}{2}(x + \sqrt{3}y) \\ &\quad + \mu[(y_☽ - \dot{x}_☽)\, p_x - (x_☽ + \dot{y}_☽)\, p_y] + V, \end{aligned}$$

where p_x, p_y have been defined for convenience as the quantities $\partial\mathscr{L}/\partial\dot{x} + \frac{1}{2}\sqrt{3}$, $\partial\mathscr{L}/\partial\dot{y} - \frac{1}{2}$, which are small for motions remaining near L_4.

H is next expanded in ascending powers of x, y, up to 6th degree. The terms linear in x, y, but independent of $e_☽$ and m, naturally drop out. The resulting H is re-scaled and separated as follows:

$$\frac{H}{m^2} = H_0 + \sum_{q=1}^{4} \frac{m^q}{q!} H_q,$$

in which H_0, the 'unperturbed' Hamiltonian, includes for convenience *all* terms quadratic in $x/m, y/m, p_x/m, p_y/m$ with constant coefficients. (The corresponding terms from the solar gravity-gradient potential were left in the perturbing Hamiltonian in [1].)

The convention adopted for the perturbation terms H_q, $q \geqq 1$, is as follows: Since H involves the mass-ratio μ in its coefficients both by itself and in the combination $(1-2\mu)$, while H_0 involves only the combination $(1-2\mu)$, this combination is treated as a numerical value of order zero, while μ is treated as a numerical multiple of m, as is $e_☽$, the ratio $(e_☽/m)$ being retained, however, as an adjustable parameter for later comparison with [5] for which it is zero. Finally, the quantity $1/\bar{r}_\odot$ is expressed as a multiple of m^2. $H_q (q=1, \ldots, 4)$ are thus functions of μ/m, $e_☽/m$, $1/\bar{d}_\odot m^2$, and time t as well as of the re-scaled $x/m, y/m, p_x/m, p_y/m$.

A canonical transformation is performed to the action and angle variables A_i and B_i of H_0:

$$\begin{pmatrix} x/m \\ y/m \\ p_x/m \\ p_y/m \end{pmatrix} = \begin{pmatrix} \\ 4 \times 4 \\ \text{matrix} \\ \end{pmatrix} \begin{pmatrix} \sqrt{A_1}\sin B_1 \\ \sqrt{A_2}\sin B_2 \\ \sqrt{A_1}\cos B_1 \\ \sqrt{A_2}\cos B_2 \end{pmatrix},$$

in which the matrix agrees with J in Equations (5), (6) of [2], except for the influence of additional solar terms in H_0, the A_i's being related to the 'energies' α_i there by $\omega_i A_i = \alpha_i$, while

$$\begin{cases} B_1 = \omega_1(t + \beta_1) \\ B_2 = \omega_2(-t + \beta_2) \end{cases}.$$

The unperturbed Hamiltonian takes the simple form:

$$H_0 = \omega_1 A_1 - \omega_2 A_2,$$

its non-definite form being associated with the fact mentioned earlier that L_4 is a maximum rather than a minimum of $[-U^*(x, y, 0)]$.

In contrast to [1] and [2], however, the terms in H linear in x/m, etc., with coefficients involving m^2 (and μe_y), have been retained in H_1. Note that the frequencies ω_1 and ω_2 have been slightly affected by the additional solar terms included in H_0. The revised values are:

$$\omega_1 = 0.949\,313, \quad \omega_2 = 0.300\,684.$$

A further canonical transformation, in powers of the small parameter m, to 'slow variables' $\bar{A}_i$, $\bar{B}_i$ is now introduced with the object of removing all trigonometric terms from the resulting Hamiltonian K except those with 'slow' arguments $\bar{B}_1 + 3\bar{B}_2$, $\bar{B}_1 - \omega_\odot t$ and combinations thereof. The algebra for carrying this out has been described in a recent paper by one of the present authors (Kamel [6]), and is performed by symbolic manipulation on a digital computer.

Since H_1 has no slow terms but only short-period terms, these formulas to 4th order take on a simpler form than in [6]:

$$\begin{cases} K_0 = H_0 \\ K_1 = H_1 - \dfrac{DW_1}{Dt} = 0 \\ K_2 = H_2 + (H_1;\, W_1) - \dfrac{DW_2}{Dt} \\ K_3 = H_3 + (H_2 + 2K_2;\, W_1) + 2(H_1;\, W_2) - \dfrac{DW_3}{Dt} \\ K_4 = H_4 + (H_3 + 3K_3;\, W_1) + 3(H_2 + K_2;\, W_2) + 3(H_1;\, W_3) \\ \qquad - 3((K_2;\, W_1);\, W_1) - \dfrac{DW_4}{Dt}, \end{cases}$$

where the W_i's are chosen successively to remove short-period parts of K_i, DW_i/Dt denoting $\partial W_i/\partial t - (H_0;\, W_i)$.

The transformation on each of the original coordinates (x, y, p_x, p_y) is given up to 2nd order by

$$\xi = \xi_0 + m\xi_1 + (m^2/2)\,\xi_2, \quad (\xi = x, y, p_x, p_y),$$

where

$$\xi_1 = (\xi_0;\, W_1) \quad \text{and} \quad \xi_2 = (\xi_0;\, W_2) + (\xi_1;\, W_1).$$

The resulting long-period Hamiltonian K is expressed for convenience in terms of the canonical variables $\bar{A}_i$, B_i^*, where $B_1^* = \bar{B}_1 - \omega_\odot t + \varphi_1$, $B_2^* = \bar{B}_2 + \frac{1}{3}\omega_\odot t + \varphi_2$. The unperturbed Hamiltonian becomes, of course,

$$K_0^* = K_0 - \omega_\odot \bar{A}_1 + \tfrac{1}{3}\omega_\odot \bar{B}_1 = (\omega_1 - \omega_\odot)\,\bar{A}_1 - (\omega_2 - \tfrac{1}{3}\omega_\odot)\,\bar{A}_2.$$

The 2nd-order part has the same general form as in [1]:

$$K_2 = C_{11}\bar{A}_1^2 + 2C_{12}\bar{A}_1\bar{A}_2 + C_{22}\bar{A}_2^2 + C_1\bar{A}_1 \cos 2B_1^* + C_2\bar{A}_1^{1/2}\bar{A}^{3/2} \cos(B_1^* + 3B_2^*),$$

if phases φ_1 and φ_2 have been suitably chosen.

The 3rd-order part has the form

$$K_3 = C_3\bar{A}_1^{1/2} \cos(B_1^* + \phi_3) + C_4\bar{A}_1 \cos(2B_1^* + \phi_4) + (e_☽/m)^2 (C_5\bar{A}_1 + C_6\bar{A}_2).$$

The 4th order part K_4, neglecting $(e_☽/m)$, includes secular terms of 3rd as well as 2nd degree in $\bar{A}_i$, and trigonometric terms with arguments $B_1^*, 2B_1^*, B_1^* + 3B_2^*$ and $B_1^* - 3B_2^*$.

The values of the constants C_{ij}, C_i, etc., will appear shortly in the Ph.D. dissertation by Kamel [7].

It should be pointed out that additional rather slow arguments, introduced along with $e_☽$, have been removed from H along with the other short-period terms. This includes, for example, the argument $\bar{B}_1 - f_☽$, where $\dot{f}_☽$ the lunar anomalistic mean motion is $1 - \frac{3}{4}m^2 - \frac{245}{32}m^3 \ldots$. The justification for removal of these additional trigonometric terms, suggested by the imitative response to $e_☽$ when $m = 0$, lies in the non-appearance of excessively large coefficients of $(e_☽/m)^j$ in the resulting computations.

It should be added, however, that, due to limitations on computer storage, the inclusion of $(e_☽/m)$ was terminated at $(e_☽/m)^3$ and at K_3 instead of K_4.

5. Resulting Periodic Orbits

Periodic orbits, in the absence of lunar eccentricity $e_☽$, are found by searching for equilibria in the $(\bar{A}_i, \bar{B}_i^*)$ phase-space, any such equilibrium representing a three-month periodic orbit if $\bar{A}_2 \neq 0$, and to a one-month periodic orbit if $\bar{A}_2 = 0$.

The search for these equilibria is accomplished more easily in terms of the related variables $P_i = \sqrt{2\bar{A}_i} \cos B_i^*$, $Q_i = \sqrt{2\bar{A}_i} \sin B_i^*$, $(i = 1, 2)$. Because of the term in K_3 linear in P_1, Q_1, arising from the higher-order solar perturbation term with factor $1/\bar{r}_\odot$, the origin is no longer an equilibrium but a nearby equilibrium exists, whose stability is essentially that of the Hamiltonian system $(\omega_1 - \omega_\odot)\bar{A}_1 + C_3\bar{A}_1 \cos 2B^*$. Since, as in [1], the coefficient C_3 is somewhat larger than $\omega_1 - \omega_\odot$, this Hamiltonian is not positive-definite, as a quadratic form in P_1, Q_1, so that mode 1 is parametrically excited and the periodic motion corresponding to this small $\bar{A}_1$ is necessarily unstable. The monthly position fluctuation due to this response to the higher-order solar perturbation is comparable in size to the direct gravity-gradient twice-monthly fluctuation.

The inclusion of higher powers of $\bar{A}_1$ in K leads to the possibility as in [1] of a stable size for mode 1 in the presence of the sun. The corresponding equilibrium points, with $P_2 = Q_2 = 0$, have been calculated in the form

$$(P_1, Q_1) = \pm(,) + m(,) + m^2(,).$$

The first term results from K_2 alone and yields a stable monthly motion in phase or 180° out of phase with the sun. The corresponding semi-axis a_1 is approximately 10% rather than 33% lower than that in [5]. The difference with [1] must arise from the inclusion of some m^2-terms in H_0.

The inclusion of the 1st power of m in the calculation of (P_1, Q_1) yields a substantially improved agreement with [5]. The $\pm$ sign, depending on motion in or out of phase with the sun, now gives two slightly different sizes for a_1, both only about 3% larger than in [5]. The variables x, y have been computed as functions of time, including the influence of W_1 and W_2, and the resulting periodic orbits plotted. Allowing for the 3% difference in size, the resulting orbits (see Figure 1) were quite indistinguishable by eye from those in Figure 4 of [5].

The inclusion of m^2, however, in the calculation of (P_1, Q_1) *failed* to yield further improved agreement with [5], but yielded instead orbits about 3% too small. The computations, then, have failed to show good convergence in powers of m.

The existence of slightly larger *unstable* orbits, as in [1], is now more questionable. Including m but not m^2, unstable equilibria (P_1, Q_1, P_2, Q_2) were found with mode 1

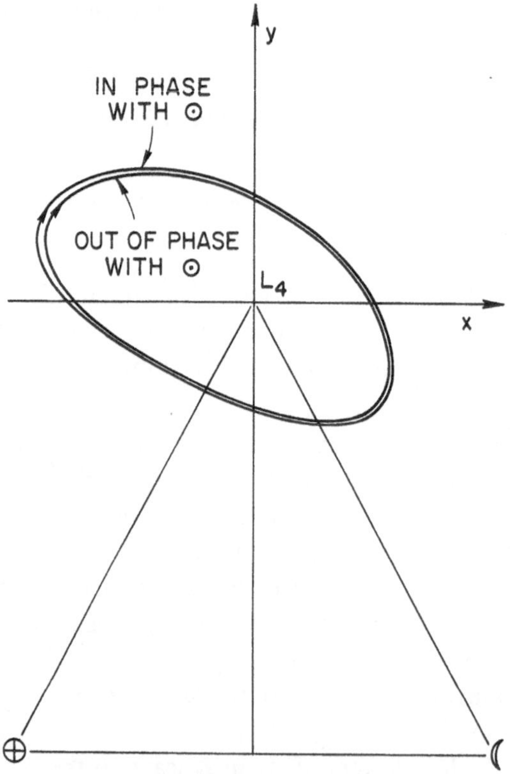

Fig. 1. One-month stable periodic orbits.

again in or out of phase with the sun but with a_1 about 25% larger than for the stable periodic orbits. The values of (P_2, Q_2) moreover, corresponded to a semi-axis a_2 of about 20 000 km. The inclusion of m^2, however, yielded a totally unreasonable change of equilibrium point. If, nevertheless, the results up to 1st power of m are taken seriously, a very rough idea of the extent of a region of stability around the stable orbits can be obtained by setting P_1 and Q_1 equal to their stable equilibrium values and expanding $K_0 + (m^2/2) K_2$ up to 3rd degree in P_2, Q_2. The resulting Hamiltonian takes the form (ignoring an additive constant):

$$K^{\ddagger} = C_7(P_2^2 + Q_2^2) - \tfrac{2}{3}C_8(P_2^3 - 3P_2Q_2^2).$$

The associated contours in the $P_2 - Q_2$ plane have, in addition to the stable equilibrium at the origin, unstable equilibria at $(C_7/C_8, 0)$ and $(-\frac{1}{2}C_7/C_8, \pm\frac{1}{2}\sqrt{3}C_7/C_8)$ forming the vertices of an equilateral triangle. These three vertices correspond to the same unstable three-month periodic orbit re-started at one-month intervals. The straight lines through those vertices are the separatrices corresponding to $K^{\ddagger} = \frac{1}{3}C_7^3/C_8^2$, so that only the interior of this triangle leads to bounded (P_2, Q_2) motion. This approximate analysis has not taken into account the fact that the unstable (P_1, Q_1) equilibrium values are larger than the stable values. Nevertheless, some idea is obtained about how small $(a_2 \sim 10\,000$ km) initial mode 2 must be to avoid eventual wild divergence of the orbit.

Finally, the effect of $(e_☽/m)$ on the *stable* orbits was computed. The contribution of $(e_☽/m)^2$ to the equilibrium values was small, yielding an increase of less than 5% in a_1. The eccentricity appears, of course, also in the removed short-period parts of the Hamiltonian, and the resulting stable orbits are no longer periodic but only quasi-periodic.

References

[1] Schechter, H. B.: 1968, *AIAA J.* **6**, No. 7.
[2] Breakwell, J. V. and Pringle, R.: 1966, *Progr. Astron.* **17**.
[3] Deprit, A., Henrard, J., and Rom, A. R. M.: 1967, *Icarus* **6**.
[4] Deprit, A.: 1969, *Celes. Mech.* **1**, 12.
[5] Kolenkiewicz, R. and Carpenter, L.: 1968, *AIAA J.* **6**, No. 7.
[6] Kamel, A. A.: 1969, *Celes. Mech.* **1**, 190.
[7] Kamel, A. A.: Ph.D. Dissertation, Stanford University, to appear.

Discussion

J. Kevorkian: I would like to report on an alternate approach to this problem undertaken by Mr. L. Mohn and myself leading us to the conclusion that the general philosophy of expanding the solution in the neighborhood of the triangular libration points is inappropriate for the case of solar perturbation because most orbits in this case do not remain close to the libration points. I believe that the remarkable agreement between theoretical and numerically integrated results reported by Prof. Breakwell is only true for the few cases of stable 'periodic' orbits. Since these orbits correspond to stationary points in the phase-plane of the averaged Hamiltonian, the solution calculated by canonical perturbation theory can be carried out to very high orders with relative ease for these special cases. Numerical integration confirms that most initial conditions near the triangular libration points lead to large deviations from these points for the case of solar perturbations and non-zero lunar eccentricity. This general behavior is exhibited in our literal theory by the divergence of the asymptotic expansions for the coordinates as functions of time for all but a few exceptional initial conditions.

Based on this fact, I believe that if one is interested in the general evolution of motion from the libration points, one must use a modified planetary theory whereby the dominant force is the earth's attraction, and the lunar as well as solar attractions are assumed small. In this approach we need not restrict ourselves to orbits that remain near the libration points.

J. V. Breakwell: As I understand it, your method introduces initial conditions into the relation between the original variables and the 'slow variables'. The behavior of the latter, then, is influenced in your analysis by initial conditions. That most initial conditions then lead to divergence (large motions away from L_4) is in agreement with my estimate of a rather small domain of stability around my fairly large stable orbits.

MOTION NEAR SUN PERTURBED EARTH-MOON COLLINEAR EQUILIBRIUM POINTS*†

Status Report

G. E. O. GIACAGLIA

University of São Paulo and University of Campinas, Brasil

and

L. N. F. FRANÇA

University of São Paulo, Brasil

The scope of this research is to study the evolution of the orbit of a small mass, P, near the points L_1 or L_2 or the earth-moon system.

More specifically, the goals of the present work are:

(1) A definition of L_1 and L_2 when solar perturbations are taken into account and a sufficiently precise theory of the moon is considered.

(2) A presentation of the equations of the motion of P, including terms up to second order in some small parameters.

Let us suppose that, at a given instant, the sun's perturbations are eliminated and the moon describes an elliptic orbit around the earth.

Furthermore, let us consider a frame of reference, with the origin at the center of mass of the earth-moon system and rotating with the moon.

As it is known, in this reference system, there exist five stationary solutions, which reduce to the equilibrium Lagrangian points $L_1, L_2, \ldots, L_5$, when the eccentricity of the moon is set equal to zero.

All these solutions are segments of straight line, which are described by the particle, in phase with the rotation of the primaries.

Such stationary solutions can also be called equilibrium points of the elliptic restricted problem, because they are the equilibrium-points of a system of differential equations which represent the problem in pulsating coordinates.

Let us go back to the proposed problem with which we are concerned.

Let us first consider the equations of motion in frame $(0, \xi, \eta, \zeta)$ with the origin at the center of mass of the earth-moon system and with inertial axes.

These equations may be written

$$\ddot{\mathbf{r}} = \operatorname{grad}_{\mathbf{r}} U$$

where

$$U = \frac{GE}{r_e} + \frac{GM}{r_m} + \frac{GS}{r_s} - \frac{GS}{E+M}\left(\frac{E\mathbf{r}_{es}}{r_{es}^3} + \frac{M\mathbf{r}_{em}}{r_{em}^3}\right)\cdot\mathbf{r},$$

* Presented by Dr. França.

† This work was partially supported by ONR Contract N00014-67-C-0347.

G. E. O. Giacaglia (ed.), Periodic Orbits, Stability and Resonances, 91–95.

where G is the gravitational constant; S, E and M are the masses of the sun, the earth and the moon, respectively. The other quantities are indicated in Figure 1.

Expanding the function U, by retaining terms larger or equal to 10^{-3} (relative magnitude), we find

$$U = \frac{GS}{r_s}\left[\frac{E}{S}\frac{r_s}{r_e} + \frac{M}{S}\frac{r_s}{r_m} + \left(\frac{r}{r_s}\right)^2 P_2(\cos\theta)\right].$$

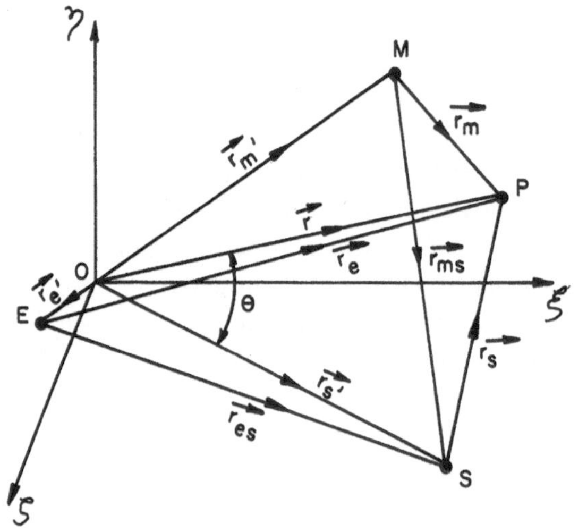

Fig. 1. Geometry of the problem.

Taking as a unit, $E/S \cdot r_s/r_e$, the second term inside brackets is approximately equal to 10^{-1}, and the third term will be, approximately 5×10^{-3} if P is considered sufficiently close to the points L_1 and L_2 of the corresponding circular restricted problem.

Within the approximation sought, it is correct to consider the sun as moving in an elliptic orbit around the center of mass of the earth-moon system.

Let us introduce, now, another reference system, $Oxyz$, which will be called the 'mean moon system' and is described in Figure 2.

The origin remains as before. The equatorial plane is the plane of the ecliptic.

The angle Ω is the secular part of the longitude of the moon's ascending node, that is

$$\Omega = (1 - g)\,\lambda + \Omega_0,$$

where g is defined by Hill's equation of the node.

λ is the moon's mean longitude,

$$\lambda = nt + \lambda_0,$$

and n is the mean motion in mean longitude.

The angle v is the secular part of the argument of the latitude of the moon, that is

$$v = g\lambda + v_0.$$

The angle I is the constant of the inclination of the moon.

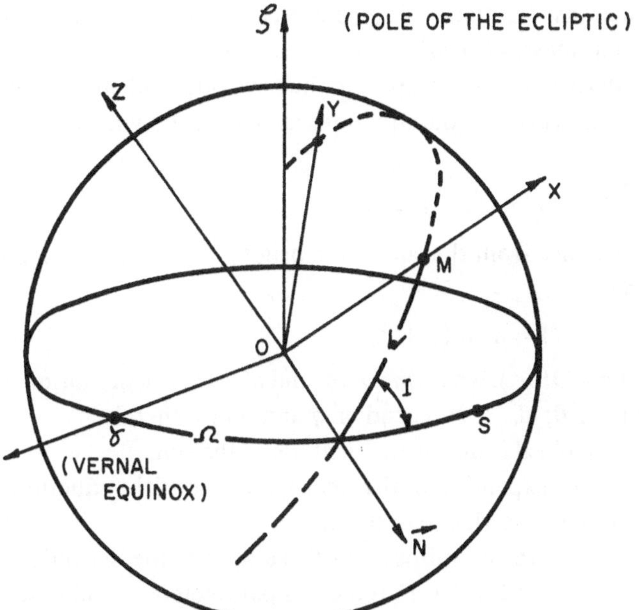

Fig. 2. The reference system.

We define the mean moon system, $Oxyz$, as being obtained from the system $O\xi\eta\zeta$ through the Eulerian angles Ω, I and v.

The equations of motion of P, in this system, are:

$$\begin{aligned}\ddot{x} - 2\dot{y} = {} & \left\{1 + A + B\cos 2v + \frac{v}{a'^3}\left[\frac{1}{2} + \frac{3}{2}\cos 2(\lambda - \lambda')\right]\right\} x \\ & + \left[- B\sin 2v - \frac{3}{2}\frac{v}{a'^3}\sin 2(\lambda - \lambda')\right] y + C(\sin v)\, z \\ & + D\dot{y} + E(\cos v)\,\dot{z} - (1+\sigma)\left[\frac{1-\mu}{r_e^3}(x - x_e) + \frac{\mu}{r_m^3}(x - x_m)\right].\end{aligned}$$

$$\begin{aligned}\ddot{y} + 2x = {} & \left[- B\sin 2v - \frac{3}{2}\frac{v}{a'^3}\sin 2(\lambda - \lambda')\right] x \\ & + \left\{1 + A - B\cos 2v + \frac{v}{a'^3}\left[\frac{1}{2} - \frac{3}{2}\cos 2(\lambda - \lambda')\right]\right\} y \\ & + C(\cos v)\, z - D\dot{x} - E(\sin v)\,\dot{z} - (1+\sigma) \\ & \times \left[\frac{1-\mu}{r_e^3}(y - y_e) + \frac{\mu}{r_m^3}(y - y_m)\right].\end{aligned}$$

$$\begin{aligned}\ddot{z} = {} & F(\sin v)\, x + F(\cos v)\, y + \left(2B - \frac{v}{a'^3}\right) z - E(\cos v)\,\dot{x} \\ & + E(\sin v)\,\dot{y} - (1+\sigma)\left[\frac{1-\mu}{r_e^3}(z - z_e) + \frac{\mu}{r_m^3}(z - z_m)\right].\end{aligned}$$

In these equations, the units adopted are: Length $=a$: the semi-major axis of the orbit of the moon; mass $=E+M$; and time $=\lambda$.
Dots represent derivatives with respect to the new independent variable λ.

The constants appearing in the equations have the following meaning

$$\nu = \frac{S}{E+M}, \qquad \mu = \frac{M}{E+M},$$

a' is the mean distance from the sun to the origin. σ is defined by Kepler's law for the perturbed moon

$$G(E+M) = n^2 a^3 (1+\sigma)$$

it is known that $\sigma = \mathrm{O}(m^2)$, where $m = n'/n$, and n' is the mean motion of the sun.

The constants $A, B, C, \ldots, F$ depend on g and on $\gamma = \sin I/2$.

Finally, the variable λ' is the mean longitude of the sun.

The next step is the expansion of the right-hand side of the equations, corresponding to the attractions of the earth and of the moon.

In the expansion of the coordinates of these bodies, the periodic perturbations are considered only to second order, in the small parameters m and e, where e is the constant of the eccentricity of the moon.

Before saying how this expansion is carried on, let us consider the stationary solutions L_1 and L_2, of the elliptic restricted problem, already mentioned.

The distances, to the center of mass, of the points which describe these solutions, are

$$X_i = k_i \varrho = k_i \frac{1-e^2}{1+e\cos f}, \quad \text{for} \quad L_i, \quad (i=1,2).$$

The distances k_1 and k_2 are, approximately equal to 1.17 and 0.85, respectively, taking as a unit the mean distance between the primaries.

The angle f is the true anomaly of the small primary (i.e. of the moon).

Expansion of ϱ gives

$$\varrho = 1 - e\cos l + \mathrm{O}(e^2),$$

where l is the mean anomaly of the moon.

This expansion has led us to consider *two points*, lying on the line joining the earth to the moon whose position is calculated taking into account the perturbations, up to second order, already mentioned.

We will define these points, as having, at any instant, distances to the origin given by

$$k_1(1 - e\cos l) \quad \text{and} \quad k_2(1 - e\cos l)$$

respectively.

From here on, when we mention L_1 and L_2 we will refer to these points just defined.

In the mean moon system, the coordinates of L_1, for example, are:

$$\begin{aligned} x_1 &= k_1(1 - e\cos l)\cos(f-l) \\ y_1 &= k_1(1 - e\cos l)\sin(f-l) \\ z_1 &= 0. \end{aligned}$$

Let us consider the new variables:

$$\xi = x - x_1$$
$$\eta = y - y_1$$
$$\zeta = z - z_1 .$$

Let h_1 denote the distance $\overline{L_1 M}$; h the distance $\overline{L_1 P}$, and ϕ the angle $PL_1 M$. We have

$$\frac{1}{r_m} = \frac{1}{h_1}\left[1 + \frac{h}{h_1} P_1(\cos\phi) + \left(\frac{h}{h_1}\right)^2 P_2(\cos\phi) + \cdots\right]$$

At this point, we would like to make a remark:

In the expansion of the function U, we should keep terms inside brackets, larger or equal to 10^{-3} (relative magnitude) as we already said.

In order to satisfy this condition, and considering P sufficiently close to L_1 or L_2, it follows that we would keep, in the expansion of $1/r_m$, terms larger or equal to 10^{-1}, and in the expansion of $1/r_e$, terms larger or equal to 8×10^{-4}.

Let us suppose, for example, that we are interested to study arcs of orbits, which start near L_1, and such that the ratio h/h_1 is smaller than $\frac{1}{10}$.

In this case it is easy to show that, in the expansions of $1/r_m$ and $1/r_e$ it is sufficient to keep only two terms, i.e., to write

$$\frac{1}{r_m} = \frac{1}{h_1}\left(1 + \frac{h}{h_1}\cos\phi\right) \quad \text{and} \quad \frac{1}{r_e} = \frac{1}{H_1}\left(1 + \frac{h}{H_1}\cos\psi\right),$$

where H_1 is the distance $\overline{L_1 E}$ and ψ is the angle $PL_1 E$.

For larger values of the ratio h/h_1, one needs to consider a larger number of terms in those developments.

In any case, to avoid a collision with the moon, the distance $h(1+e)$ must remain smaller than

$$(1-e)(k_1 - 1 + \mu) \quad \text{or} \quad (1-e)(1-\mu-k_2)$$

according to when P starts near L_1 or near L_2, respectively.

While point P is describing the considered arc of orbit, the equations of its motion, derived in the way it was described, will be correct including terms of second order in the small quantities e and m.

To verify this result, it should be noted that all the constants, $A, B, \ldots, F$, include the factor $(1-g)$ which is of order m^2.

It is worth noting that the coefficient ν/a'^3 is approximately equal to 56×10^{-4}, which is smaller than m^2.

These equations, written in the variables ξ, η, ζ are too long, to be shown here.

In the example previously given, in which two terms were needed, in the expansions of $1/r_m$ and $1/r_e$, those equations will contain terms up to ξ^4 and η^4.

It is important to observe that L_1 and L_2 are indeed approximate solutions of these equations.

In a more precise sense:

The point $\xi = \eta = \zeta = 0$, verifies these equations with an error of second order in the parameters e and m.

RESONANCES IN THE ELLIPTIC RESTRICTED PROBLEM*

G. E. O. GIACAGLIA

University of São Paulo and University of Campinas, Brasil

and

PAUL E. NACOZY

The University of Texas at Austin, U.S.A.

Abstract. A preliminary study of some cases of resonance in the planar elliptic restricted problem of three bodies is presented. The eccentricity of the primaries varies from 0 to 0.7. The long-term motion of minor planets in resonance with Jupiter ($e'=0.048$) is also given. We give special attention to effects in long period due to a change in the eccentricity of the primaries.

The method of solution applied is by now considered well known and can be named as Bohlin's Method. The equations of motion in terms of critical (resonance) argument are due to Poincaré (1902). The numerical averaging procedure that we use to eliminate short period terms has been applied since Schwarzschild (1903) to more recent works by Schubart (1968), Hori and Giacaglia (1967) and Giacaglia (1968).

The results of this study, shown graphically in the various figures, give an overall and qualitative solution of five cases of resonance. The Trojan, the Hecuba, the Thule and the Hilda groups and the case of the Pluto-Neptune system.

The method gives a qualitative behavior of orbits, mainly the long-period variations in eccentricity and perihelion, but does not produce solutions explicitly as functions of time.

1. Equations of Motion

Let the sun and Jupiter be the primaries with mass m and m' respectively and let e' be the eccentricity of Jupiter's orbit, which is supposed Keplerian. We choose a heliocentric sidereal coordinate system. Let ω' be the longitude of Jupiter's perihelion from the x-axis of this system, the plane of Jupiter's orbit being in the (x, y) plane. The equations of motion of an asteroid of negligible mass, restricted to move in that plane are

$$\begin{aligned} \dot{l} &= -K_L, & \dot{\omega} &= -K_G, & \dot{\lambda}' &= -K_T \\ \dot{L} &= K_l, & \dot{G} &= K_\omega, & \dot{T} &= K_{\lambda'} \end{aligned} \tag{1}$$

where

$$K = \mu^2/2L^2 - n'T + R$$

and

$$R = k^2 m' \left[\frac{1}{\Delta} - \frac{r\cos\psi}{r'^2}\right]. \tag{2}$$

In the above equations l=mean anomaly of asteroid; λ'=mean longitude of Jupiter $=l'+\omega'$; ω=longitude of perihelion of asteroid; L, G, T=momenta conjugate to l, ω, λ', respectively; n'=mean motion of Jupiter; and $\mu=k^2(m+m')$, (k is the gaussian constant). In Equation (2), R is the disturbing function, r and r' the distances sun-

* Presented by Dr. Nacozy.

G. E. O. Giacaglia (ed.), Periodic Orbits, Stability and Resonances, 96–127.

asteroid and sun-Jupiter respectively, ψ the elongation of the asteroid from Jupiter and Δ the distance Jupiter-asteroid, that is $\Delta = |\mathbf{r} - \mathbf{r}'|$. We have also

$$\mu = k^2(m + m') = n'^2 a'^3 ,$$

where a' is Jupiter's semi-major axis. The small parameter of the problem is $\varepsilon = m'/(m+m')$ so that $k^2 m' = \varepsilon n'^2 a'^3$ and

$$R = \varepsilon n'^2 a'^3 \left(\frac{1}{\Delta} - \frac{r \cos \psi}{r'^2} \right). \tag{3}$$

The disturbing function can formally be represented by the D'Alembert series

$$R = \varepsilon \sum_{\binom{p'_1, p_1, p_2, p_3}{k, k'}} A^{(p,k)} n'^2 a^2 \left(\frac{a}{a'} \right)^{p'_1} e^{|p_2| + 2k} e'^{|p_3| + 2k'} \times \cos \left[p_1 (\lambda - \lambda') + p_2 l + p_3 l' \right],$$

where $A^{(p,k)}$ is a numerical coefficient, and e and e' denote the eccentricities of the asteroid and Jupiter respectively. For large values of e, e' and a/a' such a series is meaningless but is useful to indicate the character of the argument of cosines in a process of numerical procedure.

We consider the resonance problem defined by

$$|pn - qn'| \leqslant n\varepsilon^{1/2} ,$$

where p and q are mutually prime integers.

We perform a canonical transformation to variables $(x_0, x_1, x_2; y_0, y_1, y_2)$ defined by

$$\begin{aligned} y_0 &= l , & x_0 &= L + pT/q \\ y_1 &= pl - ql' + q(\omega - \omega') , & x_1 &= - T/a \\ y_2 &= \omega - \omega' , & x_2 &= G + T \end{aligned}$$

and the Hamiltonian becomes

$$K' = \frac{\mu^2}{2} (x_0 + px_1)^2 + qn' x_1 + R(x; y)$$

with R expressible as a triple Fourier series of general argument

$$p_1 (\lambda - \lambda') + p_2 l + p_3 l' = \left[p_3 + p_2 + (p_3 - p_1) \frac{p}{q} \right] y_0 - (p_3 - p_1) \frac{1}{q} y_1 + p_1 y_2 .$$

We perform another canonical transformation $(x, y) \rightarrow (\xi, \eta)$ such that the new Hamiltonian $\Phi(\xi, \eta)$ should not contain short period argument η_0 (corresponding to y_0). By Bohlin's procedure, the result can be written formally as

$$\phi = \phi_0 + \phi_{1/2} + \phi_1 + \phi_{3/2} + \mathrm{O}(\varepsilon^2),$$

where

$$\phi_0 = \frac{\mu^2}{2}(\xi_0 + p\xi_1)^{-2} + qn'\xi_1$$

$$\phi_{1/2} = 0$$

$$\phi_1 = \frac{1}{2\pi q}\int_0^{2\pi q} R \, \mathrm{d}y_0 \Big|_{\substack{x=\xi \\ y=\eta}} \tag{4}$$

$$\phi_{3/2} = 0$$

and subscripts indicate order of magnitude with respect to ε, that is, for example, $\phi_{1/2} = O(\varepsilon^{1/2})$. Our solution is limited to terms $O(\varepsilon^{3/2})$ so that we neglect terms $O(\varepsilon^2)$ altogether. The new equations of motion are

$$\dot{\xi}_j = \phi_{\eta_j}, \quad \dot{\eta}_j = -\phi_{\xi j} \quad (j = 1, 2) \tag{5}$$

and η_0 is given by quadrature (ξ_0=const.), all with an error $O(\varepsilon^2)$. We have a system of two degrees of freedom where Φ depends on $\xi_0, \xi_1, \xi_2, \eta_1, \eta_2$.

Using a method already introduced elsewhere (Hori and Giacaglia, 1967; Giacaglia, 1968, 1969) we perform a canonical transformation $(\xi, \eta) \to (X, Y)$ such that the new Hamiltonian $F(X, Y)$ should not contain the critical angle Y_1 (corresponding to y_1). Assuming the development

$$\phi_0 + \phi_1 = F_0 + F_{1/2} + F_1 + F_{3/2} + \cdots \tag{6}$$

and the canonical transformation generated by

$$S(X, \eta) = X_0\eta_0 + X_1\eta_1 + X_2\eta_2 + S_{1/2}(X_0, X_1, X_2; \eta_1, \eta_2) + \cdots \tag{7}$$

we obtain

$$F_0 = \frac{\mu^2}{2}(X_0 + pX_1)^{-2} + qn'X_1$$

$$F_{1/2} = 0$$

F_1 and $S_{1/2}$ are connected by the equation

$$F_1(X; \eta_2) = \phi_1(X; \eta_1, \eta_2) + (qn' - pn^{**})\, S_{1/2, \eta_1} + \frac{3p^2 n^{**}}{L^{**}} S^2_{1/2, \eta_1}, \tag{8}$$

where double asterisks indicate averaged variables over y_0 and η_1. Equation (8) is solved by the minimum principle, that is, if $\eta_1 = \eta_1^0(X; \eta_2)$ is such that

$$\frac{\partial \phi_1}{\partial \eta_1}\Big|_{\eta_1 = \eta_1{}^0} = 0$$

and it is a minimum, then

$$F_1(X, \eta_2) = \phi_1(X; \eta_1^0, \eta_2) \tag{9}$$

and therefore

$$S_{1/2,\eta_1} = \frac{1}{2B}[-A \pm (A^2 - 4B\cdot\Delta\Phi)^{1/2}], \tag{10}$$

where

$$A = qn' - pn^{**}; \quad B = \frac{3p^2 n^{**}}{2L^{**}}$$

and

$$\Delta\phi = \Phi_1(X;\eta_1,\eta_2) - \Phi_1(X;\eta_1^0,\eta_2) = \Delta\phi(X;\eta_1,\eta_2).$$

With an error $O(\varepsilon^{3/2})$ the Hamiltonian is reduced to

$$F = F_0(X_0, X_1) + F_1(X_0, X_1, X_2; Y_2) \tag{11}$$

which is a system with a single degree of freedom:

$$\dot{X}_2 = F_{Y_2}, \quad \dot{Y}_2 = -F_{X_2} \tag{12}$$

It is important to note that in Equation (12) the small parameter is lost since F_0 does not depend on X_2 or Y_2 and it is actually a constant (error $O(\varepsilon^{3/2})$).

2. Semi-Analytic Procedure

In order to obtain a solution without restrictions on e and e' (less than unity) we approach the problem in a semi-analytic manner. The Hamiltonian Φ can be computed by considering that

$$\phi_1 = \frac{1}{2\pi q}\int_0^{2\pi q} R(p,q,e,e',a',y_0,y_1,y_2)\,\mathrm{d}y_0\Big|_{\substack{x\to\xi\\ y\to\eta}} \tag{13}$$

Choosing a', a defined by exact resonance (which gives an error less than the precision of the solution), and specific values of p, q, e, e', y_1 and y_2, we can numerically integrate R with respect to y_0, that is, find the average of R over y_0. The result will give Φ, error $O(\varepsilon^2)$, as a table of six parameters: p, q, e, e', η_1, η_2. We then proceed to Equation (11) to determine F and with an error $O(\varepsilon^{3/2})$,

$$F = F_0 + \Phi_1(X_0, X_1, X_2; \eta_1^0(X, Y_2), Y_2).$$

We proceed to find η_1^0 such that Φ_1 has a minimum at $\eta_1 = \eta_1^0$. Again we use a numerical approach. For each set of the six parameters we determine the numerical value of η_1^0. The result is that the Hamiltonian F is given as a table of five parameters: $F = F(p, q, e, e', y_2)$.

For a fixed value of the pair (p, q) and fixed e' and e, F may be plotted versus Y_2. Choosing different values of e, and holding p, q, e' fixed, gives a one parameter (e) family $F = \text{const}\,(e, Y_2)$.

Five different sets of (p, q) were studied:
(1, 1) – The Trojan group; (1, 2) – the Hecuba group; (3, 4) – the Thule group; (2, 3) – the Hilda group, and (3, 2) – the Pluto-Neptune case. For each set, the eccentricity

of Jupiter was given the values: $e'=0.001$, 0.01, 0.048, 0.075, 0.1, 0.2, 0.3, 0.4, 0.5, 0.6, 0.7. Values of e' greater than 0.7 were not included, due to prohibitive calculation times. For each set (p, q) and each e', the eccentricity of the asteroid varied from $e=0$ to $e=0.9$ by steps of 0.1 or finer steps when necessary to reveal details which were not clear for that grid size of e. And for each of the above sets (p, q, e', e), Y_2 was varied from 0 to π in 16 steps (the functions Φ and F are periodic in 2π and symmetric about π).

To find the average of the disturbing function, y_0 was varied from 0 to $2\pi q$ in 50 steps for e' less than 0.4, in 100 steps for e' 0.4 to 0.6, and in some cases in 200 steps for e' 0.6 and 0.7.

The minima of the function with respect to Y_1 (or for $Y_1=\eta_1^0$) were found by fitting a second degree polynomial to three neighboring points in the vicinity of the minimum of Φ. The minimum of the polynomial was found and the process repeated until the minimum of Φ was found to the desired accuracy. This minimum is F, the doubly averaged Hamiltonian.

At this stage we possess F, defined by Equation (11) as a table of p, q, e', e, Y_2. This system could be integrated but the essential features being geometric, they are easily found by isoenergetic curves in the (e, Y_2) plane. In order to obtain these contour curves, for different values of $F=$constant we obtain the corresponding pairs (e, Y_2). This is done for every value of e' and of the pair (p, q). Figures 1 to 54 show these contours indicating the locations of saddle points (S), and centers corresponding to minimum energy (L) or maximum energy (H). For each case, 20 equally spaced values of F were chosen to obtain the contour curves in each of these figures.

Along a trajectory in the (e, Y_2) plane one can obtain the corresponding value of Y_1 at any point. This is necessary to verify the possibility of a close approach since, near to such a situation, the present development cannot be valid.

Considering that

$$y_1 = pl - ql' + q(\omega - \omega')$$

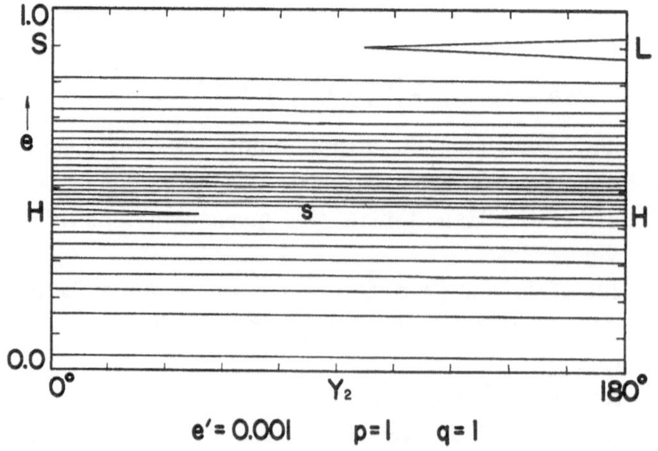

Fig. 1.

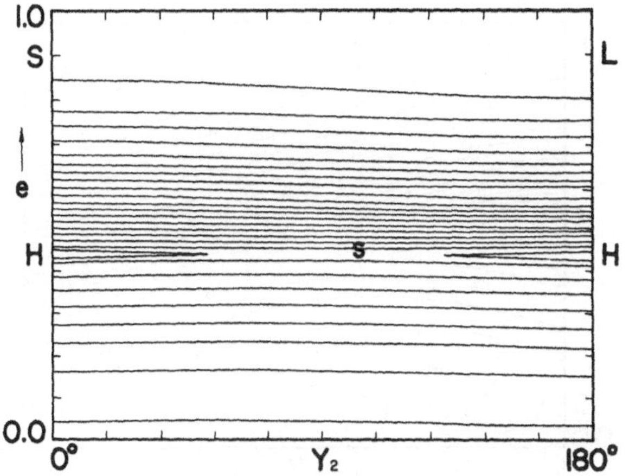

e′ = 0.01 p = 1 q = 1

Fig. 2.

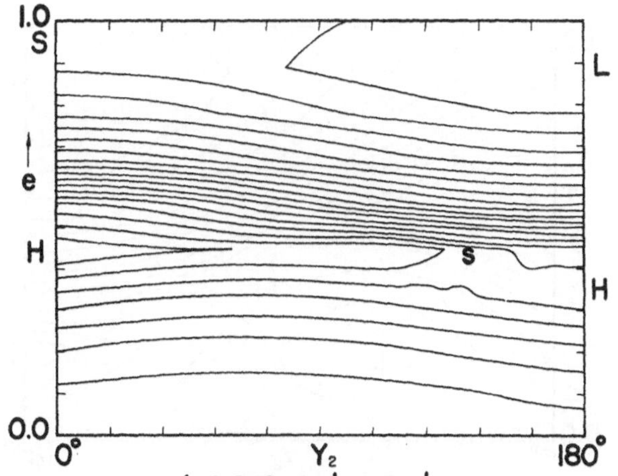

e′ = 0.048 p = 1 q = 1

Fig. 3.

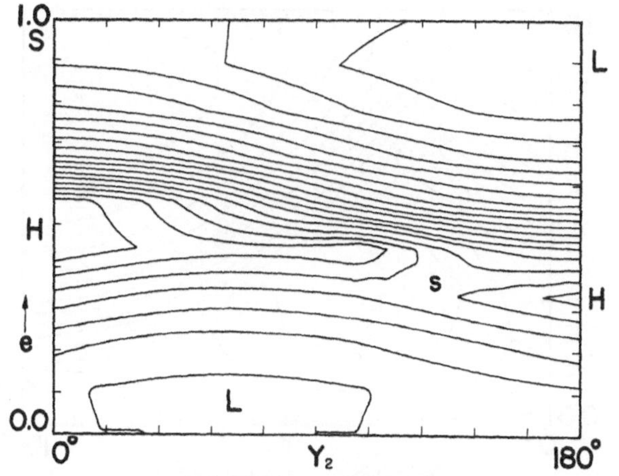

e′ = 0.075 p = 1 q = 1

Fig. 4.

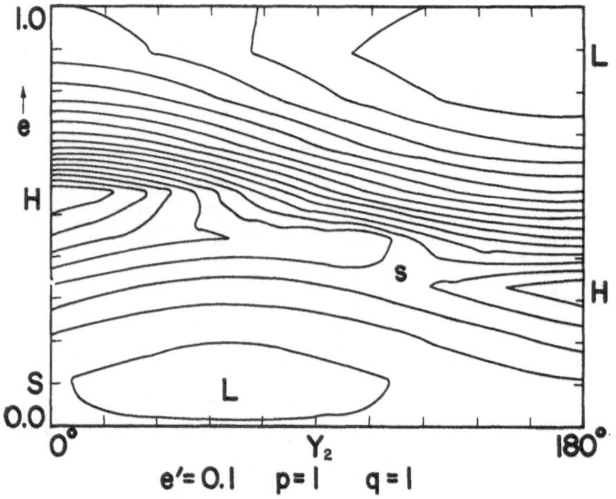

Fig. 5.

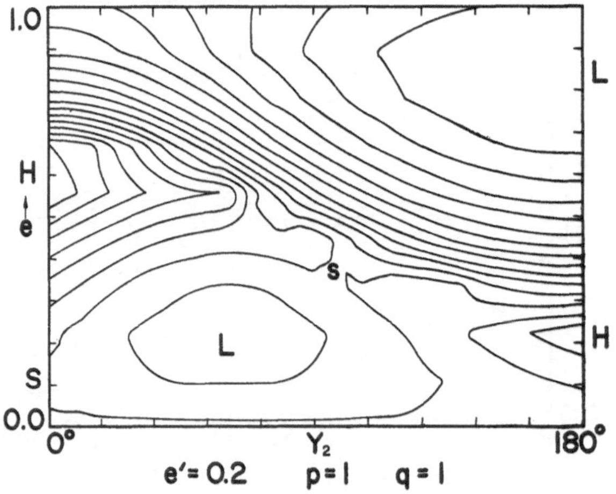

Fig. 6.

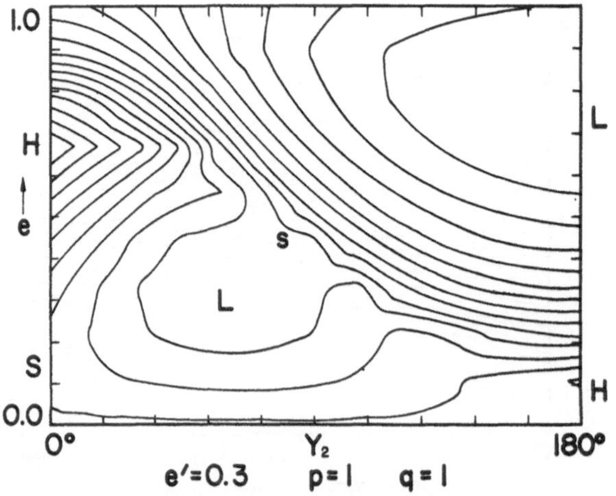

Fig. 7.

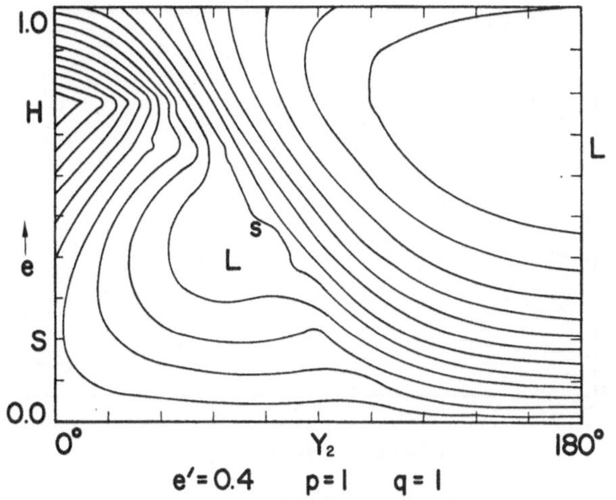

Fig. 8.

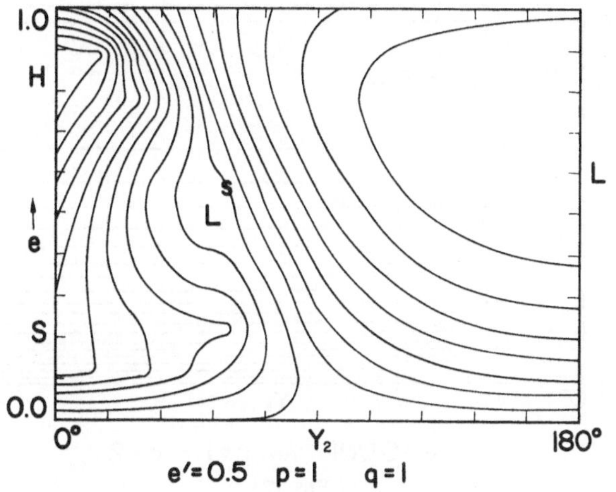

Fig. 9.

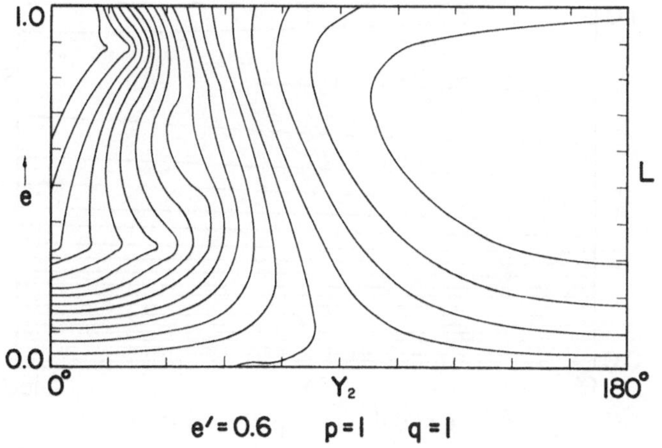

Fig. 10.

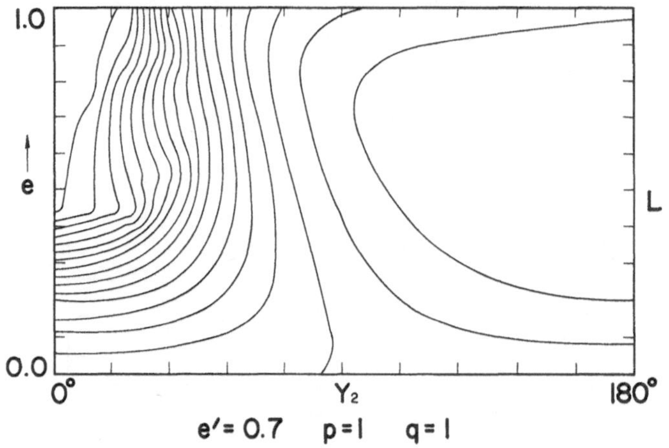

Fig. 11.

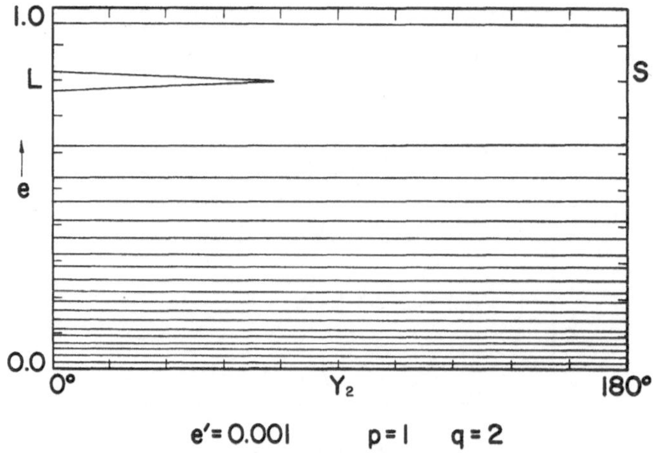

Fig. 12.

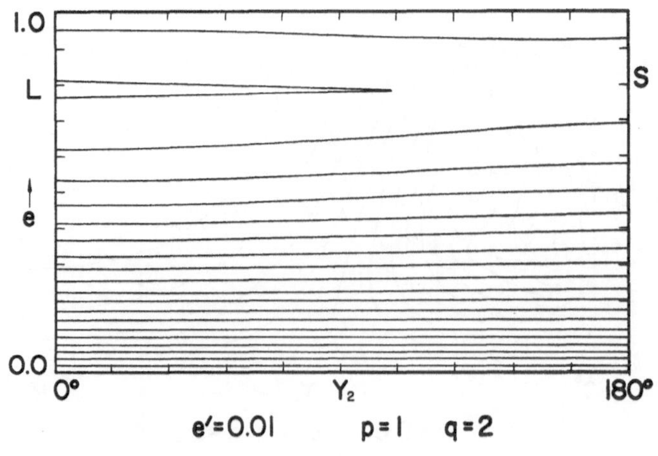

Fig. 13.

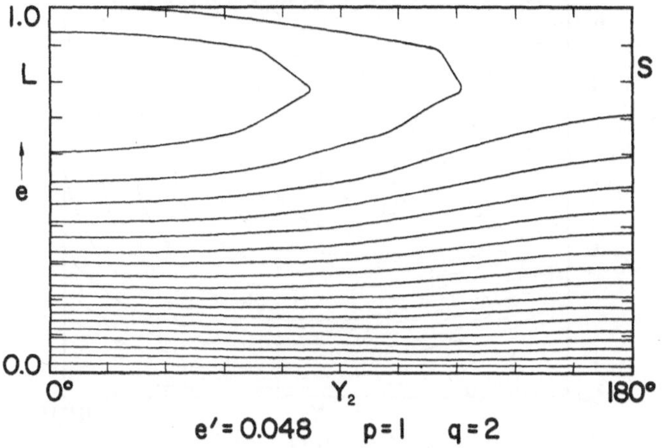

Fig. 14.

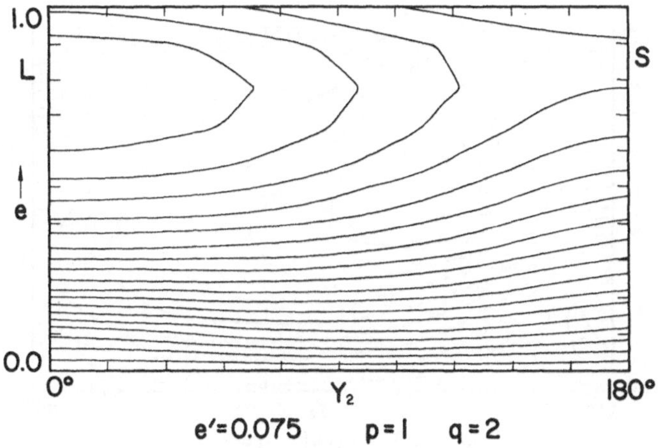

Fig. 15.

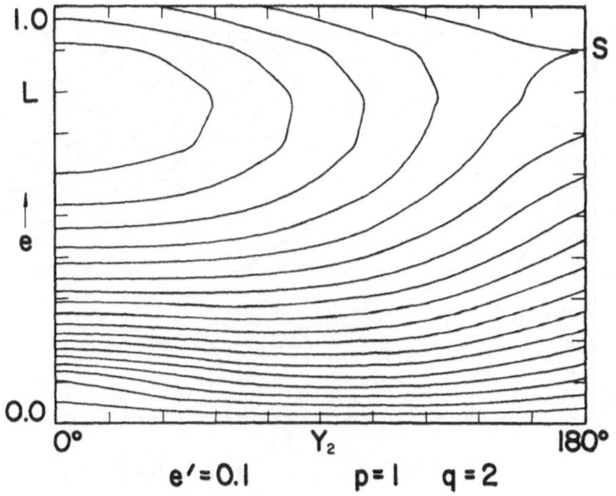

Fig. 16.

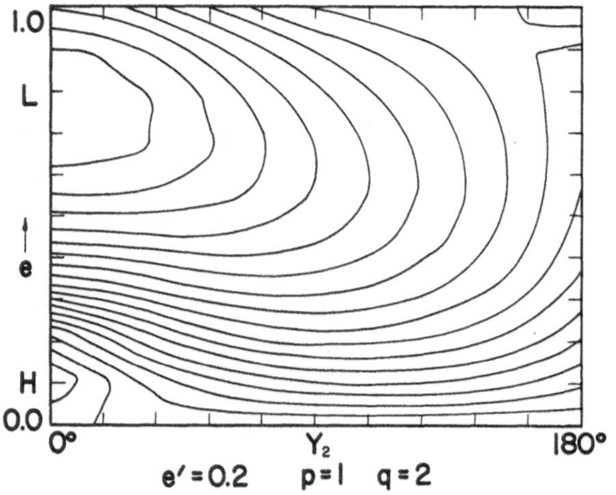

Fig. 17.

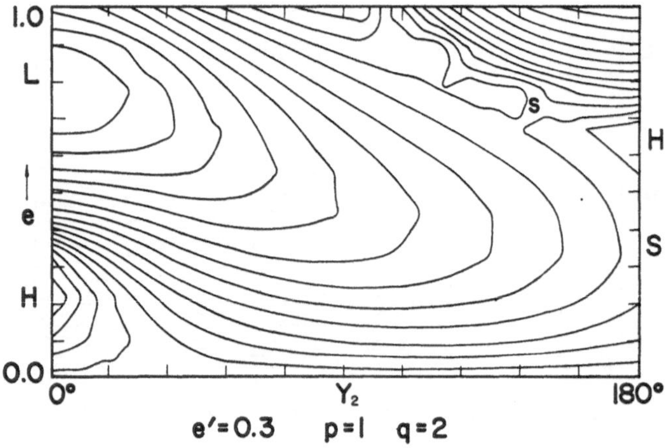

Fig. 18.

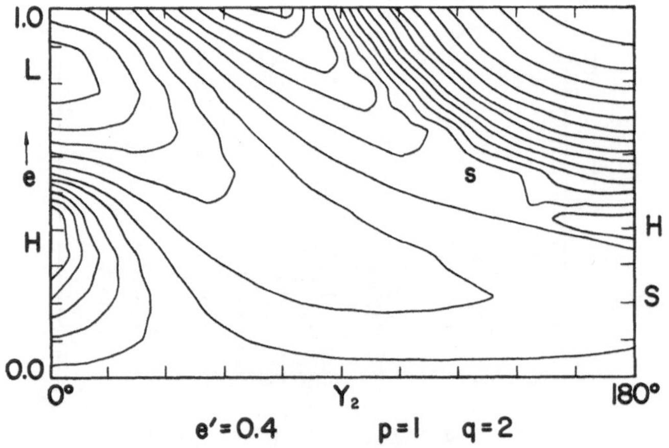

Fig. 19.

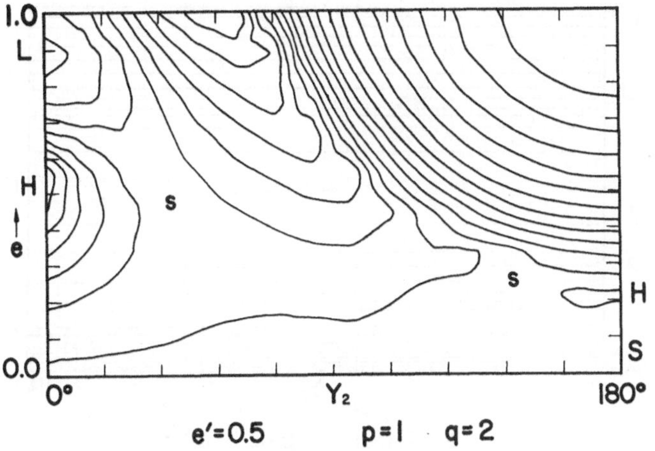

Fig. 20.

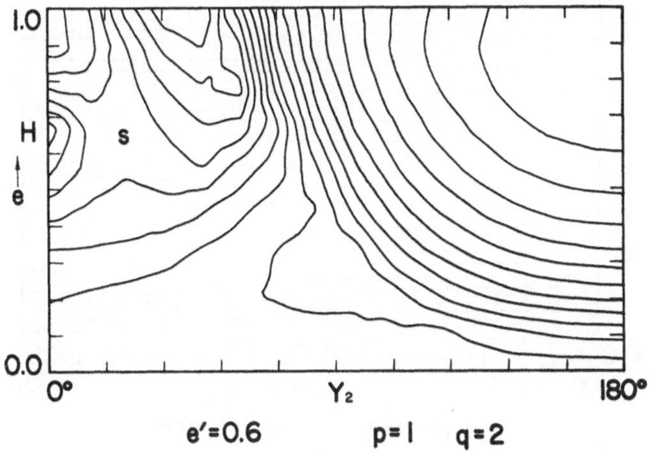

Fig. 21.

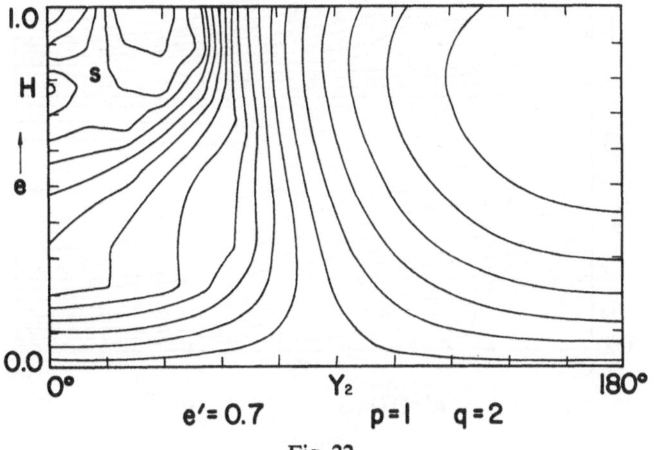

Fig. 22.

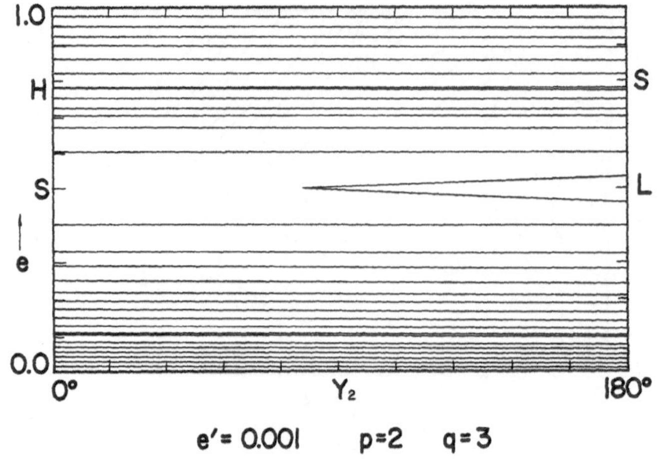

Fig. 23.

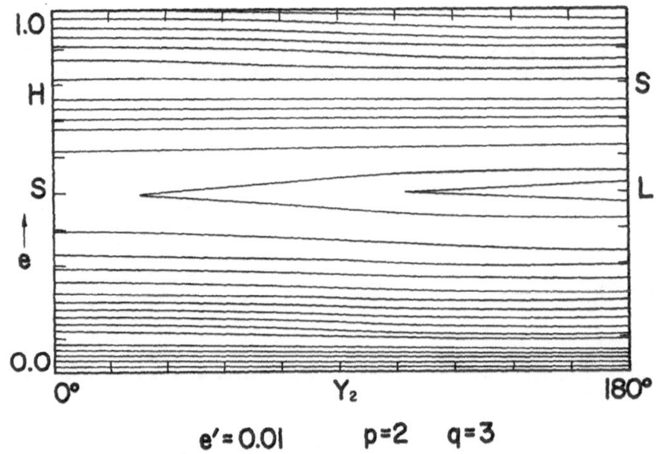

Fig. 24.

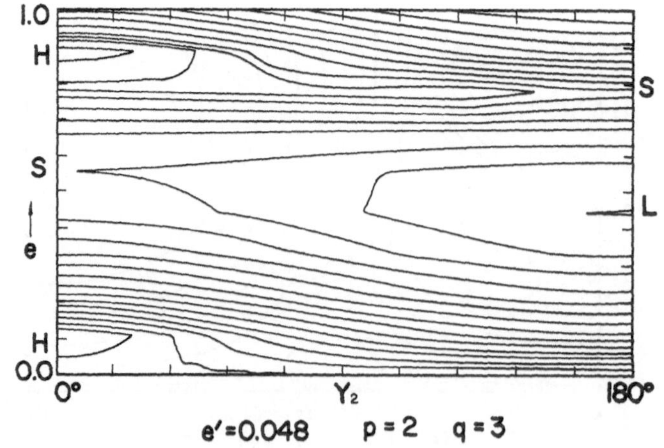

Fig. 25.

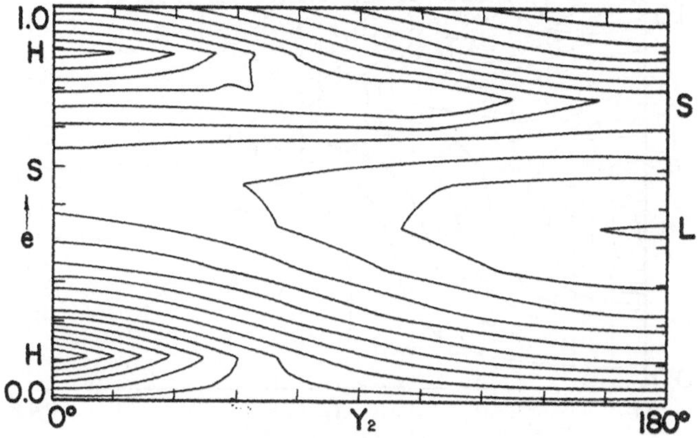

e′ = 0.075 p = 2 q = 3

Fig. 26.

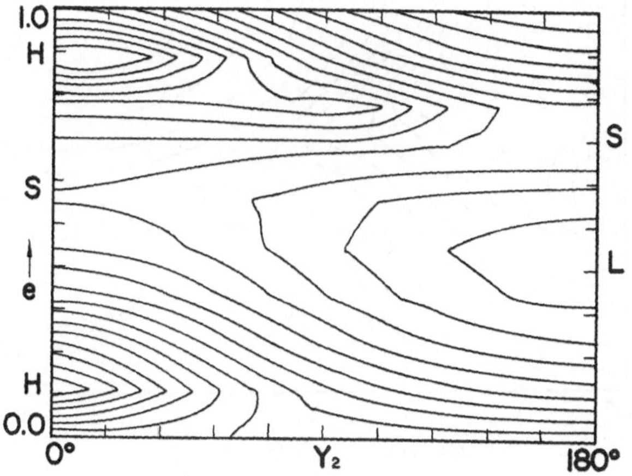

e′ = 0.1 p=2 q=3

Fig. 27.

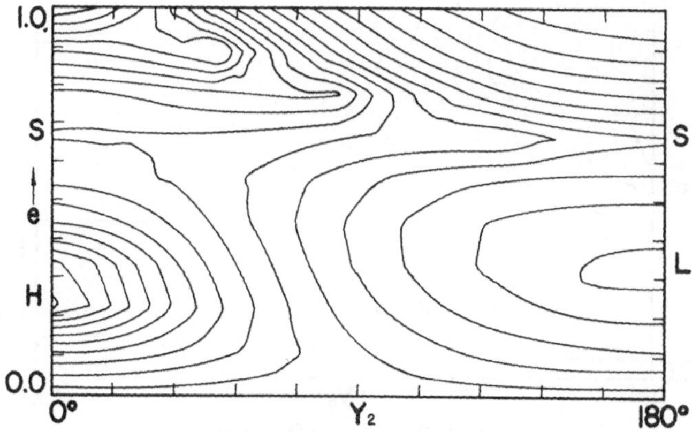

e′=0.2 p = 2 q = 3

Fig. 28.

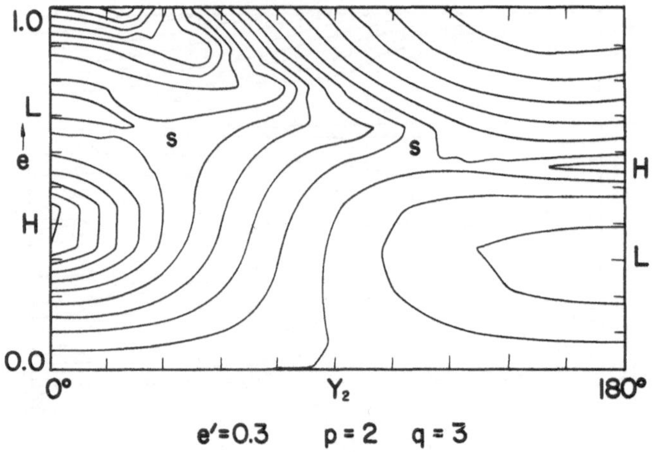

Fig. 29.

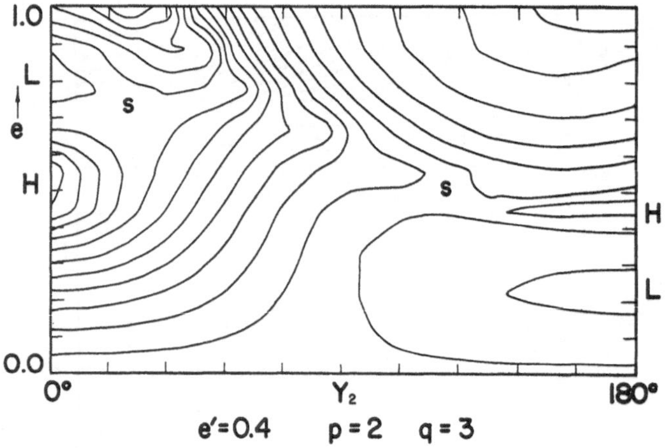

Fig. 30.

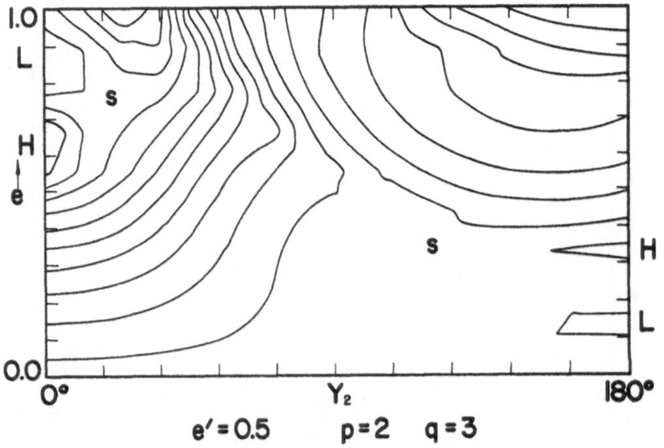

Fig. 31.

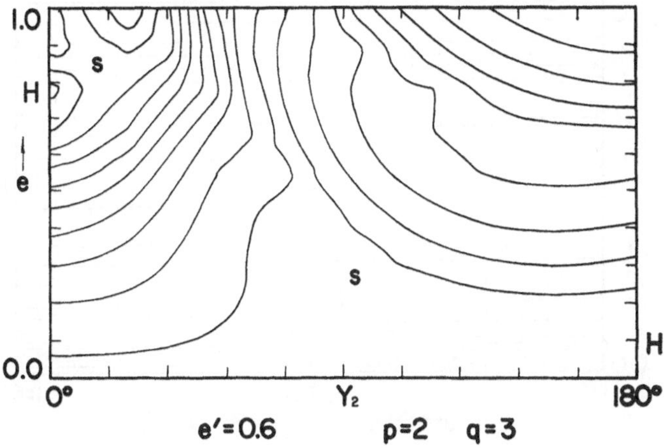

Fig. 32.

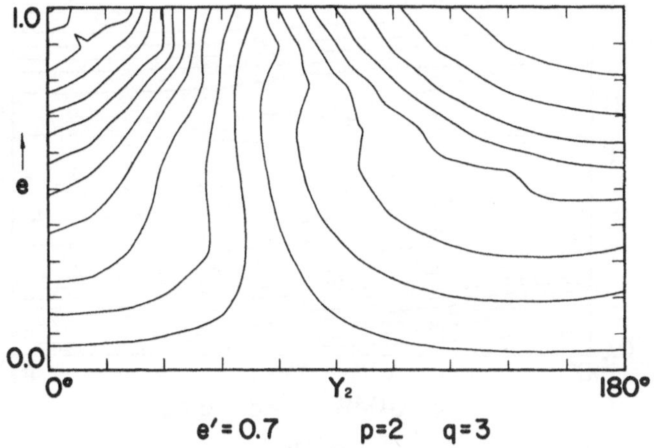

Fig. 33.

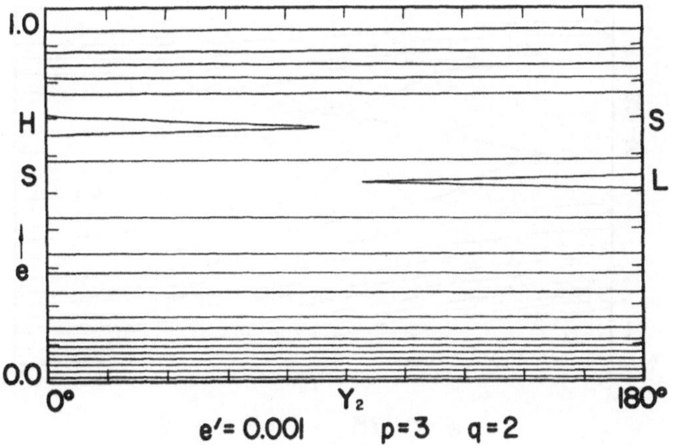

Fig. 34.

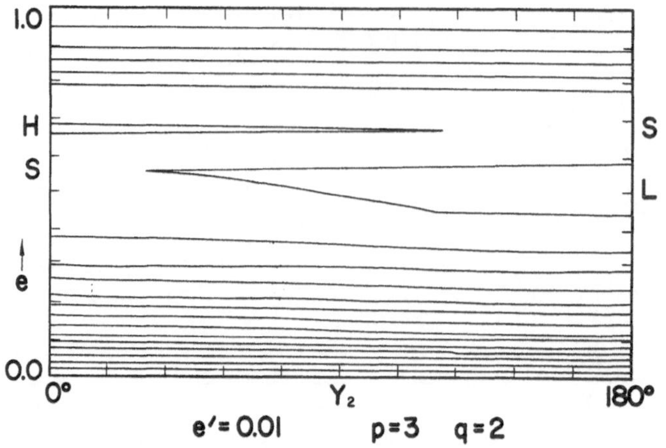

Fig. 35.

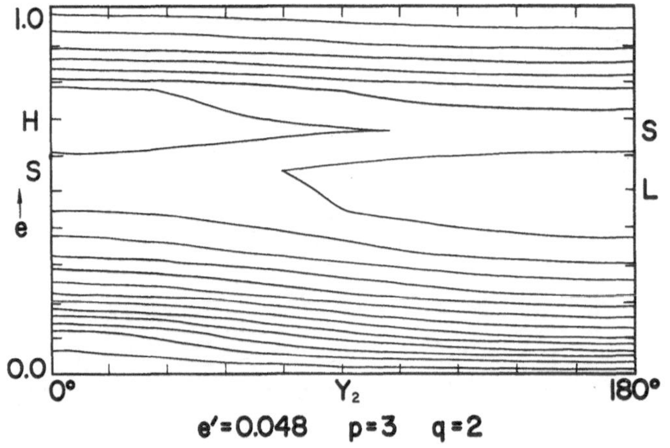

Fig. 36.

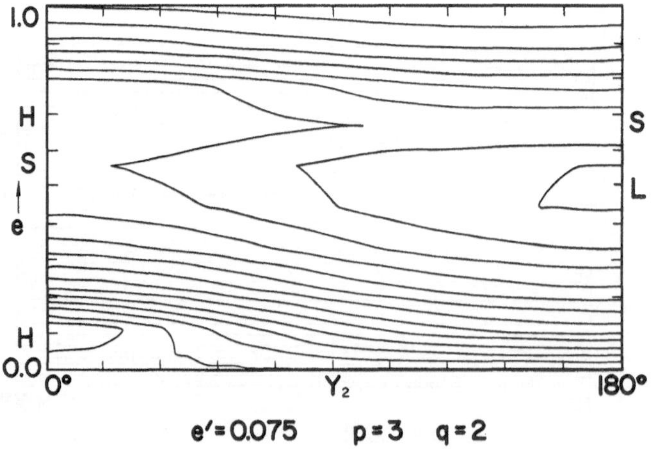

Fig. 37.

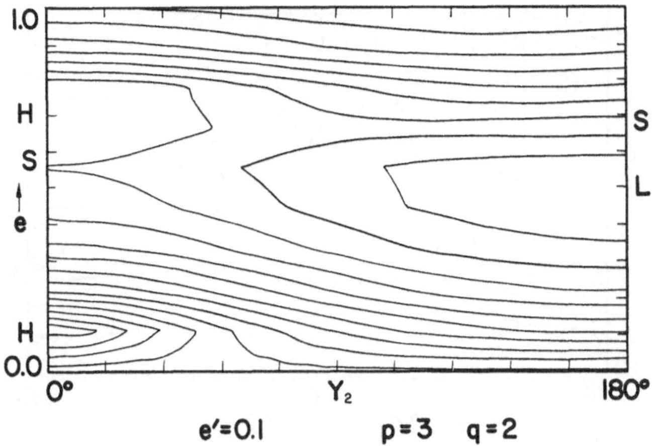

Fig. 38.

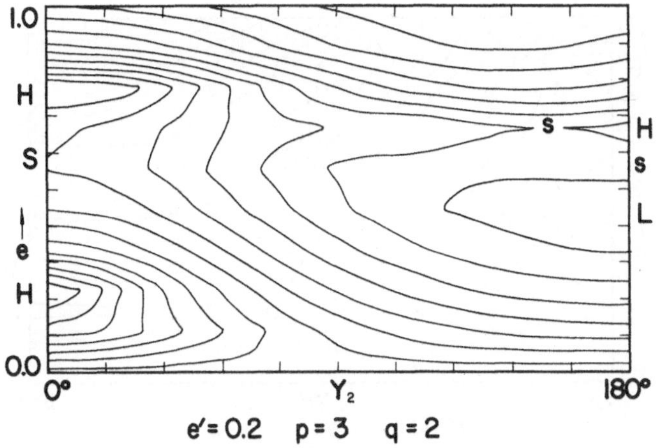

Fig. 39.

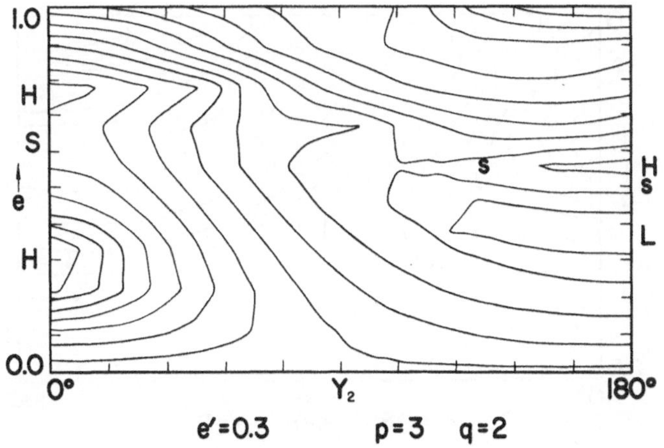

Fig. 40.

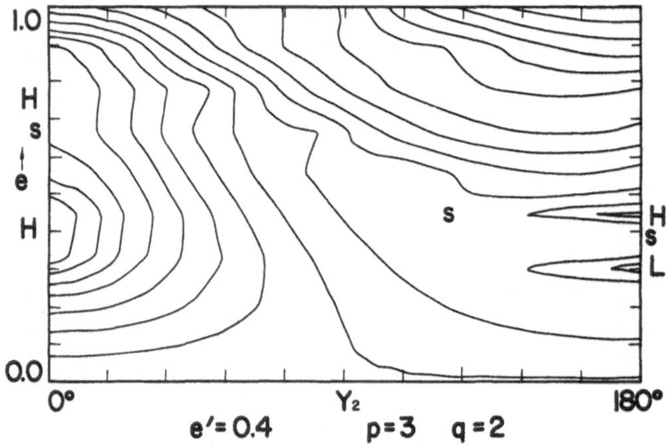

Fig. 41.

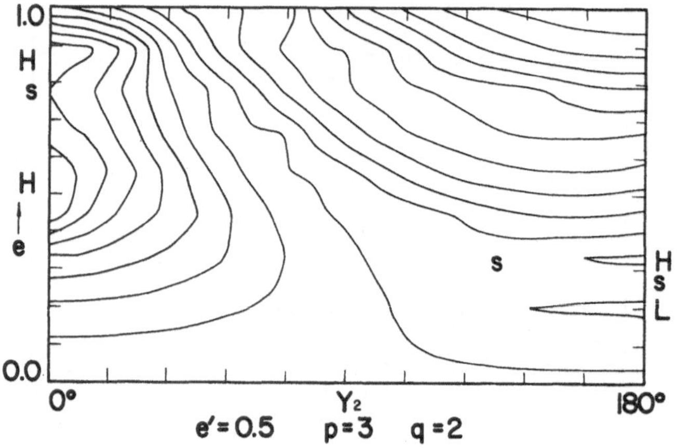

Fig. 42.

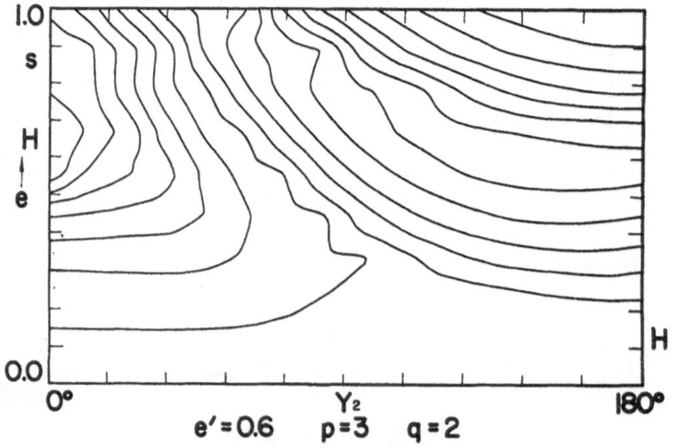

Fig. 43.

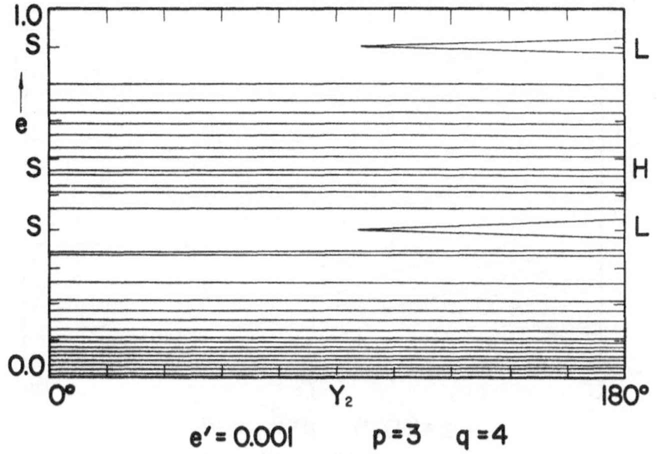

Fig. 44.

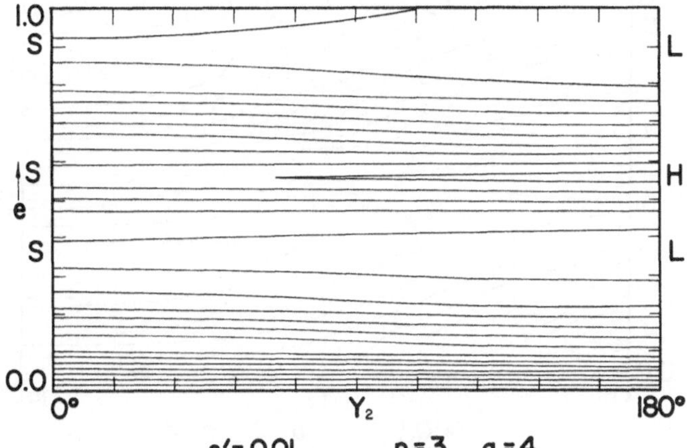

Fig. 45.

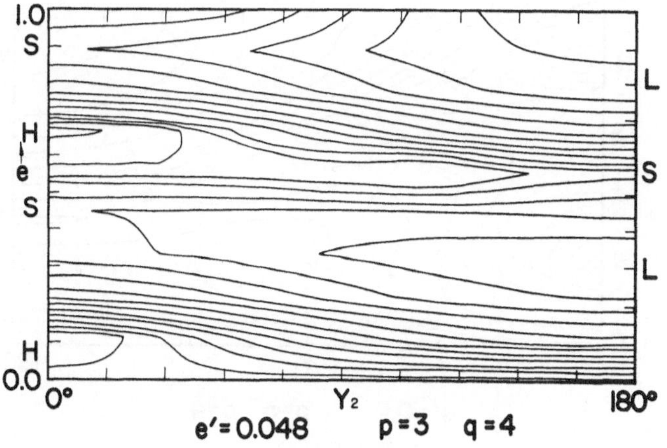

Fig. 46.

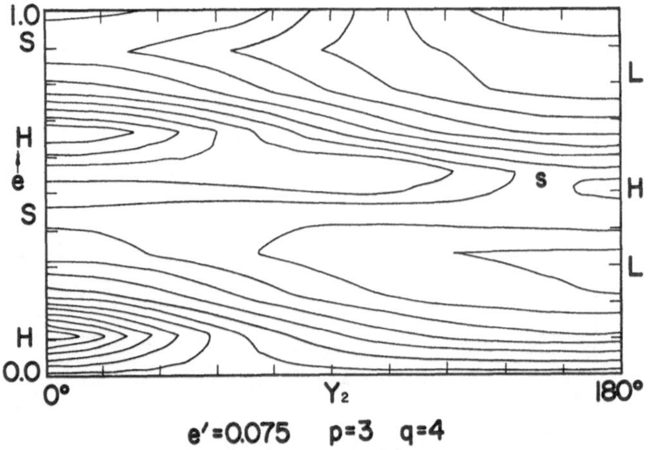

Fig. 47.

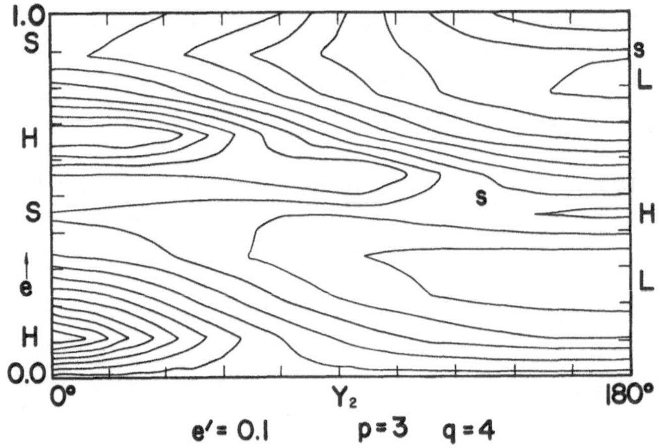

Fig. 48.

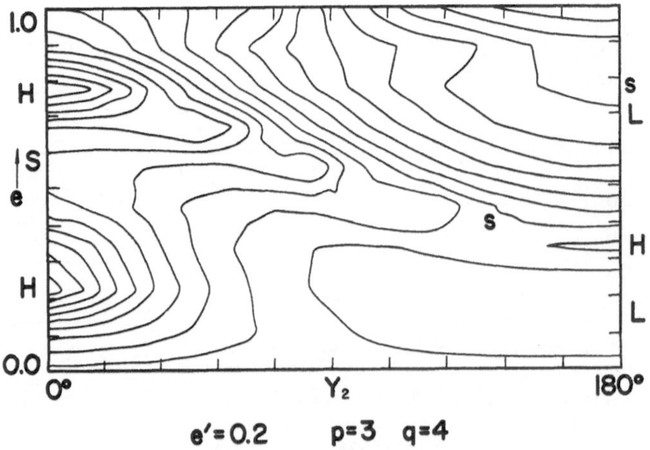

Fig. 49.

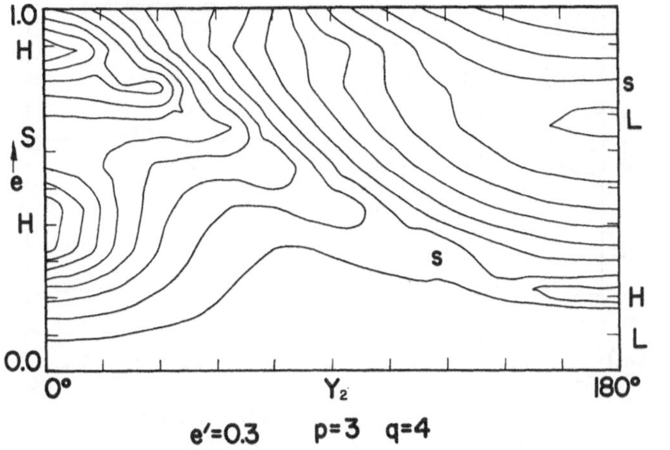

Fig. 50.

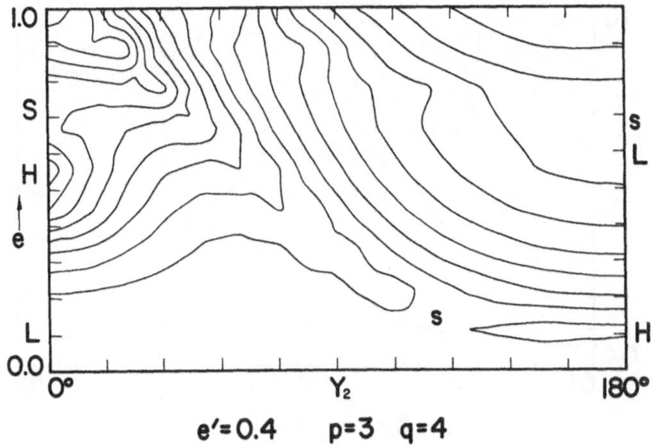

Fig. 51.

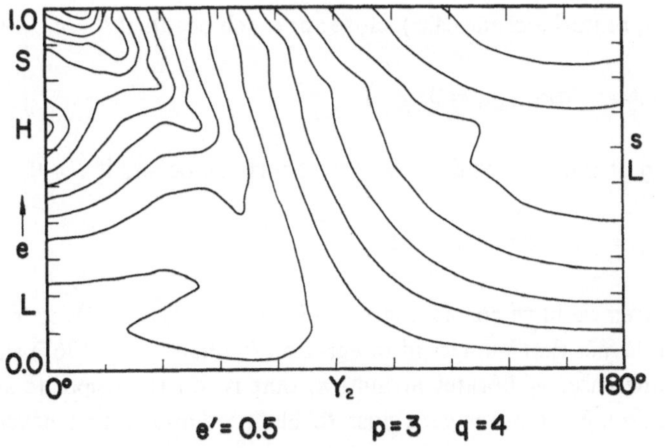

Fig. 52.

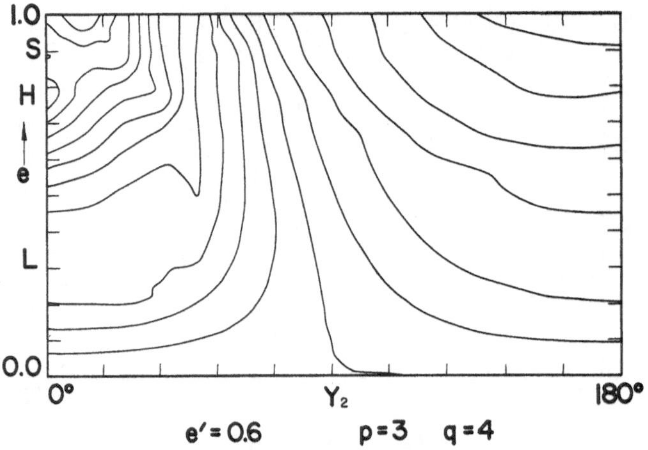

Fig. 53.

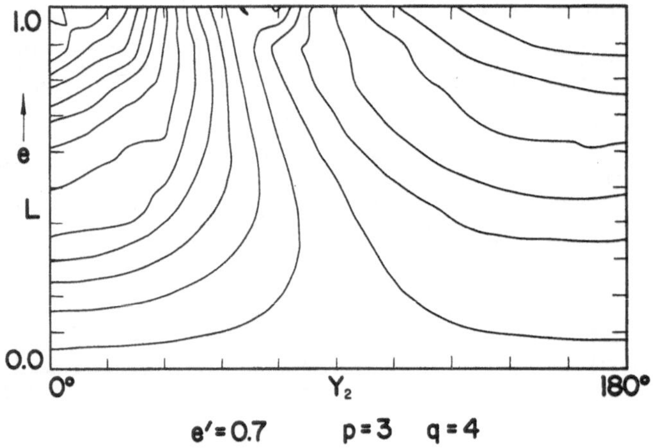

Fig. 54.

if $p>q$ ($n'>n$, Pluto-Neptune case), close approach occurs if $l'=\pi$, $l=0$ and $\omega-\omega'=\pi$ that is

$$y_1 = -q\pi + q\pi = 0 .$$

When $p \leqslant q$ ($n' \leqslant n$, Asteroid case), close approach occurs if $l'=0$, $l=\pi$, $\omega'-\omega=\pi$, that is

$$y_1 = p\pi - q\pi = (p-q)\,\pi$$

and in the cases we have considered when $|p-q|=1$, $|y_1|=\pi$. It is to be remembered the remarkable fact that numerical integration (Cohen *et al.*, 1967) shows that in the Pluto-Neptune case, y_1 librates around π, that is, on the opposite side of the close approach region. Pluto never gets close to Neptune and in fact never as close as to Uranus.

3. The behavior of e and Y_2

For each set of (p, q), Jupiter's eccentricity e' has been given the values 0.001, 0.01, 0.048, 0.075, 0.1, 0.2, 0.3, 0.4, 0.5, 0.6, 0.7, except for the set (p, q) equal (3, 2) where $e'=0.7$ has been omitted due to the excessive time of calculation required to produce sufficient accuracy.

By varying Jupiter's eccentricity one is able to determine the progressive shift of singular points (S, L, H) due to that variation. Since the planetary theory predicts a reasonable change in the eccentricity of the planets, it will be of interest to determine how this change would influence the resonance process affecting the motion of perihelion.

The contour plots for all cases (p, q) considered and for all the eccentricities of Jupiter studied here are shown in Figures (1) through (54). Along the abscissa axis of the figures is plotted the angle $Y_2=\omega-\omega'$, varying from 0 to π. The plots are symmetric about $Y_2=\pi$, with period 2π. Along the ordinate axis is plotted the eccentricity (mean mean) of the asteroid varying from zero to unity. Three essential types of long term behavior of the eccentricity and perihelion of the asteroid can be distinguished in the figures.

The first type of behavior is circulation of the argument of perihelion when a curve begins at $Y_2=0°$, at some value of the eccentricity and ends at $Y_2=180°$ at another value of the eccentricity. The second type of behavior is libration of the argument of perihelion when the curve is closed. The point all the concentric curves enclose is a libration center, that is, both the eccentricity and the argument of perihelion are stationary. The librational behavior may be separated into two cases. (1) A closed curve enclosing a high point or maximum of the Hamiltonian F. This case is denoted in the figures by the letter H. (2) A closed curve enclosing a low point or minimum of the Hamiltonian F. This case is denoted in the figures by the letter L. The third type of behavior is a point in the contour figures where there is neither libration about the point nor circulation through the point. This point representing a stationary solution is a saddle point and is denoted in the figures by the letter S. Aside from determining the long-term behavior of the motion of the asteroid for a given value of the eccentricity of Jupiter, it is of interest to determine the dependence and movement of the stationary points of the eccentricity and perihelion (as well as the critical argument) as Jupiter's eccentricity is varied. For this purpose, Figures (55) through (59) for the five cases of p and q studied here show the movement of the stationary points. The figures have Y_2 as abscissa and the eccentricity of the asteroid as ordinate. The numbers are placed approximately within the plot to denote the position of the stationary points. The values of the numbers refers to the value of the eccentricity of Jupiter at which the stationary point occurs. Maximum (H) and minimum (L) stationary points are not distinguished in the Figures (1) through (5). Saddle points are denoted by asterisks beside the numbers.

In the following, for each specific set of p's and q's, we will discuss the long-term behavior of the eccentricity of the asteroid for various initial eccentricities of the

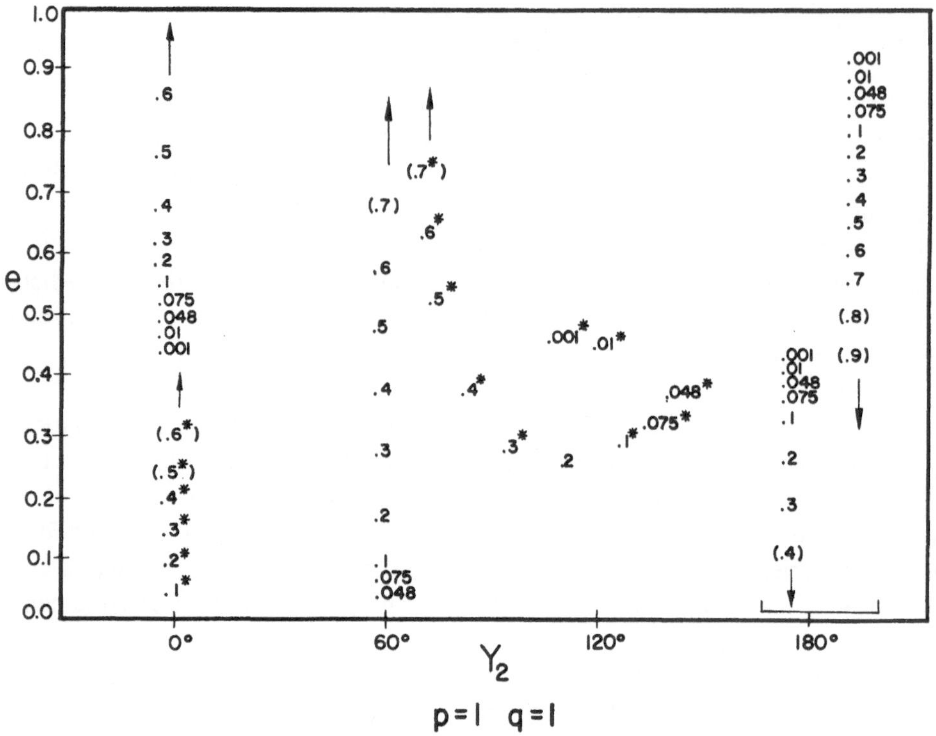

Fig. 55. Development of the stationary points.

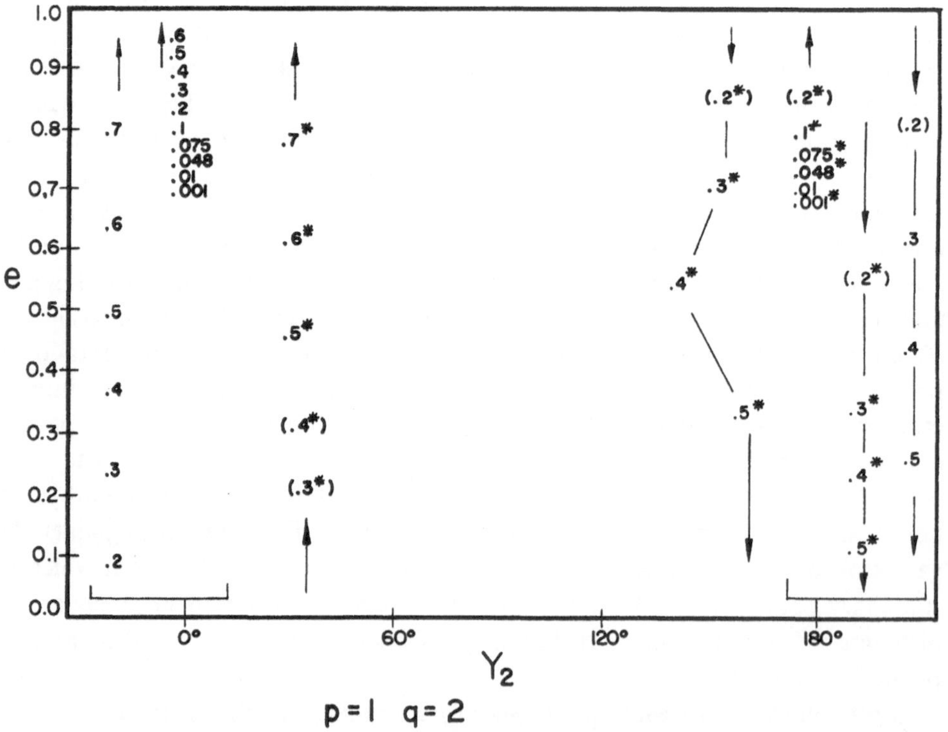

Fig. 56. Development of the stationary points.

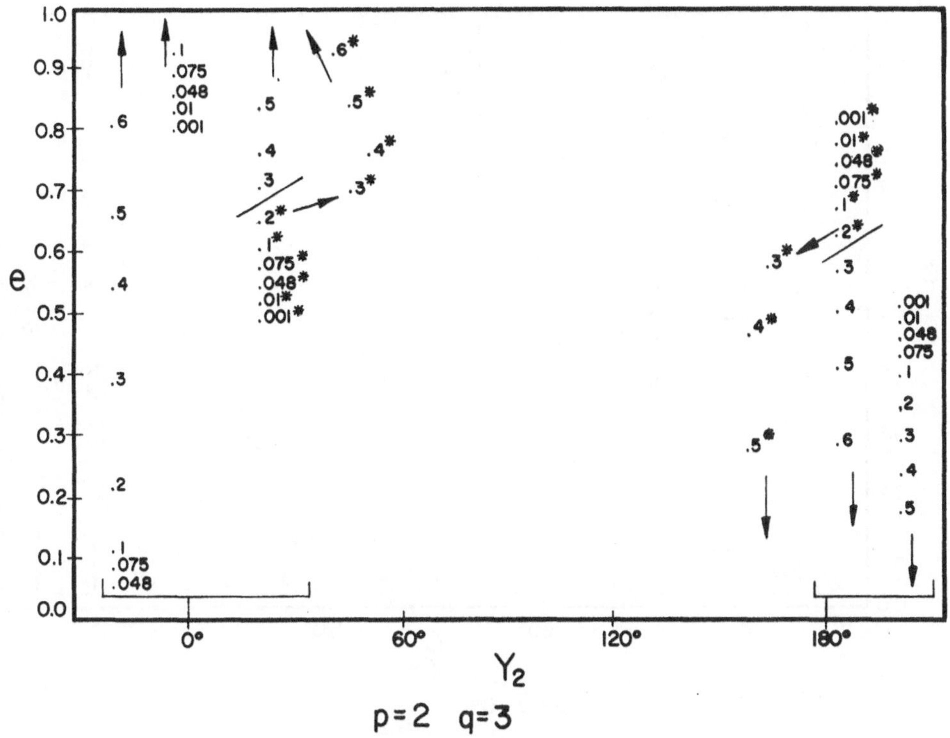

Fig. 57. Development of the stationary points.

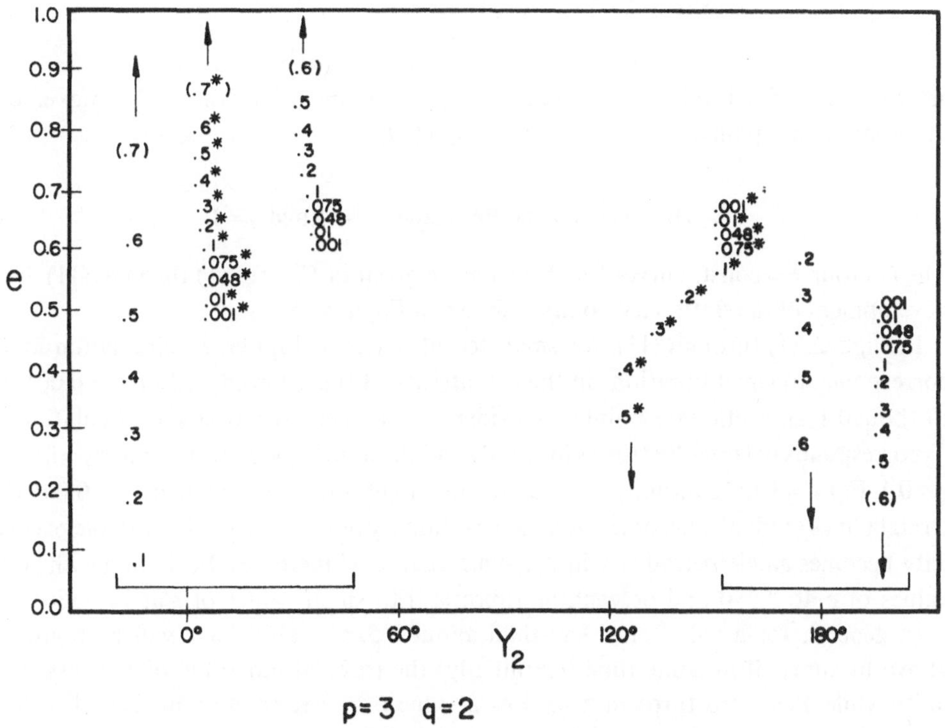

Fig. 58. Development of the stationary points.

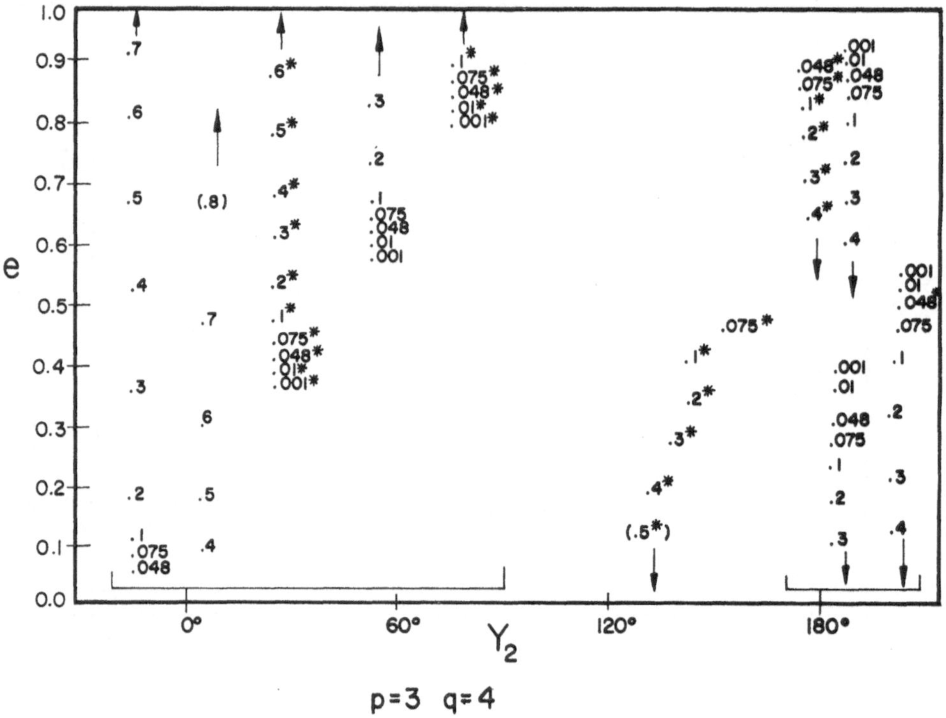

Fig. 59. Development of the stationary points.

asteroid for a fixed value of the eccentricity of Jupiter and discuss the movement of the stationary points of Figures (55) through (59).

4. The Trojan-Type Resonance: $p=1$ and $q=1$

The contour $F=$const. curves for this case are given in Figures (1) through (11). The development of the stationary points is shown in Figures 55.

In Figures (1) through (11), for small eccentricities of Jupiter, e', circulation in Y_2 corresponds to small librations in the eccentricity of the asteroid, e. For $e'=0.001$ up to about 0.1, six stationary points are evident. As e' increases, past 0.1, circulation in Y_2 corresponds to larger librations in e, and a saddle point begins to appear at $Y_2=0°$ and $e\sim 0.1$. For $e'=0$ the contour plots would be straight lines. As e' increases to 0.001 the straight lines gradually become curves and stationary points appear. The double periodicity becomes single periodicity in 2π, symmetric as e' increases, beginning with large values of e at $Y_2=0$ and descending (librating) to small values of e at $Y_2=180°$.

In general, for a value of e' less than about 0.5, all values of e will be bounded above by unity, if at some time (or initially) the (osculating) value of e is less than unity while Y_2 passed through zero. For a value of e' less than about 0.2, all values of e will be bounded above by unity, if at some time the value of e is less than about

0.7 while Y_2 passed through 180°. As the eccentricity of Jupiter, e', increases to 0.2, 0.3, and 0.4, values of e will be bounded above by unity, if at some time the value of e is less than 0.4, 0.25, and 0.1, respectively, while Y_2 passes through 180°. For e' exceeding about 0.4, all values of e will exceed unity at a later time if, initially, they possessed a non-zero eccentricity while Y_2 passed through zero.

In Figure (55) the triangular libration point begins at $e=0$, $e'=0$, $Y_2=60°$, and moves upwards. The movement of other stationary points are evident from Figure (55): beginning at $e \sim 0.45$, $e' \to 0.0$, $Y_2=0°$, moving upwards, exceeding unity in e at $e' \sim 0.6$; beginning at $e \sim 0.9$, $e' \to 0.0$, $Y_2=180°$, moving downwards, beginning at $e \sim 0.45$, $e' \to 0.0$, $Y_2=180°$, moving downwards, approaching $e=0$ at $e' \sim 0.35$; a saddle point beginning at $e \to 0.0$, $e' \sim 0.12$, $Y_2=0°$, moving upwards; and a saddle point beginning at $e \sim 0.45$, $e' \to 0.0$, $Y_2 \sim 90°$, moving circularly and approaching unity near $e'=0.7$.

The triangular libration points (for $e' \neq 0$, elliptic stationary orbits) are evidently defined by $e=e'$, $Y_2=60°$ or $-60°$, as seen in the contour plots.

5. The Hecuba-Type Resonance: $p=1$ and $q=2$

The contour curves for this case are given in Figures (12) through (22). The development of the stationary points is shown in Figure (56).

In Figures (12) through (22), circulation in Y_2 corresponds to small librations in e, for small values of e', and corresponds to larger librations in e, for larger values of e'. Again, as e' increases from zero the contour lines become curves, with single periodicity in 2π, and stationary points appear. As e' increases, the trend continues, with large values of e at $Y_2=0°$ librating to small values of e at $Y_2=180°$.

The boundedness of e is somewhat more complex than for the case $p=1$ and $q=1$. In general, e is bounded above by unity for all initial values of e, no matter what the value Y_2 initially has, except for a gap between about $e=0.4$ and $e=0.6$ at $Y_2=0°$, and for e greater than about 0.6 at $Y_2=180°$ and for larger values of e'.

For small eccentricities of Jupiter, e', in the vicinity of 0.048 for instance, only two stationary points are evident, and these occurring for large values of the eccentricity of the asteroid. As e' increases past 0.2, stationary points appear for small values of e. In Figure (20), for $e'=0.5$, at about $e=0.1$ and $Y_2=140°$ and 160°, an examination of the data gave a slight indication of two stationary points, one a maximum the other a minimum and a saddle point was indicated at about $e=0.05$ and $Y_2=180°$. More accurate calculations were run, calculating many more points for this area. The stationary points remained and it may be tentatively stated that they exist. The fact that there are five stationary points for $e'=0.5$, in a small area bounded by $e=0.0$ and $e=0.3$ and $Y_2=135°$ and 180°, is interesting.

6. The Hilda-Type Resonance: $p=2$ and $q=3$

The plots for this case are given in Figures (23) through (33). The development of the stationary points is shown in Figure (57).

In Figures (23) through (33), circulation in Y_2 is possible for values of e less than about 0.4 and for values of e' less than about 0.15. As e' exceeds 0.2, no circulation in Y_2 is possible for any value of e. For larger values of Jupiter's eccentricity, e', no orbits have bounded eccentricities except for the stationary solutions.

For the asteroid Hilda, with eccentricity, $e \sim 0.25$ at $Y_2 = 0°$, with the actual eccentricity of Jupiter, $e' = 0.048$, Figure (25) shows circulation in Y_2 and a variation in e, from 0.25 at $Y_2 = 0°$ to 0.1 at $Y_2 = 180°$. This is identical to the result given for Hilda by J. Schubart (1968). Schubart obtained his result by numerically integrating equations similar to Equations (5), given here.

In Figure (57), three stationary points appear at $Y_2 = 0°$ and at $Y_2 = 180°$, two appear, for small eccentricities of Jupiter, e'. For the value of e' equal to 0.3, a stationary point appears at $Y_2 = 0°$ and $e \sim 0.7$ and at $Y_2 = 180°$ and $e \sim 0.6$. As the two points appear at $e' = 0.3$, the two saddle points at $Y_2 = 0°$ and 180° move to higher and lower values, respectively, of Y_2.

7. The Planar Pluto-Neptune Resonance: $p = 3$ and $q = 2$

The plots for this case are given in Figures (34) through (43). The development of the stationary points is shown in figure (58).

For small eccentricities of Neptune, less than about 0.2, and for moderately small eccentricities of Pluto, less than about 0.3, circulation is possible for the angle Y_2. Exceeding $e' = 0.3$, no circulation is possible for Y_2.

The eccentricity of Pluto, e, is bounded by unity for all values of e', at least less than 0.6, if the (mean mean) osculating eccentricity of the asteroid is bounded by unity when $Y_2 = 0°$. In this case, libration in Y_2 about 0° is possible.

It is interesting to notice that for small values of the eccentricity of Neptune, the argument of perihelion of Pluto, Y_2, circulates. Also, the value of the critical argument Y_1 librates about 180°. This result is identical to the result obtained by the numerical integration of the five outer planets by Cohen and Hubbard (1964) and the semi-analytical procedure of Hori and Giacaglia (1967). Hori and Giacaglia's Pluto-Neptune model is a circular three dimensional restricted problem. Hence, it appears that the libration of the critical argument of Pluto about 180° (and the impossibility of a Pluto-Neptune close approach) exists if either the inclination is zero or Neptune's eccentricity is zero.

In Figure (58), three stationary points appear at $Y_2 = 0°$ and three points appear at $Y_2 = 180°$, for small eccentricities of Neptune. All disappear at approximately the same value of the eccentricity of Neptune.

8. The Thule-Type Resonance: $p = 3$ and $q = 4$

The contour plots for this case are given in Figures (44) through (54). The development of the stationary points is shown in Figure (59).

In Figures (44) through (54), circulation in Y_2 exists only in small ranges of the eccentricity of the asteroid, e, even for small values of the eccentricity of Jupiter, e'. As e' increases past 0.1 libration in Y_2 occurs for most values of e. For values of e that allow circulation in Y_2, variation in e has large amplitudes of about 0.3. As e' exceeds about 0.3 circulation in Y_2 is possible only for values of e less than about 0.1. For values of e' greater than about 0.5, e is not bounded above by unity except for several stationary solutions.

There appears to exist many more stationary solutions for the case $p=3$, $q=4$ than exist for the other sets of p and q studied. For $e'=0.048$, for instance, there exist four stationary solutions at $Y_2=0°$ and three at $Y_2=180°$. In Figure (59), at $Y_2=0°$, there are four stationary solutions appearing for small eccentricities of Jupiter, e', and one appearing at $e' \sim 0.4$.

At $Y_2=180°$, four appear for small e'.

9. Pairing of the Stationary Solutions

On examining the figures of the developments of the stationary solutions, Figures (55) through (59), it may be noticed that, in general, stationary solutions exist in pairs, appearing and disappearing at the same values of the eccentricity of Jupiter, e'. One member of the pair occurs at $Y_2=0°$ and as e' increases, the solution moves upwards, to larger values of the eccentricity of the asteroid, e. The other member of the pair occurs at $Y_2=180°$ and as e' increases, the solution moves downwards, to smaller values of e.

The paired solutions have values of e exceeding unity or vanishing through zero at approximately the same value of e'.

Occasionally some solutions are not paired. Whether the calculations were not thorough enough to discover the other member of the pair or whether the other member does not exist is not known for certain.

References

Cohen, C. J. *et al.*: 1967, *Astron. J.* **72**, 973.

Giacaglia, G. E. O.: 1968, SAO Special Rep. 278.

Giacaglia, G. E. O.: 1969, *Astron. J.* **74**, 1254.

Hori, G. and Giacaglia, G. E. O.: 1967, Res. Cel. Mech. Diff. Eqo., Univ. São Paulo, CEMC-IPM-USP 67/68, No. 1, 4.

Schubart, J.: 1968, *Astron. J.* **73**, 99.

Poincaré, H.: 1902, *Méthodes nouvelles de la mécanique céleste*, Vol. II, Gauthier-Villars.

Discussion

Y. Kozai: Do your results depend on values of critical argument?

P. E. Nacozy: The contour curves of the eccentricity of the asteroid versus the angle Y_2 do not depend on the actual values of the critical argument. We have obtained separately the librational character of the critical argument. Knowledge of the long-term behavior of the asteroid requires examination of both the contour plots and librational character of the critical argument. The contour plots tell us whether the argument of perihelion librates or circulates and the corresponding amplitude of oscillation of the eccentricity of the asteroid. The librational character of the critical argument tells us whether the critical argument stays near 0° or 180°, or for the Trojan case, 60°. For an asteroid whose orbit is within Jupiter's orbit close approach is not possible if Y_2 librates about 0°. Close approach is possible if Y_2 librates about 180°. For an asteroid outside of Jupiter (or the Pluto-Neptune system) close approach is not possible if Y_2 librates about 180°. And close approach is possible if Y_2 librates about 0°.

J. V. Breakwell: Do any of your periodic orbits, for enlarged eccentricity of Jupiter, yield unusually small closest distances to Jupiter?

P. E. Nacozy: Yes. This is indicated whenever, in the asteroidal case, Y librates around 180° and, in the Pluto-Neptune case, around 0°.

In order to give precise information in this respect, curves Y_1 vs. Y_2 along the $F=$const. orbits should be given. This we plan to do in the near future.

E. Rabe: Do your results give any information on the behavior of the librations about L_4 and L_5, in the Trojan case? For instance, their dependence on libration amplitude?

P. E. Nacozy: This would require additional calculations in order to obtain the generating function for the double averaged variables.

P. E. Nacozy: (Comment to Mr. Rabe's question by G. Hori) When we want to obtain the librational motion of Trojan asteroids, the determining function $S_{1/2}$ should be obtained numerically and its derivatives with respect to the variables should be estimated numerically. Nobody ever tried but worth while to do.

J. Schubart: How do your results agree with the numerical results obtained by Colombo, Munford and Franklin (*Astron. J.* **73**, 1968) for periodic solutions of the Hilda-type in the case of the elliptic restricted three-body problem with Jupiter's mass and eccentricity?

P. E. Nacozy: Colombo, Munford, and Franklin obtained periodic orbits in the circular restricted problem. They did compute a few orbits in the elliptic restricted problem. Two of their periodic orbits are stationary solutions and have an eccentricity of Jupiter equal to 0.05. In one of the orbits the asteroid is at perihelion when Jupiter is at aphelion, corresponding to a difference in the arguments of perihelion of 180°. At that moment, the osculating eccentricity of the asteroid is 0.42. In the other orbit the asteroid is at perihelion when Jupiter is at perihelion, corresponding to a perihelion difference of 0°. At that moment, the osculating eccentricity of the asteroid is 0.49.

Corresponding to these cases, our results show similar stationary solutions of e (the mean eccentricity of the asteroid) $= 0.51$, $Y_2=0°$, and $e=0.43$, $Y_2=180°$.

Another, and more direct comparison of my results may be made with your study of the motion of the asteroid (153) Hilda (J. Schubart, *Astron. J.* **73**, 1968). Both of our studies perform a numerical average, eliminating the short-period terms, in the planar elliptic restricted problem. After the averaging, your procedure utilized numerical integration whereas mine used the Hori-Giacaglia-von Zeipel transformation. Our results should be equivalent.

For the asteroid Hilda, you found that the argument of perihelion circulates. You found the values of Hilda's mean osculating eccentricity to be approximately 0.24 when $\tilde{\omega}=0°$ and 0.11 when $\tilde{\omega}=180°$. My results are identical with yours for this case. I find circulation in Hilda's perihelion and oscillation in its eccentricity also between 0.11 and 0.24.

NUMERICAL STUDIES OF SOLAR INFLUENCED PARTICLE MOTION NEAR THE TRIANGULAR EARTH-MOON LIBRATION POINTS

B. E. SCHUTZ* and B. D. TAPLEY**

The University of Texas at Austin, Austin, Tex., U.S.A.

Abstract. The equations of motion of the restricted problem of four bodies were numerically integrated using the Jet Propulsion Laboratory Ephemeris Tapes DE3 to provide the positions of the three primaries, viz., earth, moon, and sun. Using initial conditions for the particle which satisfied the elliptic restricted problem of three bodies (earth, moon, and particle), the numerical results of the restricted problem of four bodies indicate that a particle placed at L_5 on Julian Ephemeris Date 2 439 796.735 will follow a libration-point-centered motion for 2500 days. The envelope of the particle's motion about L_5 expands and contracts with a period of approximately 650 days. Additional numerical computations indicate that the motion will persist for a period in excess of 5000 days. For the same initial date, a near lunar encounter occurs at 579 days for a particle placed at L_4 with initial conditions satisfying the elliptic restricted problem of three bodies.

1. Introduction

In 1961 K. Kordylewski announced the observation of faint, cloudlike satellites near the L_4 and L_5 points of the earth-moon system. These observations and those made by other observers in subsequent years are discussed by Simpson (1967). However, the existence of the clouds has not been firmly established as evidenced by the discussions of Roosen *et al.* (1967, 1968), and Wolff *et al.* (1967).

Motivated partly by the observations of Kordylewski, the solar influenced motion of a particle near the triangular points of the earth-moon system has been studied using various models of the restricted problem of four bodies. Surveys of the literature on this problem are given by De Vries (1963), Steg and De Vries (1966), Szebehely (1967), and Schutz (1969). The simplest model used in studies of the solar influenced motion is the model proposed by Huang (1960) in which the effect of the sun on the earth-moon system is excluded. Thus the earth and moon move in a circular orbit around their center of mass which in turn describes a coplanar circular orbit around the sun. This model is referred to as the 'very restricted four-body' (VRFB) model. Using a variation of this model (i.e., the lunar orbital inclination was included) Tapley *et al.* (1964, 1965) found by numerical integration that if the particle was initially at L_4 with zero relative velocity, the particle's envelope of motion expanded and contracted with a period of approximately 1460 days. The earth, moon, and sun were initially collinear and the motion was studied for 2500 days. Feldt and Shulman (1966) extended the investigation of Tapley to 5000 days and found that the expansion-contraction of the envelope of motion did not persist due to a lunar encounter at

* Assistant Professor, Dept. of Aerospace Engineering and Engineering Mechanics.
** Professor and Chairman, Dept. of Aerospace Engineering and Engineering Mechanics.

G. E. O. Giacaglia (ed.), Periodic Orbits, Stability and Resonances, 128–142.

approximately 4000 days. However, Tapley and Schutz (1968) discuss the effect of the constants used in the model and found that if more accurate values were used, an expansion-contraction motion persisted for over 8000 days.

The VRFB models neglect the important indirect effect of the sun, i.e., the gravitational effect on the motion of the earth and the moon. Hence, the results obtained from the simplified VRFB model cannot be used to infer motion in the real world. That is, based on VRFB results, it is not known whether the expansion-contraction of the envelope of motion which exists in the VRFB models also exists in the real world. The purpose of the investigation reported in this paper was to study the nature of the solar influenced particle motion near the earth-moon triangular points by numerically integrating the equations of motion over a period of several years using a model which closely represents the real world. The data are presented using orbital elements since they are somewhat more indicative of general trends than the trajectory plots used in the previous investigations.

2. Model and Procedure

The equations of motion of the particle using the model based on ephemeris data for the earth, moon and sun were expressed in a nonrotating, rectangular coordinate system, the origin of which was at the earth-moon barycenter. This model of the restricted problem of four bodies will be referred to as the 'Ephemeris Model'. The equations of motion were integrated numerically using a computer program developed by Lastman (1964). The numerical integration procedure utilizes a fourth order Runge-Kutta to obtain three starting values in addition to the given initial conditions. Control is then switched to a fourth-order Adams-Bashforth predictor followed by one application of the fourth-order Adams-Moulton corrector. The program uses partial double precision arithmetic, i.e., all calculations are performed in single precision with the exception that the dependent variables are stored in double precision to control round-off error. This partial double precision program is referred to subsequently as single precision. All computations were performed on the Control Data Corporation (CDC) 6600 which uses a 60-bit single precision word or 14 digit accuracy. As discussed in a later section, complete double precision computer runs were made to evaluate the round-off error in the single precision runs.

Ephemeris information was obtained from the Jet Propulsion Laboratory (JPL) Ephemeris Tapes DE 3 (see Peabody *et al.*, 1964a, b). These magnetic tapes were modified to be compatible with the CDC 6600. The JPL tapes provide the position and velocity of the moon, earth-moon barycenter, and the major planets in a heliocentric rectangular coordinate system oriented such that the X-axis is in the direction of the Vernal Equinox of Epoch 1950, the Y-axis is in the Equatorial Plane of Epoch 1950, and the Z-axis is mutually perpendicular. The second and fourth differences are stored on the tapes for interpolation purposes. All computations performed in the subroutine which determines the positions of the sun, moon, and earth by interpolation were made in double precision.

The constants used were those given by Clarke (1964) with the exception that Peabody *et al.* (1964a, b) suggest a slightly different value for converting earth radii to kilometers for use with the ephemeris tapes. Of particular importance is the earth-moon mass ratio given by Clarke, viz., $m_{\oplus}/m_{☽}=81.3015\pm0.0033$. The value of 81.3015 was used in all computations reported in this paper.

3. Initial Conditions

In all cases investigated, the particle was assumed to be initially at the L_4 or L_5 point. Furthermore, the triangular points were assumed to lie in the plane perpendicular to the angular momentum vector of the earth-moon system relative to the earth-moon barycenter. Two sets of initial conditions were used for computer runs at both L_4 and L_5. In Case I, the velocity requirements for the elliptic restricted problem of three bodies were satisfied. That is, if the effects of the sun were neglected, the earth and moon would move in elliptic orbits and the particle would have the proper velocity to maintain the equilateral triangle configuration. In Case II, the radial component of the velocity relative to the earth-moon barycenter, as determined by the initial conditions used in Case I, is set equal to zero. The other conditions remained the same.

In addition to selecting initial conditions for the particle, the initial date must be chosen also. The initial date was chosen to yield a relative orientation of earth, moon, and sun which closely approximated that used in previous VRFB model studies. The orientation used by Tapley *et al.* (1964, 1965, 1968) in VRFB studies was such that the sun, moon, and earth were collinear. This is approximated in the Ephemeris Model by new moon; however, it is more closely approximated during the occurrence of a solar eclipse in which case the earth center is relatively close to the sun-moon line. In addition, the moon was chosen to be near the descending node since this was the case in Tapley's model. An initial date which satisfied these requirements was Julian Ephemeris Date (JED) 2 439 796.735 or November 2, 1967, $5^{h}.6$ GMT. The perpendicular distance from the earth center to the sun-moon line at this time was slightly less than one earth radii.

For JED 2 439 796.735, the radial velocity component in Case I was 318 km/day and the tangential component was 94 700 km/day. The earth-moon distance was 356 994 km and the moon was slightly past perigee.

4. Results

The position and velocity history of the particle in the earth-moon barycentered coordinate system which was obtained from the numerical integration was transformed into an earth-centered system and then into orbital elements relative to the earth. The data, referred to the non-rotating barycentered system and also referred to a rotating coordinate system with origin at the triangular point of interest, were shown by Schutz (1966). The gravitational parameter used in computing the orbital elements was

$$\mu = k^2(m_{\oplus} + m_{☽})$$

and the inclination and longitude of the node are referenced to the Ecliptic of Epoch 1950. Also, the numerical results are displayed as a function of elapsed time from the initial date JED 2 439 796.735. The mean anomaly is not shown since it is essentially linear.

The results for Case I and Case II in which the particle was initially at the L_4-point are shown in Figures 1 to 5. The data for the moon are included for comparison purposes. The semi-major axis shown in Figure 1 shows a sudden change for both Case I and Case II at approximately 578–580 days and 462–464 days respectively. Examination of the rectangular coordinates reveals that a near lunar encounter occurs at these times. The eccentricity, inclination, longitude of the node, and argument of perigee also show sudden changes during the lunar encounter in both cases. The dashed vertical line in Figure 5 is used to plot ω in the range 0 to 2π. It should be noted that the lunar encounter occurs over 100 days later in Case I than in Case II, i.e., the initial conditions for the elliptic restricted problem of three bodies yields a longer libration-point-centered motion than the other set of initial conditions used.

A near lunar encounter does not occur with either Case I or Case II initial conditions for L_5 during the 2500 days studied. An expansion-contraction of the envelope of motion about L_5 for Case I can be seen in Figure 6 (displacement from L_5 with a period of approximately 650 days. Thus, for 2500 days the particle remains on a

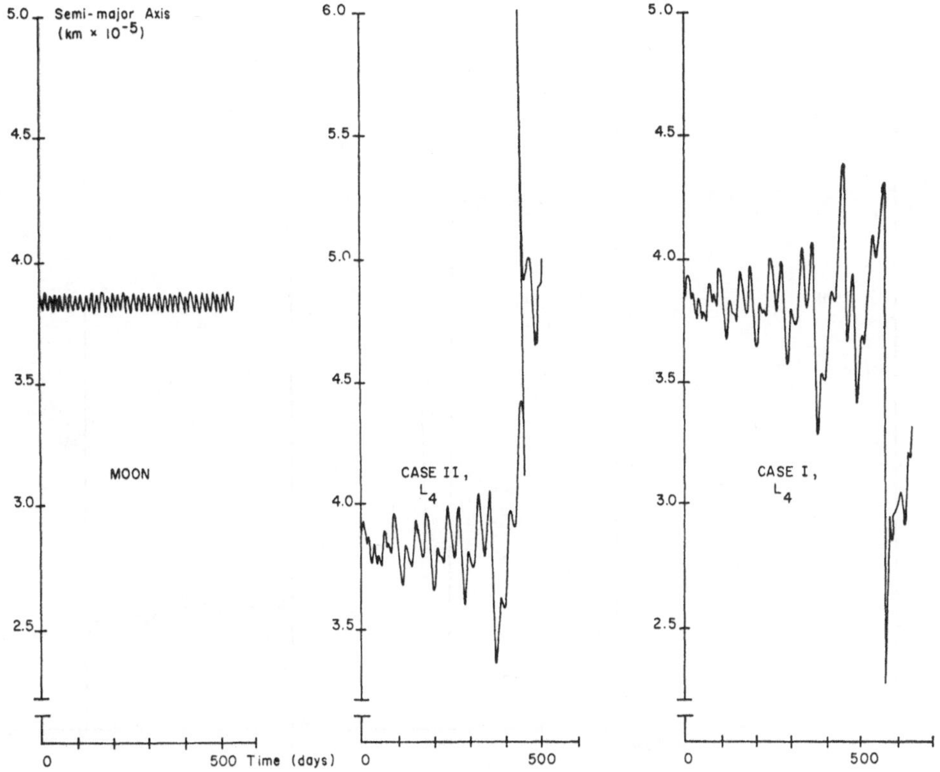

Fig. 1. Semi-major axis vs. time in Ephemeris model.

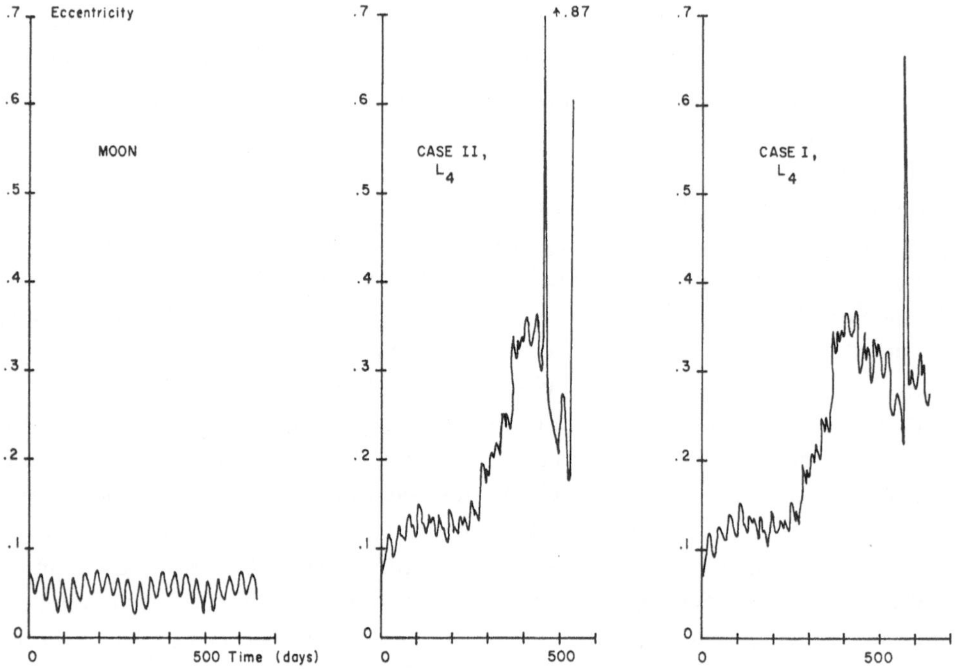

Fig. 2. Eccentricity vs. time in Ephemeris model.

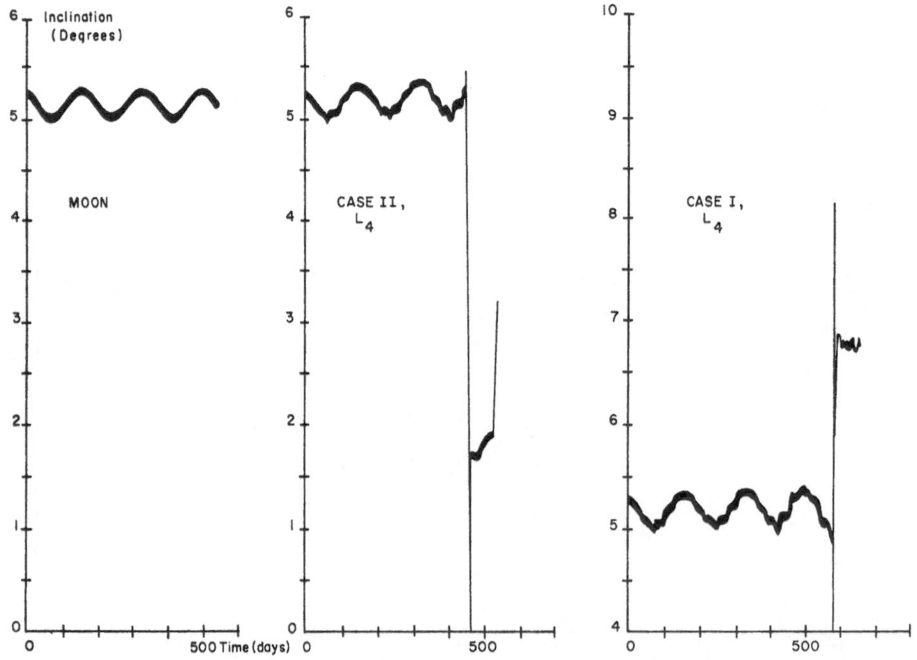

Fig. 3. Inclination vs. time in Ephemeris model.

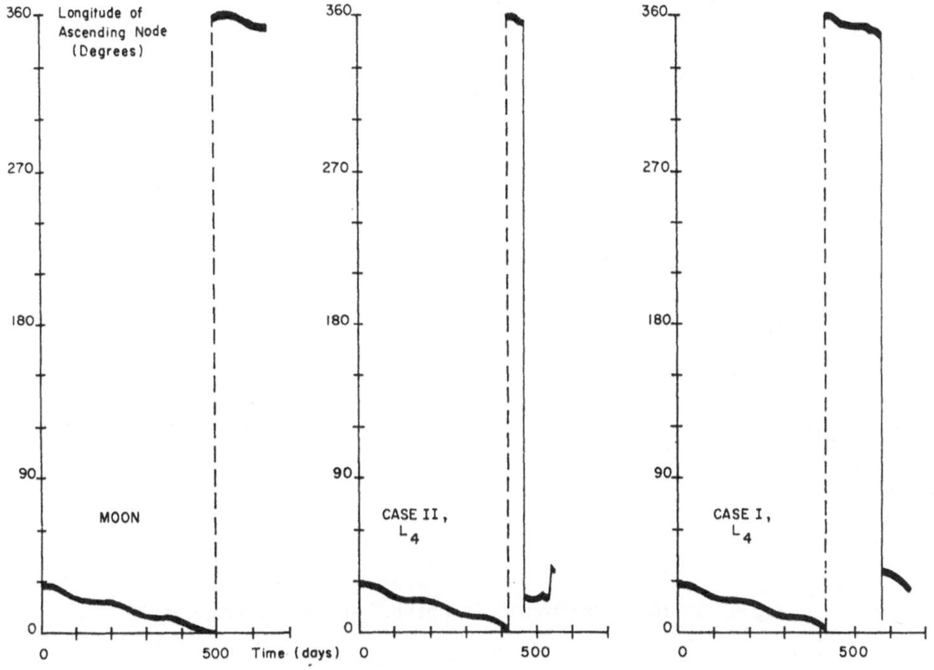

Fig. 4. Longitude of ascending mode vs. time in Ephemeris model.

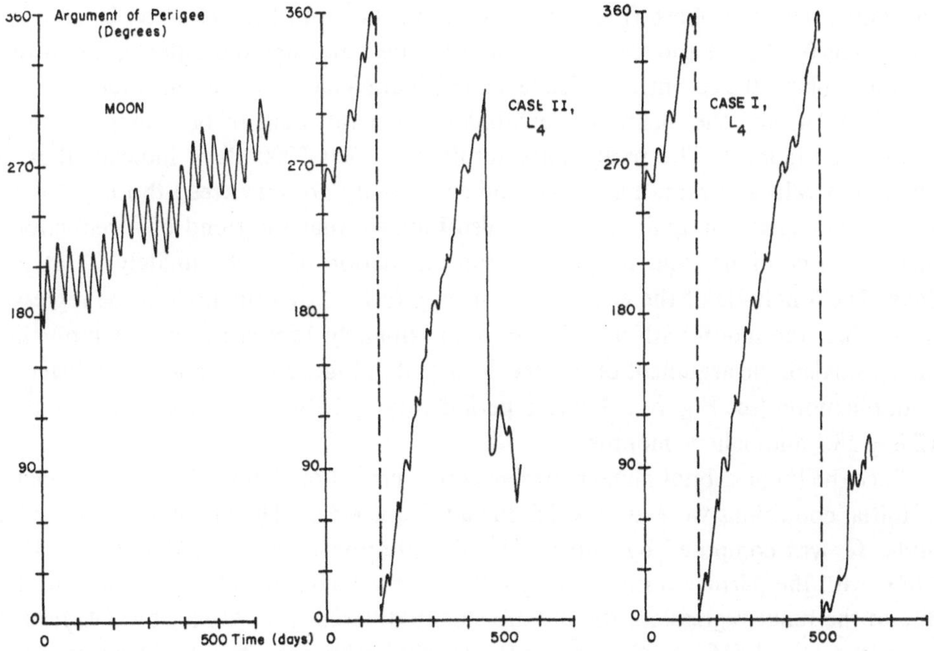

Fig. 5. Argument of perigee vs. time in Ephemeris model.

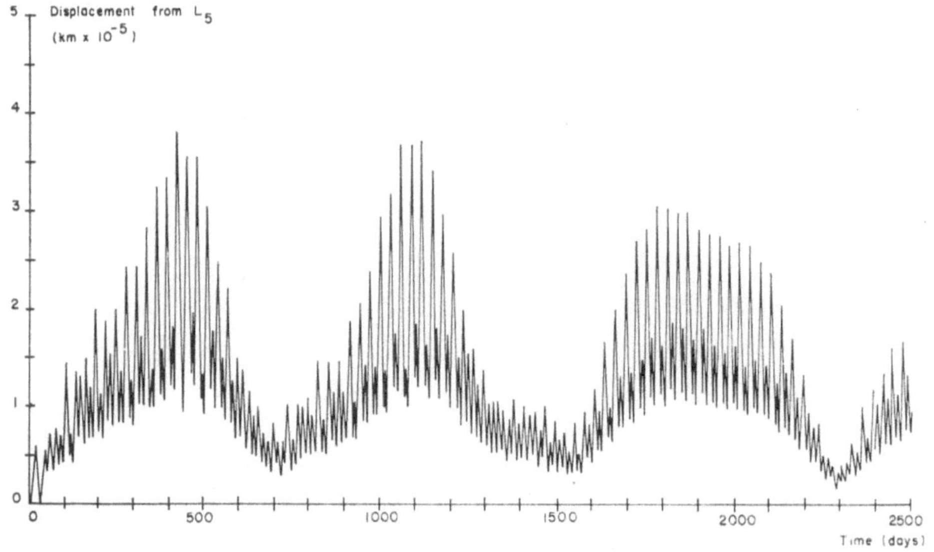

Fig. 6. Displacement from L_5 vs. time in Ephemeris model, Case I.

libration-point-centered motion. Preliminary studies of Case I for a period of 5000 days indicate that the particle continues on a libration-point-centered motion during this period. Numerical experiments to determine the accuracy of the computations during the first 2 500 days were made and are discussed in the next section. The orbital elements for Case I, Case II, and the moon are shown in Figures 7 to 12. The semi-major axis for Case I and Case II shown in Figure 7 appears to oscillate about a value of 380 000 km., the eccentricity displayed in Figure 8 has a maximum value of slightly less than 0.4, and the inclination exhibits the general trend of becoming smaller as shown in Figure 9. The preliminary results for 2500–5000 days indicate that the maximum values of semi-major axis and eccentricity do not exceed those of the first 2500 days. These preliminary results also indicate that the trend of a reduction in inclination continues and reaches a mean inclination of approximately 2° at 5000 days. The longitude of the node shown in Figure 10 shows the node to be regressing faster than the moon with a period of approximately 15.9 years or 210 anomalistic months. Also, the argument of perigee displayed in Figure 12 advances at a faster rate than the moon (see Figure 11), viz., a period varying between 350 days to 500 days or 12.8 to 18.1 anomalistic months.

Since the lunar orbital plane regresses with a period of 18.6 years, an additional set of initial conditions were used which included this effect. The angular velocity of the node, $\dot{\Omega}$, was computed by numerical differentiation to be -1.255×10^{-4} rad/day. Thus, with the particle initially at L_5 but with a velocity including the motion of L_5 due to the lunar regression, the initial velocity differed from Case I by: $(\Delta\dot{x})=0.796$ km/day, $(\Delta\dot{y})=1.515$ km/day, and $(\Delta\dot{z})=18.499$ km/day, or $\Delta v=18.5$ km/day. Using a step size of 0.025^{d}, the numerical integration results were compared with the results

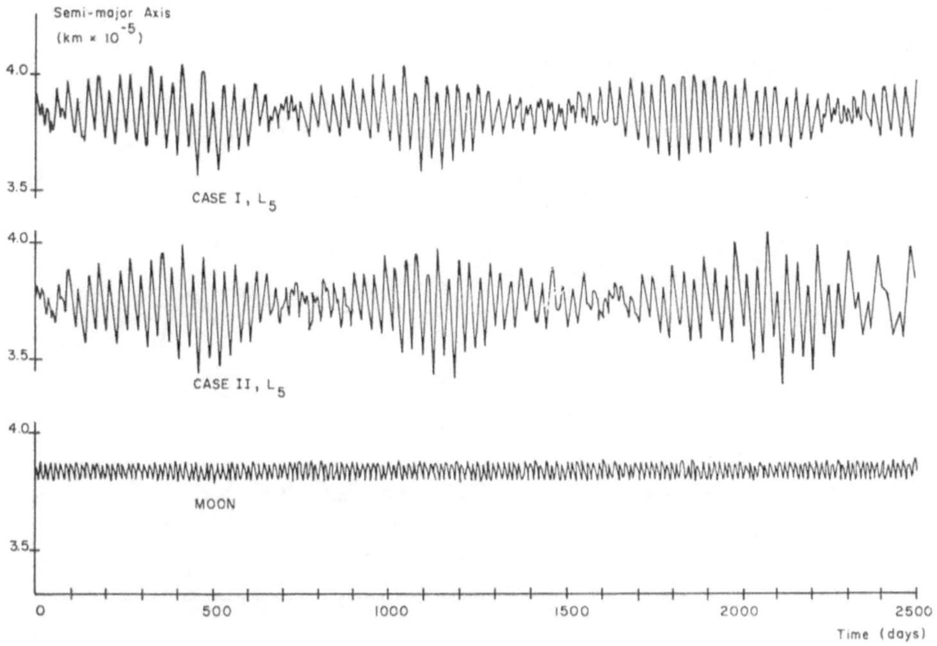

Fig. 7. Semi-major axis vs. time in Ephemeris model.

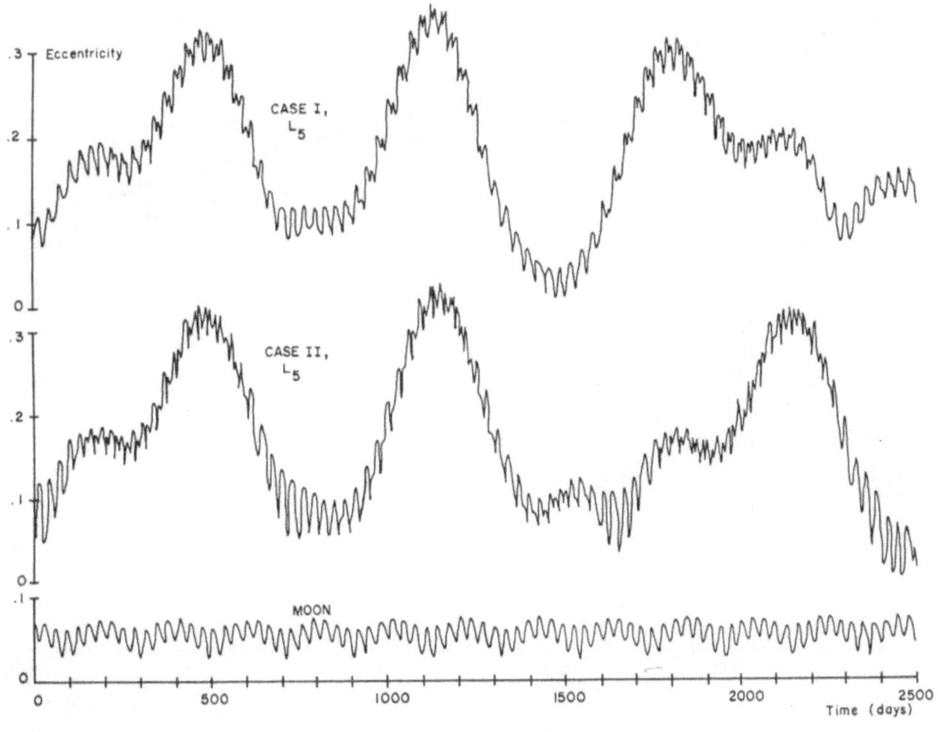

Fig. 8. Eccentricity vs. time in Ephemeris model.

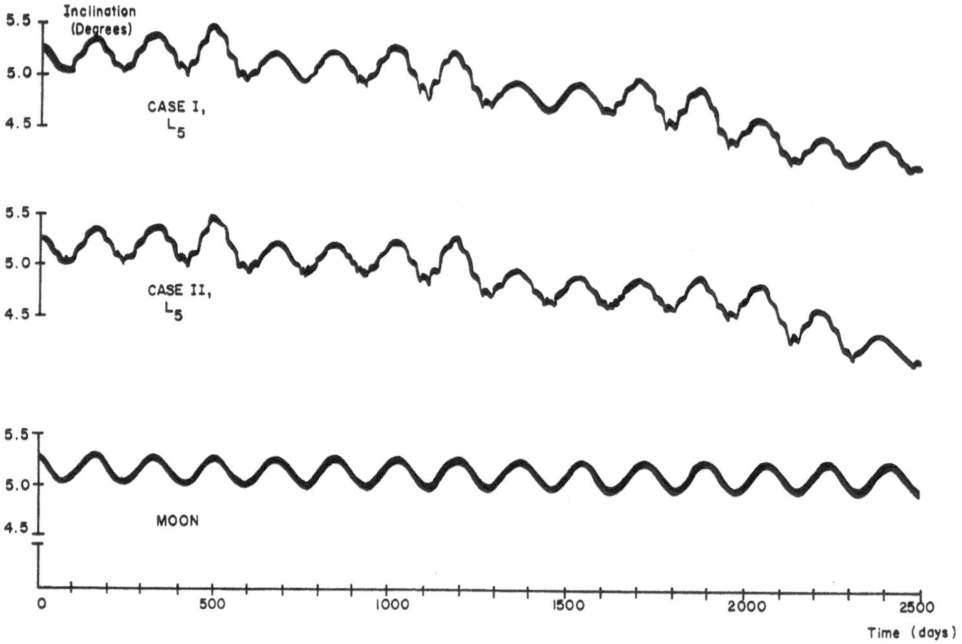

Fig. 9. Inclination vs. time in Ephemeris model.

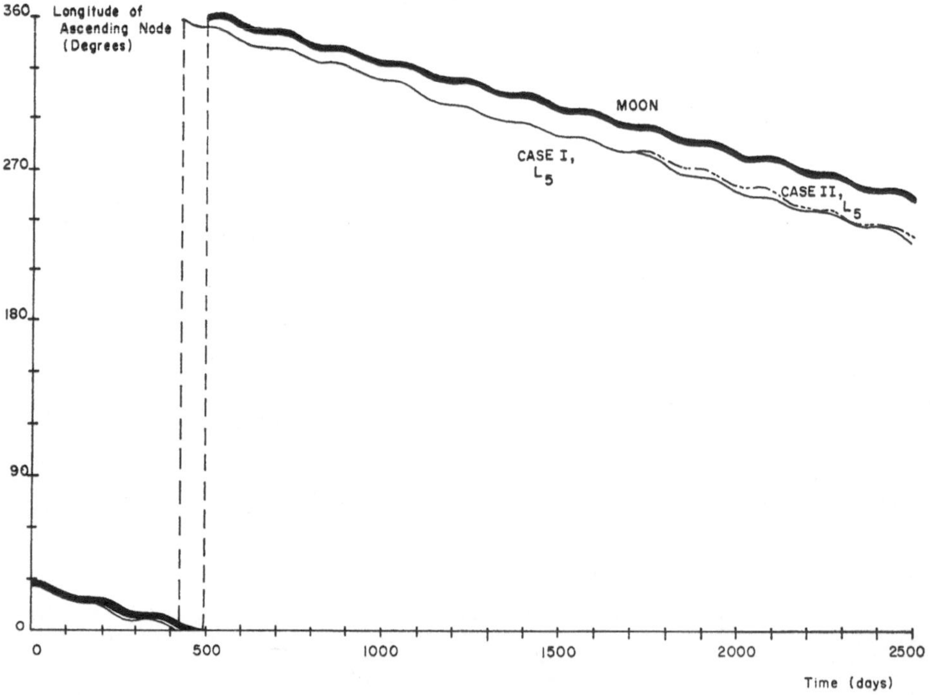

Fig. 10. Longitude of ascending node vs. time in Ephemeris model.

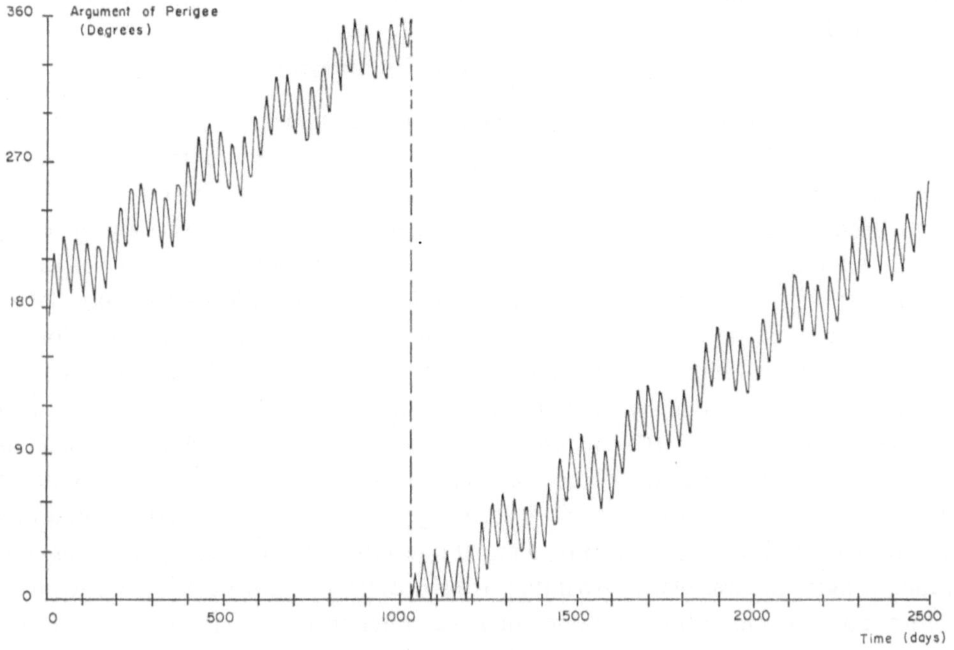

Fig. 11. Argument of perigee vs. time for the moon in Ephemeris model.

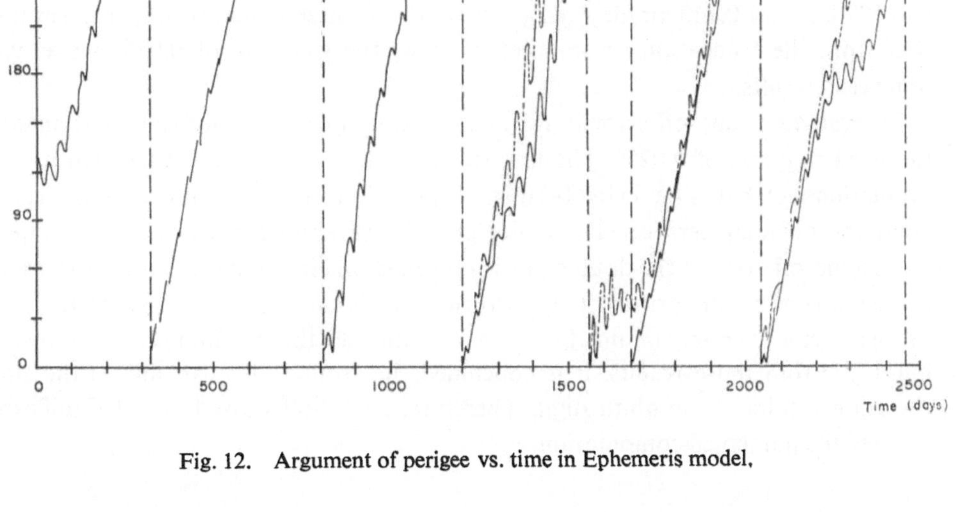

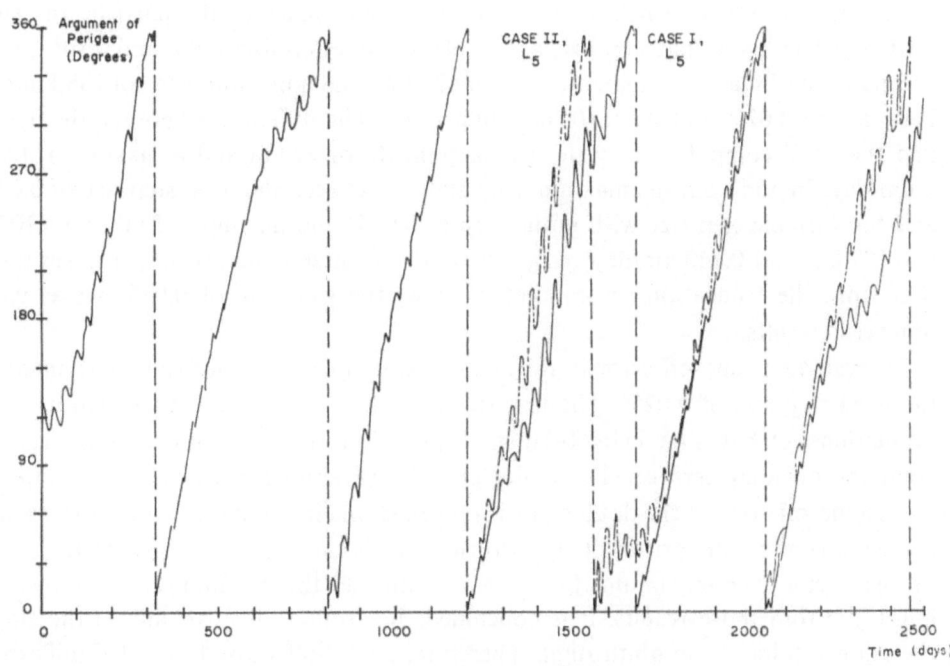

Fig. 12. Argument of perigee vs. time in Ephemeris model.

of Case I. The maximum displacement between the two cases was 107 km during the first 2 500 days.

5. Accuracy of the Numerical Integration

Since round-off error is present in all finite-precision numerical calculations, a very small step size will result in large round-off error if a large number of integration steps are used. On the other hand, a large step size will result in large truncation error after many steps. Hence, numerical experiments must be performed to determine a proper step size for an integration of a particular set of differential equations over a given range.

Since the L_5 results required a greater number of integration steps than the L_4 results, the discussion will be concerned with numerical results computed for the L_5-point, although similar computer runs were made for L_4. A variable step size computer run was originally made to compensate for rapid changes in positions in the event of a lunar or earth encounter. The initial step size was chosen to be 0.20^d with the Single Step Error (SSE) required to be in the range $1 \times 10^{-5} \leqslant \text{SSE} \leqslant 5 \times 10^{-14}$. The integration was performed for a 2500 day period. Using a smaller initial step size of 0.05^d and same SSE requirements, the results of the two computer runs were found to agree to three places. An initial step size of 0.025^d gave much better agreement with the 0.05^d computation. Since no lunar encounter occurred in these computer runs, fixed step size computations were made using step sizes of 0.10^d, 0.05^d, and 0.025^d. The latter step size results were then compared with the other two by computing at each step

$$\Delta \varrho = [(x_h - x_{0.025})^2 + (y_h - y_{0.025})^2 + (z_h - z_{0.025})^2]^{1/2}$$

$$\Delta v = [(\dot{x}_h - \dot{x}_{0.025})^2 + (\dot{y}_h - \dot{y}_{0.025})^2 + (\dot{z}_h - \dot{z}_{0.025})^2]^{1/2}$$

where (x, y, z) are the earth-moon barycenter coordinates of the particle, the subscript 0.025 refers to that step size, and h refers to the step sizes 0.05^d and 0.10^d. These computations indicated a maximum $\Delta\varrho$ of 624 km, and a maximum Δv of 138 km/day between the 0.025^d and the 0.10^d computer runs. The differences between the 0.025^d and the 0.05^d computations yield a maximum $\Delta\varrho$ of 27 km and a maximum Δv of 6 km/day. In addition, the maximum $\Delta\varrho$ and Δv between the fixed step size of 0.025^d and the variable step size with initial step of 0.025^d and maximum SSE of 1×10^{-7}, is 0.507 km and 0.140 km/day respectively. From these comparisons, it is surmised that from the truncation error point of view, the step size of 0.025^d yields valid numerical results.

To evaluate round-off error in the computation, two computer runs were made at the fixed step size of 0.025^d. The first run was the aforementioned truncation error evaluation which was made in 60-bit single precision. In the second computer run all computations were performed in double precision. Making the reasonable assumption that round-off error in the double precision run is smaller than the error for the single precision run, a comparison of the two runs yields the round-off error of the single precision run. The maximum $\Delta\varrho$ is 0.001 36 km and the maximum Δv is 0.000 372 km/day. From those results, it is concluded that round-off error affects the single precision results in the ninth digit. Therefore, round-off error has not significantly affected the numerical computation.

As a further check on the accuracy, the differential equations of motion for the particle, moon, and sun were expressed in an earth-centered rectangular coordinate system resulting in 18 first order differential equations which were numerically integrated. The initial position and velocity of the moon and sun were obtained from the JPL Ephemeris Tapes. The results were transformed to an earth-moon barycentered system and compared with those of the Ephemeris Model and found to agree through three digits throughout the 2500 day study for L_5. The discrepancy can be attributed to the following:

(1) The initial velocity specification for the moon and sun may not have been sufficiently accurate to be consistent with the results given by the theories from which the positions of these bodies were generated. The velocity data stored on the Ephemeris Tapes was generated by numerical differentiation of the position data; the velocity for the given initial date was then obtained by interpolation. A similar comparison of the lunar position obtained by numerical integration to that stored on the Ephemeris Tapes shows 3 place agreement.

(2) The indirect effect of the planets was not included in the integration of the 18 differential equations of motion for the restricted problem of four bodies, but was partially included in the Ephemeris Model through the Ephemeris Tapes (see Section 6).

(3) Since 18 first order ordinary differential equations were integrated simultaneously, round-off error may have affected the results in the fifth or sixth place instead of the ninth place as was the case in the Ephemeris Model.

The discrepancy, it should be noted, cannot be discerned on the orbital element plots or the x-y trajectory plots.

These numerical experiments indicate that at least three place accuracy and possibly five place accuracy was obtained on the L_4 and L_5 computer runs. Even on the computer runs of 100 000 steps, round-off error affected the results only in the ninth place on the single precision computations.

6. Effect of the Planets

Although the Ephemeris Model includes only the sun, moon, and earth the indirect effect of the planets is partially included since the planets are included in the generation of the ephemeris data for the JPL Ephemeris Tapes. However, the indirect effect of the planets is not included in the differential equations used in the Ephemeris Model.

A study of the effect of the planets was made by including both the direct and indirect effects in the differential equations of motion for the particle in the Ephemeris Model. In order to avoid unnecessary computations, the magnitude of the acceleration of the particle in the earth-moon barycenter frame of reference was computed assuming the particle to be continuously at L_4. It was found that Uranus, Neptune, and Pluto could be excluded in the evaluation of the effect of the planets. The numerical integration was performed for L_5 in single precision with a fixed step size of 0.025^d and compared with the results of the integration neglecting the planets at the same step size. The maximum computed difference in $\Delta\varrho$ is 488 km and the maximum Δv is 112 km/day. While the planets do not significantly affect the motion during the 2500 days considered here, it appears that their effect can be quite significant for motion which persists for longer periods.

7. Other Initial Conditions

Using an initial date of JED 2 439 501.0, and an initial particle position of L_4, additional numerical studies were made. This particular date was chosen because of its location on the JPL Ephemeris Tape, viz., it is the first date on the second tape. The results of this study were similar to the L_4 results of initial date JED 2 439 796.735, i.e., the particle experienced a close lunar encounter at 730 days for Case I and approximately 490 days for Case II. The initial position of L_5 was not used.

8. Conclusions

Starting the particle at L_5 on JED 2 439 796.735 with an initial velocity satisfying the requirements for the elliptic restricted problem of three bodies, numerical computations indicate that the particle will remain in a libration orbit or libration-point-centered orbit for over 2500 days (7 years). During this period, the envelope of motion expands and contracts with a period of approximately 650 days. Preliminary results indicate that it continues in this orbit for over 5000 days (14 years). Using an initial velocity in which the radial component of velocity (318 km/day) for the aforementioned starting date was set to zero, the particle's envelope of motion expands and contracts with approximately the same period for 2500 days and possibly longer. In addition, a slight change in the Case I velocity of 18.5 km/day yielded an orbit which deviated from Case I by a maximum of 107 km during the 2500 days.

For the same initial date a particle placed at rest relative to L_4 will experience a lunar encounter after approximately 579 days and the libration orbit is destroyed. Similar results have been obtained at L_4 for another initial date, viz., JED 2 439 501.0.

Although it has been found that a particle could have a libration orbit about L_5 for a period of time approaching at least 14 years and possibly longer, this does not confirm the observations made by Kordylewski, Simpson and others. Very special initial conditions were used which may be difficult to achieve by a natural body. In addition, although the particle moves on a libration orbit, the amplitudes of motion are quite large, e.g., the angle between the barycenter-particle line and the barycenter-L_5 line can be as large as 60° behind L_5 and 30° ahead of L_5. The observational reports have not indicated observing a cloud that far from L_5.

Note added in proof. The numerical integration has been repeated with the more recent JPL Ephemeris Tape DE 19. The differences in the results cannot be discerned on the orbital elements plots of this paper.

Acknowledgement

This research was sponsored by the U.S. Navy, Office of Naval Research under Contract Number N00014-67-A-0126-0007.

References

Clarke, V. C., Jr.: 1964, JPL Tech. Report 32-604.
Feldt, W. T. and Shulman, Y.: 1966, *AIAA J.* **4**, 1501.
Huang, S.-S.: 1960, NASA Technical Note D-501.
Lastman, G. J.: 1964, University of Texas Computation Center UTD2-03-046.
Peabody, P. R., Scott, J. F., and Orozco, E. G., 1964a, JPL Tech. Memo. 33-167.
Peabody, P. R., Scott, J. F., and Orozco, E. G., 1964b, JPL Tech. Report 32-580.
Roosen, R. G.: 1968, *Icarus* **9**, 429.
Roosen, R. G., Harrington, R. S., and Jefferys, W. H.: 1967, *Physics Today* **20**, 9.
Schutz, B. E.: 1966, Engineering Mechanics Research Lab TR-1002, University of Texas. (Also NASA-CR-65677.)
Schutz, B. E.: 1969, Dissertation, University of Texas.
Simpson, J. W.: 1967, *Physics Today* **20**, 39.
Simpson, J. W. and Miller, R. G.: 1967, *Physics Today* **20**, 11.
Steg, L. and de Vries, J. P.: 1966, *Space Sci. Rev.* **5**, 210.
Szebehely, V. G.: 1967, *Theory of Orbits*, Academic Press, New York.
Tapley, B. D. and Lewallen, J. M.: 1964, *AIAA J.* 728.
Tapley, B. D. and Schutz, B. E.: 1965, *AIAA J.* 1954.
Tapley, B. D. and Schutz, B. E.: 1968, *AIAA J.* 1405.
De Vries, J. P.: 1963, General Elect. Space Sciences Lab R63SD99.
Wolff, C., Dunkelmann, L., and Haughney, L.: 1967, *Science* **157**, 427.

Discussion

J. V. Breakwell: Does not your apparent difference between L_4 and L_5 depend really on your choice of initial time of month? (There being no essential difference here.)

I am unhappy about your gradual divergence of the plane. The difference, however, in the period of your long-period fluctuations, when you included the indirect effect of the Sun through the moon's motion, is explicable in that this period should depend essentially logarithmically on the reciprocal of the initial amplitude of the faster mode, which in your case of zero initial conditions must cancel out any sun-forced motion.

B. D. Tapley: I think that the answer to your question would be "yes". Although we have not actually found an orientation for the earth, moon and sun for which the motion at L_4 is persistent, the results shown previously do not imply that such a motion does not exist. I would expect that if a careful study of the effect of the initial conditions was made, that an orientation of the earth, moon and sun could be determined for which the motion at L_4 would be persistent. However, I would expect that, for such an orientation, the L_5 motion would persist for only a short time; as did the L_4-point motion in the case presented here. This would seem to suggest that there is a difference in the character of the two points in the real world model; since in the very restricted four body model, if the motion at L_4 is persistent for a given sun orientation, then it is persistent at L_5 for the same orientation. As pointed out in the previous results, this characteristic does not carry over to the real world model. The decrease in the inclination is a surprising trend which I cannot explain at this time. However, the computation has been carried out through 5000 days and it appears that the inclination reaches a lower limiting value of about 2° at this time.

J. Kevorkian: I am very gratified to see the enormous difference in the solutions by numerical integration using the very restricted model on the one hand and a dynamically consistent model on the other hand. L. Mohn and I derived equations of motion incorporating de Pontecoulant's lunar theory for the motion of the moon. Using these equations of motion we have derived analytic formulas for the coordinates as functions of time, and were not able to confirm your earlier numerical work for the very restricted problem. I am happy to confirm your present result that for motion starting from L_4 with zero velocity, motion reaches a large distance very soon. This result is evidenced in our theory by the divergence of our asymptotic results for this case.

B. D. Tapley: I am happy that we are in agreement on the behavior of the divergent motion at L_4. It would be interesting to determine how well your results describe the persistent motion at L_5 where the amplitude of motion initially grows to over 200 000 miles before the expansion ceases and a contraction mode is initiated.

ON USING MINOR PLANETS CLOSE TO THE 2:1, 3:2, 4:3 COMMENSURABILITIES TO DETERMINE THE MASS OF JUPITER

W. J. KLEPCZYNSKI

U.S. Naval Observatory, Washington, D.C., U.S.A.

In recent years there have been several attempts to determine the mass of Jupiter using minor planets in the 2:1 commensurability range which were suggested by Hill (1873) as being especially useful for this purpose. Table I lists the results. Minor planets 10, 24, 31, 52 were studied by Klepczynski (1969), 48 by Zielenbach (1968), 57 by Fiala (1968) and 65 by O'Handley (1967). Taking all things into consideration, I believe that these investigations are giving the best results that have so far been obtained using minor planets. By best results, I mean about five and a half significant figures in the reciprocal mass. However, before any further significant improvement can be obtained from dynamical analysis using minor planets, we must be able to identify objects which are considerably more useful for this goal. It is the purpose of this paper to present such a list of objects.

Laplace (1829) in *Mécanique Céleste* developed the first order perturbations in longitude of a minor planet. The periodic term which contains the first power of the eccentricity of the disturbing body and the disturbed body as factors is given by:

$$\delta v = \frac{m'n}{n - i(n - n')} \times \left\{eF^{(i)} \sin\left[(i-1)\,\lambda + i\lambda' - \pi\right] + e'G^{(i)} \sin\left[(i-1)\,\lambda + i\lambda' - \pi'\right]\right\}. \tag{1}$$

If we let $X = (i-1)\,\lambda - i\lambda'$
and

$$\gamma = \frac{n}{i(n - n') - n} = \frac{n}{(i-1)\,n - in'}.$$

TABLE I

Values for mass of Jupiter

Minor Planet	Reciprocal mass	Mean error
10	1047.351	0.006
24	1047.359	0.010
31	1047.372	0.006
52	1047.337	0.027
48	1047.340	0.024
57	1047.350	0.004
65	1047.387	0.004

G. E. O. Giacaglia (ed.), Periodic Orbits, Stability and Resonances, 143–150.

we can then write Equation (1) as

$$\delta v = m'\gamma \{eF^{(i)} (\sin X \cos \pi + \cos X \sin \pi) + e'G^{(i)} (\sin X \cos \pi' + \cos X \sin \pi')\}. \tag{2}$$

where $F^{(i)}$ and $G^{(i)}$ are functions of Laplace Coefficients.

By writing Equation (2) in the form

$$\delta v = m'\gamma \{\sin X (eF^{(i)} \cos \pi + e'G^{(i)} \cos \pi') + \cos X (eF^{(i)} \sin \pi + e'G^{(i)} \sin \pi')\} \tag{3}$$

and introducing the unknowns K and β by the relations

$$K \cos \beta = eF^{(i)} \cos \pi + e'G^{(i)} \cos \pi'$$
$$K \sin \beta = eF^{(i)} \sin \pi + e'G^{(i)} \sin \pi'$$

we can then derive an equation similar to the one given by Hill (1873),

$$\delta v = m'\gamma K \{\sin X \cos \beta + \cos X \sin \beta\}$$

or

$$\delta v = L \sin \{(i - 1) \lambda - i\lambda' + \beta\} \tag{4}$$

where

$$L = m'\gamma K.$$

Equation (4) is more general than that given by Hill. Hill considered only the case $i=2$ or 2:1 commensurability. In the form given here we can easily extend it to the 3:2 commensurability for $i=3$, and the 4:3 commensurability for $i=4$.

If we presume the relative distance from the resonant frequency for each of the different commensurabilities to be the same, Hill made the following important observation about this expression. Since we have here only the first power of the eccentricity in the coefficient, this term will be much larger for the 2:1, 3:2, 4:3 commensurabilities than the corresponding resonant term for the 3:1 or 5:2 or higher commensurabilities. For the 3:1 case, the eccentricity will be squared; for the 5:2 case, the eccentricity will be cubed. Thus, the resonant coefficient would normally be expected to be smaller for these cases. For cases when the 3:1 coefficients would be of equal or greater magnitude, their periods would be considerably longer than those of the 2:1, 3:2, or 4:3 cases and prohibit the practical use of those objects.

Table II lists the coefficient of the long-period term and its period found by Hill for the minor planets in the 2:1 commensurability range known at his time. The studies which have so far been made to determine the mass of Jupiter have used a representative sample of these objects; ranging from one with a large coefficient and long period, to one with a small coefficient and short period.

Since the time of Hill we have found more minor planets in the 2:1 range and even have discovered some in the 3:2 and 4:3 ranges. In view of this, the question arises whether there may be more objects useful for this purpose.

At this point I would like to point out some of the results found by Klepczynski

TABLE II

Hill's list of minor planets

Minor planet	Coefficient	Period
10 Hygeia	14676″	98
24 Themis	14606	92
31 Euphrosyne	28997	99
52 Europa	6584	68
48 Doris	5087	72
57 Mnemosyne	12956	103
65 Cybele	13145	94
49 Pales	11639	62
62 Erato	13655	83
76 Freia	32244	121
86 Semele	10861	65
90 Antiope	28568	104

(1969) in his determination of the mass of Jupiter. Figure 1 is a plot of the difference between observations and a geometric ephemeris of the minor planet 10 Hygeia which was generated using a reciprocal mass of Jupiter of 1047.355, the currently adopted value. The points represent unweighted means of the $(O-C)$'s. Next, the thick solid line represents the difference in heliocentric longitude between an ephemeris utilizing a disturbing mass of 1047.390 fitted to observations and an ephemeris based on 1047.355 fitted to observations. The thin solid line represents the difference in right ascension between the two ephemerides. If the true value for mass of Jupiter were

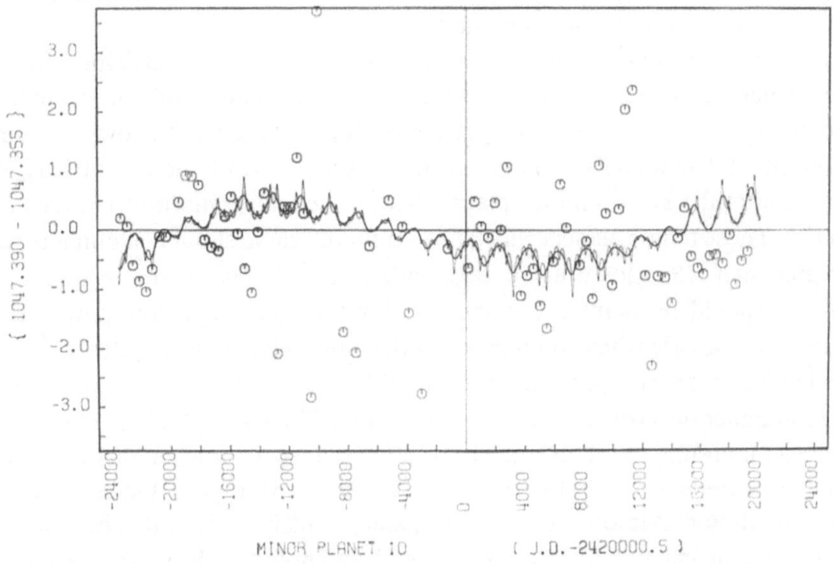

Fig. 1.

1047.390, then the $(O-C)$'s with respect to the integration based on 1047.355 should reflect the thin solid line.

There are several important items to be seen from this figure. First, the difference in heliocentric longitude between two integrations, each utilizing different disturbing masses for Jupiter and fitted to observations, is closely represented by the difference obtained on evaluating the resonant term with two different masses of Jupiter. Furthermore, the value obtained for the phase angle, β, corresponds closely to the nodal points which designate the closest approach of the minor planet to Jupiter. Secondly, the difference in longitude is magnified in right ascension due to close approaches of the minor planet to earth. Thirdly, it is of importance to point out that the mean error of unit weight of the observations of the minor planets used in these investigations so far has turned out to be about 1$''$.5.

Keeping in mind the fact that the resonant term reflects to a large degree the main perturbation caused by an error in the mass of Jupiter, that the mean error of unit weight of a minor planet observation is about 1$''$.5, and the range in values obtained for the mass of Jupiter from the investigations already cited, the following procedure was adopted to help identify those minor planets whose motion would be especially sensitive to a change in the mass of Jupiter.

The resonant term was evaluated for two different masses of Jupiter and a number of fictitious minor planets with varying ranges of eccentricity, mean motion and longitude of perihelion. The different reciprocal masses of Jupiter used were 1 047.355 and 1 047.390. This corresponds to a difference of 0.035 or 0.003%. The longitudes of perihelia were restricted to values such that the point of closest approach of the minor planet to Jupiter occurred when Jupiter was at perihelion and the minor planet was somewhere between its perihelion and aphelion. The eccentricities ranged from 0.00 to 0.95. The mean motion ranged 50$''$/d on each side of the resonant value for the 2:1, 3:2, and 4:3 commensurabilities.

Limiting curves were constructed for those values of the eccentricities and mean motions which produced a difference of 1$''$.5 in the evaluation of the resonant term corresponding to the two different masses of Jupiter. Figure 2 shows the limiting curve for the 2:1 commensurability together with a plot of the eccentricity versus mean motion of the known minor planets which have mean motions between 550$''$/d and 650$''$/d. Those minor planets which fall within the curve should exhibit a perturbation greater than 1$''$.5 in longitude corresponding to a change in the reciprocal mass of 0.035. It should be pointed out that the limiting curve is a lower limit since it corresponds to the case when conjunction of the minor planet and Jupiter occurs when the two bodies are at their perihelia. Note that the abscissa has the period of the long-period term under the corresponding mean motion. There are 42 minor planets which fall within the limiting curve. Note the lack of minor planets at the resonant mean motion.

A similar procedure was followed for the 3:2 case and Figure 3 shows the results of that computation. Almost all the minor planets which fall into this case are useful in determining the mass of Jupiter. There are 21 and note the close aggregation about the resonant frequency.

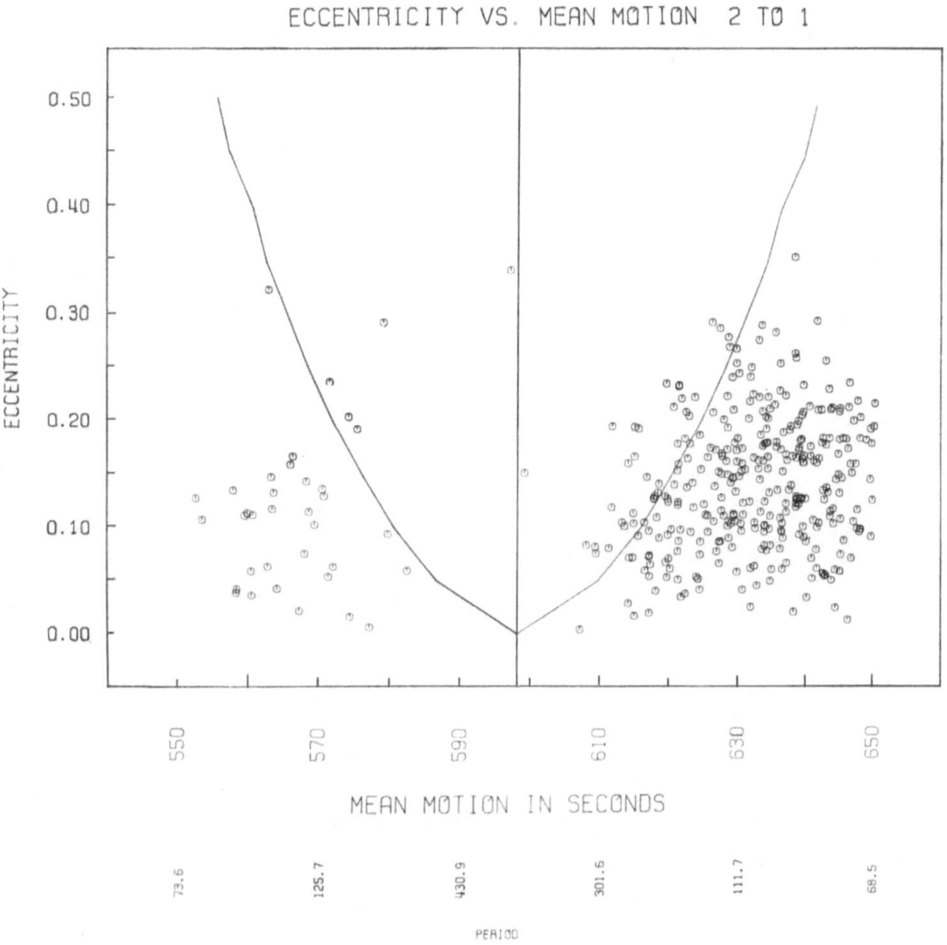

Fig. 2.

A similar figure for the 4:3 group is not necessary since there is only one member in it, 279 Thule, which upon computation is found to be an object sensitive to a change in the disturbing mass of Jupiter.

Table III is a listing of the minor planets for the two groups which fell within the limiting curves. Note that for the 2:1 case only 76 Freia is common to Hill's list and this list.

Using the osculating elements published in the 1968 ephemerides of minor planets, the resonant term was evaluated for each of the 64 minor planets sensitive to a small change in the mass of Jupiter and inspected. Table IV lists the 14 minor planets which would be considered as representative members of their respective groups.

The first 10 minor planets are in the 2:1 commensurability range. The first three were chosen from among those minor planets having a period less than 150 yr for the long-period inequality and a sensitivity greater than 1″.5.

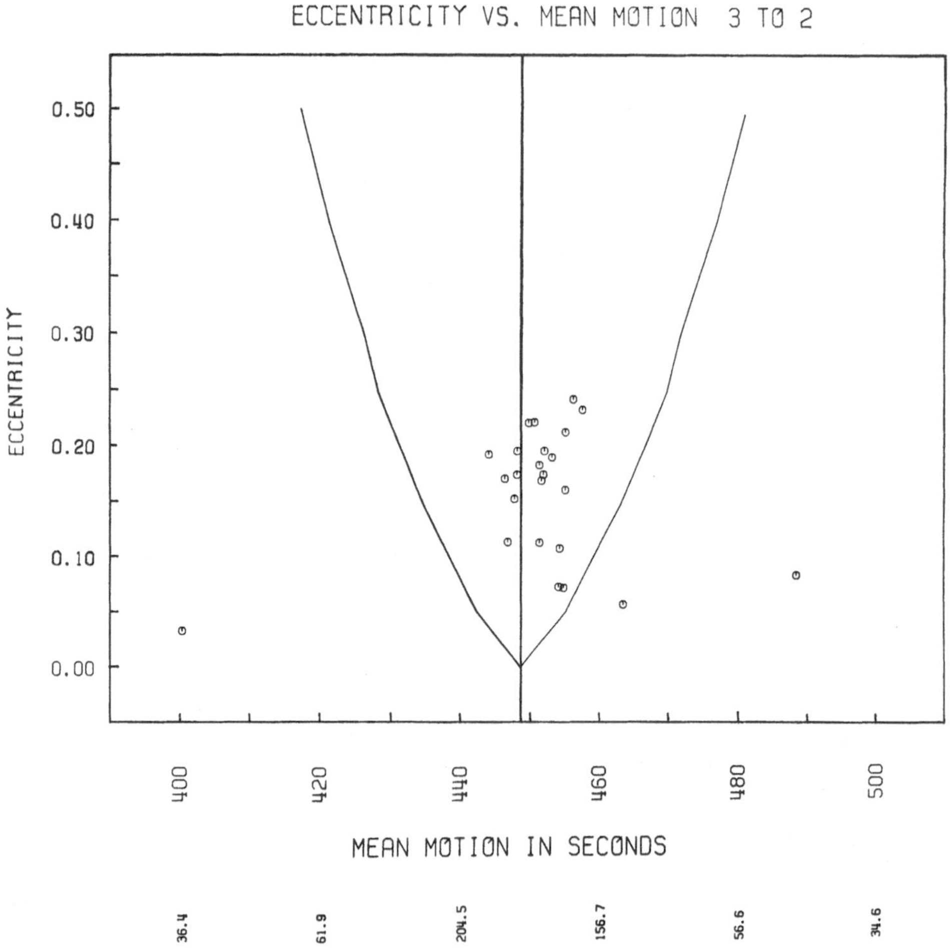

Fig. 3.

TABLE III

Minor planets useful in determining the mass of Jupiter

2:1						3:2		
76	692	781	927	1070	1309	153	1162	1345
175	696	805	959	1072	1362	190	1180	1439
225	745	814	973	1101	1438	361	1202	1512
319	756	842	978	1227	1456	499	1212	1529
530	758	859	1000	1229	1477	748	1256	1578
580	777	892	1042	1283	1491	958	1268	1746
667	778	895	1051	1303	1672	1038	1269	1748

TABLE IV

Minor planet	Ω	π	e	i	n	L	ΔL	T	Disc.
76 Freia	210°	95°	0.20	2°	574″/d	53652″	1.8	149 yr	1862
319 Leona	189	43	0.24	11	572	47347	1.6	134	1891
778 Theobalda	324	88	0.27	13	628	49228	1.6	118	1914
225 Henrietta	199	299	0.29	21	579	122265	4.1	188	1882
696 Leonora	300	46	0.23	13	621	69173	2.3	155	1910
814 Tauris	89	26	0.29	22	626	60472	2.0	128	1916
175 Andromache	209	342	0.20	3	612	164651	5.5	265	1877
530 Turandot	130	333	0.20	8	615	108787	3.6	214	1904
1101 Clematis	203	329	0.15	18	599	26318247	879.5	3861	
1362 Griqua	121	24	0.34	24	597	41603486	1390.5	3193	
153 Hilda	228	278	0.15	8	448	14870547	496.9	1735	1875
190 Ismene	177	98	0.17	6	452	1989034	66.5	624	1878
361 Bononia	19	91	0.21	13	455	508769	17.0	287	1893
279 Thule	75	275	0.03	2	400	2255838	75.4	794	1888

The next three minor planets were chosen from those minor planets with a sensitivity greater than 2″ and a period less than 200 yr. The next two minor planets have a sensitivity greater than 3″.

The minor planets 1101 and 1362 are so close to commensurability that their long-period term would most like'y correspond to a secular term of 0.9″/yr and 1.7″/yr, respectively.

In the 3:2 region, the three earliest discovered are representative of the groups sensitivity to a change in the mass of Jupiter. There is only one member of the 4:3 group and it is also sensitive to small changes in the mass of Jupiter.

In order to discuss the feasibility of using these objects to determine a correction to the mass of Jupiter, we must first look at the dates of discoveries of these objects and the period of the long-period term. We immediately see that none of the objects have observational histories which exceed the period of the long-period term. The minor planet 76 Freia, the most favorable, will not have sufficient observations until 2011. Further, due to the limited observational history of 1101 and 1362, their quasi-secular perturbations will be absorbed by a correction to the mean motion or semi-major axis by the orbit correction process.

Hopefully, there may be one consideration which will increase the feasibility of using some of these objects to determine the mass of Jupiter. If the geometry of the problem is favorable and if we have two nodal points within the span of observations, then hopefully the orbit correction process may correctly determine a correction to the mass of Jupiter provided we have correctly modeled our equations of condition. Under this circumstance, the minor planets 76 Freia and 319 Leona may prove helpful in adding to our knowledge of the mass of Jupiter before the first fly-by of an artificial planetoid.

Some additional comments still remain. First, we should keep in mind that we have used osculating elements for the minor planets in evaluating the resonant term not mean elements as is rigorously required. For this reason we have probably omitted some useful minor planets from the list.

A comment should be made concerning the sensitivity of the coefficient to the difference between the longitude of perihelia of the minor planet and Jupiter. For the cases investigated, the coefficient is not sensitive to this quantity until one gets extremely close to resonance with eccentricities in excess of 0.5. Thus, the orientation of the orbit of the minor planet with respect to the orientation of the orbit of Jupiter does not significantly influence the results. The mutual inclination of the two orbits should have more of an effect. The greater the inclination the more diluted the effect.

References

Fiala, A. D.: 1968, 'Determination of the Mass of Jupiter from a Study of the Motion of 57 Mnemosyne', Dissertation, Yale University.

Hill, G. W.: 1873, *On the Derivation of the Mass of Jupiter from the Motion of Certain Asteroids*, Collected Mathematical Works, Vol. I, p. 105, Privately printed, 1907, Johnson Reprint Corp., 1965.

Klepczynski, W. J.: 1969, 'The Motion of the Minor Planets 10, 24, 31, 52 and the Mass of Jupiter', Dissertation, Yale University.

Laplace, P. S.: 1829, *Mécanique céleste* (transl. by N. Bowditch), Hilliard, Gray, Little, and Wilkins, Publishers.

O'Handley, D. A.: 1967, 'Determination of the Mass of Jupiter from the Motion of 65 Cybele', Dissertation, Yale University.

Zielenbach, J. W.: 1968, 'Determination of the Mass of Jupiter from the Motion of 48 Doris', Dissertation, Georgetown University.

Discussion

J. Schubart: What is the definition of the period T, which appears in one of your tables? I am interested in this, since your value of T for (153) Hilda is very large in comparison with the period of libration, which I get as about $2-\frac{1}{2}$ centuries.

W. J. Klepczynski: T is the period of the resonant term which has frequency $(i-1)\,n-in'$, where n and n' are the osculating mean motions of the minor planet and Jupiter, respectively.

J. Schubart: Why did you exclude (334) Chicago from the table containing your results? We at Heidelberg are working on a determination of Jupiter's mass from selected planets of the Hilda and Thule type. We got the impression that Chicago is superior in this connection to both the planets Hilda and Thule.

B. Marsden: Following up Dr Schubart's question, I take it that you did not distinguish between librating and non-librating planets?

W. J. Klepczynski: That is true.

B. Marsden: The situation for the determination of the mass of Jupiter from the librating Hilda planets and the librating planets Griqua and Clematis could well be much worse than you say because these planets are prevented from making close approaches to Jupiter. That is why (334) Chicago, a non-librating Hilda planet, that can pass 1.1 AU from Jupiter, seems more promising.

ON THE RELATIONSHIP BETWEEN COMETS AND MINOR PLANETS

B. G. MARSDEN

Smithsonian Astrophysical Observatory, Cambridge, Mass., U.S.A.

Although the majority of the minor planets evidently arose as the result of fragmentation of larger planetary objects through collisions, there is a certain amount of evidence that *some* of them may be defunct cometary nuclei. Öpik (e.g. 1963) has long felt that the Apollo asteroids, with their perihelia inside the orbit of the earth, are especially likely to be ex-comets. Their aphelia are well inside the orbit of Jupiter (unlike those of the short-period comets), but this could be explained by the action of nongravitational forces when they were comets, just as the secular acceleration of P/Encke has apparently caused that comet's aphelion distance to decrease to the present 4.1 AU. The radii of these objects, 1 to 5 km or so, are quite comparable with those of cometary nuclei (Roemer, 1966); further, Wetherill (1968, 1969) has shown that large meteoroids, and in particular chondritic meteorites, apparently come from typically cometary orbits with perihelia near the orbit of the earth and aphelia near that of Jupiter.

But the best indication that comets turn into minor planets is that some comets actually *look* like minor planets. The two comets P/Neujmin 1 and P/Arend-Rigaux appear as very sharp condensations accompanied by the most tenuous of comas; not only do these condensations move completely in accordance with gravitational theory, but the light variation – of P/Arend-Rigaux at any rate – is precisely what would be expected of a single solid mass shining by reflecting sunlight. The motions of other comets are affected in various degrees by nongravitational forces, and it is found that these forces invariably become smaller with time (Marsden, 1968, 1969, 1970), presumably as the nuclei of the comets – regarded in terms of the Whipple (1950) icy-conglomerate model – become depleted of their volatile materials.

If P/Arend-Rigaux and P/Neujmin 1 really are objects in a transition phase between comet and asteroid, many of the asteroids must have been comets rather recently, and some of them should have orbits indistinguishable from those of Jupiter's comet family. Of course, the great majority of known minor planets have aphelion distances considerably smaller than those of the Jupiter comets, so we can concentrate here on the 3% that have aphelion distances in excess of 4.0 AU.

As is well known, the orbits of the Jupiter comets are continually being disturbed owing to encounters with Jupiter. In fact, half of these comets have passed within half an astronomical unit of Jupiter at some time during the past half century.

On the other hand, the minor planets seem to go to great lengths to avoid Jupiter. The Trojans, Thule (Takenouchi, 1962) and even the Hilda planets (Schubart, 1968) are captured in librations that appear to make it impossible for them to pass within 1.1 AU of Jupiter at the very least, forcing us to conclude that if P/Neujmin 1 and

G. E. O. Giacaglia (ed.), Periodic Orbits, Stability and Resonances, 151–163. *All Rights Reserved.*

P/Arend-Rigaux are indeed transition objects, the 'minor planets' they are to become will have lifetimes that cannot be much more than 10^3 or 10^4 years; by different reasoning Whipple has arrived at a figure of 10^4 for the lifetimes of the Apollo asteroids. On the other hand, the lifetime of a conventional minor planet must be some 10^9 years.

We have attempted to verify that these librations are stable even if perturbations by planets other than Jupiter are taken into account, and completely rigorously.

We have paid particular attention to the motion of (1362) Griqua, an asteroid which appears to be librating about the 2:1 mean motion resonance with Jupiter (Schubart, 1966; Sinclair, 1969; Schweizer, 1969); since the orbit is rather eccentric, the aphelion distance can be as high as 4.4 AU. Since our long-term calculations include perturbations by the planets Jupiter to Pluto, we took pains to derive from the observations a starting orbit in which perturbations by these five planets alone were considered. The residuals were noticeably systematic, the mean value being 2".13, while a solution in which perturbations by the inner planets were included as well gave a mean residual of only 1".25. (Parenthetically we remark that a fit in which perturbations only by Jupiter – assumed to move in a fixed ellipse – were considered gives a terrible fit, with a mean residual of 19"!)

Our numerical integration for Griqua extends from the year −3534 to the year +7417, during which time Griqua's line of apsides regressed just one complete revolution. The top portion of Figure 1 shows a small section of this perihelion

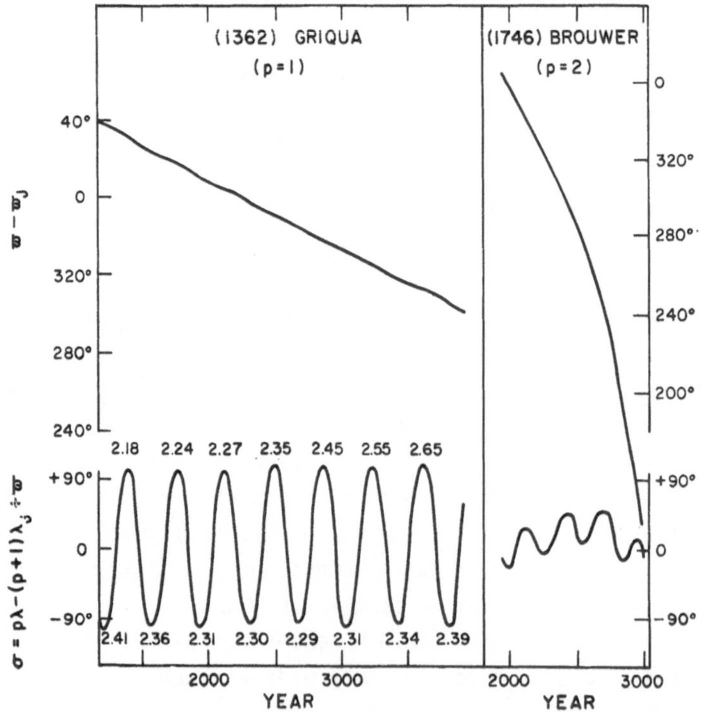

Fig. 1.

regression (actually it shows $\tilde{\omega}-\tilde{\omega}_J$, where $\tilde{\omega}$ and $\tilde{\omega}_J$ are the longitudes of perihelion of Griqua and Jupiter). The lower portion shows that the argument $\sigma=\lambda-2\lambda_J+\tilde{\omega}$ (where λ and λ_J are the mean longitudes of Griqua and Jupiter) does indeed librate about zero with amplitude 100° to 120° in a period of close to 400 years. The numbers at the maxima and minima of this curve show the least distances from Jupiter (in AU), which in fact occur when the libration is near its extremes (and the mean motion passes through the resonant value). The mean motion and eccentricity have their maxima midway up the rising branches and their minima on the falling branches.

The essentially closed curve in Figure 2 shows the variations of the eccentricity e

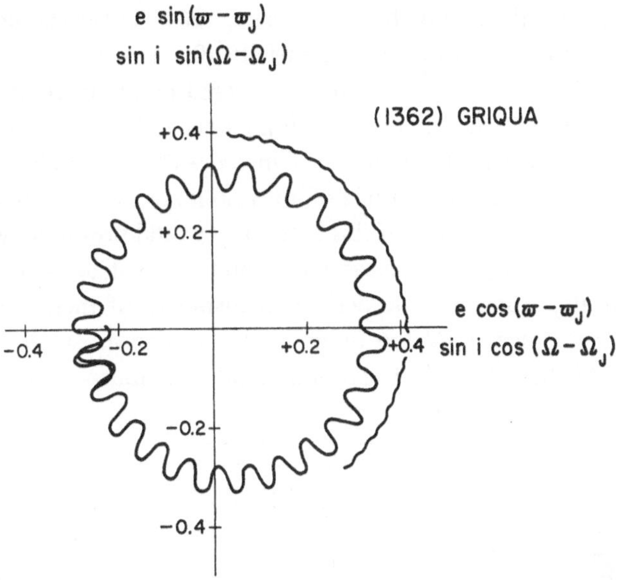

Fig. 2.

and the difference $\tilde{\omega}-\tilde{\omega}_J$ during the complete 11 000 year period of regression of the latter. The undulations show the variations during the 400-year libration periods. If we smooth out these undulations, we obtain a curve that is very close to circular, just as Schubart (1968) found in the case of the 3:2 resonance planets. If we move the ordinate-axis so that it passes through the center of the circle, at the point whose polar coordinates are $e=0.04$, $\tilde{\omega}-\tilde{\omega}_J=0$, we can describe the long-term motion in terms of a 'proper eccentricity' of 0.30 and an essentially uniformly (about the new origin) regressing 'longitude of proper perihelion' (less $\tilde{\omega}_J$). The additional variations in e during the libration period have amplitude 0.03, so the total range in e is from 0.23 to 0.37. The arc in the right-hand part of Figure 2 shows the variation of the inclination (to the ecliptic) i and the difference $\Omega-\Omega_J$ between the longitudes of the ascending nodes (again on the ecliptic) of Griqua and Jupiter. There is little physical significance in $\Omega-\Omega_J$, but it is a convenient quantity to tabulate. Small variations are evident in i having the period of the libration.

The libration of Griqua is certainly stable for one whole perihelion period, and we can infer that there is no evidence for its dissolution at any time in the past or future. That is, on the assumption that nongravitational forces are not acting, or unless a very long-term gravitational interaction should take place between the revolution period of the line of apsides and that of the line of nodes. The nodal period is evidently some 34 000 years, rather close to being in 3:1 resonance with the perihelion period.

Figure 1 also shows that (1746) Brouwer, a recently numbered member of the Hilda group, is also in libration. Figure 3 shows the discovery by Schweizer (1969) and Sinclair (1969), that (887) Alinda is librating about the 3:1 resonance with Jupiter, to be confirmed. The mean value of the libration argument $\lambda - 3\lambda_J + 2\tilde{\omega}$ is apparently 159°, not the theoretical value of 180° (Schubart, 1964), but it should be noted that, on account of Alinda's high orbital eccentricity (0.52 to 0.58), $\tilde{\omega} - \tilde{\omega}_J$ changes by only 4° during the whole 1400 year interval covered by the integration. The broken curves in Figure 3 refer to the single apparition planet 1953 EA, and it might be noted that the mean value of σ is closer to 180° (since $\tilde{\omega} - \tilde{\omega}_J$ is also closer to 180° this is perhaps to be expected). The amplitude of σ is also larger, which could mean that the two objects had a common origin; if this is so, they would presumably have separated from each other when they had the same common value of $\tilde{\omega}$, perhaps 10000 years ago. We should also note that the perihelion distances of these asteroids (at least during the interval covered by the integration) become as small as 1.03 AU for Alinda and 0.91 AU for 1953 EA. The results could be seriously affected by encounters

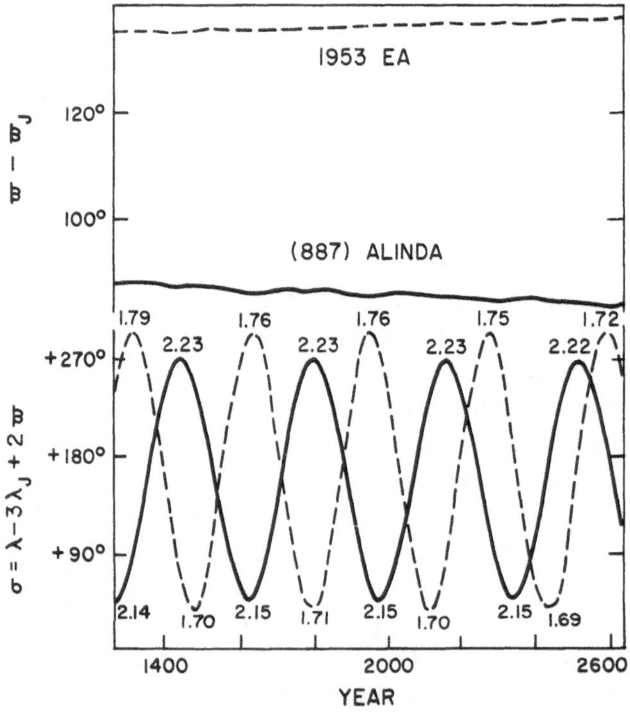

Fig. 3.

with the earth, all the more so since a 3:1 resonance with Jupiter implies approximate 1:4 resonance with the earth.

The 3:1 resonance seems to be the only one of second or higher order that is involved with librating planets. If the minor planets librate in order to avoid Jupiter, the 5:3 resonance would seem to be another one that could contain librating planets, and possibly also the nearby third order 7:4 and 8:5 resonances. There are three numbered minor planets with mean motions in the vicinity of these resonances, but none of them appears to be in libration. At first sight one planet, (522) Helga, appears to be librating about the 12:7 resonance, the argument $7\lambda - 12\lambda_J + 5\tilde{\omega}$ exhibiting pronounced periodic variations: there is a secular trend also, however, and the variations are evidently of little dynamical significance.

There is another kind of libration that can be instrumental in preventing encounters of minor planets with Jupiter. Discovered by Kozai (1962), it arises from interaction between the eccentricity and the inclination, causing the argument of perihelion ω to librate about 90° or 270°, with the result that the aphelion cannot be near a node, and hence near Jupiter's orbit plane. Kozai noted that (1373) Cincinnati is affected by this libration, and as far as we have been able to ascertain, it is the *only* such planet. We find that the range of variation of ω is from 73° to 106°, about half the range established by Kozai. The period of the libration is about 11000 years. The eccentricity and inclination have their mean values at the extremes of ω. When ω passes through 90° and is increasing e attains its maximum (0.53) and i its minimum (30°), and when ω is decreasing e attains its minimum (0.29) and i its maximum (40°).

Kozai also established that even though ω may make complete revolutions e tends to be a maximum and i a minimum when $\omega = 90°$ or 270°, and vice versa when $\omega = 0$ or 180°. This also is effective at preventing encounters between minor planets and Jupiter. As noted by Kozai, it applies in the case of (1036) Ganymed; it does so also for (1006) Lagrangea and (225) Henrietta.

All the minor planets avoid Jupiter, that is, except for (944) Hidalgo. This object has the largest mean distance (5.8 AU) of any minor planet, and its orbital eccentricity (0.66) is exceeded only by that of (1566) Icarus. In 1922, two years after its discovery, Hidalgo passed 0.89 AU from Jupiter. In 1827 the least separation was 0.84 AU, while in 1673 Hidalgo went within 0.38 AU of Jupiter. That and earlier encounters meant that while the previous orbit was still very eccentric, the mean distance was more comparable with that of Jupiter itself. Clearly, the behavior is more that of a comet than an asteroid, although nobody has ever reported any diffuseness about its appearance. Actually, since it has not been observed at less than 1.2 AU from the earth, this is not too surprising, for P/Neujmin 1 and P/Arend-Rigaux (assuming them to be the same kind of object) have always looked stellar when as far away as that.

Hidalgo has been particularly well observed, 94 reliable measurements of its position having been obtained at nine oppositions covering a span of 44 years. If it is in fact cometary, there would seem to be a fair chance of detecting the effects of nongravitational forces on its motion. A purely gravitational fit gives a mean residual of 1.″95, which is rather high for modern photographic observations of a comet, let alone an object that looks like an asteroid. There are several systematic trends to 3″ and more. If one solves for nongravitational parameters as well, however, in the manner tried in the case of several comets (Marsden, 1969, 1970), the mean residual drops to only 1.″36, and the systematic trends largely disappear. The nongravitational forces are one to two orders of magnitude smaller than found for more typical comets, but the transverse component does seem to be determined to some 15 times its mean error. Icarus is another object that is especially felt to be an old comet nucleus. In this case the transition from comet to minor planet is evidently complete, for Icarus did not show any diffuseness when only 0.1 AU from the earth, and its motion seems to be completely unaffected by nongravitational forces. The existence of Icarus is in fact the main reason for believing that the lifetimes of completely exhausted comet nuclei are not negligibly brief.

So the bona fide minor planets all avoid Jupiter, if necessary by means of librations, and apparently have existed in their present orbits for a very long time. Because of their frequent encounters with Jupiter, we should not expect the motions of the short-period comets to be affected by librations. The 2:1 resonance plays a role in their motions, but mainly in the sense that a close approach to Jupiter may by chance set a comet into resonance, while the second approach that inevitably follows 12 years later throws it out again. The most famous example that shows this phenomenon is P/Lexell, while P/Oterma illustrates the corresponding phenomenon with the 3:2 resonance.

On the other hand, there appear to be a few cases where comets have been influenced by resonances to the extent that they *temporarily* librate, sometimes for several oscillations of the libration argument. As shown in Figure 4, P/Pons-Winnecke is librating about the 2:1 resonance, the mean value of the argument $\lambda - 2\lambda_J + \tilde{\omega}$

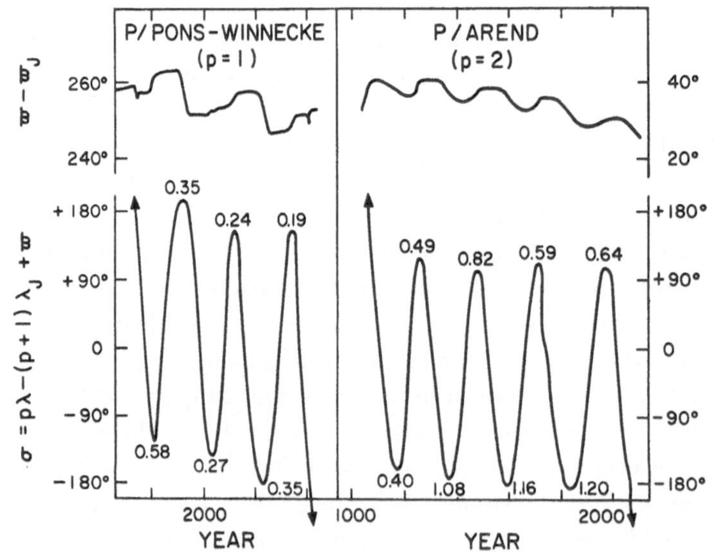

Fig. 4.

again being zero and the closest approaches to Jupiter occurring at the extremes. But the minimum approaches are considerably smaller than for the minor planets, only 0.35 AU in 1918 and 0.27 AU in 2037. The comet will continue to librate with a period of 200 years until finally the approaches to Jupiter at the beginning of the 25th century will cause the period to increase to 7.2 years and the comet to escape from the influence of the 2:1 resonance entirely. Likewise, the comet had a longer period prior to the 18th century. But the libration *does* persist for $3\frac{1}{2}$ periods of the libration argument.

The only other more-than-one-appearance comet that is now temporarily librating about the 2:1 resonance is P/Tempel-Swift. The oscillations have only just begun, but they should persist for $2\frac{1}{2}$ libration periods. (Of course, nongravitational forces have not been allowed for in any of these computations, and they could affect the results to some extent, particularly the precise number of oscillations.) Unfortunately, now that the period of P/Tempel-Swift is larger than the critical value, the perihelion distance has increased also, and the comet is now too faint for observation.

The single comet P/Arend is temporarily librating about the 3:2 resonance. The mean motion of the comet is just now passing its critical value, with the minimum separation from Jupiter being 0.64 AU. The libration has existed for about a millennium in the past, although it is just now reaching a turning value for the last time. P/Arend remains in libration for a total of $4\frac{1}{2}$ libration periods of 200 years each; see Figure 4.

In order to study the relationship between comets and minor planets we need to ask the question: "Are there any short-period comets, librating or not, that *cannot* make close approaches to Jupiter?"

Certainly, because of its relatively small aphelion distance, P/Encke cannot pass too close to Jupiter, but it does approach it to somewhat less than 0.9 AU, and on account of a weak 18:5 resonance, it does so every 59 years. Actually, of the 47 comets of more than one appearance and revolution periods less than twice that of Jupiter, 45 have passed within 0.9 AU of Jupiter during the past 200 years. Most of them have of course passed closer, and on several occasions.

The two comets that have been the most successful at avoiding Jupiter are P/Neujmin 1 and P/Arend-Rigaux – the two that look like minor planets. Neither comet has in fact approached Jupiter within 0.9 AU for about 900 years.

P/Neujmin 1 is especially interesting, because it is prevented from approaching Jupiter on account of a libration about the 2:3 resonance with Jupiter, i.e. its revolution period is 18 years, half as long again as Jupiter's. The argument $2\lambda_J - 3\lambda + \tilde{\omega}$ librates about a mean value of zero. In the otherwise similar Neptune-Pluto case the libration is about a mean value of 180°; likewise, Schubart (1964) discussed this particular libration only when the mean value is 180°. Apparently, the orbit of P/Neujmin 1 is so eccentric that conjunctions are excluded when the comet is at aphelion, rather than perihelion, even though the aphelion distance is 12 AU. The motion is obviously influenced by Saturn as well as by Jupiter. In fact, it was an encounter with Saturn (to within 0.61 AU) in the year 773, followed by another one (to within

0.34 AU) in 1038, that set the comet into libration with Jupiter in the first place. The libration is extraordinarily successful at keeping the comet away from Jupiter (see Figure 5), although there are perhaps indications in the latter part of the fifth millennium that the libration is about to break. We should certainly expect that Saturn would upset it eventually.

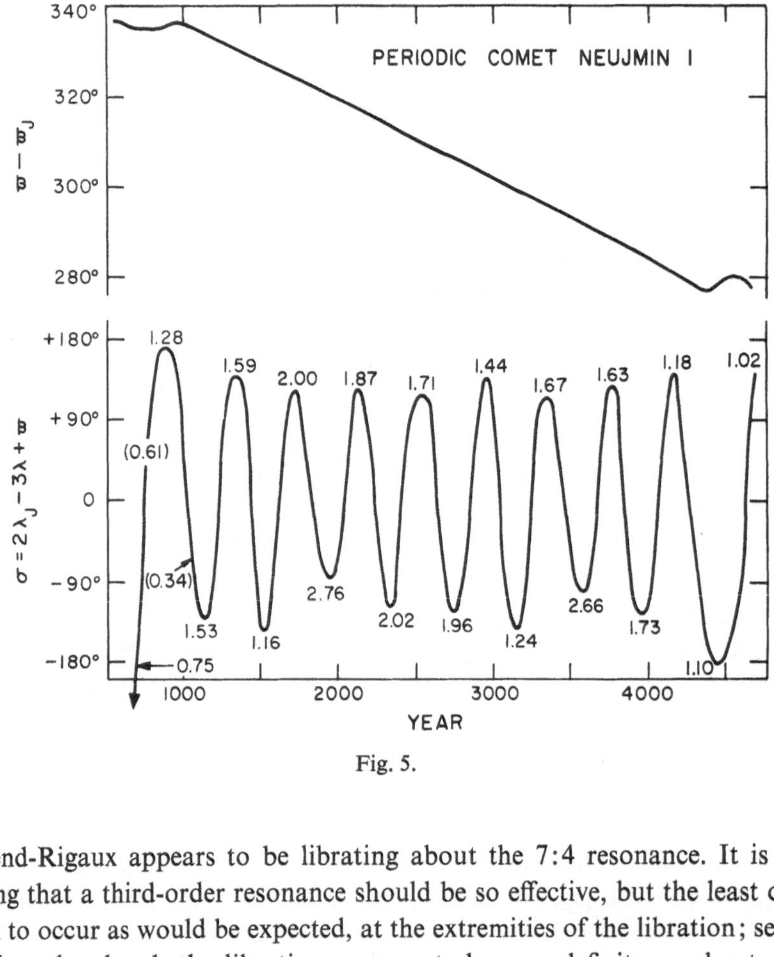

Fig. 5.

P/Arend-Rigaux appears to be librating about the 7:4 resonance. It is perhaps surprising that a third-order resonance should be so effective, but the least distances do seem to occur as would be expected, at the extremities of the libration; see Figure 6. On the other hand, the libration argument shows a definite secular trend, and the oscillations are thus probably artificial, as with (522) Helga and the 12:7 resonance. Actually, the motion of this comet is influenced more by the argument of perihelion effect noted earlier. About the year 1500, ω had a value of 270°, e then attaining its maximum of 0.63 and i its minimum of 10°; this would indeed have prevented the comet from passing within about 0.9 AU of Jupiter. On the other hand, when ω was 200°, about the year 900, a close approach to Jupiter could and did occur; and in the 23rd century, when ω is about 340°, another approach will take place.

In order to illustrate further how complicated the motion of a comet can be, even when there are relatively few close approaches to Jupiter, we mention the case

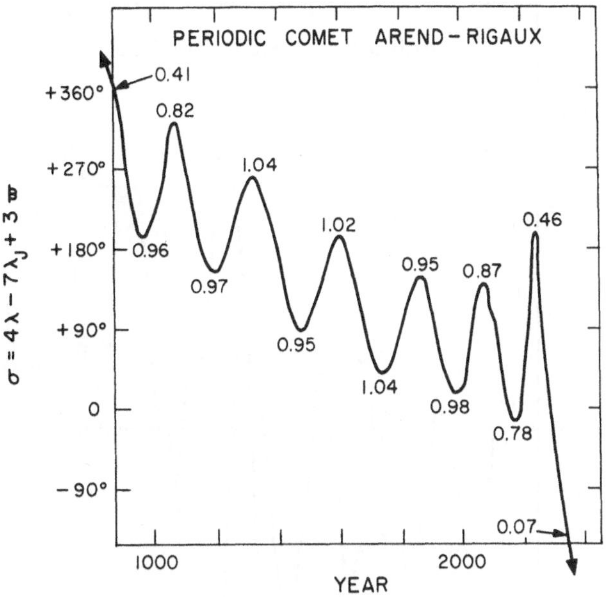

Fig. 6.

of P/Tempel 2. With an aphelion distance of only 4.7 AU, this comet has been relatively undisturbed for several centuries. As shown in the left-hand part of Figure 7, it is possible, although rather improbable, that the 9:4 resonance has been influential in keeping the comet reasonably far from Jupiter. However, an approach within 0.63 AU in 1943 caused the comet to move under the influence of the 11:5 resonance instead, as shown in the right-hand part of the figure. P/Tempel 2 is certainly very much a comet, although the influence of nongravitational forces on its motion is very small.

So perhaps there *is* a tendency for comets that look like minor planets to try to avoid Jupiter. Even Hidalgo has not been significantly less than 0.9 AU from Jupiter now for 300 years. It is not involved in any libration, but since its mean motion is not greatly different from that of Jupiter, this is perhaps not surprising. It is rather surprising that anything (i.e. Thule) should librate about the 4:3 resonance, and libration about first order resonances closer to the zeroth order 1:1 resonance would be most improbable.

P/Neujmin 1, P/Arend-Rigaux and Hidalgo, as well as P/Encke, P/Tempel 2 and P/Reinmuth 2, have been rather successful at avoiding Jupiter. These comets are now all subject to little or no nongravitational activity and perhaps approaching the ends of their lives as comets. The rate at which an icy short-period comet ages must depend rather strongly on its perihelion distance. There will have been little change in the perihelion distance of a comet that has not had any serious encounter with Jupiter, so these comets have more or less continuously been ridding themselves of their volatile materials, certainly for centuries, probably for a millennium or two – which must be a substantial fraction of their lifetimes. P/Pons-Winnecke and P/Arend should

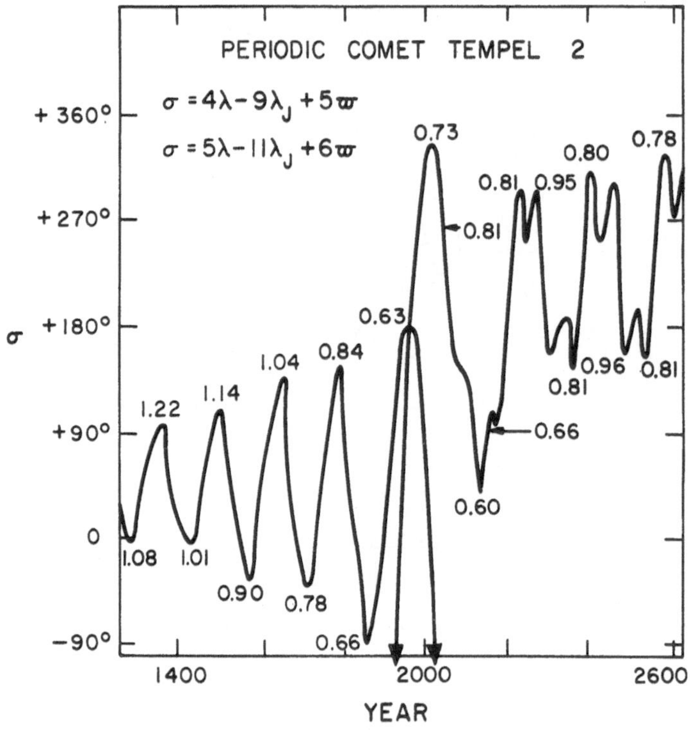

Fig. 7.

perhaps be added to the list, for their librations make it impossible for the perihelion distances to have recently been *very* much larger.

On the other hand, the more active comets – ones affected by large nongravitational forces – such as P/Schwassmann-Wachmann 2, P/Wirtanen, P/Daniel, P/d'Arrest, P/Perrine-Mrkos, P/Schaumasse and P/Honda-Mrkos-Pajdušáková, have recently and frequently passed near Jupiter. It is impossible to say what their orbits were like before several such encounters, but there is a good chance that a number of these comets would have had much larger perihelion distances not too many centuries ago. If their perihelion distances were larger than 2.5 to 3.0 AU, they could have been preserved almost indefinitely as frozen, primordial comets. P/Whipple, P/Oterma and P/Schwassmann-Wachmann 1, the motions of which are little or not at all affected by nongravitational forces, but which have large perihelion distances, should also be included in this category of 'new' comets.

If there is indeed a tendency for a dying comet to librate, it is worth while to examine the question whether the librating asteroids are particularly likely to be ex-comets. Nongravitational forces acting on them as comets could provide a mechanism for getting them into permanent libration; after they ceased to be comets, the nongravitational forces would no longer be acting, and they would be captured. But this mechanism could not easily produce the majority of the librating planets – the Trojans, Thule and the Hildas – because the motions of comets are at best only very slightly affected by such forces at the necessary heliocentric distances. The mechanism is more attractive in the cases of Griqua and Cincinnati, which after all do have rather 'cometary' values for their orbital eccentricities, inclinations and aphelion distances. Griqua is in many respects similar to P/Schwassmann-Wachmann 2, although any nongravitational forces acting on it now are at least 100 times smaller than those acting on the comet. But assuming that the forces would decrease at the same rate as those acting on the comet, it follows that 400 years ago the forces on Griqua *could* have been as large as those on the comet now. In that case, Griqua could not have remained in libration very long. Of course, this figure of 400 years is an *extreme* lower bound, but the point is that it is *possible* for Griqua to have been captured as a comet, even if it happened 40 000 years ago.

But this previous paragraph must be regarded as very speculative, for there is no real evidence that a temporarily librating comet could, with the help of nongravitational forces, become a permanently librating minor planet. All the evidence points the opposite way. In the nineteenth century, the mean motion of P/Pons-Winnecke was greater than twice that of Jupiter, and there was a nongravitational secular acceleration; now that the mean motion is less than the critical value, there is a secular deceleration! On both sides of the resonance the nongravitational forces have tended to draw the comet further from the resonance! As far as observations are concerned, this comet is unique, and if the change of sign of the nongravitational forces when a comet passes through a first-order resonance should be the rule, this would indeed be most curious; but as it stands, alone, P/Pons-Winnecke must be regarded as a curiosity.

References

Kozai, Y.: 1962, *Astron. J.* **67**, 591.
Marsden, B. G.: 1968, *Astron. J.* **73**, 367.
Marsden, B. G.:1969, *Astron. J.* **74**, 720.
Marsden, B. G.: 1970, *Astron. J.* **75**, 75.
Öpik, E. J.: 1963, *Advan. Astron. Astrophys.* **2**, 219.
Roemer, E.: 1966, *Mem. Soc. Roy. Sci. Liège Sér. 5* **12**, 23.
Schubart, J.: 1964, Smithsonian Astrophys. Obs. Spec. Rpt. No. 149.
Schubart, J.: 1966, private communication.
Schubart, J.: 1968, *Astron. J.* **73**, 99.
Schweizer, F.: 1969, *Astron. J.* **74**, 779.
Sinclair, A. T.: 1969, *Monthly Notices Roy. Astron. Soc.* **142**, 289.
Takenouchi, T.: 1962, *Ann. Tokyo Obs.* **7**, 191.
Wetherill, G. W.: 1968, *Science* **159**, 79.
Wetherill, G. W.: 1969, *Trans. Am. Geophys. Union* **50**, 224.
Whipple, F. L.: 1950, *Astrophys. J.* **111**, 375.

Discussion

W. H. Jefferys: I was interested to hear about the temporary resonance you found for Comet Pons-Winnecke, for in my integrations of the planar, circular, restricted problem orbits have been found which librate for several centuries before a too close approach to Jupiter destroys the libration.

J. Schubart: I think that we can congratulate Dr. Marsden on the discovery of a new type of libration, since he found that Comet Neujmin 1 shows libration of the critical argument around zero, although the comet is close to an outer commensurability with respect to Jupiter. I want to ask whether you believe that one can find, on the basis of the restricted three-body problem, a corresponding periodic solution with a vanishing critical argument.

B. G. Marsden: I think this would be a possibility. The situation certainly deserves further study.

J. Schubart: Did the orbital elements of Comet Neujmin 1 change very much when it was captured into libration by the attraction of Saturn?

B. G. Marsden: Not terribly much. The perihelion distance was formerly only slightly larger, about 1.6 AU. The revolution period was 18.8 years, about one year larger than the resonant value. Of course, before that the comet was able to approach Jupiter more closely, and prior to an approach (to within 0.75 AU) in the year 701 the period was 19.2 years.

G. Colombo: Do you think that Comet Neujmin 1 has been captured in the libration? In this case the periodic solution (if there is one) would not be of the same type as that of the Pluto orbit (relative to Neptune). The conjunction occurs at the perihelion of the comet, whereas at its conjunction Pluto is at aphelion. It seems to me very interesting to see if this periodic solution is stable, as in the Pluto case. In some sense this would contradict intuition.

B. G. Marsden: I think it rather certain that the libration of Comet Neujmin 1 will ultimately be destroyed, although it certainly seems to exist for a remarkably long period of time. Sooner or later Saturn is bound to upset the situation, and the libration may be destroyed even if the influence of Saturn is ignored. Maybe we should not regard this conjunction at perihelion too seriously. The mean value of the libration argument is certainly near zero, but the avoidance probably takes place when the comet is near Jupiter's mean distance – near neither perihelion nor aphelion. The value zero could be due somewhat to chance.

G. Colombo: Can you give the latest data on Icarus' size, shape and angular velocity?

B. G. Marsden: I think the radius is about 700 m, the shape is apparently remarkably close to spherical, and the rotation period is rather well determined at 2 hours 16 min.

G. Colombo: The high angular velocity is not inconsistent with asteroidal origin. If Icarus was the product of a collision between large asteroids, it had quite a good chance of acquiring such an angular spin rate. Do you think that at any perihelion passage the asteroid can lose mass by melting and escaping from the surface? I do not see any need to assume a high density (iron) asteroid because of rotational instability. Even assuming it is a fluid, a period of 2.25 hours could mean a density as low as 2.5.

B. G. Marsden: Icarus is not observable at perihelion, but there has been no indication of any 'coma' five weeks later. The motion shows no indication of nongravitational forces. If Icarus was indeed cometary, it is now in a much more advanced state of evolution than the transition objects I have discussed. Rotational stability depends upon rigidity as well as density. It was my impression that a typical stone object would break up at Icarus' angular velocity and that polarization measurements are not consistent with an iron object. But a relatively 'loose' low density comet nucleus could rotate in 2.25 hours without breaking up.

ABSOLUTE ORBITS AND JUPITER'S GREAT SATELLITES

Progress Report

S. FERRAZ-MELLO

ITA Astronomical Observatory, S. José dos Campos, Brasil

The purpose of this report is to present some results of the researches carried out, at the Astronomical Observatory of the Technological Institute of Aeronautics, on the problem of construction of absolute orbits in connection with the problem of the motion of the four great satellites of Jupiter.

The features of the galilean system of Jupiter's satellites are such that the problem of construction of a theory of the motion, is a problem of construction of absolute orbits. Indeed, the precision required in such a theory is much smaller than the precision required in planetary theories; we remember that the best observations made so far, are those by Alden, at Johannesburg, in 1925, which lead to precisions equal to 30″ in jovicentric longitudes, in the most favorable cases. On the other hand, the time scale of the motion of these satellites is such that the classical methods of Leverrier and Newcomb cannot be used. These methods lead to secular terms which limit strongly the period of validity of the theory.

So, the problem is to construct orbits which are not too precise, but have a long period of validity with this lower precision, or, as we call them, absolute orbits.

1. Outline of the Method

In preceding papers we have presented the ideas we are developing. We remember here the main ones [1].

The equations of motion are taken in a moving frame, and the variables used are almost equal to those used by Hill. These equations are:

$$(D+m)^2 u_i = \lambda_i \frac{a_i^3}{r_i^3} u_i + R_i$$

$$(D-m)^2 s_i = \lambda_i \frac{a_i^3}{r_i^3} s_i + T_i$$

$$D^2 z_i = \lambda_i \frac{a_i^3}{r_i^3} z_i + V_i$$

where the λ_i are constants depending on the constant of gravitation and masses.

R_i, T_i, V_i are functions of the perturbative forces.

The reference solutions are the circular ones, with mean motions equal to those observed.

The new variables

$$\hat{u}_i = u_i - \exp(i\gamma_i t)$$
$$\hat{s}_i = s_i - \exp(-i\gamma_i t)$$

G. E. O. Giacaglia (ed.), Periodic Orbits, Stability and Resonances, 164–167.

are introduced and the perturbation equations obtained. These equations will be non-linear.

At this point an essential modification is made. We introduce additional couples of variables P_i and Q_i, by the relations:

$$(D + m)u_i = \kappa_i P_i \zeta^{-(\kappa_i - m)} + \kappa_i \frac{a_i}{r_i} u_i + \tfrac{1}{2} A_{2i} u_i + W_{2i}$$

$$(D - m)s_i = - \kappa_i Q_i \zeta^{\kappa_i - m} - \kappa_i \frac{a_i}{r_i} s_i - \tfrac{1}{2} A_{2i} s_i - W_{2i}^*,$$

where κ_i are the mean motions in convenient units, ζ is a time exponential, and A_{2i}, W_{2i}, W_{2i}^* second-order functions of $\hat{u}_i$, $\hat{s}_i$, z_i, Dz_i. These relations are equivalent, up to the second-order, to Laplace's first integral of the two-body problem. These relations are introduced to transform the $2N$ equations of second order in $4N$ equations of first order.

The additional equations are non-disturbed equations and can be solved formally. Thus the system is reduced to $2N$ equations of first-order, in P_i and Q_i. The mean feature here is that there are no quadratic terms in the equations for P_i and Q_i, except those which came from perturbations, and which have disturbing masses as factors.

Actually, the solutions of the first $2N$ equations involve the primitive functions of P_i and Q_i, and the final equations are integro-differential equations.

For the z-equations, the reduction to first-order equations is made by introducing the variables k_i and h_i, by the relations

$$\kappa_i \zeta^{-m} k_i = u_i \, Dz_i - z_i (D + m)\, u_i$$
$$\kappa_i \zeta^{m} h_i = - s_i \, Dz_i + z_i (D - m)\, s_i$$

The variables k_i, h_i are associated to the Area Integral in the same form as the variables P_i, Q_i are associated with the Laplace Integral [2].

2. Some Results

The researches on the construction of a second-order theory has led to the following preliminary results.

(1) The equations for plane perturbations do not depend on space variables.

(2) The equations for the space variables do not depend on plane perturbations.

(3) The equations obtained are linear.

(4) The solutions present the main features of the motion; coupled natural vibrations in longitude and latitude, and bounded (no secular terms).

Nevertheless, we may notice that numerical applications had shown the existence of some small errors in algebraic computations. These errors may not change the main features of the results.

3. Jupiter's Satellites

The specific problem of the motion of the great satellites of Jupiter is now being studied by J. L. Sagnier [3]. He has preferred to maintain the variables $\hat{u}_i$ and $\hat{s}_i$, and to eliminate the quadratic terms by simple algebraic relations.

The device proposed for the study of the critical terms is very simple.

In an only qualitative approximation, the equations are

$$\kappa\, DP + \kappa^2 P = \psi$$
$$\kappa\, DQ - \kappa^2 Q = \psi^*$$

and the longitudes are connected to P_i and Q_i by

$$D^2\theta_i = \frac{\kappa_i}{2i} D(P+Q) - \frac{\kappa_i^2}{2i}(P-Q).$$

Critical terms of the form

$$k\zeta^{\Theta}$$

$(\Theta = \theta_1 - 3\theta_2 + 2\theta_3)$, are integrated as constants and give

$$P_i = \frac{k_i}{\kappa_i^2}\zeta^{\Theta}$$
$$Q_i = \frac{k_i}{\kappa_i^2}\zeta^{-\Theta}$$

The equation for Θ is easily obtained, and will be of the form

$$D^2\Theta = K \sin\Theta$$

All correlated discussions are classical.

4. Researches on the Method

There are, in the method presented here, many points to be discussed. Some researches are in progress about it. S. B. Rissi is making an analysis of non-disturbed motion. We remember that in a second-order theory, non-linearities arise only from non-disturbed terms. This analysis must decide on essential and non-essential features of the method. He is working also on the corrections of the errors mentioned above. C. M. Rodrigues, who is finishing a study on old observations of Jupiter's satellites, is going to apply the method to the study of quasi-stationary orbits of artificial satellites; this study has as objective an analysis of the domain of applicability of the method. R. V. Moraes is beginning the study of 3rd order problems. The algebra of the series involved will be performed on an IBM 1130 computer. The results, except for 3rd order, must be available for the next General Assembly of the International Astronomical Union.

5. Acknowledgements

These researches are being supported partly by the Researches Foundation of São Paulo (FAPESP), grants 68/864, 69/373, and 69/377.

References

[1] Ferraz-Mello, S.: 1969, *Compt. Rend.* **268**, 198.
[2] Ferraz-Mello, S.: 1969, *Compt. Rend.* **268**, 985.
[3] Sagnier, J. L.: ITA. Obs. Ast. NT 08/69.

Discussion

B. G. Marsden: Have you resolved or do you have hopes of resolving the problem of the equation of the center for the third satellite?

I mean the fact that Wargentin's eighteenth century eclipse observations seem to be inconsistent with the modern astrometric observations.

S. F. Mello: No. This problem is not solved so far.

G. Colombo: How did you choose the reference system?

S. F. Mello: There are several possibilities of choice for the reference frame in the problem of Galilean Satellites. The frame we had chosen was proposed by De Sitter. The choice is made in view of a more simple form for the exponents.

G. Colombo: What is the effect that would be introduced in your results by tidal torques?

S. F. Mello: The problem of tidal effects has not been considered so far. We intend to consider it only in a later stage. From the ephemeris point of view, tidal effects do not seem to be a problem.

EXISTENCE OF PERIODIC SOLUTIONS OF DIFFERENTIAL EQUATIONS OF SECOND ORDER

MARKO ŠVEC
Universidade Federal da Bahia, Brasil

Consider the differential equation

$$y'' = g(x, y) \tag{1}$$

and assume that the function $g: R^2 \to R$ is continuous in x, y and periodic in x with the period $T(>0)$. The existence of a periodic solution of (1) is investigated. Instead of solving directly the problem of the existence a periodic solution of (1), we solve the equivalent problem, i.e., the problem of the existence of the solution of the periodic boundary value problem

$$y'' + a^2 y = f(x, y), \quad f(x, y) = a^2 y + g(x, y) \tag{2}$$
$$y(0) = y(T), \quad y'(0) = y'(T) \tag{3}$$

where $a > 0$ is a suitable constant. It is evident that each periodic solution of (1) is a solution of (2), (3) and the periodic continuation of each solution of (2), (3) is also a periodic solution of (1).

Suppose that $aT \neq 2k\pi, \quad k = 0, 1, \ldots$
Then to the linear differential expression

$$Ly = y'' + a^2 y$$

and the condition (3) corresponds the Green's function

$$G(x, t) = \frac{1}{2a(1 - \cos aT)} \begin{cases} \sin[a(T - x + t)] + \sin[a(x - t)], & 0 \leqslant t \leqslant x \leqslant T, \\ \sin[a(T + x - t)] + \sin[a(t - x)], & 0 \leqslant x \leqslant t \leqslant T, \end{cases} \tag{4}$$

which is continuous and symmetric on $[0, T] \, x \, [0, T]$. Its eigenvalues are

$$\lambda_0 = a^2, \quad \lambda_{2k-1} = \lambda_{2k} = a^2 - \frac{4k^2\pi^2}{T^2}, \quad k = 1, 2, \ldots \tag{5}$$

and the corresponding eigenfunctions are

$$\phi_0(x) = \frac{1}{\sqrt{T}}, \quad \phi_{2k-1}(x) = \sqrt{\frac{2}{T}} \sin\left(\frac{2k\pi}{T} x\right),$$
$$\phi_{2k}(x) = \sqrt{\frac{2}{T}} \cos\left(\frac{2k\pi}{T} x\right),$$
$$k = 1, 2, \ldots \tag{6}$$

which form an orthonormal system.

G. E. O. Giacaglia (ed.), Periodic Orbits, Stability and Resonances, 168–175.

We will suppose, in what follows, that

$$aT < 2\pi . \tag{7}$$

Then $\lambda_0 = a^2 > 0$ and $\lambda_s < 0, s = 1, 2, \ldots$.

The boundary value problem (2), (3) is equivalent to the integral equation of Hammerstein's type

$$y(x) = \int_0^T G(x, t)\, f(t, y(t))\, \mathrm{d}t . \tag{8}$$

The problem of the existence of periodic solution of (1) is now reduced to proving the existence of a solution of (8).

There exist several proofs of the existence of a solution of (8) using some additional assumptions concerning the function $f(x, y)$. For example, if we suppose that $f(x, y)$ satisfies uniformly a Lipschitz condition

$$|f(x, y_1) - f(x, y_2)| < \alpha(x)\, |y_1 - y_2|$$

and

$$\int_0^T A^2(x)\, \alpha^2(x)\, \mathrm{d}x < 1$$

where

$$A^2(x) = \int_0^T G^2(x, t)\, \mathrm{d}t ,$$

the method of successive approximations gives a unique solution of (8).

Also, if

$$|f(x, y)| \leqslant d\,|y| + c$$

and we impose additional conditions the use of Schauder-Tichonov fixed point theorem leads to at least one periodic solution of (1).

In our consideration we will follow the basic idea of Hammerstein [1]. Assume that the solution $y(x)$ of (8) exists and that $h(x) = f(x, y(x)) \in L_2[0, T]$. Then we have

$$y(x) = \int_0^T G(x, t)\, h(t)\, \mathrm{d}t$$

and by the Hilbert-Schmidt theorem $y(x)$ can be expressed as a series

$$y(x) = \sum_{m=0}^{\infty} c_m \phi_m(x) \tag{9}$$

which converges on $[0, T]$ absolutely and uniformly.

The constants c_m are expressed by

$$c_m = \int_0^T y(x)\, \phi_m(x)\, \mathrm{d}x = \frac{1}{\lambda_m} \int_0^T f(t, y(t))\, \phi_m(t)\, \mathrm{d}t \tag{10}$$

$$m = 0, 1, 2, \ldots$$

using the fact that $\phi_m(x)$ are eigenfunctions of the kernel $G(x, t)$.

The problem of solving (8) is now reduced to that of solving the system of infinite equations with infinite number of unknowns

$$c_m = \frac{1}{\lambda_m} \int_0^T f\left(t, \sum_{s=0}^{\infty} c_s \phi_s(t)\right) \phi_m(t)\, dt, \quad m = 0, 1, 2, \ldots \tag{11}$$

Instead of this we shall consider the finite sums

$$y_n(x) = \sum_{s=0}^{n} c_{n,s} \phi_s(x) \tag{12}$$

where the coefficients $c_{n,s}$ have to satisfy the equations

$$c_{n,k} = \frac{1}{\lambda_k} \int_0^T f\left(t, \sum_{s=0}^{n} c_{n,s} \phi_s(t)\right) \phi_k(t)\, dt, \quad k = 0, 1, \ldots, n. \tag{13}$$

We shall prove the existence of a solution of (13) by assuming that

$$F(x, y) \geqq Dy^2 + C \tag{14}$$

where

$$F(x, y) = \int_0^y f(x, u)\, du \tag{15}$$

and D, C are constants such that

$$a^2 < 2D. \tag{16}$$

Consider the function

$$H(x_0, x_1, \ldots, x_n) = \sum_{k=0}^{n} \lambda_k x_k^2 - 2 \int_0^T F\left(t, \sum_{s=0}^{n} x_s \phi_s(t)\right) dt. \tag{17}$$

Its partial derivatives are

$$\frac{1}{2\lambda_k} \frac{\partial H}{\partial x_k} = x_k - \frac{1}{\lambda_k} \int_0^T f\left(t, \sum_{s=0}^{n} x_s \phi_s(t)\right) \phi_k(t)\, dt, \quad k = 0, 1, \ldots, n.$$

If we can prove that $H(x_0, x_1, \ldots, x_n)$ has an extremum at a point $(x_0', x_1', \ldots, x_n')$, then it is evident that

$$c_{n,0} = x_0', \quad c_{n,1} = x_1', \quad \ldots, \quad c_{n,n} = x_n'$$

is a solution of (13).

From (17), (14) and from the fact that $\phi_n(x)$ form an orthonormal system we get

$$H(x_0, x_1, \ldots, x_n) \leqslant \sum_{k=0}^{n} (\lambda_k - 2D)\, x_k^2 - 2CT. \tag{18}$$

From (16) it follows that

$$\lambda_k - 2D < 0 \tag{19}$$

Therefore $H(x_0, x_1, \ldots, x_n)$ has an absolute maximum at a point $(\bar{x}_0, \bar{x}_1, \ldots, \bar{x}_n)$ and

$$c_{n,0} = \bar{x}_0, \quad c_{n,1} = \bar{x}_1, \quad \ldots, \quad c_{n,n} = \bar{x}_n \tag{20}$$

is a solution of (13).

Denote

$$h_n = H(c_{n,0}, c_{n,1}, \ldots, c_{n,n}), \quad S_n = \sum_{k=0}^{n} \lambda_k c_{n,k}^2$$

$$S_n^* = \sum_{k=0}^{n} |\lambda_k| \, c_{n,n}^2.$$

Because $\lambda_0 = a^2$, $\lambda_k < 0$, $k = 1, 2, \ldots$, we have

$$0 \leqslant S_n^* = 2a^2 - S_n. \tag{21}$$

From the fact that $H(x_0, x_1, \ldots, x_n) = H(x_0, x_1, \ldots, x_n, 0)$ it follows that

$$h_n \leqslant h_{n+1}, \quad n = 0, 1, 2, \ldots \tag{22}$$

and from this and (18) we find

$$h_0 + 2CT \leqslant h_n + 2CT \leqslant S_n - 2D \sum_{k=0}^{n} c_{n,k}^2 \leqslant S_n$$

and with respect to (21)

$$S_n^* = 2a^2 - S_n \leqslant 2a^2 - (h_0 + 2CT) = K. \tag{23}$$

Now we can prove that the functions $y_n(x)$, $n = 0, 1, 2, \ldots$ are uniformly bounded. In fact, we have

$$y_n(x) = \sum_{k=0}^{n} c_{n,k}\phi_k(x) = \sum_{k=0}^{n} \sqrt{|\lambda_k|} \, c_{n,k} \frac{\phi_k(x)}{\sqrt{|\lambda_k|}}.$$

Using the Schwarz inequality for sums we get

$$y_n^2(x) \leqslant \sum_{k=0}^{n} |\lambda_k| \, c_{n,k}^2 \cdot \sum_{k=0}^{n} \frac{\phi_k^2(x)}{|\lambda_k|} \leqslant S_n^* \sum_{k=0}^{\infty} \frac{\phi_k^2(x)}{|\lambda_k|} \tag{24}$$

But our kernel $G(x, t)$ satisfies the conditions of Mercer's theorem. Therefore

$$\sum_{k=0}^{\infty} \frac{\phi_k^2(x)}{\lambda_k} = G(x, x) = \frac{\sin aT}{2a(1 - \cos aT)}$$

and

$$\sum_{k=0}^{\infty} \frac{\phi_k^2(x)}{|\lambda_k|} = 2\frac{\phi_0^2(x)}{\lambda_0} - \sum_{k=0}^{\infty} \frac{\phi_k^2(x)}{\lambda_k} = 2\frac{\phi_0^2(x)}{\lambda_0} - G(x, x)$$

From (23) and (24) we have finally

$$y_n^2(x) \leqslant K\left[2\frac{1}{a^2T} - \frac{\sin aT}{2a(1-\cos aT)}\right] = R^2. \tag{25}$$

In the following step we shall prove that the functions

$$\psi_n(x) = -y_n(x) + \int_0^T G(x,t)\, f(t, y_n(t))\, dt$$

converge uniformly to zero for $n \to \infty$. In fact, from Hilbert-Schmidt's theorem we get

$$\begin{aligned}
\psi_n(x) &= -\sum_{k=0}^{n} c_{n,k}\phi_k(x) + \sum_{s=0}^{\infty} \phi_s(s) \int_0^T \phi_s(\tau)\, d\tau \int_0^T G(\tau,u)\, f(u, y_n(u))\, du \\
&= -\sum_{k=0}^{n} c_{n,k}\phi_k(x) + \sum_{s=0}^{\infty} \phi_s(x) \int_0^T f(u, y_n(u))\, du \int_0^T \phi_s(\tau)\, G(\tau,u)\, d\tau \\
&= -\sum_{k=0}^{n} c_{n,k}\phi_k(x) + \sum_{s=0}^{n} \frac{\phi_s(x)}{\lambda_s} \int_0^T \phi_s(u)\, f(u, y_n(u))\, du \\
&\quad + \sum_{s=n+1}^{\infty} \frac{\phi_s(x)}{\lambda_s} \int_0^T \phi_s(u)\, f(u, y_n(u))\, du \\
&= \sum_{s=n+1}^{\infty} \frac{\phi_s(x)}{\lambda_s} \int_0^T \phi_s(t)\, f(t, y_n(t))\, dt = \sum_{s=n+1}^{\infty} \frac{\phi_s(x)}{\lambda_s} k_s
\end{aligned}$$

where

$$\int_0^T \phi_s(t)\, f(t, y_n(t))\, dt = k_s$$

is the Fourier coefficient of the function $f(t, y_n(t))$. Using the Schwarz inequality for sums we get

$$\psi_n^2(x) \leqq \sum_{s=n+1}^{\infty} \lambda_s^{-2}\phi_s^2(x) \cdot \sum_{s=n+1}^{\infty} k_s^2.$$

Now from Bessel's inequality

$$\sum_{s=0}^{\infty} k_s^2 \leqq \int_0^T f^2(t, y_n(t))\, dt$$

and (25) we have

$$\sum_{s=n+1}^{\infty} k_s^2 \leqq \int_0^T f^2(t, y_n(t))\, dt \leqq \max_{[0,T] x [-K,K]} f(t,z) \cdot T = P. \tag{26}$$

Therefore

$$\psi_n^2(x) \leqq \sum_{s=n+1}^{\infty} \lambda_s^{-2} \phi_s^2(x) \cdot P. \tag{27}$$

However, we know that the series $\sum_{s=0}^{\infty} \lambda_s^{-2} \phi_s^2(x)$ converges uniformly to $G_2(x, x) = \int_0^T G(x, \tau) G(\tau, x)\, d\tau = \int_0^T G^2(x, t)\, dt$. Hence $\{\psi_n(x)\}$ converges uniformly to zero as $n \to \infty$.

Next we consider the sequence

$$\omega_n(x) = \psi_n(x) + y_n(x) = \int_0^T G(x, t)\, f(t, y_n(t))\, dt.$$

Using the Schwarz inequality and (26) we have

$$\omega^2(x) \leqq \int_0^T G^2(x, t)\, dt \int_0^T f^2(t, y_n(t))\, dt \leqq P \int_0^T G^2(x, t)\, dt$$

which means that $\omega_n(x)$ are uniformly bounded. We shall prove that they are also equicontinuous. In fact, by the Schwarz inequality,

$$\begin{aligned}
[\omega_n(x_1) - \omega_n(x_2)]^2 &= \left[\int_0^T [G(x_1, t) - G(x_2, t)]\, f(t, y_n(t))\, dt\right]^2 \\
&\leqq \int_0^T [G(x_1, t) - G(x_2, t)]^2\, dt \int_0^T f^2(t, y_n(t))\, dt \\
&\leqq P\,[G_2(x_1, x_1) - 2G_2(x_1, x_2) + G_2(x_2, x_2)] \\
&= P\,[G_2(x_1, x_1) - G_2(x_1, x_2)] + P[G_2(x_2, x_2) - G_2(x_1, x_2)].
\end{aligned}$$

From the continuity of the iterated kernel $G_2(x, t)$ follows now the equicontinuity of $\omega_n(x)$, $n=0, 1, 2, \ldots$. Using the Ascoli-Arzela's theorem we can extract from the sequence $\{\omega_n(x)\}$ a subsequence $\{\omega_{n_k}(x)\}$ which converges uniformly to a continuous function $\phi(x)$. Now, because

$$y_{n_k}(x) = \omega_{n_k}(x) - \psi_{n_k}(x)$$

and $\omega_{n_k}(x)$ converges uniformly to $\phi(x)$ and $\psi_{n_k}(x)$ to zero as $n_k \to \infty$, we have that $y_{n_k}(x)$ converges uniformly to $\phi(x)$ as $n \to \infty$. Therefore from the equation

$$\psi_{n_k}(x) = -\, y_{n_k}(x) + \int_0^T G(x, t)\, f(t, y_{n_k}(t))\, dt$$

and continuity of f we get

$$0 = -\, \phi(x) + \int_0^T G(x, t)\, f(t, (t))\, dt.$$

Thus, we have proved the

THEOREM: Let $g(x, y): R^2 \to R$ be a continuous function, periodic in x with the period T. Let

$$\text{(A)} \qquad f(x, y) = a^2 y + g(x, y)$$

be such that

$$\text{(B)} \qquad F(x, y) = \int_0^y f(x, u)\, du \geqq Dy^2 + C,$$

$$aT < 2\pi, \quad a^2 < 2D.$$

then the equation

$$y'' = g(x, y) \tag{1}$$

has at least one periodic solution of the period T.

COROLLARY 1: If $g(x, y): R^2 \to R$ is continuous, periodic in x with the period T and

$$\int_0^T g(x, u)\, du \geqq \alpha^2 y^2 + C, \quad \alpha \neq 0 \tag{28}$$

Then (1) has at least one periodic solution.

PROOF: From (28) and if a is such that $aT < 2\pi$ we have

$$F(x, y) = \tfrac{1}{2} a^2 y^2 + \int_0^y g(x, u)\, du \geqq (\tfrac{1}{2} a^2 + \alpha^2)\, y^2 + C$$

If we put $\frac{1}{2}a^2 + \alpha^2 = D$, we have $a^2 < 2D$.

COROLLARY 2: Let $g(x, y): R^2 \to R$ be continuous and periodic in x with the period T. Let

$$g(x, y) \geqq \alpha_1 y + \beta_1 \quad \text{for} \quad y \geqq 0, \quad \alpha_1 > 0$$
$$g(x, y) \leqq \alpha_2 y + \beta_1 \quad \text{for} \quad y < 0, \quad \alpha_2 > 0$$

Then (1) has at least one periodic solution of the period T.

PROOF: We have

$$F(x, y) \geqq (\tfrac{1}{2} a^2 + \alpha_1)\, y^2 + \beta_1 y \quad \text{for} \quad y \geqq 0$$
$$F(x, y) \geqq (\tfrac{1}{2} a^2 + \alpha_2)\, y^2 + \beta_1 y \quad \text{for} \quad y \leqq 0$$

or

$$F(x, y) \geqq (\tfrac{1}{2} a^2 + \alpha)\, y^2 + \beta_1 y \quad \text{for all } y \quad \text{where } \alpha = \min(\alpha_1, \alpha_2).$$

We can find such $D > 0$ and C that

$$F(x, y) \geqslant Dy^2 + C.$$

It is sufficient to choose D, C such that

$$\beta_1^2 + 4(\tfrac{1}{2} a^2 + \alpha - D)\, C < 0, \quad \tfrac{1}{2} a^2 + \alpha - D > 0.$$

This requires that C be negative. Now, if $\frac{1}{2}a^2+\alpha-D>0$, D can be chosen such that $\frac{1}{2}a^2-D<0$.

Reference

[1] Hammerstein, A.: 1930, 'Nichtlineare Integralgleichungen nebst Anwendungen', *Acta Math.* **54**, 117–176.

SETS OF COLLISION PERIODIC ORBITS IN THE RESTRICTED PROBLEM

G. BOZIS

University of Thessaloniki, Greece

Abstract. The evolution of 16 collision periodic orbits established for the Copenhagen problem ($\mu = 0.50$) is numerically studied for various values of the mass ratio μ.

Some conclusions are drawn regarding the interconnection of the known families of periodic orbits for $\mu = 0.50$ as well as families studied by various investigators for fixed values of μ.

1. Introduction

A family of periodic orbits is associated with a fixed value of the mass ratio μ and is effectively described by its continuous (but not necessarily closed) characteristic curve C-x, i.e. the Jacobian constant C vs. the ordinate x to which a perpendicular crossing of definite direction with the x-axis occurs.

Families of periodic orbits have been established by G. Darwin for $\mu = 0.909\,091$, F. R. Moulton (1920) for $\mu = 0.80$ and $\mu = 0.50$, Broucke (1962, 1968) for the earth-moon system and others. E. Strömgren and his collaborators found 15 families for $\mu = 0.50$. Bartlett (1964) studied extensively the case of two equal masses, reproduced Strömgren's families and found six new ones, which, however, include orbits as complicated as those of family g. Independently from Bartlett, Hénon (1965a, b) recomputed periodic orbits of the Copenhagen problem, studied their stability and found seven new families, thus bringing their total number up to 22.

The discontinuous character of the collision periodic orbits as regards their launching velocity at the primary P_1 in the physical rotating system Oxy, compared to the velocity of the neighboring orbits of the same family, is evident. However, when regularization is performed, a collision orbit exhibits its main feature: it becomes a means of changing direction of the velocity in the neighborhood of the primary. It is worth noticing that if a family possesses a collision orbit, separating its direct and retrograde members, the normal crossing of the x-axis, close to P_1, takes place to the same side of the primary. Strömgren's families k and n are mentioned as examples.

The choice of a member of a family as family-representing orbit can be arbitrary. However, if a family includes a collision periodic orbit, preference should be given to it because of its features described above. Thus, Bartlett and Wagner (1965) in their study of some selected classes of periodic orbits present graphs and tables showing the dependence of the energy constant on the mass ratio for certain fixed values of the ordinate. Their work also includes the extension of seven collision periodic orbits (denoted by an asterisk in our Table I). On the other hand Bozis (1968), in his study of figure 8 orbits, presents a graph to describe the extension in μ of the collision periodic orbits k and j.

G. E. O. Giacaglia (ed.), Periodic Orbits, Stability and Resonances, 176–191.

This paper brings into a whole the above-mentioned results and makes an extension for all possible values of μ of seven of the remaining collision periodic orbits of the Copenhagen category. The other free parameter is the constant of energy C. In a manner similar to that used in constructing C-x curves, the μ-C curves for these orbits are traced. A certain meaning to the direction of the velocity at collision is given in the Birkhoff's regularized plane. As long as the basic characteristics of the collision periodic orbits are preserved their values of μ and C are placed on the corresponding μ-C curve. Any two collision periodic orbits belonging to the same μ-C curve as well as all members of the families to which they give rise are considered as belonging to the same set of periodic orbits.

2. Comments on Birkhoff's Transformation

Let x, y be the cartesian coordinates of the moving point in the conventional uniformly rotating frame Oxy and u, v be the regularized coordinates. Let also $z=x+iy$ and $w=u+iv$. Birkhoff's conformal transformation is

$$z = \frac{w^2 + \mu(1-\mu)}{2w + 1 - 2\mu} \qquad w \neq w_3 = \mu - \tfrac{1}{2}. \tag{1}$$

The regularized time $\bar{t}$ is related to the time t by

$$\mathrm{d}t = \frac{\varrho_1^2 \varrho_2^2}{4\varrho_3^4}\,\mathrm{d}\bar{t} \tag{2}$$

where ϱ_1, ϱ_2 and ϱ_3 give the distances of P from the primaries $P_1(\mu, 0)$, $P_2(\mu-1, 0)$ and the midpoint $P_3(\mu-\frac{1}{2}, 0)$ of P_1P_2 in the uv plane. The equations of motion are the same with those used by Bozis (1968).

The point $P_3(u=u_3,\ v=0)$ is mapped to the infinity of the xy-plane. The semi-axis $u>u_3$ goes to the semi-axis $x>\mu$ and the $u<u_3$ to $x<\mu-1$. The circumference

$$|w - w_3| = \tfrac{1}{2}, \tag{3}$$

centered at P_3 with radius $\sqrt{2}$ is mapped to the segment P_1P_2 of the x-axis. Further the lower half of the circle (3) and the upper half of the w-plane (excepting the upper half of the circle) go to positive y's and the remaining part of the w-plane goes to negative y's, as shown in Figure 1 for $\mu=0.50$.

The roots of

$$\mathrm{d}z/\mathrm{d}w = 0$$

are $w_1=\mu$ and $w_2=\mu-1$. Therefore, the angle formed by any two intersecting curves at any point of the w-plane (except for the primaries) is equal to the angle of the images of these curves in the z-plane. Further, since at the primaries it is

$$\mathrm{d}^2z/\mathrm{d}w^2 \neq 0$$

angles are doubled at these points as they are transformed from the w- to the z-plane

(Smirnov, 1964, p. 94, also Szebehely *et al.*, 1964). One implication of this fact is the following: if a collision orbit in the w-plane, associated with a certain value of μ and C, is periodic, then the collision orbit with opposite sign of the launching velocity at P_1 is also periodic for the same pair of μ and C; both orbits in the w-plane are mapped to the same orbit in the z-plane.

Thus the collision periodic orbit P_1bP_1 of family j for $\mu=0.50$, $C=3.4708$ is the image of the orbits $P_1b_1P_1$ and $P_1b_2P_1$, both calculated in the uv-plane the same value of C but opposite signs of their normal velocities at P_1.

On the other hand, the collision periodic orbit P_1cP_1 of family g for $\mu=0.50$, $C=3.736\,05$ is the image of only the orbit $P_1c_1P_1c_2P_1$ in the uv-plane which starts at P_1 perpendicularly to the u-axis with $dv/d\tilde{t}<0$, crosses perpendicularly the circumference (3) at the point c_1, comes back to P_1 perpendicularly with $dv/d\tilde{t}>0$ and then follows the symmetric to the u-axis path $P_1c_2P_1$. Clearly only the branch P_1c_1 needs to be calculated in order to assert periodicity. This one-to-one correspondence of closed curves in both planes is always the case if, in the uv-system, one of the normal crossings is with the u-axis and the other with the circumference (3).

Collision periodic orbits of which one normal crossing of the x-axis occurs between the primaries are images of asymmetric with respect to the u-axis periodic orbits in the uv-plane. Thus the orbit P_1aP_1 (of family n with $C=2.5549$) in the xy-plane is the image of the orbit $P_1a_1P_1$ in the uv-plane as well as of its symmetric with respect to the u-axis (not shown in Figure 1) which is obtained by taking the initial velocity with the opposite sign. Obviously all orbits of the family intersect the x-axis perpendicularly at two points between the primaries.

3. Sets of Collision Periodic Orbits

The value of C and the family to which 16 known collision periodic orbits for $\mu=0.50$ belong are listed in Table I.

The family g contains six collision periodic orbits. The orbits k and C are periodic with collision with both primaries.

From the remarks of the previous section it follows that μ and C are not sufficient to specify the initial conditions of a collision periodic orbit. In addition one must know whether the initial velocity in the uv plane is perpendicular to the u-axis ($v=0$) or to the circumference (3) ($K=0$), the sign of this velocity being immaterial. This information is therefore given in Table I. In Figure 3 the initial conditions of the orbits starting normally to the u-axis are given in heavy lines and of the orbits starting normally to (3) by the dashed lines.

We shall now comment on the evolution of all the collision periodic orbits included in Table I.

Set a

Family a of the Copenhagen School is generated from infinitesimal elliptic retrograde orbits around L_3 (Szebehely, 1967). From our point of view its collision orbit for

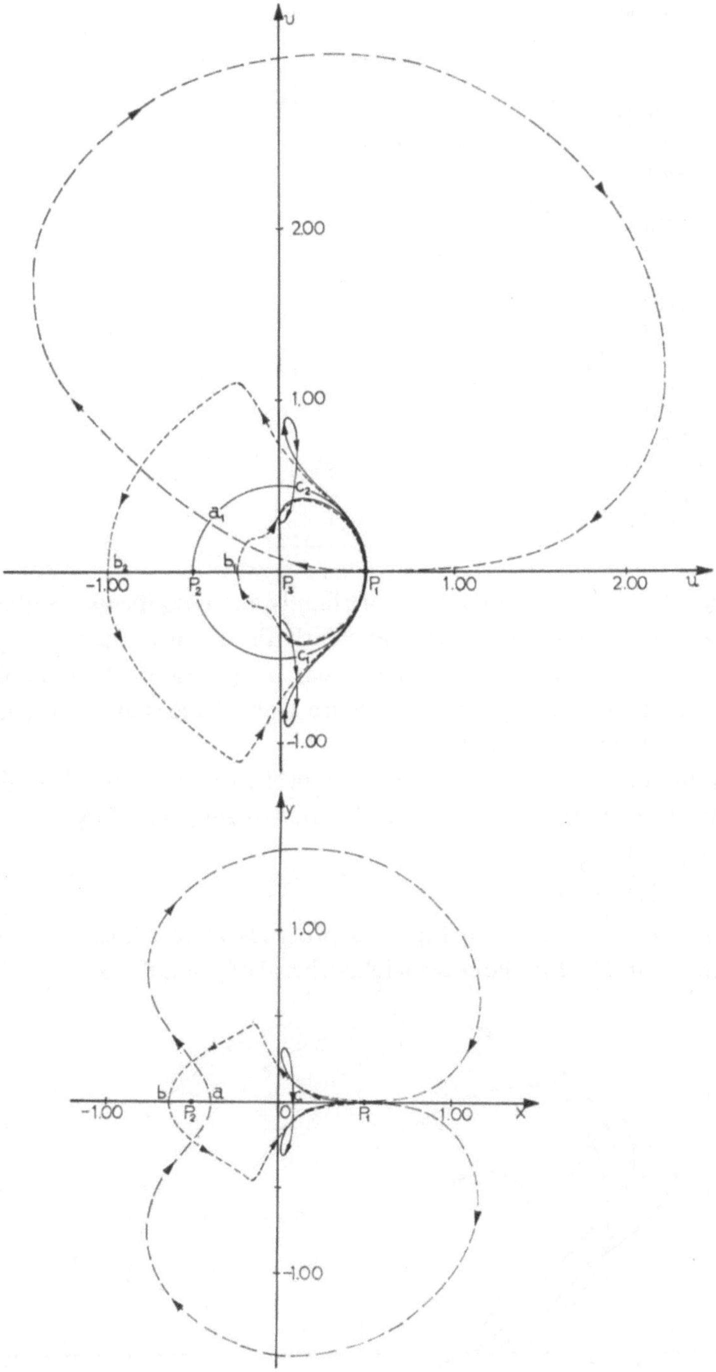

Fig. 1. Mapping of collision periodic orbits from the regularized plane *uv* to the physical plane *xy* (mass ratio $\mu = 0.50$)

TABLE I

Constant C	Family	Normal crossings			
		Starting		Recrossing	
		$v=0$	$K=0$	$v=0$	$K=0$
*3.736 05	g_a	+			+
3.669	k	+		+	
3.470 8	j	+		+	
3.088	v–u	+		+	
3.053 8	$z-y$	+			+
3.035 9	t–s	+			+
2.995	g_c		+	+	
*2.907	g_b	+			+
*2.787 9	g_d	+			+
2.664 7	g_d		+	+	
2.643 8	g_d		+	+	
*2.554 9	n		+		+
*2.503	a	+		+	
*2.432 9	c		+		+
*2.074	c*		+		+
2.044	f		+	+	

$C=2.503$ may be used to generate the entire family and describe its members. Indeed, in the uv-plane, this orbit crosses perpendicularly the u-axis at the primary P_1 and at a point lying between P_3 and P_1, its path remaining inside the circumference (3). Then the Birkhoff's transformation and Figure 1 reveal the shape of the neighboring periodic orbits in the physical plane.

The evolution of the collision orbit with respect to μ was studied by Bartlett and Wagner (1965) who use the symbols γ and T related to our μ and C by

$$\begin{aligned} \mu &= \tfrac{1}{2}(1+\gamma) \\ C &= \tfrac{1}{4}(1+T). \end{aligned} \tag{4}$$

The a-type μ-C curve given in Figure 3 proceeds almost linearly (especially for $\mu<0.50$) from $\mu=0$, $C=1$ to the close neighborhood of $\mu=1$, $C=3$.

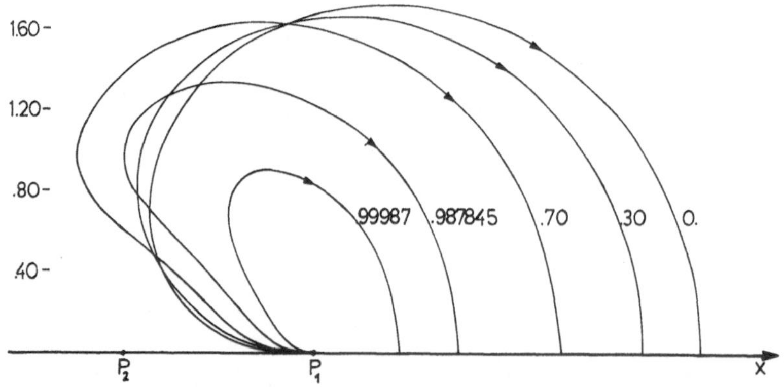

Fig. 2. Evolution of collision periodic orbits of family a as μ increases. Initial conditions: $\mu=0.$, $C=1.$; $\mu=0.30$, $C=1.969\,5$; $\mu=0.70$, $C=2.902\,6$; $\mu=0.987\,845$, $C=2.744\,564$; $\mu=0.999\,87$, $C=2.90$. Only the upper halves of the orbits are shown in the xy-system. The length $(P_1P_2)=1$ and the origin is at a distance from P_1 equal to μ.

The orbits $\mu=0.012\,155$, $C=1.042\,93$ (with collision with the earth) and $\mu=0.987\,845$, $C=2.744\,566$ (with collision with the moon) were found by Broucke (1962, 1968) and generate his families J_1 and I correspondingly. These orbits as well as Moulton's (1920) orbit $\mu=0.80$, $C=2.862$ belong to the set a.

Five sample collision orbits are given in Figure 2 in the xy-system for various values of μ. As $\mu\to 1$, the orbits decrease rapidly in size.

Sets k and j

For $\mu=0.50$ Strömgren's (1931) family k consists of symmetric with respect to both axes direct and retrograde (in the vicinity of P_1) periodic orbits around the primaries, crossing the x-axis at two or at four points correspondingly.

Hénon's family j (1965b) consists of retrograde figure 8 orbits looping around P_1 and P_2 and direct periodic orbits covering both primaries. From the collision periodic orbit of family j, shown in Figure 1, it can be seen that its neighboring periodic orbits starting perpendicularly to the left and to the right of P_1 follow the above description.

The μ-C curves for the k- and j-type orbits were given by Bozis (1968) and are included in Figure 3; Hénon's and Strömgren's orbits are linked through μ and, as a consequence, they belong to the same set. The curves terminate at the same point S^* ($\mu=0.4755$, $C=3.$) spiraling around it. The point S^* itself represents an asymptotic to L_4 and L_5 periodic branch, similar to that established by Strömgren (1931) for $\mu=0.50$.

Initial conditions for the generating orbits of the sets k and j are given in Tables II and III.

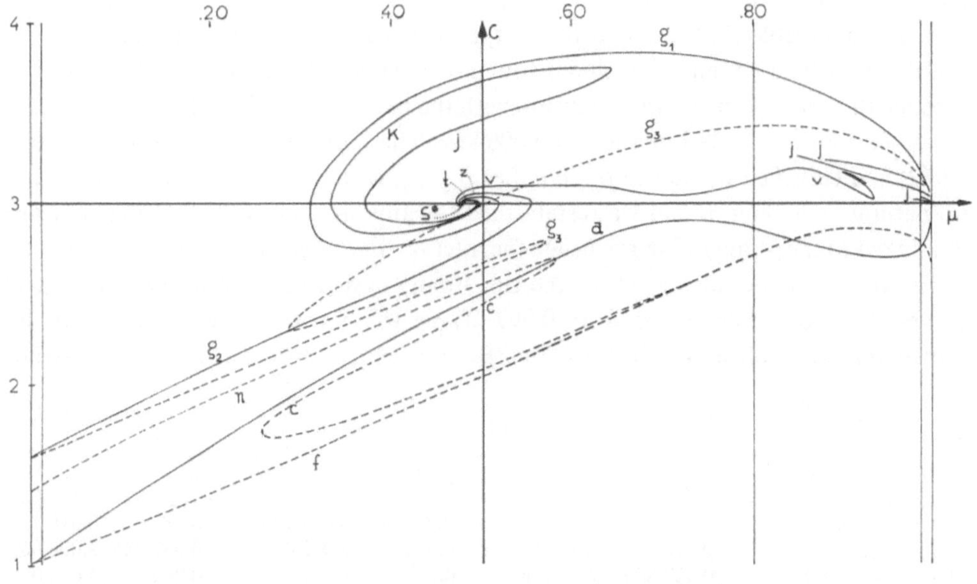

Fig. 3. Energy-mass ratio diagrams for collision periodic orbits. All orbits start at P_1 perpendicularly either to the u-axis (heavy lines) or to the circumference (3).

TABLE II

μ	C	μ	C	μ	C
0.50	3.668 8	0.343 16	2.90	0.497 45	3.
0.46	3.602 2	0.35	2.875 5	0.495 93	3.01
0.43	3.535 7	0.37	2.842 84	0.49	3.018 24
0.40	3.450 2	0.39	2.836 1	0.472 81	3.
0.38	3.378 3	0.42	2.849 9	0.475 857	3.
0.36	3.286 25	0.49	2.958 154	0.475 481	3.
0.333 28	3.	0.497 125	2.99	0.475 528 05	3.

TABLE III

μ	C	μ	C	μ	C	μ	C
0.50	3.470 77	0.64	3.731 7	0.43	3.308 6	0.49	2.967 5
0.52	3.506 35	0.60	3.748 8	0.40	3.200 9	0.495 7	3.
0.54	3.539 25	0.58	3.742 7	0.375 34	3.	0.49	3.015 01
0.56	3.570 23	0.54	3.715 17	0.378 37	2.95	0.48	3.013 73
0.60	3.629 65	0.52	3.694 44	0.40	2.890 57	0.473 2	3.
0.62	3.660 88	0.50	3.470 8	0.43	2.886 91	0.475 803	3.
0.63	3.678 9	0.48	3.431 58	0.46	2.913 0	0.475 488	3.
0.64	3.703 5	0.46	3.387 62	0.47	2.926 8	0.475 527 20	3.

The ranges of μ for which there exist collision with P_1 periodic orbits generating k and j families are $0.333<\mu\leqslant 0.50$ and $0.375<\mu<0.642$, correspondingly. This does not imply that outside these ranges families k or j do not exist. In fact such families with no collision orbits may be generated in a continuous manner by those which include collision orbits (Bozis, 1968).

It is also possible that other distinct j-type μ-C curves exist in other parts of the μ-C diagram. Thus, Figure 3 includes four μ-C curves (tabulated in Table IV) going through the periodic orbits with collision with the moon ($\mu=0.987\,845$): $C=3.120\,252$, $C=3.068\,146$, $C=3.029\,926$ and $C=3.006\,869$. These orbits were found by Broucke (1962) and classified by him to the families E_6, E_6, E_4 and E_5 respectively. Although no natural beginning or end was established for any of those we know that they are not linked to the j-type μ-C curve inside the interval $0.375<\mu<0.642$.

The above families also exist for the sun-Jupiter system; in fact the two lower μ-C curves (E_4 and E_5) persist until $\mu=0.999$ (Table IV). On the other hand it can be shown that the collision orbits E_6 and E_4, for which Broucke did not construct

TABLE IV

μ	C	μ	C	μ	C	μ	C
0.88	3.247 6	0.85	3.262	0.987 845	3.029 926	0.98	3.018 7
0.90	3.227 4	0.93	3.138	0.995	3.032 4	0.987 845	3.006 869
0.93	3.198 5	0.987 845	3.068 146	0.998 3	3.025	0.994	3.008 0
0.987 845	3.120 252	0.998	3.041 5	0.999 1	3.020	0.998 8	3.007 5
0.992 95	3.10	0.998 7	3.035	0.999 75	3.01	0.999 6	3.005 0

characteristic curves, actually belong to the same family, the C-x curve of which is shown in Figure 4.

The branch E_6abE_4 consists of direct (in the vicinity of the moon) simple periodic orbits covering both primaries whereas E_6dcE_4 consists of retrograde figure 8 orbits looping around the earth and the moon.

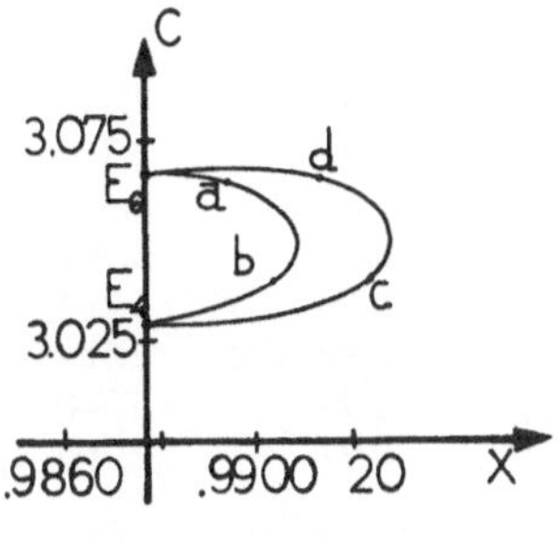

Fig. 4. Characteristic curve for the earth-moon system ($\mu = 0.987\,845$) joining the two collision periodic orbits $E_6(C = 3.068\,146)$ and $E_4(C = 3.029\,926)$. Initial conditions for sample orbits are: $a{:}x = 0.989\,394$, $C = 3.065\,431$; $b{:}x = 0.990\,278$, $C = 3.04$; $c{:}x = 0.992\,367$, $C = 3.04$; $d{:}x = 0.991\,280$, $C = 3.066\,825$.

Set f

Family *f*, studied by Möller (1935) for $\mu = 0.50$, includes two collision periodic orbits with $C = 2.044$ and $C = 1.7403$. Only the first of these is here extended to other values of μ.

The trajectory in the regularized plane needed to establish this orbit is shown in Figure 5; it starts perpendicularly to the circumference (3) and its first intersection with the u-axis is again perpendicular at a point a_1 between P_2 and P_3.

In extending the trajectory P_1a_1 for various values of μ it was observed that it remained inside the upper half of (3). Therefore the neighboring members of the corresponding families f in the physical plane, which start perpendicularly to the x-axis and to the left of P_1, are not complicated. The retrograde members are simple periodic orbits around P_1 and, for large values of C, they tend to Poincaré's infinitesimal circular orbits (Szebehely, 1967). The direct members loop first around P_1 and then around both primaries, intersecting the x-axis at three points.

The μ-C curve describing the evolution of the collision orbit $\mu = 0.50$, $C = 2.044$ is shown in Figure 3. Starting from the point $\mu = 0.$, $C = 1.$, the Jacobian constant $C = C(\mu)$ is almost linearly increasing with increasing μ till $\mu \simeq 0.92$. From this point on C is decreasing and as $\mu \to 1$, $C \to -\infty$.

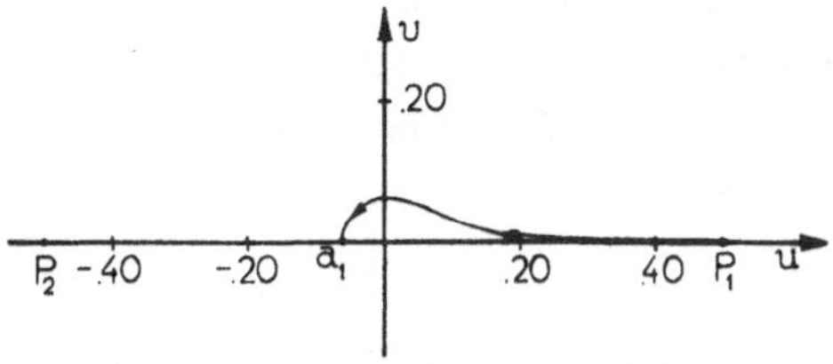

Fig. 5. f-type trajectory in the uv-plane for the collision periodic orbit $\mu = 0.50$, $C = 2.044$.

For $\mu > 0.50$ the μ-C curve goes very close and in fact intersects the μ-C curve of the c-type orbits which are discussed below.

Initial conditions for the f-type orbits are given in Table V.

TABLE V

μ	C	μ	C
0.50	2.044	0.53	2.111 8
0.40	1.819 9	0.60	2.270 1
0.35	1.709 5	0.70	2.493
0.30	1.601 5	0.80	2.701
0.20	1.389 8	0.90	2.861 5
0.10	1.188 1	0.987 845	2.784 9
0.05	1.092	0.995	2.740
0.0	1.	0.999	2.700

Sets c and n

It was shown by Bartlett and Wagner (1965) that the collision orbits c ($C = 2.4329$) and n ($C = 2.5549$) of the Copenhagen category belong to the same set (Figure 3). In constructing their μ-C curves they also found a new periodic orbit for $\mu = 0.50$ with $C = 2.074$, denoted by c^* in Table I.

An n-type collision periodic orbit is shown in Figure 1. In the uv-plane the trajectory P_1a_1 suffices to establish periodicity. As we move on the μ-C diagram from $\mu = 0$, $C = 1.401\,876$ to the maximum $\mu \simeq 0.565$ and back to $\mu = 0.50$, $C = 2.4329$, the trajectory P_1a_1 remains inside the upper half of (3). For the rest of the curve P_1a_1 intersects the u-axis between P_2 and P_3 before cutting normally the circumference (3). As $\mu \to 1$, the trajectories become even more complicated, intersecting the u-axis at two points on the segments P_2P_3 and P_3P_1 before they cross normally the circumference.

Samples of such trajectories are shown in Figure 6 in the regularized plane. The analysis of Section 2 may help in understanding the complicated shape of the neighboring periodic orbits in the physical system.

From Figure 6 it is concluded that between the orbits 5 and 6 an orbit recrossing normally the circumference at the point P_1 must exist. Of course, such an orbit should also cross the u-axis on the segment P_2P_3 perpendicularly, namely it should belong to the set f. Additional numerical calculations are necessary to indicate whether the c and f-type μ-C curves have one or more common points.

One final remark concerns orbit 1 of Figure 6. This is a consecutive collision orbit for the two-body problem going asymptotically from P_1 to P_2 which, in this case, is identified with Lagrange's point L_1. Orbit 4 is also with consecutive collision for $\mu = 0.50$ and is symmetric to both axes.

Set v-u

For $\mu = 0.50$ family v, established by Hénon (1965a), includes non-symmetric with

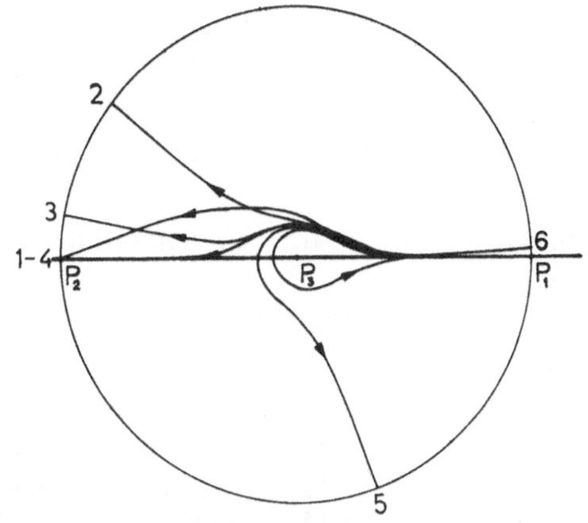

Fig. 6. Samples of c- and n-type trajectories in the uv-plane with collision at P_1. Initial conditions: $1{:}\mu=0.$, $C=1.401\,876$; $2{:}\mu=0.43$, $C=2.42$; $3{:}\mu=0.532$, $C=2.54$; $4{:}\mu=0.50$, $C=2.432\,913$; $5{:}\mu=0.27$, $C=1.70$ and $6{:}\mu=0.90$, $C=2.8597$.

respect to the y-axis periodic orbits; family u is generated by v in view of the mirror properties of the case $\mu=0.50$. The members of the set v are topologically similar to those of the set a, in the sense that, in the uv-plane, a v-type collision periodic orbit remains inside the circumference (3) and crosses the u-axis perpendicularly at a point between P_1 and P_3.

The μ-C curve of the set v, shown in Figure 3, goes from the point $\mu=0.50$, $C=3.088$ to larger values of μ, slightly fluctuating, till $\mu\simeq0.935$ where it assumes a maximum μ; then it turns back abruptly and, with decreasing values of μ goes near the point $\mu\simeq0.895$, $C\simeq3.17$. From this point on μ is again increasing. An inspection of the orbits themselves shows that, as we move on the μ-C curve, they create loops which come closer and closer to the x-axis. This is indicated in Figure 7 where a sequence of orbits is shown with initial conditions taken successively along the v-type μ-C curve. In view of this figure it is conjectured that the μ-C curve of the set v for $\mu>0.50$

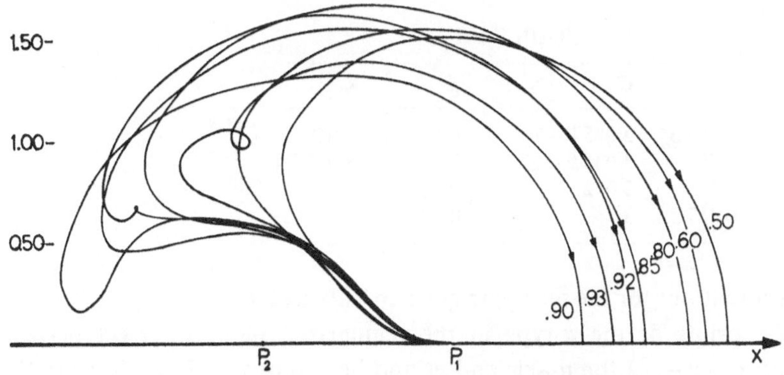

Fig. 7. The upper halves of v-type collision periodic orbits for various values of μ in the physical plane. Initial conditions are given in Table VI.

eventually terminates for some value of μ and for $C=3$ to a collision periodic branch, asymptotic to the collinear Lagrangian point L_1 (Szebehely and Williams, 1964).

For $\mu<0.50$ the μ-C curve assumes a minimum $\mu=0.469$ at $C=3$, then μ increases and the curve is spiraling around the point S^* to which the μ-C curves of the sets k and j also tend.

Table VI gives the initial conditions for the generating orbits of the families v, including those of Figure 7.

TABLE VI

μ	C	μ	C	μ	C
0.475 553	3.	0.70	3.042 5	0.935	3.040
0.459 5	3.	0.80	3.122 7	0.930	3.070
0.47	3.044 8	0.85	3.196 0	0.92	3.113
0.50	3.088	0.90	3.104 5	0.90	3.169
0.55	3.095	0.93	3.028 4	0.92	3.121
0.60	3.100 7	0.935	3.022 5	0.93	3.117

Sets g, t-s and z-y

Strömgren's family g for $\mu=0.50$ starts with simple direct periodic orbits around P_1 and, according to Szebehely (1967), it evolves in seven phases including more and more complicated orbits. (The subscripts of g in Table I correspond to these phases.) The evolution of the family was at times pursued by a number of investigators (Burrau, Fischer-Petersen) until the problem of its termination was settled by Bartlett (1964) who verified Strömgren's conjecture that the family should terminate to asymptotic orbits with respect to L_4 and L_5. Szebehely and Nacozy (1967) furnished additional initial conditions with increased accuracy near the termination of the family.

For the Copenhagen problem family g includes six collision orbits, given in Table I. One might expect a single continuous μ-C curve to link the corresponding points on the μ-C diagram. However, as it is seen in Figure 3 there exist three distinct g-type μ-C curves denoted by g_1, g_2 and g_3.

The sets g_1 and g_2 were computed by Bartlett and Wagner (1965) but not to their end. Tables of initial conditions may be constructed for these with the aid of Equations (4) and our complementary Table VII which starts with the last values of Bartlett and Wagner.

TABLE VII

μ	C	μ	C
0.503 34	3.035 2	0.531 31	3.049 7
0.50	3.035 9	0.50	3.053 8
0.48	3.022	0.48	
0.467 5	3.	0.476 14	3.

Initial conditions for the set g_3 are given in Table VIII.

Periodic orbits of the g-type in the regularized plane intersect perpendicularly at a point $u_0>\mu-\sqrt{2}$ the u-axis and at another point the circumference (3). As it is seen from Table I and is schematically shown in Figure 8 the normal crossing at collision occurs with the u-axis for g_1 and g_2 and with the circumference for g_3.

Only the branches necessary to ascertain periodicity are shown in Figure 8. Our calculations of trajectories g_1 and g_2 have indicated that for all values of μ they remain inside the lower half of the circle before they cross it perpendicularly. This remark must be taken into account when considering the termination of the $\mu-C$ curves for the g_1 and g_2 orbits.

TABLE VIII

μ	C	μ	C	μ	C
0.000	1.587 5	0.575	2.784	0.40	2.728
0.10	1.809	0.575	2.802	0.50	2.995
0.20	2.021	0.55	2.756	0.55	3.105
0.30	2.235	0.50	2.664 7	0.60	3.200
0.40	2.442	0.41	2.495	0.70	3.345
0.50	2.643 8	0.35	2.384	0.80	3.422
0.55	2.735	0.30	2.308	0.909 091	3.380
0.571	2.78	0.30	2.375	0.987 845	3.147 6

The g_3-type trajectories may or may not cross the circumference (3) before their first perpendicular intersection with the u-axis (Figure 8).

For $\mu=0.50$ g_3 and g_1 or g_2 orbits must be linked through the ordinate x, since they belong to the same family g. In the xy-system this amounts to an abrupt change of the direction of velocity at collision. Szebehely (1967, p. 469) states that the directional properties of the g-type collision periodic orbits "show possible ambiguity". The transition from g_1 to g_3 is easily described in the uv-system with the help of the intermediate orbit u_1a_1 of Figure 8. As $u_1 \to u_0$, $a_1 \to P_1$ and the symmetric branch of g_3 with respect to the u-axis is obtained.

Families t-s and z-y were discovered by Hénon for $\mu=0.50$. Their collision orbits are shown in Figure 9 in the regularized and the physical plane. They resemble topologically the collision periodic orbits g_1 and g_2 and in fact it is indicated that they belong to the corresponding μ-C curves (Figure 3). This was missed by Bartlett and Wagner probably because they stopped tracing for $\mu>0.50$ the μ-C curves going

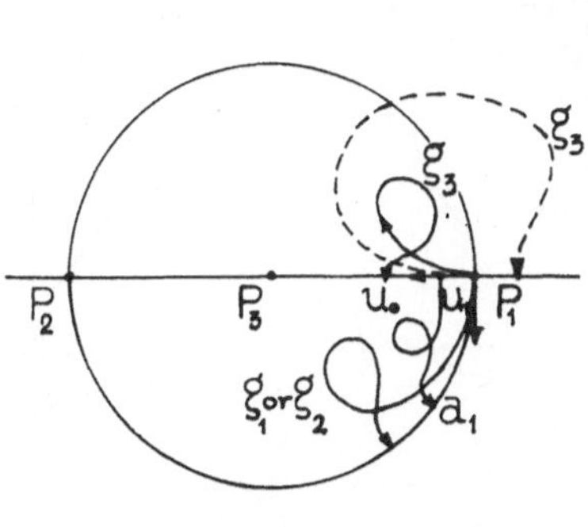

Fig. 8. Periodic orbits of the sets g_1 and g_3 shown schematically in the regularized plane. Orbits g_2 are topologically similar to g_1.

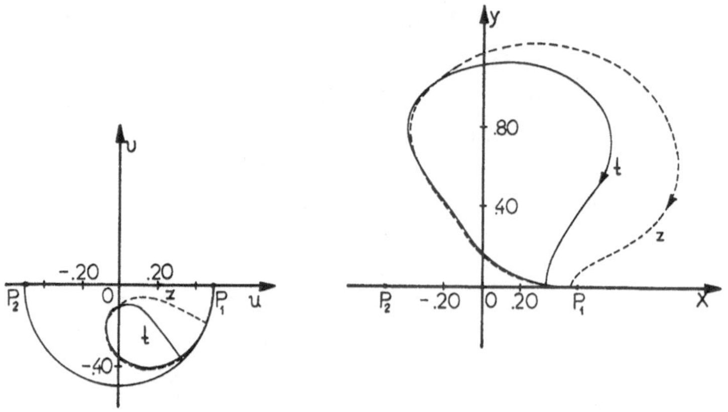

Fig. 9. Collision periodic orbits of the families t and z for $\mu=0.50$ in the uv- and xy-systems.

through the points $\mu=0.50$, $C=2.907$ and $\mu=0.50$, $C=2.7879$ before they intersected again the $\mu=0.50$ axis and, instead, they focused their attention on the existence of g_1 and g_2-type orbits for large values of μ.

The μ-C curves for Hénon's orbits t and z for $\mu<0.50$ are shown in Figure 3 as continuations of g_1 and g_2 respectively. Evidence is provided that they are spiraling around the point S^* by Figure 10. A comparison of Figure 10 with Figure 9 shows that as the initial conditions are chosen from more advanced points of the μ-C curves towards the end the orbits themselves present loops around L_4 and eventually tend to the collision asymptotic branch which corresponds to the limiting point S^* (Bozis, 1968).

The g_1 μ-C curve to the right of Strömgren's orbit $\mu=0.50$, $C=3.736\,05$ goes smoothly to the close vicinity of $\mu=1$, $C=3$ and passes through Darwin's orbit $\mu=0.909\,091$, $C=3.531\,83$, generating his family A of satellites and Broucke's orbit $\mu=0.987\,845$, $C=3.168\,4$, generating his family H_1. The similarity of the families H_1 and g is also pointed out by Broucke (1962).

The μ-C curve of the set g_3 intersects the $\mu=0.50$ axis of Figure 3 at three points,

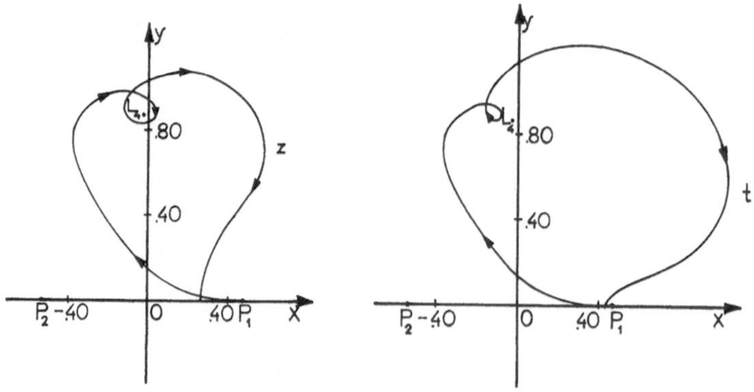

Fig. 10. Collision periodic orbits in the xy-plane of the sets $t(\mu=0.4675,\ C=3)$ and $z(\mu=0.476\,14,\ C=3)$ close to the end of their μ-C curves.

i.e. it includes three, out of the six, known collision periodic orbits of the family g. Starting from $\mu=0$, $C=1.587\,5$ the curve goes almost linearly to $\mu=0.50$ $C=2.6438$, proceeds to a maximum $\mu\simeq0.58$ and, with decreasing μ, comes back to $\mu=0.50$, $C=3.6647$. From this point on μ is decreasing almost linearly, it assumes a minimum $\mu\simeq0.275$ on the g_2 μ-C curve and, with increasing μ the curve goes to the point $\mu=0.50$, $C=2.995$. Hereafter the curve goes smoothly to values of μ close to unity passing through Darwin's orbit $\mu=0.909\,091$, $C=3.38$.

As regards the shape of the g_3-type orbits we have observed that from $\mu=0.$, $C=1.5875$ up to the first maximum μ of the μ-C curve their perpendicular crossing of the μ-axis occurs to the right of P_1 (dashed trajectory of Figure 8). The Birkhoff transformation now suggests that a neighboring to such a collision periodic orbit is rather complicated in the xy-system presenting three or four points of intersection with the x-axis. The shape of the collision periodic orbits for the rest of the g_3 μ-C curve is similar to the full trajectory g_3 of Figure 8 and the orbits generated by them are simpler.

In its way back and forth the g_3-type μ-C curve intersects at certain points the g_1 and g_2 μ-C curves. At these points two collision periodic orbits exist for the same pair of μ and C. In the xy-system the infinite velocity at collision with P_1 for a given g_1 or g_2-type orbit is pointing towards P_2, the orbit goes at first to positive y's and recrosses normally the x-axis at a point between P_1 and P_2. For a g_3-type orbit the initial velocity at collision is pointing towards the opposite direction of P_2, the orbit goes to negative y's and recrosses the x-axis at a point to the right of P_1.

An interesting case appears for the two-body problem ($\mu=0$) and for $C=1.5875$. At this point of the μ-C diagram the g_2 and g_3 μ-C curves come together and for orbits of both curves the recrossing point of the x-axis is again P_1. Therefore the two trajectories shown in Figure 11 are actually combined into one periodic orbit.

4. Concluding Remarks

In the present paper we consider some relatively simple collision periodic orbits (16 of the Copenhagen category and 4 of the earth-moon case) and describe their evolution with respect to the mass ratio μ and the shape of the members of the families to which they give rise.

This work is closely related to the last paper of the Copenhagen series by Bartlett and Wagner which deals with the overall topological structure of some selected classes of periodic orbits of the restricted problem and to a paper by Bozis which discusses the structure of the families k and j for various values of μ. It is also related to work on periodic orbits performed by a number of investigators for fixed μ. It either connects known families for different values of μ or reveals the existence of certain families for an arbitrary mass ratio.

The main results may be summarized as follows:

(a) The sets v, z, t, g_3 and f of collision periodic orbits are added in the μ-C diagram to the previously known sets g_1, g_2, a, c and n (Bartlett and Wagner, 1965) and also k and j (Bozis, 1968).

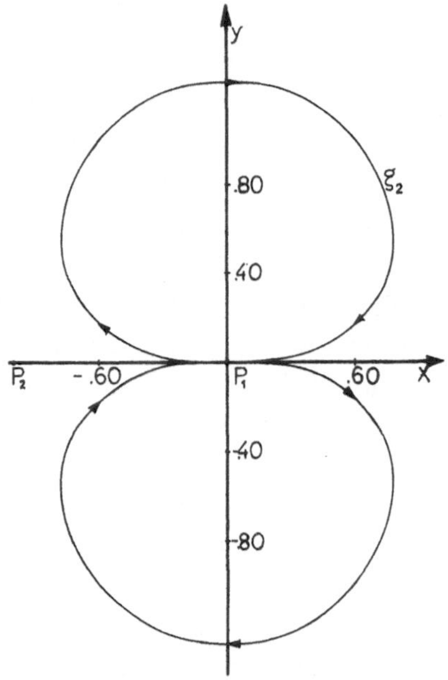

Fig. 11. Two collision with P_1 trajectories (g_2 and g_3) with opposite direction of the initial velocity at P_1 for the two-body problem ($\mu = 0$) and common $C = 1.5875$. These trajectories are combined into one periodic orbit.

(b) The Birkhoff transformation was proven very efficient in describing the shape of the members of the families generated for various values of μ by the collision periodic orbits of the above sets. In order to establish periodicity it is required that the trajectory in the regularized plane shows two normal crossings with the u-axis (k, j, v and a) or the circumference centered at P_3 with radius $\sqrt{2}$ (c and n) or both (g_1, g_2, g_3, z, t and f). Other intermediate non-perpendicular intersections with either $v=0$ or $K=0$ may occur (as in the sets k, c, f and part of g_3) or may not as it actually happens for all other sets.

(c) Hénon's families t-s and z-y belong to the same set with Strömgren's family g. Other interconnections were also found of the families of the Copenhagen School with Broucke's ($\mu=0.012\,155$ and $\mu=0.987\,845$), Moulton's ($\mu=0.80$) and Darwin's ($\mu=0.909\,091$) families.

(d) Evidence was provided that the μ-C curves of the sets c and f intersect each other and, as a consequence, they may be considered as one set.

(e) The μ-C curves of the sets k, j, t, z and v are eventually spiraling around the same point $S^*(\mu \simeq 0.475\,5, C=3)$ of the μ-C diagram. This point represents an asymptotic to L_4 and L_5 periodic branch with collision at P_1 at which Strömgren's asymptotic orbits I and II coalesce. It is of interest to note that the characteristic curves of the families k, j, t, z and v for $\mu=0.50$ are also spiraling around one point.

The coincidence of the branches I and II at S^* was also verified by Mullius (1969) who made an extension in μ of all Strömgren's and Bartlett's asymptotic branches known for $\mu = 0.50$. Evidently the asymptotic orbits can also be used, instead of the collision orbits, to follow the evolution of certain families.

(f) The μ-C curves g_1, g_3, c, f and a go to the close vicinity of $\mu = 1$. The case $\mu = 1$ requires a separate study. We remark at this point that Hénon (1969) in his study of the Hill's problem establishes five families of simple periodic orbits which, in Strömgren's nomenclature, are those mentioned above i.e. families g_1, g_3, c, f and a.

(g) It is possible that two collision periodic orbits belong to the same family but their μ-C curves are not linked as in the case of some collision orbits of the family g; it may also happen that collision orbits of the same μ-C curve do not belong to the same family for a certain value of μ as in the case of the sets k and j.

Using the method described in this paper, Zikides (1969) is now studying collision periodic orbits for the elliptic problem and for the case of two equal masses. His results may reveal possible interconnections of the known families of the circular problem through the eccentricity e.

It is hoped that all these results will contribute to the question of the classification of the periodic orbits of the restricted problem.

References

Bartlett, J. H.: 1964, 'The Restricted Problem of Three Bodies', *Mat.-Fys. Skr. Dan. Vid. Selsk.* **2**, No. 7.

Bartlett, J. H. and Wagner, C. A.: 1965, *Publ. Copenh. Observ.* **183**.

Bozis, G.: 1968, *Astron. J.* **73**, 616.

Broucke, R. A.: 1962, 'Recherches d'orbites périodiques dans le problème restreint plan, système terre-lune', Doctoral Dissertation, Univ. of Louvain, Belgium.

Broucke, R. A.: 1968, NASA Tech. Rep. 32–1168.

Carpenter, L. and Stumpff, K.: 1968, *Astron. Nachr.* **291**, No. 1.

Hénon, M.: 1965a, *Ann. Astrophys.* **28**, 499.

Hénon, M.: 1965b, *Ann. Astrophys.*, **28**, 992.

Hénon, M.: 1969, *Astron. Astrophys.* **1**, 223.

Huang, S. S.: 1962, *Astron. J.* **67**, 304.

Möller, J. P.: 1935, *Publ. Copenh. Observ.* **99**.

Mullius, L. J.: 1969, private communication.

Smirnov, V. I.: 1964, *A Course of Higher Mathematics*, Vol. 3, Pergamon Press.

Strömgren, E.: 1931, *Publ. Copenh. Observ.* **80**.

Strömgren, E.: 1935, *Publ. Copenh. Observ.* **100**.

Szebehely, V.: 1967, *Theory of Orbits*, Academic Press, New York.

Szebehely, V. and Nacozy, P.: 1967, *Astron. J.* **72**, 184.

Szebehely, V. and Williams, C.: 1964, *Astron. J.* **69**, 460.

Szebehely, V., Pierce, D., and Standish, E. M.: 1964, *Celestial Mechanics and Astrodynamics*, Progress in Astronautics and Aeronautics, Vol. 14, Academic Press, New York and London, p.35.

Zikides, M.: 1969, private communication.

Discussion

V. Szebehely: Are there any consecutive collision periodic orbits included in your μ-C diagram for $\mu \neq \frac{1}{2}$?

G. Bozis: No. Except for the consecutive collision periodic orbits of the families K and C of the Copenhagen problem and a trivial n-type orbit for the two-body problem, no other such orbit was found.

PERIODIC ORBITS IN TRIGONOMETRIC SERIES

LLOYD CARPENTER

Goddard Space Flight Center, Greenbelt, Md, U.S.A.

Abstract. A method is given for the study of families of periodic motion using trigonometric series to represent the individual solutions. The continuous deformations of a periodic orbit along a family are represented by the variations of the trigonometric coefficients with respect to a parameter.

For the applications the series are truncated, while the coefficients and their variations are determined numerically. In this form, the continuation with respect to the parameter is given by a mapping $f : R^n \to R^n$ of the space of coefficients into itself. The values of the coefficients are then improved by another mapping which is a contraction operator in some neighborhood of the fixed point representing the solution.

The method is applied to the natural families defined by Wintner and to the families of the first and second kinds of Poincaré in the restricted problem of three bodies.

1. Introduction

Trigonometric series have been used very extensively in the study of periodic orbits and families of periodic motion in the restricted problem of three bodies. Many of these studies depend on the presence of small parameters in the problem, and the results consist of trigonometric terms whose coefficients are power series in the small parameters. That is to say, the results are obtained in the form of Poisson series. The same techniques can be applied using numerical values for the parameters, and this is referred to as a semi-analytic development. However, many families of periodic orbits have not yielded to these methods and have thus far been explored by numerical integration.

The purpose of the present study is to obtain semi-analytic results for some of these more difficult families of periodic orbits and to place the continuation of these families on a sound basis when trigonometric series with numerical coefficients are used in a method of successive approximations.

Cases of near and exact resonance are of special interest and are handled without difficulty using the present technique. The periodic matrix of the variational equations plays an essential role in the method, and the linear stability analysis is done in the usual way.

A good discussion of the method, as applied to isolated periodic solutions has been given by Urabe (1965). Numerical examples are given by Urabe and Reiter (1966). More recent results and references are given by Stokes (1969). The successive approximations are similar to those of Bennett and Palmore (1968) in that each step yields a periodic function which is an approximate solution of the original equations of motion. The variations of the series coefficients with respect to a local parameter give a representation of the functions $x_1(t)$ and $y_1(t)$ discussed by Deprit and Henrard (1967) (p. 160).

The method described in this study has been applied in the computation of periodic

G. E. O. Giacaglia (ed.), Periodic Orbits, Stability and Resonances, 192–209.

orbits in the restricted problem of four bodies by Kolenkiewicz and Carpenter (1967 and 1968) and in the restricted problem of three bodies by Carpenter and Stumpff (1968). Further applications are being made by Deprit and Carpenter for locating elements of many new families or orbits in the problem of three bodies.

2. The Iteration Process for Fixed Period

The method is applicable to a wide range of problems, and a general formulation is possible. However, each class of problems has its own interesting features, and it is usually possible to greatly improve the efficiency by taking advantage of the particular forms of the equations and the solutions.

The formulation will be given for the restricted problem of three-bodies considering symmetric orbits which lie in the plane of motion of the primaries. Furthermore, the method will be developed as a modification of an iterative general perturbations technique, so a two-body reference orbit is used as the starting point. For convenience the reference orbit is circular with respect to the central primary whose mass is put equal to M (referred to as the sun).

Let a and n be the radius and mean motion respectively of the circular reference orbit of the infinitesimal particle (referred to as the minor planet), and let k be the gaussian constant so that

$$n^2a^3 = k^2M.$$

Let the true position of the minor planet be given by

$$\mathbf{r} = (1 + \alpha)\,\mathbf{r}_0 + \beta\mathbf{w},$$

where $\mathbf{r}_0$ is the position in the reference orbit,

$$\mathbf{w} = \frac{1}{n}\frac{\mathrm{d}\mathbf{r}_0}{\mathrm{d}t}$$

is the vector of length a in the direction of the velocity in the reference orbit, and the quantities α and β will represent the periodic deviations from the circular motion.

Let a' and n' be the radius and mean motion respectively of the circular motion of Jupiter, whose mass is denoted by m, so that

$$n'^2a'^3 = k^2(M + m) \quad \text{or} \quad n'^2a'^3 = k^2M(1 + m') \quad \text{with} \quad m' = m/M$$

and

$$a'/a = (1 + m')^{1/3}\nu^{2/3},$$

where

$$\nu = n/n'.$$

For the cases where $n > n'$, the relative reference motions will be periodic with period

$$T_r = \frac{2\pi}{n - n'}$$

and the orbit to be determined will have the period

$$T = \frac{2\pi}{N}$$

with

$$N = \frac{n - n'}{l}$$

for some positive integer value of l. The integer l is the winding number or index of the orbit with respect to the sun in the synodic coordinate system. Thus the trigonometric argument is

$$\theta = Nt.$$

For symmetric orbits the epoch of time is chosen such that the minor planet crosses the sun-Jupiter line at $t=0$.

With respect to the mean anomaly, $g=nt$, the equations of motion for the minor planet are

$$\frac{d^2\alpha}{dg^2} - 2\frac{d\beta}{dg} - 3\alpha = \Omega_\alpha$$

$$\frac{d^2\beta}{dg^2} + 2\frac{d\alpha}{dg} = \Omega_\beta .$$

In terms of the present coordinates

$$\Omega = \frac{1}{2}\left(\frac{r}{a}\right)^2 + \frac{a}{r} - \frac{3}{2}\alpha^2 + m'\left\{\frac{a}{\varrho} - \left(\frac{a}{a'}\right)^2 [(1+\alpha)\cos l\theta - \beta \sin l\theta]\right\},$$

where

$$\frac{r}{a} = \sqrt{(1+\alpha)^2 + \beta^2}$$

and

$$\frac{a}{\varrho} = \left\{(1+\alpha)^2 + \beta^2 + \left(\frac{a'}{a}\right)^2 - 2\left(\frac{a'}{a}\right)[(1+\alpha)\cos l\theta - \beta \sin l\theta]\right\}^{-1/2},$$

ϱ being the distance from the minor planet to Jupiter. Thus

$$\Omega_\alpha = (1+\alpha)\left[1 - \left(\frac{a}{r}\right)^3\right] - 3\alpha$$
$$+ m'\left\{\left(\frac{a}{\varrho}\right)^3 \left[\left(\frac{a'}{a}\right)\cos l\theta - (1+\alpha)\right] - \left(\frac{a}{a'}\right)^2 \cos l\theta\right\}$$

and

$$\Omega_\beta = \beta\left[1 - \left(\frac{a}{r}\right)^3\right] + m'\left\{-\left(\frac{a}{\varrho}\right)^3 \left[\left(\frac{a'}{a}\right)\sin l\theta + \beta\right] + \left(\frac{a}{a'}\right)^2 \sin l\theta\right\}.$$

Later we will also need expressions for the partial derivatives of Ω_α and Ω_β with respect to α, β, m' and ν. These are

$$\Omega_{\alpha\alpha} = -2 - \left(\frac{a}{r}\right)^3 + 3(1+\alpha)^2\left(\frac{a}{r}\right)^5 + m'\left\{-\left(\frac{a}{\varrho}\right)^3 + 3\left(\frac{a}{\varrho}\right)^5 \times\left[\left(\frac{a'}{a}\right)\cos l\theta - (1+\alpha)\right]^2\right\}$$

$$\Omega_{\alpha\beta} = \Omega_{\beta\alpha} = 3(1+\alpha)\,\beta\left(\frac{a}{r}\right)^5 - 3m'\left(\frac{a}{\varrho}\right)^5\left[\left(\frac{a'}{a}\right)\cos l\theta - (1+\alpha)\right]\left[\left(\frac{a'}{a}\right)\sin l\theta + \beta\right]$$

$$\Omega_{\beta\beta} = 1 - \left(\frac{a}{r}\right)^3 + 3\beta^2\left(\frac{a}{r}\right)^5 + m'\left\{-\left(\frac{a}{\varrho}\right)^3 + 3\left(\frac{a}{\varrho}\right)^5\left[\left(\frac{a'}{a}\right)\sin l\theta + \beta\right]^2\right\}$$

$$\Omega_{\alpha\nu} = \frac{2}{3}\frac{a'}{a}\frac{m'}{\nu}\left\{3\left(\frac{a}{\varrho}\right)^5\left[(1+\alpha)\cos l\theta - \beta\sin l\theta - \left(\frac{a'}{a}\right)\right]\left[\left(\frac{a'}{a}\right)\cos l\theta - (1+\alpha)\right] + \left[\left(\frac{a}{\varrho}\right)^3 + 2\left(\frac{a}{a'}\right)^3\right]\cos l\theta\right\}$$

$$\Omega_{\beta\nu} = -\frac{2}{3}\frac{a'}{a}\frac{m'}{\nu}\left\{3\left(\frac{a}{\varrho}\right)^5\left[(1+\alpha)\cos l\theta - \beta\sin l\theta - \left(\frac{a'}{a}\right)\right]\left[\left(\frac{a'}{a}\right)\sin l\theta + \beta\right] + \left[\left(\frac{a}{\varrho}\right)^3 + 2\left(\frac{a}{a'}\right)^3\right]\sin l\theta\right\}$$

$$\Omega_{\alpha m'} = \frac{1}{2}\frac{\nu}{1+m'}\,\Omega_{\alpha\nu} + \left\{\left(\frac{a}{\varrho}\right)^3\left[\left(\frac{a'}{a}\right)\cos l\theta - (1+\alpha)\right] - \left(\frac{a}{a'}\right)^2\cos l\theta\right\}$$

$$\Omega_{\beta m'} = \frac{1}{2}\frac{\nu}{1+m'}\,\Omega_{\beta\nu} - \left\{\left(\frac{a}{\varrho}\right)^3\left[\left(\frac{a'}{a}\right)\sin l\theta + \beta\right] - \left(\frac{a}{a'}\right)^2\sin l\theta\right\}$$

The dependence on ν and part of the dependence on m' comes from the relation

$$\frac{a'}{a} = (1+m')^{1/3}\nu^{2/3}.$$

The argument, θ, is treated as an independent quantity, because these expressions are used for computing the variations of the coefficients in the series expansions.

For the symmetric periodic orbits, α and Ω_α are even functions of θ while β and Ω_β are odd so that

$$\alpha = \sum_{k=0}^{\infty} \alpha_k \cos k\theta$$

$$\beta = \sum_{k=1}^{\infty} \beta_k \sin k\theta$$

$$\Omega_\alpha = \sum_{k=0}^{\infty} c_k \cos k\theta$$

$$\Omega_\beta = \sum_{k=1}^{\infty} s_k \sin k\theta$$

for non-collision orbits. All such series are truncated in the actual computations

Putting

$$N_k = k\frac{N}{n}$$

and substituting the series expressions into the equations of motion gives

$$\left.\begin{aligned}(3 + N_k^2)\,\alpha_k + 2N_k\beta_k &= -\,c_k\\ 2N_k\alpha_k + N_k^2\beta_k &= -\,s_k\end{aligned}\right\}\ k = 0, 1, 2, \ldots \tag{A}$$

by equating coefficients. If c_k and s_k were known, the solution would be obtained from the formulas

$$\alpha_0 = -\,c_0/3$$

$$\left.\begin{aligned}\alpha_k &= -\frac{1}{N_k^2 - 1}\,c_k + \frac{2}{N_k(N_k^2 - 1)}\,s_k\\ \beta_k &= \frac{2}{N_k(N_k^2 - 1)}\,c_k - \frac{3 + N_k^2}{N_k^2(N_k^2 - 1)}\,s_k\end{aligned}\right\}\ k = 1, 2, 3, \ldots \tag{B}$$

Since Ω_α and Ω_β depend on α and β, the coefficients c_k and s_k are not known beforehand. However, in many practical cases the solution may be obtained by the iterative general perturbations technique. This consists of starting with $\alpha=\beta=0$, for example, computing approximate values for c_k and s_k by harmonic analysis of the expressions for Ω_α and Ω_β, and using these values to solve for approximate values of α_k and β_k. These approximations for α and β are then used in a new harmonic analysis of Ω_α and Ω_β to give the next approximations for α_k and β_k etc.

This simple process fails in some very interesting cases such as resonance. For example, when

$$\frac{n}{n'} = \frac{p}{q}$$

for integer values of p and q, and

$$l = p - q$$

the formula gives

$$N_p = 1\,.$$

In this case the formulas for α_k and β_k have zero divisors at $k=p$, so the method does not work, although the original equations could have a solution with $c_p=2s_p$.

The simple iteration process has been modified to handle cases such as the one just described by taking into account the partial derivatives of c_k and s_k with respect to α_j and β_j and then computing corrections to an approximate solution. This modified process yielded, for example, a linearly stable orbit for $n/n'=\frac{3}{2}$ (the Hilda resonance). This particular orbit will be discussed later.

The partial derivatives of c_k and s_k with respect to α_j and β_j are obtained from the coefficients in the expansions of $\Omega_{\alpha\alpha}$ $\Omega_{\alpha\beta}$, $\Omega_{\beta\beta}$. These functions can be represented in the form

$$\Omega_{\alpha\alpha} = \sum_{k=0}^{\infty} x_k \cos k\theta$$

$$\Omega_{\alpha\beta} = \sum_{k=1}^{\infty} y_k \sin k\theta$$

$$\Omega_{\beta\beta} = \sum_{k=0}^{\infty} z_k \cos k\theta .$$

Numerical values of the coefficients are computed by harmonic analysis. Then

$$\frac{\partial c_k}{\partial \alpha_j} = \frac{x_{|k-j|} + x_{k+j}}{2}$$

$$\frac{\partial c_k}{\partial \beta_j} = \frac{-\eta_{k,j} y_{|k-j|} + y_{k+j}}{2}$$

$$\frac{\partial s_k}{\partial \alpha_j} = \frac{\eta_{k,j} y_{|k-j|} + y_{k+j}}{2}$$

$$\frac{\partial s_k}{\partial \beta_j} = \frac{z_{|k-j|} - z_{k+j}}{2},$$

where

$$\eta_{k,j} = \begin{cases} 1 & \text{for} \quad k \geqslant j \\ -1 & \text{for} \quad k < j . \end{cases}$$

These variations are the quantities needed for improving the iteration formulas which can now be written as

$$\left.\begin{aligned} (3 + N_k^2)\, \delta\alpha_k + 2N_k\, \delta\beta_k + \sum_{i=0}^{\infty} \frac{\partial c_k}{\partial \alpha_i}\, \delta\alpha_i + \sum_{i=1}^{\infty} \frac{\partial c_k}{\partial \beta_i}\, \delta\beta_i = \varepsilon_k \\ 2N_k\, \delta\alpha_k + N_k^2\, \delta\beta_k + \sum_{i=0}^{\infty} \frac{\partial s_k}{\partial \alpha_i}\, \delta\alpha_i + \sum_{i=1}^{\infty} \frac{\partial s_k}{\delta \beta_i}\, \delta\beta_i = \delta_k \end{aligned}\right\} k = 0, 1, 2, \ldots, \qquad \text{(C)}$$

where

$$\left.\begin{aligned} \varepsilon_k &= -(3 + N_k^2)\, \alpha_k - 2N_k\beta_k - c_k \\ \delta_k &= -2N_k\alpha_k - N_k^2\beta_k - s_k \end{aligned}\right\} k = 0, 1, 2, \ldots$$

The quantities $\delta\alpha_k$ and $\delta\beta_k$ to be computed are corrections to the approximate values α_k and β_k.

In the applications the series appearing in this system of equations can be truncated at a relatively low order depending on the rate of convergence for the particular orbit being computed. The remaining equations are then uncoupled in pairs as before.

One must also consider the possibility that the symmetric matrix associated with this system of linear equations may be singular. The places where this has occurred in the applications is at those points on a natural family of orbits where the period becomes stationary. These cases occur near a resonance and are easily treated by holding the eccentricity fixed rather than the period as in the following discussion.

3. The Eccentricity of the Orbits

In many cases, such as resonance, the eccentricity is a convenient parameter to be used in following a family of orbits. For two-body motion the position vector may be written as

$$\mathbf{r} = a\mathbf{P}(\cos E - e) + a\sqrt{1-e^2}\,\mathbf{Q}\sin E,$$

where $\mathbf{P}$ and $\mathbf{Q}$ are the usual unit vectors, a is the semi-major axis, e is the eccentricity, and E is the eccentric anomaly. The deviations from circular motion may be expressed in α and β by considering

$$\mathbf{r} = (1+\alpha)\,\mathbf{r}_0 + \beta\mathbf{w},$$

where

$$\mathbf{r}_0 = a\mathbf{P}\cos g + a\mathbf{Q}\sin g$$

$$\mathbf{w} = -\,a\mathbf{P}\sin g + a\mathbf{Q}\cos g$$

with g as the mean anomaly. Equating the coefficients of $\mathbf{P}$ in the two expressions for $\mathbf{r}$, it follows that

$$\cos E - e = (1+\alpha)\cos g - \beta\sin g\,.$$

When $\cos E$ is expanded in a cosine series in g, the constant term is $-e/2$. Now

$$\alpha = \sum_{k=0}^{\infty} \alpha_k \cos k\theta$$

$$\beta = \sum_{k=1}^{\infty} \beta_k \sin k\theta\,,$$

and, for the motion to be periodic in the rotating system

$$\theta = g/p$$

for some positive integer p. Therefore

$$e = (\beta_p - \alpha_p)/3\,.$$

This two-body formula may be used to define the mean eccentricity of a perturbed orbit. For elliptic motion all the coefficients in the expansions of α and β can be computed from e using the Bessel functions, and in many cases this is a good initial approximation for the perturbed periodic orbit.

The eccentricity may now be prescribed, and this eliminates one equation in the system by putting

$$\beta_p = 3e_p + \alpha_p\,.$$

The subscript is placed on e to identify the coefficients with which it is associated. The equation which has been eliminated is replaced by another for determining the period or the mass ratio as in the following section.

4. The General Predictor-Corrector Formulas

The Equations (A) relating the coefficients can be written in matrix form as

$$DX = f(X)$$

where X is the vector of coefficients α_k and $\beta_k, f(X)$ is the vector of coefficients $-c_k$ and $-s_k$ which were seen to depend on X, and D is the matrix of coefficients multiplying α_k and β_k in Equations (A). It is assumed that the sequence of equations is truncated at some appropriate value of k. Note that the matrix D depends only on the ratio, $\nu = n/n'$, of mean motions for fixed l. However, f depends on m' as well as ν. Allowing first order corrections, the equations become

$$\left(D - \frac{\partial f}{\partial X}\right)\delta X + \left(\frac{\partial D}{\partial \nu} X - \frac{\partial f}{\partial \nu}\right)\delta\nu - \frac{\partial f}{\partial m'}\delta m' = f - DX .$$

The components of $\partial f/\partial \nu$ and $\partial f/\partial m'$ are simply the coefficients in the trigonometric expansions of $\Omega_{\alpha\nu}$, $\Omega_{\beta\nu}$, $\Omega_{\alpha m'}$ $\Omega_{\beta m'}$. With ν and m' fixed, the iteration (or corrector) formulas become

$$\delta X = \left(D - \frac{\partial f}{\partial X}\right)^{-1}(f - DX). \tag{D}$$

From an established orbit, the predictor formulas which give the variations of X with respect to the parameters are

$$\frac{\partial X}{\partial \nu} = \left(D - \frac{\partial f}{\partial X}\right)^{-1}\left(\frac{\partial f}{\partial \nu} - \frac{\partial D}{\partial \nu} X\right) \tag{E}$$

and

$$\frac{\partial X}{\partial m'} = \left(D - \frac{\partial f}{\partial X}\right)^{-1}\frac{\partial f}{\partial m'} \tag{F}$$

so that the same symmetric matrix is to be inverted in each case.

The eccentricity, e_p, is introduced as a parameter using the relation

$$\delta\beta_p = 3\,\delta e_p + \delta\alpha_p$$

wherever $\delta\beta_p$ occurs in the equations. Then, when δe_p is specified, either $\delta\nu$ or $\delta m'$ is obtained as part of the solution. This is accomplished by simple manipulations of the vectors and matrices in the above formulas. After the solution, the new value of β_p is computed from

$$\beta_p = 3\,e_p + \alpha_p .$$

5. The Linear Stability Analysis

Putting

$$\gamma = \mathrm{d}\alpha/\mathrm{d}g$$

and

$$\delta = \mathrm{d}\beta/\mathrm{d}g$$

the equations of motion can be written as

$$\begin{aligned}
\mathrm{d}\alpha/\mathrm{d}g &= \gamma \\
\mathrm{d}\beta/\mathrm{d}g &= \delta \\
\mathrm{d}\gamma/\mathrm{d}g &= 3\alpha + 2\delta + \Omega_\alpha \\
\mathrm{d}\delta/\mathrm{d}g &= -2\gamma + \Omega_\beta
\end{aligned}$$

so that the variational equations are

$$\mathrm{d}\Phi/\mathrm{d}g = A\Phi$$

with

$$A = \begin{bmatrix} 0 & 0 & 1 & 0 \\ 0 & 0 & 0 & 1 \\ 3 + \Omega_{\alpha\alpha} & \Omega_{\alpha\beta} & 0 & 2 \\ \Omega_{\alpha\beta} & \Omega_{\beta\beta} & -2 & 0 \end{bmatrix}$$

The functions $\Omega_{\alpha\alpha}$, $\Omega_{\alpha\beta}$ and $\Omega_{\beta\beta}$ are available in series form from the iteration process, so the variational equations can be integrated directly. For the present this is being done numerically using power series.

6. The Applications

Previous applications were mentioned in the introduction. Some additional orbits and families of orbits are given here as illustrations. These orbits all close after one revolution around the sun in the synodic system ($l=1$ in the definition of the trigonometric argument).

A typical stable orbit of the first kind is shown in the synodic coordinate system in Figure 1a. Such orbits are nearly circular and very easy to compute. The deviations from the circular reference motion are shown in the α, β plot of Figure 1b, and the series coefficients are given in Table I. The values are given to six decimals for purposes of illustration. The orbits were all computed to an accuracy of about 12 decimals.

As the ratio of mean motions, $\nu = n/n'$, is decreased toward the resonance value, $\nu=2$, the deviations from circular motion increase as shown in Figure 2a for the stable orbit at $\nu=2.001$. The α, β plot of Figure 2b for this orbit consists of two loops which are nearly identical. The series coefficients, given in Table II, are much larger now, but the rate of convergence is still good. This orbit is best described as a slowly rotating perturbed ellipse with eccentricity $e_2=0.388\,739$. The coefficients with odd subscripts would all be zero if the orbit were a true ellipse.

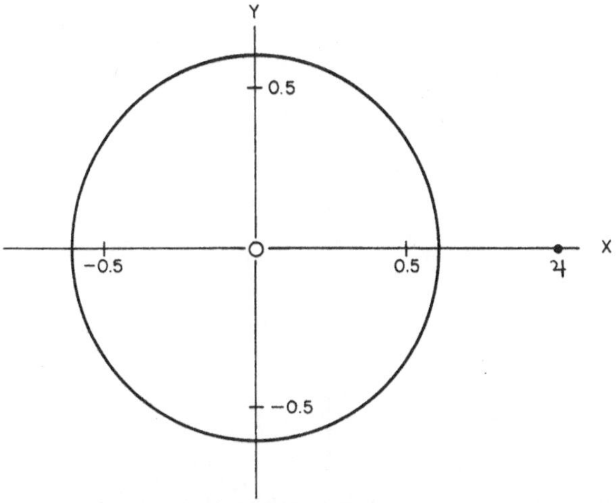

Fig. 1a. Nearly circular stable periodic orbit of the first kind with mean motion slightly greater than that of Hecuba resonance. $m' = 1/1\,047.35$, $\nu = n/n' = 2.1$, $e_2 = 0.011\,905$. (Synodic coordinate system with the sun at the origin and Jupiter at $X = 1$.)

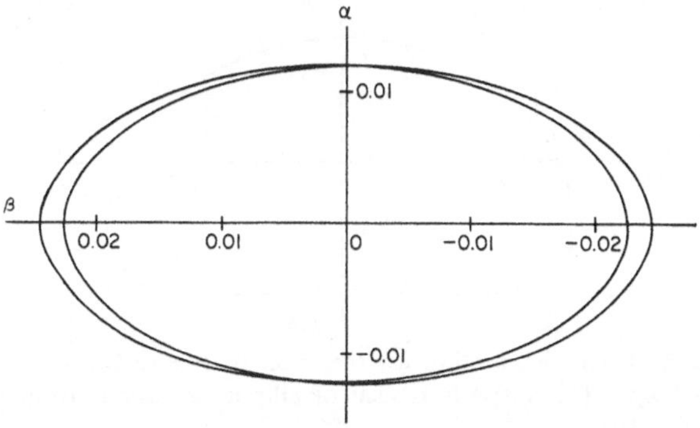

Fig. 1b. Deviations from circular motion for the orbit of Figure 1a. The two loops cross at the top of the figure, while at the bottom they are nearly tangent but do not cross. Uniform circular motion would give $\alpha = \beta = 0$ throughout the orbit. The sun is at $\alpha = -1$ while Jupiter moves on the circle $(1 + \alpha)^2 + \beta^2 = (a'/a)^2$.

TABLE I

Coefficients of the trigonometric series for α and β in the orbit of Figures 1a and 1b

k	$\alpha_k \cdot 10^6$	$\beta_k \cdot 10^6$
0	−122	0
1	595	−2 028
2	−12 158	23 556
3	−491	719
4	−60	202
5	−43	57
6	−18	23
7	−8	10
8	−4	4
9	−2	2
10	−1	1
11	−1	1

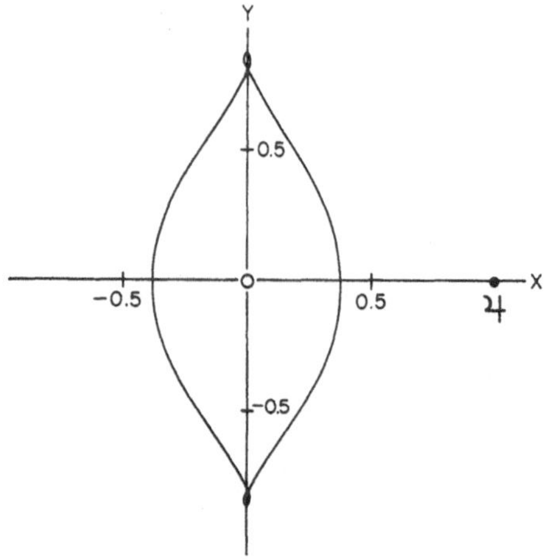

Fig. 2a. Synodic motion of a stable periodic orbit with mean motion very near resonance. $m' = 1/1047.35$, $\nu = 2.001$, $e_2 = 0.388\,739$.

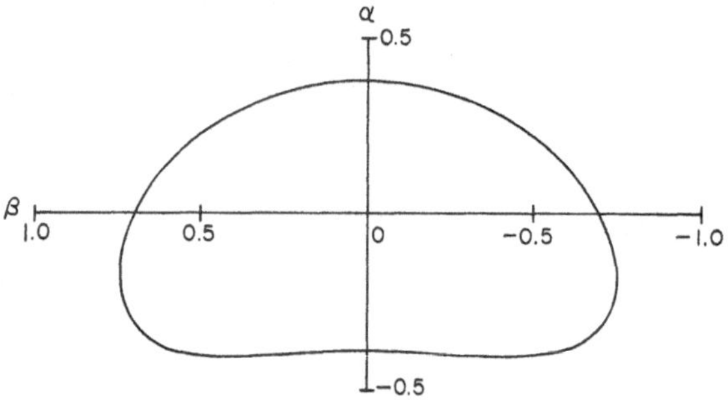

Fig. 2b. Deviations from circular motion for the orbit of Figure 2a. The figure consists of two nearly identical loops. The shape is typical of elliptic deviations from circular motion.

There are more periodic orbits which are nearly circular for $\frac{3}{2} < \nu < \frac{2}{1}$. The deviations from circular motion for one such orbit are shown in Figure 3 where $\nu = 1.6$. This orbit is also stable and has the series coefficients shown in Table III.

As the ratio of mean motions is reduced toward the $\frac{3}{2}$ resonance value, the eccentricity increases again. There is a stable orbit at the exact resonance shown in Figure 4a. Now the α, β plot, Figure 4b, consists of three nearly identical loops. The dominant coefficients (with subscripts which are multiples of 3) are given in Table IV. This orbit is a slightly perturbed ellipse with $e_3 = 0.453\,692$ and no secular motion of the perihelion.

Each of the above orbits was computed as a member of a natural family with $m' = 1/1047.35$ and using $\nu = n/n'$ as the parameter. For the orbit of Figure 4 this would not have been possible without using the partial derivatives in the iteration because of the zero divisors. For the orbits near the resonances, $\nu = \frac{2}{1}$ and $\nu = \frac{3}{2}$, the eccentricity, e_2 or e_3, would serve as a better parameter.

Each of the orbits except the one of Figure 4 has also been computed starting from circular two-body motion ($m' = 0$) and then increasing m' to the value for

TABLE II

Coefficients of the trigonometric series for the orbit of Figures 2a and 2b

k	$\alpha_k \cdot 10^6$	$\beta_k \cdot 10^6$
0	−75 926	0
1	−1 433	2 179
2	−410 408	755 808
3	198	334
4	68 305	28 584
5	149	128
6	18 652	13 566
7	72	65
8	6 429	5 342
9	35	33
10	2 429	2 137
11	17	16
12	968	879
13	8	8
14	400	370
15	4	4
16	170	159
17	2	2
18	73	69
19	1	1
20	32	31
22	14	14
24	6	6
26	3	3
28	1	1
30	1	1

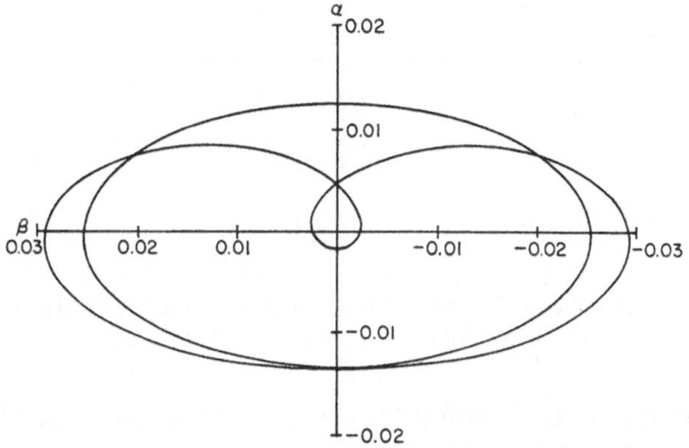

Fig. 3. Deviations from near circular motion of a stable periodic orbit between the $^2/_1$ and $^3/_2$ resonances. $m' = 1/1047.35$, $v = 1.6$, $e_2 = -0.008\,292$, $e_3 = 0.007\,791$.

Jupiter while holding v fixed. This is the technique for orbits of the first kind (Poincaré). The variations of e_2 with m' for three families of this type are shown in Figure 5.

Each of the orbits has also been computed by starting from elliptic two-body motion ($m' = 0$) and then increasing m' to the value for Jupiter while holding the

TABLE III

Coefficients of the trigonometric series for the orbit of Figure 3

k	$\alpha_k \cdot 10^6$	$\beta_k \cdot 10^6$
0	−235	0
1	1 842	−8 567
2	7 202	−17 674
3	−8 168	15 207
4	−1 234	1 935
5	−522	585
6	−185	271
7	−100	128
8	−56	66
9	−32	37
10	−19	21
11	−11	13
12	−7	8
13	−4	5
14	−3	3
15	−2	2
16	−1	1
17	−1	1
18	−1	1

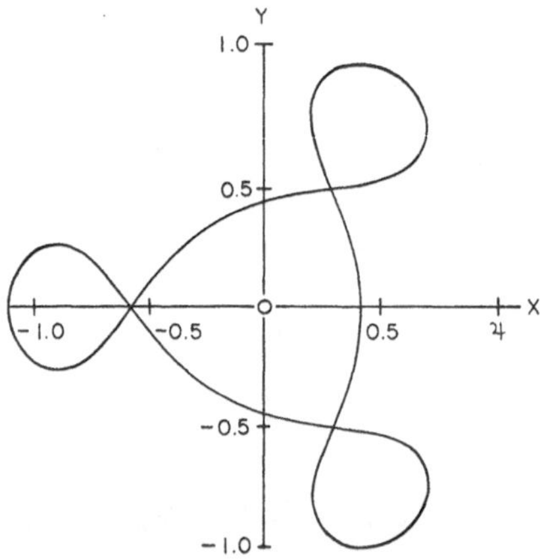

Fig. 4a. Synodic motion of a stable periodic orbit at the Hilda resonance. $m' = 1/1047.35$, $\nu = 1.5$, $e_3 = 0.453\ 692$.

eccentricity e_p fixed as in the method for orbits of the second kind (Poincaré). In this approach $p=2$ for the orbits of Figures 1 and 2 while $p=3$ for the orbit of Figure 4. As a numerical experiment the orbit of Figure 3 was computed by continuation from two different two-body elliptic orbits, one starting from $\nu=2$ and holding e_2 fixed, and the other starting from $\nu=\frac{3}{2}$ and holding e_3 fixed. The variations of ν with respect to m' for these two families along with three others are shown in Figure 6.

In following families of the types illustrated in Figures 5 and 6 there will be cases where the curves have vertical tangents (the mass m' is stationary at a point on the

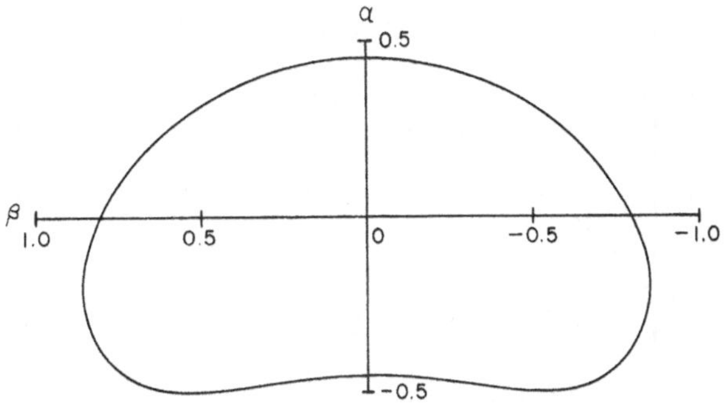

Fig. 4b. Deviations from circular motion for the orbit of Figure 4a. The figure consists of three nearly identical loops similar to those of Figure 2b except that the eccentricity is now larger.

TABLE IV

Dominant coefficients of the trigonometric series for the orbit of Figures 4a and 4b

k	$\alpha_k \cdot 10^6$	$\beta_k \cdot 10^6$
0	−103 710	0
3	−487 824	873 251
6	89 846	32 529
9	27 907	19 671
12	11 018	8 978
15	4 778	4 144
18	2 187	1 963
21	1 038	952
24	506	471
27	252	236
30	127	120
33	65	62
36	34	32
39	18	17
42	9	9
45	5	5
48	3	3
51	1	1
54	1	1

family). The mass m' is not a suitable parameter in a neighborhood of such a point, and the matrix involved in the computation becomes singular. In such cases the ratio of mean motions, ν, and the eccentricity, e_p, are specified, while m' is obtained in the solution along with the other coefficients. This is somewhat akin to prescribing a function and then searching for the problem which it solves, but in this case the procedure is necessary and justified for the purpose of continuing a family of orbits.

The relationship between the α, β and synodic coordinate systems is indicated in Figure 7.

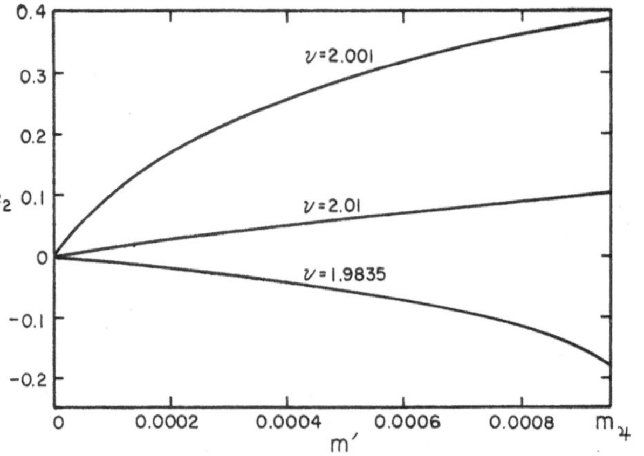

Fig. 5. Three families of the first kind. These families start from circular motion at $m'=0$ and are continued up to $m'=1/1047.35$ holding the ratio of mean motions, $\nu=n/n'$, at a fixed value. These curves show the variations of the eccentricity e_2 for three values of ν near 2. The cases $\nu=p/(p-1)$ for integral $p>1$ are singular for continuations of this type.

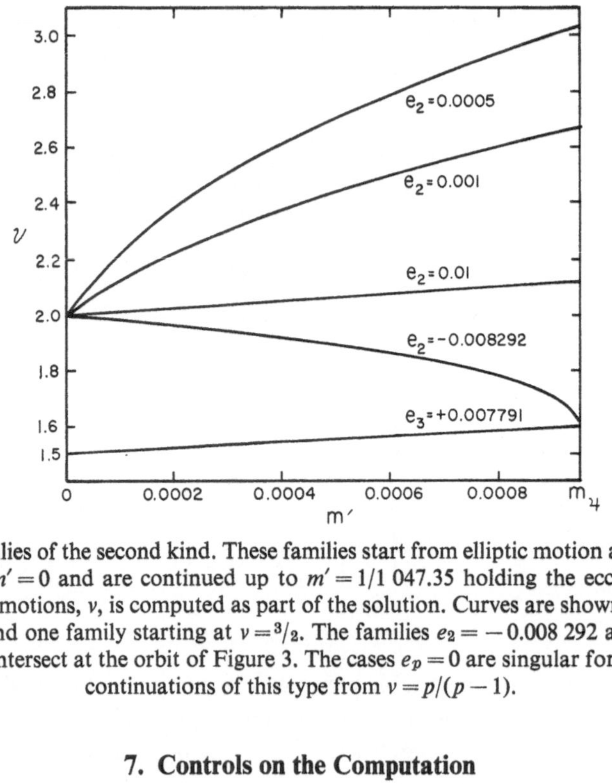

Fig. 6. Five families of the second kind. These families start from elliptic motion at exact resonance $\nu=p/(p-1)$ for $m'=0$ and are continued up to $m'=1/1\,047.35$ holding the eccentricity e_p fixed. The ratio of mean motions, ν, is computed as part of the solution. Curves are shown for four families starting at $\nu=2$ and one family starting at $\nu={}^3/_2$. The families $e_2=-0.008\,292$ and $e_3=0.007\,791$ intersect at the orbit of Figure 3. The cases $e_p=0$ are singular for continuations of this type from $\nu=p/(p-1)$.

7. Controls on the Computation

The Jacobi constant plays no role in the computational scheme used in this study. Its value is monitored as an indicator of numerical inaccuracies or difficulties.

The iteration process is continued until the residuals in the differential equations fall below an acceptable tolerance. The number of terms kept in the series and the

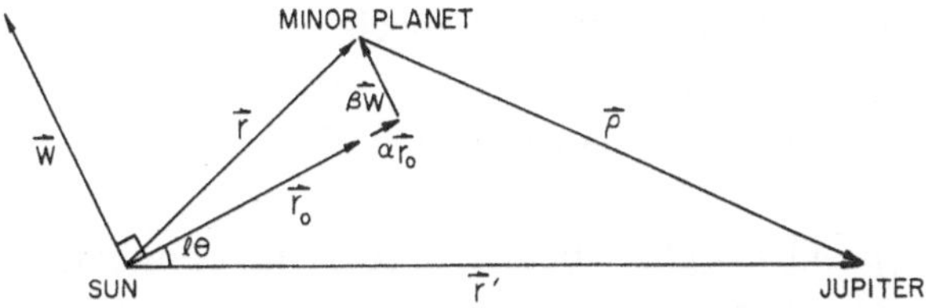

Fig. 7. Geometry of the coordinate systems.
$|\mathbf{r}_0| = a$, $|\mathbf{w}| = a$, $|\mathbf{r}| = a[(1+\alpha)^2 + \beta^2]^{1/2}$, $|\mathbf{r}'| = a'$, $l\theta = (n - n')\,t$,
$|\varrho| = a\{(1+\alpha)^2 + \beta^2 + (a'/a)^2 - 2(a'/a)\,[(1+\alpha)\cos l\theta - \beta\sin l\theta]\}^{1/2}$.

order of the matrix used in the iteration are determined for each case so as to give the required accuracy and convergence. Generally it was required that the last ten terms computed in the series be less than 0.5×10^{-12}.

When the iteration matrix becomes ill-conditioned, a change is made to a different independent parameter. No cases have been encountered where this presents any serious difficulty.

Several cases have been checked by comparison with numerical integration, and the expressions for the variations have been checked by numerical differencing of the series coefficients.

8. Comparison with Numerical Integration

The author is aware of no numerical integration program which has the facility for continuing families of orbits in all of the ways discussed here, or which gives the variations with respect to a parameter of a representation of the entire orbit such as those given in equations (E) and (F). On the other hand, the present technique is not suitable when the series convergence becomes too slow, nor is it useful for computing orbits which are not periodic. The necessary modifications for regularization have not yet been made.

It should be mentioned that, although the restricted problem of three bodies is extremely rich in natural families (Wintner) of periodic orbits, elements of many of these families have been discovered only by accident. Continuation from two-body motion as suggested by Poincaré and applied here provides a basis for the systematic exploration of many of these families of orbits.

The present method can be used in conjunction with a good numerical integration program taking full advantage of the special features of each.

9. Conclusion

The present study demonstrates the strength of a semi-analytic development in cases where small parameter methods fail.

The method described here has been justified mathematically for isolated periodic solutions (see Urabe (1965) and Stokes (1969)), but further work is needed for singular solutions (periodic orbits which are members of a natural family) as in this study. The indeterminancies are removed here by applying constraints. The value of a suitable parameter is specified, and the epoch of time is defined to occur at a perpendicular crossing of the axis for symmetric orbits. For non-symmetric orbits one might require one of the leading coefficients in the series to be zero to tie down the epoch.

In the restricted problem of four-bodies, once the motions of the three primaries are determined, the periodic solutions for the infinitesimal body are isolated, so these difficulties are not encountered. Nor would they be present in the reduced (or elliptic) problem of three-bodies.

There are many possible applications of this method in celestial mechanics problems, and several are being explored.

Acknowledgments

The author expresses his appreciation to Prof. K. Stumpff whose interests lead to the present study, to R. Kolenkiewicz for his cooperation in applying the method in the problem of four bodies, to E. Goodrich for the use of his numerical integration program in checking, to Prof. A. Stokes for many interesting discussions of previous works and current research, and to Prof. A. Deprit for using the results in an extensive joint survey of periodic orbits of the minor planet type.

References

Bennett, A. and Palmore, J.: 1968, 'A New Method for Constructing Periodic Orbits in Nonlinear Dynamical Systems', AAS Paper No. 68-085.

Carpenter, L. and Stumpff, K.: 1968, 'Über periodische Bahnen im eingeschränkten Dreikörperproblem und in der Nähe der Kommensurabilitäten vom Typus $(k+1)/k$', *Astron. Nachr.* **291**, No. 1.

Deprit, A. and Henrard, J.: 1967, 'Natural Families of Periodic Orbits', *Astron. J.* **72**.

Kolenkiewicz, R. and Carpenter, L.: 1967, 'Periodic Motion Around the Triangular Libration Point in the Restricted Problem of Four Bodies', *Astron. J.* **72**.

Kolenkiewicz, R. and Carpenter, L.: 1968, 'Stable Periodic Orbits about the Sun Perturbed Earth-Moon Triangular Points', *AIAA J.* **6**, No. 7.

Stokes, A.: 1969, 'On The Approximation of Periodic Solutions of Differential Equations', to appear in the *Proceedings of the Fifth International Conference on Nonlinear Oscillations*.

Urabe, M.: 1965, 'Galerkin's Procedure for Nonlinear Periodic Systems', *Arch. Rational Mech. Anal.* **20**.

Urabe, M. and Reiter, A.: 1966, 'Numerical Computation of Nonlinear Forced Oscillations by Galerkin's Procedure', *J. Math. Anal. Appl.* **14**, 107–140.

Discussion

G. Colombo: Did you analyze in the case of eccentric restricted three body problem the stability of periodic orbits with periods $\frac{1}{3}$, $\frac{3}{5}$, $\frac{1}{4}$, $\frac{2}{5}$ of the period of the primaries?

You should find instability for the case $\frac{1}{3}$, $\frac{3}{5}$ even when the mass ratio is very small and the eccentricity of the second body is very small.

A typical example is the case of Saturn-Mimas where the $\frac{1}{3}$ case corresponds to the location of the internal boundary of ring B, the $\frac{3}{5}$ case corresponds to the location of the Encke division.

L. Carpenter: In the elliptic problem I have so far computed families with the eccentricity of the primaries as the parameter using the sun-Jupiter mass ratio at the resonances $\frac{3}{2}$, $\frac{3}{1}$ and $\frac{5}{2}$. (My notation is the reciprocal of yours.) For the cases $\frac{3}{2}$ and $\frac{5}{2}$ orbits of both types, stable and unstable, were obtained. For the $\frac{3}{1}$ case all orbits I have obtained are of the unstable type. The work on the elliptic problem is being continued, and these are only the initial results.

E. Rabe: You found the existence of some selected periodic orbits in the elliptic restricted problem, for appropriate values of e' and e_s. Are these all for the same value of the mass ratio μ?

If yes, then I assume μ can be varied, leading to an infinite number of periodic orbits for μ-values adjacent to the one which you actually used.

L. Carpenter: The orbits I have shown were all for the Sun-Jupiter mass ratio.

Yes, they could be continued to different mass ratios. I have done this in the case of the $\frac{3}{2}$ resonance continuing the mass ratio up to the value for the earth-moon system and obtaining a stable periodic orbit for the elliptic problem for this case.

J. M. A. Danby: For your unstable periodic orbit were all the characteristic exponents of the unstable type?

L. Carpenter: One of the exponents was of the unstable type and the other was of the stable type. On the orbits of the elliptic problem which I have called unstable at least one of the characteristic exponents was of this type.

BIFURCATION LIMITS FOR THE EXISTENCE OF PERIODIC ORBITS*

SAMUEL PINES**

Analytical Mechanics Associates, Inc., Jericho, N.Y., U.S.A.

Abstract. A theory is presented for determining the limiting bounds for the existence of periodic orbits in the neighborhood of a known existing periodic orbit. A linear perturbation analysis is presented for finding the number of periodic orbits and their periods. Bifurcation limits are determined beyond which no periodic orbits can exist. The number of periodic orbits is shown to depend on the number of perturbing bodies, and the bifurcation limits are shown to result from a frequency coalescence. The results are general for dynamical systems with real masses and restoring potential forces.

1. Introduction

The existence of almost periodic orbits is an observed fact in celestial mechanics, and has led many investigators to analyze the conditions under which such orbits exist and the stability of these orbits. Techniques have been developed for finding strictly periodic solutions for several restricted problems by numerical iteration methods [1, 2, 3] and by analytical asymptotic expansion methods of the Hamilton-Jacobi equations [4, 5]. The literature in this single area of investigation is so extensive that one is apt to commit an error of omission in listing even principal researchers. I propose to comment on only a few, by way of illustration, as an introduction to the theory being presented here.

2. The Equations of Motion

We choose, as an inertial reference frame, the center of mass of the planetary system of N masses. Let the three-dimensional Cartesian coordinate position of the ith planet be R_i, and its equations of motion be given by

$$\ddot{R}_i = -\sum_{j\neq i}^{N} \mu_j \frac{R_{ji}}{r_{ji}^3}; \qquad i = 1, 2, \ldots, N, \tag{1}$$

where $$R_{ji} = R_i - R_j \tag{1a}$$

and $$r_{ji} = |R_{ji}| = r_{ij}. \tag{1b}$$

We take as the position vector of the point mass, whose periodic orbit is under study, R_v. Then the equations of motion of the point mass with respect to each one of the N planets is given by

$$\ddot{R}_{iv} = -\mu_i \frac{R_{iv}}{r_{iv}^3} - \sum_{j\neq i}^{N} \mu_j \left(\frac{R_{jv}}{r_{jv}^3} - \frac{R_{ji}}{r_{ji}^3}\right); \qquad i = 1, 2, \ldots, N. \tag{2}$$

* This research was performed under Contract NAS 5-11738 with NASA Goddard Space Flight Center, Greenbelt, Maryland.

** President.

G. E. O. Giacaglia (ed.), Periodic Orbits, Stability and Resonances, 210–215. All Rights Reserved.

The relative position vector of the ith planet from the jth planet can be given in terms of the relative position vector of the point mass from the ith and the jth planets.

$$R_{ji} = R_{jv} - R_{iv} \tag{3}$$

and

$$r_{ji} = |R_{jv} - R_{iv}|\,. \tag{3a}$$

Thus, each of the N-vector differential equations can be expressed entirely in terms of the relative position vectors of the point mass with respect to each of the N planets.

$$\ddot{R}_{iv} = -\left(\frac{\mu_i}{r_{iv}^3} + \sum_{j \neq i}^{N} \frac{\mu_j}{r_{ji}^3}\right) R_{iv} - \sum_{j \neq i}^{N} \mu_j \left(\frac{1}{r_{jv}^3} - \frac{1}{r_{ji}^3}\right) R_{jv}; \quad i = 1, 2, \ldots, N. \tag{4}$$

We are dealing with a set of N second-order vector differential equations in the N-vector variables R_{iv}. To find periodic solutions to these equations, we have recourse to a corresponding problem in linear vector differential equations with constant scalar coefficients.

3. The Equivalent Linear System

Consider a set of N second-order vector differential equations with constant coefficients

$$\ddot{P}_i = \sum_{j=1}^{N} a_{ij} P_j\,; \quad i = 1, 2, \ldots, N \tag{5}$$

with a_{ij} real scalar constants and either

$$a_{ij} = a_{ji} \quad \text{or} \quad a_{ij} \neq a_{ji}\,. \tag{5a}$$

Such a system has N distinct periodic solutions if, and only if, the matrix of the

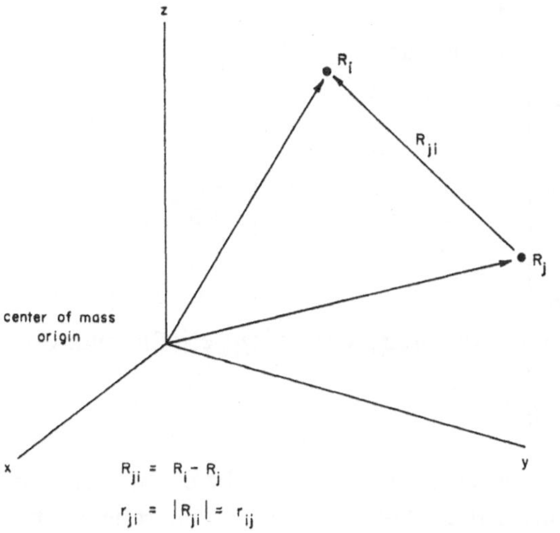

Fig. 1.

coefficients is diagonalizable. Thus, if the matrix, A, of coefficients, a_{ij}, has a Jordan canonical form of the type

$$AT = T\Lambda, \tag{6}$$

where

$$\Lambda = \begin{bmatrix} \omega_1^2 & 0 & 0 & & \\ 0 & \omega_2^2 & 0 & & \\ 0 & 0 & \omega_3^2 & & \\ & & & \ddots & \\ 0 & 0 & 0 & & \omega_N^2 \end{bmatrix} \tag{6a}$$

then the N ω_i^2 are the squares of the periodic frequencies and the eigenvalues of the matrix A. The matrix T consists of the set of distinct eigenvectors of the matrix A. Furthermore, T^{-1} exists. To obtain the corresponding periodic solution of the vectors R_{iv} in Cartesian coordinates, we proceed as follows.

Consider the solution of the single linear second-order vector differential equation

$$\ddot{Q}_i = \omega_i^2 Q_i \tag{7}$$

with initial conditions Q_{i_0} and $\dot{Q}_{i_0}$. Then

$$\begin{aligned} Q_i(t) &= \cos\omega_i t Q_{i_0} + \frac{\sin\omega_i t}{\omega_i}\dot{Q}_{i_0} \\ \dot{Q}_i(t) &= -\omega_i \sin\omega_i t Q_{i_0} + \cos\omega_i t \dot{Q}_{i_0}\,. \end{aligned} \tag{8}$$

For every set of vector values Q_{i_0} and $\dot{Q}_{i_0}$, there exists a periodic solution $Q_i(t)$ and $\dot{Q}_i(t)$ given by Equation (8).

The matrix T is an $N \times N$ matrix of the eigenvectors of A, where each vector t_i satisfies the matrix equation

$$At_i = \omega_i^2 t_i\,. \tag{9}$$

The elements of the N-vector t_i are

$$t_i = \begin{bmatrix} t_{1i} \\ t_{2i} \\ t_{3i} \\ \vdots \\ t_{Ni} \end{bmatrix}. \tag{9a}$$

The relationship of the vectors P_i to the N Q_i periodic solutions is given by

$$P_i = \sum_{j=1}^{N} t_{ij} Q_j \tag{10}$$

In the above, it is not necessary for the ω_i^2 to be distinct; it is only necessary that the matrix A be diagonalizable. Thus, we could theoretically have periodic *resonant* frequencies which are still stable.

Instability occurs when A is not diagonalizable, and the Jordon form is of the type

$$\Lambda = \begin{bmatrix} \omega_1^2 & 1 & 0 & & \\ 0 & \omega_1^2 & 0 & & \\ 0 & 0 & \omega_2^2 & & \\ & & & \ddots & \\ & & & & \omega_N^2 \end{bmatrix} \tag{11}$$

In this case, the N Q_i solutions are not separable, and two of them are irreducibly coupled. For the nondiagonalizable case, we have

$$\begin{aligned} \ddot{Q}_1 &= \omega_1^2 Q_1 + Q_2 \\ \ddot{Q}_2 &= \omega_1^2 Q_2 \end{aligned} \tag{12}$$

The solutions are nonperiodic and grow in time like $t \sin \omega_1 t$. In this case, the energy from one mode is absorbed in the second, and the motion of the second grows beyond bounds while the first vanishes. This phenomenon can occur in physical systems, and is commonly called bifurcation due to frequency coalescence. A. Deprit [4] gives evidence of this abrupt disappearance of the periodic solution in his description of the behavior of the Trojan manifold in the Earth-Moon system.

4. Some Inferences and Conclusions

If we investigate the conditions under which bifurcation can occur, we may arrive at several conclusions.

(1) We know that every symmetric matrix is diagonalizable. It follows that, if $a_{ij} = a_{ji}$, we can never run into this mode of instability.

In examining the condition of the coefficients in our case, we find that the corresponding values of a_{ij} are given by

$$\begin{aligned} a_{ij} &= -\mu_j \left(\frac{1}{r_{jv}^3} - \frac{1}{r_{ji}^3} \right) \\ a_{ji} &= -\mu_i \left(\frac{1}{r_{iv}^3} - \frac{1}{r_{ij}^3} \right) \end{aligned} \tag{13}$$

and, in general,

$$a_{ij} \neq a_{ji} \tag{14}$$

Thus, diagonalizability is not assured, and the bifurcation phenomenon can occur with its corresponding disappearance of periodic orbits.

(2) When periodic orbits do exist, that is, when the matrix A is diagonalizable, then the number of solutions is N. This means that, in general, we should be able to find N distinct periodic orbits corresponding to each set of the coefficients a_{ij}.

The significant item in this regard is that it points out a serious limitation in the Hamilton-Jacobi perturbation technique, which uses periodic orbits of a reduced problem of $N-1$ planetary bodies in order to start searching for periodic orbits in the N-body case. It is plain that such a method [5] can, at best, produce only $N-1$ perturbed periodic solutions, whereas, in general, we must expect N solutions to exist.

In what sense can we regard the linear approximation given by Equation (5) as representative of our Equation (4)? The correspondence is arrived at thus:

Imagine that we have a periodic solution of Equation (4). Then, we can replace the coefficients of R_{iv} by the average values over the period τ_k as follows:

$$a_{ij}(\tau_k) = \frac{\int_0^{\tau_k} - \mu_j \left(\frac{1}{r_{jv}^3} - \frac{1}{r_{ji}^3} \right) \mathrm{d}t}{\tau_k}$$

$$a_{ii}(\tau_k) = \frac{\int_0^{\tau_k} - \left(\frac{\mu_i}{r_{iv}^3} + \sum_{j \neq i}^{N} \frac{\mu_j}{r_{ji}^3} \right) \mathrm{d}t}{\tau_k} \tag{15}$$

Furthermore, since the coefficients, a_{ij}, are to be considered as constants over each period τ_k, and since there are N distinct periods, it follows that we must expect the nonlinear functions, whose integrals are to form the equivalent constants a_{ij} over the N periods, themselves to be oscillatory and bounded. We have the conditions

$$\frac{\int_0^{\tau_k} - \mu_j \left(\frac{1}{r_{jv}^3} - \frac{1}{r_{ji}^3} \right) \mathrm{d}t}{\tau_k} = \frac{\int_0^{\tau_p} - \mu_j \left(\frac{1}{r_{iv}^3} - \frac{1}{r_{ji}^3} \right) \mathrm{d}t}{\tau_p} \tag{16}$$

over each period τ_k.

To understand the process through which the bifurcation limits are reached, we may imagine a search for periodic solutions in the neighborhood of a known, existing, periodic solution. Let R_{iv} be a known, existing, periodic solution. We add a small vector L to R_{iv}. Then the new periodic solution will be R'_{iv}.

$$R'_{iv} = R_{iv} + L\,; \qquad i = 1, 2, \ldots, N \tag{17}$$

and

$$r'_{iv} = |R_{iv} + L|\,. \tag{17a}$$

This variation causes a change in the N^2 a_{ij} elements. In the limit, as L increases, we achieve a condition where the matrix A loses its diagonalizability property, and we may expect two of the periodic solutions to vanish simultaneously. Periodic orbits for these two solutions beyond this limiting L vector will no longer exist.

The author is presently engaged in investigating numerical techniques for utilizing the ideas presented herein.

References

[1] Schechter, H. B.: 1967, 'Three-Dimensional Nonlinear Stability Analysis of the Sun-Perturbed Earth-Moon Equilateral Points', Paper No. 67-566, *AIAA Symposium.*
[2] Kolenkiewicz, R. and Carpenter, L.: 1967, 'Periodic Motion Around the Triangular Libration Point in the Restricted Problem of Four Bodies', *Astron. J.* **72**, 180–183.
[3] Steffensen, J. F.: 1956, *Kgl. Danske Videnskab. Selskab, Mat.-Fys. Medd.*, **30**, No. 18.
[4] Deprit, A. *et al.*: 1967, 'The Trojan Manifold in the System Earth-Moon', *Monthly Notices Roy. Astron. Soc.* **137**, 311–335.
[5] Breakwell, J. V. and Pringle, R., Jr.: 1966, *Progress in Astronautics*, Vol. 17, Academic Press, Inc., New York, pp. 55–73.

ON THE ACCURACY IN THE NUMERICAL COMPUTATION OF ORBITS*

PEDRO E. ZADUNAISKY
Instituto Torcuato Di Tella, Buenos Aires, Argentina

1. Introduction

Any system of ordinary differential equations of N-th order with initial conditions given at the point $t=a$, can be written in the general form

$$\begin{aligned} x' &= f(t, x) \\ x(a) &= x_0, \end{aligned} \tag{1}$$

where x and f are vectors of N functions and x_0 is a vector of N constants. Putting $t_n = a + nh$ we assume that to solve the system (1) it is used a recursive algorithm of the general form

$$x_{n+1} = \phi(x_n, x_{n-1}, \ldots, x_{n-k}; h). \tag{2}$$

Applying this algorithm one obtains numbers $\tilde{x}_n$ and after n steps the accumulated error is the difference

$$\varepsilon_n^{(0)} = x(t_n) - \tilde{x}_n. \tag{3}$$

It can be written in the form

$$\varepsilon_n^{(0)} = [x(t_n) - x_n] + [x_n - \tilde{x}_n], \tag{4}$$

where the first difference is the *truncation* error ε_n and the second difference is the *round-off* error r_n.

In an earlier paper (Zadunaisky, 1966) we have presented a method for estimating the *total* accumulated errors $\varepsilon_n^{(0)}$ which consists in the following:

First, we may determine a vector $P(t)$ of empirical functions $P_i(t)$, $i=1, 2, \ldots, N$ that can be, for instance, polynomials or other simple functions adjusted in order to represent in the best possible way the respective components of $\tilde{x}_n$. Now we can establish a new system of equations, that we have called the 'Pseudo-Problem', of the following form:

$$\begin{aligned} z' &= f(t, z) + P'(t) - f(t, P(t)) \\ z(a) &= P(a). \end{aligned} \tag{5}$$

Evidently the solution of this problem is the function vector $z = P(t)$. If we integrate numerically the system (5) we shall obtain again numbers $\tilde{z}_n$ and the accumulated error after n steps will be, exactly,

$$\varepsilon_n^{(1)} = P(t_n) - \tilde{z}_n. \tag{6}$$

* Some parts of the present investigation were developed while the author was working at the Laboratory for Theoretical Studies of the Goddard Space Flight Center in Greenbelt, Md. as a Senior Research Associate of the National Academy of Sciences of the U.S.A.

G. E. O. Giacaglia (ed.), Periodic Orbits, Stability and Resonances, 216–227.

If the interpolating functions $P(t)$ are properly adjusted then the Pseudo-Problem should not differ too much from the original problem and the behaviour of the propagated errors in both problems should be rather similar. In other words $\varepsilon_n^{(1)}$ can be adopted as a good estimation of $\varepsilon_n^{(0)}$. In our earlier paper we have found that the conditions under which this assertion is true are

$$\left.\begin{array}{l} \|\tilde{x}_n - P(t_n)\| \\ \|x_n^{(p+1)} - P^{(p+1)}(t_n)\| \end{array}\right. \leqslant \delta = \mathrm{O}(h^2) \tag{7}$$

where $\tilde{x}_n^{(p+1)}$ is a numerical approximation for the $(p+1)$-th derivative of x.

Our proof was based on the results of the asymptotic theory (when $h \to 0$) of error propagation which can be outlined as follows (see P. Henrici, 1963). First, it is necessary to make some assumptions about the local behaviour of errors that can be summarized in the following form:

The *local truncation error* is expressed in the form

$$T(x) = -C\frac{\mathrm{d}^{(p+1)}x}{\mathrm{d}t^{(p+1)}}, \tag{8}$$

where C is a constant.

The *expected value of the local round-off* error is given by

$$E[\varepsilon_n] = \mu Q(t_n), \tag{9}$$

where μ = constant and $Q(t)$ is a given function, and the *covariance matrix of the local round-off error* is

$$\operatorname{covar}(\varepsilon_n) = \sigma^2 Q(t_n) \tag{10}$$

where δ is a constant and $C(t)$ is also a given function.

Under these hypotheses the accumulated errors may be estimated as follows: First, it is necessary to establish the *fundamental system of variational equations*

$$\begin{array}{lll} \text{(I)} & w' = G(t)\,w + T(x), & w(a) = 0 \\ \text{(II)} & m' = G(t)\,m + mG^*(t) + Q(t), & m(a) = 0 \\ \text{(III)} & V_\nu' = \lambda_\nu G(t)\,V_\nu + V_\nu\lambda_\nu G^*(t) + C(t), & V(a) = 0 \\ & (\nu = 1, 2, \ldots, M) & \end{array} \tag{11}$$

where $G(t)$ is the Jacobian Matrix $(\partial f(t, x(t))/\partial x)$ and $G^*(t)$ its conjugate transpose and the parameters λ_ν and the number M depend fundamentally on the recursive algorithm (2). Those variational equations are of the non-homogeneous linear type; to solve them it is necessary to solve first the corresponding homogeneous equations

$$\begin{aligned} u' &= G(t)\,u \\ u_\nu' &= \lambda_\nu G(t)\,u_\nu \end{aligned} \tag{12}$$

and then obtain the unknown functions w, m and V_ν by quadratures:

$$\text{(1)} \qquad w = \int_a^t u(t, \tau)\, T(\tau)\, \mathrm{d}\tau$$

$$\text{(II)} \qquad m = \int_a^t u(t,\tau)\, Q(\tau)\, u^*(t,\tau)\, \mathrm{d}\tau \tag{13}$$

$$\text{(III)} \qquad V_\nu = \int_a^t u_\nu(t,\tau)\, C(\tau)\, u_\nu^*(t,\tau)\, \mathrm{d}\tau$$

Finally, for the errors accumulated after n steps of integration of the original problem the following estimations are obtained:

Truncation error:

$$\varepsilon_n = h^p w(t_n) + \mathrm{O}(h^{p+1})$$

Primary component of the round-off error:

(a) Expected Value $E[r_n] = \dfrac{\mu}{kh}[m(t_n) + \mathrm{O}(h \log h)]$

(b) Covariance Matrix $\operatorname{covar}(r_n) = \dfrac{\sigma^2}{h}\left[\sum_{\nu=1}^{M} CV_\nu(t_n) + \mathrm{O}(h \log h)\right].$

In the application of this procedure appears the difficulty that the actual solution $x(t)$ of the problem, which figures as an argument in the matrix $G(t)$, is, of course, not known. Neither it is known the $(p+1)$-th derivative of x that appears in (7). What can be done is to substitute for $x(t)$ and

$$\frac{\mathrm{d}^{(p+1)}x}{\mathrm{d}t^{(p+1)}}$$

their numerical approximations $\tilde{x}_n$ and $\tilde{x}_n^{(p+1)}$ obtained in the course of the numerical integration of the problem and then to apply also numerically the method of the variational equations.

The essential result we obtained in our earlier paper is that the method of the 'Pseudo-Problem' is entirely equivalent to the method of the variational equations provided that the interpolating function $P(t)$ fulfils the conditions (7). However the 'Pseudo-Problem' method is much simpler and it does not require any assumptions like (8), (9) and (10), about the local behaviour of errors.

We have applied successfully our method and in what follows we give a detailed description of several examples. In two of these examples the actual results were known analytically so that we could find the real errors accumulated in the process of numerical integration and check the errors estimated by our method.

2. Examples

EXAMPLE 1. JACOBIAN ELLIPTIC FUNCTIONS

The original problem is here:

$$\begin{aligned} x_1' &= x_2 x_3, & x_1(0) &= 0 \\ x_2' &= -x_1 x_3, & x_2(0) &= 1 \\ x_3' &= -\tfrac{1}{2} x_1 x_2, & x_3(0) &= 1. \end{aligned}$$

The actual solution of this problem is known to be the set of Jacobian Elliptic Functions

$$x_1 = sn(t, k)$$
$$x_2 = cn(t, k)$$
$$x_3 = dn(t, k)$$

for the modulus $k = 2^{-1/2}$. These functions are given in tables or can be calculated by series with any desired accuracy.

For the numerical integration of the original problem we used Runge-Kutta's method, where the truncation error is $T = O(h^5)$, carrying 16 significant digits. Subtracting the numerical results from the corresponding values calculated from the analytic formulas we obtained the real errors.

To apply our method of estimation of errors we applied first Bessel's formula of interpolation to find three polynomials $P_1(t)$, $P_2(t)$ and $P_3(t)$ of 14th degree which represented the numerical results $\tilde{x}_1$, $\tilde{x}_2$ and $\tilde{x}_3$ on 15 equidistant points of the total interval of integration. With these polynomials we formed the Pseudo-Problem

$$z_1' = z_2 z_3 + P_1'(t) - P_2(t) P_3(t), \qquad z_1(0) = P_1(0)$$
$$z_2' = - z_1 z_3 + P_2'(t) + P_1(t) P_3(t), \qquad z_2(0) = P_2(0)$$
$$z_3' = - \tfrac{1}{2} z_1 z_2 + P_3'(t) + \tfrac{1}{2} P_1(t) P_2(t), \quad z_3(0) = P_3(0).$$

Applying again Runge-Kutta's method we integrated this system obtaining numerical results $(\tilde{z}_1)_n$, $(\tilde{z}_2)_n$ and $(\tilde{z}_3)_n$, and finally estimated the errors accumulated in the numerical solution of the original problem by the differences:

$$(\varepsilon_1)_n = P_1(t_n) - (\tilde{z}_1)_n$$
$$(\varepsilon_2)_n = P_2(t_n) - (\tilde{z}_2)_n$$
$$(\varepsilon_3)_n = P_3(t_n) - (\tilde{z}_3)_n$$

The results are given in Figure 1 and Table I.

We can observe that in all cases the error was correctly estimated not only in its order of magnitude and sign but in two or three figures of the error itself. This offers the possibility of using these estimations of the errors to correct the results of the numerical integration. This is precisely what we have done in the following example.

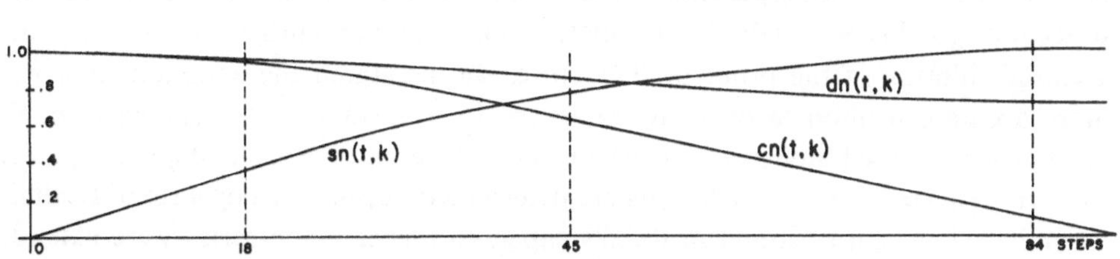

Fig. 1.

TABLE I

Jacobian elliptic functions

	$dn(t, k)$		
Error	Est. -0.585×10^{-10}	-0.425×10^{-10}	$+0.5383 \times 10^{-9}$
	Real -0.586	-0.428	$+0.5379$
	$cn(t, k)$		
Error	Est. -0.2969×10^{-9}	-0.1125×10^{-8}	-0.1861×10^{-8}
	Real -0.2963	-0.1126	-0.1862
	$sn(t, k)$		
Error	Est. $+0.949 \times 10^{-9}$	$+0.1125 \times 10^{-8}$	$+0.1940 \times 10^{-9}$
	Real $+0.950$	$+0.1126$	$+0.1948$

EXAMPLE 2. A PERIODIC SOLUTION OF THE RESTRICTED PROBLEM OF THREE BODIES

The original problem is here

$$p'' = 2Nq' - Mp(r^{-3} - 1) - (p - 1)(s^{-3} - 1) = \phi_1(p, q)$$
$$q'' = -2Np' - Mq(r^{-3} - 1) - q(s^{-3} - 1) = \phi_2(p, q)$$
$$M = 1/1047.355 \qquad N^2 = 1 + M$$

It is worth noting that this system has not the form of (1). However, we have found that also for this case the method of the Pseudo-Problem is valid.

These equations belong to the case where the primaries are Jupiter and the Sun and the motion of the asteroid is referred to a rotating system centered at Jupiter.

Following Rabe (1961), Goodrich (1966) found that the initial conditions

$$p(0) = 0.4 \qquad q(0) = 1.039\ 2304\ 8454\ 1326$$
$$p'(0) = 0.3734\ 1677\ 1713\ 7303, \quad q'(0) = 1.072\ 0524\ 2989\ 4477$$

correspond to a periodic solution where the period $T \cong 6.3$.

Deprit and Palmore (1966) have found that this orbit belongs to a family of the stable type. However there is a close approach to the Sun for $5 < t < 6$ (see Figure 2) where the numerical process of integration becomes unstable and it should require the necessity of regularizing time and coordinates at least for a certain interval. In the present case we did not regularize the problem letting the errors grow freely in order to see how good is our method of estimation. On the other hand, we do not have here a closed solution of the problem. Therefore, we integrated the equations first by a more accurate method in order to use its results as a standard for comparison. For this purpose we used Steffensen's method where the coordinates p and q are expanded as Taylor's Series and the coefficients are determined at each step by a set of recursive formulas which can be found in Rabe's paper. In this case the series were extended to 16 terms and the calculations were made carrying 16 significant figures. Further-

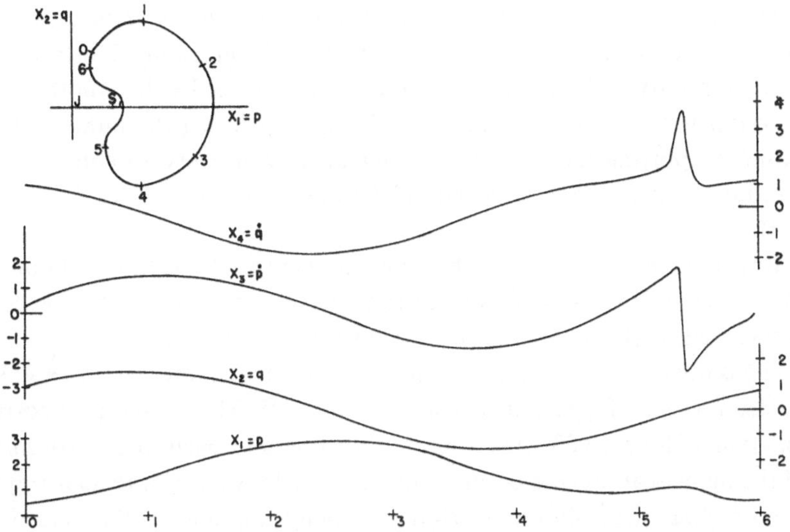

Fig. 2. A periodic solution of the restricted problem of three bodies.

more we applied our method and estimated that in no place along a complete period, the errors were larger than 10^{-12}.

For the second and less accurate method of integration we reduced first the original problem to the equivalent one

$$x_1' = x_3 \quad x_3' = 2Nx_4 - Mx_1(r^{-3} - 1) - (x_1 - 1)(s^{-3} - 1) = \phi_3(x_1 x_2 x_3 x_4)$$
$$x_2' = x_4 \quad x_4' = -2Nx_3 - Mx_2(r^{-3} - 1) - x_2(s^{-3} - 1) = \phi_4(x_1, x_2, x_3, x_4)$$
$$r^2 = x_1^2 + x_2^2 \quad s^2 = (x_1 - 1)^2 + x_2^2$$

where evidently

$$x_1 = p \quad x_2 = q \quad x_3 = p' \quad x_4 = q'$$

and which has now the form of (1).

The corresponding Pseudo-Problem is

$$z_1' = z_3 + P_1'(t) - P_3(t), \quad z_3' = \phi_3(z_1, z_2, z_3, z_4) + P_3'(t) - \phi_3(P_1, P_2, P_3, P_4)$$
$$z_2' = z_4 + P_2'(t) - P_4(t), \quad z_4' = \phi_4(z_1, z_2, z_3, z_4) + P_4'(t) - \phi_4(P_1, P_2, P_3, P_4)$$
$$z_1(0) = P_1(0), \quad z_2(0) = P_2(0), \quad z_3(0) = P_3(0), \quad z_4(0) = P_4(0)$$

For the numerical integration we applied Runge-Kutta's method carrying 16 significant digits. Besides, we improved systematically the solution by the following procedure.

After calculating the first 15 steps we applied our method of error estimation and correspondingly corrected all the results obtained from Runge-Kutta's formulas. After that we restarted the numerical solution and proceeded for another 15 steps. Then we estimated the errors in this second stage, corrected the numerical solution and continued applying the procedure systematically at every 15 consecutive steps. The procedure is illustrated graphically in Figure 3 and our results are summarized in Table II.

What we call there a 'real' error is the difference between the solution obtained by the first accurate method (Steffensen's method) and the solution obtained by the second and less accurate method (Runge-Kutta plus systematic corrections).

The table is divided in two parts; the first one covers the interval $0 \leqslant t \leqslant 4.95$ where the integration was performed at regular steps $h = 0.05$. The second part corresponds to the interval $4.95 < t \leqslant 5.50$ where occurs the close approach of the asteroid to the Sun and the numerical process becomes unstable; here we used a regular step 5 times smaller but anyway the solution deteriorates rapidly and at $t = 5.50$ practically all the accurate figures are lost.

The estimation of errors is very good in the first part. In the second part it is generally good except for a few places where the errors are overestimated by one or two orders of magnitude and two places where the errors were slightly underestimated.

EXAMPLE 3. LINEARIZED EQUATIONS OF THE RESTRICTED PROBLEM OF THREE BODIES

The original problem is in this case

$$x'' - 2y' = 2x$$
$$y'' + 2x' = 2y$$

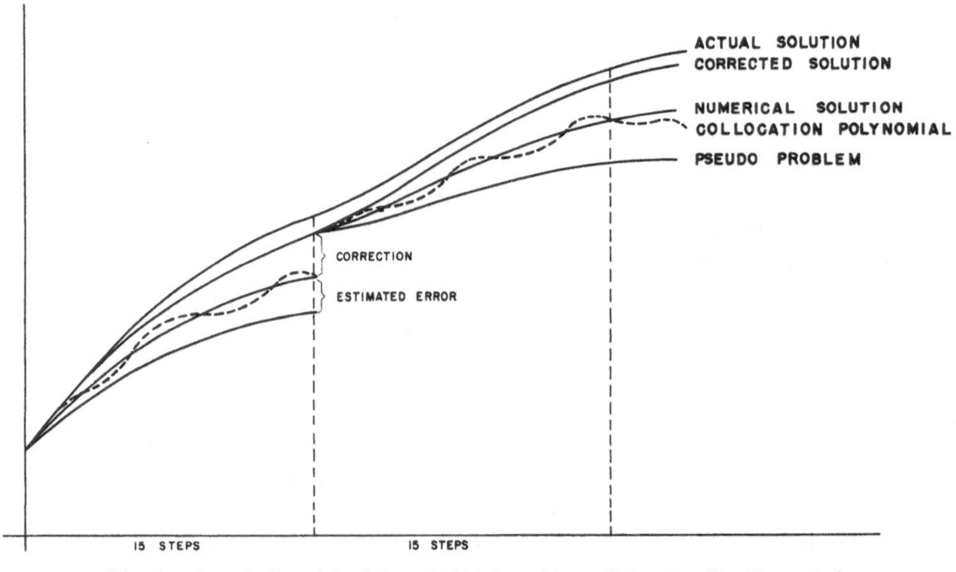

Fig. 3. A periodic orbit of the restricted problem of three bodies. Systematic corrections of the numerical solution.

TABLE II

A periodic orbit of the restricted problem of three bodies. Estimated and real accumulated errors

		$h=0.05$			$h=0.01$		
No. of steps		20	60	99	124	139	154
Time		1.00	3.00	4.95	5.20	5.35	5.50
$X_1(p)$ Error	Est.	$+0.2740\times10^{-7}$	-0.93×10^{-8}	$+0.147\times10^{-6}$	$+0.9\times10^{-9}$	$+0.1\times10^{-5}$	$+0.7\times10^{-3}$
	Real	$+0.2742$	-0.92	$+0.146$	$\times 0.4$	$+0.2$	$+0.3$
$X_2(q)$ Error	Est.	$+0.1803\times10^{-7}$	-0.87×10^{-8}	$+0.231\times10^{-6}$	$+0.8\times10^{-8}$	$+0.100\times10^{-4}$	$+0.15\times10^{-2}$
	Real	$+0.1805$	-0.86	$+0.230$	$+0.6$	$+0.004$	$+0.02$
$X_3(\dot{p})$ Error	Est.	$+0.486\times10^{-7}$	$+0.919\times10^{-8}$	$+0.797\times10^{-6}$	$+0.1\times10^{-7}$	$+0.3\times10^{-4}$	$+0.23\times10^{-2}$
	Real	$+0.488$	$+0.922$	$+0.795$	$+0.2$	$+0.9$	$+0.08$
$X_4(\dot{q})$ Error	Est.	-0.288×10^{-7}	$+0.50\times10^{-8}$	-0.1036×10^{-5}	-0.07×10^{-6}	$+0.169\times10^{-4}$	$+0.17\times10^{-1}$
	Real	-0.287	$+0.48$	-0.1035	-0.14	$+0.004$	$+0.03$

being its general solution

$$x = e^t [C_1 \sin t - C_2 \cos t] + e^{-t} [C_3 \sin t - C_4 \cos t]$$
$$y = e^t [C_1 \cos t + C_2 \sin t] + e^{-t} [C_3 \cos t + C_4 \sin t].$$

With the initial conditions

$$x(0) = -2, \quad x'(0) = 2, \quad y(0) = 0, \quad y'(0) = 2$$

we have the particular solution

$$x = -2 e^{-t} \cos t$$
$$y = 2 e^{-t} \sin t.$$

The trajectory defined by these equations is a logarithmic spiral which tends asymptotically to the origin of coordinates.

In this case the numerical process of integration is highly unstable because after the first steps the errors begin to accumulate, the numerical results step out the particular solution and the increasing exponential terms of the general solution start affecting drastically the entire calculation (see Figures 4a and 4b).

The numerical process was performed not on the original equations but on the equivalent problem

$$x_1' = x_2$$
$$x_2' = x_3$$
$$x_3' = x_4$$
$$x_4' = -4x_1$$

with the initial conditions

$$x_1(0) = 0, \quad x_2(0) = 1, \quad x_3(0) = -2, \quad x_4(0) = 2$$

and the new variables being related to the old ones by the expressions

$$x_1 = \tfrac{1}{2} y \qquad x_3 = x$$
$$x_2 = \tfrac{1}{2} y' \qquad x_4 = x'.$$

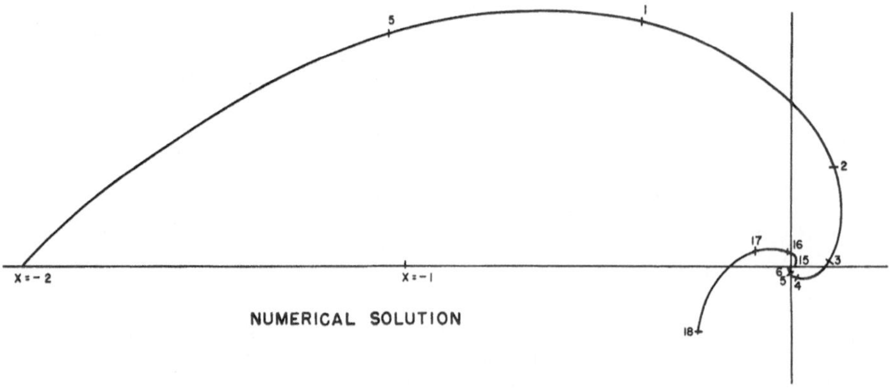

Fig. 4a.

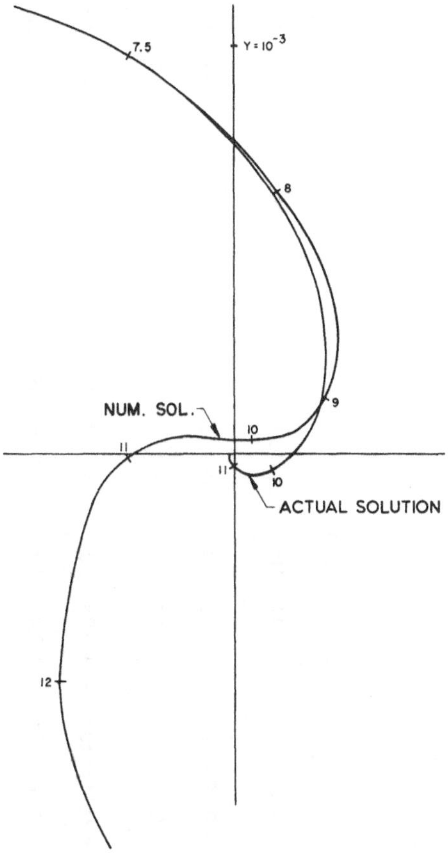

Fig. 4b.

We used three different methods of integration as follows:

Method I: Taylor's Series with 12 terms ($T = O(h^{13})$), carrying 8 significant digits.

Method II: Runge-Kutta ($T = O(h^5)$), carrying 9 significant digits.

Method III: Taylor's Series with 8 terms ($T = O(h^9)$), carrying 16 significant digits.

As interpolating functions we used in all cases collocation polynomials of fifth degree on every 6 consecutive points and the equations of the Pseudo-Problem were

$$z_1' = z_2 + P_1'(t) - P_2(t)$$
$$z_2' = z_3 + P_2'(t) - P_3(t)$$
$$z_3' = z_4 + P_3'(t) - P_4(t)$$
$$z_4' = -4z_1 + P_4'(t) + 4P_1(t)$$
$$z_1(0) = P_1(0), \quad z_2(0) = P_2(0), \quad z_3(0) = P_3(0), \quad z_4(0) = P_4(0).$$

The results obtained, shown in Table III, reveal that the solution is mostly affected by the accumulation of rounding-off errors. In fact, the results obtained with *Method II*, which has a much larger truncation error than *Method I*, but was performed with one more significant digit, are considerably better. Still much better are the results of *Method III* where the truncation error had an intermediate size but the calculations were carried with 16 significant digits.

TABLE III

Linearized equations of the restricted problem of 3 bodies. Estimated and real accumulated errors

No. of steps		80			200			320	360
Time		4.0			10.0			16.0	18.0
Method		I	II	III	I	II	III	III	III
$X_1(Y/2)$									
Error	Est.	$+0.10 \times 10^{-3}$	-0.33×10^{-7}	$+0.33 \times 10^{-13}$	$+0.41 \times 10^{-1}$	-0.28×10^{-4}	$+0.13 \times 10^{-10}$	$+0.49 \times 10^{-8}$	$+0.08 \times 10^{-8}$
	Real	$+0.13 \times 10^{-3}$	-0.89×10^{-7}	$+0.39 \times 10^{-13}$	$+0.55 \times 10^{-1}$	-0.44×10^{-4}	$+0.23 \times 10^{-10}$	$+0.82 \times 10^{-8}$	$+0.33 \times 10^{-8}$
$X_2(Y'/2)$									
Error	Est.	$+0.09 \times 10^{-3}$	$+0.85 \times 10^{-7}$	$+0.29 \times 10^{-13}$	$+0.48 \times 10^{-1}$	$+0.21 \times 10^{-4}$	$+0.15 \times 10^{-10}$	$+0.07 \times 10^{-7}$	-0.39×10^{-7}
	Real	$+0.11 \times 10^{-3}$	-0.29×10^{-7}	$+0.52 \times 10^{-13}$	$+0.60 \times 10^{-1}$	-0.22×10^{-4}	$+0.27 \times 10^{-10}$	$+0.12 \times 10^{-7}$	-0.65×10^{-7}
$X_3(X)$									
Error	Est.	-0.23×10^{-4}	$+0.31 \times 10^{-6}$	-0.08×10^{-13}	$+0.14 \times 10^{-1}$	$+0.98 \times 10^{-4}$	$+0.43 \times 10^{-11}$	$+0.47 \times 10^{-8}$	-0.08×10^{-6}
	Real	-0.48×10^{-4}	$+0.31 \times 10^{-6}$	$+0.24 \times 10^{-13}$	$+0.11 \times 10^{-1}$	$+0.44 \times 10^{-4}$	$+0.89 \times 10^{-11}$	$+0.84 \times 10^{-8}$	-0.14×10^{-6}
$X_4(X')$									
Error	Est.	-0.23×10^{-3}	$+0.26 \times 10^{-6}$	-0.07×10^{-12}	$+0.68 \times 10^{-1}$	$+0.18 \times 10^{-3}$	-0.23×10^{-11}	-0.50×10^{-8}	-0.08×10^{-6}
	Real	-0.32×10^{-3}	$+0.30 \times 10^{-6}$	-0.19×10^{-12}	-0.98×10^{-1}	$+0.13 \times 10^{-3}$	-0.45×10^{-11}	-0.79×10^{-8}	-0.14×10^{-6}
Jacobian int.		-0.69×10^{-4}			-0.37×10^{-4}				

The estimation of the accumulated errors was in all cases quite satisfactory; it is worth noting that the larger are the errors the better is their estimation. A possible explanation for this is that generally the errors are small quantities obtained from the difference between two much larger quantities, an operation which is highly affected by the rounding-off noise.

It is also interesting to note that the values of the Jacobian Constant given at the bottom of the table for Method I, which in this case should be equal to zero, fail to give a good indication of the size of the actual errors.

Acknowledgement

A part of the present work was done at the Laboratory of Theoretical Studies of Goddard Space Flight Center under a Senior Post Doctoral Associateship granted by the National Academy of Sciences of the U.S.A.

Bibliography

Deprit, A. and Palmore, J.: 1966, *Astron. J.* **71**.
Goodrich, E.: 1966, *Astron. J.* **71**, and also personal communication.
Henrici, P.: 1961, *Discrete Variable Methods for Ordinary Differential Equations*, Wiley, New York.
Rabe, E.: 1961, *Astron. J.* **66**.
Zadunaisky, P. E.: 1966, Proceedings of the IAU Symposium No. 25, Thessaloniki, Greece, 1966.

Discussion

J. V. Breakwell: Do you make pessimistic assumptions about the time-history of the sign of the round-off error in order to maximize its convoluted contribution to the terminal error?

P. E. Zadunaisky: No. One of the main advantages of the present method is that it does not require any assumptions about the local behaviour of the truncation and round-off errors. It only requires the fulfillment of conditions (7).

C. F. Peters: This method requires the extra work of integrating the Pseudo-Problem.

P. E. Zadunaisky: That is true. However any method of error estimation requires also an extra work of the same kind. For instance, the numerical integration of the system (11) of variational equations should require more work than the integration of the Pseudo-Problem. It is also true that it is necessary to find the interpolating function $P(t)$ but this can be done by simple standard procedures.

ON THE NON-EXISTENCE OF TRANSFORMATIONS TO NORMAL FORM IN CELESTIAL MECHANICS

RICHARD B. BARRAR

Mathematics Department, University of Oregon, Eugene, Oreg., U.S.A.

Kolmogorov (1954) has shown that under rather general conditions a Hamiltonian of the form:

$$H = \lambda_1 p_1 + \lambda_2 p_2 + A_0(q_1, q_2) + A_1(q_1, q_2)\, p_1 + A_2(q_1, q_2)\, p_2 + \sum_{i+j=2}^{\infty} A_{ij}(q_1, q_2)\, p_1^i p_2^j \tag{1}$$

can be reduced to the form:

$$H = \lambda_1 p_1 + \lambda_2 p_2 + \sum_{i+j=2}^{\infty} A_{ij}(q_1, q_2)\, p_1^i p_2^j \tag{2}$$

by a suitable canonical transformation, that is the constant and the linear terms periodic in q_1, q_2 can be eliminated from the Hamiltonian. This implies in particular that $p_i = 0$, $q_i = \lambda_i(t - \tau_i)$ is a solution of the canonical differential equations in terms of the new variables.

Once the differential equation is in the form (2), it is always possible if λ_1/λ_2 is irrational and for any preassigned integer n to reduce (2) to the form:

$$H = \lambda_1 P_1 + \lambda_2 P_2 + \sum_{i+j=2}^{n-1} A_{ij} P_1^i P_2^j + \sum_{i+j=n}^{\infty} A_{ij}(Q_1, Q_2)\, P_1^i P_2^j \tag{3}$$

(for notational purposes we will designate a general series of this last form as $S_n(P, Q)$). That is, it is possible to eliminate periodic terms up to an arbitrarily high order n, where, of course, in going from (1) to (2) to (3) the circle of convergence gets smaller and smaller.

The question then arises if it is possible to eliminate all periodic terms from the Hamiltonian (2) and still maintain a finite radius of convergence. Most workers in celestial mechanics would assert that in general it certainly is not possible to eliminate all periodic terms. The purpose of this paper is to outline a general procedure for treating this problem.

Before proceeding we mention that both Moser (1955) and Siegel (1954) have treated similar questions. However due to certain technical difficulties their work is not directly applicable to the Hamiltonian considered by Kolmogorov. In an effort to extend their methods, the author was led to realize the close connection between the Birkhoff fixed point theorem and the reduction to normal form discussed in this paper.

Consider a Hamiltonian as in (2) with $H(p_1, p_2, q_1, q_2) = \text{constant}$. Since $\partial H/\partial p_1 \neq 0$, we can solve for $p_1 \equiv K(p_2, q_2, q_1)$. Wintner (1941, p. 128) shows that the solution of

G. E. O. Giacaglia (ed.), Periodic Orbits, Stability and Resonances, 228–231.

the differential equations

$$\mathrm{d}p_2/\mathrm{d}q_1 = -\,\partial K/\partial q_2 \qquad \mathrm{d}q_2/\mathrm{d}q_1 = \partial K/\partial p_2$$

is equivalent to the solution of the original Hamiltonian system of differential equations with constant energy. Further the solution of this differential equation is an area preserving mapping T that maps the points (p_2°, q_2°) at $q_1=0$ into (p_2, q_2) at $q_1=2\pi$, with $p_2^\circ=0$ going into itself. Clearly for the original differential equation to have a periodic solution with q_1 changing $2\pi m$ is equivalent to the mapping $T^{(m)}$ having a fixed point.

Assume now that there is a convergent canonical transformation obtained by means of a generating function of the form $P_1q_1+P_2q_2+S_3(P, q)$ such that Equation (2) goes over to the form:

$$H(p_i, q_i) = H(P) = \lambda_1 P_1 + \lambda_2 P_2 + AP_1^2 + BP_1P_2 + CP_2^2 + \sum_{i+j=3}^{\infty} A_{ij}P_1^iP_2 \tag{4}$$

where $H(P)$ has a finite radius of convergence.

In terms of the new variables, the solution of $H(P)=0$ is:

$$P_1 = \alpha P_2 + \beta P_2^2 + \mathrm{O}(P_2^3)$$

with $\alpha=-\lambda_2/\lambda_1$, $\beta=A\alpha^2+B\alpha+C$.

The mapping $\tilde{T}$ corresponding to (P_2°, Q_2°) at $Q_1=0$ going into (P_2, Q_2) at $Q_1=2\pi$ is

$$P_2 = P_2^\circ$$
$$Q_2 = Q_2^\circ + \alpha + 4\pi\beta P_2^\circ + \mathrm{O}((P_2^\circ)^2)$$

which is a twist mapping if $\beta\neq 0$. Hence whole circles $P_2=P_2^\circ$ are fixed points of the mapping $\tilde{T}^{(m)}$. Since q_1 changing by $2\pi m$ corresponds to Q_1 changing by $2\pi m$, this readily implies that the mapping $T^{(m)}$ does not have isolated fixed points arbitrarily near the origin, but rather whole curves are fixed by the mapping $T^{(m)}$. Thus if we can show that for general transformations the fixed points near the origin are isolated, we will have shown that in general a Hamiltonian cannot be brought to the normal form (4).

Although we could treat the Hamilton differential equations directly, for purposes of clarity we will treat the case of an area preserving mapping and show that in general it has isolated fixed points arbitrarily near the origin.

Thus we assume that we are given an area preserving transformation of the form:

$$W(p_1, q) = (q+\gamma)\,p_1 + p_1^2/2 + S_3(p_1, q)$$
$$q_1 = \partial W/\partial p_1 \qquad p = \partial W/\partial q$$

so

$$p_1 = p + \mathrm{O}(p^3)$$
$$q_1 = q + \gamma + p + \mathrm{O}(p^2)$$

which we write as $(p_1, q_1)=T(p, q)$ and $(p_k, q_k)=T^{(k)}(p, q)$ for the kth iterate.

For such a transformation Siegel (1956), Section 22 in proving the Birkhoff fixed point theorem proves there is some $\delta > 0$ such that if $n > 5\pi/\delta$, there are a_n, b_n such that:

$$\begin{aligned}
&0 < a_n < b_n < \delta \\
&0 < p_k(p, q) < 5\delta/4 \quad k = 1 \ldots n, \quad \text{if} \quad 0 < p < b_n \\
&q_n(a_n, q) - q < 2g_n\pi < q_n(b_n, q) - q \quad g_n \text{ an integer} \\
&\partial q_n(p, q)/\partial p > 0 \quad \text{for} \quad a_n \leqslant p \leqslant b_n .
\end{aligned}$$

It follows from this that for fixed q there is one and only one p such that

$$q_n(p, q) - q = 2g_n\pi \quad 0 < a_n < p < b_n < \delta . \tag{5}$$

Further by the implicit function theorem the p satisfying (5) is an analytic function of q. We write it as:

$$p = k_n(q) .$$

The Birkhoff fixed point theorem easily follows from this. Since if we consider the points p, q as polar coordinates, the points on the curve $p = k_n(q)$ only move radially under the mapping $T^{(n)}$. Hence whenever the curves $k_n(q)$ and $T^{(n)}(k_n(q))$ intersect there is a fixed point. However since the mapping is area preserving they must intersect, and this insures fixed points.

However the function $k_n(q) - T^{(n)}(k_n(q))$ being analytic in q, either is identically zero or its zeros are isolated. If its zeros are isolated so are the fixed points of the mapping $T^{(n)}$ on the curve $k_n(q)$.

The function $k_n(q) - T^{(n)}(k_n(q))$ or more precisely each of its Fourier coefficients depends on the coefficients in the series of the transformation. Following Siegel (1954) if the transformation is of the form:

$$\begin{aligned}
&W(p_1, q) = (q + \gamma)\, p_1 + p_1^2/2 + S_3(p_1, q) \\
&S_3(p_1, q) = \sum_{i=3}^{\infty} (\tilde{a}_{ij}^{(1)} p_1^i \cos jq + \tilde{a}_{ij}^{(2)} p_1^i \sin jq)
\end{aligned}$$

and if $\varepsilon_{ij}^{(k)} k = 1, 2$ is any sequence of numbers $0 < \varepsilon_{ij}^{(k)} < 1$, we form the Cartesian product with the weak topology, call it C, of all the sets $|a_{ij}^{(k)} - \tilde{a}_{ij}^{(k)}| \leqslant \varepsilon_{ij}^{(k)}$, which by the Tynchoff theorem is compact.

Thus the points of C represent sets of coefficients near the coefficients of $S_3(p_1, q)$. It is possible to show, since all functions that are involved are analytic, that for any n, the set of points in C making a suitable Fourier coefficient of $k_n(q) - T^{(n)}(k_n(q))$ vanish is nowhere dense in C. Hence by the Baire theorem, if we are given any countable set of equations $k_n(q) - T^{(n)}(k_n(q))$, the set of points in C implying that none of these equations is identically zero is dense in C. It readily follows that each of the transformations corresponding to this dense set of points have isolated fixed points arbitrarily close to the origin.

This same procedure may be applied to transformations resulting from Hamiltonian differential equations, to show that in general a Hamiltonian (2) of the type treated by Kolmogorov cannot be brought to normal form in a whole neighborhood of the origin.

Acknowledgement

This research was supported in part by NSF Grant GP-9384.

References

Kolmogorov, A. N.: 1954, 'The Conservation of Conditionally Periodic Motions with a Small Change in the Hamiltonian', *Doklady Akademii Nauk SSSR* **98**, 527–30.

Moser, J.: 1955, 'Nonexistence of Integrals for Canonical Systems of Differential Equations', *Comm. Pure Appl. Math.* **8**, 409–36.

Siegel, C. L.: 1954, 'Über die Existenz einer Normalform analytischer Hamiltonischer Differentialgleichungen in der Nähe einer Gleichgewichtslösung', *Math. Ann.* **128**, 144–70.

Siegel, C. L.: 1956, *Vorlesungen über Himmelsmechanik*, Springer Verlag, Berlin.

Wintner, A.: 1941, *The Analytic Foundations of Celestial Mechanics*, Princeton University Press, Princeton.

Discussion

A. Deprit: On one side, astronomers have been convinced by mathematicians that a complete normalization does not make sense; on the other side, astronomers know from long experience that truncated normalizations do not distort appreciably the portrait of the phase space as it can be reached within a given accuracy. Can you explain the contrast of Gustavson's third integral – of no mathematical existence – justifying Hénon's portrait of a galactic potential around an equilibrium?

R. B. Barrar: I think that one of the main problems for mathematicians working in celestial mechanics at the present time is to explain the amazing accuracy that astronomers achieve with series that appear to diverge. The Kolmogorov theorem is a major step in this direction.

G. Hori: Suppose $A_{ij}(q_1, q_2) = A_{ij} + \varepsilon B_{ij}(q_1, q_2)$. Then your conclusion is that the transformation to the normal form is divergent regardless the smallness of ε and this conclusion seems to me not plausible.

R. B. Barrar: I only assert that for given $B_{ij}(q_1, q_2)$ and given $\varepsilon > 0$, there are arbitrarily small $C_{ij}(q_1, q_2)$ such that the transformation to normal form of

$$A_{ij} + \varepsilon B_{ij}(q_1, q_2) + C_{ij}(q_1, q_2)$$

will not converge in a whole neighborhood of the origin.

A UNIFIED TREATMENT OF LUNAR THEORY AND ARTIFICIAL SATELLITE THEORY

WILLIAM A. MERSMAN

NASA Ames Research Center, Moffett Field, Calif., U.S.A.

Abstract. Lunar theory and artificial satellite theory are treated by a unified method that is a generalization of the von Zeipel procedure. The technique of separation of variables is used to generate a single canonical transformation that eliminates the time and all the angle variables from the Hamiltonian. The validity of the method is established and the calculations are carried far enough to illustrate the techniques involved.

Symbols

a	semimajor axis
e	eccentricity
f	true anomaly
g	argument of perigee
h	longitude of node from moving reference line
k	gaussian gravitational constant
l	mean anomaly
m	dimensionless ratio of mean motions
n	mean motion of moon or satellite
p	angle variable
q	action variable
r	geocentric distance
t	time
u	eccentric anomaly
x	action variable
y	angle variable
B	earth-moon barycenter
$C_{n,m}$, $S_{n,m}$	dimensionless coefficients in tesseral harmonics
D_n	zonal harmonic
E	new Hamiltonian, mass of earth
F	old Hamiltonian
ϱ	distance from sun to earth-moon barycenter
ϕ	true anomaly of sun
G	$\sqrt{\mu a(1-e^2)}$
H	$G \cos I$
I	inclination angle
J_n	dimensionless coefficients in zonal harmonics
L	$\sqrt{\mu a}$
M	mass of moon
P_n	Legendre polynomial
P_n^m	associated Legendre function
R	equatorial radius of the earth
S	determining function, mass of sun
$T_{n,m}$	tesseral harmonic
W	determining function
X, Y, Z	decomposition types
β, δ	solar factors
γ	$1-\cos I$
Γ	longitude of sun's perigee
ε	eccentricity of sun's orbit
θ	geocentric latitude
λ	mean anomaly of sun
Λ	geographic longitude
μ	gravitational parameter
ν	mean motion of sun, spin velocity of earth
ψ	elongation of moon
Ω	longitude of node from inertial reference line

G. E. O. Giacaglia (ed.), Periodic Orbits, Stability and Resonances, 232–258.

1. Introduction

The theory of artificial satellite motion has been developed in recent years by methods that had not previously been considered applicable to lunar theory. The problems appear quite different, mathematically, because of the explicit appearance of the time in the lunar problem.

In the present paper the familiar von Zeipel procedure is generalized in two respects. Explicit dependence on the time is permitted, in a certain restricted form, and then the time and all the angle variables are removed from the Hamiltonian by means of a single canonical transformation. This generalization is capable of solving the complete lunar problem far more efficiently than the older methods of von Zeipel or Delaunay. It is then shown that this method can be applied directly, in a simplified, degenerate form, to the artificial satellite problem.

The familiar Delaunay variables are used, with one modification. The longitude of the node is measured from a moving, rather than from an inertial reference direction. For lunar theory the reference direction is that of the sun, for artificial satellite theory, it is the meridian of Greenwich. This device removes the explicit time dependence from the satellite problem, and drastically simplifies the lunar problem.

The generalized procedure is presented in the next section, and its application to lunar and satellite theory follow in that order.

2. The Generalized von Zeipel Transformation

Recent applications of the von Zeipel transformation have been restricted essentially to Hamiltonians in which the time does not occur explicitly [1, 2]. When it does occur, it is formally removed by the artifice of adjoining an additional pair of canonical variables ([3, pp. 530–31], [4]). A generalized method is developed here that is applicable to certain problems involving a time-dependent Hamiltonian. In the case of lunar theory this permits many of the solar terms to be expressed in closed form, thereby avoiding certain Fourier series expansions that frequently complicate the problem unnecessarily.

Consider the canonical system

$$\mathrm{d}x_i/\mathrm{d}t = \partial F/\partial y_i, \qquad \mathrm{d}y_i/\mathrm{d}t = -\,\partial F/\partial x_i$$

with 'action variables' $x=(x_1, x_2, x_3)$ and 'angle variables' $y=(y_1, y_2, y_3)$. Let the Hamiltonian be of the form

$$F = F(x, y, \lambda)$$

where λ (an angle) is a linear function of the time, with $\nu=\mathrm{d}\lambda/\mathrm{d}t$ a constant. This is the only form of time-dependence that will be considered.

Now let $q=(q_1, q_2, q_3)$ and $p=(p_1, p_2, p_3)$ be new canonical variables defined by the implicit transformation equations

$$p_i = \partial W(q, y, \lambda)/\partial q_i$$
$$x_i = \partial W(q, y, \lambda)/\partial y_i$$

with the 'determining function', W, given by

$$W = q_i y_i + S(q, y, \lambda)$$

(Einstein's summation convention will be used throughout this paper: a repeated subscript is to be summed over its range.) Thus, in vector notation,

$$x = q + \Delta q$$
$$y = p + \Delta p$$

with the increments having components given implicitly by

$$(\Delta q)_i = \partial S(q, y, \lambda)/\partial y_i$$
$$(\Delta p)_i = -\partial S(q, y, \lambda)/\partial q_i .$$

Expanding the right member of the last equation in a Taylor series near $y=p$ gives

$$(\Delta p)_i = -\frac{\partial S(q, p, \lambda)}{\partial q_i} - \frac{\partial^2 S}{\partial q_i \, \partial p_j}(\Delta p)_j - \frac{1}{2}\frac{\partial^3 S}{\partial q_i \, \partial p_j \, \partial p_k}(\Delta p)_j (\Delta p)_k \ldots .$$

This can be solved iteratively to any desired degree of accuracy, if S and its derivatives are assumed small:

First order:

$$(\Delta p)_i = -\frac{\partial S}{\partial q_i}.$$

Second order:

$$(\Delta p)_i = -\frac{\partial S}{\partial q_i} - \frac{\partial^2 S}{\partial q_i \, \partial p_j}\left(-\frac{\partial S}{\partial q_j}\right).$$

Third order:

$$(\Delta p)_i = -\frac{\partial S}{\partial q_i} - \frac{\partial^2 S}{\partial q_i \, \partial p_j}\left(-\frac{\partial S}{\partial q_j} + \frac{\partial^2 S}{\partial q_j \, \partial p_k}\frac{\partial S}{\partial q_k}\right) - \frac{1}{2}\frac{\partial^3 S}{\partial q_i \, \partial p_j \, \partial p_k}\left(-\frac{\partial S}{\partial q_j}\right)\left(-\frac{\partial S}{\partial q_k}\right).$$

Similarly,

$$(\Delta q)_i = \frac{\partial S}{\partial p_i} - \frac{\partial^2 S}{\partial p_i \, \partial p_j}\frac{\partial S}{\partial q_j} + \frac{\partial^2 S}{\partial p_i \, \partial p_j}\frac{\partial^2 S}{\partial q_j \, \partial p_k}\frac{\partial S}{\partial q_k} + \frac{1}{2}\frac{\partial^3 S}{\partial p_i \, \partial p_j \, \partial p_k}\frac{\partial S}{\partial q_j}\frac{\partial S}{\partial q_k} \ldots .$$

Thus the equations

$$x = q + \Delta q$$
$$y = p + \Delta p$$

with Δq, Δp given above, constitute an explicit solution of the implicit transformation equations.

Now, q and p are canonical variables:

$$\mathrm{d}q_i/\mathrm{d}t = \partial E/\partial p_i, \qquad \mathrm{d}p_i/\mathrm{d}t = -\partial E/\partial q_i$$

with the new Hamiltonian, E, given implicitly by

$$E = F(x, y, \lambda) - \frac{\partial W(q, y, \lambda)}{\partial t} = F - \nu \frac{\partial S(q, y, \lambda)}{\partial \lambda}$$

[5, ch. 6].

If the right member is again expanded in a Taylor series and the explicit forms of Δq and Δp are inserted, an explicit representation $E(q, p, \lambda)$ is obtained. Successive differentiations with respect to p can be used to yield the symmetric equation

$$\begin{aligned} E(q, p, \lambda) &+ \frac{\partial E}{\partial p_i}\frac{\partial S}{\partial q_i} + \frac{1}{2}\frac{\partial^2 E}{\partial p_i\,\partial p_j}\frac{\partial S}{\partial q_i}\frac{\partial S}{\partial q_j} + \frac{1}{6}\frac{\partial^3 E}{\partial p_i\,\partial p_j\,\partial p_k}\frac{\partial S}{\partial q_i}\frac{\partial S}{\partial q_j}\frac{\partial S}{\partial q_k} + \cdots \\ &= F(q, p, \lambda) - \nu\frac{\partial S}{\partial \lambda} + \frac{\partial F}{\partial q_i}\frac{\partial S}{\partial p_i} + \frac{1}{2}\frac{\partial^2 F}{\partial q_i\,\partial q_j}\frac{\partial S}{\partial p_i}\frac{\partial S}{\partial p_j} \\ &+ \frac{1}{6}\frac{\partial^3 F}{\partial q_i\,\partial q_j\,\partial q_k}\frac{\partial S}{\partial p_i}\frac{\partial S}{\partial p_j}\frac{\partial S}{\partial p_k} + \cdots \end{aligned}$$

and only the new variables q and p appear.

Now let n be the mean motion of the body being studied (moon or satellite in the present paper) and let $m = \nu/n$ be a small quantity (i.e., $m \ll 1$). Introducing series expansions

$$F = \sum_{k=0}^{\infty} F_k \quad E = \sum_{k=0}^{\infty} E_k \quad S = \sum_{k=1}^{\infty} S_k \quad \Delta q = \sum_{k=1}^{\infty} \Delta_k q \quad \Delta p = \sum_{k=1}^{\infty} \Delta_k p$$

with the subscript, k, denoting $0(m^k)$, permits the separation of the equations according to order of magnitude, that is, according to powers of m.

Thus, the explicit transformation equations become

$$\begin{aligned} (\Delta_1 p)_i &= -\frac{\partial S_1}{\partial q_i} \\ (\Delta_2 p)_i &= -\frac{\partial S_2}{\partial q_i} + \frac{\partial^2 S_1}{\partial q_i\,\partial p_j}\frac{\partial S_1}{\partial q_j} \\ (\Delta_3 p)_i &= -\frac{\partial S_3}{\partial q_i} + \frac{\partial^2 S_2}{\partial q_i\,\partial p_j}\frac{\partial S_1}{\partial q_j} + \frac{\partial^2 S_1}{\partial q_i\,\partial p_j}\frac{\partial S_2}{\partial q_j} \\ &- \frac{\partial^2 S_1}{\partial q_i\,\partial p_j}\frac{\partial^2 S_1}{\partial q_j\,\partial p_k}\frac{\partial S_1}{\partial q_k} \\ &- \frac{1}{2}\frac{\partial^3 S_1}{\partial q_i\,\partial p_j\,\partial p_k}\frac{\partial S_1}{\partial q_j}\frac{\partial S_1}{\partial q_k} \end{aligned}$$

and so forth, for Δp, with analogous equations for Δq.

The equation for the Hamiltonian separates into the 'von Zeipel equations':

$$E_0 = F_0$$

$$E_1 + \frac{\partial E_0}{\partial p_i}\frac{\partial S_1}{\partial q_i} = F_1 + \frac{\partial F_0}{\partial q_i}\frac{\partial S_1}{\partial p_i}$$

$$E_2 + \frac{\partial E_0}{\partial p_i}\frac{\partial S_2}{\partial q_i} + \frac{\partial E_1}{\partial p_i}\frac{\partial S_1}{\partial q_i} + \frac{1}{2}\frac{\partial^2 E_0}{\partial p_i \partial p_j}\frac{\partial S_1}{\partial q_i}\frac{\partial S_1}{\partial q_j}$$

$$= F_2 - \nu\frac{\partial S_1}{\partial \lambda} + \frac{\partial F_0}{\partial q_i}\frac{\partial S_2}{\partial p_i} + \frac{\partial F_1}{\partial q_i}\frac{\partial S_1}{\partial p_i} + \frac{1}{2}\frac{\partial^2 F_0}{\partial q_i \partial q_j}\frac{\partial S_1}{\partial p_i}\frac{\partial S_1}{\partial p_j}$$

$$E_3 + \frac{\partial E_0}{\partial p_i}\frac{\partial S_3}{\partial q_i} + \frac{\partial E_1}{\partial p_i}\frac{\partial S_2}{\partial q_i} + \frac{\partial E_2}{\partial p_i}\frac{\partial S_1}{\partial q_i} + \frac{\partial^2 E_0}{\partial p_i \partial p_j}\frac{\partial S_1}{\partial q_i}\frac{\partial S_2}{\partial q_j}$$

$$+ \frac{1}{2}\frac{\partial^2 E_1}{\partial p_i \partial p_j}\frac{\partial S_1}{\partial q_i}\frac{\partial S_1}{\partial q_j} + \frac{1}{6}\frac{\partial^3 E_0}{\partial p_i \partial p_j \partial p_k}\frac{\partial S_1}{\partial q_i}\frac{\partial S_1}{\partial q_j}\frac{\partial S_1}{\partial q_k}$$

$$= F_3 - \nu\frac{\partial S_2}{\partial \lambda} + \frac{\partial F_0}{\partial q_i}\frac{\partial S_3}{\partial p_i} + \frac{\partial F_1}{\partial q_i}\frac{\partial S_2}{\partial p_i} + \frac{\partial F_2}{\partial q_i}\frac{\partial S_1}{\partial p_i} + \frac{\partial^2 F_0}{\partial q_i \partial q_j}\frac{\partial S_1}{\partial p_i}\frac{\partial S_2}{\partial p_j}$$

$$+ \frac{1}{2}\frac{\partial^2 F_1}{\partial q_i \partial q_j}\frac{\partial S_1}{\partial p_i}\frac{\partial S_1}{\partial p_j} + \frac{1}{6}\frac{\partial^3 F_0}{\partial q_i \partial q_j \partial q_k}\frac{\partial S_1}{\partial p_j}\frac{\partial S_1}{\partial p_j}\frac{\partial S_1}{\partial p_k}$$

and so forth.

It may be remarked that this representation of the transformation equations and of the Hamiltonian explicitly in terms of the new variables q and p is not the usual practice. Brouwer [1] uses the mixed set of variables q and y.

In [1], [2], and [4] the von Zeipel transformation is used successively to eliminate only one angle variable, p_i, at a time. In the applications in the present paper, a single transformation will be exhibited that eliminates λ and all three angle variables simultaneously. Thus, the partial derivatives disappear from the left members of the von Zeipel equations, giving simply the components, E_k, explicitly. The equations will be used here only in this simplified form.

These equations can be regarded as a simultaneous set of partial differential equations for the components, S_k, of the determining function, S. A technique that is similar to the classical one of separation of variables will be used to construct a recursive set of equations, each of which is a linear, first-order partial differential equation with constant coefficients. Specifically, the angle variables, p, will be the Delaunay variables (l, g, h), respectively, the mean anomaly, argument of perigee, and longitude of the node. Each term, S_k, will be decomposed into the sum of three terms

$$S_k = X_k + Y_k + Z_k$$

where X a periodic function of l; Y a periodic function of g, h, and λ, with l absent and the sum, $h+\lambda$, excluded; Z a periodic function of two variables, g and the sum $h+\lambda$, with l absent.

Similarly, each term, F_k, of the Hamiltonian will be decomposed into the sum of four terms.

$$F_k = XF_k + YF_k + ZF_k + EF_k$$

with the prefix E denoting a function that is independent of λ and all the angles variables (l, g, h), the other prefixes having the meaning assigned above. The calculations will then proceed according to the scheme:

Order 0;	E_0			
Order 1;	E_1	X_1		
Order 2;	E_2	X_2	Y_1	
Order $k \geqslant 3$;	E_k	X_k	Y_{k-1}	Z_{k-2}.

This is reminiscent of the schemes of Brouwer [1] and Hori [2] which yield short-period terms to a higher order than long-period terms. Later it will be shown that the components X, Y, Z are obtained as follows:

The component X_k will be obtained by simple quadrature:

$$X_k = \int (\text{function of type } X)\, \mathrm{d}l$$

The component Y_k will satisfy an equation of the type

$$\partial Y/\partial\lambda - \partial Y/\partial h = f(\theta)$$

$$\theta = i\lambda + jh + kg\,,\quad i \neq j, \quad i, j, k \text{ integers}$$

Clearly, the solution is

$$Y = \frac{1}{i-j} \int f(\theta)\, \mathrm{d}\theta$$

The component Z_k will satisfy an equation of the form

$$\xi \frac{\partial Z}{\partial g} + \eta \frac{\partial Z}{\partial h} = f(\theta)$$

$$\theta = ig + j(h + \lambda), \quad i, j \text{ integers}$$

and the solution is

$$Z = \frac{1}{\xi i + \eta j} \int f(\theta)\, \mathrm{d}\theta$$

where ξ and η are functions of the action variables (L, G, H) and it will be proved that the denominator never vanishes, so that no critical case occurs in the lunar theory. In satellite theory, of course, the case of critical inclination has to be excluded.

Lunar theory will be treated first, and then it will be shown that satellite theory is simply a degenerate case.

It is clear from the preceding discussion that the first task is to obtain a series representation of the Hamiltonian, in terms of the Delaunay variables, and then to exhibit the decomposition by type (X, Y, Z, E). This will be accomplished in the next section.

3. Lunar Theory

A. DEVELOPMENT OF THE HAMILTONIAN

The lunar theory to be developed here is that of the motion of the moon about the earth, under the assumption that the earth-moon barycenter (B) moves about the sun in a Keplerian orbit, the plane of which will be referred to as the ecliptic (Figure 1). Earth, moon and sun are regarded as point masses, and planetary perturbations are ignored. The Hamiltonian for the classical Delaunay variables (L, G, H) and (l, g, Ω) is

$$F' = \frac{\mu^2}{2L^2} + \frac{k^2 S}{\varrho^3} r^2 P_2(\cos\psi) + \frac{k^2 S}{\varrho^4}\frac{E-M}{E+M} r^3 P_3(\cos\psi)\ldots$$

(see [3], pp. 271 and 291), and the law of cosines of spherical trigonometry gives (see sketch):

$$\cos\psi = \cos(f+g)\cos h - \sin(f+g)\sin h \cos I$$

where h is the elongation of the node, measured from the sun:

$$h = \Omega - (\Gamma + \phi).$$

If h is used, rather than Ω, as the conjugate variable to H, then the Hamiltonian is simply

$$F = F' + H\, \mathrm{d}\phi/\mathrm{d}t$$

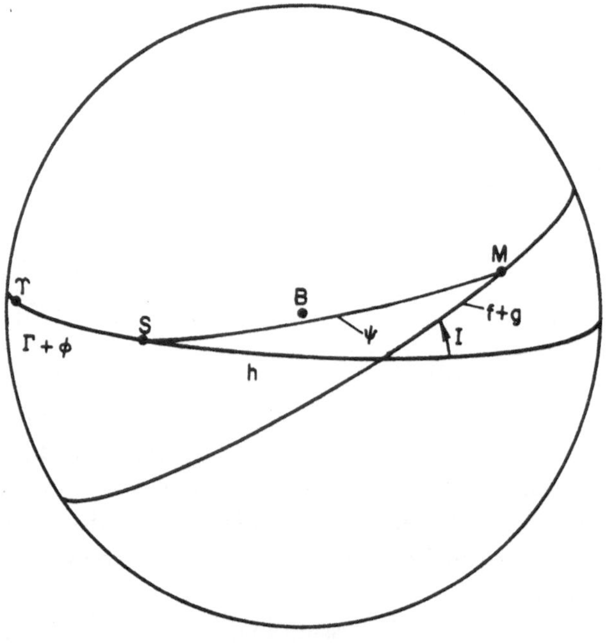

Fig. 1.

since

$$\frac{\mathrm{d}h}{\mathrm{d}t} = \frac{\mathrm{d}\Omega}{\mathrm{d}t} - \frac{\mathrm{d}\phi}{\mathrm{d}t} = -\frac{\partial F'}{\partial H} - \frac{\partial}{\partial H}\left(\frac{\mathrm{d}\phi}{\mathrm{d}t}\cdot H\right) = -\frac{\partial F}{\partial H}$$

and

$$\frac{\mathrm{d}H}{\mathrm{d}t} = \frac{\partial F'}{\partial \Omega} = \frac{\partial F'}{\partial h} = \frac{\partial F}{\partial h}.$$

This choice of variables makes ψ independent of the time, so that solar effects enter only via ϱ, the distance from the sun to the earth-moon barycenter. Many Fourier series expansions and multiplications are thus avoided.

The usual methods of the theory of elliptic motion ([3, ch. 2], and [6, ch. 3]) can be used to obtain the solar factors in closed form and in series. The only ones needed here are

$$\mathrm{d}\phi/\mathrm{d}t = \nu(1 + \delta)$$
$$k^2 S/\varrho^3 = \nu_2(1 + \beta)$$

where ν is the sun's mean motion and

$$\nu_2 = \nu\nu_1, \qquad \nu_1 = \frac{S}{S + E + M}\cdot\frac{\nu}{(1 - \varepsilon^2)^{3/2}}$$

$$\delta = \frac{\mathrm{d}(\phi - \lambda)}{\mathrm{d}\lambda} = \sum_{k=1}^{\infty} \delta_k \cos k\lambda$$

$$\beta = \frac{\mathrm{d}(\phi - \lambda + \varepsilon \sin\phi)}{\mathrm{d}\lambda} = \sum_{k=1}^{\infty} \beta_k \cos k\lambda$$

and δ_k, β_k are power series starting with ε^k; for example,

$$\delta_1 = 2\varepsilon - \tfrac{1}{4}\varepsilon^3 \ldots$$
$$\delta_2 = \tfrac{5}{2}\varepsilon^2 - \tfrac{11}{24}\varepsilon^4 \ldots$$
$$\beta_1 = 3\varepsilon - \tfrac{9}{8}\varepsilon^3 \ldots$$
$$\beta_2 = \tfrac{9}{2}\varepsilon^2 - \tfrac{13}{4}\varepsilon^4 \ldots$$

To obtain the desired form for the lunar factors, write

$$\cos\psi = \cos(f + g + h) + \gamma \sin(f + g) \sin h$$

where

$$\gamma = 1 - \cos I = 1 - H/G$$

Then the Legendre polynomials can be written as

$$P_n(\cos\psi) = \sum_{k=0}^{n} p_{n,k}\gamma^k$$

where $p_{n,k}$ is a trigonometric polynomial in f, with coefficients that are trigonometric functions of g and h; for example,

$$p_{2,0} = \tfrac{1}{4} + \tfrac{3}{4}\cos(2g+2h)\cos 2f - \tfrac{3}{4}\sin(2g+2h)\sin 2f$$
$$p_{2,1} = -\tfrac{3}{2}\sin^2 h + \tfrac{3}{2}\sin h\sin(2g+h)\cos 2f + \tfrac{3}{2}\sin h\cos(2g+h)\sin 2f$$
$$p_{2,2} = \tfrac{3}{4}\sin^2 h - \tfrac{3}{4}\sin^2 h\cos 2g\cos 2f + \tfrac{3}{4}\sin^2 h\sin 2g\sin 2f\,.$$

The Hamiltonian can now be written as a series

$$F = \sum_{k=0}^{\infty} F_k$$

where

$$F_0 = \mu^2/2L^2$$
$$F_1 = \nu H$$
$$F_2 = \nu\,\delta H + \nu_2 r^2 p_{2,0}$$
$$F_3 = \nu_2 \beta r^2 p_{2,0} + \nu_2 \gamma r^2 p_{2,1}\,.$$

and so forth, under the assumptions that $\varepsilon = 0(m)$, $\gamma = 0(m)$.

Finally, to separate each term by type (X, Y, Z, E) note that the true anomaly, f enters in the form

$$r^n \cos kf\,,\; r^n \sin kf\,,\quad 0 \leqslant k \leqslant n$$

and these can be expressed as trigonometric polynomials in u, the eccentric anomaly, the coefficients being functions of a and e (see [2] and [4]). For example,

$$r^2 = a^2(1 + \tfrac{3}{2}e^2) + A_0$$
$$r^2 \cos 2f = \tfrac{5}{2}a^2 e^2 + B_0$$
$$r^2 \sin 2f = C_0$$

where

$$A_0 = a^2[-2e(\tfrac{1}{2}e + \cos u) + \tfrac{1}{2}e^2 \cos 2u]$$
$$B_0 = a^2[-2e(\tfrac{1}{2}e + \cos u) + (1 - \tfrac{1}{2}e^2)\cos 2u]$$
$$C_0 = a^2\sqrt{1-e^2}\,(-2e\sin u + \sin 2u)$$

and each of these is periodic in l, the mean anomaly, with vanishing mean value (type X).

The desired representation of the Hamiltonian, then, is

$$F_0 = \mu^2/2L^2 \quad \text{Type } E$$
$$F_1 = \nu H \quad \text{Type } E$$
$$F_2:\quad XF_2 = \tfrac{1}{4}\nu_2[A_0 + 3B_0\cos(2g+2h) - 3C_0\sin(2g+2h)]$$
$$YF_2 = \nu\,\delta H + \tfrac{15}{8}\nu_2 a^2 e^2 \cos(2g+2h)$$
$$ZF_2 = 0$$
$$EF_2 = \tfrac{1}{4}\nu_2 a^2(1 + \tfrac{3}{2}e^2)$$
$$F_3:\quad XF_3 = \beta\cdot XF_2 + \tfrac{3}{2}\gamma\nu_2 \sin h[-A_0\sin h + B_0\sin(2g+h) + C_0\cos(2g+h)]$$

$$YF_3 = \tfrac{15}{16}\nu_2\beta_1 a^2 e^2 [\cos(2g + 2h + \lambda) + \cos(2g + 2h - \lambda)]$$
$$+ \tfrac{3}{4}\gamma\nu_2 a^2[(1 + \tfrac{3}{2}e^2(\cos 2h - \tfrac{5}{2}e^2 \cos(2g + 2h)] + \beta \cdot EF_2$$
$$ZF_3 = \tfrac{15}{8}\gamma\nu_2 a^2 e^2 \cos 2g$$
$$EF_3 = - \tfrac{3}{4}\gamma\nu_2 a^2 (1 + \tfrac{3}{2}e^2)$$
$$F_4{:} \quad ZF_4 = \tfrac{15}{16}\nu_2\beta_2 a^2 e^2 \cos(2g + 2h + 2\lambda) - \tfrac{15}{16}\gamma^2\nu_2 a^2 e^2 \cos 2g$$
$$EF_4 = \tfrac{3}{8}\gamma^2\nu_2 a^2 (1 + \tfrac{3}{2}e^2).$$

The X and Y components of F_4 will not be given, since the analysis will not be carried far enough in the present paper to require them.

It may be remarked that F_3 contains the term $\nu_2\beta r^2 p_{2,0}$, which yields, among others, the term

$$\tfrac{15}{8}\nu_2\beta a^2 e^2 \cos(2g + 2h)$$

and here the series for β must be used, giving

$$\tfrac{15}{16}\nu_2 a^2 e^2 \beta_1 [\cos(2g + 2h + \lambda) + \cos(2g + 2h - \lambda)]$$
$$+ \tfrac{15}{16}\nu_2 a^2 e^2 \beta_2 [\cos(2g + 2h + 2\lambda) + \cos(2g + 2h - 2\lambda)]$$
$$+ \dots .$$

The first line has been included in YF_3, and the first term of the second line is included in ZF_4. There are also higher order contributions that must be properly assigned in a complete theory.

B. SOLUTION OF THE LUNAR PROBLEM

The lunar problem can now be solved by means of the generalized von Zeipel transformation from the old variables

$$x = (L, G, H), \qquad y = (l, g, h)$$

to the new variables

$$q = (\bar{L}, \bar{G}, \bar{H}), \qquad p = (\bar{l}, \bar{g}, \bar{h})$$

where the bar notation is used to emphasize the fact that the new variables are the mean values of the old ones, in the usual astronomical sense. That is, the difference between each new variable and the corresponding old one is a sum of terms each of which is a periodic function of the time with vanishing mean value.

Since the new Hamiltonian contains only the action variables q, the new canonical equations are

$$\mathrm{d}q_i/\mathrm{d}t = \partial E/\partial p_i = 0$$

so that the q's $(\bar{L}, \bar{G}, \bar{H})$ are constants. Then

$$\mathrm{d}p_i/\mathrm{d}t = - \partial E/\partial q_i = \text{constant}$$

and the p's $(\bar{l}, \bar{g}, \bar{h})$ are linear functions of the time.

Since the von Zeipel equations connecting E, S, and F contain only the new variables, it is convenient to omit the bars and write simply (L, G, H) and (l, g, h) during the calculations. Of course, the bars must be restored before the explicit transformation equations and the new canonical equations can be written.

It is convenient to divide the discussion into two parts, since the equations of order 0, 1, and 2 are degenerate. It is also possible to omit many of the algebraic details by recalling that, for example, X_k is the only part of S_k that contains l, and that

$$\frac{\partial F_1}{\partial H}\frac{\partial Z_k}{\partial h} - \nu\frac{\partial Z_k}{\partial \lambda} = 0$$

since $F_1 = \nu H$ and Z_k contains h and λ only via their sum, $h + \lambda$.

1. *Initial stages*

The first von Zeipel equation is simply

$$E_0 = F_0 = \mu^2/2L^2$$

and the second is

$$E_1 = F_1 + \frac{\partial F_0}{\partial L}\frac{\partial X_1}{\partial l} = \nu H - \frac{\mu^2}{L^3}\frac{\partial X_1}{\partial l},$$

which separates immediately into $X_1 = 0$, $E_1 = \nu H$, since F_1 is entirely of type E.

The equation of second order is, in view of the result just obtained,

$$E_2 = F_2 - \frac{\mu^2}{L^3}\frac{\partial X_2}{\partial l} + \nu\left(\frac{\partial Y_1}{\partial h} - \frac{\partial Y_1}{\partial \lambda}\right).$$

Since $ZF_2 = 0$, this equation separates into three:

$$\frac{\mu^2}{L^3}\frac{\partial X_2}{\partial l} = XF_2$$

$$\nu\left(\frac{\partial Y_1}{\partial \lambda} - \frac{\partial Y_1}{\partial h}\right) = YF_2$$

$$E_2 = EF_2$$

and integrating gives

$$X_2 = \frac{1}{4}\nu_2\frac{L^3}{\mu^2}[A_1 + 3B_1\cos(2g + 2h) - 3C_1\sin(2g + 2h)]$$

$$Y_1 = (\phi - \lambda)H - \frac{15}{16}\nu_1 a^2 e^2 \sin(2g + 2h)$$

$$E_2 = \frac{1}{4}\nu_2 a^2\left(1 + \frac{3}{2}e^2\right) = \frac{1}{8}\frac{\nu_2}{\mu^2}(5L^4 - 3G^2L^2)$$

where

$$A_1 = \int A_0 \, dl = \int A_0 (1 - e \cos u) \, du$$
$$= a^2 [(\tfrac{3}{4} e^3 - 2e) \sin u + \tfrac{3}{4} e^2 \sin 2u - \tfrac{1}{12} e^3 \sin 3u]$$
$$B_1 = \int B_0 \, dl$$
$$= a^2 [(\tfrac{5}{4} e^3 - \tfrac{5}{2} e) \sin u + (\tfrac{1}{2} + \tfrac{1}{4} e^2) \sin 2u + (\tfrac{1}{12} e^3 - \tfrac{1}{6} e) \sin 3u]$$
$$C_1 = \int C_0 \, dl$$
$$= a^2 \sqrt{1 - e^2} \, [\tfrac{5}{2} e (\tfrac{1}{2} e + \cos u) - \tfrac{1}{2} (1 + e^2) \cos 2u + \tfrac{1}{6} e \cos 3u] .$$

Following Hori [2] the term $\frac{1}{2}e$ is added to $\cos u$ to annihilate the mean value with respect to l.

2. *The general case, order $k \geqslant 3$*

When the explicit equation for the new Hamiltonian was separated according to the order k, it was implicitly assumed that the order of a term is not affected by the process of differentiation. However, the present Hamiltonian, beginning with F_3 contains the factor $\gamma = 1 - (H/G)$, so that differentiation may 'lose' one power of γ, thereby reducing the order. For example, the terms

$$\frac{\partial F_3}{\partial \gamma} \frac{\partial \gamma}{\partial G} \quad \text{and} \quad \frac{\partial F_3}{\partial \gamma} \frac{\partial \gamma}{\partial H}$$

are of second order. When this is taken into account, the general equation, of order k can be written in the form

$$E_k = -\frac{\mu^2}{L^3} \frac{\partial X_k}{\partial l} + \nu \left(\frac{\partial Y_{k-1}}{\partial h} - \frac{\partial Y_{k-1}}{\partial \lambda} \right) + \left(\frac{\partial F_2}{\partial G} + \frac{\partial F_3}{\partial \gamma} \frac{\partial \gamma}{\partial G} \right) \frac{\partial Z_{k-2}}{\partial g}$$
$$+ \left(\frac{\partial F_2}{\partial H} + \frac{\partial F_3}{\partial \gamma} \frac{\partial \gamma}{\partial H} \right) \frac{\partial Z_{k-2}}{\partial h} + Q_k$$

where Q_k is a sum of terms involving F_k and previously computed quantities. The currently undetermined components X_k, Y_{k-1}, Z_{k-2} occur only in the terms exhibited here explicitly.

Recalling the separation of F_2 and F_3 into components, it can be seen that the terms

$$\left(\frac{\partial X F_2}{\partial G} + \frac{\partial X F_3}{\partial \gamma} \frac{\partial \gamma}{\partial G} \right) \frac{\partial Z_{k-2}}{\partial g} + \left(\frac{\partial X F_2}{\partial H} + \frac{\partial X F_3}{\partial \gamma} \frac{\partial \gamma}{\partial H} \right) \frac{\partial Z_{k-2}}{\partial h}$$

are entirely of type X, while the terms

$$\left(\frac{\partial Y F_2}{\partial G} + \frac{\partial Y F_3}{\partial \gamma} \frac{\partial \gamma}{\partial G} \right) \frac{\partial Z_{k-2}}{\partial g} + \left(\frac{\partial Y F_2}{\partial H} + \frac{\partial Y F_3}{\partial \gamma} \frac{\partial \gamma}{\partial H} \right) \frac{\partial Z_{k-2}}{\partial h}$$

are entirely of type Y. Since ZF_3 contains the factor e^2 as well as $\gamma\nu_2$, the terms containing $\partial ZF_3/\partial\gamma$ and Z_{k-2} can be regarded as of order $k+1$, under the assumption that $e^2=0(m)$; that is, these terms are deferred to the next stage. Since $ZF_2=(\partial EF_2/\partial H)=0$, the Z component of the equation for E_k becomes simply

$$\left(\frac{\partial EF_2}{\partial G}+\frac{\partial EF_3}{\partial\gamma}\frac{\partial\gamma}{\partial G}\right)\frac{\partial Z_{k-2}}{\partial g}+\frac{\partial EF_3}{\partial\gamma}\frac{\partial\gamma}{\partial H}\frac{\partial Z_{k-2}}{\partial h}+ZQ_k=0$$

Inserting EF_2 and EF_3 gives

$$\xi\frac{\partial Z_{k-2}}{\partial g}+\eta\frac{\partial Z_{k-2}}{\partial h}=ZQ_k$$

where

$$\xi=\frac{3}{4}\nu_2\frac{a^2}{G}\left[1-e^2+(1-\gamma)\left(1+\frac{3}{2}e^2\right)\right]$$

$$\eta=-\frac{3}{4}\nu_2\frac{a^2}{G}\left(1+\frac{3}{2}e^2\right).$$

If the typical term of ZQ_k has the form $f(\theta)$ with $\theta=ig+j(h+\lambda)$, then the corresponding term in Z_{k-2}, is

$$\frac{1}{\xi_i+\eta j}\int f(\theta)\,\mathrm{d}\theta.$$

The only case that can produce a small denominator is $j=2i$, giving

$$\xi i+\eta j=-\frac{3}{4}\nu_2\frac{a^2 i}{G}\left[\frac{5}{2}e^2+\gamma\left(1+\frac{3}{2}e^2\right)\right].$$

If γ vanishes, the orbit plane coincides with the ecliptic, the node loses its identity, and g, h, occur only in the sum $g+h$. Hence, terms of type Z must contain the single argument, $g+h+\lambda$, so that $j=i$ and the small divisor does not occur. Hence, the divisor never vanishes.

Once Z_{k-2} has been obtained, the equation for E_k separates into three:

$$\frac{\mu^2}{L^3}\frac{\partial X_k}{\partial l}=\text{sum of terms of type } X$$

$$\nu\left(\frac{\partial Y_{k-1}}{\partial\lambda}-\frac{\partial Y_{k-1}}{\partial h}\right)=\text{sum of terms of type } Y$$

$$E_k=\text{sum of terms of type } E$$

and these can be integrated by quadrature.

3. *The equations of order 3 and 4*

If the notation of the preceding section is used, the known terms of order three are

$$Q_3=F_3+\nu\left(\frac{\partial X_2}{\partial h}-\frac{\partial X_2}{\partial\lambda}\right)+\left(\frac{\partial F_2}{\partial G}+\frac{\partial F_3}{\partial\gamma}\frac{\partial\gamma}{\partial G}\right)\frac{\partial Y_1}{\partial g}+\left(\frac{\partial F_2}{\partial H}+\frac{\partial F_3}{\partial\gamma}\frac{\partial\gamma}{\partial H}\right)\frac{\partial Y_1}{\partial h}$$

First consider the terms involving F_3 and Y_1. These can be put in the form

$$\frac{1}{G}\frac{\partial F_3}{\partial \gamma}\left(\frac{\partial Y_1}{\partial g} - \frac{\partial Y_1}{\partial h}\right) - \frac{\gamma}{G}\frac{\partial F_3}{\partial \gamma}\frac{\partial Y_1}{\partial g}$$

by noting that

$$\begin{aligned} \gamma &= 1 - H/G \qquad & \partial\gamma/\partial G &= H/G^2 = (1-\gamma)/G \\ H &= G(1-\gamma) \qquad & \partial\gamma/\partial H &= -1/G\,. \end{aligned}$$

The second term, due to the reappearance of the 'lost' γ, is of fourth order, and hence can be deferred to the next stage (this is true at every stage). The first term vanishes, since Y_1 contains the single argument $g+h$ (this does not occur at later stages).

Next consider the term

$$\frac{\partial F_2}{\partial H}\frac{\partial Y_1}{\partial h} = -\frac{15}{8}\nu_2\,\delta a^2 e^2 \cos(2g + 2h)\,.$$

Again the series expansion for the solar factor δ must be used, giving

$$\begin{aligned} &-\tfrac{15}{16}\nu_2\,\delta_1 a^2 e^2\left[\cos(2g+2h+\lambda) + \cos(2g+2h-\lambda)\right] \\ &-\tfrac{15}{16}\nu_2\,\delta_2 a^2 e^2\left[\cos(2g+2h+2\lambda) + \cos(2g+2h-2\lambda)\right] \\ &+\cdots, \end{aligned}$$

The terms on the first line are of third order, while the remaining terms can be deferred to later stages. The treatment of the other terms in Q_3 is straightforward. In particular, the only term of type Z is

$$ZQ_3 = ZF_3 = \tfrac{15}{8}\gamma\nu_2 a^2 e^2 \cos 2g$$

and the general theory of the preceding section, with $i=2, j=0$, gives

$$Z_1 = \frac{\tfrac{5}{8}\gamma G e^2 \sin 2g}{1 + \tfrac{1}{4}e^2 - \tfrac{1}{2}\gamma(1 + \tfrac{3}{2}e^2)}\,.$$

Expanding the denominator by the binomial theorem, with

$$\zeta = \tfrac{1}{2}\gamma(1 + \tfrac{3}{2}e^2) - \tfrac{1}{4}e^2$$

gives

$$Z_1 = \tfrac{5}{8}\gamma G e^2 \sin 2g$$

and the remaining terms

$$Z_1(\zeta + \zeta^2 + \zeta^3 + \cdots)$$

can be assigned to the higher stages $Z_2, Z_3, \ldots$.

The process of inserting Z_1 in the terms not used in its calculation is straightforward. In particular, the deferred terms involving $\partial ZF_3/\partial\gamma$ become

$$\frac{\partial ZF_3}{\partial \gamma}\frac{\partial \gamma}{\partial G}\frac{\partial Z_1}{\partial g} = \frac{75}{64}\gamma\nu_2 a^2 e^4 (1-\gamma)(1 + \cos 4g)$$

thus making contributions of types Z and E to the fourth-order equation.

The rest of the procedure is straightforward, provided that wherever H/G occurs it is to be replaced by $1-\gamma$ and the terms with the extra γ are reassigned to the next stage. Then the third-order equation yields

$$
\begin{aligned}
X_3 = {} & \beta X_2 + \frac{3}{2}\gamma\nu_2 \frac{L^3}{\mu^2}\sin h\left(1+\frac{5}{4}e^2\cos 2g\right) \\
& \times -[A_1 \sin h + B_1 \sin(2g+h) + C_1\cos(2g+h)] \\
& - \frac{3}{2}\nu\nu_2\frac{L^6}{\mu^4}[B_2\sin(2g+2h) + C_2\cos(2g+2h)] \\
& + \frac{1}{4}\nu_2\frac{L^3}{\mu^2}\left[\frac{5}{4}\gamma e^2 G\cos 2g - \frac{15}{8}\nu_1 a^2 e^2\cos(2g+2h)\right] \\
& \times\left[\frac{\partial A_1}{\partial G} + 3\frac{\partial B_1}{\partial G}\cos(2g+2h) - 3\frac{\partial C_1}{\partial G}\sin(2g+2h)\right]
\end{aligned}
$$

$$
\begin{aligned}
Y_2 = {} & \frac{1}{4}\nu_1 a^2\left(1+\frac{3}{2}e^2\right)(\phi - \lambda + \varepsilon\sin\phi) \\
& - \frac{45}{64}\nu_1^2 a^4 e^2\frac{G}{L^2}\left[\sin(2g+2h) + \frac{5}{4}\sin(4g+4h)\right] \\
& + \frac{15}{16}\nu_1(\delta_1-\beta_1)\,a^2e^2\left[\sin(2g+2h+\lambda) + \frac{1}{3}\sin(2g+2h-\lambda)\right] \\
& - \frac{3}{8}\gamma\nu_1 a^2\left(1 - \frac{13}{8}e^2 + \frac{25}{16}e^4\right)\sin 2h \\
& + \frac{45}{128}\gamma\nu_1 a^2e^2(2-e^2)\sin(2g+2h) + \frac{15}{64}\gamma\nu_1 a^2e^2 \\
& \times\left(1+\frac{3}{2}e^2\right)\sin(2g-2h) + \frac{75}{128}\gamma\nu_1 a^2 e^2(2-e^2)\sin(4g+2h)
\end{aligned}
$$

$$
E_3 = \frac{225}{64}\nu_1\nu_2 a^4 e^2\frac{G}{L^2} - \frac{3}{4}\gamma\nu_2 a^2\left(1+\frac{3}{2}e^2\right)
$$

where

$$
B_2 = \int B_1\,dl \quad\text{and}\quad C_2 = \int C_1\,dl.
$$

Partial results from the fourth stage are

$$
\begin{aligned}
Z_2 = {} & \frac{5}{8}Ge^2(\beta_2 - \delta_2 + \delta_1^2 - \delta_1\beta_1)\sin(2g+2h+2\lambda) \\
& + \left[-\frac{5}{32}\gamma Ge^4 + \frac{5}{4}\gamma^2 Ge^2\left(1 - \frac{75}{64}e^2 + \frac{75}{256}e^4\right)\right. \\
& \left. - \frac{45}{256}\gamma\nu_1 a^2e^2(12 - 23e^2 + 11e^4)\right]\sin 2g
\end{aligned}
$$

$$+\left[\frac{25}{128}\gamma G e^4 - \frac{25}{1024}\gamma^2 G e^2 (16 - 16e^2 + e^4)\right.$$
$$\left. + \frac{225}{1024}\gamma \nu_1 a^2 e^4 (1 - e^2)\right] \sin 4g - \frac{125}{3072}\gamma^2 e^4 (4 - e^2) \sin 6g$$
$$E_4 = \frac{\nu_2^2 a^6}{L^2}\left(-\frac{49}{64} + \frac{873}{64} e^2 - \frac{4347}{512} e^4\right) + \frac{675}{1024}\nu_1^2 \nu_2 \frac{a^6 e^2}{L^2}(4 - 5\, e^2)$$
$$+ \frac{9}{512}\frac{\gamma \nu_1 \nu_2}{G} a^4 (16 - 352e^2 + 411e^4 - 75e^6) + \frac{75}{64}\gamma \nu_2 a^2 e^4$$
$$+ \frac{3}{512}\gamma^2 \nu_2 a^2 (64 - 304e^2 + 400e^4 - 25e^6).$$

This completes the discussion of the lunar theory. The validity of the method has been established, and the calculations have been carried far enough to illustrate the manipulative techniques that are required.

4. Artificial Satellite Theory

A. DEVELOPMENT OF THE HAMILTONIAN

The primary purpose here is to demonstrate that the generalized von Zeipel transformation can be applied directly, in a simple, degenerate form, to artificial satellite theory. The usual assumptions about the gravitational field of a triaxial ellipsoid give the Hamiltonian

$$F = \frac{\mu^2}{2L^2} + \nu H + \sum_{n=2}^{\infty} D_n + \sum_{n=2}^{\infty} \sum_{m=1}^{n} T_{n,m}$$

where the zonal harmonics, D_n, and the tesseral harmonics, $T_{n,m}$ are

$$D_n = -\mu J_n \frac{R^n}{r^{n+1}} P_n(\sin\theta)$$

$$T_{n,m} = \mu \frac{R^n}{r^{n+1}} (C_{n,m} \cos m\Lambda + S_{n,m} \sin m\Lambda) P_n^m(\sin\theta)$$

Here ν is the spin velocity of the earth, R is the equatorial radius of the earth, θ and Λ are the geographical latitude and longitude (measured east from Greenwich), J, C, and S are dimensionless constants, P_n is the Legendre polynomial, and P_n^m is the associated Legendre function

$$P_n^m(Z) = (1 - Z^2)^{m/2}\, \mathrm{d}^m P_n(Z)/\mathrm{d}Z^m.$$

The notation for the zonal harmonics is Brouwer's [1] and the notation for the tesseral harmonics is Izsak's [7]. As in the lunar theory the term νH is added to the Hamiltonian because the longitude of the node, h, is measured from a rotating reference line (the meridian of Greenwich).

Orders of magnitude can be defined by introducing n, the mean motion of the satellite, the dimensionless parameter $m = \nu/n \ll 1$, and the usual assumptions (Garfinkel [8])

$$\begin{aligned} J_2 &= 0(m^2) \\ J_n &= 0(J_2^2), \quad n > 2 \\ C_{n,m} &= 0(J_2^2) \\ S_{n,m} &= 0(J_2^2). \end{aligned}$$

Thus,

$$\begin{aligned} F &= F_0 + F_1 + F_2 + F_3 + F_4 \\ F_0 &= \mu^2/2L^2 \\ F_1 &= \nu H \\ F_2 &= D_2 \\ F_3 &= 0 \\ F_4 &= \text{sum of tesseral harmonics and remaining zonal harmonics}. \end{aligned}$$

To decompose F_k by types (X, Y, Z, E)

$$F_k = XF_k + YF_k + ZF_k + EF_k$$

note that, since the present Hamiltonian does not contain the time, the definitions reduce to

X: periodic in l
Y: periodic in g and h, independent of l
Z: periodic in g, independent of l and h
E: independent of all three angle variables.

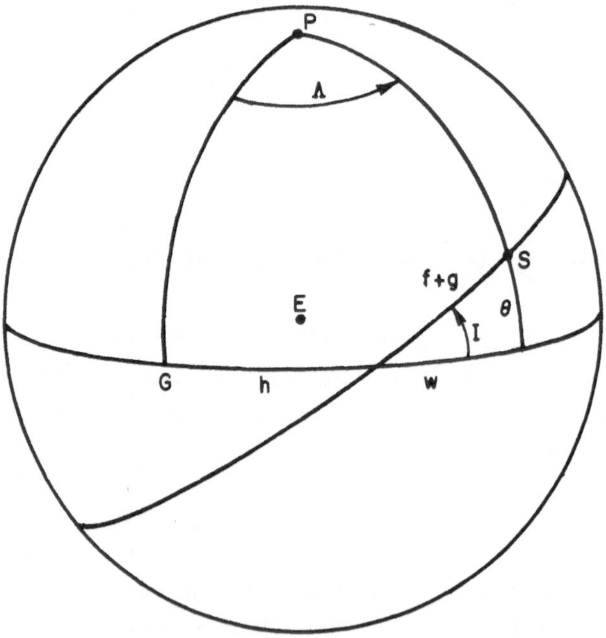

Fig. 2.

Following Brouwer [1], introduce l, the mean anomaly, implicitly via f, the true anomaly, by means of the formulas

$$r = \frac{G^2/\mu}{1 + e\cos f}$$
$$\sin\theta = \sin I \sin(f + g).$$

Then the Legendre polynomials become trigonometric polynomials in $f+g$. For example,

$$P_2(\sin\theta) = \tfrac{3}{2}\sin^2\theta - \tfrac{1}{2} = \tfrac{1}{4}(1 - 3\cos^2 I) - \tfrac{3}{4}\sin^2 I\cos(2f + 2g)$$
$$P_3(\sin\theta) = \tfrac{5}{2}\sin^3\theta - \tfrac{3}{2}\sin\theta$$
$$= \tfrac{3}{8}\sin I(1 - 5\cos^2 I)\sin(f + g) - \tfrac{5}{8}\sin^3 I\sin(3f + 3g)$$
$$P_4(\sin\theta) = \tfrac{35}{8}\sin^4\theta - \tfrac{15}{4}\sin^2\theta + \tfrac{3}{8}$$
$$= \tfrac{3}{64}(3 - 30\cos^2 I + 35\cos^4 I)$$
$$- \tfrac{5}{16}\sin^2 I(1 - 7\cos^2 I)\cos(2f + 2g)$$
$$+ \tfrac{35}{64}\sin^4 I\cos(4f + 4g)$$

and the associated Legendre functions can be treated similarly, since

$$P_n^m(\sin\theta) = \cos^m\theta\,\frac{d^m P_n(\sin\theta)}{d(\sin\theta)^m}.$$

Thus, for example,

$$P_2^2(\sin\theta) = 3\cos^2\theta$$
$$P_3^1(\sin\theta) = \cos\theta(\tfrac{15}{2}\sin^2\theta - \tfrac{3}{2}) = \tfrac{3}{4}\cos\theta$$
$$\times [3 - 5\cos^2 I - 5\sin^2 I\cos(2f + 2g)]$$
$$P_3^2(\sin\theta) = 15\cos^2\theta\sin\theta = 15\cos^2\theta\sin I\sin(f + g)$$
$$P_3^3(\sin\theta) = 15\cos^3\theta.$$

The factor $\cos^m\theta$ will disappear by cancellation once the longitudinal factors have been obtained. To show this, use the trigonometric formulas [9]

$$w = \Lambda - h$$
$$\sin w = \frac{\sin(f + g)\cos I}{\cos\theta}$$
$$\cos w = \frac{\cos(f + g)}{\cos\theta}$$

to obtain

$$\sin\Lambda = \frac{c_1\sin h + s_1\cos h}{\cos\theta}$$
$$\cos\Lambda = \frac{c_1\cos h - s_1\sin h}{\cos\theta}$$
$$c_1 = \cos(f + g)$$
$$s_1 = \cos I\sin(f + g)$$

and hence, by induction,

$$\sin m\Lambda = \frac{c_m \sin mh + s_m \cos mh}{\cos^m \theta}$$

$$\cos m\Lambda = \frac{c_m \cos mh - s_m \sin mh}{\cos^m \theta}$$

$$c_{m+1} = c_m c_1 - s_m s_1$$
$$s_{m+1} = c_m s_1 + s_m c_1 .$$

so that, for example,

$$c_2 = \tfrac{1}{2} \sin^2 I + \tfrac{1}{2}(1 + \cos^2 I) \cos(2f + 2g)$$
$$s_2 = \cos I \sin(2f + 2g)$$
$$c_3 = \tfrac{3}{4} \sin^2 I \cos(f + g) + \tfrac{1}{4}(1 + 3 \cos^2 I) \cos(3f + 3g)$$
$$s_3 = \tfrac{3}{4} \sin^2 I \cos I \sin(f + g) + \tfrac{1}{4}(3 + \cos^2 I) \cos I \sin(3f + 3g).$$

These yield the desired expressions for the tesseral harmonics, typical examples being

$$T_{2,2} = 3\mu \frac{R^2}{r^3} \{(C_{2,2} \cos 2h + S_{2,2} \sin 2h) [\tfrac{1}{2} \sin^2 I + \tfrac{1}{2}(1 + \cos^2 I) \times \cos(2f + 2g)] + (S_{2,2} \cos 2h - C_{2,2} \sin 2h) \cos I \sin(2f + 2g)\}$$

$$T_{3,1} = \frac{3}{4} \mu \frac{R^3}{r^4} [(C_{3,1} \cos h + S_{3,1} \sin h) \cos(f + g) + (S_{3,1} \cos h - C_{3,1} \sin h) \cos I \sin(f + g)] \times [3 - 5 \cos^2 I - 5 \sin^2 I \cos(2f + 2g)].$$

Thus, both zonal and tesseral harmonics can be expressed as trigonometric polynomials in $f + g$, multiplied by negative powers of r, and the coefficients are functions of I and h. To effect the decomposition by types, introduce the mean values with respect to l (Brouwer, [1]):

$$\bar{D}_n = \frac{1}{2\pi} \int_0^{2\pi} D_n \, dl$$

$$\bar{T}_{n,m} = \frac{1}{2\pi} \int_0^{2\pi} T_{n,m} \, dl$$

so that the components of type X are simply

$$XD_n = D_n - \bar{D}_n$$
$$XT_{n,m} = T_{n,m} - \bar{T}_{n,m} .$$

Perform the integrations by changing to the true anomaly, f, as the variable of integration, using the formulas

$$\mathrm{d}l = \frac{\mu^2 r^2}{GL^3}\,\mathrm{d}f\,, \qquad r = \frac{G^2/\mu}{1 + e\cos f}\,.$$

Then the negative powers of r become positive powers of $1 + e\cos f$, and the integrands become trigonometric polynomials in f.

The zonal harmonics have no components of type Y, and the tesseral harmonics have no components of type Z or E. Hence,

$$\bar{D}_n = ZD_n + ED_n$$
$$\bar{T}_{n,m} = YT_{n,m}$$

and the separation of $\bar{D}_n$ is done by inspection. Typical results are:

$$F_2 = -\mu J_2 \frac{R^2}{r^3} P_2(\sin\theta) = \frac{1}{4}\mu J_2 \frac{R^2}{r^3}$$
$$\times\,[3\cos^2 I - 1 + 3\sin^2 I\cos(2f + 2g)]$$
$$EF_2 = \bar{F}_2 = \frac{1}{4} J_2 \frac{\mu^4 R^2}{G^3 L^3}(3\cos^2 I - 1)$$
$$YF_2 = ZF_2 = 0\,, \qquad XF_2 = F_2 - EF_2\,.$$

The third zonal harmonic yields

$$ZD_3 = \bar{D}_3 = \frac{3}{8} J_3 \frac{\mu^5 R^3 e}{G^5 L^3}\sin I(5\cos^2 I - 1)\sin g$$
$$YD_3 = ED_3 = 0\,, \qquad XD_3 = D_3 - ZD_3$$

and the fourth gives

$$ZD_4 = \frac{15}{64} J_4 \frac{\mu^6 R^4 e^2}{G^7 L^3}\sin^2 I(1 - 7\cos^2 I)\cos 2g$$
$$ED_4 = -\frac{3}{64} J_4 \frac{\mu^6 R^4}{G^7 L^3}(1 + \tfrac{3}{2}e^2)(3 - 30\cos^2 I + 35\cos^4 I)$$
$$YD_4 = 0\,, \qquad XD_4 = D_4 - (ZD_4 + ED_4)\,.$$

The first two tesseral harmonics give

$$YT_{2,2} = \frac{3}{2}\frac{\mu^4 R^2}{G^3 L^3}\sin^2 I(C_{2,2}\cos 2h + S_{2,2}\sin 2h)$$
$$XT_{2,2} = T_{2,2} - YT_{2,2}$$

and

$$YT_{3,1} = \frac{3}{8}\frac{\mu^5 R^3 e}{G^5 L^3}[(C_{3,1}\cos h + S_{3,1}\sin h)(1 - 5\cos^2 I)\cos g$$
$$+ (S_{3,1}\cos h - C_{3,1}\sin h)(11 - 15\cos^2 I)\sin g]\,.$$

It is clear that every harmonic can be decomposed in this fashion.

B. SOLUTION OF THE ARTIFICIAL SATELLITE PROBLEM

1. *Initial stages*

The generalized von Zeipel procedure gives, for the first three stages, essentially the first-order results obtained by Brouwer [1]

$$E_0 = F_0 = \mu^2/2L^2$$

$$E_1 = F_1 - \frac{\mu^2}{L^3}\frac{\partial X_1}{\partial l} = F_1 = \nu H$$

and, as in the lunar theory, $X_1 = 0$. The second-order equation is

$$E_2 = XF_2 + EF_2 - \frac{\mu^2}{L^3}\frac{\partial X_2}{\partial l} + \nu \frac{\partial Y_1}{\partial h}.$$

Separating and integrating gives

$$Y_1 = 0$$

$$E_2 = EF_2 = \frac{1}{4} J_2 \frac{\mu^4 R^2}{G^3 L^3}\left(3\frac{H^2}{G^2} - 1\right)$$

$$X_2 = \frac{L^3}{\mu^2}\int XF_2 \, dl = \frac{L^3}{\mu^2}\left(\int F_2 \, dl - E_2 l\right)$$

$$= \tfrac{1}{4} J_2 \frac{\mu^2 R^2}{G^3}\{(3\cos^2 I - 1)(f - l + e\sin f)$$

$$+ 3\sin^2 I\left[\tfrac{1}{2}\sin(2f + 2g) + \tfrac{1}{2}e\sin(f + 2g) + \tfrac{1}{6}e\sin(3f + 2g)\right]\}.$$

2. *The general case, order $k \geqslant 3$*

Since the quantities λ and γ of the lunar theory do not occur in the satellite problem, the von Zeipel equations do not suffer any loss of order due to differentiation. Hence, the general equation becomes

$$E_k = -\frac{\mu^2}{L^3}\frac{\partial X_k}{\partial l} + \nu\frac{\partial Y_{k-1}}{\partial h} + \frac{\partial F_2}{\partial G}\frac{\partial Z_{k-2}}{\partial g} + Q_k$$

where Q_k is a sum of known terms. The component Z_{k-2} is obtained from

$$\frac{\partial EF_2}{\partial G}\frac{\partial Z_{k-2}}{\partial g} + ZQ_k = 0$$

that is,

$$\tfrac{3}{4}J_2 \frac{\mu^4 R^2}{G^4 L^3}(5\cos^2 I - 1)\frac{\partial Z_{k-2}}{\partial g} = ZQ_k$$

which can be integrated by quadrature. The general equation then becomes

$$E_k = -\frac{\mu^2}{L^3}\frac{\partial X_k}{\partial l} + \nu\frac{\partial Y_{k-1}}{\partial h} + \frac{\partial XF_2}{\partial G}\frac{\partial Z_{k-2}}{\partial g} + Q_k - ZQ_k$$

which separates into three:

$$\frac{\mu^2}{L^3}\frac{\partial X_k}{\partial l} = XQ_k + \frac{\partial XF_2}{\partial G}\frac{\partial Z_{k-2}}{\partial g}$$

$$-\nu\frac{\partial Y_{k-1}}{\partial h} = YQ_k$$

$$E_k = EQ_k.$$

Notice that the partial differential equations of lunar theory have degenerated to ordinary equations here.

3. *Equations of order 3 and 4*

The third-order equation is completely degenerate since the known terms are

$$Q_3 = F_3 + \nu\frac{\partial X_2}{\partial h} + \frac{\partial F_2}{\partial G}\left(\frac{\partial X_1}{\partial g} + \frac{\partial Y_1}{\partial g}\right) + \frac{\partial F_2}{\partial L}\frac{\partial X_1}{\partial l} + \frac{\partial^2 F_0}{\partial L^2}\frac{\partial X_1}{\partial l}\frac{\partial X_2}{\partial l} = 0$$

giving

$$E_3 = X_3 = Y_2 = Z_1 = 0$$

For the fourth-order equation the known terms are

$$Q_4 = F_4 + \frac{\partial F_2}{\partial L}\frac{\partial X_2}{\partial l} + \frac{\partial F_2}{\partial G}\frac{\partial X_2}{\partial g} + \frac{1}{2}\frac{\partial^2 F_0}{\partial L^2}\left(\frac{\partial X_2}{\partial l}\right)^2.$$

Since this expression is quite lengthy the components of type X will not be exhibited. The zonal harmonics give the components of types Z and E which yield Brouwer's results [1]

$$Z_2 = \frac{1}{32}\frac{\mu^2 R^2 e^2}{G^3}\frac{\sin^2 I}{5\cos^2 I - 1}\left[J_2(1 - 15\cos^2 I) + \frac{5J_4}{J_2}(1 - 7\cos^2 I)\right]\sin 2g - \frac{1}{2}\frac{J_3}{J_2}\frac{\mu Re}{G}\sin I\cos g$$

$$E_4 = \frac{3}{128}J_2^2\frac{\mu^6 R^4}{G^7 L^3}\left[\frac{G^2}{L^2}(5\cos^4 I - 18\cos^2 I + 5) + 4\frac{G}{L}(3\cos^2 I - 1)^2 + 5(7\cos^4 I + 2\cos^2 I - 1)\right] + \frac{3}{128}J_4\frac{\mu^6 R^4}{G^7 L^3}\left(3\frac{G^2}{L^2} - 5\right)(35\cos^4 I - 30\cos^2 I + 3).$$

The first two tesseral harmonics yield the typical contribution to Y_3:

$$Y_3 = -\frac{3}{4}\frac{\mu^4 R^2}{\nu G^3 L^3}\sin^2 I(C_{2,2}\sin 2h - S_{2,2}\cos 2h) + \frac{3}{8}\frac{\mu^5 R^3 e}{\nu G^5 L^3}[(5\cos^2 I - 1)\cos g(C_{3,1}\sin h - S_{3,1}\cos h) + (15\cos^2 I - 11)\sin g(S_{3,1}\sin h + C_{3,1}\cos h)].$$

Again, the development can be terminated with the remark that the validity of the method has been established, provided the critical inclination is avoided.

5. Concluding Remarks

The generalized von Zeipel transformation presented here is clearly capable of producing a highly efficient lunar theory. The major simplification resides in the partial uncoupling of solar and lunar factors. This occurs in the initial development of the Hamiltonian and in the final development of the determining function. At both stages many Fourier series expansions are avoided, the result being a product of two finite expressions. This is achieved primarily because the elimination of the short period terms is effected by solving an ordinary differential equation, in which solar factors play the role of constants.

If the inclination angle is not restricted to small values, the same technique can be applied to eliminate the terms of types X and Y, with terms of type Z remaining in the new Hamiltonian. The resulting system might well be a fruitful subject for future research.

Appendix. Explicit Recursive Algorithms for the von Zeipel Transformation

The basic, implicit equations of the von Zeipel transformation are

$$x_i = q_i + \partial S(q, y)/\partial y_i$$
$$p_i = y_i + \partial S(q, y)/\partial q_i$$

where, for present purposes, the dependence on t, via λ, is irrelevant. Let $\phi(x, y)$ be an arbitrary function and let

$$\psi(q, p) = \phi(x, y).$$

That is, this is an identity in either set of variables under the transformation $(x, y) \leftrightarrow (q, p)$. Eliminating p in the left member and x in the right member gives

$$\psi\left\{q, \left[y_i + \frac{\partial S(q, y)}{\partial q_i}\right]\right\} = \phi\left\{\left[q_i + \frac{\partial S(q, y)}{\partial y_i}\right], y\right\}$$

an identity in (q, y). Expanding each member in Taylor's series gives

$$\psi(q, y) + \sum_{k=1}^{\infty} \frac{1}{k!} \frac{\partial^k \psi(q, y)}{\partial y_{i_1} \partial y_{i_2} \cdots \partial y_{i_k}} \frac{\partial S}{\partial q_{i_1}} \frac{\partial S}{\partial q_{i_2}} \cdots \frac{\partial S}{\partial q_{i_k}}$$
$$= \phi(q, y) + \sum_{k=1}^{\infty} \frac{1}{k!} \frac{\partial^k \phi(q, y)}{\partial q_{i_1} \partial q_{i_2} \cdots \partial q_{i_k}} \frac{\partial S}{\partial y_{i_1}} \frac{\partial S}{\partial y_{i_2}} \cdots \frac{\partial S}{\partial y_{i_k}}$$

with the repeated subscripts $i_1, i_2, \ldots, i_k$ satisfying the summation convention. Since this is an identity in the variables (q, y), the variables are dummies, and can be replaced by (q, p) or by (x, y). Introducing the series expansion

$$\phi = \sum_{n=0}^{\infty} \phi_n$$

$$\psi = \sum_{n=0}^{\infty} \psi_n$$

$$S = \sum_{n=1}^{\infty} S_n$$

and collecting terms of equal order gives, with (q, y) replaced by (q, p):

$$\psi_0(q, p) = \phi_0(q, p)$$

and, for $n \geqslant 1$,

$$\psi_n(q, p) + \sum_{k=1}^{n} \frac{1}{k!} \sum_{\alpha} \frac{\partial^k \psi_\alpha(q, p)}{\partial p_{i_1} \partial p_{i_2} \ldots \partial p_{i_k}} \frac{\partial S_{\alpha_1}(q, p)}{\partial q_{i_1}} \cdots \frac{\partial S_{\alpha_k}(q, p)}{\partial q_{i_k}}$$
$$= \phi_n(q, p) + \sum_{k=1}^{n} \frac{1}{k!} \sum_{\alpha} \frac{\partial^k \phi_\alpha(q, p)}{\partial q_{i_1} \partial q_{i_2} \ldots \partial q_{i_k}} \frac{\partial S_{\alpha_1}(q, p)}{\partial p_{i_1}} \cdots \frac{\partial S_{\alpha_k}(q, p)}{\partial pi_k}$$

where $\sum_\alpha$ denotes the sum over all combinations of $\alpha, \alpha_1, \alpha_2, \ldots, \alpha_k$ satisfying

$$\alpha \geqslant 0, \qquad \alpha_j \geqslant 1$$
$$\alpha + \sum_{j=1}^{k} \alpha_j = n.$$

This can be put in the explicit form

$$\psi_n + \sum_{k=1}^{n} \frac{1}{k!} \sum_{\alpha_1=1}^{r_1} \sum_{\alpha_2=1}^{r_2} \cdots \sum_{\alpha_k=1}^{r_k} \frac{\partial^k \psi_\alpha}{\partial p_{i_1} \partial p_{i_2} \ldots \partial p_{i_k}} \frac{\partial S_{\alpha_1}}{\partial q_{i_1}} \cdots \frac{\partial S_{\alpha_k}}{\partial q_{i_k}}$$
$$= \phi_n + \sum_{k=1}^{n} \frac{1}{k!} \sum_{\alpha_1=1}^{r_1} \sum_{\alpha_2=1}^{r_2} \cdots \sum_{\alpha_k=1}^{r_k} \frac{\partial^k \phi_\alpha}{\partial q_{i_1} \partial q_{i_2} \ldots \partial q_{i_k}} \frac{\partial S_{\alpha_1}}{\partial p_{i_1}} \cdots \frac{\partial S_{\alpha_k}}{\partial p_{i_k}}$$

where

$$r_1 = n + 1 - k$$
$$r_{j+1} = r_j - \alpha_j + 1 \quad \text{for} \quad j = 1, 2, \ldots, k-1$$
$$\alpha = r_k - \alpha_k .$$

Thus, for example,

$$\psi_0 = \phi_0$$
$$\psi_1 + \frac{\partial \psi_0}{\partial p_i} \frac{\partial S_1}{\partial q_i} = \phi_1 + \frac{\partial \phi_0}{\partial q_i} \frac{\partial S_1}{\partial p_i}$$

$$\psi_2 + \frac{\partial \psi_1}{\partial p_i}\frac{\partial S_1}{\partial q_i} + \frac{\partial \psi_0}{\partial p_i}\frac{\partial S_2}{\partial q_i} + \frac{1}{2}\frac{\partial^2 \psi_0}{\partial p_i \partial p_j}\frac{\partial S_1}{\partial q_i}\frac{\partial S_1}{\partial q_j}$$
$$= \phi_2 + \frac{\partial \phi_1}{\partial q_i}\frac{\partial S_1}{\partial p_i} + \frac{\partial \phi_0}{\partial q_i}\frac{\partial S_2}{\partial p_i} + \frac{1}{2}\frac{\partial^2 \phi_0}{\partial q_i \partial q_j}\frac{\partial S_1}{\partial p_i}\frac{\partial S_1}{\partial p_j}$$

and so forth, a recursive system. Thus, if $\phi_n(x, y)$ are given, then $\psi_n(q, p)$ are obtained recursively. Similarly, if these same equations are written with (x, y) in place of (q, p), then $\phi_n(x, y)$ can be obtained recursively when $\psi_n(q, p)$ are given. The equations for transforming the variables (x, y) and (q, p), in either direction, now follow immediately as special cases, by taking, successively

$$\phi_0(x, y) = x_\beta, \quad \phi_\alpha = 0 \quad \text{for} \quad \alpha \neq 0$$
$$\phi_0(x, y) = y_\beta, \quad \phi_\alpha = 0 \quad \text{for} \quad \alpha \neq 0$$
$$\psi_0(q, p) = q_\beta, \quad \psi_\alpha = 0 \quad \text{for} \quad \alpha \neq 0$$
$$\psi_0(q, p) = p_\beta, \quad \psi_\alpha = 0 \quad \text{for} \quad \alpha \neq 0.$$

The results are

$$x_\beta = \sum_{n=0}^{\infty} q_{\beta, n}$$

where

$$q_{\beta, 0} = q_\beta$$
$$q_{\beta, 1} = \partial S_1 / \partial p_\beta$$

and, for $n \geqslant 2$,

$$q_{\beta, n} = \frac{\partial S_n}{\partial p_\beta} - \sum_{k=1}^{n-1} \frac{1}{k!} \sum_{\alpha_1 = 1}^{t_1} \cdots \sum_{\alpha_k = 1}^{t_k} \frac{\partial^k q_{\beta, \alpha}}{\partial p_{i_1} \ldots \partial p_{i_k}} \frac{\partial S_{\alpha_1}}{\partial q_{i_1}} \cdots \frac{\partial S_{\alpha_k}}{\partial q_{i_k}}$$

where

$$t_1 = n - k$$
$$t_{j+1} = t_j - \alpha_j + 1 \quad \text{for} \quad j = 1, 2, \ldots, k - 1$$
$$\alpha = t_k - \alpha_k + 1.$$

Thus, for example,

$$q_{\beta, 2} = \frac{\partial S_2}{\partial p_\beta} - \frac{\partial q_{\beta, 1}}{\partial p_i}\frac{\partial S_1}{\partial q_i}$$
$$q_{\beta, 3} = \frac{\partial S_3}{\partial p_\beta} - \frac{\partial q_{\beta, 1}}{\partial p_i}\frac{\partial S_2}{\partial q_i} - \frac{\partial q_{\beta, 2}}{\partial p_i}\frac{\partial S_1}{\partial q_i} - \frac{1}{2}\frac{\partial^2 q_{\beta, 1}}{\partial p_i \partial p_j}\frac{\partial S_1}{\partial q_i}\frac{\partial S_1}{\partial q_j}$$

and so forth, where $S_k = S_k(q, p)$.

The transformation equation for y is

$$y_\beta = \sum_{n=0}^{\infty} p_{\beta, n}$$

where

$$p_{\beta,0} = p_\beta$$
$$p_{\beta,1} = -\partial S_1/\partial q_\beta$$

and, for $n \geqslant 2$,

$$p_{\beta,n} = -\sum_{k=1}^{n-1} \frac{1}{k!} \sum_{\alpha_1=1}^{r_1} \cdots \sum_{\alpha_k=1}^{r_k} \frac{\partial^k p_{\beta,\alpha}}{\partial p_{i_1} \ldots \partial p_{i_k}} \frac{\partial S_{\alpha_1}}{\partial q_{i_1}} \cdots \frac{\partial S_{\alpha_k}}{\partial q_{i_k}}$$

Thus, for example,

$$p_{\beta,2} = -\frac{\partial S_2}{\partial q_\beta} - \frac{\partial p_{\beta,1}}{\partial p_i}\frac{\partial S_1}{\partial q_i}$$

$$p_{\beta,3} = -\frac{\partial S_3}{\partial q_\beta} - \frac{\partial p_{\beta,1}}{\partial p_i}\frac{\partial S_2}{\partial q_i} - \frac{\partial p_{\beta,2}}{\partial p_i}\frac{\partial S_1}{\partial q_i} - \frac{1}{2}\frac{\partial^2 p_{\beta,1}}{\partial p_i \partial p_j}\frac{\partial S_1}{\partial q_i}\frac{\partial S_1}{\partial q_j}$$

and so forth, again with $S_k = S_k(q, p)$.

The reverse transformation is

$$q_\beta = \sum_{n=0}^{\infty} x_{\beta,n}$$

where

$$x_{\beta,0} = x_\beta$$
$$x_{\beta,1} = -\partial S_1/\partial y_\beta$$

and, for $n \geqslant 2$,

$$x_{\beta,n} = -\sum_{k=1}^{n-1} \frac{1}{k!} \sum_{\alpha_1=1}^{r_1} \cdots \sum_{\alpha_k=1}^{r_k} \frac{\partial^k x_{\beta,\alpha}}{\partial x_{i_1} \ldots \partial x_{i_k}} \frac{\partial S_{\alpha_1}}{\partial y_{i1}} \cdots \frac{\partial S_{\alpha_k}}{\partial y_{i_k}}.$$

Thus, for example,

$$x_{\beta,2} = -\frac{\partial S_2}{\partial y_\beta} - \frac{\partial x_{\beta,1}}{\partial x_i}\frac{\partial S_1}{\partial y_i}$$

$$x_{\beta,3} = -\frac{\partial S_3}{\partial y_\beta} - \frac{\partial x_{\beta,1}}{\partial x_i}\frac{\partial S_2}{\partial y_i} - \frac{\partial x_{\beta,2}}{\partial x_i}\frac{\partial S_1}{\partial y_i} - \frac{1}{2}\frac{\partial^2 x_{\beta,1}}{\partial x_i \partial x_j}\frac{\partial S_1}{\partial y_i}\frac{\partial S_1}{\partial y_j}$$

and so forth, where now $S_k = S_k(x, y)$.

Finally, the transformation equation for p is

$$p_\beta = \sum_{n=0}^{\infty} y_{\beta,n}$$

where

$$y_{\beta,0} = y_\beta$$
$$y_{\beta,1} = \partial S_1/\partial x_\beta$$

and, for $n \geqslant 2$,

$$y_{\beta,n} = \frac{\partial S_n}{\partial x_\beta} - \sum_{k=1}^{n-1} \frac{1}{k!} \sum_{\alpha_1=1}^{t_1} \cdots \sum_{\alpha_k=1}^{t_k} \frac{\partial^k y_{\beta,\alpha}}{\partial x_{i_1} \ldots \partial x_{i_k}} \frac{\partial S_{\alpha_1}}{\partial y_{i_1}} \cdots \frac{\partial S_{\alpha_k}}{\partial y_{i_k}}$$

Thus, for example,

$$y_{\beta,2} = \frac{\partial S_2}{\partial x_\beta} - \frac{\partial y_{\beta,1}}{\partial x_i}\frac{\partial S_1}{\partial y_i}$$

$$y_{\beta,3} = \frac{\partial S_3}{\partial x_\beta} - \frac{\partial y_{\beta,1}}{\partial x_i}\frac{\partial S_2}{\partial y_i} - \frac{\partial y_{\beta,2}}{\partial x_i}\frac{\partial S_1}{\partial y_i} - \frac{1}{2}\frac{\partial^2 y_{\beta,1}}{\partial x_i \partial x_j}\frac{\partial S_1}{\partial y_i}\frac{\partial S_1}{\partial y_j}$$

and so forth, again with $S_k = S_k(x, y)$.

It may be noted that the authors of recent papers on the Lie transformation [10, 11] have remarked that the von Zeipel transformation is unsatisfactory because of its implicit nature. Such remarks are no longer valid in view of the present explicit algorithms for effecting the transformation of coordinates and of arbitrary functions in either direction.

References

[1] Brouwer, D.: 1959, 'Solution of the Problem of Artificial Satellite Theory Without Drag', *Astron. J.* **64**, 378–97.

[2] Hori, G. I.: 1963, 'A New Approach to the Solution of the Main Problem of the Lunar Theory', *Astron. J.* **68**, 125–46.

[3] Brouwer, D. and Clemence, G. M.: 1961, *Methods of Celestial Mechanics*, Academic Press, New York.

[4] Kovalevsky, J.: 1966, 'Sur la théorie du mouvement d'un satellite à fortes inclinaison et excentricité', in *The Theory of Orbits in the Solar System and in Stellar Systems*, Academic Press, New York, pp. 326–44.

[5] Szebehely, V. G.: 1967, *Theory of Orbits*, The Restricted Problem of Three Bodies, Academic Press, New York.

[6] Smart, W. M.: 1960, *Celestial Mechanics*, Longmans, London.

[7] Izsak, I. G.: 1964, 'Tesseral Harmonics of the Geopotential and Corrections to Station Coordinates', *J. Geophys. Res.* **69**, 2621–30.

[8] Garfinkel, B.: 1965, 'Tesseral Harmonic Perturbations of an Artificial Satellite', *Astron. J.* **70**, 784–86.

[9] Garfinkel, B.: 1965, 'The Disturbing Function for an Artificial Satellite', *Astron. J.* **70**, 699–704.

[10] Hori, G. I.: 1966, 'Theory of General Perturbations with Unspecified Canonical Variables', *Publ. Astron. Soc. Japan* **18**, 287–96.

[11] Deprit, A.: 1968, Canonical Transformations Depending on a Small Parameter, Mathematical Note No. 574, Mathematics Research Laboratory, Boeing Scientific Research Laboratories.

Discussion

J. Kovalevsky: Although the procedure you describe does appear to be simple in a presentation, it has the drawback that the known quantities that you called Q are more and more complicated when the order increases. As in von Zeipel or Hori method, these is no simple way – by recurrence or otherwise – to compute it. At the Bureau des Longitudes, J. Chapront and L. Mangeney have devised quite a different approach to the lunar theory, where the formulae are the same for any order, permitting an iterative programmation of the computation. This method is published in the August 1969 issue of *Astronomy and Astrophysics*.

W. Mersman: Several recursive versions of the Lie transformation have been devised since the symposium. The technique of separating the variables can be applied using any of these recursive algorithms.

STABILITY OF FREE ROTATION OF A RIGID BODY*

JOHN P. VINTI

M.I.T. Measurement Systems Laboratory, Cambridge, Mass., U.S.A.

Abstract. The paper shows that a greater excursion of initial conditions, than that previously derived by the author, permits stability of free rotation of a rigid body about either of its stable principal axes. The results are exhibited by means of contour plots, in a new diagram that represents the dynamical properties of any rigid body.

For time intervals short compared with geological eras, the paper also shows that, if the earth were truly rigid, its pole of rotation would move about its pole of figure in an ellipse of small eccentricity, dependent only on the principal moments of inertia. Because of non-rigidity the observed path of the pole of rotation is very different from this. The question is raised whether the data on polar motion could ever be taken so accurately and so frequently that a sort of osculating polar ellipse, corresponding to an instantaneously rigid earth, could be fitted reliably at closely spaced intervals. If so, the path of the center of the ellipse would give the path of the pole of figure, which may have a seasonal period of one year.

1. Introduction

A recent paper by the author, henceforth referred to as V1969, derived some new quantitative results concerning an old problem, the stability of free rotation of a rigid body. The method used only the integrals of motion and the physical assumption that the body's angular velocity $\boldsymbol{\omega}$ is a continuous function of time.

The present paper deals with the same topic, but yields as its main results sufficient conditions for stability of rotation that are much less restrictive. It then depicts these conditions by means of contour lines in a new diagram that classifies all rigid bodies dynamically. Finally, it improves the earlier discussion of polar wandering in the earth.

For a rigid body rotating freely or in a uniform gravitational field, if T is its kinetic energy of rotation and $\mathbf{L}$ its angular momentum, both taken relative to the center of mass as origin, their constancy results in the equations

$$2T = A\omega_1^2 + B\omega_2^2 + C\omega_3^2 = A\omega_{10}^2 + B\omega_{20}^2 + C\omega_{30}^2 \tag{1}$$

$$\mathbf{L}^2 = A^2\omega_1^2 + B^2\omega_2^2 + C^2\omega_3^2 = A^2\omega_{10}^2 + B^2\omega_{20}^2 + C^2\omega_{30}^2 . \tag{2}$$

Here the principal moment of inertia are to be so denoted that $A \leqq B \leqq C$, the respective components of angular velocity about the principal axes are ω_1, ω_2, and ω_3, and ω_{k0} denotes $\omega_k(0)$, the value of ω_k at time $t=0$. Unit vectors along the principal axes will be denoted by $\mathbf{1}_A$, $\mathbf{1}_B$, and $\mathbf{1}_C$.

Then

$$\begin{aligned} \mathbf{L}^2 - 2TA &= B(B-A)\,\omega_2^2 + C(C-A)\,\omega_3^2 \\ &= B(B-A)\,\omega_{20}^2 + C(C-A)\,\omega_{30}^2 \end{aligned} \tag{3}$$

$$\begin{aligned} 2TC - \mathbf{L}^2 &= A(C-A)\,\omega_1^2 + B(C-B)\,\omega_2^2 \\ &= A(C-A)\,\omega_{10}^2 + B(C-B)\,\omega_{20}^2 . \end{aligned} \tag{4}$$

* This paper was prepared under the sponsorship of the Electronics Research Center of the National Aeronautics and Space Administration through NASA Grant NGR 22-009-311.

G. E. O. Giacaglia (ed.), Periodic Orbits, Stability and Resonances, 259–271.

As mentioned in V1969, the graphs of (3) and (4) are ellipses, suggesting the well known stability of rotation about the axes $\mathbf{1}_A$ and $\mathbf{1}_C$. By forming $2TB-\mathbf{L}^2$, we should find a hyperbola, suggesting the familiar instability of rotation about the intermediate axis $\mathbf{1}_B$.

Since my purpose is to obtain quantitative results, I use the language of Liapounov stability, with an ε and a δ, and try to find the ratio δ/ε. I speak only of the 'language' of Liapounov stability (hereafter called L-stability), because the first order property of the usual Eulerian equations

$$\dot{\mathbf{L}} = 0 \tag{5}$$

does not permit any other kind of stability. This is in contradistinction to the case of the orbital motion of a particle, where the second order property of the differential equations permits other kinds of stability of motion or of equilibrium.

As applied to rotational equilibrium, L-stability exists at $\boldsymbol{\omega}=\boldsymbol{\Omega}$ if for arbitrarily small ε there exists a $\delta \leqq \varepsilon$ such that $|\boldsymbol{\omega}(t)-\boldsymbol{\Omega}|<\varepsilon$ for all t, positive or negative, whenever $|\boldsymbol{\omega}(0)-\boldsymbol{\Omega}|<\delta$. In considering stability about $\mathbf{1}_A$, e.g., we should let

$$\boldsymbol{\Omega} \equiv \boldsymbol{\Omega}_A = \mathbf{1}_A\omega_0 . \tag{6}$$

Then in

$$\boldsymbol{\omega}(0) = \mathbf{1}_A\omega_{10} + \mathbf{1}_B\omega_{20} + \mathbf{1}_C\omega_{30} , \tag{7}$$

if the above inequalities hold when $\omega_{10}\neq\omega_0$, we speak of *general* L-stability. If they hold for $\omega_{10}=\omega_0$, we speak of *restricted* L-stability. In the restricted case we add only perpendicular components to $\boldsymbol{\Omega}$ and see what happens; in the general case we also change the main component. When we also change the main component, it is clear that speeding it up makes the rotation more stable and slowing it down makes it less stable. In the mathematics this intuitive result corresponds to the necessity of imposing an *upper* limit on ε, viz. $\varepsilon<|\omega_{10}|$, in order to avoid a change of sign of the main component ω_1.

In a way it is unnecessary to treat the general case, because we could always begin with $\boldsymbol{\Omega}$ as equal to the changed value of the main component. The previous paper (V1969), however, gave a proof in the general case and the same proof applies here. If the result for the restricted case is $\delta/\varepsilon=F_k^{-1}$, the result for the general case is $\delta/\varepsilon=(1+F_k)^{-1}$. The present paper derives smaller values for F_1 and F_3, thus establishing restricted L-stability for a greater excursion of initial conditions than did V 1969. A correspondingly greater excursion then holds for general L-stability.

2. Free Rotation about $\mathbf{1}_A$

If $\boldsymbol{\omega}(0)=\mathbf{1}_A\omega_0$, then $\omega_{20}=\omega_{30}=0$ and (3) then shows that $\omega_2(t)=\omega_3(t)=0$, so that $\boldsymbol{\omega}(t)=\boldsymbol{\omega}(0)$ for all t. Thus $\boldsymbol{\Omega}_A\equiv(\omega_0, 0, 0)$ is a point of equilibrium in $\boldsymbol{\omega}$-space. To show that it has restricted L-stability we let $\omega_{10}=\omega_0$, but choose $\omega_{20}^2+\omega_{30}^2\neq 0$. Our task is then to find a $\delta\leqq\varepsilon$ such that, for all t,

$$(\omega_1-\omega_0)^2+\omega_2^2+\omega_3^2<\varepsilon^2 \tag{8}$$

whenever

$$\omega_{20}^2 + \omega_{30}^2 < \delta^2 , \tag{9}$$

Since we shall work only with the integral relations (3) and (4) which hold 'for all t', we may henceforth omit that phase.

To place an upper limit on $\omega_2^2 + \omega_3^2$, we note that (3) is an ellipse

$$\frac{\omega_2^2}{\beta_1^2} + \frac{\omega_3^2}{\gamma_1^2} = 1 , \tag{10}$$

where

$$\beta_1^2 \equiv \omega_{20}^2 + \frac{C(C-A)}{B(B-A)} \omega_{30}^2 \tag{11}$$

$$\gamma_1^2 \equiv \frac{B(B-A)}{C(C-A)} \omega_{20}^2 + \omega_{30}^2 . \tag{12}$$

Since $A \leqq B \leqq C$, it follows that

$$\frac{C(C-A)}{B(B-A)} \geqq 1 \geqq \frac{B(B-A)}{C(C-A)} . \tag{13}$$

Thus $\beta_1^2 \geqq \gamma_1^2$, so that β_1 is the semi-major axis and

$$(\omega_1^2 + \omega_2^2)_{\max} = \beta_1^2 . \tag{14}$$

Then

$$\omega_1^2 + \omega_2^2 \leqq \beta_1^2 \leqq \frac{C(C-A)}{B(B-A)} (\omega_{20}^2 + \omega_{30}^2) , \tag{15}$$

by (11), (13), and (14).

The next step is to place an upper limit on $(\omega_1 - \omega_0)^2$ and this is more difficult. The work is the same as in V1969, however, so that I shall omit most of it, except for a few remarks. The procedure is this. First write

$$|\omega_1 - \omega_0| \equiv \frac{|\omega_1^2 - \omega_0^2|}{|\omega_1 + \omega_0|} . \tag{16}$$

Next use the integral relations (3) and (4), along with the continuity of $\omega_1(t)$, to prove that

$$\omega_0^2 > \frac{B(C-B)}{A(C-A)} \omega_{20}^2 + \frac{C(C-B)}{A(B-A)} \omega_{30}^2 \tag{17}$$

is a sufficient condition that ω_1 shall not change sign. Then use (3) and (4) again to establish the inequalities

$$-\frac{C(C-B)}{A(B-A)} \omega_{30}^2 \leqq \omega_1^2 - \omega_0^2 \leqq \frac{B(C-B)}{A(C-A)} \omega_{20}^2 . \tag{18}$$

Now if a number of c of unknown sign satisfies

$$-c_1 \leqq c \leqq c_2 , \tag{19.1}$$

where c_1 and c_2 are non-negative numbers and where $c_1 - c_2$ is of unknown sign, we can say only that

$$|c| \leqq c_1 + c_2 . \tag{19.2}$$

Application of (19) to (18) then shows that

$$|\omega_1^2 - \omega_0^2| \leqq \frac{B(C-B)}{A(C-A)} \omega_{20}^2 + \frac{C(C-B)}{A(B-A)} \omega_{30}^2 . \tag{20}$$

Next impose the condition (17), so that

$$|\omega_1 + \omega_0| > |\omega_0| . \tag{21}$$

Insertion of (21), (20), and (17) into (16) then yields

$$(\omega_1 - \omega_0)^2 < \frac{B(C-B)}{A(C-A)} \omega_{20}^2 + \frac{C(C-B)}{A(B-A)} \omega_{30}^2 . \tag{22}$$

But since $A \leqq B \leqq C$, we have

$$\frac{B(C-B)}{A(C-A)} \leqq \frac{C(C-B)}{A(B-A)}, \tag{23}$$

and (22) and (23) then give the kind of upper limit that we need, viz,

$$(\omega_1 - \omega_0)^2 \leqq \frac{C(C-B)}{A(B-A)} (\omega_{20}^2 + \omega_{30}^2) . \tag{24}$$

Addition of (15) and (24) then shows that

$$(\boldsymbol{\omega} - \boldsymbol{\Omega}_A)^2 \equiv (\omega_1 - \omega_0)^2 + \omega_2^2 + \omega_3^2 < (\omega_{20}^2 + \omega_{30}^2)\, F_1 \tag{25.1}$$

$$F_1^2 \equiv \frac{C(C-B)}{A(B-A)} + \frac{C(C-A)}{B(B-A)}, \tag{25.2}$$

provided that (17) is satisfied. That is,

$$(\boldsymbol{\omega} - \boldsymbol{\Omega}_A)^2 < \varepsilon^2 , \tag{26.1}$$

if

$$[\boldsymbol{\omega}(0) - \boldsymbol{\Omega}_A]^2 \equiv \omega_{20}^2 + \omega_{30}^2 < \delta^2 , \tag{26.2}$$

with

$$\delta = \varepsilon F_1^{-1} , \tag{26.3}$$

if (17) is satisfied.

We may easily satisfy (17), however, by placing an upper limit on ε, viz.

$$\varepsilon < |\omega_0| \tag{27}$$

To show this, multiply (26.2) by F_1^2 and impose (27) on the result, with use of (26.3). Then

$$(\omega_{20}^2 + \omega_{30}^2)\, F_1^2 < \varepsilon^2 < \omega_0^2 . \tag{28}$$

Inspection of (28), (25.2), (23), and (17) then shows that (27) guarantees fulfillment of (17).

Thus given an arbitrarily small $\varepsilon<|\omega_0|$ we have

$$|\boldsymbol{\omega}(t)-\boldsymbol{\Omega}_A|<\varepsilon<|\omega_0| \quad \text{if} \quad |\boldsymbol{\omega}(0)-\boldsymbol{\Omega}_A|<\delta=\varepsilon F_1^{-1}.$$

This completes the proof of restricted L-stability of $\boldsymbol{\Omega}_A\equiv(\omega_0, 0, 0)$.

3. Free Rotation about $\mathbf{1}_C$

If $\boldsymbol{\omega}(0)=\mathbf{1}_C\omega_0$, then $\omega_{10}=\omega_{20}=0$ and (4) shows that $\omega_1(t)=\omega_2(t)=0$, so that $\boldsymbol{\omega}(t)=\boldsymbol{\omega}(0)$. Thus $\boldsymbol{\Omega}_C=(0, 0, \omega_0)$ is a point of equilibrium in $\boldsymbol{\omega}$-space. To investigate its restricted L-stability we let $\omega_{30}=\omega_0$, but choose $\omega_{10}^2+\omega_{20}^2\neq 0$. Then we try to find a $\delta\leqq\varepsilon$ such that

$$(\omega_3-\omega_0)^2+\omega_1^2+\omega_2^2<\varepsilon^2, \tag{29}$$

whenever

$$\omega_{10}^2+\omega_{20}^2<\delta^2 \tag{30}$$

Just as we did in the $\mathbf{1}_A$ case, we may place an upper limit on $\omega_1^2+\omega_2^2$ by noting that (4) is an ellipse

$$\frac{\omega_1^2}{\alpha_3^2}+\frac{\omega_2^2}{\beta_3^2}=1, \tag{31}$$

where

$$\alpha_3^2\equiv\omega_{10}^2+\frac{B(C-B)}{A(C-A)}\omega_{20}^2 \tag{32}$$

$$\beta_3^2\equiv\frac{A(C-A)}{B(C-B)}\omega_{10}^2+\omega_{20}^2 \tag{33}$$

Now

$$\frac{A(C-A)}{B(C-B)}\equiv 1+\frac{(B-A)(A+B-C)}{B(C-B)} \tag{34}$$

and for any rigid body it is easy to show that

$$A+B-C\geqq 0, \tag{35}$$

the equality sign holding in the limiting case of a flat plate of finite mass and zero thickness. Then, from (34) and (35)

$$\frac{A(C-A)}{B(C-B)}\geqq 1\geqq\frac{B(C-B)}{A(C-A)} \tag{36}$$

Thus $\beta_3^2\geqq\alpha_3^2$, so that β_3 is the semi-major axis of the ellipse and

$$\omega_1^2+\omega_2^2\leqq\beta_3^2\leqq(\omega_{10}^2+\omega_{20}^2)\frac{A(C-A)}{B(C-B)} \tag{37}$$

To place an upper limit on $(\omega_3-\omega_0)^2$, we proceed as we did for $(\omega_1-\omega_0)^2$.

From (3) and (4) and the continuity of $\omega_3(t)$, show that

$$\omega_0^2 > \frac{A(B-A)}{C(C-B)}\omega_{10}^2 + \frac{B(B-A)}{C(C-A)}\omega_{20}^2 \tag{38}$$

is a sufficient condition that ω_3 shall not change sign. Then write

$$|\omega_3 - \omega_0| \equiv \frac{|\omega_3^2 - \omega_0^2|}{|\omega_3 + \omega_0|} \tag{39}$$

Imposing the condition (38) then gives

$$(\omega_3 + \omega_0)^2 > \omega_0^2 \tag{40}$$

Now with (3) and (4) establish the inequalities

$$-\frac{A(B-A)}{C(C-B)}\omega_{10}^2 \leqq \omega_3^2 - \omega_0^2 \leqq \frac{B(B-A)}{C(C-A)}\omega_{20}^2, \tag{41}$$

so that

$$|\omega_3^2 - \omega_0^2| \leqq \frac{A(B-A)}{C(C-B)}\omega_{10}^2 + \frac{B(B-A)}{C(C-A)}\omega_{20}^2 \tag{42}$$

Then (38), (39), (40), and (42) lead to the result

$$(\omega_3 - \omega_0)^2 < \frac{B-A}{C}\left(\frac{A}{C-B}\omega_{10}^2 + \frac{B}{C-A}\omega_{20}^2\right) \tag{43}$$

But

$$\frac{A}{C-B} \equiv \frac{B}{C-A} + \frac{(B-A)(A+B-C)}{(C-B)(C-A)} \geqq \frac{B}{C-A} \tag{44}$$

Relations (43) and (44) then lead to the appropriate inequality

$$(\omega_3 - \omega_0)^2 < (\omega_{10}^2 + \omega_{20}^2)\frac{A(B-A)}{C(C-B)} \tag{45}$$

Addition of (37) and (45) finally gives

$$(\boldsymbol{\omega} - \boldsymbol{\Omega}_C)^2 \equiv (\omega_3 - \omega_0)^2 + \omega_1^2 + \omega_2^2 < (\omega_{10}^2 + \omega_{20}^2)F_3^2, \tag{46.1}$$

where

$$F_3^2 \equiv \frac{A(C-A)}{B(C-B)} + \frac{A(B-A)}{C(C-B)}, \tag{46.2}$$

provided that (38) is satisfied. That is

$$(\boldsymbol{\omega} - \boldsymbol{\Omega}_C)^2 < \varepsilon^2 \tag{47.1}$$

if

$$[\boldsymbol{\omega}(0) - \boldsymbol{\Omega}_C]^2 \equiv \omega_{10}^2 + \omega_{20}^2 < \delta^2, \tag{47.2}$$

where

$$\delta = \varepsilon F_3^{-1}, \tag{47.3}$$

with the proviso (38).

Just as for $\mathbf{1}_A$ we may easily guarantee satisfaction of (38) by placing an upper limit on ε, viz

$$\varepsilon < |\omega_0| \tag{48}$$

To show this, multiply (47.2) by F_3^2 and require (48), to find

$$(\omega_{10}^2 + \omega_{20}^2)\, F_3^2 < \varepsilon^2 < \omega_0^2 \tag{49}$$

But from (44) and the relations $A \leqq B \leqq C$

$$\frac{A(C-A)}{B(C-B)} \geqq 1 \geqq \frac{B(B-A)}{C(C-A)} \tag{50}$$

and from (46.2) and (50)

$$F_3^2 \geqq \frac{A(B-A)}{C(C-B)} + \frac{B(B-A)}{C(C-A)} \tag{51}$$

Relations (49) and (51) then show that (38) is satisfied. Thus, for an arbitrarily small $\varepsilon < |\omega_0|$, it follows that $|\boldsymbol{\omega}(t) - \boldsymbol{\Omega}_C| < \varepsilon < |\omega_0|$ if $|\boldsymbol{\omega}(0) - \boldsymbol{\Omega}_C| < \delta = \varepsilon F_3^{-1}$. This completes the proof of the restricted L-stability of $\boldsymbol{\Omega}_C \equiv (0, 0, \omega_0)$.

4. A Rigid Body Diagram

To illustrate the above results it is convenient to plot contour lines of δ/ε in the diagram of Figure 1. If we put

$$x \equiv A/C \qquad y \equiv B/C \tag{52}$$

then since $A \leqq B \leqq C$ and since $A + B - C \geqq 0$, we find that any rigid body is represented by some point either within or on the boundary of the cross-hatched right triangle of Figure 1.

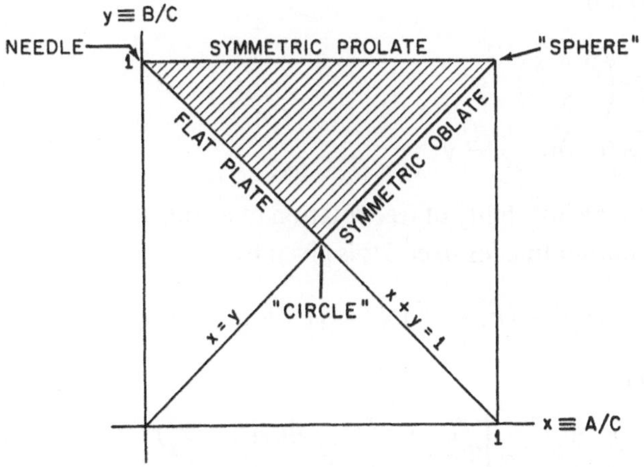

Fig. 1. Classification of rigid bodies.

The hypotenuse $y=1$ represents all symmetric prolate bodies, varying from a needle at (0,1) to a dynamical sphere at (1,1). (For uniform density a 'dynamical sphere', characterized by $A=B=C$, is exemplified by a sphere or by any of the five regular polyhedra, among other possibilities.) The right leg $x=y$ represents all symmetric bodies, varying from a 'dynamical sphere' at (1,1) to a 'dynamical circle' with $A=B=C/2$ at $(\frac{1}{2}, \frac{1}{2})$. (For uniform surface density a 'dynamical circle' is exemplified by a circle, any regular polygon, or any flat plate whose appearance is invariant to a rotation through any angle $2\pi m/n$, where n is an integer greater than 2 and m is an integer less than n.) The left leg $x+y=1$ represents all flat plates, varying from a 'dynamical circle' at $(\frac{1}{2}, \frac{1}{2})$ to a needle at (0,1).

5. Contour Lines of δ/ε for Stability of Free Rotation about $\mathbf{1}_A$

If δ is an excursion of $\boldsymbol{\omega}(0)$ from $(\omega_0, 0, 0)$ that will permit an excursion ε of $\omega(t)$ satisfying $\varepsilon < |\omega_0|$, then for this $\mathbf{1}_A$ case

$$\frac{\varepsilon^2}{\delta^2} = \frac{1-y}{x(y-x)} + \frac{1-x}{y(y-x)} = \frac{1}{k_1}, \tag{53}$$

by (25.2), (26.3) and (52). Then

$$\delta/\varepsilon = \sqrt{k_1} \tag{54}$$

The value of ε^2/δ^2 in (53) is less than that in V1969 by unity, so that for a given body (x, y) we now obtain a larger estimate of the permissible initial excursion for a given excursion ε of $\boldsymbol{\omega}(t)$.

For a symmetric oblate body $(x=y)$ we obtain $\delta/\varepsilon=0$, as before, corresponding to the known instability of free rotation of such a body about any axis perpendicular to its axis of symmetry. For a symmetric prolate body $(y=1)$ we obtain $\delta/\varepsilon=1$, instead of the previous value $\sqrt{2}/2$. This means that for free rotation of such a body the excursion of $\boldsymbol{\omega}(t)$ can never exceed an initial excursion δ if $\delta < |\omega_0|$. For a flat plate $(x+y=1)$ we obtain

$$\delta/\varepsilon = \left(\frac{y-x}{2}\right)^{1/2} \tag{55.1}$$

$$= 0 \quad \text{for} \quad x = y, \tag{55.2}$$

corresponding to the instability of free rotation of a coin about an axis in its plane.

From (53) a contour line for fixed δ/ε is given by

$$(k_1 + x)\, y^2 - (k_1 + x^2)\, y - k_1(x - x^2) = 0, \tag{56}$$

with the solution

$$2y = \frac{k_1 + x^2}{k_1 + x} + \left[\left(\frac{k_1 + x^2}{k_1 + x}\right)^2 + \frac{4k_1(x - x^2)}{k_1 + x}\right]^{1/2} \tag{57}$$

Its minimum is at

$$x_m = (k_1 + k_1^2)^{1/2} - k_1, \quad y_m = (1 + 2^{1/2})\, x_m \tag{58}$$

If $k_1 < \frac{1}{2}$, its intersection with $x+y=1$ is at

$$x_i = \tfrac{1}{2} - k_1, \quad y_i = \tfrac{1}{2} + k_1 \tag{59}$$

Its boundary slopes are

$$y'(0) = 1 - 1/k_1, \quad y'(1) = (1 - k_1)(1 + k_1)^{-1} \tag{60}$$

Equations (59) show that $k_1 = \frac{1}{2}$ is the separating contour between those contours that go all the way from $x=0$ to $x=1$ and those that begin at the flat plate line $x+y=1$. For this separating contour the boundary slopes are -1 and $\frac{1}{3}$.

Some typical contours are shown in Figure 2. Inspection of the figure shows that

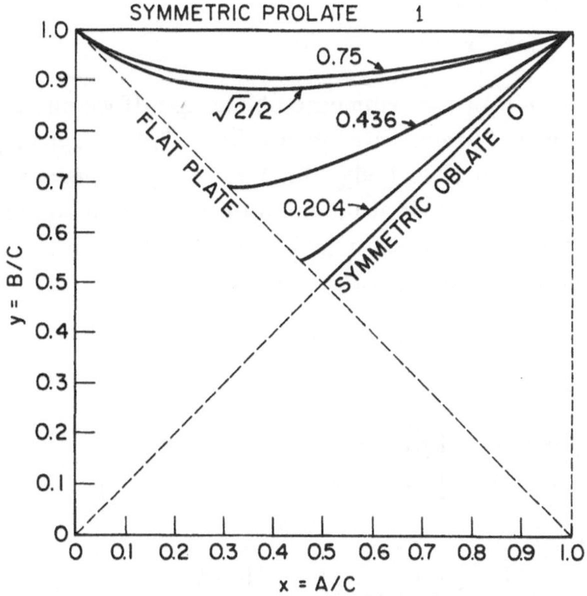

Fig. 2. Contour lines of δ/ε for stability of rotation about $\mathbf{1}_A$.

the contour lines crowd together at the singular points (0, 1) and (1, 1). Thus the whole picture of rotational stability about $\mathbf{1}_A$ changes rapidly if the body is close to being a needle or a 'dynamical sphere'. Use of (53) and (61) near (1, 1) may be useful in designing approximately spherical bodies for stability of free rotation.

6. Contour Lines of δ/ε for Stability of Free Rotation about $\mathbf{1}_C$

The introductory discussion here is the same as that for $\mathbf{1}_A$, but now

$$\varepsilon^2/\delta^2 = \frac{x(1-x)}{y(1-y)} + \frac{x(y-x)}{1-y} = \frac{1}{k_3}, \tag{61}$$

by (46.2), (47.3), and (52). Then

$$\delta/\varepsilon = \sqrt{k_3} \tag{61.1}$$

This expression for ε^2/δ^2 is much simpler than that in V1969 and smaller in important cases, so that again larger excursions of $\boldsymbol{\omega}(0)$ from $(0, 0, \omega_0)$ are permissible for a given excursion ε of $\boldsymbol{\omega}(t)$.

For a symmetric prolate body ($y=1$) we obtain $\delta/\varepsilon=0$, as in V1969, corresponding to the familiar instability of free rotation of a toy top about any axis perpendicular to its axis of symmetry. For a symmetric oblate body ($x=y$), we obtain $\delta/\varepsilon=1$, instead of the value $\sqrt{3}/3$ as in V1969. This means, for example, that for a symmetric rigid planet the excursion of $\boldsymbol{\omega}(t)$ cannot exceed an initial excursion $\delta<|\omega_0|$. For a flat plate ($x+y=1$), we obtain

$$\delta/\varepsilon = [2(1-x)]^{-1/2} \tag{62.1}$$

$$= 1 \quad \text{for} \quad x = \tfrac{1}{2}, \tag{62.2}$$

agreeing with the result for the symmetric oblate case. If we put $x=0$ in this expression, corresponding to a needle, we obtain $\delta/\varepsilon=\sqrt{2}/2$, disagreeing with the result $\delta/\varepsilon=0$ for a symmetric prolate body. But this kind of behavior is expected, since the contour lines crowd together at the point (0, 1). It means simply that we have to be careful near such a singular point in making any idealizations.

From (61) a contour line for fixed δ/ε is given by

$$(1 + k_3 x)\, y^2 - (1 + k_3 x^2)\, y + k_3 (x - x^2) = 0, \tag{63}$$

with the solution

$$2y = \frac{1 + k_3 x^2}{1 + k_3 x} + \left[\left(\frac{1 + k_3 x^2}{1 + k_3 x}\right)^2 - \frac{4k_3 (x - x^2)}{1 + k_3 x}\right]^{1/2} \tag{64}$$

Its minimum is at

$$x_m = \frac{1}{1 + \sqrt{1 + k_3}}, \quad y_m = \frac{1 + \sqrt{1 - k_3}}{1 + \sqrt{1 + k_3}} \tag{65}$$

Its intersection with $x+y=1$ is at

$$x_i = 1 - (2k_3)^{-1}, \quad y_i = (2k_3)^{-1} \tag{66}$$

If $k_3 > \frac{1}{2}$, its boundary slopes are

$$y'(0) = -2k_3, \quad y'(1) = 2k_3 (1 + k_3)^{-1} \tag{67}$$

Equation (66) shows that $k_3 = \frac{1}{2}$ is again the separating contour between those contours that go all the way from $x=0$ to $x=1$ and those that begin at the flat plate line $x+y=1$. For this separating contour the initial and final slopes are -1 and $\frac{2}{3}$.

The contour map for rotational stability about $\mathbf{1}_C$ is given in Figure 3. Again the contour lines crowd together at the singular points (0, 1) and (1, 1). Figures 2 and

3 exhibit a resemblance, but it is only qualitative. One cannot obtain one figure from the other by merely relabeling the curves. E.g., the contour lines for $k_1 = \frac{1}{2}$ and for $k_3 = \frac{1}{2}$ are both separating curves, which would have to be identical if relabeling would work. Although their slopes $y'(0)$ are both -1, their slopes $y'(1)$ are respectively $\frac{1}{3}$ and $\frac{2}{3}$.

7. Polar Wandering of a Rigid Oblate Planet

Let O be the center of mass of the planet, Q its 'pole of figure', and P its pole of rotation. Q is then the intersection of the planet's surface by the principal axis along $\mathbf{1}_C$ and P its intersection by $\boldsymbol{\omega}$. For the earth the wandering of P about Q is restricted to so small a region that we may regard the latter as flat. If we adopt rectangular coordinates X and Y in this tangent plane, with Q as origin, QX parallel to $\mathbf{1}_A$ and QY parallel to $\mathbf{1}_B$ and let $OQ \equiv R_0$ and $OP \equiv R$, then

$$R = (R_0^2 + X^2 + Y^2)^{1/2} \approx R_0 \left(1 + \frac{X^2 + Y^2}{2R^2}\right). \tag{68}$$

Plots of polar motion do not show Q, but suggest that $(X^2 + Y^2)^{1/2}$ does not exceed 30 m. Since $R_0 \approx (6)\,10^6$ m, we see that R remains equal to R_0, within one part in 10^{11}.

For the earth $|\boldsymbol{\omega}| \equiv \omega$ diminishes secularly by about one part in 10^{10} per year and varies periodically by not more than one part in 10^8 per year (Woolard and Clemence, 1966). Thus, since

$$X = \frac{R}{\omega}\,\omega_1 \approx \frac{R_0}{\omega}\,\omega_1\,, \quad Y = \frac{R}{\omega}\,\omega_2 \approx \frac{R_0}{\omega}\,\omega_2\,, \tag{69}$$

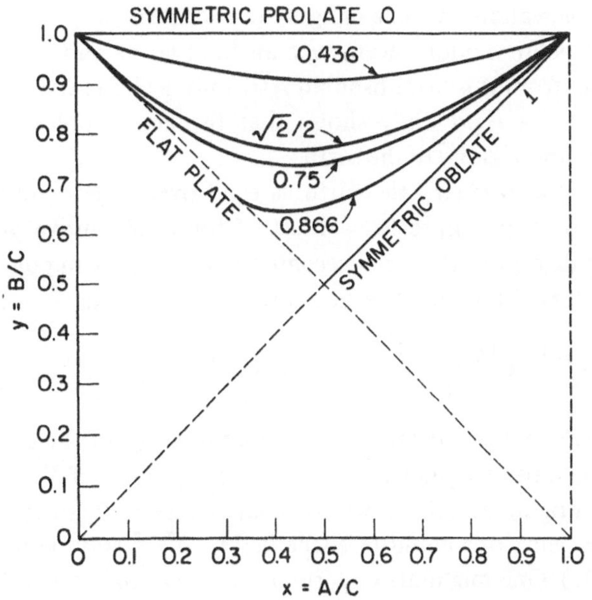

Fig. 3. Contour lines of δ/ε for stability of rotation about $\mathbf{1}_C$.

we may assume, in following polar motion for a century, say, that

$$X = K\omega_1 \quad Y = K\omega_2, \tag{70}$$

where K is constant, to one part in 10^8.

Except for a change of scale, the curve of X versus Y is the same as that of ω_1 versus ω_2. By (31) this would be an ellipse if the earth were rigid. Then by (32) and (33) the semi-axes α_3 and β_3 would be related by

$$B(C-B)\beta_3^2 = A(C-A)\alpha_3^2 \tag{71}$$

and the eccentricity e would be given by

$$e^2 = 1 - \frac{\alpha_3^2}{\beta_3^2} = \frac{(B-A)(A+B-C)}{A(C-A)}, \tag{72}$$

independently of the initial conditions.

If M is the mass of the earth and R_e its equatorial radius, then

$$\frac{C-(A+B)/2}{MR_e^2} = J_2 \approx 10^{-3}; \quad \frac{B-A}{4MR_e^2} = (C_{2,2}^2 + S_{2,2}^2)^{1/2} \approx 10^{-6};$$
$$\frac{C-A}{A} \approx \frac{1}{305}. \tag{73}$$

It then follows from (72) and (73) that

$$e^2 \approx \frac{B-A}{C-A} \approx (4)\,10^{-3} \quad e \approx 0.06. \tag{74}$$

Thus, if the earth were rigid, the path of the pole of rotation P would be an ellipse of eccentricity ≈ 0.06 about the pole of figure Q as center.

The observed polar motion, even over an interval as short as eight years, is in fact very different from this (Smithsonian Astrophysical Observatory 1966). Thus the stability theory for a rigid body shows that the actual polar motion is strongly influenced by the non-rigidity of the earth.

Suppose one were to regard the earth as rigid over a short interval of time. One might then try to fit an ellipse to each set of four consecutive points of the polar path and thus to determine the mean eccentricity and the mean pole of figure over this short interval of time. This would require solving four simultaneous equations

$$\frac{(\xi_i - \xi_0)^2}{a^2} + \frac{(\eta_i - \eta_0)^2}{b^2} = 1, \quad (j = 1, 2, 3, 4) \tag{75}$$

for the semi-axes a and b and for the coordinates ξ_0, η_0 of the center of the ellipse. Here ξ and η would be rectangular coordinates in the plane of X and Y, with some arbitrary point as origin. By using first the polar points $P_1\ P_2\ P_3\ P_4$, then $P_2\ P_3\ P_4\ P_5$, then $P_3\ P_4\ P_5\ P_6$, etc., one could test whether $1 - a^2/b^2$ stays constant and equal to e^2 as given by (72). One might also interpret (ξ_0, η_0) as the pole of figure and attempt to find how it moves in a year.

The work might be reduced to the solution of three simultaneous equations by using

$$1 - a^2/b^2 = e^2 , \tag{76}$$

with e^2 given by (72). A still further simplification would be to neglect e^2 and then simply fit circles successively to $P_1 P_2 P_3$, $P_2 P_3 P_4$, $P_4 P_5 P_6$, etc. To fit a circle to the points P_1, P_2, P_3 one has simply to construct the perpendicular bisectors of the straight line segments $P_1 P_2$ and $P_2 P_3$. To solve the set (75) by successive approximations, one might reverse this procedure, by first determining the circles, then the ellipses with the aid of (76), and then the ellipses from (75).

I doubt that such a project is presently feasible, but I wish to raise the following question. Could the data on polar motion ever be taken so accurately and so frequently that a sort of osculating polar ellipse, corresponding to an instantaneously rigid earth, could be fitted reliably at closely spaced intervals? If so, one could follow the small variations in time of $B-A$ and thus of $J_{2,2} \equiv (C_{2,2}^2 + S_{2,2}^2)^{1/2}$. One could also follow the pole of figure and see whether its changes of position could be correlated with the seasons.

References

Smithsonian Astrophysical Observatory 1966, Special Report 200, Vol. 1, p. 34.
Vinti, J. P.: 1969, *Celes. Mech.* **1**, 59–71.
Woolard, E. W. and Clemence, G. M.: 1966, *Spherical Astronomy*, Academic Press, New York, pp. 362–363.

WILD DYNAMICAL SYSTEMS, AND THE ROLE OF TWO OR MORE SMALL DIVISORS

J. M. A. DANBY

North Carolina State University, Raleigh, N.C., U.S.A.

Abstract. Numerical results are presented describing conditions in two dynamical systems on the borderline of wildness. The first refers to motion in the field of an oblate planet, and the second to a system due to Walker and Ford in which the role of resonance can be recognized more specifically. In particular, a wild asymptotic surface is investigated in some detail. The work and the observations on it are intended to be clinical in nature, referring to systems having certain parameters much larger than those that can be covered under existing mathematical theories.

Most dynamical systems are non-integrable. There is a school of thought that they should be called 'non-integrated', with the implication that all systems may eventually succumb to integration given sufficient human ingenuity. But the first statement expresses a genuine property proved by Siegel, where 'integrability' is associated with the convergence of Birkhoff's normalization. For me, as an astronomer, the word 'series' is apt to produce a picture of the type of series that are used in generating planetary or lunar ephemerides. These do not satisfy the equations of motion, although they may satisfy an integrable approximation to these equations. They are merely the start of formal series solutions of the equations of motion – and of formal series that almost certainly diverge. The problem of divergence, or of non-integrability, has not been acute in classical celestial mechanics because the approximations have done a quite remarkably good job. Leverrier, for instance, could detect the gravitational discrepancy in the motion of the line of apsides in Mercury's orbit. Concern about convergence has arisen most readily when questions are asked about the long-term properties of solutions; the principal worries being the presence of secular and mixed secular terms (so that the quasi-periodic nature and stability of the actual motion must be questioned), and the appearance of small divisors.

Mathematically speaking, we can say that in general the quasi-periodic solutions of the astronomers are wrong. But this need not be because of secular terms that result in instability; methods for avoiding these secular terms are known, although in general impractical to apply. But what is the nature of the error, and how serious is it? Is it a detail in the fine print of mathematical theory, or is it something that must be reckoned with in practice? Some of these questions can be answered through the application of numerical experiments. These experiments are not mathematically acceptable, but they can provide an extrapolation from situations where the current mathematical theorems can be applied to actual physical problems. The approach here is not to examine situations where the wildness can only be seen through a microscope, but where the break-up of a system is actually taking place. The work is descriptive and clinical in nature. Where extrapolation seems to be reasonable, reference is made to basic theory.

G. E. O. Giacaglia (ed.), Periodic Orbits, Stability and Resonances, 272–285.

An illuminating discussion of the role of small divisors is contained in a paper by Walker and Ford. They consider in turn the three Hamiltonian systems defined by

$$H_1 = H_0(J_1, J_2) + \alpha J_1 J_2 \cos(2\phi_1 - 2\phi_2) \tag{1}$$

$$H_2 = H_0(J_1, J_2) + \beta J_1 J_2^{3/2} \cos(2\phi_1 - 3\phi_2), \tag{2}$$

and

$$H_3 = H_0(J_1, J_2) + \alpha J_1 J_2 \cos(2\phi_1 - 2\phi_2) + \beta J_1 J_2^{3/2} \cos(2\phi_1 - 3\phi_2), \tag{3}$$

where

$$H_0(J_1, J_2) = J_1 + J_2 - J_1^2 - 3J_1 J_2 + J_2^2 .$$

The action-angle variables (J_i, ϕ_i) are related to 'cartesian' variables (q_i, p_i) by

$$\begin{aligned} q_i &= (2J_i)^{1/2} \cos\phi_i , \\ & \qquad\qquad i = 1, 2 . \\ p_i &= -(2J_i)^{1/2} \sin\phi_i . \end{aligned} \tag{4}$$

In each case (1) and (2) give integrable systems. In fact the Hamiltonian is an integral in both cases, and $(2J_1 + 2J_2)$ and $(3J_1 + 2J_2)$ are respectively integrals for (1) and (2).

These systems can be investigated by Poincaré's (1899) method of consequents or, in alternative terminology, the computation of surfaces of section. Consider orbits having a given energy (where the integral $H = h$ is, as is customary, called the integral of energy); one coordinate, say p_1, can be eliminated using the integral of energy, and the orbits considered in a reduced three-dimensional phase space. Take as a surface of section the plane $q_1 = 0$. Let an orbit cross this plane at P_0; if it next crosses the plane, moving in the same direction, at P_1, then $P_1 = TP_0$ is the consequent of P_0, T representing a mapping. The computation of successive consequents can indicate whether or not the orbit lies on a torus in the reduced phase space. This search for tori is important. Their existence, and persistence when the constant of energy is slightly changed are related to questions of stability, and their existence is also related to the existence of quasi-periodic solutions.

If the orbit lies on a torus, then the consequents will lie on an invariant curve or 'island'. In the cases of systems (1) and (2), the islands are continuously packed, and typical surfaces of section are shown in Figures 1 and 2. In the case of H_2, the triple archipelago does not appear for very low energies; it first appears *at the origin* and then expands. If $T^n P_0 = P_0$ for some n, then P_0 is a fixed point of a periodic orbit. For the integrable case, any other point lying on the invariant curve on which P_0 lies will also be a fixed point of a periodic orbit. So on this invariant curves there are fixed points of infinitely many periodic orbits; all have the same period and energy and all have characteristic exponents zero. These are ordinary periodic orbits. The fixed point at the center of an island system gives a periodic orbit that is singular; in general its characteristic exponents will not all vanish.

Each of the systems (1) and (2) has a 'resonance zone'. That for H_1 resembles in some ways certain sections in the problem of the motion of a satellite of an oblate

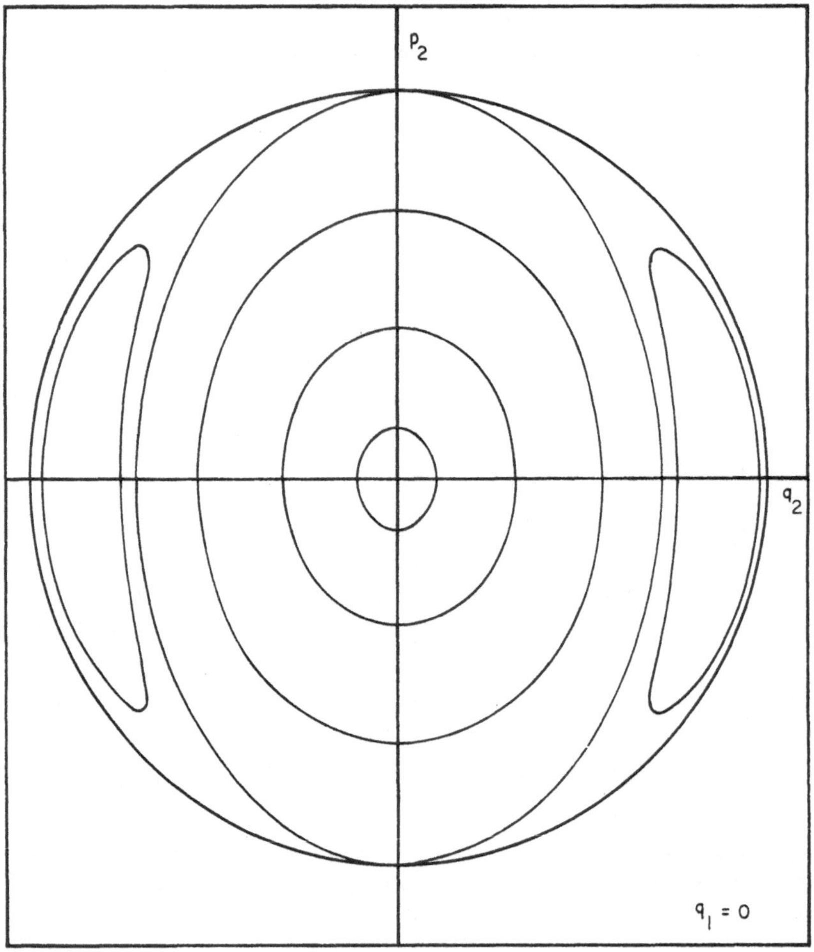

Fig. 1. Island systems for H_1.

planet (Danby, 1968, Figure 4). There the resonance zone corresponds to librational motion of the line of apsides close to the critical inclination.

The system given by H_3 is in general non-integrable. Suppose that we are concerned with initial conditions that are such that one of the resonance zones will be unimportant; such low energy, say, that the triple archipelago will not appear. Then we can use the usual canonical methods for eliminating the angle $(2\phi_1 - 3\phi_2)$. But in doing so we introduce into the new Hamiltonian a trigonometric Fourier series having arguments $[n(2\phi_1 - 2\phi_2) + m(2\phi_1 - 3\phi_2)]$, and we suddenly have on our hands an infinity of potential donors of small divisors (and consequently, resonance zones). These are called 'secondary resonances' by Walker and Ford. In an integrable case, these would furnish the ordinary periodic orbits referred to above. But in a non-integrable case there are no ordinary periodic orbits, and where, plotting consequents, we find $T^n P_0 = P_0$, then P_0 gives us a singular periodic orbit; if it is of the stable type then it is likely in its turn to be surrounded by island systems forming an archipelago.

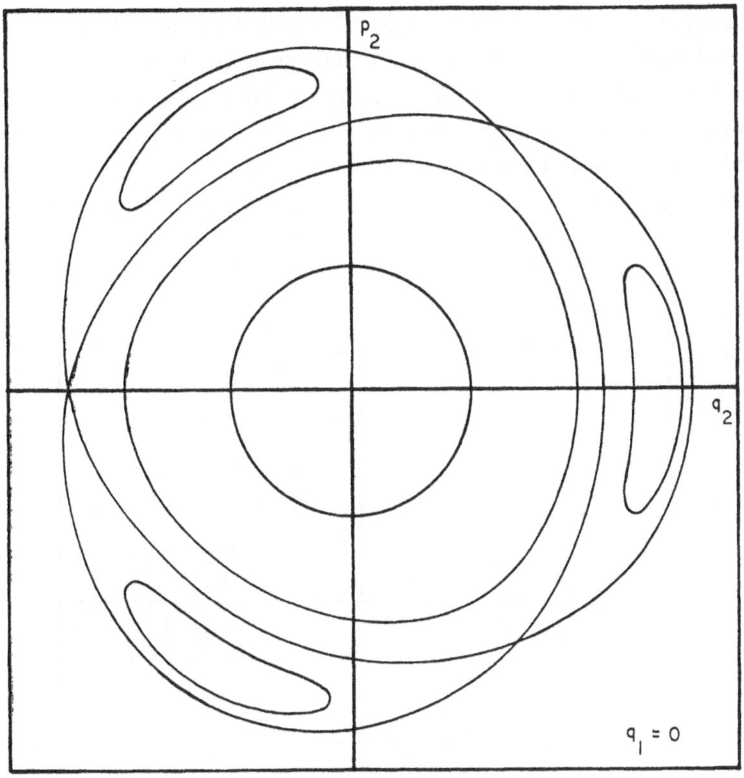

Fig. 2. Island systems for H_2.

If we consider β in H_3 to be a small parameter, then for $\beta=0$ the system is integrable. For sufficiently small β, from Kolmogorov's theorem we can expect most of the islands of Figure 1 to survive. But an island that for $\beta=0$ contains ordinary periodic orbits will not survive; instead, we find a stable-unstable pair of singular periodic orbits (these are orbits of the *deuxième genre*, in the terminology of Poincaré): the island is replaced by a thin archipelago which is a section of what in three dimensions will seem like a system of winding tapeworms in the reduced phase space. In this case every tapeworm will in its turn have tapeworms, and so on.

Mathematical theory is necessary for a description of what happens for small β, but it is of no help to us as the parameter increases. There is no way of considering the continuation of a torus (or a quasi-periodic orbit) in terms of a parameter, so one can only speak loosely here. Let us use rotation number to identify tori having different values of a parameter. For a time most tori seem to survive, but it does appear that some disappear, as our brain cells do as we get older, and that eventually there is usually a cataclysm when, over quite a small change of the parameter all of the large surviving tori perish. The tapeworms can survive this event, but in turn we find that the larger of these become progressively decimated. A major result of Walker and Ford is that for H_3 the principal break-up occurs when the energy (which they use as a parameter) reaches a value for which the resonance zones for H_1 and H_2

considered separately, meet. Figure 3, taken from their paper, illustrates this situation.

The calculation of a torus numerically can never be definitive. We can never be sure, for instance, that the torus is real, and not a seductive disguise like that assumed by Zeus in his pursuit of Europa. Two possibilities must in particular be considered. The island might turn out under close examination to be 'fuzzy'. This need not be serious, for we might be in a narrow region imprisoned between two tori, and study of

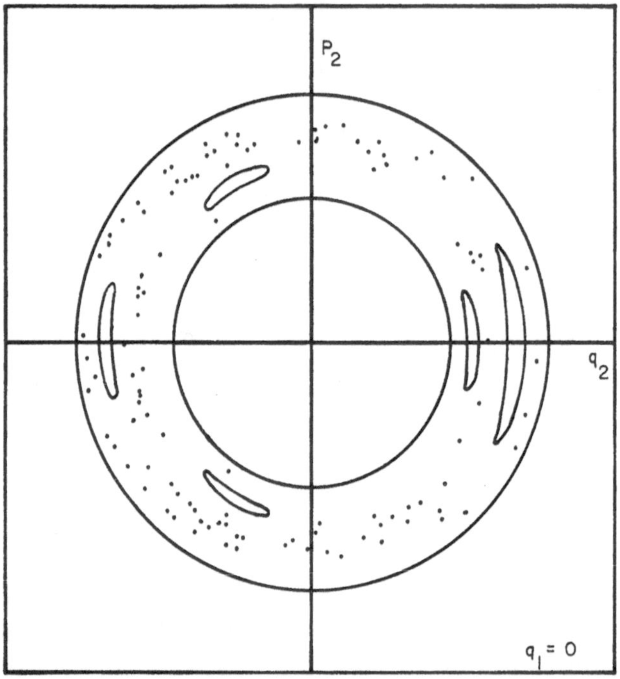

Fig. 3. Surface of section for H_3, with $q_1 = 0$. The energy constant is 0.209 5.

the consequents can be as useful as that for consequents on a true island. More serious, consequents may eventually leave completely the 'island' that they seem for a time to be defining. This possibility can never be excluded; however, when it happens, it appears that the progressive mapping of the consequents is irregular, and shows signs of clustering.

Numerical methods have been most successful in work on periodic orbits. These can provide a useful framework for the consideration of non-periodic motion, and there is adequate theory to cover their continuation and, in many cases, their existence. The most helpful existence theorem for the present discussion is that due to Birkhoff concerning the existence of periodic orbits of arbitrarily long periods, arbitrarily close to a given periodic orbit of the stable type. These are orbits of the deuxième genre, occurring in stable-unstable pairs. Figure 4a shows the sort of pattern of fixed points that we can expect; here the fixed point of an instable orbit is marked by a cross. In Figure 4b the fixed points of the stable type have been clothed in islands, and one can readily see the tapeworm sections; note that this is very similar to the picture

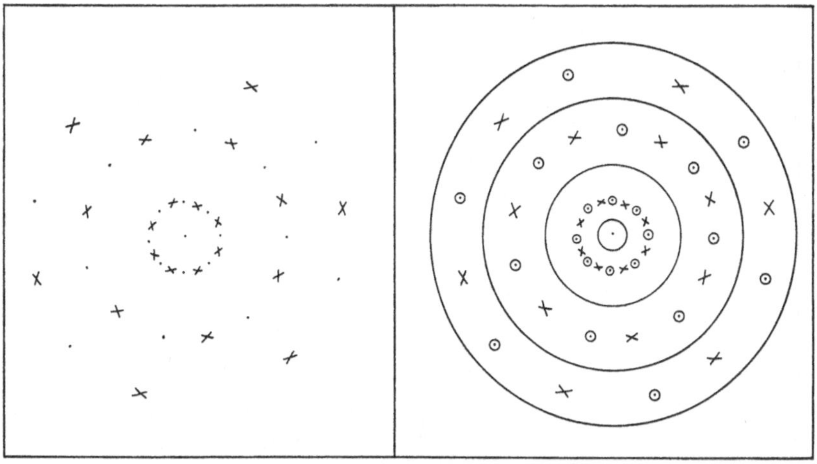

Fig. 4a. Fixed points for orbits of the *deuxième genre* of the type predicted by Birkhoff's theorem.
Fig. 4b. The fixed points of Figure 4a, with surrounding islands added.

obtaining when a small parameter destroys integrability. The islands in Figure 4b are not guaranteed by theory; but they are to be expected, and are usually found in computations when sought after. Where an island (or quasi-periodic orbit) exists, we can expect it to be bracketed as closely as desired by periodic orbits, rather as an irrational number can be approximated by a rational. If a torus is characterized by rotation number, then that number is irrational; while the rotation number for a periodic orbit is rational. The orbits bracketing a torus can be continued; they cannot suddenly appear or disappear; but not so with the torus.

Before discussing the system (3) further, I shall discuss some of the phenomena of break-up in a more complicated system, the one considered in Danby (1968). The potential is that for an oblate planet with only the second zonal harmonic, having the usual coefficient J_2. Since the longitude is ignorable, the problem can be reduced to a conservative one having two degrees of freedom, and motion is followed in the meridian plane that contains the satellite. Figures 5a, b, c show the evolution of surfaces of section with increasing parameter J_2, while the parameters for energy and angular momentum about the axis of symmetry remain fixed. Between 5b and 5c a periodic orbit has disappeared; instead of having two stable and one unstable, we have remaining a stable-unstable pair. This illustrates the phenomenon of branching in a family of periodic orbits; notice that when an orbit disappears by branching, its characteristic pattern in the surface of sections shrinks to a point and vanishes. Later, for higher J_2 than that for 5c the remaining unstable orbit disappears leaving a stable orbit that can apparently be continued indefinitely. Again, the figure 8 of the asymptotic section in 5c shrinks to a point.

The sections of asymptotic surfaces appearing in Figure 5 do not show any sign of the customary wildness found in non-integrable systems, and none could be found numerically. There is plenty of evidence for the non-integrability of this problem, but it happens that on and within this asymptotic surface, none could be found. (For

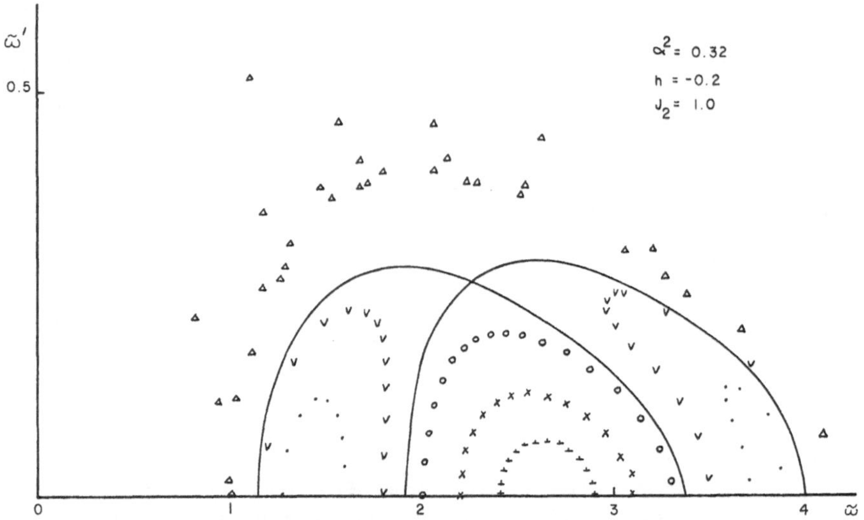

Fig. 5a. Sections for the problem of motion of a satellite of an oblate planet.

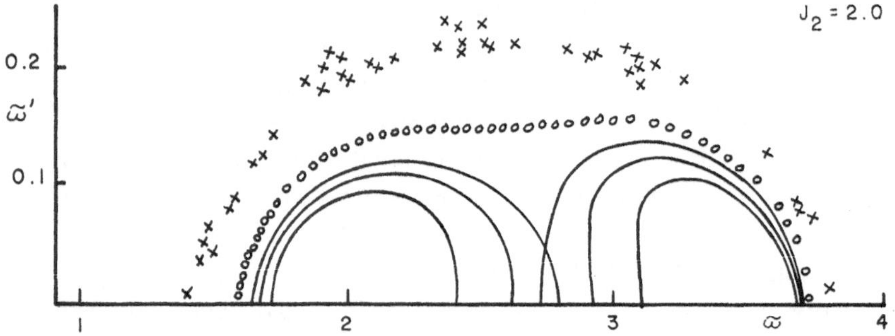

Fig. 5b.

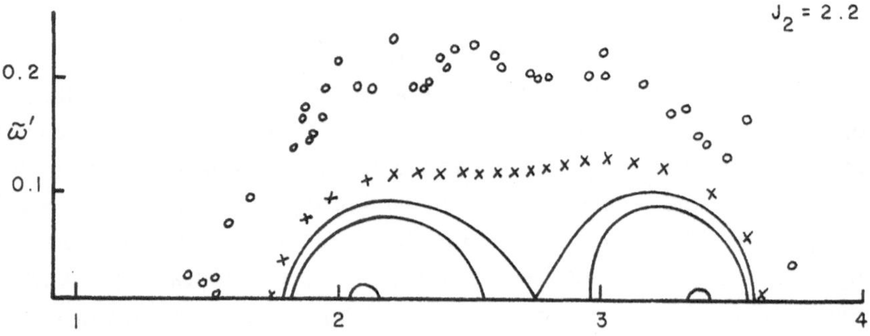

Fig. 5c.

instance, characteristic exponents of periodic orbits of the *deuxième genre* that were investigated differed from zero only in the eleventh place of decimals, and this was also the possible error of the calculation.)

The usual behavior of an asymptotic section is shown in Figure 11 of Danby (1968); this is reproduced here in Figure 6. The features here include at top a region containing

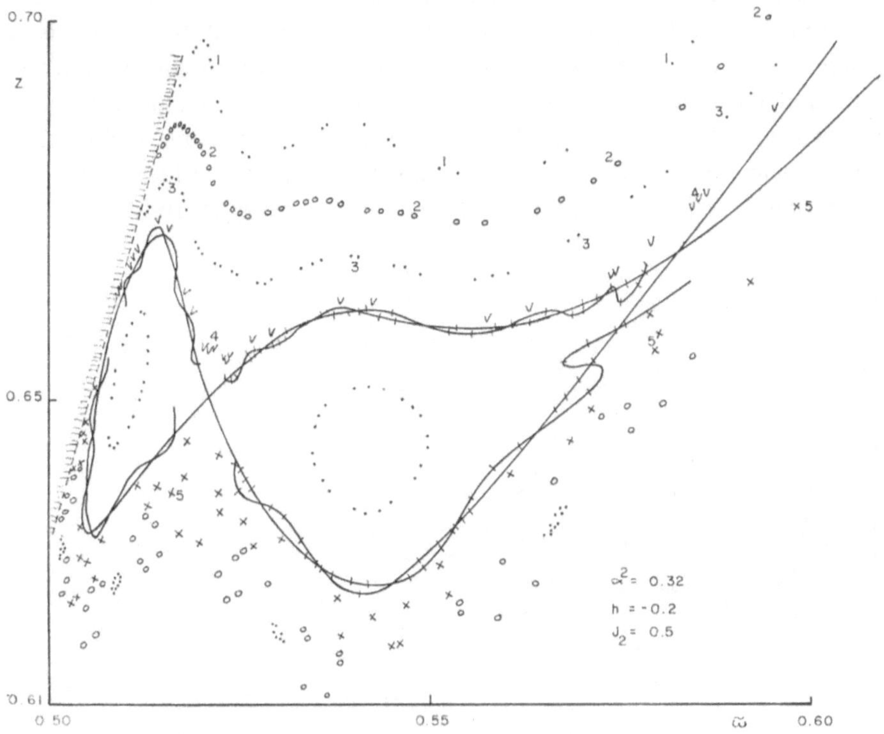

Fig. 6. Section for the motion of the satellite of an oblate planet.

apparently well defined invariant curves (with shapes radically influenced by the bulging archipelago system below); these have between them very thin archipelago systems corresponding to resonant periodic orbits. On the other side of the figure there is a region having no apparent organization. Adjectives such as 'chaotic' or 'ergodic' have been used to describe motion in such a region (which also appears in Figure 3). This judgement seems to me to be premature. We may not have yet tried the right method for looking for 'order' in this situation. The word 'ergodic' carries with it properties that go beyond an apparent lack of organization. In this latter region archipelago systems, very thin, persist, one appearing in the figure. In between these regions there is a border zone where wildness is taking over. The resonance corresponding to the archipelago system has some reason been stimulated, and has greatly expanded, and with this there appears very clearly the conventional wild behavior of the asymptotic sections that emanate from the fixed points of the unstable periodic orbit that is the pair with the stable orbit having fixed points at the center of the expanded island system. We see very clearly that a secondary resonance is apparently at the root of the breakdown; but there is no immediate explanation as to why it should be stimulated in this way.

Figures 7a, b, c still refer to this problem. Each contains a central region which, on and inside an apparently non-wild asymptotic section showed no evidence of non-integrability. Then there is a region that contains some apparently well defined invariant curves. Finally a wild region. In Figure 7a part of an asymptotic section is shown emanating from the fixed points that are circled. Figure 7b shows another such section, and these two are combined in Figure 7c. These figures pose, in summary, many questions. Can a dynamical system be integrable in some region and non-integrable in another? In what manner can a wild asymptotic surface vanish when its parent periodic orbit is absorbed into another branch of periodic orbits? To what extent

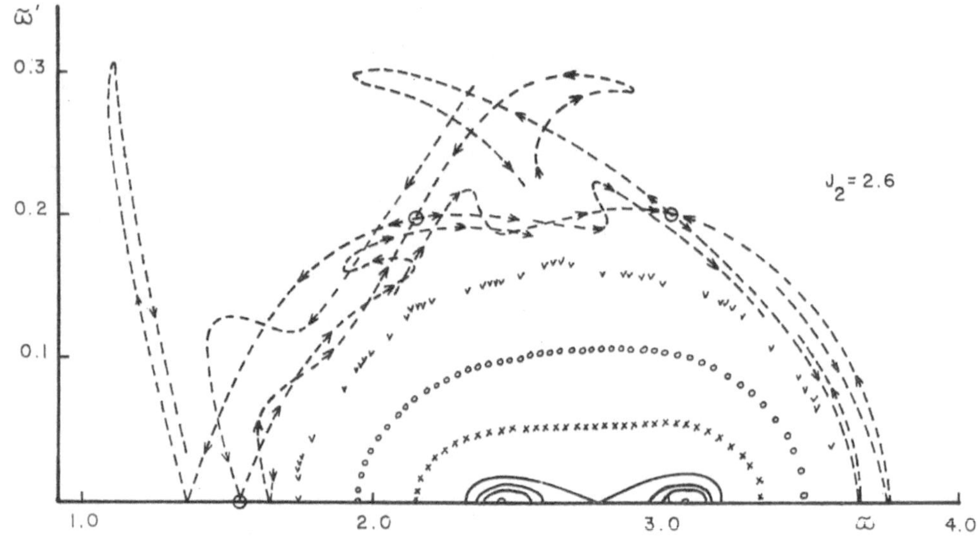

Fig. 7a.

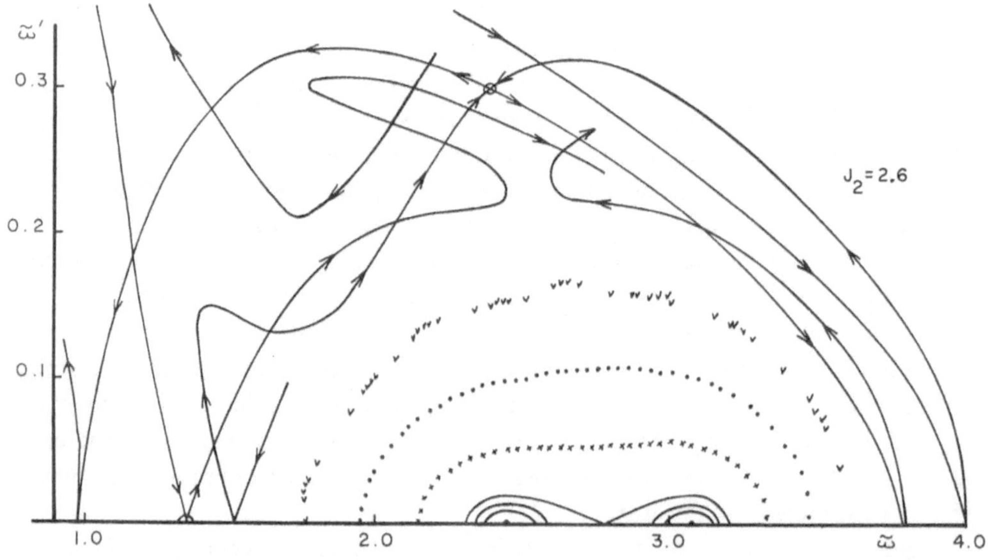

Fig. 7b.

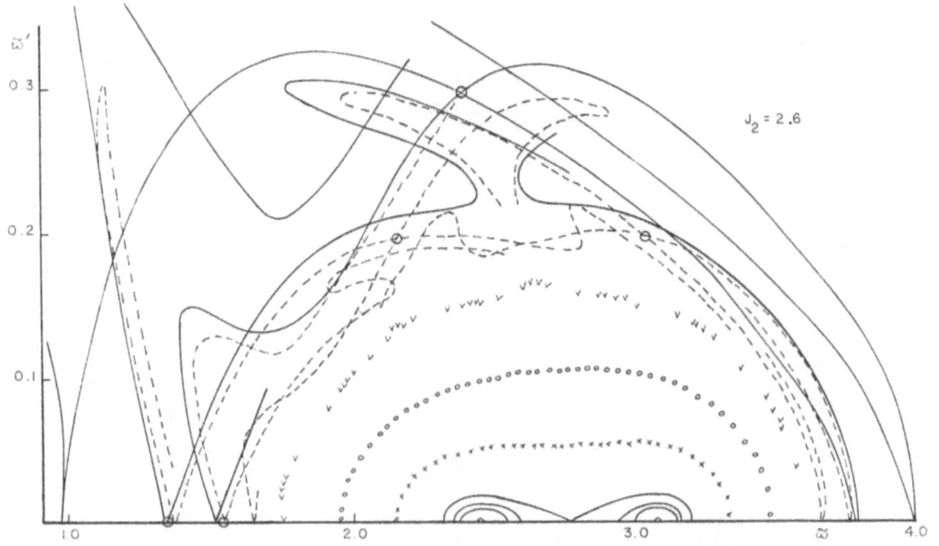

Fig. 7c.

are adjectives like 'chaotic' or 'ergodic' justified in regions that I prefer at present simply to call 'wild'? Can the immediate cause of break-up be traced to a resonance, or small divisor, and what role does the increasing parameter play?

The potential of an oblate planet involves singularities for which no known regularization is available, and the consequent numerical problems of accurate integration were too great for a more detailed investigation than that described. The problem posed by H_3 has, however, no singularity, and it is easier to discus the role of resonances here. Accordingly this problem was integrated numerically using Steffensen's method in double precision; the constant of energy was monitored, and never varied by more than the fourteenth place of decimals. The parameters used were those considered by Walker and Ford, $\alpha=\beta=0.02$, and attention was confined to the energy 0.209 5, for which a section is given in Figure 3. The islands shown in Figure 3 have at their centers fixed points of stable periodic orbits. The corresponding unstable periodic orbits were found, and then, using the methods of Danby (1968), the directions of the asymptotic sections emanating from these. These sections were then followed until the first signs of wildness appeared. The results are shown in Figure 8. The sections for the 2-2 and 2-3 cases do not meet, and, indeed, over much longer integration, no meeting was found. This must be a fortunate coincidence, for with slightly larger energy, the sections should surely intersect. In the calculations of these sections I felt confident that I could depend on at least eight good figures.

Checks were made of the symmetry of some homoclinic and heteroclinic points about the q_2-axis; a lack of symmetry would have shown errors of integration or errors in the initial conditions used to find points on the asymptotic sections. Those asymptotic sections from the 2-2 orbit that are turned out, and those from the 2-3 orbit that are turned inward toward the origin yielded scarcely any signs of wildness.

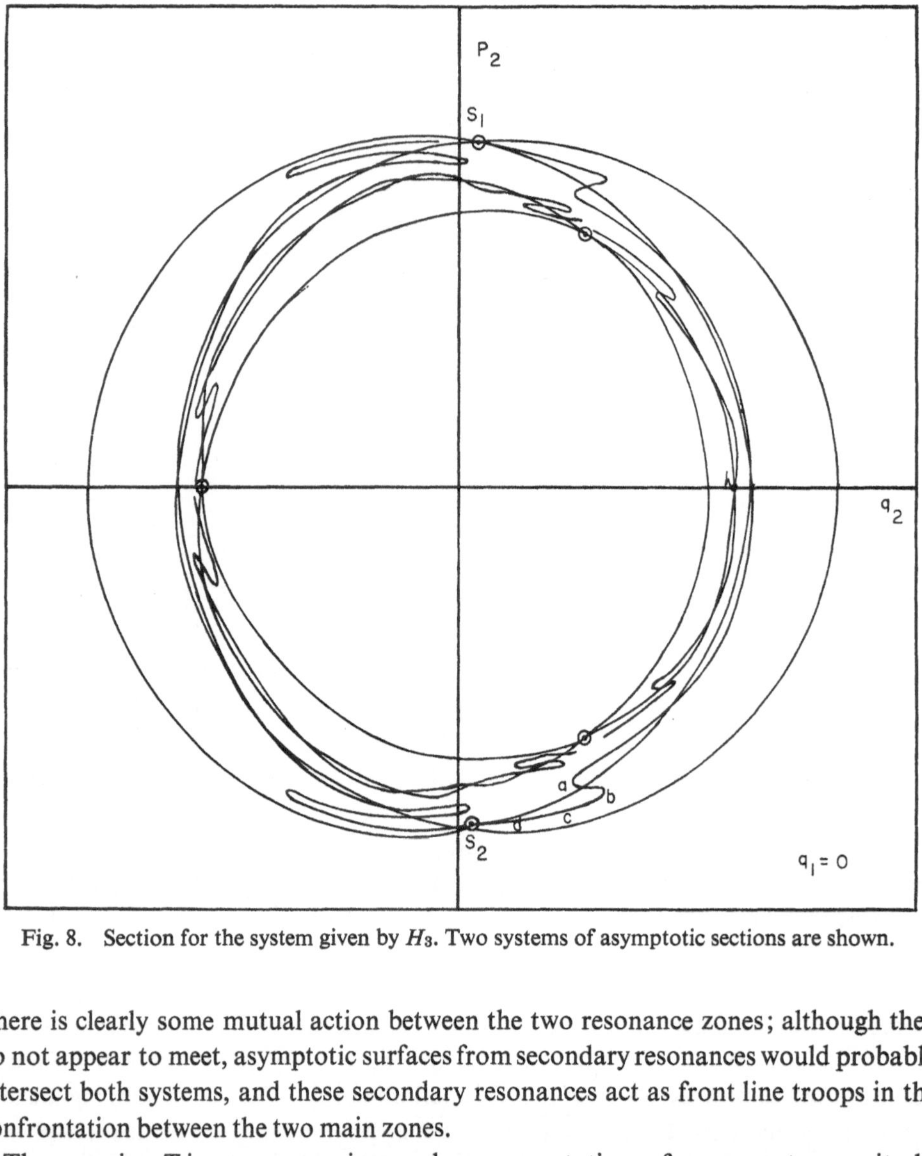

Fig. 8. Section for the system given by H_3. Two systems of asymptotic sections are shown.

There is clearly some mutual action between the two resonance zones; although they do not appear to meet, asymptotic surfaces from secondary resonances would probably intersect both systems, and these secondary resonances act as front line troops in the confrontation between the two main zones.

The mapping T is area preserving, and an asymptotic surface cannot cross itself. These conditions alone suggest an extremely complicated behavior of an asymptotic section of which Figures such as 6 or 8 give no clue; Figure 7 gives some idea of what to expect but the details could not be followed accurately. I found it of interest to trace out a section much further than is done in Figure 8. The result is shown in Figure 9. Numbers are included to try to make it easier to follow the curve. The same number before a decimal point gives points lying on the consequent of a loop such as abcd in Figure 8. The curve plotted is continuous up to the numbers 10.... The numbers 11... refer to a later portion of the same curve. Also the curve has been deliberately distorted in places to try to make the structure clearer; even so, the necessary thickness of a line swamps much of the detail; some of the 'lines' are in fact

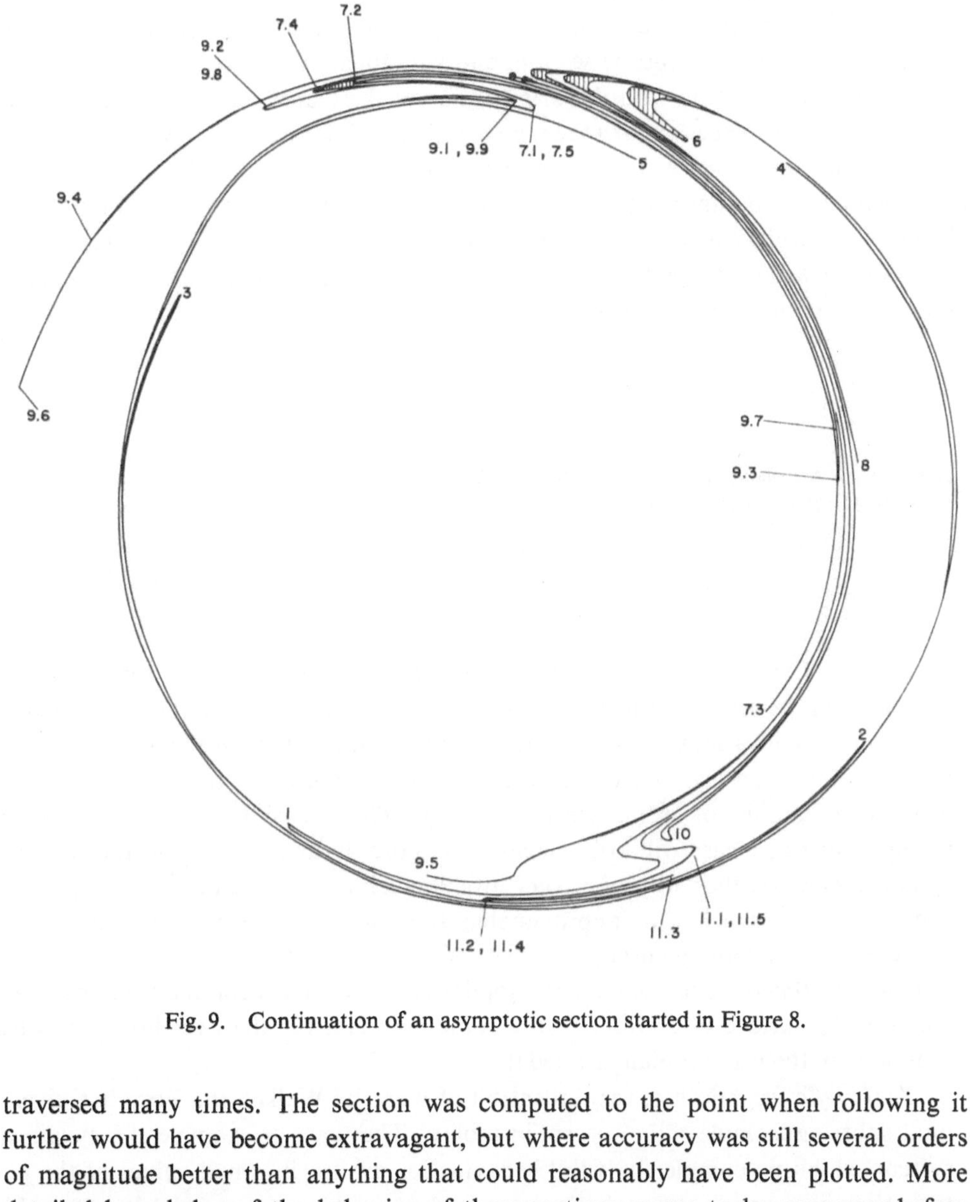

Fig. 9. Continuation of an asymptotic section started in Figure 8.

traversed many times. The section was computed to the point when following it further would have become extravagant, but where accuracy was still several orders of magnitude better than anything that could reasonably have been plotted. More detailed knowledge of the behavior of these sections seems to be necessary before statements about possible ergodicity can be made about orbits like the one represented by the scattered dots in Figure 3. Since ergodicity in a region involves an orbit returning arbitrarily close to a point in the region, this might imply that an asymptotic surface should fill the region densely.

A fixed point for an unstable orbit has asymptotic surfaces of stable and unstable types; surfaces of the same type cannot intersect. Surfaces of opposite type intersect in homoclinic points; these are points of homoclinic orbits which are doubly asymptotic to the original periodic orbit. Sections from two different orbits intersect in heteroclinic points, which correspond to heteroclinic orbits which are asymptotic to the two different given orbits at different extremes of time. The reflection of the curve of Figure 9 is another asymptotic section, and will apparently have an infinite number of points of intersection with the original curve; these are heteroclinic points. Now consider a part of a section, like the segment abcd in Figure 8. This becomes stretched without limit in its successive consequents, and there appears to be no limit to the heteroclinic points that it contains; if this is so, then there is an infinite number of heteroclinic orbits asymptotic to the periodic orbits with fixed points marked S_1 and S_2 in Figure 8. This suggests the possibility that heteroclinic points, and perhaps also homoclinic points, are dense on an asymptotic section.

References

Danby, J. M. A.: 1968, 'Motion of a Satellite of a Very Oblate Planet', *Astron. J.* **73**, 1031.
Poincaré, H.: 1899, *Méthodes nouvelles de la mécanique céleste*, Tome 3.

Discussion

O. Godart: The quasi invariant features mainly near the center of the surface of section might be envisaged if two crossings of the surface of section in two different directions give invariant curves for their respective mappings. In the integrable case, their translation on the same origin will give superposable lines and as a consequence definition of families of ordinary periodic orbits. The extension of those mappings in the non integrable case will not coincide any more. Near the center in the case of a certain symmetry, they will differ very little in such a way that the singular periodic orbits will appear at first on approximately the same invariant lines crossing of two of the constant mappings curves.

J. M. A. Danby: The theory only applies to consequents for the same direction crossing the surface of section. I only combined data for the two directions when symmetry of the Hamiltonian justified it.

W. H. Jefferys: Another, very recent bit of work by Walker and Ford has given a quite sobering example of the onset of instability. They gave an example of a dynamical system exhibiting completely 'wild' behavior for arbitrarily small values of the perturbation parameter, although the unperturbed problem is completely integrable. This happened because changing the size of the perturbation produced exactly the same orbit but with different time scales. This example should be kept in mind by workers in this area.

A. Deprit: Your very extensive study of an asymptotic manifold impresses me very much. Indeed in the summer of 1967, Jacques Henrard and I tried to do it for an unstable periodic orbit emanating from L_3; the figure, though, looked so intricate that we gave up. Concerning periodic orbits clustering around homoclinic orbit, I found in collaboration with Lloyd Carpenter that many bridges of symmetric periodic orbits in the Restricted Problem appear to accumulate for some part of their length against the family of unstable periodic orbits emanating from L_1.

J. M. A. Danby: Periodic orbits with fixed points close to homoclinic or heteroclinic points are easy to find, but I have not found them discussed by Birkhoff. Perhaps we need to introduce more 'genres' in order to discuss the variety that seems to exist.

THE PLANAR MOTION OF A TROJAN ASTEROID

J. KEVORKIAN

University of Washington, Seattle, Wash., U.S.A.

Abstract. The planar motion of an asteroid in an orbit close to that of Jupiter is considered within the framework of the restricted circular three-body problem. The solution is derived asymptotically to second order using the two-variable expansion procedure for the case of small Jupiter-sun mass ratio. It is shown that the solution derived under the assumption of small Jupiter-sun mass ratio becomes singular for orbits that approach Jupiter. The nature of the singularity is exhibited as a guide for future work valid for this case. The results are given in explicit form for the coordinates as functions of time including both short and long periodic terms. Finally, the present solution is specialized to the family of long periodic Trojan orbits about the sun-Jupiter libration points.

1. Introduction

This paper concerns a class of planar orbits in the restricted circular three-body problem. It is assumed that μ, the mass of the smaller primary (Jupiter) divided by the total mass of the system (sun-Jupiter), is very small. In addition, the orbit in phase-space of the particle of negligible mass (asteroid) is taken initially at some point very close to the curve describing the orbit of Jupiter.

The Trojan asteroids are found clustered in neighborhoods of the sun-Jupiter triangular libration points. Thus, the present model is only an approximation for the case of the Trojan asteroids in the sense that the orbital eccentricity of Jupiter ($e=0.05$) is ignored, motion is assumed to occur in the sun-Jupiter orbital plane, and the perturbation of Saturn is not considered. It is pointed out, however, that no assumption is made regarding the closeness of the asteroids to the sun-Jupiter triangular libration points.

The basic features of the motion for the case of the Trojan asteroids was first discussed by Brown (1911a, b) and appears again in more generality in Brown and Shook (1933). Hertz (1943) gives some details of the latter work for the case of motion near the libration points. Unfortunately, the archaic method of solution in the foregoing treatments and the lack of explicit formulas for the coordinates provide little insight into the nature and general applicability of the results. A case in point is the fact that solution of the problem by any method in which the effect of Jupiter is neglected in the first approximation cannot be valid for the case of close approach to Jupiter. The nature of the singularity in the solution for the case of close approach to Jupiter is not discussed in the foregoing references perhaps because none of the known Trojan asteroids depart from the sun-Jupiter triangular libration points by more than 10° of arc (i.e., none come closer than 50° of arc to Jupiter).

Rather than presenting another theory for the motion of the Trojan asteroids, the goals of the present work are a systematic derivation of the solution for the simplified problem, clarification of the nature of the singularity in the solution for the case of close approach to Jupiter, and specialization of the results to periodic motion.

G. E. O. Giacaglia (ed.), Periodic Orbits, Stability and Resonances, 286–303.

The solution is developed asymptotically in the limit of small μ using the two-variable expansion procedure (see Kevorkian (1966) for a discussion of the method and Eckstein and Shi (1969) for a recent application of the technique to a satellite problem). Williams (1966) also used the two-variable expansion procedure to study the motion of asteroids perturbed by Jupiter. However, his work specifically excludes the conditions approximating the Trojan asteroids.

To insure ease of applicability of the results, all constants of integration are expressed in terms of initial conditions and the solution is exhibited explicitly including short as well as long periodic terms. In addition, as a partial check of the calculations, the results are substituted into the Jacobi integral; and it is shown that the present results are consistent with the constancy of the Jacobi integral to the highest order calculated.

Particular attention is focused on initial conditions which lead to a close approach of the asteroid to Jupiter as it is hoped that the present study will form the basis for later work concerning the details of the evolution of the entire orbit for this case.

Finally, it is shown that the solution derived here includes as a special case the family of long periodic orbits about the sun-Jupiter triangular libration points. This family was calculated numerically by Rabe (1961, 1962) for $\mu = 9.5388 \times 10^{-4}$.

2. Problem Formulation

Consider the planar, circular restricted three-body problem where two particles of mass m_1 and m_2 $(m_1 > m_2)$ move in circular orbits about their center of mass, and a third body of negligible mass moves in the orbital plane of the two primaries. Lengths are nondimensionalized by dividing by the distance between m_1 and m_2 and the time is normalized by the reciprocal angular velocity of the primaries. Let x and y denote the Cartesian coordinates of the massless particle referred to a non-rotating frame centered at m_1 and let r and θ be the corresponding polar coordinates (see Figure 1). The

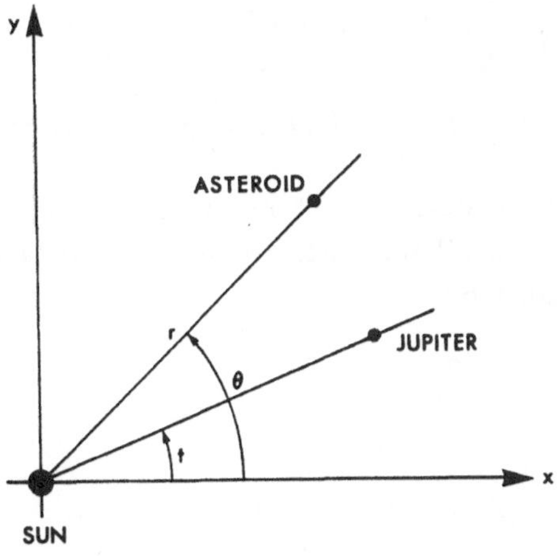

Fig. 1. Geometry.

equations of motion in polar coordinates are:

$$\frac{d^2 r}{dt^2} - r\left(\frac{d\theta}{dt}\right)^2 = -\frac{(1-\mu)}{r^2} + \mu\left\{\frac{\cos(\theta - t) - r}{[1 + r^2 - 2r\cos(\theta - t)]^{3/2}} - \cos(\theta - t)\right\} \tag{1a}$$

$$r\frac{d^2\theta}{dt^2} + 2\left(\frac{dr}{dt}\right)\left(\frac{d\theta}{dt}\right) = \mu\sin(\theta - t)\left\{1 - \frac{1}{[1 + r^2 - 2r\cos(\theta - t)]^{3/2}}\right\}. \tag{1b}$$

For application of the two-variable expansion procedure it is convenient to adopt $u = 1/r$ and the time t as the dependent variables and let θ be the independent variable. Under the above transformation, Equations (1) become

$$\frac{d^2 u}{d\theta^2} + u - u^4\left(\frac{dt}{d\theta}\right)^2 = \mu\left\{u\left(\frac{dt}{d\theta}\right)^2\left[u\cos(\theta - t) - \frac{du}{d\theta}\sin(\theta - t)\right]\right\} + \mu u^4\left(\frac{dt}{d\theta}\right)^2\left\{\frac{1 - u\cos(\theta - t) + \frac{du}{d\theta}\sin(\theta - t)}{[1 + u^2 - 2u\cos(\theta - t)]^{3/2}} - 1\right\} \tag{2a}$$

$$\frac{d}{d\theta}\left(u^2\frac{dt}{d\theta}\right) = \mu\left(u\frac{dt}{d\theta}\right)^3\sin(\theta - t)\left\{\frac{1}{[1 + u^2 - 2u\cos(\theta - t)]^{3/2}} - 1\right\}. \tag{2b}$$

The well-known Jacobi integral of the system (1)

$$\frac{1}{2}\left[\left(\frac{dr}{dt}\right)^2 + r^2\left(\frac{d\theta}{dt}\right)^2\right] - r^2\frac{d\theta}{dt} + \mu r\cos(\theta - t) - \frac{\mu}{[1 + r^2 - 2r\cos(\theta - t)]^{1/2}} - \frac{(1-\mu)}{r} = \frac{\mu^2}{2} - \frac{C}{2} = \text{const}. \tag{3}$$

transforms to

$$\frac{1}{2}u^{-2}\left(\frac{dt}{d\theta}\right)^{-2}\left[u^{-2}\left(\frac{du}{d\theta}\right)^2 + 1\right] - u^{-2}\left(\frac{dt}{d\theta}\right)^{-1} + \mu u^{-1}\cos(\theta - t) - (1-\mu)u - \mu u[1 + u^2 - 2u\cos(\theta - t)]^{-1/2} = \frac{\mu^2}{2} - \frac{C}{2} \tag{4}$$

and will be used in the subsequent calculations as a partial check of the results.

At time $t = 0$ it is assumed that Jupiter is at $\theta = 0$ and the coordinates and velocities of the asteroid are specified by

$$u = 1 + \mu^{1/2}\alpha \tag{5a}$$

$$du/d\theta = \mu^{1/2}\beta \tag{5b}$$

$$\theta = \theta_0 \tag{5c}$$

$$dt/d\theta = 1 + \mu^{1/2}\gamma \tag{5d}$$

where α, β and γ are constants not very large compared to unity in absolute value.

The fact that for $\mu=0$ the above corresponds to a circular orbit at a unit distance from the sun is immediately apparent, and the reason for chosing $\mu^{1/2}$ as a measure of the deviation from this condition will be elaborated in the next section.

Use of Equations (5) in (4) gives the following expansion for the Jacobi constant C in terms of the initial conditions.

$$-\frac{C}{2} = -\frac{3}{2} + \mu\left\{\frac{\beta^2}{2} - \frac{3}{2}\alpha^2 - \frac{\gamma^2}{2} + \cos\theta_0 + 1 - [2(1-\cos\theta_0)]^{-1/2}\right\}$$
$$+ \mu^{3/2}\left\{2\alpha^2 - (\alpha+\gamma)(2\beta^2+\gamma^2) + \alpha(1-\cos\theta_0) - \frac{\alpha}{2}[2(1-\cos\theta_0]^{-1/2}\right\}. \tag{6}$$

In the usual application of the two-variable expansion procedure one assumes that the leading terms of the asymptotic expansions for u and t are the Keplerian solutions (resulting from setting $\mu=0$ in Equations (2)) with slowly varying elements. Williams (1966) showed that for orbits of small eccentricity and values of the semi-major axis, a, close to the resonances

$$a = ((n-m)/n)^{2/3} = a_{\text{res}} \tag{7}$$

the resulting expressions for the higher-order terms in u_1 and t_1 contain small divisors. In Equation (7) n and m are relatively prime non-negative integers with $n>m$. This phenomenon is quite common in celestial mechanics and was discussed in connection with the asteroid problem by Brown and Shook (1933). Guided by the above work and translating Brown and Shook's procedure to the present notation, Williams showed that for near resonant values of a the Keplerian elements in the leading term of the solution must be assumed in the form

$$a = ((n-m)/n)^{2/3} + \mu^{1/2}\bar{a} \tag{8a}$$
$$e = \mu^{1/2}\,\bar{e} = \text{eccentricity} \tag{8b}$$
$$\omega = \mu^{1/2}\,\bar{\omega} = \text{longitude of periapsis} \tag{8c}$$
$$\tau = \mu^{1/2}\,\bar{\tau} = \text{time at periapsis} \tag{8d}$$

where the barred quantities depend on the slow variable

$$\tilde{\theta} = \mu^{1/2}(\theta - \theta_0). \tag{9}$$

Actually, if one assumed in Equations (8) and (9) that the exponents of μ are unknown, one can show that only the choice $\mu^{1/2}$ leads to a consistent asymptotic expansion. Let $\mu^{1/2}=\varepsilon$.

Clearly, for the case $a_{\text{res}}=1$, the initial conditions of Equations (5) together with the following expansions for u and t are equivalent to the choice of the elements in the form of Equations (8).

$$u = 1 + \sum_{n=1}^{N} \varepsilon^n u_n(\theta, \tilde{\theta}) + 0(\varepsilon^{N+1}) \tag{10a}$$

$$t = \theta + \sum_{n=1}^{N} \varepsilon^n t_n(\theta, \tilde{\theta}) + 0(\varepsilon^{N+1}). \tag{10b}$$

Due to the perturbation of Jupiter, the mean motion of the asteroid is not identically equal to t. Thus, the expansion for t according to Equation (10b) will contain secular terms in the t_n violating the uniformity of the results for times of order ε^{-1}. To avoid this inconsistency it is necessary to use a modified fast variable φ defined by the transformation

$$d\varphi/d\theta = 1 + \sum_{n=1}^{N} \varepsilon^n f_n(\tilde{\theta}) + 0(\varepsilon^{N+1}) \tag{11a}$$

where the f_n are unknown functions of $\tilde{\theta}$ to be determined by requiring the expansion for $t - \varphi$ to be bounded. Integrating Equation (11a) defines φ in terms of θ in the form

$$\varphi = \theta - \theta_0 - \sum_{n=1}^{N} \varepsilon^{n-1} F_n(\tilde{\theta}) + 0(\varepsilon^{N+1}) \tag{11b}$$

where

$$F_n(\tilde{\theta}) = -\int_{\theta_0}^{\tilde{\theta}} f_n(s)\, ds. \tag{11c}$$

Now, instead of Equations (10), u and t are taken in the form

$$u = 1 + \sum_{n=1}^{N} \varepsilon^n u_n(\varphi, \tilde{\theta}) + 0(\varepsilon^{N+1}) \tag{12a}$$

$$t = \varphi + \sum_{n=1}^{N} \varepsilon^n t_n(\varphi, \tilde{\theta}) + 0(\varepsilon^{N+1}). \tag{12b}$$

Equation (11a) is equivalent to the definition of a modified variable denoted by w_1' in Equation (2) of Section 9.5 of Brown and Shook (1933). Use of such a transformation is also not new in a two-variable procedure, as discussed, for example, by Kuzmak (1959) (see also Cole (1968) for a discussion of Kuzmak's method.)

In the present work u and t are calculated to $0(\varepsilon^2)$, hence the differential equations resulting from substitution of Equations (12) into Equations (2) must be obtained to $0(\varepsilon^3)$. Straightforward but lengthy calculations lead to the following equations for u_1, t_1, u_2, t_2, u_3, and t_3.

$$\frac{\partial^2 u_1}{\partial \varphi^2} - 3u_1 - 2\frac{\partial t_1}{\partial \varphi} - 2f_1 = 0 \tag{13a}$$

$$\frac{\partial}{\partial \varphi}\left(2u_1 + \frac{\partial t_1}{\partial \varphi}\right) = 0 \tag{13b}$$

$$\begin{aligned} &\frac{\partial^2 u_2}{\partial \varphi^2} - 3u_2 + 2f_1\frac{\partial^2 u_1}{\partial \varphi^2} + 2\frac{\partial^2 u_1}{\partial \varphi\, \partial \tilde{\theta}} - 2\frac{\partial t_2}{\partial \varphi} \\ &- 2f_1\frac{\partial t_1}{\partial \varphi} - 2f_2 - 2\frac{\partial t_1}{\partial \tilde{\theta}} - \left(\frac{\partial t_1}{\partial \varphi} + f_1\right)^2 \\ &- 8u_1\left(\frac{\partial t_1}{\partial \varphi} + f_1\right) - 6u_1^2 = \cos\xi_1 - 1 + 2^{-3/2}(1 - \cos\xi_1)^{-1/2} \end{aligned} \tag{14a}$$

$$\frac{\partial}{\partial\varphi}\left[2u_2 + u_1^2 + 2u_1\frac{\partial t_1}{\partial\varphi} + 4u_1 f_1 + \frac{\partial t_2}{\partial\varphi} + 2f_1\frac{\partial t_1}{\partial\varphi} + \frac{\partial t_1}{\partial\tilde{\theta}}\right]$$
$$= -\frac{\partial}{\partial\tilde{\theta}}\left(2u_1 + \frac{\partial t_1}{\partial\varphi} + f_1\right) + [2(1-\cos\xi_1)]^{-3/2}\sin\xi_1 - \sin\xi_1, \quad (14b)$$

$$\frac{\partial^2 u_3}{\partial\varphi^2} - 3u_3 + 2f_1\frac{\partial^2 u_2}{\partial\varphi^2} + (2f_2 + f_1^2)\frac{\partial^2 u_1}{\partial\varphi^2} + 2\frac{\partial^2 u_2}{\partial\varphi\,\partial\tilde{\theta}}$$
$$+ 2f_1\frac{\partial^2 u_1}{\partial\varphi\,\partial\tilde{\theta}} + \frac{\mathrm{d}f_1}{\mathrm{d}\tilde{\theta}}\frac{\partial u_1}{\partial\varphi} + \frac{\partial^2 u_1}{\partial\tilde{\theta}^2} - 12u_1u_2 - 8u_2\frac{\partial t_1}{\partial\varphi}$$
$$- 8u_2f_1 - 12u_1^2\frac{\partial t_1}{\partial\varphi} - 12f_1u_1^2 - 8u_1\frac{\partial t_2}{\partial\varphi}$$
$$- (4f_2 + 2f_1^2)\left(2u_1 + \frac{\partial t_1}{\partial\varphi}\right) - 16f_1u_1\frac{\partial t_1}{\partial\varphi} - 2\frac{\partial t_3}{\partial\varphi}$$
$$- 4f_1\frac{\partial t_2}{\partial\varphi} - 2f_3 - 2\frac{\partial t_2}{\partial\tilde{\theta}} - 4u_1^3 - 4u_1\left(\frac{\partial t_1}{\partial\varphi}\right)^2$$
$$- 2\frac{\partial t_1}{\partial\varphi}\frac{\partial t_2}{\partial\varphi} - 2f_1\left(\frac{\partial t_1}{\partial\varphi}\right)^2 - 2\frac{\partial t_1}{\partial\varphi}\frac{\partial t_1}{\partial\tilde{\theta}} - 2f_1\frac{\partial t_1}{\partial\tilde{\theta}} - 2f_1f_2$$
$$- 8u_1\frac{\partial t_1}{\partial\tilde{\theta}} = 2\left(u_1 + \frac{\partial t_1}{\partial\varphi} + f_1\right)\cos\xi_1 - \left(F_2 - t_1 + \frac{\partial u_1}{\partial\varphi}\right)\sin\xi_1$$
$$- 2\left(2u_1 + \frac{\partial t_1}{\partial\varphi} + f_1\right) + 2^{-3/2}(1-\cos\xi_1)^{-1/2}\left[-\frac{3}{2}u_1\right.$$
$$- \frac{3}{2}\frac{\sin\xi_1}{(1-\cos\xi_1)}(F_2 - t_1) + 2\frac{\partial t_1}{\partial\varphi} + 2f_1 + 4u_1 - u_1\frac{\cos\xi_1}{1-\cos\xi_1}$$
$$\left. + \frac{\sin\xi_1}{1-\cos\xi_1}\left(F_2 - t_1 + \frac{\partial u_1}{\partial\varphi}\right)\right], \quad (15a)$$

$$\frac{\partial}{\partial\varphi}\left[2u_3 + \frac{\partial t_3}{\partial\varphi} + 2u_1u_2 + 2u_2\frac{\partial t_1}{\partial\varphi} + f_1u_2^2 + 4f_1u_2\right]$$
$$+ 2f_1\frac{\partial t_2}{\partial\varphi} + 2u_1\frac{\partial t_2}{\partial\varphi} + u_1^2\frac{\partial t_1}{\partial\varphi} + f_1u_1^2 + \frac{\partial t_2}{\partial\tilde{\theta}}$$
$$+ 4f_1u_1\frac{\partial t_1}{\partial\varphi} + 2u_1\frac{\partial t_1}{\partial\tilde{\theta}} + (2f_2 + f_1^2)\left(2u_1 + \frac{\partial t_1}{\partial\varphi}\right) + f_1\frac{\partial t_1}{\partial\tilde{\theta}}$$
$$= \frac{\partial}{\partial\tilde{\theta}}\left[2u_2 + u_1^2 + 2u_1\frac{\partial t_1}{\partial\varphi} + f_1\left(2u_1 + \frac{\partial t_1}{\partial\varphi}\right) + \frac{\partial t_2}{\partial\varphi} + \frac{\partial t_1}{\partial\tilde{\theta}}\right]$$
$$+ [2(1-\cos\xi_1)]^{-3/2}\left\{-\frac{3}{2}\left[u_1 + (F_2 - t_1)\frac{\sin\xi_1}{1-\cos\xi_1}\right]\sin\xi_1\right.$$
$$\left. + (F_2 - t_1)\cos\xi_1 + 3\left(2u_1 + \frac{\partial t_1}{\partial\varphi} + f_1\right)\sin\xi_1\right\}$$
$$- 3\left(u_1 + \frac{\partial t_1}{\partial\varphi} + f_1\right)\sin\xi_1 - (F_2 - t_1)\cos\xi_1. \quad (15b)$$

In Equations (14) and (15) the following notation has been introduced:

$$\xi_1(\tilde{\theta}) = \theta_0 + F_1(\tilde{\theta}) \tag{16}$$

and the meaning of ξ_1 is discussed in Section 3.

The initial conditions at $\theta = \theta_0$ that result from applying the conditions (5) to Equations (12) and their derivatives are:

$$u_1 = \alpha \tag{17a}$$

$$\partial u_1/\partial\varphi = \beta \tag{17b}$$

$$t_1 = 0 \tag{17c}$$

$$\partial t_1/\partial\varphi = \gamma - f_1(0) \tag{17d}$$

$$u_2 = 0 \tag{18a}$$

$$\frac{\partial u_2}{\partial\varphi} = -f_1(0)\beta - \frac{\partial u_1}{\partial\tilde{\theta}}(0,0) \tag{18b}$$

$$t_2 = 0 \tag{18c}$$

$$\frac{\partial t_2}{\partial\varphi} = -f_1(0)\gamma + f_1^2(0) - f_2(0) - \frac{\partial t_1}{\partial\tilde{\theta}}(0,0). \tag{18d}$$

Substitution of the expansions (12) and (6) into the Jacobi integral (4) leads to the requirements

$$\begin{aligned}
&\frac{1}{2}\left[\left(\frac{\partial u_1}{\partial\varphi}\right)^2 + \left(\frac{\partial t_1}{\partial\varphi}\right)^2\right] + f_1\frac{\partial t_1}{\partial\varphi} + \frac{1}{2}f_1^2 - \frac{3}{2}u_1^2 + (1 + \cos\xi_1)\\
&- [2(1-\cos\xi_1)]^{-1/2} = \frac{\beta^2}{2} - \frac{3}{2}\alpha^2 - \frac{\gamma^2}{2} + 1 + \cos\theta_0\\
&- [2(1-\cos\theta_0)]^{-1/2} = \text{const}.
\end{aligned} \tag{19a}$$

$$\begin{aligned}
&\frac{1}{3}f_1\frac{\partial t_2}{\partial\varphi} - 2f_2u_1 - \frac{1}{3}f_1f_2 - 2u_1\frac{\partial t_2}{\partial\varphi} + \frac{14}{3}f_1u_1^2 - 2u_1\frac{\partial t_1}{\partial\tilde{\theta}}\\
&- \frac{1}{3}f_1\frac{\partial t_1}{\partial\tilde{\theta}} + \frac{13}{27}f_1^3 + 2u_1^3 + \frac{50}{9}u_1f_1^2 + 3u_1u_2 + \frac{\partial u_1}{\partial\varphi}\frac{\partial u_2}{\partial\varphi}\\
&+ \frac{4}{3}f_1\left(\frac{\partial u_1}{\partial\varphi}\right)^2 + \frac{\partial u_1}{\partial\varphi}\frac{\partial u_1}{\partial\tilde{\theta}} - (F_2 - t_1)\sin\xi_1 + u_1(1-\cos\xi_1)\\
&- \frac{u_1}{2}[2(1-\cos\xi_1)]^{-1/2} + [2(1-\cos\xi_1)]^{-3/2}(F_2 - t_1)\sin\xi_1\\
&= 2\alpha^2 - (\alpha+\gamma)(2\beta^2 + \gamma^2) + \alpha(1-\cos\theta_0)\\
&- \frac{\alpha}{2}[2(1-\cos\theta_0)]^{-1/2} = \text{const}.
\end{aligned} \tag{19b}$$

After the solution for u and t is calculated to each order, it is substituted into (19) and the constancy of the latter to each order is used as a partial check of the calculations.

3. First Order Solution

It follows from Equation (13b) that

$$2u_1 + \partial t_1/\partial\varphi = p_1(\tilde{\theta}). \tag{20}$$

Upon substitution of Equation (20) into Equation (13a) the following solution for u_1 can be calculated

$$u_1 = \varrho_1 \cos(\varphi + \psi_1) + 2(p_1 + f_1) \tag{21}$$

where p_1, ϱ_1 and ψ_1 are functions of $\tilde{\theta}$ to be determined together with $f_1(\tilde{\theta})$ by requiring that the expansion (12) be consistent.

Using Equation (21) in Equation (20) gives

$$\partial t_1/\partial\varphi = -(3p_1 + 4f_1) - 2\varrho_1 \cos(\varphi + \psi_1). \tag{22}$$

Integration of Equation (22) with respect to φ would result in a secular term in t_1 unless the first term is set equal to zero, and this gives the first relation governing the four unknown functions

$$p_1 = -\tfrac{4}{3} f_1. \tag{23}$$

Equation (22) is now integrated to give

$$t_1 = -2\varrho_1 \sin(\varphi + \psi_1) + q_1 \tag{24}$$

where q_1 is a constant. In view of Equation (23), u_1 becomes

$$u_1 = \varrho_1 \cos(\varphi + \psi_1) - \tfrac{2}{3} f_1. \tag{25}$$

It is shown next that there is no loss of generality in taking q_1 constant as long as f_2 is allowed to depend on $\tilde{\theta}$. Consider the expression for φ in terms of θ and $\tilde{\theta}$ given in Equation (11b).

Since the leading term in the expansion for t is φ, the final expression for t in terms of θ involves q_1 only through the difference $(q_1 - F_2)$. In fact, to any order ε^n in the expansion for t, only combinations of the form $(q_n - F_{n+1})$ appear. It therefore follows that allowing the q_n to vary will merely change the definitions of the F_{n+1} with no net effect on the final result.

Applying the initial conditions to the solutions for u_1 and t_1 leads to the following initial values (at $\theta = \theta_0$) for the four unknown functions and q_1.

$$f_1(0) = -3(2\alpha + \gamma) = -\tfrac{3}{4} p_1(0) \tag{26a}$$

$$\varrho_1(0) \sin\psi_1(0) = -\beta \tag{26b}$$

$$\varrho_1(0) \cos\psi_1(0) = -(3\alpha + 2\gamma) \tag{26c}$$

$$q_1 = -2\beta. \tag{26d}$$

Consider next the expression for $\partial t_2/\partial\varphi$ and u_2 governed by Equation (14b). Substituting for u_1 and t_1 into the second bracketed term and use of Equation (23) gives

$$\frac{\partial}{\partial\varphi}\left[2u_2 + u_1^2 + 2u_1\frac{\partial t_1}{\partial\varphi} + 4f_1u_1 + \frac{\partial t_2}{\partial\varphi} + 2f_1\frac{\partial t_1}{\partial\varphi} + \frac{\partial t_1}{\partial\tilde{\theta}}\right]$$
$$= \frac{1}{3}\xi_1'' + [2(1-\cos\xi_1)]^{-3/2}\sin\xi_1 - \sin\xi_1 \tag{27}$$

where the prime denotes differentiation with respect to $\tilde{\theta}$. In deriving Equation (27), Equations (11c) and (16) have been used to set

$$f_1' = -\xi_1''. \tag{28}$$

In order that the asymptotic expansion for u and t be consistent, u_2 and t_2 cannot contain terms proportional to φ. Therefore, the right hand side of Equation (27) must be set equal to zero; and this leads to the following differential equation governing ξ_1.

$$\xi_1'' + 3\{1 - [2(1-\cos\xi_1)]^{-3/2}\}\sin\xi_1 = 0. \tag{29}$$

In view of Equations (11b) and (12b), ξ_1 is the leading term in the mean motion of the asteroid in a frame rotating with Jupiter. This follows from the fact that the mean value of $\theta - t$ to $O(1)$ is merely ξ_1.

Equation (29) possesses the following integral which defines the evolution of the mean motion of the asteroid in the phase-plane of ξ_1 and ξ_1'.

$$\tfrac{1}{2}\xi_1'^2 + 3\{[2(1-\cos\xi_1)]^{-1/2} - \cos\xi_1\} = E = \text{const}. \tag{30a}$$

where the constant E may be expressed in terms of the initial parameters in the form

$$E = \tfrac{9}{2}(\gamma + 2\alpha)^2 + 3\{[2(1-\cos\theta_0)]^{-1/2} - \cos\theta_0. \tag{30b}$$

Equation (30a) is in agreement with the first un-numbered equation in Section 9.14 of Brown and Shook (1933).

Figure 2 is a scale drawing of curves of constant E in the phase plane ξ_1, ξ_1'. In view of the symmetry of Equation (30a) only the quadrant $\xi_1' > 0$, $0 < \xi_1 < \pi$ need be exhibited. As expected, the equilateral libration points (L_4, L_5) appear as centers and the saddle corresponds to the straight-line equilibrium point (L_3) at opposition with Jupiter.

For the special case of motion near the three equilibrium points, Equation (29) provides the leading term in the linearized long-periodic result. In particular, consider the motion near L_4, i.e., $\xi_1 \approx \pi/3$. Equation (29) may be linearized to

$$\xi_1'' + \tfrac{27}{4}\xi_1 = 0 \tag{31a}$$

according to which

$$\xi_1 = \xi_1^* \sin\left(\frac{\sqrt{27}}{2}\tilde{\theta} + \text{const.}\right) \tag{31b}$$

valid for sufficiently small ξ_1^*. Since, to first order $\theta = t$, Equation (31b) shows that

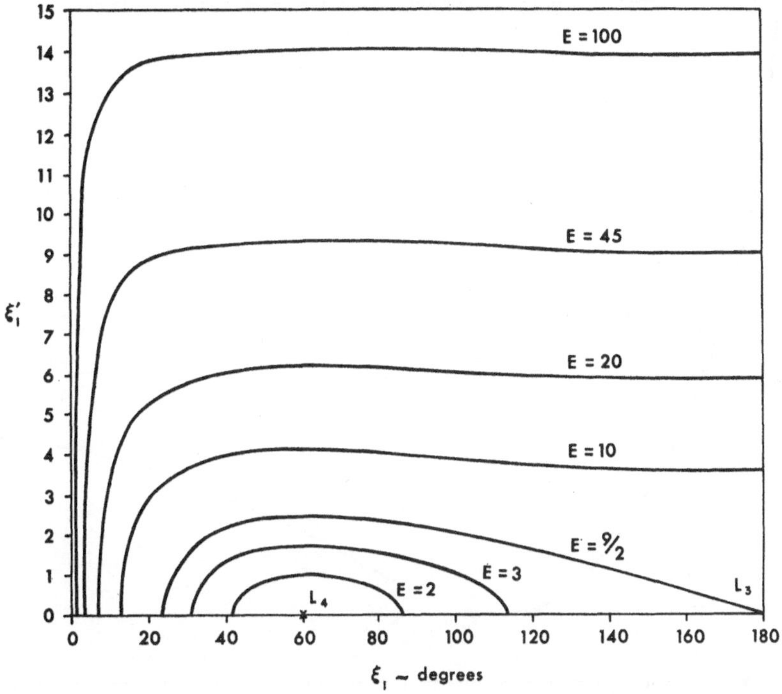

Fig. 2. Phase-plane solution for mean motion of asteroid relative to Jupiter.

near L_4, the long periodic librations have the frequency $(\sqrt{(27\mu)}/2)$ in agreement with the leading term in the expansion for the frequency ω in the linear solution

$$\omega = \left\{\frac{1 - [1 - 27\mu(1-\mu)]^{1/2}}{2}\right\}^{1/2} = \left[\frac{27\mu}{4}\right]^{1/2} [1 + 0(\mu)]. \tag{32}$$

An integral representation for ξ_1 in terms of $\tilde{\theta}$ follows immediately from Equation (30); however, this is of little computational value. As suggested by Brown and Shook (1933) one can construct the Fourier cosine series representation for ξ_1 in terms of $\tilde{\theta}$ easily as long as ξ_1 is near L_4 or L_5. This follows from the fact that for motion near L_4 or L_5 the coefficients of the Fourier series for ξ_1 proceed in decreasing powers of the amplitude in the phase plane (ξ_1, ξ_1') measured from L_4 and L_5. Clearly, for initial values such that E begins to approach $\frac{9}{2}$ (i.e., near or outside the separatrix) the above procedure fails and one must use the exact representation for ξ_1 to numerically evaluate the coefficients and period of the appropriate Fourier series.

Of course, even the exact result predicted by Equation (30) is not actually valid for large values of E, since as ξ_1 approaches zero or 2π (i.e., as the asteroid approaches Jupiter) the expansions for u and t break down, as will be shown in Section 5. As an indication of the closeness of ξ_1 to zero with increasing E, consider Figure 3 drawn to scale using Equation (30a) and showing $E - \frac{1}{2}(\xi_1')^2$ as a function of ξ_1 on the interval $0 < \xi_1 < \pi$. For example, the separatrix $(E = \frac{9}{2})$ in the phase-plane intercepts the ξ_1 axis at the two points with $\cos\xi_1 \approx 0.9$ which implies that the asteroid comes to within 24°

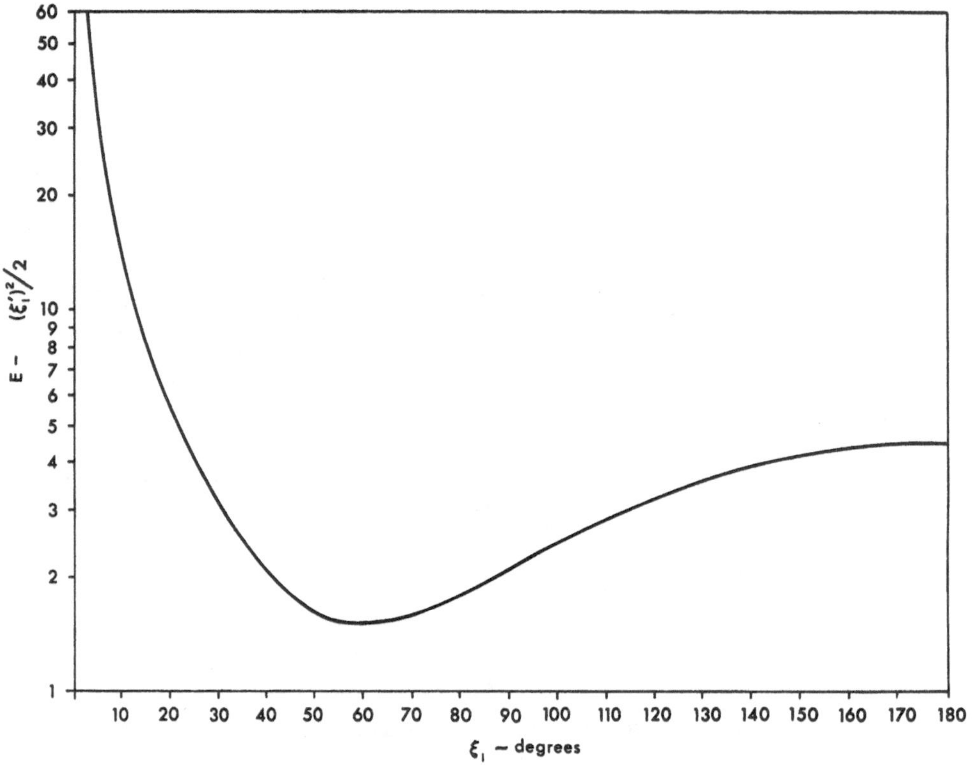

Fig. 3. Energy integral for mean motion of asteroid.

of arc of Jupiter. Clearly, the assumption in this case that Jupiter's gravitational attraction is very small compared to the sun's is not strictly valid. This question will be elaborated in Section 5.

To complete the solution to first order, it is necessary to evaluate ϱ_1 and ψ_1 by requiring that u_2 be bounded. With the right hand side equal to zero, Equation (27) integrates to

$$2u_2 + u_1^2 + 2u_1 \frac{\partial t_1}{\partial \varphi} + 4u_1 f_1 + \frac{\partial t_2}{\partial \varphi} + 2f_1 \frac{\partial t_1}{\partial \varphi} + \frac{\partial t_1}{\partial \varphi} = p_2(\tilde{\theta}) \tag{33}$$

where p_2 is an unknown function of $\tilde{\theta}$.

Substituting the value of $\partial t_2/\partial\varphi$ given by Equation (33) into Equation (14a), and use of the forms for u_1 and t_1 according to Equations (21) and (24) gives

$$\frac{\partial^2 u_2}{\partial \varphi^2} + u_2 = 2\varrho_1 (f_1 + \psi_1') \cos(\varphi + \psi_1) + 2\varrho' \sin(\varphi + \psi_1) + \tfrac{25}{9} f_1^2$$
$$+ 2f_2 + 2p_2 - (1 - \cos\xi_1) + 2^{-3/2}(1 - \cos\xi_1)^{-1/2}. \tag{34}$$

In order to avoid mixed-secular terms in u_2, which would render the expansion for u inconsistent, one must set

$$\varrho_1' = 0 \tag{35a}$$
$$\psi_1' + f_1 = 0. \tag{35b}$$

Equation (35) can be solved immediately to give

$$\varrho_1 = \text{const.} = [\beta^2 + (3\alpha + 2\gamma)^2]^{1/2} \tag{36a}$$

$$\psi_1(\tilde{\theta}) = \xi_1(\tilde{\theta}) - \theta_0 + \psi_0 \tag{36b}$$

where ψ_0 is defined in Equations (26b, c).

Substitution of the above solution for u_1 and t_1 into the left hand side of Equation (19a) shows that all the short periodic terms cancel *identically* and that the long periodic terms also cancel identically if ξ_1 obeys Equation (30). The remaining constant terms on the left hand side of Equation (19a) then add up precisely to the value required by the right hand side of Equation (19a). Thus, the Jacobi integral provides a check for the accuracy of the calculation to first order. This completes the solution to $0(\varepsilon)$.

4. Second Order Solution

The procedure for calculating the solution to second order parallels that used in the preceding section and will be presented here without repeating detailed arguments.

The solution of Equation (34) subject to Equation (35) gives

$$\begin{aligned} u_2 &= \varrho_2 \cos(\varphi + \psi_2) + \tfrac{25}{9} f_1^2 + 2f_2 + 2p_2 \\ &\quad + \cos\xi_1 - 1 + 2^{-3/2}(1 - \cos\xi_1)^{-1/2} \end{aligned} \tag{37}$$

where ϱ_2 and ψ_2 are undetermined functions of $\tilde{\theta}$.

The above is used in Equation (32) to calculate $\partial t_2/\partial\varphi$ in the form

$$\begin{aligned} \partial t_2/\partial\varphi &= \{-3p_2 - 4f_2 - \tfrac{10}{3} f_1^2 + \tfrac{3}{2}\varrho_1^2 \\ &\quad + 2(1 - \cos\xi_1) - [2(1 - \cos\xi_1)]^{-1/2}\} - 2\varrho_2 \cos(\varphi + \psi_2) \\ &\quad - \tfrac{10}{3}\varrho_1 f_1 \cos(\varphi + \psi_1) + \tfrac{3}{2}\varrho_1^2 \cos 2(\varphi + \psi_1). \end{aligned} \tag{38}$$

Thus, boundedness of t_2 requires that

$$-3p_2 - 4f_2 - \tfrac{10}{3} f_1^2 + \tfrac{3}{2}\varrho_1^2 + 2(1 - \cos\xi_1) - [2(1 - \cos\xi_1)]^{-1/2} = 0 \tag{39}$$

which defines p_2 in terms of f_2 and known quantities. Equation (38) can now be integrated to give t_2 in the form

$$\begin{aligned} t_2 &= -2\varrho_2 \sin(\varphi + \psi_2) - \tfrac{10}{3}\varrho_1 f_1 \sin(\varphi + \psi_1) \\ &\quad + \tfrac{3}{4}\varrho_1^2 \sin 2(\varphi + \psi_1) + q_2 \end{aligned} \tag{40}$$

where q_2 is constant.

The initial conditions to $0(\varepsilon^2)$ given by Equations (18) imply that q_2 and the four unknown functions at $\tilde{\theta} = 0$ have the following values

$$\begin{aligned} f_2(0) &= \tfrac{69}{2}\alpha^2 + 36\alpha\gamma + 12\gamma^2 + \tfrac{3}{2}\beta^2 \\ &\quad + 2(1 - \cos\theta_0) - [2(1 - \cos\theta_0)]^{-1/2} \end{aligned} \tag{41a}$$

$$p_2(0) = -\tfrac{163}{2}\alpha^2 - 80\alpha\gamma - 24\gamma^2 - \tfrac{3}{2}\beta^2$$

$$- 2(1 - \cos\theta_0) + [2(1 - \cos\theta_0)]^{-1/2} \tag{41b}$$

$$\varrho_2(0)\cos\psi_2(0) = -6\alpha^2 - 8\alpha\gamma - \gamma^2 + (1 - \cos\theta_0) - 2^{-3/2}(1 - \cos\theta_0)^{-1/2} \tag{41c}$$

$$\varrho_2(0)\sin\psi_2(0) = 2\{[2(1 - \cos\theta_0)]^{-3/2} - 1\}\sin\theta_0 \tag{41d}$$

$$q_2 = 4\{[2(1 - \cos\theta_0)]^{-3/2} - 1\}\sin\theta_0 + \beta(\tfrac{31}{2}\alpha + 7\gamma). \tag{41e}$$

Determination of the four unknown functions follows from the consistency of the solution to $0(\varepsilon^3)$ and this is considered next.

The right hand side of Equation (15b) is made up entirely of terms of order ε^2 and lower. If this right hand side is denoted by Q one calculates

$$\begin{aligned} Q = {} & \tfrac{1}{3}f_2' + \tfrac{5}{3}\{1 - [2(1 - \cos\xi_1)]^{-3/2}\}\,\xi_1'\sin\xi_1 - 2^{-5/2}(1 - \cos\xi_1)^{-3/2} \\ & \times (F_2 - q_1)(3 + \cos\xi_1) - (F_2 - q_1)\cos\xi_1 - [2(1 - \cos\xi_1)]^{-3/2} \\ & \times \{\tfrac{3}{2}\varrho_1\sin\xi_1\cos(\varphi + \psi_1) + (3 + \cos\xi_1)\,\varrho_1\sin(\varphi + \psi_1)\} \\ & + 3\varrho_1\sin\xi_1\cos(\varphi + \psi_1) - 2\varrho_1\cos\xi_1\sin(\varphi + \psi_1). \end{aligned} \tag{42a}$$

Boundedness of the bracketed quantity on the left hand side of Equation (15b) requires that the terms independent of φ in Equation (42) vanish, i.e.,

$$\begin{aligned} & \tfrac{1}{3}f_2' + \tfrac{5}{3}\{1 - [2(1 - \cos\xi_1)]^{-3/2}\}\,\xi_1'\sin\xi_1 - 2^{-5/2}(1 - \cos\xi_1)^{-3/2} \\ & \times (F_2 - q_1)(3 + \cos\xi_1) - (F_2 - q_1)\cos\xi_1 = 0. \end{aligned} \tag{42b}$$

Equations (42b) and (39) define f_2 and p_2 completely. Let ξ_2 denote the term of order ε for the mean-motion of the asteroid in the rotating frame, i.e.,

$$\xi_2 = F_2 - q_1 \tag{43a}$$

$$f_2 = -\xi_2'. \tag{43b}$$

Eliminating p_2 from Equation (42b) gives

$$\tfrac{1}{3}\xi_2'' + \tfrac{5}{9}\xi_1'\xi_1'' + \{2^{-5/2}(1 - \cos\xi_1)^{-3/2}(3 + \cos\xi_1) + \cos\xi_1\}\,\xi_2 = 0. \tag{44}$$

Again, it is noted that for small ξ_2 the homogeneous part of Equation (44) predicts harmonic oscillations with frequency $\sqrt{(27\mu)/2}$. This is in agreement with the linearized result since (32) has no term of order μ.

Using the following identity

$$\{[2(1 - \cos\xi_1)]^{-3/2}\sin\xi_1\}' = -2^{-5/2}(1 - \cos\xi_1)^{-3/2}(3 + \cos\xi_1) \tag{45}$$

and some algebra provides the solution of Equation (44) in the form

$$\xi_2 = -\tfrac{5}{9}\xi_1\xi_1' + C_2\xi_1' + C_1\xi_1'\int_0^{\tilde{\theta}} \frac{ds}{\xi_1'^2(s)} \tag{46}$$

where the constants C_1 and C_2 are related to the initial conditions by

$$C_1 = \xi_1'(0)\xi_2'(0) - \xi_2(0)\xi_1''(0) + \tfrac{5}{9}\xi'^3(0) \tag{47a}$$

$$C_2 = \frac{\xi_2(0)}{\xi_1'(0)} + \tfrac{5}{9}\xi_1(0). \tag{47b}$$

The first two terms of ξ_2 are bounded periodic functions of $\tilde{\theta}$. To show that the third term is also bounded and periodic it is necessary to demonstrate that it is not singular as $\xi_1'(0)\to 0$. According to Equation (30a) $\xi_1' = 0[(\xi_1-\xi_0)^{1/2}]$ as $\xi_1\to\xi_0$, where ξ_0 is a zero of ξ_1'. It then follows that the third term in Equation (47) tends to a finite value as $\tilde{\theta}$ tends to one of the zeros of ξ_1'. Thus, since ξ_2 has the same period as ξ_1, the mean motion of the asteroid is periodic to $0(\varepsilon)$ and qualitatively dominated by the behavior of ξ_1.

To calculate ϱ_2 and ψ_2 one must require u_3 to be bounded. After solving for t_3 from Equation (15b) with Q given by Equations (42), the resulting expression is used to define Equation (15a). Again, the homogeneous solution for u_3 is simple harmonic in φ with unit frequency. Therefore, all terms with unit frequency in the forcing function must be eliminated. After some tedious but rather straightforward calculations, one finds the following conditions for the boundedness of u_3

$$2[\varrho_2 \cos(\psi_2-\psi_1)]' - k(\tilde{\theta})\,\varrho_2 \sin(\psi_2-\psi_1) = g(\tilde{\theta}) \tag{48a}$$

$$2[\varrho_2 \sin(\psi_2-\psi_1)]' + k(\tilde{\theta})\,\varrho_2 \cos(\psi_2-\psi_1) = h(\tilde{\theta}) \tag{48b}$$

where

$$k(\tilde{\theta}) = -\tfrac{20}{9}f_1^3 + \tfrac{8}{3}f_1 f_2 - 4\varrho_1^2 f_1 - \tfrac{4}{3}f_1(1-\cos\xi_1) + \tfrac{2}{3}^{1/2}(1-\cos\xi_1)^{-1/2} f_1 \tag{48c}$$

$$g(\tilde{\theta}) = -12\varrho_1 f_1 q_1 + \varrho_1 f_1' \tag{48d}$$

$$h(\tilde{\theta}) = 8\varrho_1^3 + \tfrac{22}{3}\varrho_1 f_1^2 - 2\varrho_1 f_2 - 2^{-5/2}(1-\cos\xi_1)^{-3/2} \times \varrho_1(9+5\cos\xi_1) - 2\varrho_1\cos\xi_1 - 4\varrho_1 q_1^2 . \tag{48e}$$

The solution of (48) for ϱ_2 and ψ_2 gives

$$\varrho_2 \sin(\psi_2-\psi_1) = -A\sin\sigma + B\cos\sigma + \int_0^{\sigma} \frac{[h(s)-g(s)]\sin\sigma + [h(s)+g(s)]\cos\sigma}{2k(s)}\,\mathrm{d}s \tag{49a}$$

$$\varrho_2 \cos(\psi_2-\psi_1) = A\cos\sigma + B\sin\sigma + \int_0^{\sigma} \frac{[h(s)+g(s)]\sin\sigma - [h(s)-g(s)]\cos\sigma}{2k(s)}\,\mathrm{d}s \tag{49b}$$

where

$$\sigma = \int_0^{\tilde{\theta}} \frac{k(s)}{2}\,\mathrm{d}s \tag{49c}$$

$$A = \varrho_2(0)\cos[\psi_2(0) + f_1(0)] \tag{49d}$$

$$B = \varrho_2(0)\sin[\psi_2(0) + f_1(0)] . \tag{49e}$$

The lengthy details of the verification of the above results to $0(\varepsilon^2)$ by substitution in Equation (19b) have been carried out and will be omitted. As noted previously, all the short periodic terms cancel identically while the long periodic terms vanish as long as ξ_2 satisfies Equation (46). Finally, the constant part remaining on the left hand side of Equation (19b) reduces to the value on the right hand side.

Although the solution by this method can, in principle, be carried out routinely to any order in ε, hand calculations become extremely laborious; and one must use a formula manipulation scheme on the computer to continue the solution.

5. Discussion of Results

Before considering specialization of the foregoing results to the particular initial values for periodic orbits, it is necessary to derive transformation relations for expressing the solution as a function of time.

If the quantity $\tilde{\varphi}=\varepsilon\varphi$ is introduced temporarily, Equation (11b) can easily be inverted to $0(\varepsilon^2)$ in the form

$$\tilde{\theta} = \tilde{\varphi} + \varepsilon F_1(\tilde{\varphi}) + \varepsilon^2 [F_2(\tilde{\varphi}) + F_1(\tilde{\varphi}) F_1'(\tilde{\varphi})] + 0(\varepsilon^3). \tag{50}$$

Thus, the expansion for t, given by Equation (12b) becomes

$$t = \varphi + \varepsilon t_1(\varphi, \tilde{\varphi}) + \varepsilon^2 \left[t_2(\varphi, \tilde{\varphi}) + F_1(\tilde{\varphi}) \frac{\partial t_1}{\partial \tilde{\theta}} (\varphi, \tilde{\varphi}) \right] + 0(\varepsilon^3) \tag{51}$$

Equation (51) can now be inverted to give

$$\varphi = t - \varepsilon t_1(t, \tilde{t}) + \varepsilon^2 \left[- t_2(t, \tilde{t}) + t_1(t, \tilde{t}) \frac{\partial t_1}{\partial \varphi} (t, \tilde{t}) - F_1(\tilde{t}) \frac{\partial t_1}{\partial \tilde{\theta}} (t, \tilde{t}) \right] + 0(\varepsilon^3) \tag{52}$$

where $\tilde{t}=\varepsilon t$.

In the above t_1, t_2, and F_1 are the functions of φ and $\tilde{\theta}$ already defined; and the notation is self-explanatory.

If Equation (52) is multiplied by ε and the resulting expression for $\tilde{\varphi}$ in terms of t and $\tilde{t}$ is used in Equation (50), one obtains the following relation between $\tilde{\theta}$ and the time.

$$\tilde{\theta} = \tilde{t} + \varepsilon F_1(\tilde{t}) + \varepsilon^2 [F_2(\tilde{t}) - t_1(t, \tilde{t}) + F_1(\tilde{t}) F_1'(\tilde{t})] + 0(\varepsilon^3). \tag{53}$$

Equations (52) and (53) can now be used to write the solution as a function of time. The result for $\theta(t)$ is immediate and follows from substituting Equation (52) and (53) into Equation (11b).

$$\begin{aligned}\theta = \theta_0 &+ t + F_1(\tilde{t}) + \varepsilon [F_2(\tilde{t}) + F_1(\tilde{t}) F_1'(\tilde{t}) \\ &- t_1(t, \tilde{t})] + \varepsilon^2 \left[F_3(\tilde{t}) + t_1(t, \tilde{t}) \frac{\partial t_1}{\partial \varphi} (t, \tilde{t}) \right. \\ &\left. - F_1(\tilde{t}) \frac{\partial t_1}{\partial \tilde{\theta}} (t, \tilde{t}) - t_2(t, \tilde{t}) \right] + 0(\varepsilon^3).\end{aligned} \tag{54}$$

As can be seen from above, knowledge of F_3 is needed to define θ completely to $0(\varepsilon^2)$ as a function of time. This not surprising result is an inherent property of the solution and is often overlooked when only formal solutions are obtained. It is only in deriving explicit formulae for the coordinates as functions of time that one encounters the above situation. As pointed out earlier in this paper and discussed in detail by Eckstein *et al.* (1966), one can use a known exact integral of the motion (such as the Jacobi integral in this case) to calculate F_3 without considering the solution to the next order.

To calculate u as a function of time, one substitutes the expressions for φ and $\tilde{\theta}$ given by Equations (52) and (53) into the solution to obtain

$$u = 1 + \varepsilon u_1(t, \tilde{t}) + \varepsilon^2 \left[u_2(t, \tilde{t}) + F_1(\tilde{t}) \frac{\partial u_1}{\partial \tilde{\theta}}(t, \tilde{t}) - t_1(t, \tilde{t}) \frac{\partial u_1}{\partial \varphi}(t, \tilde{t}) \right] + 0(\varepsilon^3). \tag{55}$$

Equations (54) and (55) define the solution completely as a function of time.

A necessary and sufficient condition for an orbit to be periodic in a coordinate system rotating with Jupiter is that both u and $\theta - t$ be periodic functions of time with the same period. Consider first the solution to order ε calculated earlier and expressed here as a function of time.

$$u = 1 + \varepsilon \{\varrho_1 \cos[t + \xi_1'(\tilde{t})] + \tfrac{2}{3}\xi_1'(\tilde{t})\} \tag{56a}$$

$$\theta - t = \xi_1(\tilde{t}) + \varepsilon \{\xi_2(\tilde{t}) - \xi_1'(\tilde{t})[\xi_1(\tilde{t}) - \theta_0] + 2\varrho_1 \sin[t + \xi_1'(\tilde{t})]\}. \tag{56b}$$

Clearly, if $\varrho_1 = 0$, the functions above are periodic with the same period, since both ξ_1 and ξ_2 have the same period. The condition $\varrho_1 = 0$ implies that $\beta = 0$ and $\gamma = -(\frac{3}{2})\alpha$. Thus, for varying values of α one has a one parameter family of long periodic orbits which are the continuations of the family of long periodic orbits close to the triangular libration points.

Since $r = 1 - \varepsilon u_1 + 0(\varepsilon^2)$, the coordinates r and $\theta - t$ for the periodic solution are defined by

$$r = 1 - \tfrac{2}{3}\varepsilon\xi_1'(\tilde{t}) + 0(\varepsilon^2) \tag{57a}$$

$$\theta - t = \xi_1(\tilde{t}) + \varepsilon\{\xi_2(\tilde{t}) - [\xi_1(\tilde{t}) - \theta_0]\,\xi_1'(\tilde{t})\} + 0(\varepsilon^2). \tag{57b}$$

In view of the behavior of ξ_1 depicted in Figure 2, the above formulas are in qualitative agreement with Rabe's (1961) numerical solutions. In fact, the evolution of periodic orbits from the two lenticular shapes near L_4 and L_5 to coalescence at L_3 and ultimately to the horseshoe shaped orbits enclosing L_3, L_4 and L_5 follows immediately from Equation (57).

It would be interesting to also compare quantitatively the above analytic approximation with the corresponding numerically integrated solutions and to derive the conditions for periodicity to $0(\varepsilon^2)$. In this connection, it is pointed out that the conditions $\beta = 0$, $\gamma = -(\frac{3}{2})\alpha$ only apply to $0(\varepsilon)$; and one must develop α, β and γ in powers of

ε to obtain further conditions for periodicity for the higher-order terms. Since a complete definition of the solution to $0(\varepsilon^2)$ also requires knowledge of F_3 obtained by using the Jacobi integral to $0(\varepsilon^4)$, further work by hand calculation becomes impractical and is left to investigators having access to automatic formula manipulation programs.

Consider now the behavior of the solution for values of the initial parameters such that ξ_1 approaches zero (or 2π). It is important to note that the condition of eventual close approach to Jupiter occurs, according to Equation (30a), even if θ_0 is not small as long as E is sufficiently large.

For example, if $\theta_0=\pi$ and $\alpha=\gamma=1$, one calculates $E=45$ from Equation (30b); and according to Figure 3, $\xi_1'=0$ occurs when $\xi_1\approx 3.5°$ in this case. The choice $\alpha=\gamma=1$ is certainly consistent with the order of magnitude assumed for the initial conditions in Equation (5). However, the results predicted for this case are dynamically impossible. To show this one merely need consider the motion relative to Jupiter resulting from the initial conditions near the turning point $\xi_1=3.5°$, $\xi_1'=0$. Since the asteroid is within a distance of order $\mu^{1/2}$ of Jupiter, the gravitational attraction of Jupiter is now dominant and the orbit of the asteroid should describe at least one loop around Jupiter, contrary to the results given by Equations (56). A more direct evidence of the breakdown of the solution for large values if E is the occurrence of small divisors proportional to powers of $(1-\cos\xi_1)$ in the solution for μ_2 and t_2 in Equations (37) and (40). Thus, the asymptotic expansions in the form of Equations (12) are not uniformly valid for large E. This loss of validity for the case of close approach is to be expected since the expansions (12) are constructed with the assumption that the attraction of Jupiter is of order μ and hence small regardless of the distance between Jupiter and the asteroid. A similar but somewhat simpler problem was first solved using matched asymptotic expansions by Lagerstrom and Kevorkian (1963) for the case of close approach to the moon in earth-to-moon trajectories. Unfortunately, matching of the present solution with one valid close to Jupiter is not possible. The procedure required is currently being investigated and a discussion of the question is beyond the scope of this paper.

Acknowledgement

This research was sponsored by the Air Force Office of Scientific Research, Office of Aerospace Research, U.S. Air Force, under AF-AFOSR Grant 68-1456.

References

Brown, E. W.: 1911a, *Monthly Notices Roy. Astron. Soc.* **71**, 438.
Brown, E. W.: 1911b, *Monthly Notices Roy. Astron. Soc.* **71**, 492.
Brown, E. W., and Shook, C. A.: 1933, *Planetary Theory*, (Cambridge University Press, New York and London), reprinted by Dover Publications, New York.
Cole, J. D.: 1968, *Perturbation Methods in Applied Mathematics*, Blaisdell Publishing Company, Waltham, Mass.
Eckstein, M. C., Shi, Y. Y., and Kevorkian, J.: 1966, *Astron. J.* **71**, 301.
Eckstein, M. C. and Shi, Y. Y.: 1969, *Astron. J.* **74**, 551.

Hertz, H. B.: 1943, *Astron. J.* **50**, 121.
Kevorkian, J.: 1966, 'The Two Variable Expansion Procedure for the Approximate Solution of Certain Nonlinear Differential Equations', *Space Math.* **3**, A. M. S.
Kuzmak, G. E.: 1959, *P.M.M.* **23**, 515.
Lagerstrom, P. A. and Kevorkian, J.: 1963, *J. Mécan.* **2**, 189.
Rabe, E.: 1961, *Astron. J.* **66**, 500.
Rabe, E.: 1962, *Astron. J.* **67**, 382.
Williams, R. R.: 1966, Application of the Two Variable Expansion Procedure to the Commensurable Planar Restricted Three-Body Problem, Ph.D. Thesis, California Institute of Technology, Pasadena, Calif.

Discussion

E. Rabe: For initial conditions representative of the actual Trojans, the approximate theory published by Thüring in 1931 or 50 (*Astron. Nachr.*) proceeds along lines similar to your approach, so that Thüring's theory should be obtainable as a special application of your more general treatment.

J. V. Breakwell: I would like to share with Prof. Kevorkian an optimism about the possibility of application of expansion-matching to orbits with close approaches to Jupiter. I am less optimistic about his suggestion for the sun-perturbed earth-moon equilateral libration points, mainly because $\sqrt{\mu}$ there is not very small.

J. Kevorkian: Regarding the possibility of studying the sun-perturbed earth-moon libration point problem by a generalization of the method I have discussed, I am aware of the fact that $\sqrt{\mu}$ there is about 0.1, and not very small. However, this will probably mean, as in the problem of earth to moon trajectories, that one has to derive expansions to order $\mu^{3/2}$ for accurate results. This is not too high a price to pay for relinquishing the requirement of motion close to the libration points.

ON THE LONG-TERM EVOLUTION OF LUNAR SATELLITE ORBITS

JURIS VAGNERS

Aeronautics and Astronautics, University of Washington, Seattle, Wash., U.S.A.

Abstract. The long-term behavior of close, low-eccentricity lunar orbiters is studied according to the Poincaré theory of the critical points. Lunar gravity effects are described by a spherical harmonic expansion through $J_{4,4}$ and the earth is assumed to be a point mass moving in an eccentric orbit in the lunar equational plane. After averaging out the short-period terms and terms dependent on the lunar longitude, the remaining slowly varying Hamiltonian defines a single degree of freedom system. Conclusions concerning the long-term evolut ion of the orbits are drawn from analysis of the critical points of the averaged equations of motion. It is shown that the evolution of the equi-energy contours in the eccentricity, argument of pericenter plane with varying semi-major axis and angular momentum is most easily determined from study o f the critical points.

1. Introduction

As is well known, the theory of an artificial earth satellite can be developed in analytical series valid for long time intervals by application of successive approximation techniques. The accuracy of such developments is governed, in general, by the order of the terms carried in the solution. The equations of motion can be reduced to trivial quadratures with the order of accuracy restricted only by the excessive algebra required for determining high order terms.

In attempting to apply successive approximation techniques to the lunar orbiter problem, it was soon determined (Kozai, 1963; Giacaglia, 1965; Oesterwinter, 1965) that a complete solution was not possible except under quite restrictive assumptions. This result is a consequence of two facts; the near equality, within a factor of 10, of all the low order and degree lunar gravitational harmonics $J_{n,m}$, and the greatly amplified influence of the earth perturbation. Nevertheless, the short-period terms and terms dependings on the lunar longitude may still be determined by the conventional techniques. The long-period and secular motion defined by the slowly varying Hamiltonian must, in general, be then determined from numerical integration of the slow equations of motion (Giacaglia, 1969) or from analysis of equi-energy contours in an appropriate slow variable phase plane (Kozai, 1963; Vagners, 1967).

The drawbacks of numerical integration for the study of long-term evolution of satellite orbits are self-evident; the disadvantages of equi-energy contour analysis lie in the density and number of contour maps required to completely classify the motion. In the lunar orbiter problem, the nature of the long-term motion depends on two parameters; the semi-major axis and the polar component of angular momentum, or effectively, the inclination. Hence, one must study a two parameter family of contour maps. It is the purpose of this paper to show that a more productive approach is through analysis of the critical points of the slowly varying equations of motion. Such an approach is based on the Poincaré theory of the critical points (Cesari, 1963).

G. E. O. Giacaglia (ed.), Periodic Orbits, Stability and Resonances, 304–313.

It is well known that the number, position and type of critical points will completely determine the behavior of the solution as a whole i.e. the contour maps.

In modeling the lunar orbiter problem, we make the following assumptions:

(a) The earth moves in an eccentric orbit in the lunar equatorial plane; the actual orbit is inclined about 6° 41′ (±9′) to the lunar equator.

(b) The earth may be treated as a point mass.

(c) Only harmonics through $J_{4,4}$ are retained in the lunar potential.

(d) Physical librations of the moon are ignored.

(e) Solar radiation pressure and solar gravity perturbations are ignored.

For close arbiters, the neglected terms contribute $O(10^{-7})$ perturbations; the included perturbations are of $O(10^{-4})$ to $O(10^{-5})$. (A close orbiter is defined as having a semi-major axis $a<5$ mean lunar radii.) We note that under certain circumstances, the earth orbit inclination can cause the lunar satellite orbit plane to remain stationary or to perform very slow librations about a stationary value (Dasenbrock, 1969). For such orbit configurations, the analysis of this paper should be re-examined, for here it is proposed that all of the ignored effects can be treated as (small) perturbations of the present results.

2. The Slow Hamiltonian and Equations of Motion

We will omit the derivation of the slowly varying Hamiltonian as the procedure (e.g. von Zeipel) is well known. The details may be found in Giacaglia (1969). For compactness of notation, we will employ mixed Keplerian and Delaunay variables with the understanding that the Kepler elements are used as a convenient short-hand for their more involved canonical counterparts. Thus, with $\mathfrak{J}$ equal to the negative of the Hamiltonian, we have

$$\begin{aligned}\mathfrak{J} = {} & \frac{\mu^2}{2L^2} + n_1 H + \frac{1}{4} J_2 \frac{\mu^3}{L^3 G^3}\left(1 - 3\frac{H^2}{G^2}\right) + \frac{n_1^2 L^4}{16\mu^2(1-e_1^2)^{3/2}} \\ & \times \left[\left(5 - 3\frac{G^2}{L^2}\right)\left(3\frac{H^2}{G^2} - 1\right) + 15\left(1 - \frac{G^2}{L^2}\right)\left(1 - \frac{H^2}{G^2}\right)\right] - \frac{3J_4\mu^6}{64G^7L^3} \\ & \times \left[(35\sin^4 i - 40\sin^2 i + 8)\left(1 + \frac{3}{2}e^2\right)\right. \\ & \left. - 5(1 - 7\cos^2 i)\, e^2 \sin^2 i\right] - \frac{3J_3\mu^5}{8G^5L^3}\sin i(1 - 5\cos^2 i)\, e\sin g \\ & - \left[\frac{15}{32}\frac{J_4\mu^6}{G^7L^3}\sin^2 i(1 - 7\cos^2 i) + \frac{15}{8}\frac{n_1^2L^4}{\mu^2(1-e_1^2)^{3/2}}\right. \\ & \left. \times\left(1 - \frac{H^2}{G^2}\right)\right] e^2\sin^2 g \end{aligned} \tag{1}$$

where

$$L = (\mu a)^{1/2},\ G = L(1-e^2)^{1/2},\ H = G\cos i,\ g = \omega$$

and n_1, e_1 denote the earth orbit mean motion and eccentricity respectively. The Kepler elements have their conventional meaning and the unit of distance is the mean lunar radius. It is understood that all quantities of Equation (1) are slowly varying. Since $\mathfrak{J}$ is independent of the time, it is a constant of the motion. Furthermore, since the mean anomaly l and longitude of the node, $h=\Omega-\Omega_1$, do not appear in $\mathfrak{J}$, we have L=constant and H=constant.

The slow Hamiltonian $\mathfrak{J}$ defines a single degree of freedom system described by the differential equations

$$\begin{aligned} \dot{g} &= -\,\partial\mathfrak{J}/\partial G \\ \dot{G} &= \partial\mathfrak{J}/\partial g \end{aligned} \tag{2}$$

with a known integral $\mathfrak{J}(g, G)$=constant. Since for our investigation we have assumed that J_2, J_3, J_4 and the earth perturbation are all of (roughly) the same order of magnitude, the variable part of $\mathfrak{J}$ is factored by small parameters and we cannot solve Equations (2) by successive approximation. The general solution of (2) requires an integral of a non-separable Hamilton-Jacobi partial differential equation which at present does not appear possible. In this respect, the lunar satellite problem is similar to other 'resonance problems', such as the secular motion of resonant asteroids studied by Giacaglia (1968), wherein the slow variables obey their own 'resonance dynamics'.

We have then basically three methods for studying the long-term motion: numerical integration of Equations (2), computation of the equi-energy contours $\mathfrak{J}(g, G)$= constant, and study of the critical points of the differential Equations (2). The results of previous studies utilizing equi-energy contours (Kozai, 1963; Vagners, 1967) have indicated that the eccentricity can vary from zero up to quite large values. Consequently, in order that a comprehensive picture of the overall behavior be obtained, a set of elements which are non-singular for zero eccentricity must be introduced. We abandon the canonical formalism at this point in favor of the more easily visualized combinations

$$\begin{aligned} A &= e\cos g \\ B &= e\sin g\,. \end{aligned} \tag{3}$$

The variables A, B are simply the rectangular components of the polar phase plane e, g where e is the radius and g the polar angle.

The differential equations for these elements are written as follows:

$$\begin{aligned} \dot{A} &= \frac{G\cos g}{L(1-G^2/L^2)^{1/2}}\,\dot{G} - B\dot{g} \\ \dot{B} &= \frac{G\cos g}{L(1-G^2/L^2)^{1/2}}\,\dot{G} + A\dot{g} \end{aligned} \tag{4}$$

or, in explicit form:

$$\frac{4L}{3\mu}\,\dot{A} = \frac{J_3}{2a^4\eta^4}\,(1-\beta\eta^{-2})^{1/2}(1-5\beta\eta^{-2}) + \frac{5Ba^2}{KR^3\eta(1-e_1^2)^{3/2}}$$

$$\times (\eta^2 - \beta - \beta\eta^{-2}B^2) - B\left\{\frac{J_2}{a^3\eta^4}(1 - 5\beta\eta^{-2}) + \frac{2a^2\eta}{KR^3(1-e_1^2)^{3/2}}\right.$$
$$\left. - \frac{5J_4}{32a^5\eta^{12}}[9\eta^6 - (50\beta - 7)\eta^4 - (7\beta^2 + 126\beta)\eta^2 - 231\beta^2]\right\}$$
$$- \frac{B^2J_3}{2a^4\eta^7}\left\{-5(1 - 5\beta\eta^{-2})(1 - \beta\eta^{-2}) + \frac{\beta(11\eta^2 - 15\beta)}{\eta^4(1 - \beta\eta^{-2})^{1/2}}\right\}$$
$$+ \frac{5B^3J_4}{8a^5\eta^{12}}(7\eta^4 - 72\beta\eta^2 + 77\beta^2) \tag{5a}$$

$$\frac{4L}{3\mu}\dot{B} = A\left\{\frac{J_2}{a^3\eta^4}(1 - 5\beta\eta^{-2}) + \frac{2a^2\eta}{KR^3(1-e_1^2)^{3/2}} - \frac{5J_4}{32a^5\eta^{12}}\right.$$
$$\left.\times [\eta^6 + 7(2\beta + 1)\eta^4 - 63(\beta^2 + 2\beta)\eta^2 - 231\beta^2]\right\} + \frac{ABJ_3}{2a^4\eta^5}$$
$$\times\left\{-\frac{5}{\eta^2}(1 - 5\beta\eta^{-2})(1 - \beta\eta^{-2})^{1/2} + \frac{\beta(11\eta^2 - 15\beta)}{\eta^4(\eta^2 - \beta)^{1/2}}\right\}$$
$$+ \frac{5AB^2}{\eta^3}\left[\frac{a^2\beta}{KR^3(1-e_1^2)^{3/2}} - \frac{J_4}{8a^5\eta^9}(7\eta^4 - 72\beta\eta^2 + 77\beta^2)\right] \tag{5b}$$

where $\eta = [1-(A^2+B^2)]^{1/2} = (1-e^2)^{1/2}$; $\beta = H^2/L^2 = \eta^2\cos^2 i = \text{constant}$; K = ratio of mass of moon/mass of earth; R = apparent earth orbit semi-major axis measured in lunar radii.

The singular points A_0, B_0 are then defined by the solutions to the equations $\dot{A}=0$, $\dot{B}=0$ which, in general, must be determined numerically. Note that the Hamiltonian in terms of A, B variables is symmetric in A, i.e. aside from $e\sin g$, $e^2\sin^2 g$ terms, we have only terms involving e^2 or equivalently A^2+B^2 (cf. Equation (1)). Therefore, if we were to determine equi-energy contours in the A, B plane, the contours would be symmetrical about the B axis. Similarly, symmetry of $\mathfrak{J}$ implies that the singular point distribution must be symmetrical about the B axis. We remark that some of the features of the critical point behavior in the range of eccentricity between $e=0.1$ and $e=e_{\text{crit}}$ are available in the paper by Vagners (1967). The value e_{crit} is defined by $a(1-e)=1$ as the maximum allowable eccentricity for a given semi-major axis such that the radius of pericenter remains greater than the radius of the moon. Behavior for $e<0.1$ is unclear in the quoted reference due to the variables employed.

3. Low Eccentricity Equations of Motion

In this paper we are concerned specifically with the low eccentricity behavior, hence the equations of motion may be expanded on e up to, and including, e^2 terms, in order to obtain more tractable expressions. The resulting equations are:

$$\begin{aligned}\dot{A} &= k_1 + k_2B + k_3B^2 + k_4A^2\\ \dot{B} &= A[k_5 + k_6B]\end{aligned} \tag{6}$$

where the k_i are defined as follows:

$$\begin{aligned}
k_1 &= \frac{3\mu}{4L}\left[\frac{J_3}{2a^4}(1-\beta)^{1/2}(1-5\beta)\right]\\
k_2 &= \frac{3\mu}{4L}\left[\frac{n_1^2a^2(3-5\beta)}{\mu(1-e_1^2)^{3/2}} - \frac{J_2}{a^3}(1-5\beta) + \frac{5J_4}{16a^5}(8-88\beta-119\beta^2)\right]\\
k_3 &= \frac{3\mu}{4L}\left[\frac{J_3}{4a^4(1-\beta)^{1/2}}(14-117\beta+115\beta^2)\right]\\
k_4 &= \frac{3\mu}{4L}\left[\frac{J_3}{4a^4(1-\beta)^{1/2}}(4-35\beta+35\beta^2)\right]\\
k_5 &= \frac{3\mu}{4L}\left[\frac{J_2}{a^3}(1-5\beta) + \frac{2n_1^2a^2}{\mu(1-e_1^2)^{3/2}} - \frac{5J_4}{16a^5}(4-56\beta-147\beta^2)\right]\\
k_6 &= -\frac{3\mu}{4L}\left[\frac{J_3}{2a^4(1-\beta)^{1/2}}(5-41\beta+40\beta^2)\right].
\end{aligned} \tag{7}$$

For fixed semi-major axis and inclination ($\beta=(1-e^2)^{1/2}\cos^2 i\approx\cos^2 i$) values the k_i are constants. Also, we have $k_3=k_4-k_6$. If we form the (time) reduced equation

$$\frac{\mathrm{d}A}{\mathrm{d}B} = \frac{k_1+k_2B+k_3B^2+k_4A^2}{A(k_5+k_6B)} \tag{8}$$

we can determine the integral curves in the A, B plane near the origin under certain conditions on k_6 and the ratio $2k_4/k_6$. These integral curves are the appropriate low-e approximations of the slow Hamiltonian written as a function of A and B. Since these expressions are necessary in attempting to determine A and B as functions of time, we give their form here for reference purposes.

$$\begin{aligned}
&A^2+B^2+p_1B+p_2(k_5+k_6B)^{2k_4/k_6} = p_3; \quad k_6\neq 0,\quad 2k_4/k_6\neq 1\\
&A^2+B^2+p_4B-\exp[(-2k_4/k_5)B] = p_5; \quad k_6=0\\
&A^2+B^2+p_6B+p_7(k_5+k_6B)\log(k_5+k_6B) = p_8;\\
&\qquad\qquad\qquad\qquad\qquad\qquad k_6\neq 0,\quad 2k_4/k_6=1\\
&k_5A^2-k_2B^2-2k_1B = p_9; \quad k_3=k_4=k_6=0
\end{aligned} \tag{9}$$

where $p_j=p_j(\beta)$.

In this paper, however, we are not concerned with the determination of A and B as explicit functions of time.

4. Critical Points of the Low-e Equations

If we write Equations (6) in the form

$$\begin{aligned}
\dot{A} &= f_1(A, B, a, \beta)\\
\dot{B} &= f_2(A, B, a, \beta)
\end{aligned} \tag{10}$$

the critical points A_0, B_0 are defined by

$$\begin{aligned} f_1(A_0, B_0, a, \beta) &= 0 \\ f_2(A_0, B_0, a, \beta) &= 0 . \end{aligned} \tag{11}$$

The nature of a critical point A_0, B_0 will be determined by the value of the Jacobian

$$\Delta = \left. \begin{vmatrix} \dfrac{\partial f_1}{\partial A} & \dfrac{\partial f_1}{\partial B} \\ \dfrac{\partial f_2}{\partial A} & \dfrac{\partial f_2}{\partial B} \end{vmatrix} \right|_{\substack{A = A_0 \\ B = B_0}} . \tag{12}$$

If $\Delta < 0$, the critical point is a center or stable point; if $\Delta > 0$, then a saddle or unstable point. Since the Jacobian is a function of two parameters a, β, the nature and value of a given critical point will change as these parameters are varied. At the value $\Delta = 0$, we have transition from a stable to unstable point or vice versa. The exact behavior *at* $\Delta = 0$ is governed by the higher order terms in e.

The simplified form of Equations (6) allows us to write two conditions for A_0, B_0:

(1) $$\begin{aligned} A_0 &= 0 \\ B_0 &= \frac{-k_2 \pm [k_2^2 - 4k_3k_1]^{1/2}}{2k_3} \neq -\frac{k_5}{k_6} \neq 0 . \end{aligned} \tag{13}$$

These critical points lie on the $\pm B$ axis, i.e. $g = \pm 90°$. If $B_0 = -k_5/k_6 \neq 0$, we have $\Delta = 0$ and the transition condition is

$$\begin{aligned} &\frac{k_5}{k_6} + \left[\frac{-k_2 \pm [k_2^2 - 4k_3k_1]^{1/2}}{2k_3}\right] = 0 \\ &k_j = k_j(a, i) \end{aligned} \tag{14}$$

(2) $$\begin{aligned} A_0 &= \pm \left\{-\frac{1}{k_4}\left[k_1 - \frac{k_2k_5}{k_6} + \frac{k_3k_5^2}{k_6^2}\right]\right\}^{1/2} \neq 0 \\ B_0 &= -k_5/k_6 \neq 0 . \end{aligned} \tag{15}$$

These critical points occur for $|g| \neq 90°$ except at two points for a given semi-major axis where $|g| = 90°$. If $A_0 = 0$, $B_0 = k_5/k_6 \neq 0$, then $\Delta = 0$ and the transition condition becomes

$$-k_1 + \frac{k_2k_5}{k_6} - \frac{k_3k_5^2}{k_6^2} = 0 . \tag{16}$$

Of course, we only consider the real solutions of Equations (13) through (16). Under certain conditions, a critical point may exist at the origin of the A, B plane ($e = 0$): when the inclination is zero or 63.4°. Thus we see that the 'critical inclination' found in the earth satellite problem is preserved in this more complicated dynamical system for zero eccentricity. However, the nature of the singularity may be stable if $k_2/k_5<0$ or unstable if $k_2/k_5>0$ with $k_2=0$ defining the transition. With the value of β fixed at 0.2, the existence of a transition is governed by the semi-major axis value and the magnitudes (and signs) of J_2 and J_4 (cf. Equation (7)).

Note that when the J_n are taken as zero – corresponding to higher values of the semi-major axis – we recover the results of Lidov (1963). The low eccentricity equations reduce to

$$\begin{aligned} \dot{A} &= k_2' B \\ \dot{B} &= k_5' A \end{aligned} \tag{17}$$

with

$$\begin{aligned} k_2' &= \frac{3\mu}{4L} \frac{n_1^2 a^2}{\mu(1-e_1^2)^{3/2}} (3-5\beta) \\ k_5' &= \frac{3\mu}{4L} \frac{2n_1^2 a^2}{\mu(1-e_1^2)^{3/2}} . \end{aligned} \tag{18}$$

The e^2 terms drop out and we have the condition that, for $e\neq 0$, $\beta>0.6$ yields a stable and $\beta<0.6$ yields an unstable solution with $\beta=0.6$ the condition for the transition.

5. Critical Point Evolution

The critical point conditions (13) and (15) define, for a fixed value of the semi-major axis, curves in the eccentricity – $\beta=(1-e^2)\cos^2 i$ plane. Equations (13) define two solutions for $g=-90°$ and one solution for $g=+90°$. From (15) we obtain values of $|g|\neq 90°$, but such values must be symmetrically distributed about $|g|=90°$. The evolution of the critical point curves with increasing semi-major axis is shown in Figure 1. We note that the $|g|\neq 90°$ curves exist only for inclinations greater than about 60° and that there are no such solutions for $e<0.1$ for $a>1.73$. The points of intersection of the $|g|\neq 90°$ and the $g=\pm 90°$ curves correspond to transition points from center to saddle points or vice versa. A clearer understanding may be gained from Figure 2 where we present (typical) results for $a=1.7$. (The behavior for large eccentricities is conjectured and will be verified by further analysis.) The variable g curve departs from $g=-90°$ with increasing β and the g values change from $-90°$ to $+90°$ as eccentricity and β increase. One conjectures that as g goes to $+90°$, the variable g curve must intersect the g-$+90°$ curve with the intersection defining again a point of transition from unstable to stable point for $g=+90°$.

With the aid of figures such as Figure 2 for other values of the semi-major axis, one can easily sketch out the evolution of the equi-energy contours in the *A*-*B* plane with changing β. Closed contours that encircle the origin correspond to circulation of the periapse; closed contours *not* encircling the origin correspond to libration of the periapse. The critical points correspond to a stationary periapse analogous to the critical inclination in the earth satellite problem. A typical evolution with increasing β for $a=1.7$ is sketched in Figure 3. The behavior for large eccentricity at this point is not clear and hence the sketches are incomplete, however, the pattern of the

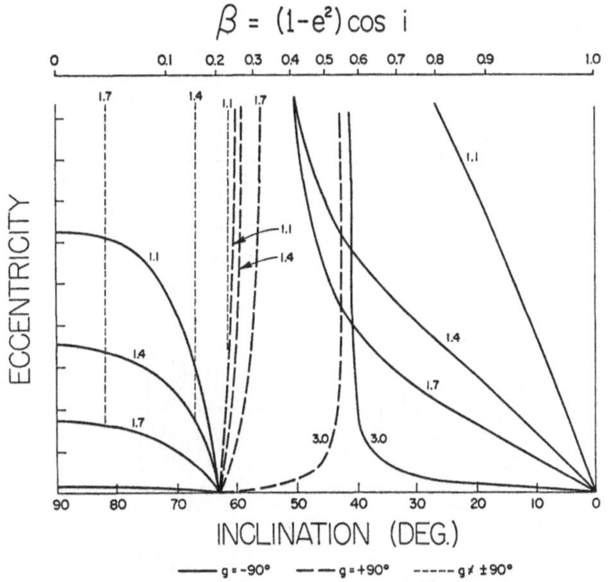

Fig. 1. Critical point evolution with increasing a.

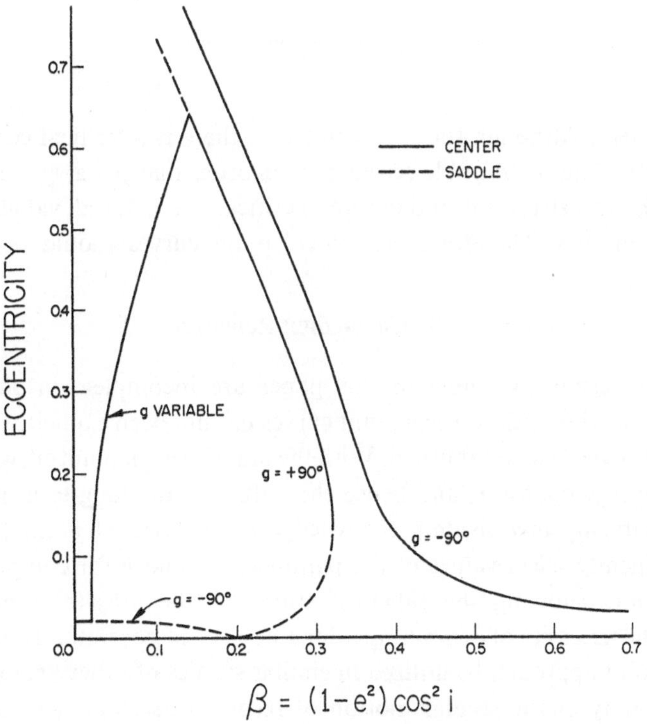

Fig. 2. Critical point distribution, $a = 1.7$.

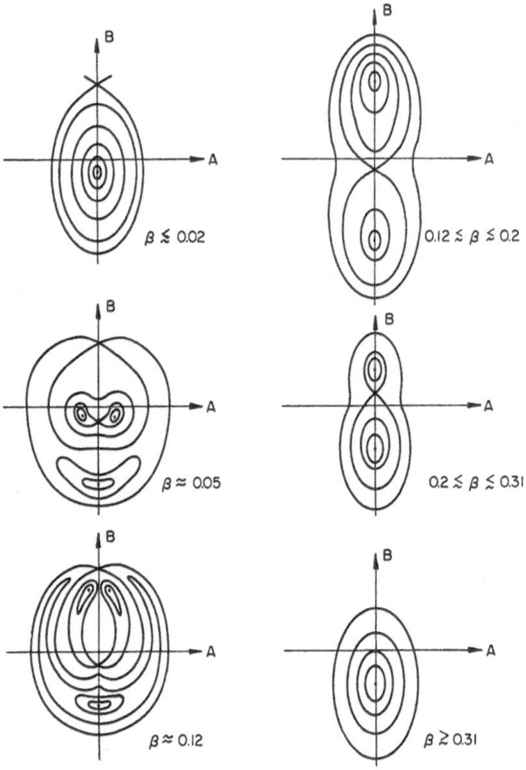

Fig. 3. Equi-energy contour evolution, $a = 1.7$.

evolution is clear. Although for each value of a there is a 'critical eccentricity circle' defining the maximum allowable eccentricity in order that periapses remain above the lunar surface, the distribution and nature of critical points for all values of e determine the overall contours. Therefore, the critical point curves should be defined for all values of e.

6. Concluding Remarks

Although the results presented in this paper are incomplete insofar as the large eccentricity behavior of the critical point curves has not been explicitly determined, the general trends have been established. With the aid of critical point curves, the evolution of the equi-energy contours, and hence the nature of the long-term motion, may be more systematically investigated. Knowledge of the critical point behavior allows one to intelligently select values of the parameters a and β for computation of equi-energy contours showing the salient features of the long-term motion, thereby eliminating the necessity of computing a high density of contours. It is suggested that the critical point approach be utilized in similar studies of other analogous problems, such as the study of the secular motion of resonant asteroids (cf. Giacaglia, 1968; Schubart, 1964).

Acknowledgement

This research was sponsored by NASA grant NGR-48-002-074.

References

Cesari, L.: 1963, *Asymptotic Behavior and Stability Problems in Ordinary Differential Equations*, Academic Press, New York, Ch. III, p. 156.

Dasenbrock, R.: 1969, Third Complication of Papers on Trajectory Analysis and Guidance Theory, NASA PM 81, pp. 231–51.

Giacaglia, G.E.O. *et al.*: 1965, The Motion of a Satellite of the Moon, Goddard Space Flight Center Report X-547-65-218, Greenbelt, Md.

Giacaglia, G.E.O.: 1968, Secular Motion of Resonant Asteroids, SAO Special Report 278.

Giacaglia, G.E.O.: 1969, A Semi-Analytical Theory for the Motion of a Lunar Satellite, NASA Technical Note, NASA TN-D-5043.

Kozai, Y.: 1963, *J. Astron. Soc. Japan* **15**.

Lidov, M. L.: 1963, Russian Supplement, *AIAA J.* **1**.

Oesterwinter, C.: 1965, The Motion of a Lunar Satellite, Ph.D. Dissertation presented to Yale University.

Schubart, J.: 1964, Long-period Effects in Nearly Commensurable Cases of the Restricted Three-body Problem, SAO Special Report No. 149.

Vagners, J.: 1967, Some Resonant and Non-resonant Perturbations of Earth and Lunar Orbiters, Ph.D. Dissertation presented to Stanford University.

TRANSIENT ANNULAR STRUCTURES IN EXPLODING GALAXIES

J. L. SÉRSIC
Observatorio Astronómico, Córdoba, Argentina

1. Introduction

The purpose of this paper is to show that the explosion and fragmentation of the nucleus of a galaxy may produce circumstances theoretically resembling a generalization of a Restricted Problem in Celestial Mechanics.

Most astronomers agree today on the decisive role played by the nucleus in the violent events observed in galaxies [1]. We shall here be particularly concerned with giant ellipticals and lenticulars known to be associated with radioemission (otherwise called radiogalaxies), and also with certain peculiar galaxies whose relationship to radiosources is still a matter of discussion.

The radiogalaxies themselves are a whole class of objects with definite characteristics [2], mainly high luminosity, high mass, and double structure of the associated radiosource [3]. Some radiogalaxies show peculiarities in their optical structures, like chains of absorbing clouds, dust belts and other features suggesting that radioemission is accompanied or followed by a drastic redistribution of stars and dust (Figures 1 and 2).

Within the wealth of queer structures observed in peculiar galaxies [4], some of them (Figure 3) are associated with E-SO galaxies, whose possible evolutive connection with radiogalaxies is proposed later in this paper as a late stage of a galaxy explosion.

2. Dynamics of a Radiogalaxy

It is fair to say that our knowledge of nuclei of radiogalaxies is so scarce that we could term them as 'unobservable entities'. This is mostly because the radiogalaxies themselves are so scarce (only 6% of the E-SO galaxies are radiogalaxies [5]) that their average mutual distance is 60 Mpc. The closest specimen is NGC 5128 (= Cen A) which is only 4 Mpc away [6], but a dust belt that screens its nuclear region prevents observing it. The second closest radiogalaxy is NGC 4486 (= M87 = Vir A) which is 15 Mpc away in the Virgo Cluster, a distance large enough to preclude resolution of its nuclear region. Confronted with these facts, we are only left to make 'reasonable guesses' regarding the properties of nuclei:

We shall assume that the nucleus of a going-to-be radiogalaxy is a high density, high mass object, which means a large central concentration of mass toward the center. Thus, we could think of the stars filling the observable volume of the galaxy as moving in a point-mass gravitational field, as the effect of collisions and close approaches are negligible everywhere except in the nuclear region.

An explosive event means, on the other hand, a change of the effective gravitational

G. E. O. Giacaglia (ed.), Periodic Orbits, Stability and Resonances, 314–321.

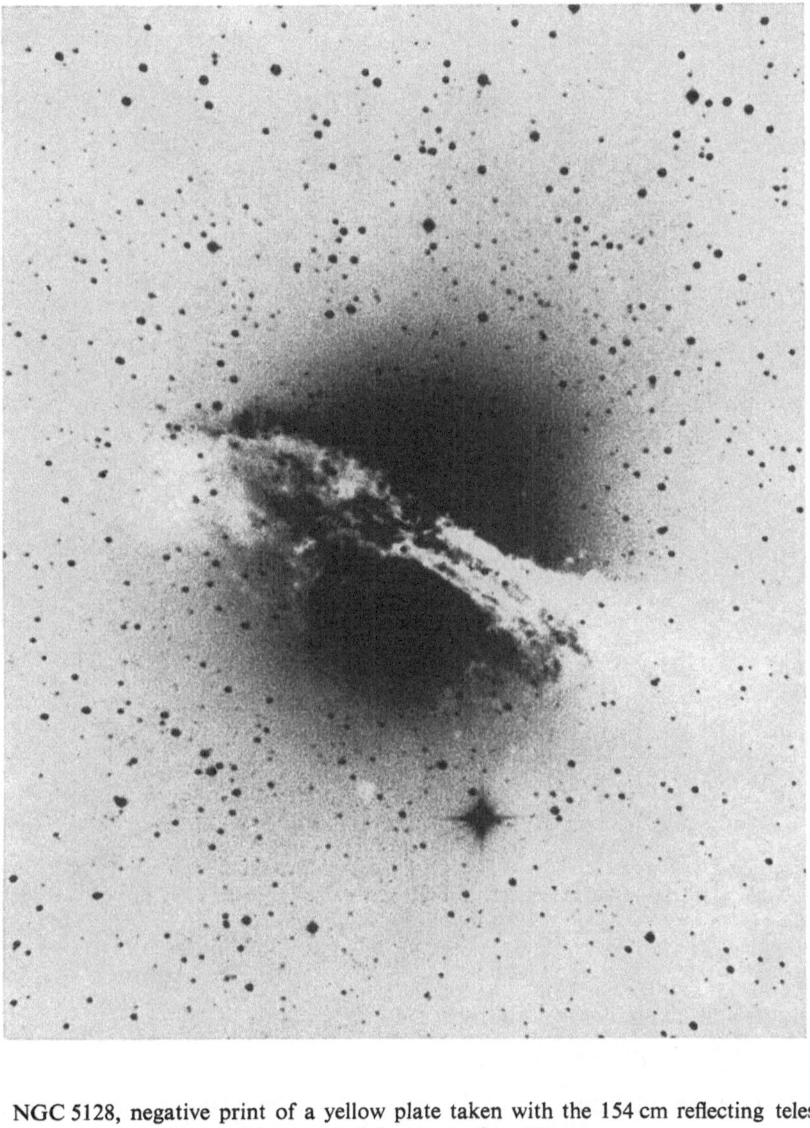

Fig. 1. NGC 5128, negative print of a yellow plate taken with the 154 cm reflecting telescope at Bosque Alegre, Córdoba, Argentina, 60^{m} exposure.

mass of the nucleus because of its dependence on the binding energy. As an explosion is essentially a radial rearrangement of the concerned masses, the binding energy will change and also the effective gravitational mass. Thus, we are due to accept the variability of the mass of the nucleus if we postulate an explosive event in it.

Lastly, let us assume the nucleus fragments in two nucleus-like objects with fractional masses $1-\mu$ and μ, while the total mass varies because of the reasons given above.

With the preceding hypothesis, we consider now the motion of an infinitesimal particle in the field

$$V(r_1, r_2, t) = -GM(t)\left[(1-\mu)/r_1 + \mu/r_2\right] \tag{1a}$$

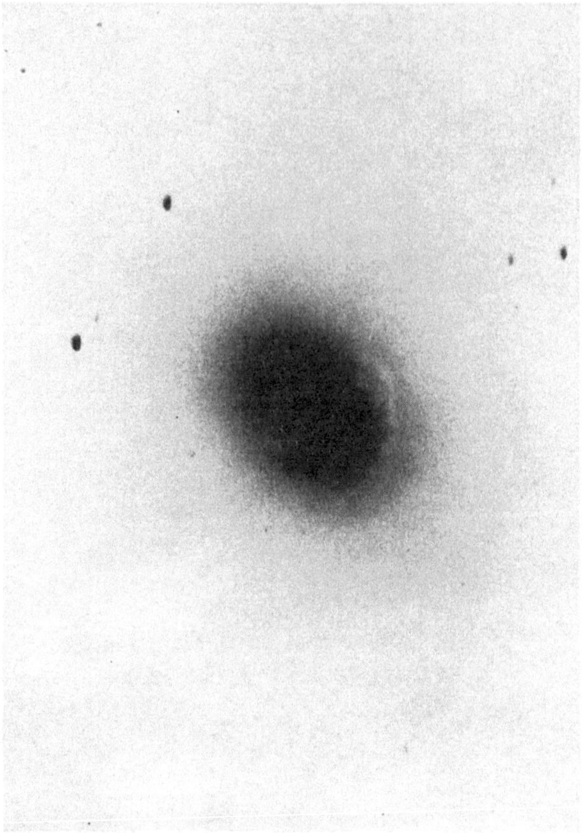

Fig. 2. NGC 1316, same as in Figure 1. Notice here, as in Figure 2, the belt of dark clouds surrounding the object.

whose motion is given by the equations

$$X_i'' = -\partial V/\partial X_i \quad i = 1, 2, 3 \tag{1b}$$

where $M(t)$ is the total mass of both nuclei, r_1, r_2 the distances of the particle from the nuclei with fractional masses $1-\mu$ and μ respectively, and G the Newtonian constant. As the problem is not conservative because of the time-dependent potential $V(r_1, r_2, t)$, we carry out the following space-time transformation

$$X_i = \xi_i/\phi \qquad \mathrm{d}\tau/\mathrm{d}t = \phi^2(\tau) \tag{2a, b}$$

known as Schürer's [7]. The function $\phi(\tau)$ is arbitrary and will be selected after imposing certain conditions. The new independent variable τ replaces the time t and the derivatives respect of τ will be written with dots. It is easy to see that (1) becomes

$$\ddot{\xi}_i = -\partial\Omega/\partial\xi_i; \quad \Omega = -\frac{1}{2}\frac{\ddot{\phi}}{\phi}\varrho^2 - \frac{GM(t)}{\phi}\left(\frac{1-\mu}{\varrho_1} + \frac{\mu}{\varrho_2}\right)$$

after using (2).

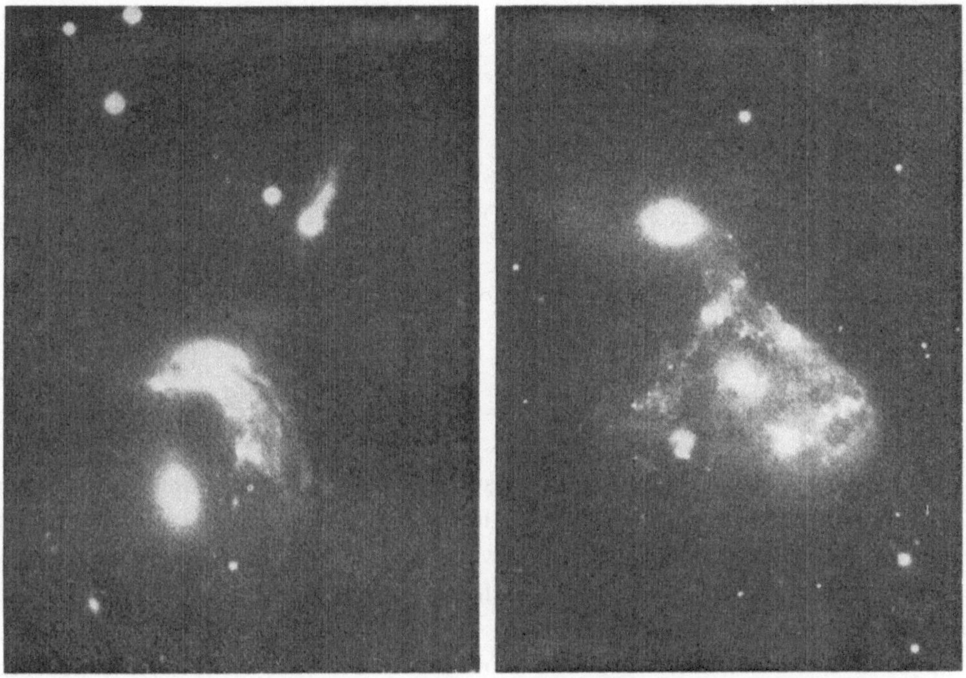

Fig. 3. Peculiar galaxies with structures which could be interpreted as Lagrangian rings (see text).

We shall select ϕ in order to have the new potential Ω independent of τ, so that we write

$$\phi(\tau) = GM(t) \qquad \ddot{\phi} = k^2 \phi . \tag{3}$$

The function ϕ is now defined, as well as the possible laws for temporal variation of the mass and the link between τ and t through (2b).

We are now concerned with the motion of the infinitesimal particle by means of the equations

$$\ddot{\xi}_i = - \partial\Omega/\partial\xi_i \quad i = 1, 2, 3 \tag{4a}$$

in the conservative potential

$$\Omega(\varrho_1, \varrho_2) = - \tfrac{1}{2} k^2 \varrho^2 - \frac{1-\mu}{\varrho_1} - \frac{\mu}{\varrho_2} . \tag{4b}$$

The 'Jacobian' integral of the problem is now

$$V^2 + C + 2\Omega(\varrho_1, \varrho_2) = 0 \tag{5}$$

where $v^2 = \dot{\xi}_i \cdot \dot{\xi}_i$ is twice the kinetic energy of the particle, C is Jacobi's constant and $\Omega(\varrho_1, \varrho_2)$ the potential.

Let P be the distance between the two centers (nuclei), ϱ, ϱ_1 and ϱ_2 are the distances of the particle from the barycenter and from each nucleus respectively.

We have then

$$\varrho^2 = (1 - \mu)\,\varrho_1^2 + \mu\varrho_2^2 + \mu(1 - \mu)\,P^2$$

so that the potential may be written in bipolar coordinates as

$$-2\Omega(\varrho_1, \varrho_2) = (1 - \mu)\,(k^2\varrho_2^2 + 2/\varrho_1) + \mu(k^2\varrho_2^2 + 2/\varrho_2) \qquad (6)$$

after suppressing the term $-(\frac{1}{2})k^2\mu(1-\mu)P^2$, which only means a change in the zero of the energies.

The motion of the particle will be real (possible) if $v^2 \geqslant 0$, so the surfaces

$$-2\Omega(\varrho_1, \varrho_2) = C \qquad (7)$$

define – for a given Jacobian constant C – the frontier of the space where the motion is allowed.

As it can be seen, the theoretical formulation of the problem after the transformation (2) and the relations (3) is analogous to the Restricted Problem of Three Bodies but taking into account now that the Hill curves in the plane of the three bodies are here the meridian sections of the surfaces (7), while the symmetry axis is here the line joining both nuclei.

There is however a difference, since in our problem there is no *a priori* relationship between the distance between nuclei P and the constant k, while in the Restricted Problem $k^2P^3=1$. This introduces an additional parameter (P) which changes the metric and the topology of the surfaces (7).

As in the case of the Restricted Problem, we also have here the singular Eulerian points L_1, L_2 and L_3, which are not fixed points now but whose positions vary with P. The main topological feature we have is that the Lagrangian triangular points L_4, L_5 now degenerate in a circumference equidistant from both nuclei, whose distance from them is ϱ_0, so that $k^2\varrho_0^3=1$. This last relationship means that ϱ_0 depends on the time scale of the mass variation process, so that the distance of the Lagrangian ring (as we shall call hereafter the L_4-L_5 circumference) from the nuclei is fixed for a given mass variation process.

Figure 4 shows the position of L_1, L_2 and L_3 relative to those of $1-\mu$ and μ as a function of P. The unit of length was taken as to make $k=1$, which also means $\varrho_0=1$. Notice that for $P=2$, $L_1(\mu)$ lies half way from both nuclei and the Lagrangian ring disappears, because $P=2$ means that only the central point satisfies the condition of being at unit distance from both nuclei. For $P>2$ the topology of the surfaces (7) changes, as there is no more Lagrangian ring, neither L_1 is any more a saddle point, now becoming an ordinary minimum. The existence of the Lagrangian ring will be conditioned to the separation P of the nuclei and becomes in principle, a transient feature.

The time-evolution of the surface (7) may be followed through the time variation of P, given by the equation

$$\ddot{P} = k^2P - 1/P^2$$

some of whose solutions in the (P, τ) space are given in Figure 5 after numerical integrations for several values of the Jacobian constant C.

If $C > 3$, we only have oscillations of the nuclei around the barycenter. In such cases P never reaches the critical value 2 and the Lagrangian ring lasts as long as the mass variation process. On the other hand, if $C < 3$, P reaches the critical value $P = 2$ after a finite τ^*. If $C = 3$ there is an asymptotic solution $P = 1$ whose stability has yet to be studied.

By reversing the transformation (2) we may recover the solution in terms of the old coordinates and physical time t. It is easy to see that the topology of the surfaces (7) is not changed and its properties should be reflected in the structures observed in galaxies with an exploding nucleus.

3. Comparison with Observed Structures in Radio- and Peculiar Galaxies

The model just developed predicts the existence of transient ring structures in galaxies suffering violent events in their nuclei accompanied by mass variation and frag-

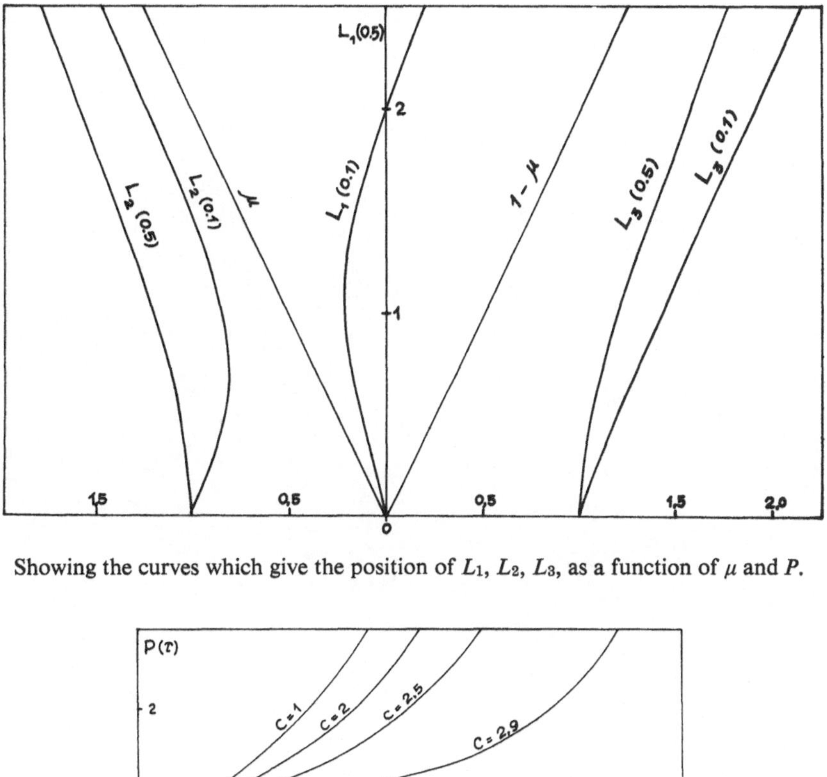

Fig. 4. Showing the curves which give the position of L_1, L_2, L_3, as a function of μ and P.

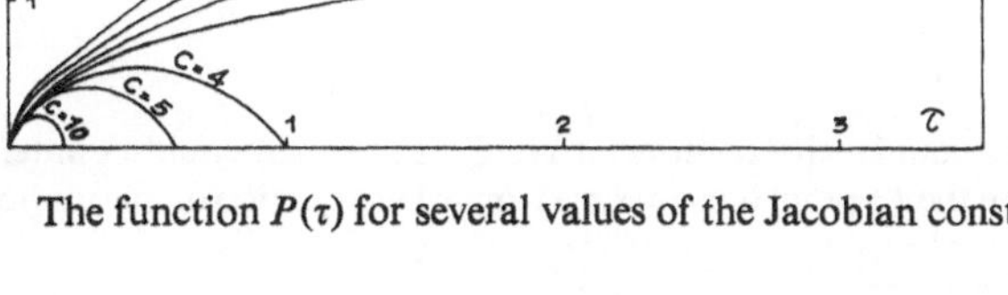

Fig. 5. The function $P(\tau)$ for several values of the Jacobian constant C.

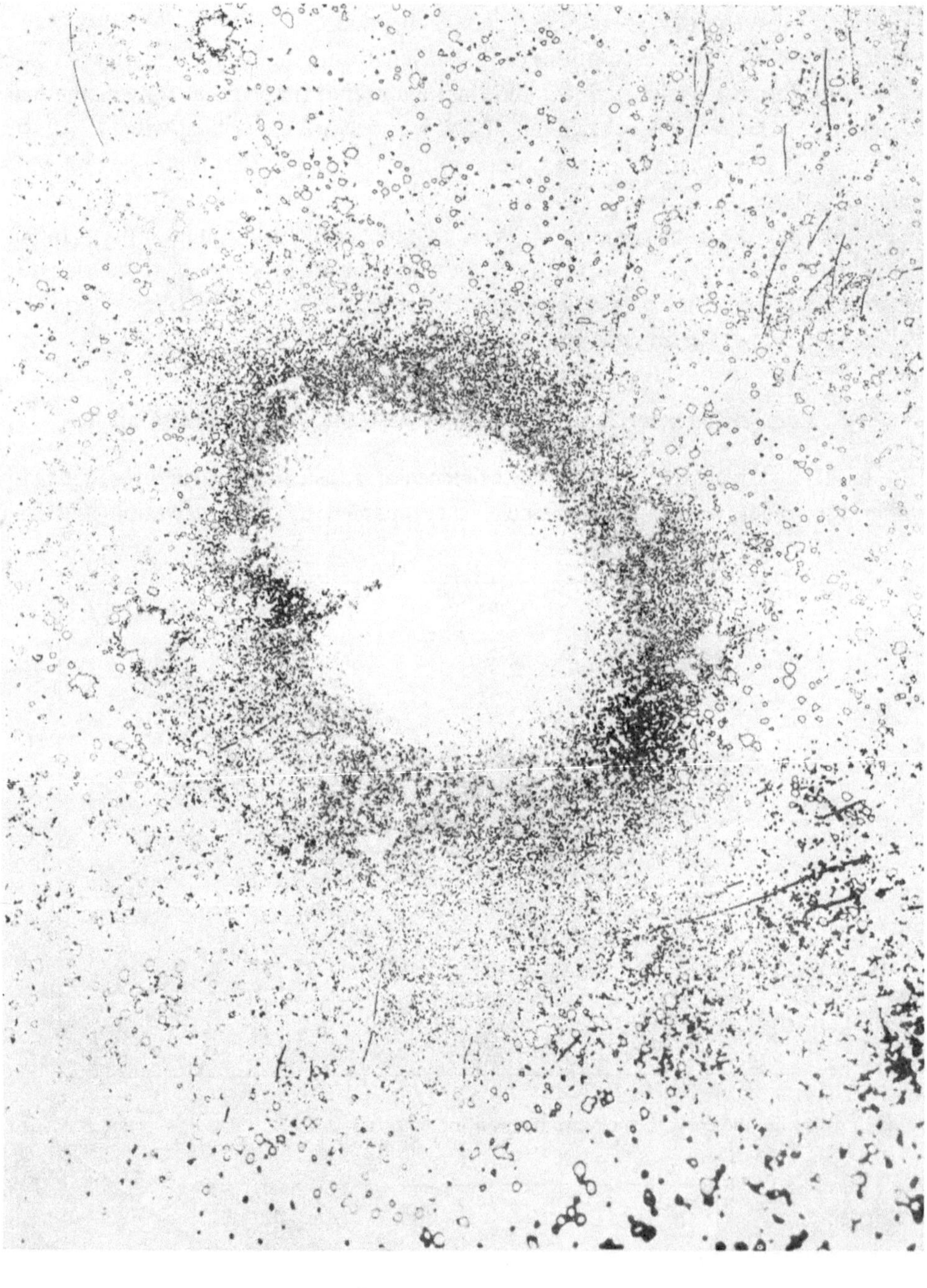

Fig. 6. Pear-shaped equidensity contour of NGC 5128 which could be interpreted as the equipotential going through the L_2 point (cusp-point) from where ejection of mass is possible. See Johnson [8].

mentation. We think this is the case in radiogalaxies like NGC 5128 NGC 1316, Cyg A, where the Lagrangian ring presents itself as a dark belt surrounding the galaxy on a plane normal to the preferential direction of the radiosource, that is to say, the direction of the explosive event.

The ring structure, where C is a minimum, controls the dynamics of the particles in that region, producing a 'dynamic filtration' of stars, gas and dust according to their energies. Only the most energetic particles populate the ring, so that we should expect a change in the velocity field of a given kind of population, for example, the stars [10].

On the other hand, the Eulerian point L_2 behaves as an escape point for particles filling the environment of μ which are sufficiently energetic. Suggestions of one-sided ejection of matter in NGC 5128 may be seen in a composite picture by Johnson [8]. The shape of the outer isophotes in this galaxy shows a cusp-point analogous to L_2 as it can be seen in [9] and also in Figure 6.

Lastly, it is well known that if $\mu(1-\mu)<\frac{1}{27}$, there are stable orbits around L_4, L_5 in the Restricted Problem. A conjecture presents itself now: is this result extended to a Lagrangian ring? If so, we should expect in case of a sensible difference of mass between the two nuclei the presence of an autonomous annular structure of stars, gas and dust midway between them. Figure 3 seems to suggest this possibility to be worth further research.

Acknowledgement

This research has been supported in part by the Consejo Nacional de Investigaciones of Argentina.

References

[1] Burbidge, G. R. and Burbidge, E. M.: 1969, *Nature* **224**, 21.
[2] Matthews, W. G., Morgan, W. W., and Schmidt, M.: 1964, *Astrophys. J.* **140**, 35.
[3] Ryle, M. and Longair, M. S.: 1967, *Monthly Notices Roy. Astron. Soc.* **136**, 123.
[4] Arp, H.: 1967, *Atlas of Peculiar Galaxies*, California Institute of Technology.
[5] Rogstad, D. H. and Ekers, R. D.: 1969, *Astrophys. J.* **157**, 481.
[6] Sérsic, J. L.: 1960, *Z. Astrophys.* **51**, 64.
[7] Schürer, M.: 1943, *Astron. Nachr.* **273**, 230.
[8] Johnson, H. M.: 1963, *Publ. Nat. Rad. Astron. Obs.* **1**, No. 15.
[9] Sérsic, J. L.: 1968, *Atlas de Galaxies Australes*, Observatorio Astronómico, Córdoba, Argentina.
[10] Sérsic, J. L.: 1969, *Nature* **224**, 253.

RESONANCE PHENOMENA IN SPIRAL GALAXIES*

G. CONTOPOULOS

University of Chicago and University of Thessaloniki

Galaxies, from the point of view of stellar dynamics, are particular cases of the n-body problem, with n so large (of the order of 10^{11}) that individual stellar encounters can be neglected during their whole history (of the order of 10^{10} years). Thus the motions of stars are governed by the 'averaged' potential V of the galaxy, where V is a function of the coordinates and, eventually, the time.

Spiral galaxies are flat stellar systems characterized by a number of spiral arms (usually two main arms emanating from a nucleus). As an idealization we usually consider spiral galaxies as two-dimensional. Their potential in the plane is assumed to be of the form

$$V = V_0 + V_1 \tag{1}$$

where V_1 is small with respect to V_0 and V_0 is a function of r only, while V_1 has a spiral form. In the general case of a two-armed spiral we have

$$V_1 = \mathrm{Re}\left[A(r)\exp\{i\left[2\Omega_s t - 2\theta + \Phi(r)\right]\}\right] \tag{2}$$

where $\Omega_s(>0)$ is the rotational velocity of the spiral pattern; $A(r)$ and $\Phi(r)$ are the amplitude and phase of the spiral field at distance r.

The spiral arms at any given time are given by the equations

$$\theta = \tfrac{1}{2}\Phi(r) + \text{const.}\,(+\,\pi)\,. \tag{3}$$

There are three kinds of problems of stellar dynamics connected with spiral arms:

(a) To find the orbits of stars in a given potential field of the form (2).

(b) Given the distribution of positions and velocities of stars at a given moment to follow the change of the distributions functions, the density and the potential in time.

(c) To find potentials of the form (2) and corresponding distributions functions that do not change in time (Ω_s, $A(r)$ and $\Phi(r)$ are time-independent).

In all cases the equations of motion are

$$\mathrm{d}r/\mathrm{d}t = \dot{r}, \qquad \mathrm{d}\theta/\mathrm{d}t = \dot{\theta} = J_0/r^2\,, \tag{4}$$

$$\mathrm{d}\dot{r}/\mathrm{d}t = -\,\partial V/\partial r + J_0^2/r^3\,, \qquad \mathrm{d}J_0/\mathrm{d}t = -\,\partial V/\partial\theta\,. \tag{5}$$

These equations form the corresponding system of the 'collisionless Boltzmann equation', satisfied by the distribution function f,

$$\frac{\partial f}{\partial t} + \dot{r}\frac{\partial f}{\partial r} + \frac{J_0}{r^2}\frac{\partial f}{\partial \theta} + \left(-\frac{\partial V}{\partial r} + \frac{J_0^2}{r^3}\right)\frac{\partial f}{\partial \dot{r}} - \frac{\partial V}{\partial \theta}\frac{\partial f}{\partial J_0} = 0\,. \tag{6}$$

In the problem of the first kind V is given; however in problems of the second and

* Presented by Dr. M. Hénon.

G. E. O. Giacaglia (ed.), Periodic Orbits, Stability and Resonances, 322–327.

third type V is connected to the density ϱ through Poisson's equation

$$\nabla^2 V = 4\pi G\varrho \,, \tag{7}$$

where

$$\varrho = \delta(z) \iint f r \, \mathrm{d}\dot{r} \, \mathrm{d}\dot{\theta} \,; \tag{8}$$

$\delta(z)$ is Dirac's delta function and the integral extends over all values of $\dot{r}$ and $\dot{\theta}$, at a given point.

The problem of the orbits is intimately connected with the problem of integrals of motion. The most general solution of Equation (6) is an arbitrary function of the integrals of motion. In general we know only one analytic integral of motion in a spiral field, rotating with angular velocity Ω_s, namely the Jacobi integral

$$H = \tfrac{1}{2}(\dot{r}^2 + J_0^2/r^2) + V - \Omega_s J_0 \,. \tag{9}$$

It is usual, in galactic dynamics, to linearize the collisionless Boltzmann equation. We write

$$f = f_0 + f_1 + f_2 + \cdots \tag{10}$$

and solve Equation (6) for f_1, assuming that f_0 and V_0 satisfy the zero order equation

$$\dot{r} \frac{\partial f_0}{\partial r} + \left(- \frac{\mathrm{d}V_0}{\mathrm{d}r} + \frac{J_0^2}{r^3} \right) \frac{\partial f_0}{\partial \dot{r}} = 0 \,. \tag{11}$$

If we omit f_2 and higher order terms, as well as terms of degree higher than the first in f_1 and V_1, we find the 'linearized collisionless Boltzmann equation'

$$\frac{\partial f_1}{\partial t} + \dot{r} \frac{\partial f_1}{\partial r} + \frac{J_0}{r^2} \frac{\partial f_1}{\partial \theta} + \left(- \frac{\mathrm{d}V_0}{\mathrm{d}r} + \frac{J_0^2}{r^3} \right) \frac{\partial f_1}{\partial \dot{r}} = \frac{\partial V_1}{\partial r} \frac{\partial f_0}{\partial \dot{r}} + \frac{\partial V_1}{\partial \theta} \frac{\partial f_0}{\partial J_0} = Q \,. \tag{12}$$

One can find a stationary solution of this equation in the rotating frame if certain non-resonance conditions (written explicitly below) are satisfied:

$$f_1 = \mathrm{Re} \left\{ \frac{\exp\left[2i(\Omega_s t - \theta)\right]}{2i \sin\left[2(\Omega_s \tau_0 - \theta_0)\right]} \int_{-\tau_0}^{\tau_0} Q \exp\left[2i(\Omega_s t' - \theta')\right] \mathrm{d}t' \right\} \tag{13}$$

(Shu, 1968), where $2\tau_0$ is the period of an unperturbed orbit (orbit in the axisymmetric field $V_0(r)$), corresponding to the same initial conditions as the real orbit, and θ_0 is the angle between the pericentron and the apocentron of the unperturbed orbit.

This relation is valid if the denominator $\sin[2(\Omega_s\tau_0 - \theta_0)]$ is not zero, or near zero. If $\sin[2(\Omega_s\tau_0 - \theta_0)]$ is zero the solution f_1 contains a secular term, and if it is very small, f_1 is larger than f_0 and the approximation scheme implied by the linearization is not valid. Thus $2(\Omega_s\tau_0 - \theta_0)$ has to be far from any multiple of π. Under this restriction f_1 can be evaluated, through Equation (13), if f_0 is a given solution of Equation (11).

Two integrals of motion of the unperturbed (axisymmetric) system are the energy

$$E_0 = \tfrac{1}{2}(\dot{r}^2 + J_0^2/r^2) + V_0(r) \tag{14}$$

and the angular momentum J_0. In fact Equation (11) is satisfied if we put $f_0 = E_0$ or $f_0 = J_0$. If we are away from resonances these are the only isolating integrals of the unperturbed system. Therefore the most general form of f_0 is

$$f_0 = f_0(E_0, J_0). \tag{15}$$

The method described above for the calculation of f_1 is the same as that used in calculating the successive terms of the 'third' integral. In fact one can calculate successively the terms of the series (10) and find f as an asymptotic expansion, which is a formal integral of motion.

The above method gives satisfactory results if the amplitude of the potential V_1 is small and if we are away from resonances. However at, or near, a resonance, this method is not applicable.

We say that we are at a Lindblad resonance if

$$2(\Omega_s \tau_0 - \theta_0) = \pm \pi \tag{16}$$

(+ for the 'outer Lindblad resonance' and − for the 'inner Lindblad resonance'). In such a case the ratio of the frequency of radial oscillation to the frequency of rotation around the center of the galaxy, measured in the rotating frame, is

$$(\pi/\tau_0)/(\theta_0/\tau_0 - \Omega_s) = \mp 2. \tag{17}$$

Thus the unperturbed orbits perform two radial oscillations, while they make a revolution around the center of the galaxy in the rotating frame, i.e. they look like ellipses with their center at the center of the galaxy.

In cases at, or near, the Lindblad resonances the integral (10) contains secular terms or small divisor terms. Therefore, as in other cases near resonance (Contopoulos, 1968), the integral f takes a quite different form. In the exact resonance case we notice that there is one more isolating integral of the unperturbed problem, besides E_0 and J_0, namely

$$S_0 = \sin\left[2(\theta_1 - \Omega_s \tau_1)\right] \quad \text{or} \quad C_0 = \cos\left[2(\theta_1 - \Omega_s \tau_1)\right], \tag{18}$$

where θ_1, τ_1 are the angle and the time of a pericentron passage. One can find then a particular function f_0 of E_0, J_0, C_0 (or S_0), such that f_1 has no secular terms.

The form of f_0 is given in a recent paper (Contopoulos, 1970b). Its main characteristic is that it depends on the angle coordinate through a term of the form

$$\cos\left[4(\Omega_s t - \theta) + \Phi(r)\right],$$

where $\Phi(r)$ is a function of r.

If we write the position angle

$$\theta' = \theta - \Omega_s t$$

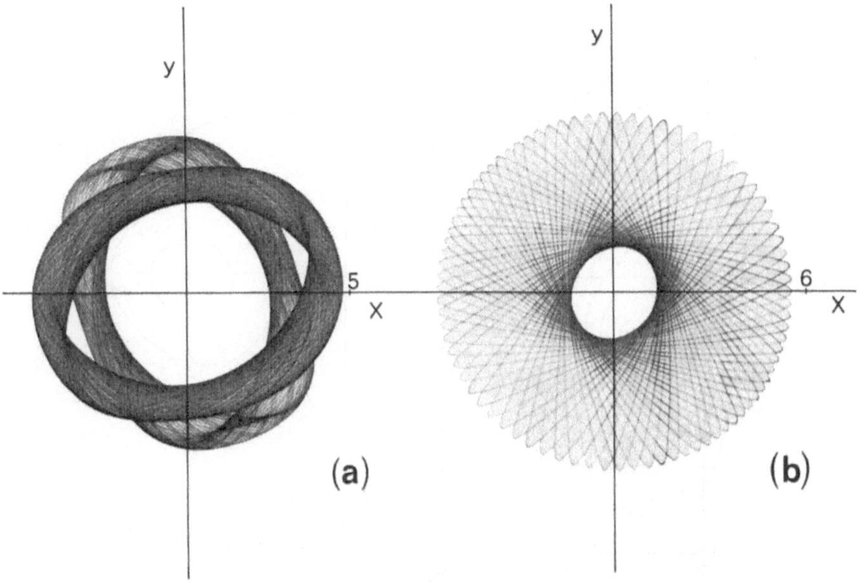

Fig. 1. Orbits in a spiral field. (a) Two tube orbits near the inner Lindblad resonance. (b) An epicyclic type orbit.

in the rotating frame, f_0 does not change if we change θ' by $\pi/2$, π or $3\pi/2$. This means that the distribution function has an (approximate) quadruple symmetry.

For a given value H of the Jacobi integral, near H_R (where H_R corresponds to a circular unperturbed orbit at the inner Lindblad resonance, which is at a distance of about 4 kpc from the center of our Galaxy), one finds only two stable periodic orbits of the resonant type. Orbits near them form tubes around one or the other of the stable periodic orbits (Figure 1a). These orbits are quite different from the usual, epicyclic type, orbits, that fill almost circular rings around the center of the galaxy (Figure 1b).

The invariant curves of the tube orbits in a diagram $(r, \dot{r})$, giving the values of r and $\dot{r}$ whenever an orbit crosses an axis $\theta = \text{const.}$ are closed curves around two resonant invariant points (Figure 2a), while in cases away from resonances there is only one invariant point, corresponding to an almost circular periodic orbit, and the rest of the invariant curves surround this point (Figure 2b).

The invariant curves near the inner Lindblad resonance are topologically similar to those of the resonance case $\frac{2}{1}$ of the potential

$$V = \tfrac{1}{2}(Ax^2 + By^2) - \varepsilon xy^2, \tag{19}$$

when $A^{1/2}/B^{1/2} = 2$ (Contopoulos and Moutsoulas, 1966) (Figure 3).

Similar results occur at the outer Lindblad resonance and at resonances of higher order, when the ratio of the frequencies of the radial and rotational motion is a rational number.

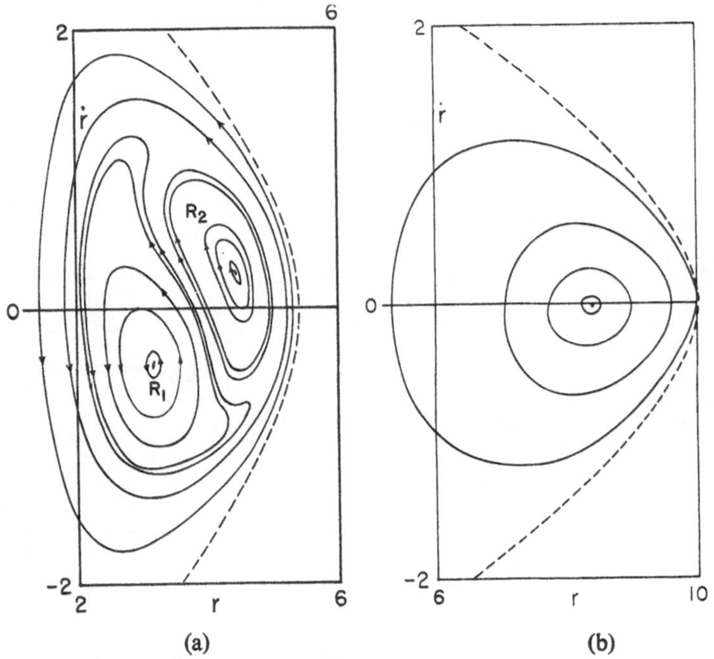

Fig. 2. Invariant curves for orbits with fixed Hamiltonian in a spiral Field. (a) $H = 0.74\ \text{kpc}^2\,(10^7\,\text{yr})^{-2}$, $\Omega_s = \Omega/2$(sun) (near the Hamiltonian of a circular unperturbed orbit at the inner Lindblad resonance); (b) $H = 3.5\ \text{kpc}^2(10^7\,\text{yr})^{-2}$, $\Omega_s = 2\Omega/3$ (sun) (far from resonance).

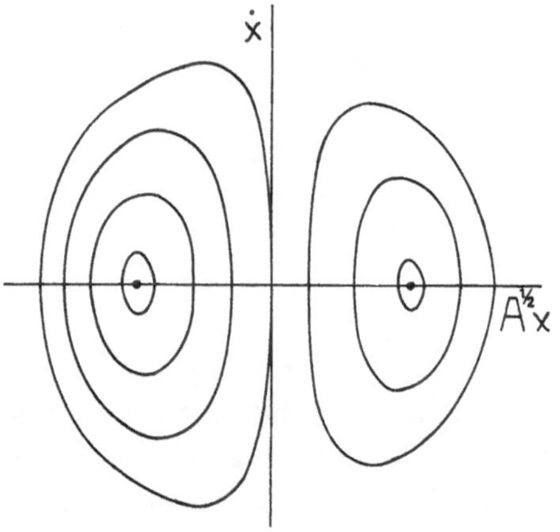

Fig. 3. Invariant curves of the potential $V = (Ax^2 + By^2)/2 - \varepsilon xy^2$, for $A^{1/2}/B^{1/2} \simeq 2$.

A different type of resonance occurs when

$$\theta_0/\tau_0 \simeq \Omega_s. \tag{20}$$

This happens near the co-rotation distance, where the angular velocity of the stars is equal to the angular veloci'y of the spiral pattern Ω_s.

In such a case the motions of stars with small velocities in the rotating frame librate

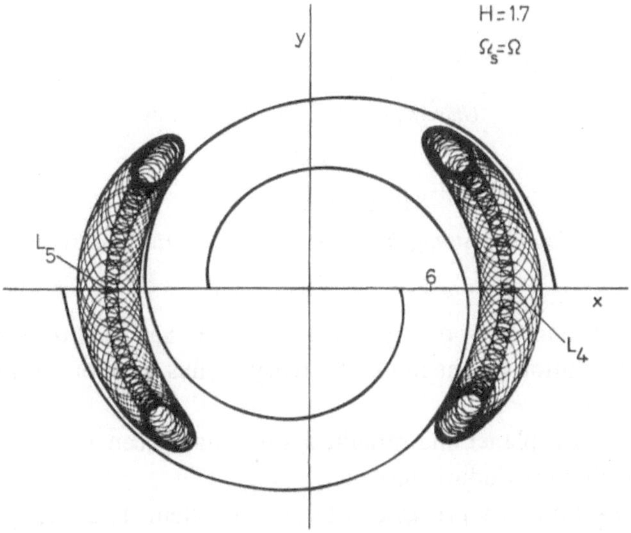

Fig. 4. Orbits in a spiral field, librating around the Lagrangian points L_4, L_5 (at the co-rotation distance); $H = 1.7$ kpc^2(10^7 yr)$^{-2}$, $\Omega_s = \Omega$ (sun).

around two points L_4 and L_5 (Barbanis, 1970) (Figure 4). These points are 90° away from the points of minimum spiral potential, where we have approximately the maximum of spiral density.

Stars with larger velocities, or starting away from the co-rotation distance, describe epicyclic orbits of the usual type (Figure 1b). The libration around the points L_4 and L_5, where we have the maximum of the potential, is completely analogous to the librations around the Lagrangian points L_4 and L_5 of the restricted three-body problem.

The implications of these resonance phenomena for the dynamics of spiral galaxies are described elsewhere (Contopoulos, 1970a; Barbanis, 1970).

References

Barbanis, B.: 1970, in *IAU Symposium* **38**, 343.
Contopoulos, G.: 1968, *Astrophys. J.* **153**, 83.
Contopoulos, G.: 1970a, in *IAU Symposium* **38**, 303.
Contopoulos, G.: 1970b, *Astrophys. J.* (to be published).
Contopoulos, G. and Moutsoulas, M.: 1966, *Astron. J.* **71**, 687.
Shu, F.: 1968, Thesis, Harvard University.

ON THE EVOLUTION OF THE SOLAR SYSTEM AND THE PLUTO-NEPTUNE CASE

G. COLOMBO
Università degli Studi, Padova, Italy
and
F. A. FRANKLIN
Smithsonian Astrophysical Observatory, Cambridge, Mass., U.S.A.

Spin-orbital synchronization, or sub-synchronization, has been observed and studied as a steady state solution for the moon, Mercury, Venus, and for several of the Jupiter and Saturn satellites.

Motions of certain planet and satellite systems have been recognized as a free or forced libration about periodic solutions.

Some of the evolutionary processes which lead systems to a steady state have also been investigated more recently as a consequence of tidal dissipation and of transfer of energy from rotational to orbital motion and vice versa. Darwin's pioneering work on the evolution of the earth-moon system was the starting point of this field of investigation which has led to important results in recent years.

The dynamical problem of the evolution of the angular velocity vector of a celestial body toward synchronization or subsynchronization of its orbital and spin periods caused by tidal dissipation has been the object of recent research, particularly in relation to the peculiar spin-orbital resonance of Mercury and Venus. There remains still a problem (even assuming a short time scale of spin evolution with respect to the time scale of the orbital evolution) with regard to the probability of capture of Mercury in the present steady state. For Venus, the spin-orbital coupling is more complicated; even the stability of the present steady state implies values of certain physical parameters of the planet which are, to some extent, unrealistic.

Tides are most probably responsible for orbital evolution which has led several satellite systems to librate about a periodic solution. Typical cases are the three inner Galilean satellites of Jupiter (the relation $n_1-3n_2+2n_3=0$ is exact up to the very high accuracy of observations), and the couples, Mimas-Tethys, Enceladus-Dione, and probably also Titan-Hyperion. These cases represent different kinds of resonance. In fact, it is well known that the Mimas-Tethys resonance involves the motion of the nodes of the two satellites, the Enceladus-Dione case involves motion of apsidal lines, and the Jupiter satellite case involves mean motion. To be more clear, the ratio 1:2 in the case of Mimas-Tethys is exact when we consider the draconian (nodal) period; for the case of Enceladus-Dione the ratio 1:2 is exact for the anomalistic period. Finally, in the Jupiter satellite case the single relation is in the mean motion. Actually we may write the relation $n_1-3n_2+2n_3=0$ in the form $n_1-n_2=2(n_2-n_3)$ and this may be interpreted as a simple relation 1 to 2 between the synodic periods of Io and

G. E. O. Giacaglia (ed.), Periodic Orbits Stability and Resonances, 328–331.

Ganymede with respect to Europa as a reference. These relations correspond to periodic solutions with respect to a rotating system, in all the three cases. The same is true for the couple Titan-Hyperion, and also for the Neptune-Pluto case. The libration of Pluto about the periodic solution with a period $\frac{3}{2}$ of Neptune's sidereal period has been recently proven. In addition, several asteroids are librating about periodic solutions. A problem arises when we try to discuss simple numerical relations which involve the mean motion of two or more bodies. Typical cases are the periods of the major planets from Jupiter to Neptune. The ratio of the mean motions of Jupiter and Saturn is close to $\frac{5}{2}$; for Saturn-Uranus we have a ratio near $\frac{3}{1}$; and for Uranus and Neptune, the ratio of mean motions is close to 2 to 1. None of these bodies seems to be librating. Somewhat more complicated numerical relations are nearly satisfied by certain satellites of the Uranus system. (The mean motion of the four large satellites of Uranus follow the relation $n_1 - n_2 - n_3 \cong n_3 - n_4$.) It is clear that observations of the satellites are sufficiently accurate to ensure that these relations are not exactly satisfied.

There are more complicated relations which have been considered in the literature but there is doubt about the value of spending time in trying to discover the meaning of these relations since they seem quite fictitious.

Certainly the simple relations among the mean motions of the largest planets are more impressive and one feels that they have some important bearing upon the origin and/or evolution of the solar system. We feel uncomfortable to believe that they are purely due to chance, but as a reasonable way of proceeding, we think that first a clear separation has to be made between those systems in which libration is certain and those in which it is absent or doubtful.

Even considering only the well-established orbital libration cases, dealing as we are with evolutionary processes, we have a very well defined problem on the evolution of the Pluto-Neptune system which certainly cannot be explained in terms of tidal processes. There is also some doubt whether the Titan-Hyperion case can be explained by the same process, but the insufficiency of tidal evolution for the Pluto-Neptune case may be easily shown. Tides may have caused Pluto to escape from Neptune's sphere of influence but the process which led to the capture of Pluto in the present dynamical situation is certainly not a tidal one. We think that the problem of capture in a librating configuration for satellite systems via tidal evolution is far from being solved, but we are also convinced that it is only a question of finding the reasonably simple model to treat a complicated mechanical system. At least we have a dissipative mechanism of interaction, in addition to the conservative gravitational interaction; the dynamical model, although complex, is clear enough to be treated mathematically in a deterministic way. With the Neptune-Pluto case we are facing a new problem which may imply a choice between a deterministic model and a statistical one.

The deterministic approach assumes that the interplanetary environment in the early stages of evolution of the solar system was quite different from the present one. In particular, we can suppose that at one time planets moved through a resisting medium associated with interplanetary particles having a higher space density than found today. The presence of such a medium would in general result in relative changes in the periods of the planets. Thus the question naturally arises: can realistic values of the drag alter orbital periods by a sufficient amount and result in capture at certain resonances as well?

The statistical approach views the solar system as having been populated with many more bodies in the distant past. Close approaches and collisions between bodies would then reduce this population as the system evolves. The planets surviving today are those that have managed to avoid close approaches either because they were initially near resonance with a massive neighbor or have managed to become so. (Note that the Pluto-Neptune case is a system which avoids close approaches by this means.) If this general viewpoint is to be useful, then clearly the region in phase space around stable resonant periodic orbits in which libration can occur should be sufficiently large. Again we have a natural field for inquiry: How large is this libration region, specifically, for example, in the Pluto-Neptune case.

We have considered therefore first the simple model of the planar restricted three body problem, starting Pluto near its present dynamical configuration and integrated backwards in time, assuming the presence of a resisting medium whose density increases with negative time. Results show that capture in the libration mode about the 2 to 3 resonance via a close approach to Neptune does not depend critically on the density distribution of the interplanetary medium assumed to move in circular orbits about the primary. A second prelimimary result has been obtained, also with a numerical analysis based upon the restricted three body problem. We consider the class of periodic orbits for the massless body, symmetric with respect to the sun-Neptune vector, with periods close to 2:3, and are looking for the size of the domain in three dimensional phase space in which the massless body can exist in a libration mode about a periodic solution. This analysis is currently in progress.

Another statistical and very general viewpoint would approach the question in terms of the n body problem. We propose using the same numerical techniques usually employed in studying the development of globular clusters, with the single exception that one body is given a much larger mass than the others. For globular clusters the masses of the bodies are assumed equal and the most probable final state is one in which only two bodies in Keplerian orbits remain. The presence of a single large mass may lead to a more complex final state with many bodies left in stable configurations. It will certainly be interesting to compare the features of the final state with the present planetary distribution in the solar system.

Discussion

Y. Kozai: In 1956 I found secular acceleration of order of $1°/(100\text{ yr})$ for Mimas but not for Tethys. If this is true, the commensurability relation for Mimas-Tethys system will be destroyed.

G. Colombo: If this acceleration in the mean motion of Mimas is confirmed using possibly recent and future observations, this would mean that the Mimas-Tethys system is not librating, or at least that Thethys is not able to keep Mimas in the resonance, feeding to Mimas the energy it is losing by some unknown mechanism.

If the two satellites decelerate together, the system remaining in a resonant situation, this would mean that we have a case where evolution toward resonance is caused by another dissipative mechanism.

FLOWS NEAR ISOLATED INVARIANT SETS IN DIMENSION 3

ROBERT EASTON

Center for Dynamical Systems, Division of Applied Mathematics, Brown University, Providence, R.I., U.S.A.

In this paper we investigate the behavior of flows on 3-manifolds near isolated invariant sets (see Definition 1). We show in particular that if the invariant set is a circle (perhaps a periodic orbit), then it is contained as a deformation retract in an isolating block (see Definition 2) which is a solid torus. Similarly we are able to treat flows near isolated invariant sets which are points or tori.

Using the present work and that of Conley [1] we show that if T is an invariant set of a volume preserving flow φ on a 3-manifold and if φ/T is conjugate to an irrational flow on the torus, then T is not isolated.

DEFINITION 1: Let φ be a flow on a metric space (X, d) and let I be a closed invariant set of φ. Then I is *isolated* if there exists an open subset U of X containing I such that I is the maximal invariant set in U.

DEFINITION 2: (Notation) Let M be a smooth manifold and let V be a C^1 vector field on M which generates a flow φ_V on M. Let N be a submanifold with boundary of M of the same dimension as M. Define

$$\begin{aligned} n &= \partial N \\ n^+ &= \{p \in n : \exists \varepsilon > 0 \ni \varphi_V(p, t) \notin N \quad \text{for} \quad -\varepsilon < t < 0\} \\ n^- &= \{p \in n : \exists \varepsilon > 0 \ni \varphi_V(p, t) \notin N \quad \text{for} \quad 0 < t < \varepsilon\} \\ \tau &= \{p \in n : V \quad \text{is tangent to} \quad n \quad \text{at} \quad p\}. \end{aligned}$$

DEFINITION 3: With the notation of Definition 2, N is an *isolating block* for φ_V if $n^+ \cap n^- = \tau$ and if τ is a smooth submanifold of n with codimension one and (as a consequence) n^+ and n^- are submanifolds of n with common boundary τ.

DEFINITION 4: Let N be an isolating block for φ_V and define

$$\begin{aligned} A^+ &= \{p \in N : \varphi_V(p, t) \in N \, \forall t \geqq 0\} \\ A^- &= \{p \in N : \varphi_V(p, t) \in N \, \forall t \leqq 0\} \\ I &= A^+ \cap A^- \\ a^+ &= A^+ \cap n = A^+ \cap n^+ \\ a^- &= A^- \cap n = A^- \cap n^- . \end{aligned}$$

If J is an invariant set of φ_V, and N is an isolating block, N is said to be an *isolating block for* J if J is the maximal invariant set in N (i.e. $A^+ \cap A^- = J$).

The main result of the paper is the following:

THEOREM 1: Let V be a C^1 vector field on an orientable smooth 3-manifold which generates a flow φ_V and suppose J is a compact isolated invariant set of φ_V. Then there exists an isolating block N for J such that the homomorphism $\bar{H}^*(N) \rightarrow$

G. E. O. Giacaglia (ed.), Periodic Orbits, Stability and Resonances, 332–336.

$\bar{H}^*(J)$ induced by the inclusion of J in N is injective. Further, if $\bar{H}^*(J)$ is finitely generated, N can be chosen so that $\bar{H}^*(N) \to \bar{H}^*(J)$ is an isomorphism, and so that $n^+ - a^+$ is homeomorphic to $\tau \times [0,1)$.

Here $\bar{H}^*(N)$ denotes the kth Alexander cohomology group of N with integer coefficients. Since N is a manifold, $\bar{H}^*(N)$ is isomorphic to $H^*(N)$, the singular cohomology of N. (See [4].)

COROLLARY 1: Under the hypothesis of Theorem 1 if J is respectively a torus, simple closed curve or a point, then there exists an isolating block N for J which is respectively homeomorphic to $J \times D^1$, $J \times D^2$, $J \times D^3$ (where D^k is the unit disk in R^k). Furthermore $n^+ - a^+$ and $n^- - a^-$ are homeomorphic to $\tau \times [0, 1)$ and consequently the inclusion of $\bar{H}^*(N)$ in $\bar{H}^*(J)$ is an isomorphism. Also J is a deformation retract of N provided J is a deformation retract of some neighborhood of J in M.

Remarks: In [1] Conley shows that if J is a periodic orbit, then n^+ must consist of annuli.

If J is a smooth submanifold of M, then J is a deformation retract of a tubular neighborhood of itself in M.

The proofs of Theorem 1 and Corollary 1 rely on the following lemma (which is false in higher dimensions).

LEMMA 1: Let M be a compact, connected, orientable 2-manifold with boundary and let C be a closed subset of the interior of M. Then given any neighborhood U of C in M, there exists a compact manifold with boundary B such that $C \subset B \subset U$ and such that $\bar{H}^*(B) \to \bar{H}^*(C)$ is an injection. If in addition $\bar{H}^*(C)$ is finitely generated, then B can be chosen so that $B - C$ is homeomorphic to $\partial B \times [0, 1)$ and consequently the inclusion $\bar{H}^*(B) \to \bar{H}^*(C)$ is an isomorphism.

Proof: The inclusion induced homomorphism $i: \bar{H}^*(M) \to \bar{H}^*(C)$ has finitely generated kernel. Thus by the continuity of the Alexander cohomology theory, there exists a neighborhood V of C in M such that if R is any manifold with boundary such that $C \subset \operatorname{int} R \subset R \subset V$, then $\ker i \subset \ker j$ where $j: \bar{H}^*(M) \to \bar{H}^*(R)$ is induced by inclusion. Equivalently, $\operatorname{kernel} \alpha \subset \operatorname{kernel} \beta$ where α and β occur in the commutative diagram

$$\begin{array}{ccc} H_*(R, \partial R) & \overset{\beta}{\leftarrow} & H_*(M, \partial M) \\ \downarrow & & \downarrow \alpha \\ H_*(R, R - C) & \approx & H_*(M, M - C). \end{array}$$

Here α is induced by inclusion and β is the composition of an inclusion β_1 with an excision isomorphism β_2,

$$H_*(M, \partial M) \overset{\beta_1}{\to} H_*(M, M - R) \overset{\beta_2}{\to} H_*(R, \partial R).$$

Choose a compact manifold with boundary B_1 such that $C \subset \operatorname{int} B_1 \subset B_1 \subset U \cap V$. B_1 can be modified by surgery to form a compact manifold with boundary $B \subset B_1$ so that the inclusion induced homomorphisms $\bar{H}^0(B) \to \bar{H}^0(C)$ and $H_0(\partial B) \to H_0(B - C)$ are injections as follows: if two components of ∂B_1 lie in the same component of $B_1 - C$, joint them with an arc in $B_1 - C$ and remove a tubular neighborhood of this arc thus obtaining a new manifold with one boundary component where there were formerly two. After finitely many surgeries of this type a compact manifold with boundary B_2 is obtained so that any component of $B_2 - C$ contains at most one component of ∂B_2. It follows that $H_0(\partial B_2)$ injects in $H_0(B_2 - C)$. Finally let B be the union of those components of B_2 which meet C. This insures that $\bar{H}^0(B)$ injects in $\bar{H}^0(C)$.

We now show that $\bar{H}^1(B)$ injects in $\bar{H}^1(C)$. By Alexander duality, it is sufficient to show that $H_1(B, \partial B)$ injects in $H_1(B, B-C)$. Consider the commutative diagram

$$\begin{array}{ccccccc} H_1(B-C) & \xrightarrow{i} & H_1(B) & \xrightarrow{j} & H_1(B, \partial B) & \xrightarrow{\partial} & H_0(\partial B) \\ & & & \searrow^{k} & \downarrow \gamma & & \downarrow \gamma' \\ & & & & H_1(B, B-C) & \xrightarrow{\partial} & H_0(B-C). \end{array}$$

Since γ' is an injection, kernel $\gamma \subset$ image $j \circ i$ (using exactness of the sequence of the pair $(B, B-C)$).

Commutativity of the diagram

$$\begin{array}{ccccccc} H_1(B-C) & \rightarrow & H_1(M) & \rightarrow & H_1(M, \partial M) & \xrightarrow{\alpha} & H_1(M, M-C) \\ & \searrow^{i} & & & \downarrow \beta & & \wr\wr \\ & & H_1(B) & \xrightarrow{j} & H_1(B, \partial B) & \xrightarrow{\gamma} & H_1(B, B-C) \end{array}$$

implies that the image of $j \circ i$ is contained in the image of β. But since $B \subset U \cap V$, it follows by choice of V that kernel $\alpha \subset$ kernel β and hence that kernel $\gamma =$ kernel $\gamma \cap$ image $\beta = 0$.

It remains to show that $\bar{H}^k(B) \rightarrow \bar{H}^k(C)$ is an injection for $k \geqq 2$. By duality, $\bar{H}^k(B)$ is isomorphic to $H_{2-k}(B, \partial B)$ which is zero for $k \geqq 2$ and thus $\bar{H}^k(B)$ injects in $\bar{H}^k(C)$ for $k \geqq 2$ and hence for all k.

Now suppose $\bar{H}^*(C)$ is finitely generated.

By duality this implies that $\bar{H}_*(M, M-C)$ is finitely generated. $H_*(M)$ is finitely generated since M is a manifold and using the exactness of the homology sequence of the pair $(M, M-C)$ it follows that $H_*(M, C)$ is finitely generated. The classification theorem for 2-manifolds (see [3]) asserts in particular that any connected orientable open 2-manifold with finitely generated homology is homeomorphic to a sphere with finitely many handles minus a finite set of points. Each component W of $(\text{int} M) - C$ is of this type and consequently there exists a compact manifold with boundary K_W contained in W such that $W - \text{int} K_W$ is homeomorphic to $\partial K_W \times [0, 1)$. Let $B' = M - \cup\{\text{int} K_W : W \text{ is a component of } (\text{int } M) - C\}$ and let B be the union of those components of B' which meet C (no component of B' meets both C and ∂M). Then $B - C$ is homeomorphic to $\partial B \times [0, 1)$ and consequently $\bar{H}_*(B-C, \partial B) \equiv 0$. Using the exact homology sequence of the triple $(B, B-C, \partial B)$ and the Alexander duality theorem it follows that $\bar{H}^*(B)$ is isomorphic to $\bar{H}^*(C)$. This completes the proof of the lemma.

Proof of Theorem 1: It is shown in [2] that any compact isolated invariant set of a C^1 flow on a smooth orientable manifold is contained as the maximal invariant set inside some isolating block. Thus let R be an isolating block for J.

Since M is orientable, so is R and hence r^+. By the lemma, choose a 2-manifold with boundary $b \subset r^+$ such that $\bar{H}^*(b) \rightarrow \bar{H}^*(a^+)$ is an injection. Use the flow φ_V to 'carry' points of $b-a^+$ across R to r^-. The set of points 'swept out' by $b-a^+$ together with those of $A^+\cup A^-$ can be modified by 'rounding off' its corners to form an isolating block N for J with $N\subset R$ and with $\bar{H}^*(n^+)\to\bar{H}^*(a^+)$ an injection. A more complete description of this process can be found in [2]. The fact that $\bar{H}^*(n^+)$ injects in $\bar{H}^*(a^+)$ is equivalent (using duality and exactness of the homology sequence of the triple (n^+, n^+-a^+, τ)) to commutativity of the diagram

$$\text{(A)}\qquad H_*(n^+-a^+,\tau)\to H_*(n^+,\tau)\to H_*(N,n).\quad\text{with } H_*(n^+-a^+,\tau)\searrow 0\nearrow H_*(N,n)$$

It is shown in [2] that for any isolating block N there exists an exact sequence

$$\text{(B)}\qquad H_*(n^+-a^+,\tau)\xrightarrow{i} H_*(N,n)\xrightarrow{j} H_*(N,N-J)\xrightarrow{\partial}$$

where i and j are induced by inclusion and ∂ is a map of degree -1. Thus it follows from the exactness of (B) and the commutativity of (A) that $H_*(N,n)$ injects in $H_*(N,N-J)$ and hence by duality that $\bar{H}^*(N)$ injects in $\bar{H}^*(J)$.

In the case that $\bar{H}^*(J)$ is finitely generated, it follows by duality that $H_*(N,N-J)$ is finitely generated. If $G'\to G\to G''$ is an exact sequence of Abelian groups with G' and G'' finitely generated, then G is also finitely generated. Using this fact it follows from the exactness of (B) that $H_*(n^+-a^+,\tau)$ is finitely generated, hence from exactness of the homology sequence of the triple (n^+, n^+-a^+, τ) that $H_*(n^+, n^+-a^+)$ is finitely generated and therefore by duality that $\bar{H}^*(a^+)$ is finitely generated. Thus Lemma 1 can be used to choose a compact manifold with boundary $b\subset n^+$ such that $b-a^+$ is homeomorphic to $\partial b\times[0,1)$. As above an isolating block $N_1\subset N$ for J can be constructed with $n_1^+-a_1^+$ homeomorphic to $\tau_1\times[0,1)$. Thus $H_*(n_1^+-a_1^+,\tau)\equiv 0$. Using the sequence (B) (with N replaced by N_1, etc.) and duality it follows that the inclusion of $\bar{H}^*(N_1)$ in $\bar{H}^*(J)$ is an isomorphism.

Proof of Corollary 1: We may assume without loss of generality that $M=R^3$. Choose an isolating block N for J by Theorem 1 so that n^+-a^+ is homeomorphic to $\tau\times[0,1)$. It follows from the exactness of the sequence (B) that $H_*(N,n)$ is isomorphic to $H_*(N,N-J)$ and thus by duality that $H_K(N,n)$ is isomorphic to $\bar{H}^{3-k}(J)$. It also follows that $H_*(N)$ is isomorphic to $H_*(J)$ and thus $H_*(n)$ can be computed using the exact sequence of the pair (N,n) once the groups $\bar{H}^*(J)$ and $H_*(J)$ are known. When J is respectively a torus, simple closed curve or a point n has the homology of $S^1\times S^1\times S^0$, $S^1\times S^1$, S^2 respectively. Since the topological type of a compact orientable 2-manifold is determined by its homology groups, n is respectively homeomorphic to $S^1\times S^1\times S^0$, $S^1\times S^1$, S^2. It follows since n is contained in R^3 and N is compact that N is respectively homeomorphic to $J\times D^1$, $J\times D^2$, $J\times D^3$.

Since n^+-a^+ is homeomorphic to $\tau\times[0,1)$ it follows that n^+ can be deformed into an arbitrarily small neighborhood of a^+. This deformation can be extended to an orbit preserving deformation of N. Following this deformation by one defined in a neighborhood of $A^+\cup A^-$ which contracts orbits towards I it is possible to deform N into an arbitrary neighborhood of J. Finally, if J is a deformation retract of some neighborhood of itself in M, it follows that J is a deformation retract of N. This completes the proof of the corollary.

A consequence of Corollary 1 and recent work of Conley [1] is the following

THEOREM 2: If φ is a C^1 volume preserving flow on a smooth 3-manifold and T is an invariant set of φ such that φ/T is conjugate to an irrational flow on the torus, then T is not an isolated invariant set of φ.

Proof: Suppose T is isolated. Then construct an isolating block N containing T such that N is homeomorphic to $T \times [-1, 1]$. This can be done even if the flow on T is not an irrational flow; but in the latter case Conley shows in [1] that $n^+ = a^+$ and $a^- = n^-$. As a consequence each component of ∂N is in either n^+ or n^- which is clearly not possible for a volume preserving flow.

Acknowledgements

This research was supported by the National Aeronautics and Space Administration under Contract No. NGL 40-002-015, the U.S. Air Force under Contract No. AF-AFOSR 67-0693A, and by the National Science Foundation under Contract No. GK 2788.

References

[1] Conley, C.: 1969, 'On the Set of Orbits Asymptotic to a Periodic Orbit', to appear.
[2] Conley, C. and Easton, R.: 1969, 'Isolated Invariant Sets and Isolating Blocks', to appear.
[3] Richards, I.: 1963, 'On the Classification of Noncompact Surfaces', *Trans. Amer. Math. Soc.* **106**, 259–69.
[4] Spanier, E.: 1966, *Algebraic Topology*, McGraw-Hill, New York.

Discussion

W. H. Jefferys: Your last demonstration of the existence of sets staying close to the invariant torus for all time in the volume-preserving case is very interesting. One can show (with a few more conditions, generally satisfied) that this is true using the Moser theorem, but your demonstration, being topological, is much nicer.

R. Easton: This is reminiscent of the fact that it is far easier and was in fact done much earlier to prove that the flow on a torus is preserved under perturbation of the differential equation when the torus is a stable or unstable manifold than in the Hamiltonian case.

W. H. Jefferys: I noticed that the periodic orbit you discussed is unstable. Is this a general situation?

R. Easton: Yes.

DYNAMICAL SYSTEMS ON MANIFOLDS

W. M. OLIVA

Escola Politécnica, University of São Paulo, Brasil

Let M be a C^∞ compact manifold, $T(M)$ its tangent bundle, I the closed interval $[-r, 0]$, $r>0$ and $C^0(I, M)$ the totality of continuous maps ϕ of I into M. The function space $C^0(I, M)$ is a C^∞ manifold modeled on a separable Banach space [1].

Let ϱ be the map $\phi \to \phi(0)$ of $C^0(I, M)$ onto M. This evaluation map is C^∞ differentiable and is a split C^∞ local fibration [2]. The map $\Sigma: M \to C^0(I, M)$ such that to each $p \in M$ associates the constant map $\Sigma(p)(\theta)=p$, $\theta \in [-r, 0]$, is a C^∞ cross section of $C^0(I, M)$ with respect to ϱ, then $\Sigma(M)$ is a compact submanifold diffeomorphic to M, transversal to the fibers of ϱ.

A C^1 retarded vector field or a functional differential equation (F.D.E.) is a C^1 map $F: C^0(I, M) \to T(M)$ such that $\pi F = \varrho$.

Let $x(t)$ be a continuous function with values on M defined for all $t \geqslant t_0 - r$, t_0 a given real number; it defines a map $t \to x_t$ for $t \geqslant t_0$, $x_t \in C^0(I, M)$ in the following way: $x_t(\theta) = x(t+\theta)$, $\theta \in [-r, 0]$. This kind of lifting with respect to ϱ enables us to introduce the notion of solution of a (F.D.E.): A solution of a C^1 (F.D.E.) with initial condition

$$(t_0, \phi) \in RxC^0(I, M)$$

is a continuous function $x(t)$, $t_0 - r \leqslant t < +\infty$, such that [1, 3]):

$$x_{t_0} = \phi \tag{i}$$
$$x(t) \text{ is } C^1 \text{ for } t \geqslant t_0 \tag{ii}$$
$$F(x_t) = (x(t), \dot{x}(t)) \text{ for } t \geqslant t_0. \tag{iii}$$

If F is a C^1 (F.D.E.) for a given (t_0, ϕ) there exists a unique solution $x(t)$ with initial condition (t_0, ϕ).

Remark that some solutions of a given C^1 (F.D.E.) cannot be defined for $-\infty < t < +\infty$. Let $x_t(\phi)$ be the solution of a C^1 (F.D.E.) associated to the initial condition $(0, \phi)$ and θ_t the flow $\theta_t: C^0(I, M) \to C^0(I, M)$ defined by $\theta_t(\phi) = x_t(\phi)$.

This flow is a C^1 map but not one to one and $\Phi_{t_1+t_2} = \Phi_{t_1} \cdot \Phi_{t_2}$ for all $t_1, t_2 \geqslant 0$.

Generic Theory

A critical point of $F \in BX^1(I, M)$ is a constant map ϕ_0 such that $F(\phi_0)=0$. Since locally $F(\phi) = (\phi(0))$ the C^1 map f has the decomposition $f(\phi) = L(\phi) + N(\phi)$ where $L(\phi)$ is the linear operator corresponding to the derivative of F at ϕ_0 (see [1, 3]). Since

$$L(\phi) = \int_{-r}^{0} [d\eta(\theta)]\, \phi(\theta)$$

G.E.O. Giacaglia (ed.), Periodic Orbits, Stability and Resonances, 337–338. All Rights Reserved.

one can consider the linear operator $\Delta(\lambda): C^n \to C^n$ by $\Delta(\lambda) = I - \int_{-r}^{0} [d\eta(\theta)]\ e^{\lambda\theta}$. λ is a characteristic root of ϕ_0 if and only if det $\Delta(\lambda) = 0$. Let G_1 be the set of all C^1 (F.D.E.) such that the characteristic roots of all critical points have non zero real part. The following result is the main result of [1]: G_1 *is open and dense in* $BX^1(I, M)$. The most important point in the proof of the density result is that in the manifold $BX^1(I, M)$ the norms of the tangent spaces are not differentiable and then we do not have bump functions.

One can define generic periodic orbits and for a fixed $T>0$ let $BX(T) \subset G_1$ be the set such that $F \in BX(T)$ implies that all periodic orbits of the (F.D.E.) F of period $\leqslant T$ are generic. It is proved that $BX(T)$ *is open in* G_1. The author believes that $BX(T)$ is also dense in G_1 but he did not find the proof.

Acknowledgement

This research was partially supported by ONR, Contract N00014-67-C-0347.

References

[1] Oliva, W. M.: 1969, 'Functional Differential Equations on Compact Manifolds and an Approximation Theorem', *J. Diff. Equations* **5**, No. 3.
[2] Eells Jr., J.: 1966, 'A Setting for Global Analysis', *Bull. Amer. Math. Soc.* **72**, 751–807.
[3] Hale, J. K.: 1969, 'Functional Differential Equations of Retarded Type', 7th Brazilian Coll. of Math., Impa, Poços de Caldas.

ON A CRITERION OF INSTABILITY FOR DIFFERENTIAL EQUATIONS WITH TIME DELAY

NELSON ONUCHIC
Escola de Engenharia, Universidade de São Paulo, São Carlos, S.P., Brasil

1. Introduction and Notation

Let C be the space of continuous functions taking the interval $[-h, 0]$, $0 \leqslant h < \infty$, into R^n. For any $\varphi \in C$ we define $\|\varphi\| = \sup_{-h \leqslant \theta \leqslant 0} |\varphi(\theta)|$, where $|\cdot|$ is any vector norm in R^n. If x is any continuous function defined in the interval $[-h, A)$, $0 < A \leqslant \infty$, x_t for $0 \leqslant t < A$ is defined by $x_t(\theta) = x(t+\theta)$, $-h \leqslant \theta \leqslant 0$. For H, $0 < H \leqslant \infty$, $C_H = \{\varphi \in C \mid \|\varphi\| < H\}$.

Let $f(\varphi) \in R^n$ be a function defined on C_H. Then

$$\dot{x}(t) = f(x_t) \tag{1}$$

is called an autonomous differential equation with time delay. For the special case in which $h = 0$, Equation (1) is an ordinary differential equation.

Let $\varphi \in C_H$. A continuous function $x(t)$ defined on $[-h, A)$, $A > 0$, is said to be a solution of (1) with initial function φ at $t = 0$ if $x_t \in C_H$, $\dot{x}(t) = f(x_t)$ for each t, $0 \leqslant t < A$ and $x_0 = \varphi$. If $f(\varphi)$ is locally Lipschitzian with respect to φ then, given $\varphi \in C_H$, there is a unique solution of (1) with initial function φ at $t = 0$. We use the notations $x(t, \varphi)$, $x(\varphi)$, $x_t(\varphi)$ to denote such a solution. It is easy to see that there is t^+, $0 < t^+ \leqslant \infty$, such that $x(\varphi)$ can be extended to $[-h, t^+)$ as a solution of (1) and $x(\varphi)$ can not be extended to any $[-h, \alpha)$, $\alpha > t^+$. Also if $\|x_t(\varphi)\| \leqslant H_1 < H$ for $0 \leqslant t < t^+$, then $t^+ = \infty$, namely, $x(t, \varphi)$ is defined in the future.

If $f(0) = 0$, then the solution $x = 0$ of (1) is said to be stable if for every $\varepsilon > 0$ there is $\delta = \delta(\varepsilon) > 0$ such that $\varphi \in C_\delta$ implies $x_t(\varphi)$ exists for $0 \leqslant t < \infty$ and $\|x_t(\varphi)\| < \varepsilon$ for all $t \geqslant 0$. If $x = 0$ is stable with $H = \infty$, that is, $C_H = C$, and $x_t(\varphi) \to 0$ as $t \to \infty$ for all $\varphi \in C$, we say that $x = 0$ is globally asymptotically stable.

A set M in C_H is called an invariant set if for any φ in M there corresponds a function $x(t) \in R^n$, defined on R, so that $\dot{x}(t) = f(t, x_t)$ for all $t \in R$ and $x_0 = \varphi$. Let $x(t, \varphi)$ be a solution of (1) with $\|x_t(\varphi)\| \leqslant H_1 < H$. The ω-limit set of $x(t, \varphi)$, denoted by $\Omega(\varphi)$, is the set of all functions ψ such that there is a sequence $\{t_m\}$, $t_m \to \infty$ as $m \to \infty$, satisfying $x_{t_m}(\varphi) \to \psi$ as $m \to \infty$. It is known that $\Omega(\varphi)$ is a nonempty, compact, connected, invariant set and $\operatorname{dist}(x_t, \Omega) \to 0$ as $t \to \infty$ (Hale, 1965, Lemma 2).

Let $V = V(\varphi)$ be a continuous scalar function defined on C_H. We define

$$\dot{V}_{(1)}(\varphi) = \overline{\lim_{r \to 0+}} \frac{1}{r} [V(x_r(\varphi)) - V(\varphi)].$$

If $U \subset C_H$, $\bar{U}$ denotes the closure of U and ∂U denotes the boundary of U with respect to C_H.

G. E. O. Giacaglia (ed.), Periodic Orbits, Stability and Resonances, 339–342.

2. Sufficient Conditions for Instability

The objective of this section is to present a criterion of instability for Equation (1). This result is more general than the one due to Hale (1965, Theorem 4).

THEOREM 1: *Let* $f(0)=0$. *Suppose that there is* $V(\varphi)$ *continuous on* C_H *and that there is an open set* U *of* C_H *such that the following conditions are satisfied:*

(i) $V(\varphi)>0$ *on* U *and* $V(\varphi)=0$ *on* ∂U
(ii) $0\in\partial U$
(iii) $\dot V(\varphi)\geqslant 0$ *on* U
(iv) *The set* $F=\{\varphi\in U \mid \dot V(\varphi)=0\}$ *contains no invariant set of* (1).

Then for each $\varphi\in U$ *there is a sequence* $\{t_m\}$ *such that* $|x(t_m,\varphi)|\to H$ *as* $m\to\infty$ *and hence, by* (ii), *the solution* $x=0$ *of* (1) *is unstable.*

Proof: Assume the contrary. Then there exist $\varphi\in U$ and H_1, $0<H_1<H$, such that $\|x_t(\varphi)\|\leqslant H_1$ for $t\geqslant 0$.

First of all let us show that $x_t(\varphi)\in U$ for all $t\geqslant 0$. Indeed, because otherwise there would be $\tau>0$ such that $x_\tau(\varphi)\in\partial U$ and $x_t(\varphi)\in U$ for $0\leqslant t<\tau$. But this would imply $V(x_t(\varphi))>0$ for $0\leqslant t<\tau$ and $V(x_\tau(\varphi))=0$, which is impossible since by (iii), $V(x_t(\varphi))$ must be nondecreasing on $[0,\tau)$. Then $x_t(\varphi)\in U$ for $0\leqslant t<\infty$.

We know that $\Omega(\varphi)$ is a nonempty and invariant set. Since $x_t(\varphi)\in U$ and $V(x_t(\varphi))\geqslant \geqslant V(x_0(\varphi))=V(\varphi)>0$ for all $t\geqslant 0$, it follows that $\Omega(\varphi)\subset U$.

Since $V(x_t(\varphi))$ is nondecreasing and $\Omega(\varphi)$ is nonempty it follows that $V(x_t(\varphi))\to\varrho$, ϱ being a real number, as $t\to\infty$. Consequently $V(\psi)$ is a constant function on $\Omega(\varphi)$, which implies $\dot V(\psi)=0$ on $\Omega(\varphi)$.

Then the nonempty and invariant set $\Omega(\varphi)$ is contained in F, which contradicts (iv). The proof is complete.

COROLLARY (Theorem of Cetaev).

Let $f(0)=0$. *Suppose that there is* $V(\varphi)$ *continuous on* C_H *and that there is an open set* U *of* C_H *such that the following conditions are satisfied:*

(i) $V(\varphi)>0$ *on* U *and* $V(\varphi)=0$ *on* ∂U
(ii) $0\in\partial U$
(iii) $\dot V(\varphi)>0$ *on* U.

Then the solution $x=0$ *of* (1) *is unstable*

3. Application

Consider the equation

$$\dot x(t) = -\frac{1}{h}\int_{-h}^{0} g(x(t+\theta))\,[h+\theta]\,d\theta\,,\quad t\geqslant 0\,, \tag{2}$$

that is, $\dot x(t)=X(x_t)$,

where

$$X(\varphi) = -\frac{1}{h}\int_{-h}^{0} g(\varphi(\theta))\,[h+\theta]\,\mathrm{d}\theta .$$

For the stability properties and the physical motivation of this equation, see Levin and Nobel (1964) and Hale (1965). It is known that if $g(x)$ is locally Lipschitzian in x $x\in R$, $xg(x)>0$ for $x\neq 0$, $G(x)=\int_0^x g(s)\,\mathrm{d}s\to\infty$ as $|x|\to\infty$, then the solution $x=0$ of (2) is stable and every solution of (2) is bounded in the future and the ω-limit set of any solution of (2) consists of the orbit of some solution $y(t)$ of

$$\ddot{y} + g(y) = 0 \tag{3}$$

with $y(t+h)=y(t)$ for all $t\in R$. Then if the only solution of (3) such that $y(t+h)=y(t)$ for all $t\in R$ is $y=0$, it follows that the zero solution of (2) is globally asymptotically stable (Hale, 1965; Levin and Nobel, 1964).

The objective of this section is to obtain a sufficient condition for the instability of the zero solution of (2) as an application of the above Theorem 1.

THEOREM 2: *Let $g(x)$ be locally Lipschitzian in x, $|x|<H$ with $g(0)=0$. Assume either*

(i) *that $g(x)<0$ for $0<x<H$, or*
(ii) *that $g(x)>0$ for $-H<x<0$.*

Then the zero solution of (2) is unstable.

Proof: Consider the case $g(x)<0$ for $0<x<H$. The treatment in the other case is similar. Define

$$\tilde{g}(x) = g(x) \quad \text{for} \quad 0 \leqslant x < H, \quad \text{and}$$
$$\tilde{g}(x) = 0 \quad \text{for} \quad -H < x \leqslant 0$$
$$V(\varphi) = -\tilde{G}(\varphi(0)) - \frac{1}{2h}\int_{-h}^{0}\left[\int_{\tau}^{0} g(\varphi(\theta))\,\mathrm{d}\theta\right]^2 \mathrm{d}\tau,$$

where $\tilde{G}(x)=\int_0^x \tilde{g}(s)\,\mathrm{d}s$. This function $V(\varphi)$ is related to the one considered in Hale (1965) and Levin and Nobel (1964). Let $U=\{\varphi\in C_H \mid V(\varphi)>0\}$. For this set U conditions (i) and (ii) of Theorem 1 are satisfied. Let us show that condition (iii) is also satisfied.

Let $\varphi\in U$. Then $\tilde{G}(\varphi(0))<0$ and consequently $\tilde{G}(\varphi(0))=G(\varphi(0))$ and

$$V(\varphi) = -G(\varphi(0)) - \frac{1}{2h}\int_{-h}^{0}\left[\int_{\tau}^{0} g(\varphi(\theta))\,\mathrm{d}\theta\right]^2 \mathrm{d}\tau .$$

A calculation shows that

$$\dot{V}(\varphi) = \frac{1}{2h}\left[\int_{-h}^{0} g(\varphi(\theta))\,\mathrm{d}\theta\right]^2 \geqslant 0$$

and then condition (iii) of Theorem 1 is satisfied. Let us show that condition (iv) also holds. For this purpose it is enough to show that there is no solution $x(t)$ of (2), bounded on R, such that $x_t \in U$ with $\dot{V}(x_t)=0$ for all $t \in R$. In fact, otherwise $0 < x(t) < H$ for $t \in R$ and as $\ddot{x}(t)+g(x(t))=(1/h)\int_{t-h}^{t} g(x(s))\,ds$, it would follow $\ddot{x}(t)+g(x(t))=0$. This would imply that $\dot{x}(t)$ is an increasing function for $t \in R$ and consequently $x(t)$ would not be bounded on R, a contradiction.

The proof is complete.

References

Hale, J. K.: 1965, 'Sufficient Conditions for Stability and Instability of Autonomous Functional-Differential Equations', *J. Diff. Equations* **1**, 452–82.

Levin, J. J. and Nobel, J.: 1964, 'On a Nonlinear Delay Equation', *J. Math. Anal. Appl.* **8**, 31–44.

THE LIBRATION CASE OF THE STELLAR PROBLEM OF THREE BODIES

DAVID FISHER

Goddard Space Flight Center, Greenbelt, Md., U.S.A.

Abstract. The methods of nonlinear mechanics are introduced to facilitate the study of the libration case of the stellar problem of three bodies. To illustrate the method, the problem of a lunar satellite perturbed by the earth and under the influence of the J_2 and J_3 terms of the moon's gravitational field is considered. By means of the energy integral and the equation of motion of the argument of pericenter, a closed curve is obtained in phase space. Analysis of the geometry of the variables in phase space yield the period of libration and expansions of the argument of pericenter and eccentricity in terms of time.

1. Introduction

In 1936 Ernest W. Brown wrote a series of four papers on the 'Stellar Problem of Three Bodies' with applications to satellite theory (Brown, 1936). The stellar problem is defined by a pair of celestial objects relatively close to each other perturbed by a third body at a considerable distance away. The motion of such a system is similar in many respects to the fundamental problem of the long period motion of a lunar satellite. Thus the moon and its satellite correspond to the close pair perturbed by the earth whose distance to the moon is large compared to the moon's distance from its satellite. In this paper the problem is expanded to include the effects of the J_2 and J_3 harmonics of the moon.

In order to solve the problems he considered, Brown developed a semi-analytic method, based on harmonic analysis, which was valid for arbitrary eccentricities and inclinations. His solution gave a great deal of insight into the nature of the perturbations and avoided a great deal of tedious computations.

In the problem considered by Brown the nature of the problem was such that the resulting motion was of the circulation type. In this paper, the initial conditions are such that libration of the argument of pericenter occurs. The differences between the two types of motion are illustrated by a pendulum which is free to rotate about a horizontal axis. When the pendulum makes complete revolutions about the axis the motion is known as circulation. However when the pendulum swings to and fro like the pendulum of a clock we say libration occurs. This concept applied to satellites says that when the argument of pericenter takes on values over the entire interval between 0 and 2π as well as integral multiples of this range circulation occurs. However, when the argument of pericenter is restricted to a smaller interval than between 0 and 2π, libration occurs.

The method described by Brown cannot be carried over to the case when libration occurs. In this paper an entirely different semi-analytic approach is developed which is a direct consequence of geometrical considerations of the problem in Poincaré phase space. Discussions of the geometry in phase space is given in books on nonlinear mechanics such as Stoker (1950).

G. E. O. Giacaglia (ed.), Periodic Orbits, Stability and Resonances, 343–348.

A brief outline of the method applied in this report which will be explained below, now follows:

A. USE OF THE ENERGY INTEGRAL

The energy constant is first evaluated from the initial conditions. In this report the initial conditions were

$$e_0 = 0.22, \qquad g_0 = 90°, \qquad \sin i_0 = 0.7,$$

and

$$a = 0.608\,3 \quad \text{decamegameters}$$

where e=eccentricity, g=argument of pericenter, i=inclination of the plane of the satellite's orbit to the moon's equator, and a=semi-major axis of the lunar satellite.

Having obtained the energy constant, a table of g for a set of values of e is obtained which satisfies the energy equation. The results for the initial conditions chosen above are given in Table I and shown in Figure 1.

B. OBTAINING THE ORDINATES NEEDED IN PHASE SPACE

From the expression of the Hamiltonian of the system, the formula for $\dot{g}$ is first derived, where $\dot{g} = \mathrm{d}g/\mathrm{d}t$.

Using the values of e and g obtained from the energy integral, corresponding values of $\dot{g}$ are found. These values are also listed in Table I. A plot of $\dot{g}$ vs. g is given in Figure 2.

C. OBTAINING THE PERIOD OF LIBRATION

The libration period is found by integrating over the curve C by the formula

$$T = \int_C \frac{\mathrm{d}g}{\dot{g}}.$$

TABLE I

e	$\pm(90° - g)$	g	$\bar{g}$
0.22	0	−0.226 7	0
0.23	5.007	−0.182 4	36
0.24	6.655	−0.141 2	51
0.25	7.300	−0.102 2	60
0.26	8.227	−0.065 1	76
0.27	8.533	−0.029 2	84
0.28	8.606	+0.005 8	90
0.29	8.466	+0.040 3	101
0.30	8.118	+0.074 6	109
0.31	7.542	+0.108 9	118
0.32	6.691	+0.143 7	128
0.33	5.442	+0.179 0	140
0.34	3.389	+0.215 3	153

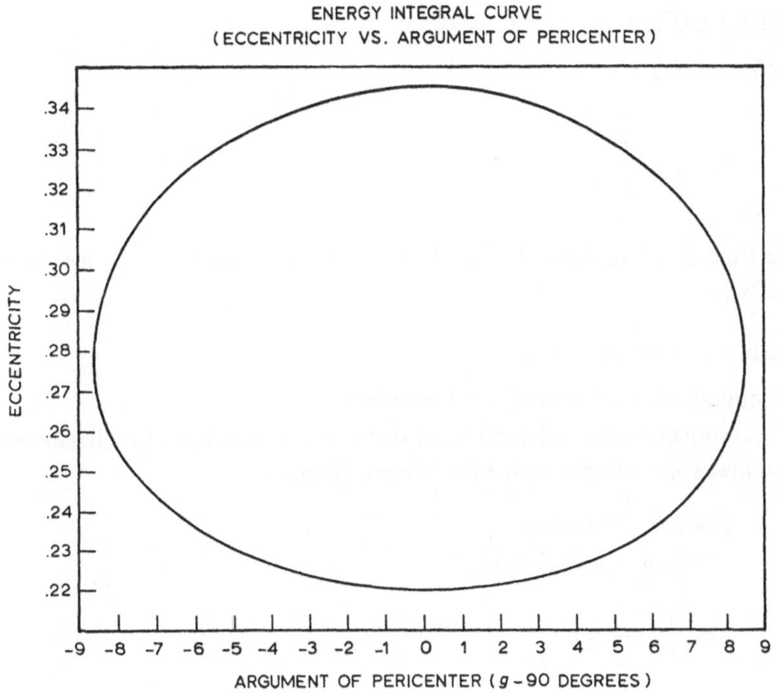
ENERGY INTEGRAL CURVE
(ECCENTRICITY VS. ARGUMENT OF PERICENTER)
ECCENTRICITY
.34
.33
.32
.31
.30
.29
.28
.27
.26
.25
.24
.23
.22
-9 -8 -7 -6 -5 -4 -3 -2 -1 0 1 2 3 4 5 6 7 8 9
ARGUMENT OF PERICENTER (g-90 DEGREES)

Fig. 1.

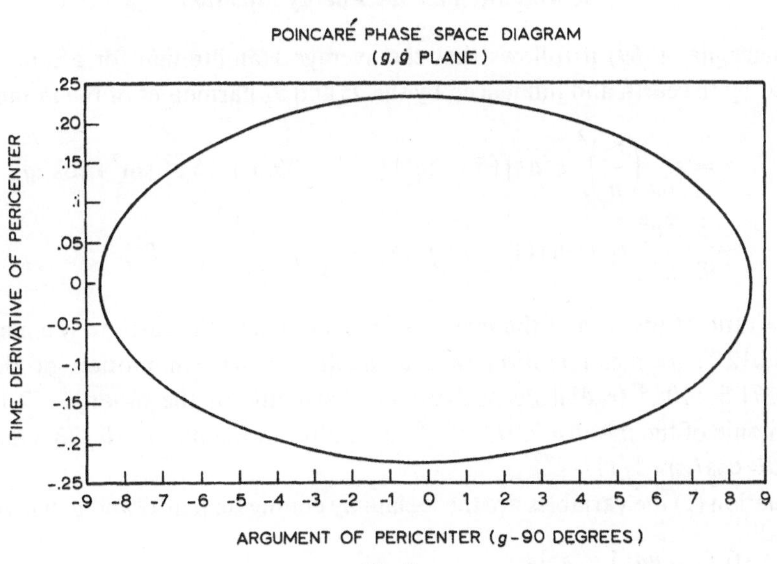
POINCARÉ PHASE SPACE DIAGRAM
(g, ġ PLANE)
TIME DERIVATIVE OF PERICENTER
.25
.20
.15
.1
.05
0
-0.5
-.1
-.15
-.2
-.25
-9 -8 -7 -6 -5 -4 -3 -2 -1 0 1 2 3 4 5 6 7 8 9
ARGUMENT OF PERICENTER (g-90 DEGREES)

Fig. 2.

The curve C is found by inspection of the geometry in phase space (Figure 2).

In this report the integration was performed numerically, using Simpson's rule. The period T was found to be 1600 days.

D. THE TIME INTEGRAL

The indefinite integral

$$t - t_0 = \int_{g_0}^{g} \frac{dg}{\dot{g}}$$

gives the time as a function of g, and by means of Table I also as a function of the eccentricity e.

E. INVERSION OF THE TIME EQUATION

The time equation given above is next inverted.

For the initial conditions described in this report it was found by harmonic analysis that good first approximations for g and e are given by

$$g = 90° - 8\overset{\circ}{.}6 \sin \bar{g}$$
$$e = 0.28 - 0.0625 \sin \bar{g}$$

where

$$\bar{g} = \frac{2\pi}{T}(t - t_0)$$

with $T = 1600$ days.

A brief description of the equations used in the development is now given.

2. Hamiltonian and Energy Equation

From Giacaglia (1969) it follows that the average Hamiltonian for a lunar satellite perturbed by the earth and influenced by the J_2 and J_3 harmonics of the moon is given by

$$F = \frac{1}{16q}\left(\frac{n_{☾}}{n}\right)^2 n^2 a^2 [(5 - 3\eta^2)(-1 + 3\theta^2) + 15\, e^2 \sin^2 i \cos 2g]$$
$$- \frac{3}{8}\frac{b^3 n^2}{\eta^5} J_3\, e \sin i (1 - 5\theta^2) \sin g - \frac{1}{4} J_2 \frac{b^2 n^2}{\eta^3}(1 - 3\theta^2) \qquad (1)$$

where q = ratio of the sum of the masses of the moon and the earth to the mass of the earth = 1.012 3; n = mean motion of the satellite; $n_{☾}$ = mean motion of the moon = $2.280\,271\,3 \times 10^{-3}$ (radians/centiday); J_3 = harmonic of the moon = -2.1×10^{-5}; J_2 = harmonic of the moon = 2.07×10^{-4}; b = radius of the moon = 0.173 8 decamegameters; $\theta = \cos i$; $\eta = \sqrt{(1 - e^2)}$.

In Equation (1) the variables are the Delaunay conjugate pair G and g, where

$$G = \sqrt{\mu a (1 - e^2)}.$$

The elements H and L are constant where

$$H = G \cos i \quad \text{and} \quad L = \sqrt{\mu a}$$

with μ=gravitational constant times the mass of the moon=3.660 1891×10^{-3} (decamegameters)^3/(centiday)^2. Since the time does not enter explicitly in the Hamiltonian we may write

$$F = \text{constant}. \tag{2}$$

Equation (2) is the Energy integral of the system.

If we utilize the fact that

$$\eta^2\theta^2 = H^2/L^2 = \text{constant}$$

we can rewrite Equation (2) in the form

$$e^2\left(1 - \tfrac{5}{2}\sin^2 i \sin^2 g\right) + \frac{\alpha_2}{\eta^3}\left(1 - 3\theta^2\right) + \frac{\alpha_3}{\eta^5} e \sin i \left(1 - 5\theta^2\right) \sin g = c \tag{3}$$

where

$$\alpha_2 = -\frac{1}{3} J_2 \frac{b^2}{a^2}\left(\frac{n}{n_☾}\right)^2 q$$

$$\alpha_3 = -\frac{1}{2} J_3 \frac{b^3}{a^3}\left(\frac{n}{n_☾}\right)^2 q.$$

Since the Delaunay elements L and H are constant, Equation (3) gives a relation between e and g since

$$\cos^2 i = \nu^2/(1 - e^2) \quad \text{where} \quad \nu^2 = H^2/L^2 = \text{const}.$$

If a plot of e vs. g derived from Equation (3) results in a closed curve as given in Figure 1 then libration occurs. This is apparent from the figure since g is bounded. In the present problem, the argument of pericenter librates about the 90° axis with a swing of 8.°6.

Felsentreger (1968) has shown the boundaries of libration and circulation when the J_2 harmonic of the moon is taken into account.

3. The Geometry in Phase Space

To obtain an expression for $\dot{g} = \mathrm{d}g/\mathrm{d}t$ we apply the rule in canonical variable theory, namely

$$\dot{g} = -\,\partial F/\partial G.$$

Performing the indicated differentiation where F is given by Equation (1) we have

$$\dot{g} = \frac{3n_☾^2}{4nq}\left\{\frac{2\eta^2 + 5\sin^2 g\,(\theta^2 - \eta^2)}{\eta} + \frac{3\alpha_2(1 - 5\theta^2)}{\eta^4} - \frac{\alpha_3 \sin g}{\eta^6}\left[\frac{(1 - 5\theta^2)(\theta^2 - \eta^2)}{e \sin i} + e \sin i\,(-5 + 35\theta^2)\right]\right\}. \tag{4}$$

Using the values of e and $\sin g$ obtained from the energy integral we obtain corresponding values of $\dot{g}$ from Equation (4). These values are also listed in Table I.

A plot of $\dot{g}$ vs. g given in Figure 2 results in an ellipse. An immediate consequence of a closed curve in phase space is that the variables are periodic. This excludes a secular growth in g and verifies the conclusion that g librates. The path in phase space also provides the contour to be followed in the process of integration to find the period from the integral

$$T = \int_C \frac{\mathrm{d}g}{\dot{g}}.$$

4. Perturbations in the Other Elements

The perturbations in the mean anomaly l, the right ascension of the node h are found from the expressions

$$\frac{\mathrm{d}l}{\mathrm{d}t} = -\frac{\partial F}{\partial L} \quad \frac{\mathrm{d}h}{\mathrm{d}t} = -\frac{\partial F}{\partial H}. \tag{5}$$

The Hamiltonian F is given by Equation (1). The values of e and g which appear in (5) may be found in Table I.

The inclination i is found from the expression

$$\cos i = v/\sqrt{1-e^2}. \tag{6}$$

5. Conclusions

A semi-analytic method for obtaining the long period perturbations of a librating lunar satellite is decribed.

For the initial conditions given here, it was found that simple sinusoidal curves describe the perturbations in e and the libration in g.

References

Brown, E. W.: 1936, 'The Stellar Problem of Three Bodies', *Monthly Notices Roy. Astron. Soc.* **97**, 56–66, 116–27, 388–95.

Felsentreger, T. L.: 1968, 'Classification of Lunar Satellite Orbits', *Planetary Space Sci.* **16**, 285–95.

Giacaglia, G. E. O. *et al.*: 1969, A Semi-Analytic Theory for the Motion of a Lunar Satellite, NASA TN D-5043, Washington, D.C.

Stoker, J. J.: 1950, *Non-Linear Vibrations*, Interscience, New York.

STABILITY OF PERIODIC ORBITS IN THE RESTRICTED PROBLEM

MICHEL HÉNON
Observatoire de Nice, France

and

MONIQUE GUYOT
Département de Mathématiques, Faculté des Sciences de Nice, France

Abstract. Critical orbits, separating stable from unstable periodic orbits, have been computed for the six main families f, g, h, i, l, m and for all values of the mass ratio μ. The results are given by tables and graphs. Domains of stability of the periodic orbits are discussed as a function of μ. The limiting cases $\mu \to 0$ and $\mu \to 1$ are studied in detail and the limiting forms of the critical orbits are explained. One particular result is that if the relative mass of one of the bodies is less than 0.0477, all retrograde orbits around that body are stable.

1. Introduction

In previous papers (Hénon, 1965a, b, 1966a, b, 1969, 1970) the general solutions of the restricted problem of three bodies have been explored numerically, for two particular values of the mass ratio (the ratio of the mass of the second body to the total mass of the system): $\mu = \frac{1}{2}$ and $\mu = 0$.

It would be desirable to extend this exploration to all values of μ; this, however, appears as a gigantic task, both in terms of computing time and of the amount of data to be displayed. As a first step, one should therefore aim at a less detailed study, giving only the essential features of the solutions, but for all values of μ.

The knowledge gained in the particular cases $\mu = \frac{1}{2}$ and $\mu = 0$ can be used to find which features are essential, and how they can be determined. It was found that the four-dimensional phase space can be approximately divided into three regions, respectively occupied by quasi-periodic orbits, semi-ergodic orbits and escape orbits; these three kinds of orbits have strikingly different properties. In many applications, the interesting orbits are those which stay in a bounded vicinity of the primaries; escape orbits are then excluded. On the other hand, the semi-ergodic region fills only a comparatively small part of phase space (Hénon, 1966b, Figure 12). Thus the main object for consideration is the quasi-periodic region.

Theoretical results (Arnold, 1963; Moser, 1962) as well as numerical computations have clarified the structure of the quasi-periodic region. Each quasi-periodic orbit lies on a two-dimensional torus in phase space; the tori are arranged in coaxial sets, and the axis of each set is a *stable periodic orbit*. Equivalently, one may look at the picture in a *surface of section*, and consider the mapping of that surface onto itself: to each quasi-periodic orbit corresponds then an invariant curve; these curves are arranged in concentric sets, and at the centre of each set lies a stable invariant point.

G. E. O. Giacaglia (ed.), Periodic Orbits, Stability and Resonances, 349–374. All Rights Reserved.

This picture is not peculiar to the restricted problem, but is found in all dynamical problems with two degrees of freedom.

Thus, stable periodic orbits, although they are particular cases among orbits in general, nevertheless play a fundamental role; to borrow a picture used by Deprit and Henrard (1969), they constitute the skeleton in the body of all the orbits.

For a given value of the parameter μ, periodic orbits are grouped in one-parameter families; each family contains a simple infinity of periodic orbits, whose properties vary continuously from one end of the family to the other. Among these properties is the *stability index a* (Hénon, 1965b); a periodic orbit is stable if $|a|<1$, unstable if $|a|>1$.* We define *critical periodic orbits* (or shortly: *critical orbits*) as those for which $|a|=1$; they separate, along the family, intervals of stable periodic orbits from intervals of unstable periodic orbits. Critical orbits are obviously of prime importance; in phase space, they mark the termination of a quasi-periodic region (Hénon, 1966b, Figure 12). Thus our first task is to find them. Accessorily, critical orbits are also interesting because in many cases they are 'branching points', where two families of periodic orbits intersect.

It was shown in previous papers that among all families of periodic orbits, six are of particular importance, and determine the main features of the picture in phase space. They are, in Strömgren's notation (1935):

Family f: retrograde orbits around the second body.
Family g: direct orbits around the second body.
Family h: retrograde orbits around the first body.
Family i: direct orbits around the first body.
Family l: direct orbits (in fixed axes) around both bodies.
Family m: retrograde orbits around both bodies.

In the present paper, we shall restrict our attention to these families, and we finally define our goal as follows: *To find the critical periodic orbits of families f, g, h, i, l, m for all values of the mass ratio* μ.

2. Method

We use the same notations as in previous papers (see Hénon, 1965a). For a given value of the mass ratio μ, critical orbits are particular, isolated orbits. If now μ is varied, each critical orbit generates a continuous one-parameter family, which can be numerically followed. Critical orbits are known for $\mu=\frac{1}{2}$ (Hénon, 1965b) and $\mu=0$ (Hénon, 1969); they can be found for $\mu=\frac{1}{11}$ and for $\mu=0.012\,15...$ by interpolation from the results of Shearing (1960) and Broucke (1968). All these orbits have been used as starting points.

Care should be taken not to confuse a *family of periodic orbits*, which depends on two parameters when μ is allowed to vary, and a *family of critical orbits*, which is a

* As pointed out by Dr. Jefferys during the discussion, periodic orbits are also unstable for the particular value $a=-\frac{1}{2}$, and may be unstable for the particular value $a=0$. These two cases correspond to Moser's resonances 3 and 4 (see Hénon, 1966a, p. 74). They will not be considered here.

subset of a family of periodic orbits and depends on one parameter. We shall always represent a family of periodic orbits by a simple letter, for example m, and a family of critical orbits by a letter followed by a number, for example $m1$.

Periodic orbits of the six main families are symmetrical with respect to the x-axis. Numerical integration of an orbit is therefore started on the x-axis, and perpendicular to it. Three quantities must be specified: the mass ratio μ, and the two initial values x_0 and $\dot{y}_0$. Instead of $\dot{y}_0$, it is in practice more convenient to specify the Jacobi constant C, from which $\dot{y}_0$ can be deduced by:

$$C = x_0^2 + \frac{2(1-\mu)}{|x_0 + \mu|} + \frac{2\mu}{|x_0 - 1 + \mu|} - \dot{y}_0^2. \tag{1}$$

On the other hand, two conditions must be satisfied: the orbit must intersect the x-axis again at right angles, in order to be periodic; it must have a stability index equal to ± 1, in order to be critical. We have thus two conditions for three unknowns, which shows again that there is a one-parameter family of solutions.

A program has been set up which, for a given value of one of the three quantities μ, x_0, C, automatically adjusts the two other quantities by trial and error, until a critical periodic orbit is found. The program gives also the sign of the variation of a along the family of periodic orbits, for a given μ, in the vicinity of the critical orbit; this determines on which side of the critical orbit there is stability.

The *types* of the critical orbits (Hénon, 1965b) were found to be very useful, because all critical orbits of a given family have the same type by continuity. This helped to guess the connections between the first fragments of families, and to disentangle different families in some crowded regions.

Thanks to the symmetry, the periodicity can be established by a computation of one half of the orbit only. The stability index and the type, on the other hand, had been defined in Hénon (1965b) by reference to a complete periodic orbit. It is possible, however, to compute these quantities too using only one half of the orbit (Deprit and Price, 1965), and thus to save a factor 2 in computing time. Let us label with 0 the starting point on the x-axis, with 1 the next intersection point, and with 2 the next one (Figure 1); points 0 and 2 coincide for a periodic orbit. In analogy with what has

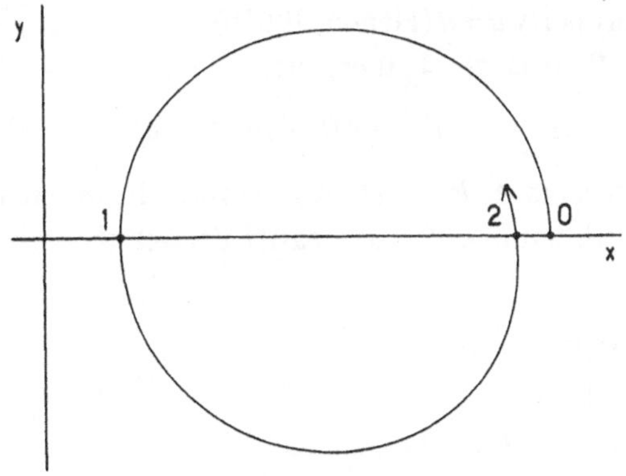

Fig. 1. Labels for intersection points.

been done in Hénon (1965b) we consider small perturbations of the periodic orbit and we write:

$$\begin{pmatrix} \Delta x_1 \\ \Delta \dot{x}_1 \end{pmatrix} = \begin{pmatrix} A & B \\ C & D \end{pmatrix} \begin{pmatrix} \Delta x_0 \\ \Delta \dot{x}_0 \end{pmatrix}, \tag{2}$$

$$\begin{pmatrix} \Delta x_2 \\ \Delta \dot{x}_2 \end{pmatrix} = \begin{pmatrix} A' & B' \\ C' & D' \end{pmatrix} \begin{pmatrix} \Delta x_1 \\ \Delta \dot{x}_1 \end{pmatrix}, \tag{3}$$

so that a, b, c, d, defined in Hénon (1965b), are given by:

$$\begin{pmatrix} a & b \\ c & d \end{pmatrix} = \begin{pmatrix} A' & B' \\ C' & D' \end{pmatrix} \begin{pmatrix} A & B \\ C & D \end{pmatrix}. \tag{4}$$

Because of the symmetry of the restricted problem, given any orbit, one finds another orbit by changing the signs of y and t. If we apply this transformation to the half-orbit from 1 to 2, it becomes the half-orbit from 1 to 0, but described in the opposite direction, i.e. from 0 to 1; thus the periodic orbit is transformed into itself. For orbits close to the periodic orbit, (3) becomes:

$$\begin{pmatrix} \Delta x_0 \\ -\Delta \dot{x}_0 \end{pmatrix} = \begin{pmatrix} A' & B' \\ C' & D' \end{pmatrix} \begin{pmatrix} \Delta x_1 \\ -\Delta \dot{x}_1 \end{pmatrix}. \tag{5}$$

Taking into account the area-preserving property:

$$AD - BC = 1\,, \qquad A'D' - B'C' = 1\,, \tag{6}$$

Equation (5) becomes:

$$\begin{pmatrix} \Delta x_1 \\ \Delta \dot{x}_1 \end{pmatrix} = \begin{pmatrix} D' & B' \\ C' & A' \end{pmatrix} \begin{pmatrix} \Delta x_0 \\ \Delta \dot{x}_0 \end{pmatrix}. \tag{7}$$

Comparing with (2), we find:

$$A' = D\,, \quad B' = B\,, \quad C' = C\,, \quad D' = A\,; \tag{8}$$

and substituting into (4):

$$a = d = AD + BC\,, \quad b = 2BD\,, \quad c = 2AC\,. \tag{9}$$

Thus the stability index can be computed from A, B, C, D. Incidentally, we have redemonstrated the property $a = d$ (Hénon, 1965b).

For a critical orbit, there is $|a| = 1$; therefore:

$$4ABCD = (AD + BC)^2 - (AD - BC)^2 = a^2 - 1 = 0\,. \tag{10}$$

One of the four coefficients A, B, C, D must be zero. From Hénon (1965b, Equation (29)), and the above Equations (9), one easily derives the correspondence with the orbit types:

$$\begin{aligned} A = 0&: \quad \text{type } 5\,. \\ B = 0&: \quad \text{type } 4\,. \\ C = 0&: \quad \text{type } 1, 2 \text{ or } 3\,. \\ D = 0&: \quad \text{type } 6\,. \end{aligned} \tag{11}$$

The numerical integration of the equations of motion has been made in double precision, in regularized coordinates, by means of Bulirsch and Stoer's (1966) algorithm. This ensured, if desired, an accuracy of the order of 10^{-14}, as measured by the variation of C. In most cases, however, such a high precision was quite unnecessary, and the integration accuracy was kept at about 10^{-6}. The conditions of periodicity and criticality were also satisfied with an accuracy of the order of 10^{-6}. The time required to find one critical orbit, with an IBM 7040, is of the order of 3 minutes.

3. Results

We remark first that if the (x, y)-plane is rotated through 180° around its origin, one finds again an instance of the restricted problem, with μ replaced by $1-\mu$. In other words: given a solution of the restricted problem, one obtains another solution by replacing x, y, μ by $-x$, $-y$, $1-\mu$. This is true in particular for periodic solutions. Family f is changed into family h, and vice versa; families i and g are similarly exchanged; family l is changed into itself, and so is family m. This is summarized as follows:

$$\frac{\mu \qquad x \quad y \quad f \; g \; h \; i \; l \; m}{1-\mu \quad -x \quad -y \quad h \; i \; f \; g \; l \; m}. \tag{12}$$

The stability index is not changed by the transformation, so a critical orbit is changed into a critical orbit.

1. *Family m*

We begin with family m, which is the simplest. Figure 2 shows, as a function of μ, the

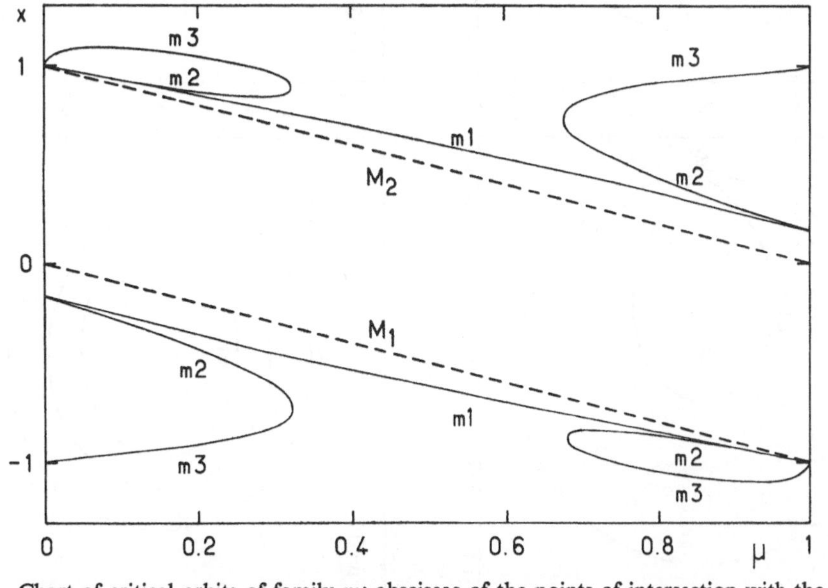

Fig. 2. Chart of critical orbits of family m: abscissas of the points of intersection with the x-axis, as a function of the mass ratio μ. Dashed lines: positions of the two primaries M_1 and M_2.

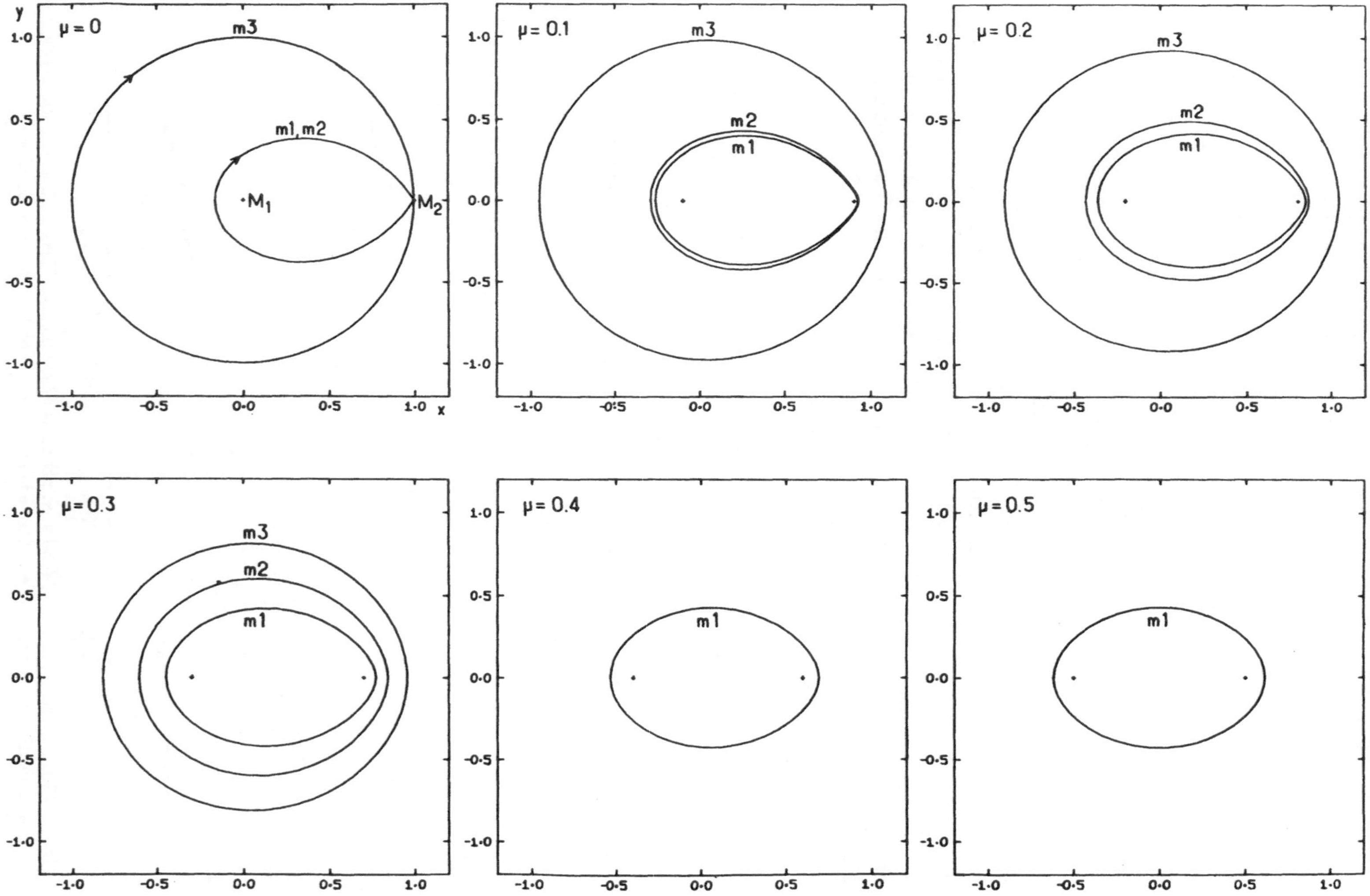

Fig. 3. Critical orbits of family m for some values of μ. Periodic orbits of family m are unstable inside $m1$ and between $m2$ and $m3$.

intersections of the critical orbits with the x-axis. Each critical orbit intersects the x-axis twice and is therefore represented by two points, situated on the same vertical line. Curves represent families of critical orbits. The straight dashed lines represent the positions of the two primaries M_1 and M_2. The picture is symmetrical with respect to the point $\mu=\frac{1}{2}$, $x=0$; this is a consequence of the transformation property described above.

Figure 3 represents the critical orbits, in the (x, y)-plane, for a few values of μ. All these orbits are retrograde. Tables at the end of the paper give the parameters for some of the computed critical orbits.

Family $m1$, which was known from the case $\mu=\frac{1}{2}$, is seen to extend continuously from $\mu=0$ to $\mu=1$. Another family has been discovered; it starts from $\mu=0$, reaches a maximal value $\mu=\mu_0=0.327...$ and comes back to $\mu=0$. A symmetrical family exists for $1-\mu_0 \leqslant \mu \leqslant 1$. It will be convenient to give two different names, $m2$ and $m3$, to the two branches separated by the extremum in μ, although Figure 2 makes clear that these two branches are two halves of a unique family.

For $\mu_0<\mu<1-\mu_0$, there is only one critical orbit, $m1$ (see last two frames of Figure 3). If one follows the family m of *periodic* orbits for a given value of μ, starting from the quasi-circular orbits of large dimensions, then the orbits are stable until $m1$ is met, and unstable afterwards. Roughly speaking: periodic orbits larger than $m1$ are stable, periodic orbits smaller than $m1$ are unstable.

For $0<\mu<\mu_0$, or $1-\mu_0<\mu<1$, there are three critical orbits (see first frames of Figure 3). If one again follows family m of periodic orbits for a given μ, starting from large orbits, one finds that the orbits are stable until the critical orbit $m3$; unstable between $m3$ and $m2$; stable between $m2$ and $m1$; and finally unstable after $m1$.

Thus, as μ becomes less than μ_0, a new zone of instability appears, and becomes progressively larger. It is a curious fact that, broadly speaking, orbits of family m appear to be less stable for smaller values of μ.

The limiting cases $\mu=0$ and $\mu=1$ will be considered in Section 4.

2. *Family l*

This family has an infinite number of critical orbits; but in practice only the first few are of importance (Hénon, 1965b). Figure 4 shows the points of intersection with the x-axis as a function of μ for three families of critical orbits: $l1$, $l'1$ and $l2$. The figure is again symmetrical with respect to the point $\mu=\frac{1}{2}$, $x=0$. (But, despite appearances, the curves are not symmetrical with respect to the vertical line $\mu=\frac{1}{2}$.) Figure 5 represents the critical orbits for three values of μ. These orbits are direct in fixed axes, retrograde in rotating axes.

If one follows family l of periodic orbits for a given μ, starting from quasi-circular orbits of large dimensions: the orbits are stable until $l1$; unstable in the narrow range between $l1$ and $l'1$; stable between $l'1$ and $l2$; and mostly unstable after $l2$, with the exception of very short intervals of stability (Hénon, 1965b).

The critical orbits $l1$ and $l'1$ become identical for $\mu=0$, $\mu=\frac{1}{2}$, $\mu=1$, and remain close to each other for all values of μ. They are nearly circular, and their radius changes

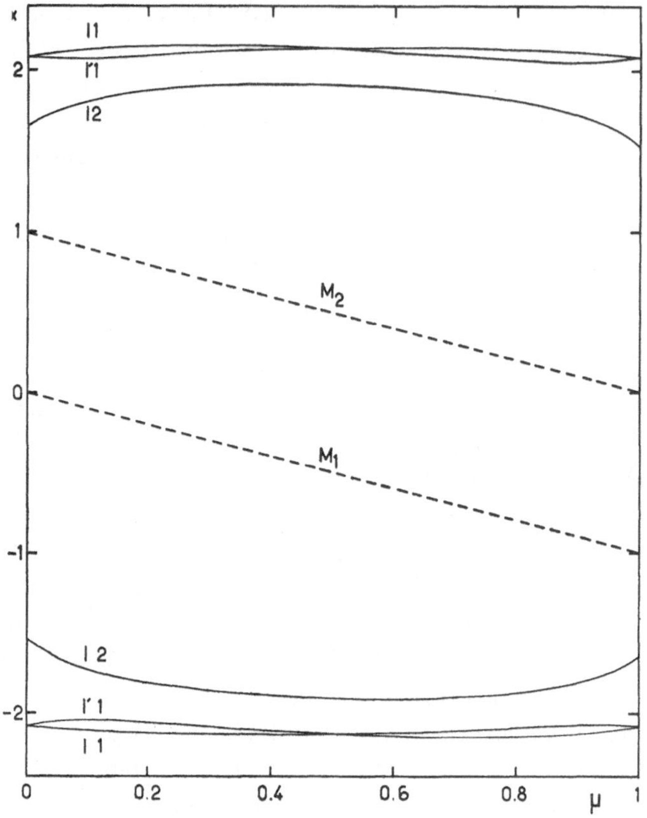

Fig. 4. Chart of critical orbits of family *l*.

remarkably little with μ (Figure 4): instability always makes its first appearance for a value of x between 2.08 and 2.16. Roughly speaking: direct orbits around both primaries are stable when their radius is larger than twice the separation of the primaries.

3. *Family h*

This family also has an infinite number of critical orbits; the first eleven have been listed in Hénon (1965b). Continuation of the first nine orbits for $\mu \neq \frac{1}{2}$ has led to the curves shown on Figure 6. The picture is much more complicated than for families m and l. Sometimes a curve started from a given critical orbit has an extremum in μ and comes back to another critical orbit for $\mu = \frac{1}{2}$. Thus critical orbits $h3$ and $h4$ belong to two branches of the same family. The same is true for the pairs $h1$ and $h5$, $h_2 6$ and $h_2 9$, $h_2 7$ and $h_2 8$. (The subscript 2 indicates double-periodic orbits.)

Branches $h_2 8$ and $h_2 9$ have not been followed to their end, and their termination is unknown.

If x and μ are changed into $-x$ and $1-\mu$, family h is changed into family f. Thus, rotating Figure 6 through 180° and reading the two scales which are now at the left and bottom, one obtains the curves describing the critical orbits for family f.

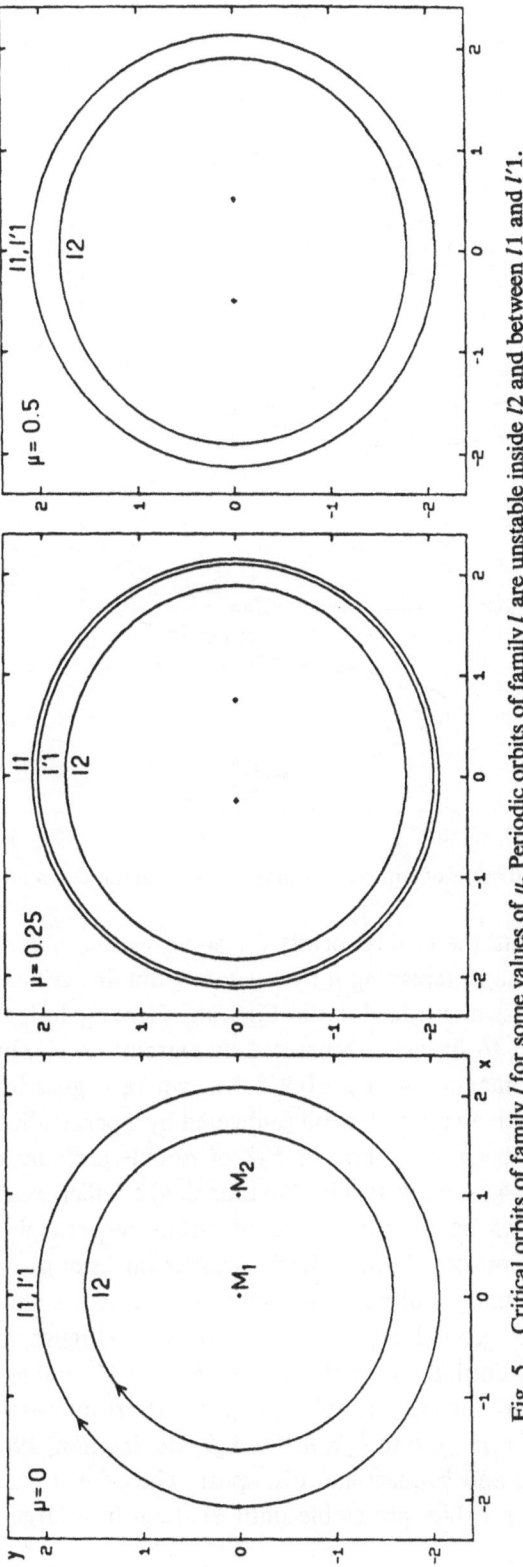

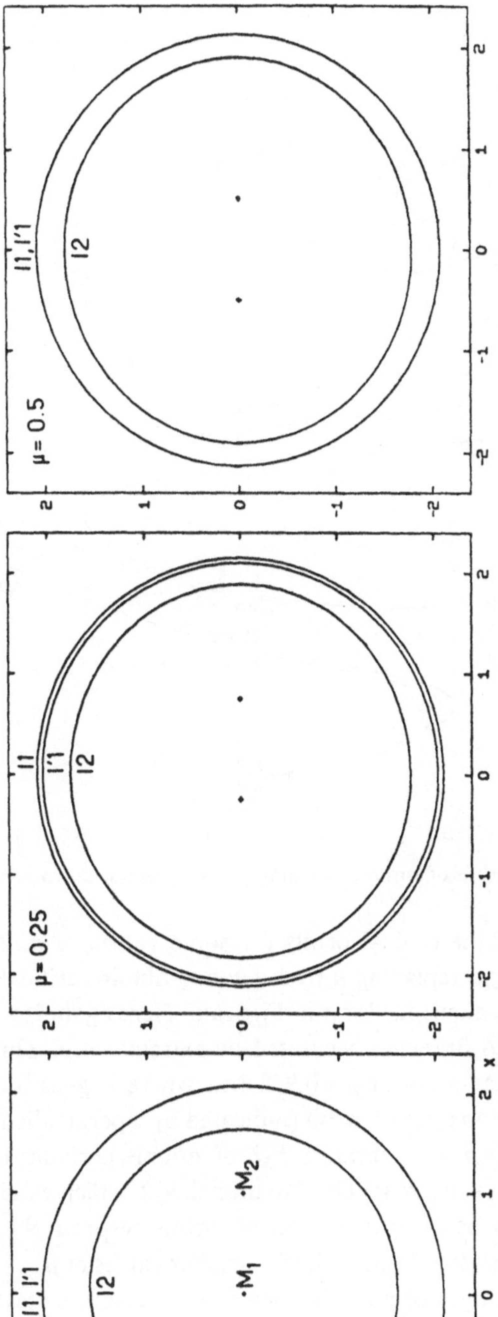

Fig. 5. Critical orbits of family *l* for some values of μ. Periodic orbits of family *l* are unstable inside *l*2 and between *l*1 and *l*′1.

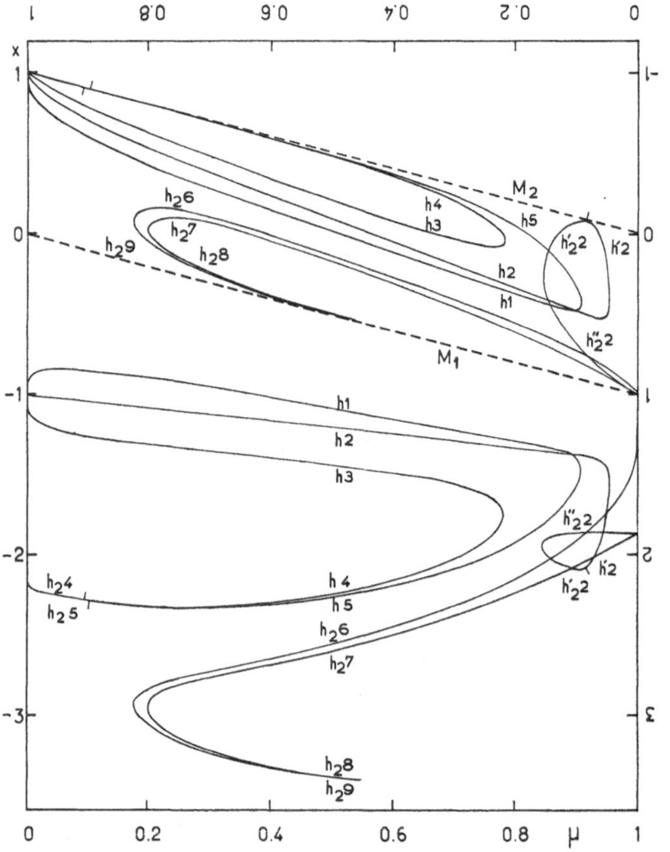

Fig. 6. Chart of critical orbits of family *h*. Turning the figure upside down gives the chart for family *f*.

Figure 7 represents the critical orbits for some values of μ. Again, rotating this figure through 180°, and replacing μ by $1-\mu$, one obtains critical orbits for family *f*.

Family *h*2 describes a curious loop in Figure 6. Here again it will be convenient to give different names to branches separated by extrema in μ. Thus, branch *h*2 starts at $\mu=0$, and has a maximum at $\mu=0.952\,3...$, where it goes into branch $h'2$. This branch passes through an ejection orbit (indicated by a perpendicular bar on Figure 6) at $\mu=0.914...$ and becomes a branch $h_2'2$ of double-periodic orbits; it reaches a minimum at $\mu=0.844...$, where it goes into branch $h_2''2$, which ends at $\mu=1$.

Families *h*4 and *h*5 pass through ejection orbits respectively at $\mu=0.094...$ and $\mu=0.102...$ and become double-periodic for smaller values of μ.

We shall follow family *h* of periodic orbits, for a given μ, starting from the small quasi-circular orbits around M_1. For $0<\mu<0.783...$ (Figure 7, first four frames), the orbits are stable until *h*1, unstable between *h*1 and *h*2, and stable between *h*2 and *h*3; after that they are mostly unstable, except for short intervals of stability such as h_24 (or *h*4) to h_25 (or *h*5), h_26 to h_27, h_28 to h_29, etc. (Hénon, 1965b). At $\mu=0.783...$, the two branches *h*3 and *h*4 meet and disappear. Thus, for $0.783...<\mu<0.844...$ (see Figure 7 for $\mu=0.8$), orbits are stable until *h*1, then in a large interval between *h*2

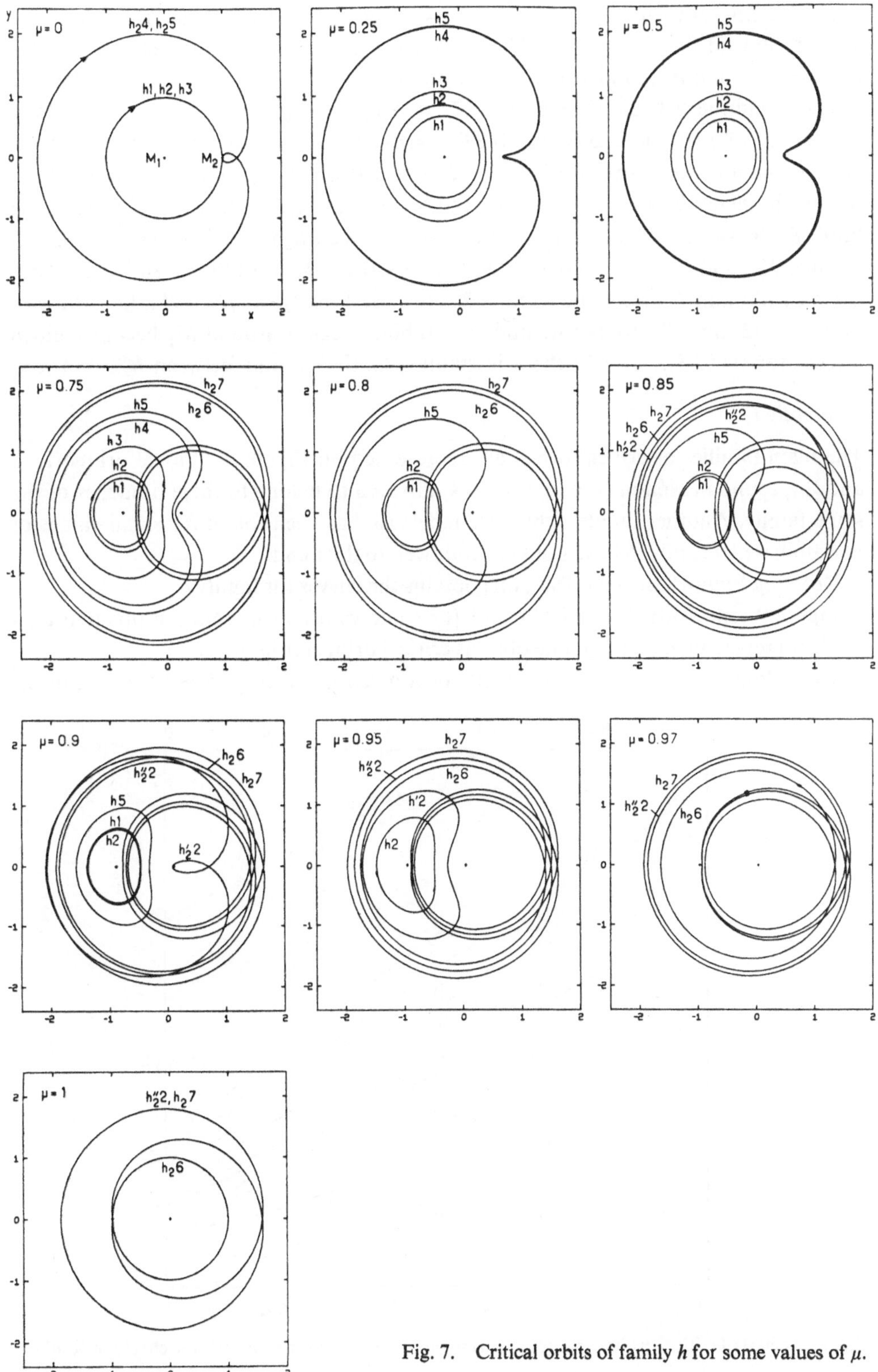

Fig. 7. Critical orbits of family *h* for some values of μ.

and $h5$, then between h_26 and h_27. At $\mu=0.844...$, the two branches $h_2'2$ and $h_2''2$ appear, and for $0.844...<\mu<0.892...$, orbits are stable until $h1$, between $h2$ and $h5$, between $h_2'2$ and $h_2''2$, and between h_26 and h_27. At $\mu=0.892...$, families $h1$ and $h2$ intersect and the critical orbits $h1$ and $h2$ exchange their places in the sequence along the family of periodic orbits; for $0.892...<\mu<0.909...$, orbits are stable until $h2$, between $h1$ and $h5$, between $h_2'2$ and $h_2''2$, and between h_26 and h_27. At $\mu=0.909...$, the two branches $h1$ and $h5$ disappear, and for $0.909...<\mu<0.921...$, orbits are stable until $h2$, between $h_2'2$ (or $h'2$) and $h_2''2$, and between h_26 and h_27. At $\mu=0.921...$, families $h_2''2$ and h_26 intersect; for $0.921...<\mu<0.952\,3...$, orbits are stable until $h2$, between $h'2$ and h_26, and between $h_2''2$ and h_27. Finally, at $\mu=0.952\,3...$, the two branches $h2$ and $h'2$ disappear, and the stability region around M_1 becomes much larger: for $0.952\,3...<\mu<1$, there is stability until h_26 and between $h_2''2$ and h_27.

4. *Family i*

The main families of critical orbits are represented on Figure 8. Critical orbits $i1$, $i2$ and $i14$, found in Hénon (1965b) for $\mu=\frac{1}{2}$, are seen to belong to three branches of the same family. Another family, which we label $i16$, has been found for small values of μ. Families $i2$, $i3$ and $i16$ have not been followed to their end.

Rotating Figure 8 through 180°, one obtains the curves for family g.

Figure 9 represents the critical orbits for some values of μ; rotating this figure by 180° and replacing μ by $1-\mu$, one obtains critical orbits for family g.

We follow family i of periodic orbits for a given μ, starting from the small orbits

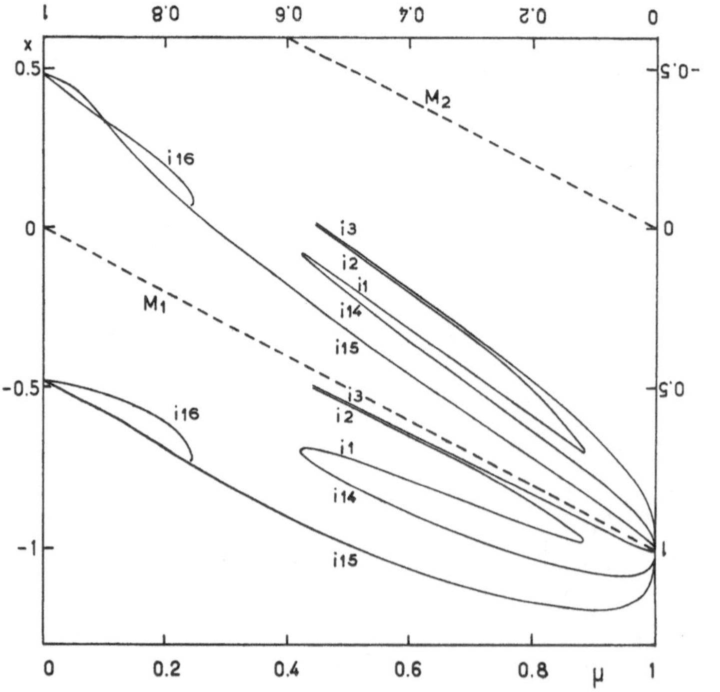

Fig. 8. Chart of critical orbits of family *i*. Turning the figure upside down gives the chart for family *g*

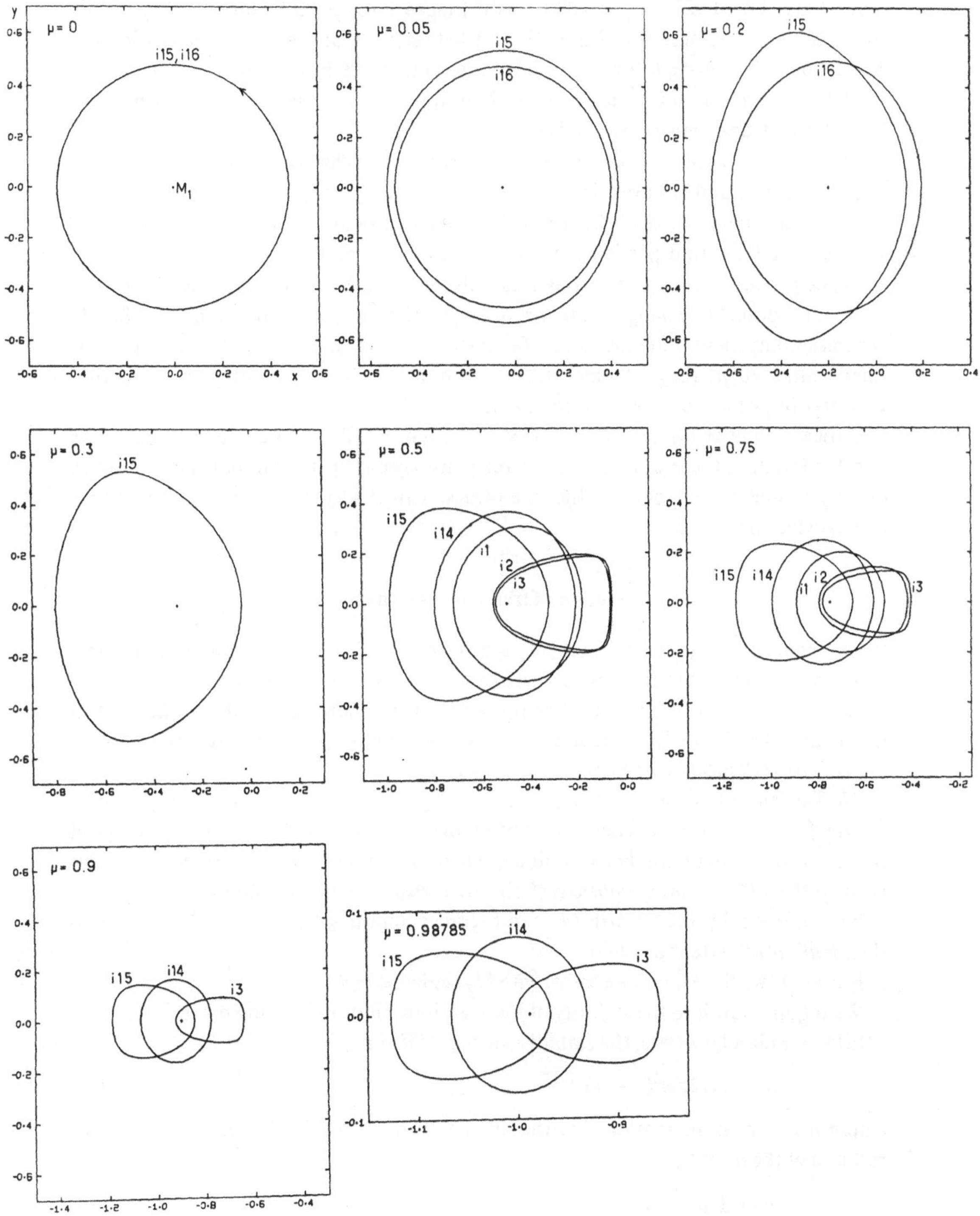

Fig. 9. Critical orbits of family i for some values of μ. The last frame which corresponds to the earth-moon case ($\mu = 0.987\,85$) has a different scale.

around M_1. For $0.423... < \mu < 0.890...$ (see Figure 9 for $\mu = 0.5$ and $\mu = 0.75$), orbits are stable until $i1$, between $i2$ and $i3$, and between $i14$ and $i15$; elsewhere they are mostly unstable, except for very short intervals of stability (Hénon, 1965b). At $\mu = 0.890...$, the two branches $i1$ and $i2$ disappear; for $0.890... < \mu < 1$, orbits are stable until $i3$, and between $i14$ and $i15$.

At $\mu = 0.423...$, the two branches $i1$ and $i14$ meet. This is unexpected, since the critical orbits $i1$ and $i14$ are widely distant along family i of periodic orbits, and separated by a number of other critical orbits. What probably happens is that family i *intersects itself* for that particular value of μ. Four branches of family i meet then at the same point. When μ goes through the value $\mu = 0.423...$, the connections between these four branches change, and the order of the critical orbits along the family becomes completely different. Thus, for $\mu < 0.423...$, orbits are stable until $i15$. For small values of μ, they become stable again after $i16$, presumably until another critical orbit, which has not been computed.

Critical orbits of family i were found to be more difficult to compute than orbits of other families: the trial-and-error search converged only if the initial guessed values of the parameters were close to the true values. This is why the picture presented here is somewhat incomplete.

4. Limiting Orbits for $\mu \to 0$ and $\mu \to 1$

When the mass of the second body M_2 tends towards zero, i.e. $\mu \to 0$, a periodic orbit tends towards some limiting form. Three cases must be distinguished:

(a) The periodic orbit does not approach M_2 at less than some finite distance. Then in the limit the effect of M_2 becomes negligible; the limiting orbit is circular or elliptic in fixed axes (Poincaré's first and second species).

(b) The closest distance of approach to M_2 tends towards zero, but the orbit retains finite dimensions. Then the limiting orbit will be unaffected by M_2 except at one point (or a finite number of points), where it will have an angle in general. This is an *orbit with consecutive collisions* (Poincaré's second species) (Hénon, 1968).

(c) The orbit shrinks towards M_2 and becomes infinitely small. In the limit, we find then *Hill's orbits* (Hénon, 1969).

For $\mu \to 1$, we have the same cases, with M_2 replaced by M_1.

We inquire now into the stability of these various kinds of limiting orbits.

(a1) For circular orbits, the stability index is (Hénon, 1969, Section 6):

$$a = \cos 2\pi\, n/(n-1)\,, \tag{13}$$

where n is the mean motion, i.e. the angular velocity in fixed axes, related to the radius ϱ of the orbit by:

$$n = \pm\, \varrho^{-3/2}\,. \tag{14}$$

Thus, if we write:

$$2n/(n-1) = k, \tag{15}$$

circular orbits are critical when k is an integer, with $a = +1$ for k even and $a = -1$ for k odd, and stable when k is not an integer. Figure 10 shows the relation between k and n.

For μ slightly different from zero, the extrema in $|a|$ along the family of periodic orbits, instead of being exactly equal to 1 as shown by (13), will be in general slightly larger or smaller. In the first case, there will be two critical orbits near to each other, with the same value of a. Thus a critical circular orbit may be the common termination of two families of critical orbits.

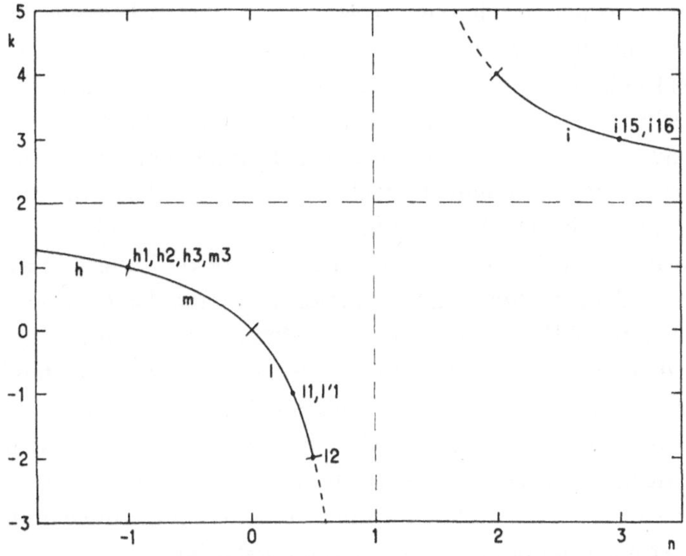

Fig. 10. Circular orbits for $\mu = 0$: relation between the mean motion n and $k = 2n/(n-1)$. Integral values of k correspond to critical orbits.

(a2) An elliptical orbit in fixed axes is periodic in rotating axes if and only if it consists of an integral number of revolutions j on the ellipse, and if during the same time the second body makes another integral number of revolutions i. In other words, the mean motion must be rational: $n = j/i$. The period in the rotating frame is $2\pi i$.

If we perturb slightly such a periodic orbit, starting from $x_0 + \Delta x_0$ and $\dot{x}_0 + \Delta \dot{x}_0$, we obtain an orbit which is still elliptical in fixed axes, and which makes nearly an integral number of revolutions before the next intersection with the x-axis. Therefore we have, to first order:

$$\Delta x_2 = \Delta x_0 , \qquad \Delta \dot{x}_2 = \Delta \dot{x}_0 . \tag{16}$$

This means that the stability index is $a = 1$: all elliptical orbits are critical.

For μ slightly different from zero, a will be slightly different from 1 in general, and the orbit will be either stable or unstable. For particular elliptical orbits it may happen that a remains equal to 1 to first order in μ; such an orbit will be the termination of a family of critical orbits. A more refined treatment is required to find analytically which orbits have this property; such a treatment has just been given by Message (1970).

(b) In general, orbits with consecutive collisions are 'infinitely unstable'; any angular displacement after a collision, however small, results in the next collision being missed, and thus in a completely different orbit after one revolution. The stability index is therefore equal to $\pm\infty$.

On the other hand, it was shown in Hénon (1965b) that an extremum in C in a family of periodic orbits, or an intersection with another family, corresponds to a critical orbit, with $a=\pm 1$. Inspection of the families of orbits with consecutive collisions (Hénon, 1968) shows that they do possess extrema in C. This seems to contradict the previous paragraph. The explanation is that, when one follows a given family, the stability index a jumps occasionally from $+\infty$ to $-\infty$, or vice versa; in so doing, it passes infinitely quickly through the values $a=+1$ and $a=-1$.

For μ slightly different from zero, the behaviour of a will be similar, but it will only reach high values instead of infinite values, and the transition from positive to negative values will take place over a finite interval, so that there will be two critical orbits near to each other, one of them with $a=+1$ and the other with $a=-1$.

Thus we may expect that an orbit with consecutive collisions corresponding to an extremum in C will be the common termination of two families of critical orbits.

(c) The stability of Hill's orbits has been studied in Hénon (1969). Family f for $\mu=0$ (or family h for $\mu=1$) is entirely stable. Family g for $\mu=0$ (or family i for $\mu=1$) has three main critical orbits, named $g1$, $g'2$ and $g'2$ in Hénon (1969) (there are two critical orbits $g'2$, one on each branch of family g'). The stability curve has a non-zero slope in the vicinity of each of these orbits; therefore we may expect that the three critical orbits still exist for μ different from zero, and that the three limiting orbits for $\mu=0$ are the terminations of three families of critical orbits.

Now we consider the fate of the families of periodic orbits and their stability for $\mu\to 0$ and $\mu\to 1$.

For $\mu=0$, family m begins with retrograde circular orbits, with a radius decreasing from ∞ to 1; n decreases from 0 to -1, and k increases from 0 to 1 (Figure 10). The stability index a decreases from 1 to -1: all orbits within this range are stable. The orbit at the end of the range, i.e. the retrograde circular orbit of radius 1, is critical, with $a=-1$; it is the termination of family $m3$ of critical orbits (Figures 2 and 3). At that particular orbit, family m of periodic orbits for $\mu=0$ branches off to family A_0 of orbits with consecutive collisions, at the point $\tau/\pi=\eta/\pi=\frac{1}{2}$, in the direction of τ decreasing (Hénon, 1968, Figures 5 and 3, Tables II and I). These orbits are unstable, with $a=-\infty$ at first. Family A_0 has an extremum in C; the parameters of the corresponding orbit are:

$$\tau/\pi = 0.213\,45\ldots,\quad x_0 = -0.162\,94\ldots,\quad C = -0.399\,13\ldots. \tag{17}$$

The numerical results indicate that this orbit is the termination of families $m2$ and $m1$, which have respectively $a=-1$ and $a=+1$, in agreement with the above predictions. At that orbit, a jumps from $-\infty$ to $+\infty$.

Family l for $\mu=0$ begins with direct circular orbits, with a radius ϱ decreasing from $+\infty$; n increases from 0 and k decreases from 0 (Figure 10). A first critical orbit is met for:

$$k = -1, \quad n = \tfrac{1}{3}, \quad \varrho = 3^{2/3} = 2.080\,08\ldots, \quad C = 3.365\,25\ldots; \qquad (18)$$

it is the termination of families $l1$ and $l'1$, which both have $a = -1$, as predicted. A second critical orbit is met for:

$$k = -2, \quad n = \tfrac{1}{2}, \quad \varrho = 2^{2/3} = 1.587\,40\ldots, \quad C = 3.149\,80\ldots. \qquad (19)$$

At this orbit, family l branches off to the family of elliptic orbits with mean motion $\frac{1}{2}$ (Guillaume, 1969). All these orbits are critical, so that any one of them can be the termination of a family of critical orbits. Extrapolation from the numerical results suggests in fact that family $l2$ terminates in one such elliptical orbit, with an eccentricity approximately equal to 0.036 5. This value was derived independently of, and is in excellent agreement with the value $e = 0.036\,52$ found by Message (1970).

Family h for $\mu = 0$ begins with retrograde circular orbits, with a radius increasing from 0 to 1; n increases from $-\infty$ to -1, and k decreases from 2 to 1 (Figure 10). Thus all the orbits within this range are stable. The retrograde circular orbit with radius 1 is critical; it is the termination of families $h1$, $h2$, $h3$. Family h then branches off to family A_0 of orbits with consecutive collisions, at the point $\tau/\pi = \eta/\pi = \frac{1}{2}$, in the direction of τ increasing; these orbits are unstable, with $a = +\infty$ at first. The family has an extremum in C for:

$$\tau/\pi = 1.997\,35\ldots, \quad x_0 = -2.174\,36\ldots, \quad C = 2.970\,94\ldots, \qquad (20)$$

where a jumps from $+\infty$ to $-\infty$. This orbit is the termination of the two families $h_2 4$ and $h_2 5$.

Family h for $\mu = 1$ begins with Hill's orbits, which are all stable. It continues with elliptic orbits of mean motion $n = 1$, through an ejection orbit (rectilinear in fixed axes), then with retrograde elliptic orbits of mean motion $n = -1$. All these orbits are critical; numerical results indicate, however, that they all become stable for $\mu \neq 1$, so none of them is the termination of a family of critical orbits. Family h then reaches a retrograde circular orbit of radius 1, described once in fixed axes, and therefore twice in rotating axes; this orbit is the termination of family $h_2 6$. At that point, family h branches off to family B_1 of orbits with consecutive collisions, at the point $\tau/\pi = \eta/\pi = 1$, in the direction of τ increasing (Hénon, 1968, Figure 7, Table V). These orbits are unstable, with $a = -\infty$ at first. Family B_1 has an extremum in C for:

$$\tau/\pi = 1.339\,62\ldots, \quad x_0 = -1.872\,12\ldots, \quad C = -1.439\,48\ldots, \qquad (21)$$

where a jumps from $-\infty$ to $+\infty$. This orbit is the termination of families $h_2'' 2$ and $h_2 7$.

Family i for $\mu = 0$ begins with direct circular orbits, with a radius increasing from 0; n decreases from $+\infty$, and k increases from 2 (Figure 10). A first critical orbit is met for:

$$k = 3, \quad n = 3, \quad \varrho = 3^{-2/3} = 0.480\,75\ldots, \quad C = 3.466\,81\ldots \qquad (22)$$

It is the termination of families $i15$ and $i16$. A second critical orbit is met for:

$$k = 4, \quad n = 2, \quad \varrho = 2^{-2/3} = 0.629\,96\ldots, \quad C = 3.174\,80\ldots. \qquad (23)$$

At this point, family i branches off to the family of elliptic orbits with mean motion 2 (Colombo *et al.*, 1968; Broucke, 1968). All these orbits are critical, and one or more of them could be the end point of a family of critical orbits. Further work is required on this point.

Finally, family i for $\mu=1$ begins with Hill's orbits, for which three main critical orbits have been found (Hénon, 1969). As predicted above, the numerical results show that these critical orbits are the end points of families $i3$, $i14$, $i15$. This is particularly obvious from a comparison of the last frame of Figure 9 with Figure 5 of Hénon (1969).

Family g for $\mu=0$ (equivalent to family i for $\mu=1$) consists of two parts, g and g', which intersect at $g1$ (Hénon, 1969, Figure 11). For $\mu\neq0$, the intersection vanishes; instead, one half of g becomes connected with one half of g', and the other half of g with the other half of g'; two new families are thus formed (Broucke, 1968, Figures 33 and 34). One of these new families begins with direct quasi-circular orbits of small dimensions; orbits of this family are stable until $i3$. The other family contains only orbits of finite dimensions, which are stable between $i14$ and $i15$.

5. Comments

(a) In the present paper, stability has been considered only with respect to perturbations in the plane. As already emphasized in Hénon (1965b), an orbit which is stable in the plane is not necessarily stable in three-dimensional space; the present study should therefore be supplemented by a study of the stability with respect to perturbations perpendicular to the plane. This has been undertaken by one of us (M.H.). The first results show that instability in the vertical direction indeed can be present, and examples have been found of orbits which are stable in the plane, but unstable in space.

(b) The prediction that retrograde orbits are more stable than direct ones in general (Jackson, 1913) is amply confirmed by our results, in particular by a comparison of Figures 2 and 4 for orbits surrounding both primaries and of Figures 6 and 8 for 'satellites' of one of the primaries.

(c) The present results also confirm the remarkable stability of retrograde satellites around the lighter body, when the mass ratio is small. This had already been found in Hill's limiting case $\mu\to0$ (Hénon, 1969). The above results for family h, transposed according to the recipe (12), show that *for*:

$$0\leqslant\mu<0.047\,7\ldots, \tag{24}$$

essentially all retrograde periodic orbits around the lighter body M_2 are stable; the interval of stability extends from the orbits of infinitely small dimensions to the orbit of ejection with the heavy body M_1 and farther, being limited only by the double-periodic critical orbit h_26 (Figure 7, last frame). This fact is of some interest since the interval (24) encompasses all instances, natural or artificial, of the restricted problem in the solar system.

(d) For direct satellites, Hill's case represents a good approximation when μ is small. Orbits are stable until *i*3, and between *i*14 and *i*15 (Figure 9, last two frames). The dimensions of these critical orbits is approximately proportional to $\mu^{\frac{1}{3}}$.

(e) It is frequently assumed that triple star systems can be stable only if they have a 'hierarchical' configuration, with two stars on a small elliptical orbit, this double system being itself engaged on a large elliptical orbit with the third star. The present results show that, at least in the case of one mass much smaller than the others, many stable configurations exist where the three distances are comparable in magnitude.

(f) Similarly, stable planetary orbits can exist in double star systems not only in the extreme cases of small radii (around one star) or large radii (around both stars), but also with dimensions comparable to the separation of the two stars.

6. Tables

In the following tables, each line corresponds to one critical orbit. x_0 and x_1 are the abscissas of the two intersections of the orbit with the x-axis. x_0 always corresponds to an intersection in the direction of y increasing; i.e., $\dot{y}_0>0$. At x_1, one may have $\dot{y}_1<0$ or $\dot{y}_1>0$, depending on whether the orbit is simple-periodic or double-periodic. C is the Jacobi constant, given by:

$$C = x^2 + y^2 + \frac{2(1-\mu)}{r_1} + \frac{2\mu}{r_2} - \dot{x}^2 - \dot{y}^2. \tag{25}$$

If μ, x_0, x_1 are changed into $1-\mu$, $-x_0$, $-x_1$, one obtains other critical orbits, belonging to another family or to the same family, as given by (12). In this transformation, types 1 to 4 are conserved, types 5 and 6 are exchanged.

The particular value $\mu=0.012\,15$ corresponds to the earth-moon case, as well as the complementary value $\mu=0.987\,85$. (The exact value used in the computation was: $\mu=\frac{1}{82.30}$, as recommended by the IAU (1966).) For $\mu=\frac{1}{2}$, the orbits have been recomputed and the values given here are more accurate than those given in Hénon (1965b).

Family *m*1

Type 1, $a=1$

μ	x_0	x_1	C
0.	−0.162 94	1.	−0.399 13
0.012 15	−0.175 09	0.990 85	−0.412 92
0.05	−0.212 83	0.961 98	−0.450 65
0.1	−0.262 19	0.923 57	−0.491 67
0.15	−0.310 73	0.885 19	−0.524 85
0.2	−0.358 24	0.846 99	−0.551 58
0.25	−0.404 57	0.808 96	−0.572 80
0.3	−0.449 61	0.771 03	−0.589 23
0.35	−0.493 31	0.733 07	−0.601 42
0.4	−0.535 71	0.694 89	−0.609 83
0.45	−0.576 89	0.656 28	−0.614 75
0.5	−0.617 02	0.617 02	−0.616 36

Family $m2$-$m3$
Type 5, $a = -1$

μ	x_0	x_1	C
0.	−0.162 94	1.	−0.399 13
0.012 15	−0.178 14	0.990 90	−0.413 04
0.05	−0.225 81	0.962 78	−0.452 60
0.1	−0.290 19	0.927 03	−0.499 76
0.15	−0.357 17	0.893 97	−0.544 76
0.2	−0.428 44	0.865 31	−0.592 16
0.25	−0.507 64	0.844 41	−0.649 97
0.3	−0.607 23	0.841 03	−0.740 80
0.32	−0.669 45	0.855 73	−0.812 71
0.32	−0.776 88	0.916 98	−0.966 96
0.3	−0.820 94	0.955 26	−1.038 78
0.25	−0.874 66	1.010 47	−1.127 55
0.2	−0.907 48	1.046 36	−1.176 72
0.15	−0.932 07	1.070 98	−1.204 12
0.1	−0.952 39	1.085 06	−1.211 22
0.05	−0.970 95	1.083 92	−1.188 46
0.02	−0.983 13	1.066 44	−1.141 25
0.012 15	−0.987 18	1.055 88	−1.116 88
0.01	−0.988 38	1.051 89	−1.108 01
0.005	−0.991 87	1.039 23	−1.080 59
0.001	−0.996 37	1.019 14	−1.038 71
0.	−1.	1.	−1.

Family $l1$
Type 5, $a = -1$

μ	x_0	x_1	C
0.	−2.080 08	2.080 08	3.365 25
0.05	−2.098 31	2.113 19	3.386 74
0.1	−2.109 87	2.132 14	3.401 26
0.15	−2.117 83	2.143 23	3.411 60
0.2	−2.123 57	2.149 28	3.419 03
0.25	−2.127 75	2.151 78	3.424 27
0.3	−2.130 72	2.151 59	3.427 73
0.35	−2.132 61	2.149 24	3.429 68
0.4	−2.133 44	2.145 03	3.430 27
0.45	−2.133 15	2.139 12	3.429 57
0.5	−2.131 86	2.131 86	3.427 72

Family $l'1$
Type 6, $a = -1$

μ	x_0	x_1	C
0.	−2 080 08	2.080 08	3.365 25
0.05	−2.049 66	2.067 55	3.364 92
0.1	−2.042 30	2.070 52	3.371 09
0.15	−2.048 55	2.080 47	3.380 40
0.2	−2.060 56	2.091 93	3.390 28
0.25	−2.074 36	2.102 64	3.399 53
0.3	−2.088 11	2.111 81	3.407 69
0.35	−2.101 01	2.119 25	3.414 61
0.4	−2.112 72	2.125 02	3.420 25
0.45	−2.123 10	2.129 25	3.424 65
0.5	−2.131 86	2.131 86	3.427 72

Family *l*2

Type 4, $a = 1$

μ	x_0	x_1	C
0.	−1.529 43	1.645 37	3.148 12
0.000 1	−1.529 82	1.645 73	3.148 27
0.001	−1.532 69	1.649 37	3.149 58
0.012 15	−1.570 29	1.682 05	3.164 35
0.05	−1.653 22	1.754 30	3.198 98
0.1	−1.721 91	1.810 58	3.228 86
0.15	−1.770 13	1.847 06	3.250 07
0.2	−1.806 73	1.872 26	3.266 01
0.25	−1.835 47	1.889 82	3.278 19
0.3	−1.858 32	1.901 65	3.287 44
0.35	−1.876 43	1.908 84	3.294 26
0.4	−1.890 49	1.912 05	3.298 94
0.45	−1.900 92	1.911 69	3.301 69
0.5	−1.907 96	1.907 96	3.302 59

Family *h*1-*h*5

Type 6, $a = -1$

μ	x_0	x_1	C
0.	−1.	1.	−1.
0.000 1	−0.966 74	0.965 17	−0.928 85
0.001	−0.932 66	0.924 81	−0.842 87
0.012 15	−0.871 26	0.822 23	−0.612 16
0.05	−0.843 80	0.695 30	−0.319 08
0.1	−0.851 45	0.588 77	−0.076 57
0.2	−0.897 37	0.422 45	0.294 28
0.3	−0.956 83	0.278 03	0.614 44
0.4	−1.021 63	0.142 84	0.917 70
0.5	−1.088 74	0.012 32	1.216 89
0.6	−1.156 44	−0.115 92	1.518 75
0.7	−1.223 46	−0.243 43	1.827 56
0.8	−1.289 67	−0.370 51	2.145 92
0.85	−1.325 33	−0.431 53	2.307 62
0.88	−1.353 60	−0.462 31	2.401 78
0.903 19	−1.400 00	−0.465 88	2.459 70
0.907 49	−1.500 00	−0.393 28	2.412 00
0.896 36	−1.600 00	−0.297 22	2.308 36
0.88	−1.689 93	−0.205 33	2.199 27
0.85	−1.803 14	−0.085 47	2.052 93
0.8	−1.927 76	0.051 24	1.900 69
0.7	−2.083 33	0.232 25	1.787 21
0.6	−2.186 64	0.367 00	1.816 97
0.5	−2.260 86	0.483 78	1.933 55
0.4	−2.310 49	0.592 26	2.109 02
0.3	−2.333 48	0.696 62	2.324 13
0.2	−2.323 68	0.798 89	2.558 45
0.15	−2.305 16	0.849 64	2.671 92
0.05	−2.254 83	0.948 57	2.845 39
0.02	−2.237 31	0.973 84	2.884 70
0.	−2.174 45	1.	2.970 99

Family $h2$
Type 5, $a = -1$

μ	x_0	x_1	C
0.	−1.	1.	−1.
0.001	−1.003 74	0.978 40	−0.956 25
0.012 15	−1.013 65	0.917 13	−0.828 07
0.05	−1.030 82	0.810 33	−0.596 94
0.1	−1.049 25	0.706 51	−0.367 56
0.2	−1.085 74	0.533 06	0.022 21
0.3	−1.124 81	0.377 74	0.377 84
0.4	−1.166 61	0.230 88	0.720 57
0.5	−1.210 36	0.088 48	1.059 41
0.6	−1.254 87	−0.051 91	1.399 65
0.7	−1.298 44	−0.192 37	1.745 21
0.8	−1.338 48	−0.335 28	2.099 70
0.9	−1.374 22	−0.480 81	2.465 27
0.93	−1.394 35	−0.515 53	2.570 73
0.95	−1.471 63	−0.480 45	2.595 80
0.952 12	−1.600 00	−0.365 50	2.489 10
0.947 20	−1.800 00	−0.171 84	2.193 96
0.932 72	−2.000 00	0.021 08	1.618 88
0.92	−2.062 83	0.076 41	1.234 26
0.9	−2.086 65	0.083 78	0.757 54
0.875	−2.060 50	0.023 18	0.278 53
0.844 71	−1.964 72	−0.200 00	−0.376 50
0.851 05	−1.906 43	−0.400 00	−0.721 24
0.9	−1.860 91	−0.688 63	−1.104 41
0.95	−1.857 93	−0.864 97	−1.302 51
0.98	−1.864 70	−0.949 89	−1.390 14
1.	−1.872 12	−1.	−1.439 48

Family $h3$-$h4$
Type 1, a = 1

μ	x_0	x_1	C
0.	−1.	1.	−1.
0.005	−1.119 41	0.977 65	−0.938 79
0.012 15	−1.153 59	0.957 11	−0.885 23
0.05	−1.221 62	0.874 75	−0.679 31
0.1	−1.261 65	0.784 61	−0.460 76
0.2	−1.311 76	0.625 06	−0.078 51
0.3	−1.353 53	0.478 43	0.274 86
0.4	−1.395 60	0.339 22	0.616 97
0.5	−1.440 99	0.205 65	0.955 55
0.6	−1.492 65	0.078 15	1.294 75
0.7	−1.561 37	−0.036 55	1.635 66
0.75	−1.619 30	−0.077 45	1.804 27
0.78	−1.706 24	−0.067 95	1.900 29
0.779 47	−1.800 00	−0.010 82	1.900 07
0.757 13	−1.900 00	0.073 43	1.857 33
0.7	−2.028 11	0.200 64	1.803 66
0.6	−2.160 40	0.354 25	1.824 67
0.5	−2.246 33	0.477 90	1.937 60
0.4	−2.301 91	0.589 44	2.111 25
0.3	−2.328 23	0.695 29	2.325 40
0.2	−2.320 61	0.798 36	2.559 10
0.1	−2.278 07	0.899 99	2.771 59
0.05	−2.253 85	0.948 82	2.845 50
0.	−2.174 45	1.	2.970 99

Family $h_2 6$-$h_2 9$
Type 6, $a = -1$

μ	x_0	x_1	C
1.	−1.	−1.	−1.
0.999 75	−1.209 42	−0.994 38	−1.179 20
0.999	−1.283 84	−0.986 74	−1.225 56
0.987 85	−1.512 54	−0.934 70	−1.300 99
0.96	−1.703 89	−0.855 61	−1.278 69
0.9	−1.921 56	−0.728 78	−1.147 31
0.8	−2.150 84	−0.556 70	−0.867 47
0.7	−2.318 84	−0.405 11	−0.547 58
0.6	−2.453 43	−0.264 84	−0.196 45
0.5	−2.563 96	−0.132 80	0.182 77
0.4	−2.654 76	−0.008 88	0.588 26
0.3	−2.730 80	0.101 75	1.017 23
0.25	−2.769 01	0.144 87	1.237 27
0.2	−2.827 41	0.158 22	1.450 91
0.180 07	−2.900 00	0.113 46	1.518 10
0.225	−3.117 21	−0.110 37	1.374 54
0.3	−3.233 11	−0.250 05	1.317 70
0.4	−3.328 40	−0.382 32	1.380 68
0.5	−3.391 34	−0.494 54	1.535 49
...			

Family $h_2 7$-$h_2 8$
Type 1, $a = 1$

μ	x_0	x_1	C
1.	−1.872 12	−1.	−1.439 48
0.987 85	−1.896 89	−0.971 33	−1.410 94
0.95	−1.976 60	−0.886 54	−1.316 00
0.9	−2.072 41	−0.784 84	−1.180 90
0.8	−2.238 29	−0.605 72	−0.884 76
0.7	−2.378 38	−0.447 01	−0.558 23
0.6	−2.498 13	−0.301 47	−0.203 70
0.5	−2.600 39	−0.166 09	0.177 43
0.4	−2.687 51	−0.041 11	0.583 92
0.3	−2.765 97	0.065 96	1.013 03
0.25	−2.812 26	0.101 34	1.232 37
0.21	−2.884 29	0.090 72	1.402 42
0.205 44	−3.000 00	−0.001 90	1.422 91
0.25	−3.138 82	−0.147 87	1.344 21
0.3	−3.218 66	−0.241 66	1.319 50
0.4	−3.321 50	−0.379 38	1.381 49
0.5	−3.387 47	−0.493 43	1.535 89
...			

Family $i14$-$i1$-$i2$
Type 1, $a = 1$

μ	x_0	x_1	C
1.	−1.	−1.	3.
0.999	−0.974 00	−1.031 80	3.040 43
0.987 85	−0.930 59	−1.066 54	3.182 67
0.95	−0.855 91	−1.081 16	3.382 19
0.9	−0.777 34	−1.070 28	3.509 64
0.8	−0.634 40	−1.023 48	3.614 69
0.7	−0.495 90	−0.962 08	3.631 27

(Continued)

Family *i*14-*i*1-*i*2
Type 1, $a=1$

μ	x_0	x_1	C
0.6	−0.356 02	−0.889 44	3.605 95
0.5	−0.209 93	−0.800 96	3.566 41
0.45	−0.131 42	−0.744 54	3.556 95
0.425	−0.087 21	−0.704 01	3.576 85
0.425	−0.082 34	−0.690 44	3.608 68
0.45	−0.113 44	−0.694 19	3.671 70
0.5	−0.181 36	−0.721 45	3.738 96
0.6	−0.317 23	−0.787 54	3.808 42
0.7	−0.451 98	−0.857 84	3.816 37
0.8	−0.587 57	−0.927 75	3.755 19
0.85	−0.656 31	−0.960 38	3.688 84
0.88	−0.696 46	−0.976 73	3.632 54
0.88	−0.677 80	−0.959 94	3.632 81
0.85	−0.611 46	−0.919 96	3.691 55
0.8	−0.503 72	−0.851 43	3.766 68
0.7	−0.343 89	−0.746 93	3.856 62
0.6	−0.201 43	−0.650 49	3.885 05
0.5	−0.064 45	−0.556 86	3.858 93
...			

Family *i*3
Type 6, $a=-1$

μ	x_0	x_1	C
1.	−1.	−1.	3.
0.999	−0.941 10	−1.005 76	3.039 33
0.987 85	−0.857 21	−1.001 60	3.184 51
0.96	−0.768 68	−0.978 71	3.363 04
0.9	−0.643 93	−0.923 67	3.575 68
0.8	−0.479 82	−0.829 11	3.762 93
0.7	−0.333 34	−0.734 16	3.853 81
0.6	−0.194 35	−0.639 89	3.882 67
0.5	−0.059 24	−0.547 13	3.856 85
...			

Family *i*15
Type 5, $a=-1$

μ	x_0	x_1	C
0.	0.480 75	−0.480 75	3.466 81
0.012 15	0.469 55	−0.491 75	3.448 84
0.05	0.434 21	−0.530 01	3.363 48
0.1	0.333 85	−0.576 52	3.247 09
0.14	0.245 66	−0.618 19	3.221 86
0.2	0.130 94	−0.689 07	3.230 20
0.3	−0.034 48	−0.803 47	3.292 38
0.4	−0.184 01	−0.902 59	3.370 64
0.5	−0.324 56	−0.987 34	3.445 85
0.6	−0.459 13	−1.059 83	3.507 90
0.7	−0.589 56	−1.120 48	3.546 26
0.8	−0.717 46	−1.166 76	3.543 72
0.9	−0.845 29	−1.188 21	3.460 28
0.95	−0.911 38	−1.176 40	3.349 80
0.987 85	−0.966 94	−1.126 85	3.170 01
0.998	−0.987 53	−1.073 17	3.058 64
1.	−1.	−1.	3.

Family *i*16

Type 6, $a = -1$

μ	x_0	x_1	C
0.	0.480 75	−0.480 75	3.466 81
0.012 15	0.460 95	−0.483 25	3.474 84
0.05	0.404 09	−0.497 38	3.487 17
0.1	0.334 64	−0.524 57	3.483 62
0.15	0.267 24	−0.559 01	3.460 28
0.2	0.194 77	−0.607 15	3.408 08
0.23	0.137 26	−0.658 04	3.353 50
0.245	0.093 34	−0.703 42	3.311 18
0.246 74	0.080 00	−0.718 32	3.294 03
0.242 49	0.070 00	−0.730 46	3.267 38
...			

References

Arnold, V.: 1963, *Russian Math. Surveys* **18**, 9, 85.
Broucke, R. A.: 1968, N.A.S.A.-J.P.L. Technical Report 32-1168.
Bulirsch, R. and Stoer, J.: 1966, *Numerische Math.* **8**, 1.
Colombo, G., Franklin, F. A., and Munford, C. M.: 1968, *Astron. J.* **73**, 111.
Deprit, A. and Henrard, J.: 1969, *Astron. J.* **74**, 308.
Deprit, A. and Price, J. F.: 1965, *Astron. J.* **70**, 836.
Guillaume, P.: 1969, *Astron. Astrophys.* **3**, 57.
Hénon, M.: 1965a, *Ann. Astrophys.* **28**, 499.
Hénon, M.: 1965b, *Ann. Astrophys.* **28**, 992.
Hénon, M.: 1966a, *Bull. Astron.* (3) **1**, fasc. 1, 57.
Hénon, M.: 1966b, *Bull. Astron.* (3) **1**, fasc. 2, 49.
Hénon, M.: 1968, *Bull. Astron.* (3) **3**, 377.
Hénon, M.: 1969, *Astron. Astrophys.* **1**, 223.
Hénon, M.: 1970, *Astron. Astrophys.*, in press.
IAU: 1966, *Bull. Astron.* (3) **1**, 16.
Jackson, J.: 1913, *Monthly Notices Roy. Astron. Soc.* **74**, 62.
Message, P. J.: 1970, this volume, p.19.
Moser, J.: 1962, *Nachr. Akad. Wiss. Göttingen*, Math.-Phys. Kl., 1.
Shearing, G.: 1960, Thesis, Manchester University.
Strömgren, E.: 1935, *Bull. Astron.* **9**, 87.

Discussion

P. J. Message: (1) Do you confirm that there are no simple bifurcations on your family l for $0<n<\frac{1}{2}$, but that they become increasingly frequent as n increases from $\frac{1}{2}$ to 1, as the commensurabilities are met?

(2) Is your theory, like the one I presented, confined to critical orbits corresponding to simple bifurcations only, and so not designed to find the location of multiple bifurcations, and corresponding isolated unstable members of the families?

M. Hénon: (1) Yes, I confirm that there are no simple bifurcations for $\mu=0$ and $0<n<\frac{1}{2}$. But at $n=\frac{1}{2}$, family l (as defined by continuity for $\mu\to0$) bifurcates from circular to elliptic orbits. Since it is this family which was followed, my present results do not include circular orbits with $\frac{1}{2}<n<1$.

(2) The computations have been so far confined to simple bifurcations ($a=+1$) and double bifurcations ($a=-1$). However, the program could be used to follow a family corresponding to any given multiple bifurcation, since such a family is characterized by a constant stability index $a=\cos 2\pi p/q$.

W. H. Jefferys: (1) The family of retrograde orbits around the smaller mass for $\mu>0.952\,3$, which you found to be stable even through a collision with the larger mass, does not agree with my calculations for the earth-moon case, for I find the retrograde orbit to become unstable before then. (Just before $C=2.9$. Figure 8 in my paper for this Symposium.) Could you suggest a reason for this?

(2) Referring to my earlier question, it may be that we are here facing a phenomenon similar to that described by Moser in a conference reported in the *Astronomical Journal* in 1958. That is, there may be a resonance between the rotation around the moon and the motion of the earth which is of low order. This would produce a single unstable orbit along the sequence of stable ones. If this hypothesis is correct, I should be able to find these stable orbits again for smaller Jacobi constants than have been tried.

M. Hénon: (1) The explanation may be that the set of invariant curves surrounding the stable periodic orbit is quite small in some cases, so that you do not find it if you do not specifically look for it, and if you do not know where it is.

(2) I quite agree that this can happen; in fact, I am sure that it happens for family h and for values of μ close to 1, because I observed it in Hill's limiting case (see Hénon, 1969, 1970). The stability index a crosses twice the value $a=-\frac{1}{2}$, which corresponds to Moser's resonance 3:1. Thus, there must exist two isolated unstable orbits inside the stability interval which I described. I did not mention this kind of unstable orbits because they are isolated and therefore of secondary importance.

I also agree that you would very probably find sets of invariant curves again for smaller Jacobi constants.

A. Deprit: Your conclusions at $\mu=0.952\,3$ might not be justified if, while the orbits of indifferent stability that you followed came to coincidence, other pairs of orbits of indifferent stability appeared behind the vanishing pair. Also I am inclined to believe that the families do not evolve continuously with the mass ratio. For the families of librations emanating from L_4, I found enough instances of singular mass ratios where the evolution broke down.

M. Hénon: This is indeed a theoretical possibility. In numerical computations of this kind, one can never guarantee that all families have been discovered. There is, however, one independent piece of evidence which supports my conclusions: for $\mu=0.987\,85...$, i.e., the earth-moon case, Broucke (1968) computed family h and found that the orbits are stable up to and beyond the ejection orbit.

J. M. A. Danby: Do you have any theory, say à la Poincaré, that guarantees the possibility of analytical continuation of the type that you discuss?

M. Hénon: No.

NEW PERIODIC ORBITS IN THE GENERAL PROBLEM OF THREE BODIES

E. MYLES STANDISH, JR.
Yale University, New Haven, Conn., U.S.A.

1. Introduction

Two years ago, Szebehely and Peters (1967) published a periodic orbit in the general problem of three bodies. Although the orbit is only two-dimensional, the word general is used here because the three masses of the system are all of the same order and the three distances separating the masses are also all of the same order.

The three masses, having the ratios, 3:4:5, all start with zero initial velocity. The subsequent motion of the system leads to a configuration in which the lightest mass comes to rest at the same time when the other two are undergoing a collision. At this point, the system reverses its motion and comes back to its initial configuration, thereby completing a full period.

This paper investigates the question as to whether this orbit is unique or whether it is a member of a family of similar periodic orbits. During the course of the study, other new periodic orbits are found and it is indicated that many more periodic orbits exist in the general problem of three bodies.

2. Points of Symmetry

The orbit of Szebehely and Peters is interesting because of the way in which it is periodic. It contains two configurations in phase space which may be called 'points of symmetry'.

An n-body system contains a 'point of symmetry' at time T if

$$\boldsymbol{\xi}(T + t) = \boldsymbol{\xi}(T - t)$$

for all time t, where $\boldsymbol{\xi}$ is a $3n$-dimensional vector representing the $3n$ position coordinates of the system. It also follows that at such a point, T,

$$\boldsymbol{\eta}(T + t) = -\boldsymbol{\eta}(T - t),$$

where $\boldsymbol{\eta}$ represents the $3n$ velocity components of the system.

It may be shown that if a system contains two points of symmetry, T_1 and T_2, then the system is periodic, with the period, P, given by

$$P = 2(T_2 - T_1),$$

and, therefore,

$$\boldsymbol{\xi}(t + P) = \boldsymbol{\xi}(t)$$

G. E. O. Giacaglia (ed.), Periodic Orbits, Stability and Resonances, 375–381. *All Rights Reserved.*

and

$$\eta(t+P)=\eta(t).$$

The two points of symmetry in the orbit of Szebehely and Peters occur at

(1) $t=T_1=0$, where $|\mathbf{v}_1|=|\mathbf{v}_2|=|\mathbf{v}_3|=0$, and

(2) $t=T_2\cong 15.8$, where $|\mathbf{v}_1|=0$ and $|\mathbf{r}_2|=|\mathbf{r}_3|$.

Arriving at each point of symmetry, the system is seen to instantaneously reverse its motion and thereby retrace its previous motion. Upon two reversals, the system has completed one period.

3. Description of a Family of Orbits

The question arises as to whether the orbit of Szebehely and Peters is a member of a family of periodic orbits or not. Perhaps the best way of visualizing the question is to examine the case of a three body system which is initially at rest and which has the property that $|\mathbf{v}_1|=0$ at some later time in the motion. Using the integral of angular momentum ($|\mathbf{L}|=0$), it may be shown that when $|\mathbf{v}_1|=0$, the following relation holds:

$$\dot{X}_2/\dot{Y}_2=\dot{X}_3/\dot{Y}_3=(X_2-X_3)/(Y_2-Y_3).$$

Therefore, when $|\mathbf{v}_1|=0$, the masses m_2 and m_3 are in one of the following four configurations:

(A) they are apart and approaching directly toward each other,

(B) they are apart and receding directly away from each other,

(C) they are apart and at rest, or

(D) they are undergoing a collision.

It is seen that both conditions (C) and (D) are points of symmetry. One is led, therefore, to create a family of orbits in which the three bodies start at rest and which have the property that $|\mathbf{v}_1|=0$ at some later time. If a member of the family satisfies either condition (C) or condition (D) at this later time, then it will be periodic.

A section of such a family has been found in this study. The masses are chosen in the ratios 3:4:5 and are scaled so that $\sum_{i=1}^{3} m_i=1$. The initial configuration is normalized so that m_3 is at the origin and m_2 is at the position $(-1, 0)$ on the x-axis. The initial conditions for each member of the family may then be completely specified by the two parameters, X_1 and Y_1. The family may be represented by a single curve in the $X-Y$ plane containing the set of pairs, (X_1, Y_1), for all members of the family, as shown in Figure 1. The numbers along the curve indicate the times for which the condition, $|\mathbf{v}_1|=0$, is fulfilled. For the sake of reference, the point labelled B is at the apex of a 3-4-5 right triangle formed with m_2 and m_3. The point labelled S–P represents the initial conditions for the orbit of Szebehely and Peters.

The orbits of the family were found by integrating the equations of motion numerically, and by differentially correcting the initial position of m_1 until the desired requirement, $|\mathbf{v}_1|=0$, was fulfilled at some later time. The equations of motion were regularized using the method of Levi-Civita as adopted by Peters (1968).

Figures 2–6 show some of the orbits of the family. The three masses are denoted by the numerals 3, 4 and 5, with the subscripts O indicating the initial positions.

It is evident from these orbits that certain segments of the curve in Figure 1 correspond to the final condition (A) from above, while other segments correspond to the condition (B). For example, the orbit shown in Figure 2 leads to a final configuration

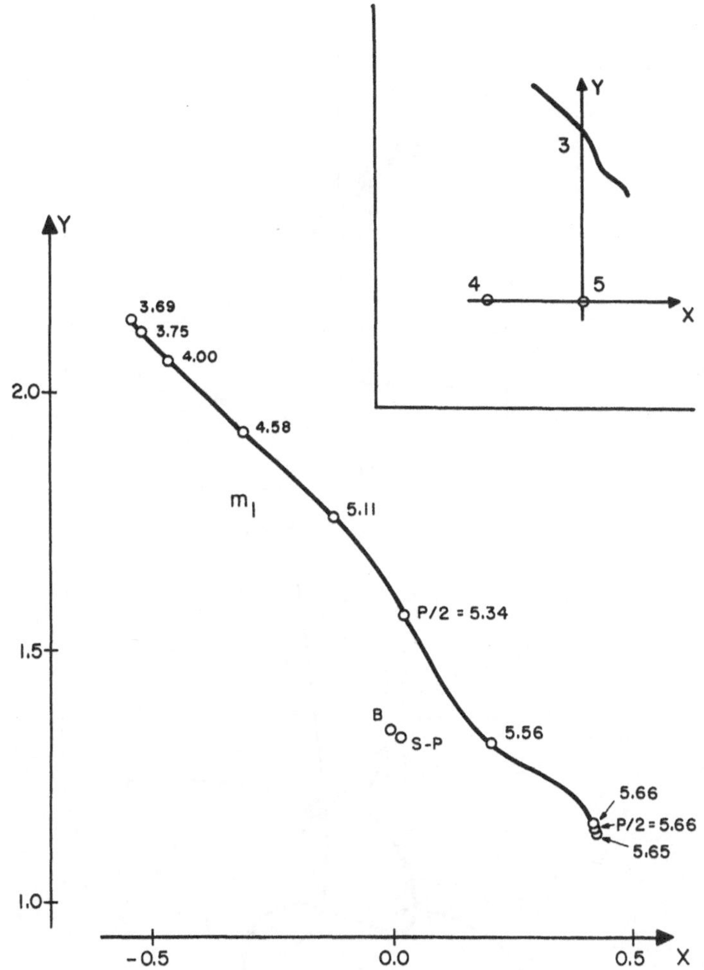

Fig. 1.

at time $t=4.00$ where the masses m_2 and m_3 are apart and are receding directly away from each other. This is an example of condition (B). The orbits in Figures 3 and 5 are also examples of condition (B). However, there is a segment of the curve in Figure 2 which contains examples of condition (A). Figures 4 and 6 are such examples.

Because of the continuous nature of the curve in Figure 1, there must be examples of condition (C) or (D) between the different segments illustrated above.

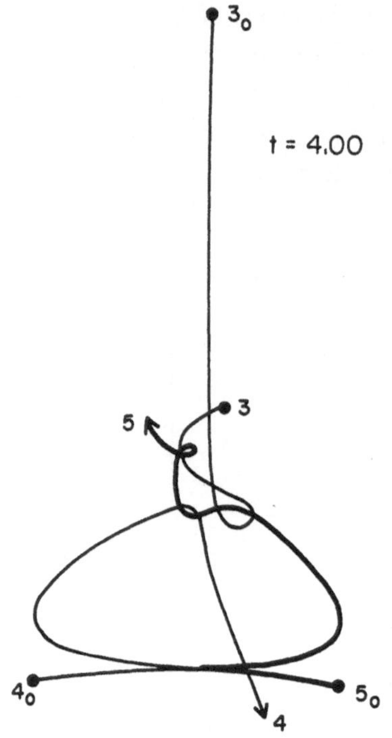

Fig. 2.

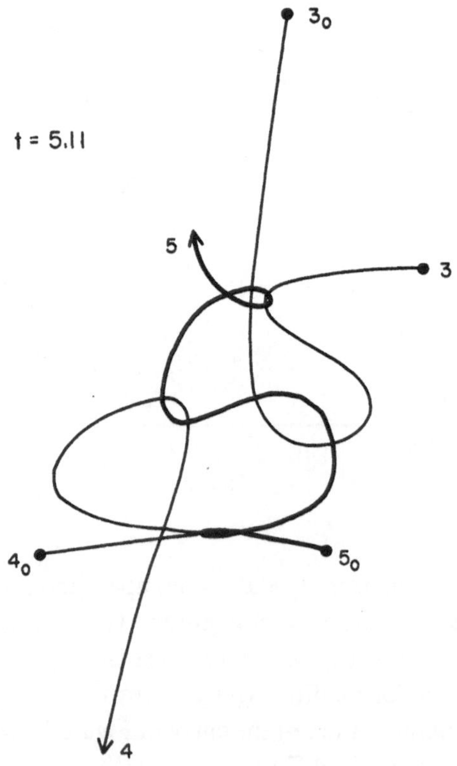

Fig. 3.

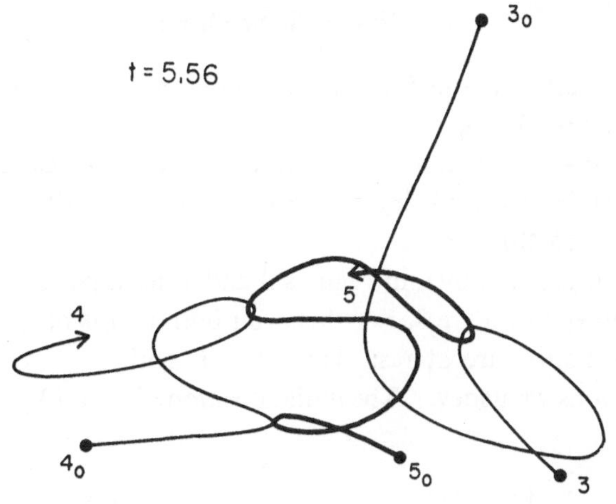

Fig. 4.

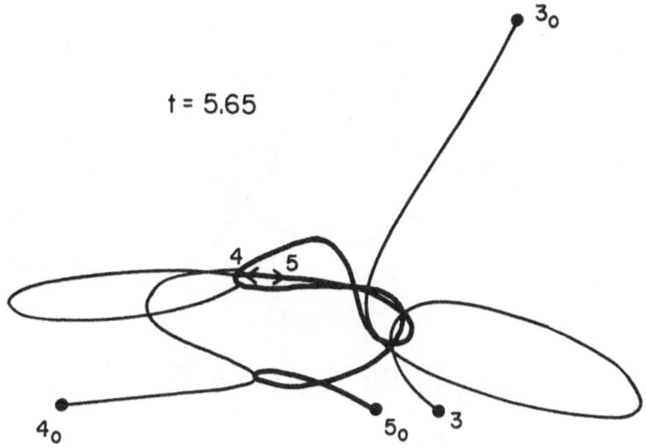

Fig. 5.

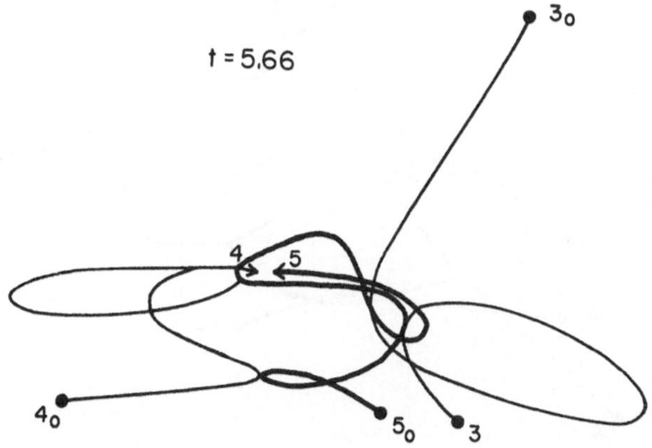

Fig. 6.

4. New Periodic Orbits

Between the two orbits given in Figures 5 and 6 lies a periodic orbit representing the condition (D) (Figure 7). The orbit is similar in nature to the orbit of Szebehely and Peters. The three masses all start at rest and the system evolves to a second point of symmetry in which $|\mathbf{v}_1|=0$ while $|\mathbf{r}_2|=|\mathbf{r}_3|=0$. The initial position of m_1 is $(X_1, Y_1)=$ $=(0.421\,433\,25, 1.185\,401\,83)$.

Between the two orbits shown in Figures 3 and 4 lies a periodic orbit representing condition (C), shown in Figure 8. For this orbit both points of symmetry are similar, because all three bodies are at rest. This orbit is entirely new, therefore, since it contains no collisions whatsoever. The initial position of m_1 is $(X_1, Y_1)=(0.028\,961\,29,$ $1.563\,064\,15)$.

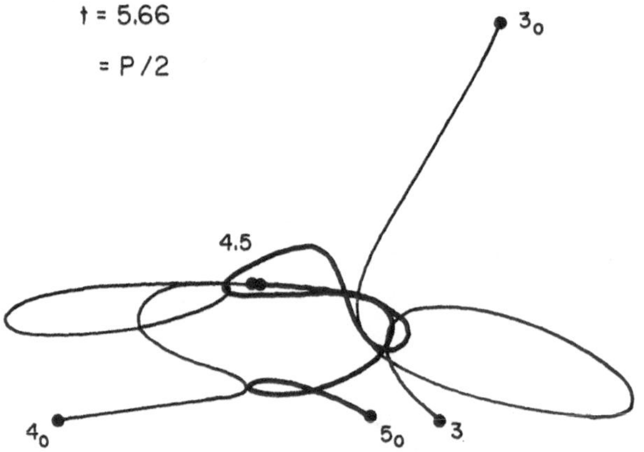

Fig. 7.

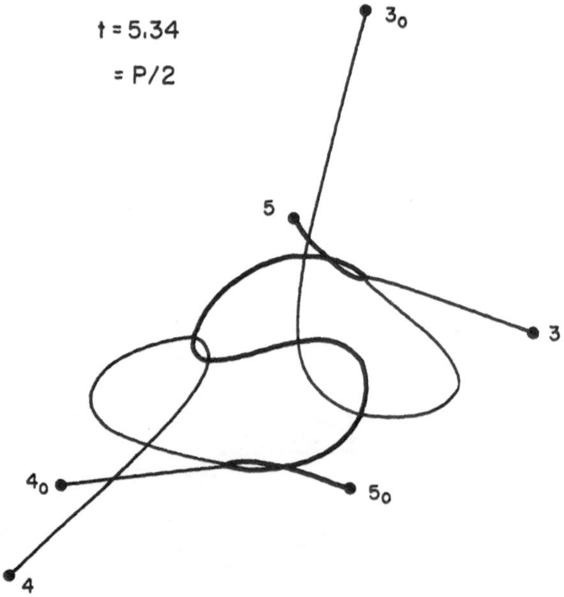

Fig. 8.

5. Discussion

The family has not been extended to its natural ends. At present, one may only conjecture as to what these may be. However, it seems quite certain that the trend of alternating A- and B-type segments will continue and that more periodic orbits lie in between these segments.

It must be assumed that the orbit of Szebehely and Peters lies on the curve of a family such as the one presented here. Possibly it is the same family.

Similar periodic orbits may be found by creating families where the condition $|\mathbf{v}_1|=0$ used here is replaced by either $|\mathbf{v}_2|=0$ or $|\mathbf{v}_3|=0$.

Finally, by not adhering to the mass ratios, 3:4:5, one may find *families* of periodic orbits as done by Szebehely (1969).

6. Conclusions

The initial question as to whether the orbit of Szebehely and Peters is a member of a family of *periodic* orbits, may be answered in the following way: No, if one retains the mass ratios, 3:4:5, and yes, if one allows these ratios to vary.

Nothing may be concluded from this study about the existence of periodic orbits for which the angular momentum is non-zero. In such a case, the two points of symmetry described above would be impossible.

Acknowledgements

This work was supported in part by the Office of Naval Research and by the Air Force Office of Scientific Research.

References

Peters, C. F.: 1968, Thesis, Yale University, New Haven.
Szebehely, V. G.: 1970, this volume, p. 382.
Szebehely, V. G. and Peters, C. F.: 1967, *Astron. J.* **72**, 1187.

Discussion

C. F. Peters: Did you continue some of the orbits in which you obtained a close triple encounter to see if one of the members escaped?

E. M. Standish: No, not in this study. However, other experience indicates that nearly all, if not all, such orbits lead eventually to the escape of one member, except, of course, the periodic orbits.

NEW FAMILIES OF PERIODIC ORBITS IN THE GENERAL PLANAR PROBLEM OF THREE BODIES

VICTOR SZEBEHELY
The University of Texas at Austin, Austin, Tex., U.S.A.

Abstract. This paper contains a preliminary report of recent results showing new periodic orbits in the two-dimensional general problem of three bodies. These periodic orbits may form families when the mass of one of the three bodies is changed. The conjecture is presented according to which these orbits may be generated from a special solution of the restricted problem of three bodies. Several numerical examples are described. These examples are based on the author's previously published results on the Pythagorean problem of three bodies.

The background of the Pythagorean problem and the pertinent literature are described in the Introduction. Then two families of new periodic orbits are described in considerable detail. This is followed by the presentation of newly discovered single periodic orbits which may be looked upon as generators for new families. Then a conjecture, furnishing analytical basis of the new families, is proposed. A short description of methods for establishing regions of the phase-space where periodic orbits may appear concludes the paper.

1. Introduction

The first investigation of the Pythagorean problem of three bodies may be traced to Burrau (1913) whose numerical integration covered the interval of the dimensionless time from zero to 3.35. Since then, integrations up to and beyond 100 dimensionless time units have been performed in several centers and with a variety of methods of integration. A detailed description of the historical aspects is offered by Szebehely (1967) and a complete numerical solution is given by Szebehely and Peters (1967a) by showing that the asymptotic solution of the Pythagorean problem (known as hyperbolic-elliptic motion) consists of the formation of a binary and of the separation and escape of the third body.

There has been a suggestion made even prior to Burrau's work according to which the Pythagorean problem possesses a periodic solution. This conjecture proved to be incorrect since the Pythagorean problem as proposed in its original form results in the above-described separation into a binary and an escaping body.

Nevertheless, it has been shown that if the initial positions of the bodies are slightly changed; that is, if the Pythagorean triangle is somewhat distorted, a periodic solution indeed may be obtained. (See Szebehely and Peters (1967b).) This solution gives rise to families of periodic solutions by changing the masses of the participating bodies. Such a family (Family of Class 5) is discussed in this paper.

Recent papers published in Russia (Agekyan and Anosova (1967, 1968)) show the dynamical breaking-up of triple star systems. These numerical integrations demonstrate that systems of three bodies with equal masses and with zero or almost zero initial velocities usually end up as a binary and an escaping star. Similar investigations using the Pythagorean problem as basis are conducted presently under the direction of this writer at The University of Texas. In this paper no results of these studies are

G. E. O. Giacaglia (ed.), Periodic Orbits, Stability and Resonances, 382–396.

discussed; nevertheless, our systematic investigations lead to the discovery of several new periodic orbits. The fact that variations of the Pythagorean problem resulted in periodic orbits with much shorter periods and consequently much greater simplicity than possessed by the periodic solution obtained by Szebehely and Peters (1967b) allowed the formulation of a conjecture regarding the origin of these orbits, as well as offered a possibility to understand the groups of families of these periodic orbits and even produce a possible analytical existence proof. According to these results, it may be expected that certain periodic orbits of the restricted problem of three bodies may be analytically continued into families of periodic orbits of the general problem of three bodies. It is also possible that some periodic solutions with symmetric properties of the general problem of three bodies can be used as generating solutions to derive families of new periodic orbits. Such solutions of the restricted problem and of the general problem have been given by Pavanini (1907), MacMillan (1911), Deprit (1953), and by Sitnikov (1960).

This introduction is concluded with the definition of the Pythagorean problem of three bodies. In this problem we consider the motion of three bodies (mass points) under their mutual gravitational attraction. The masses are 3, 4, and 5. The initial

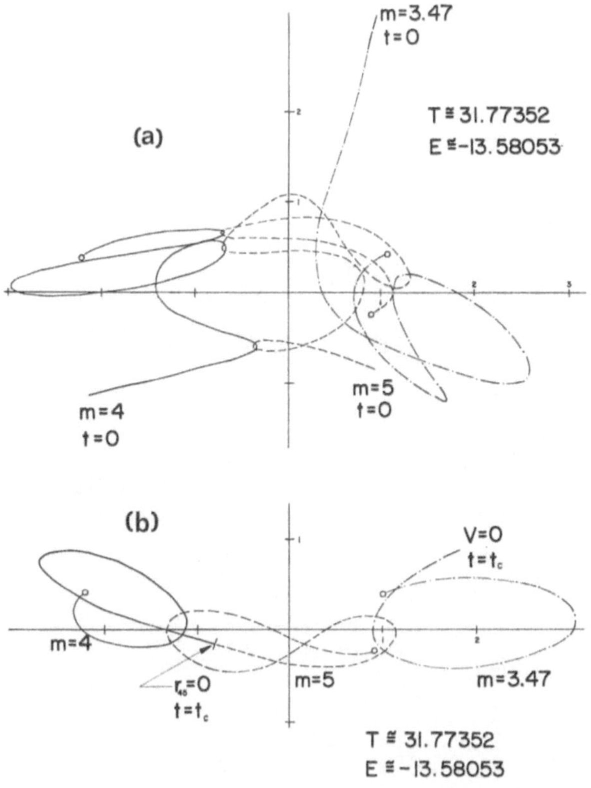

Fig. 1. Family 1 of periodic orbits; $m_1 = 3.47$.

conditions are given by zero velocities of all bodies. The bodies are placed initially at the apices of a Pythagorean triangle with sides 3, 4, and 5 so that the body with mass 3 is opposite of the side whose length is 3 and the same applies to the other two bodies. The gravitational constant is one.

2. Description of Family 1 of Periodic Orbits

As mentioned before, by changing the initial positions of the masses (but keeping the initial velocities zero), a periodic orbit was obtained two years ago by Szebehely and Peters (1967b). This orbit will be referred to as orbit S-P. A search for new initial conditions to obtain new periodic orbits in the neighborhood of the initial conditions which gave the original periodic orbit proved to be futile and the periodic orbit found in 1967 appeared to be an isolated periodic orbit. On the other hand, it was found that if the mass of one of the three bodies was changed slightly, new initial positions could be established giving a new periodic orbit of the dynamical system. If, therefore, we accept the concept of building a family of periodic orbits by changing the masses of the participating bodies, then the construction of families from the periodic orbits is a straightforward matter.

Figures 1 to 4 show characteristic members of Family 1. Attention is directed first

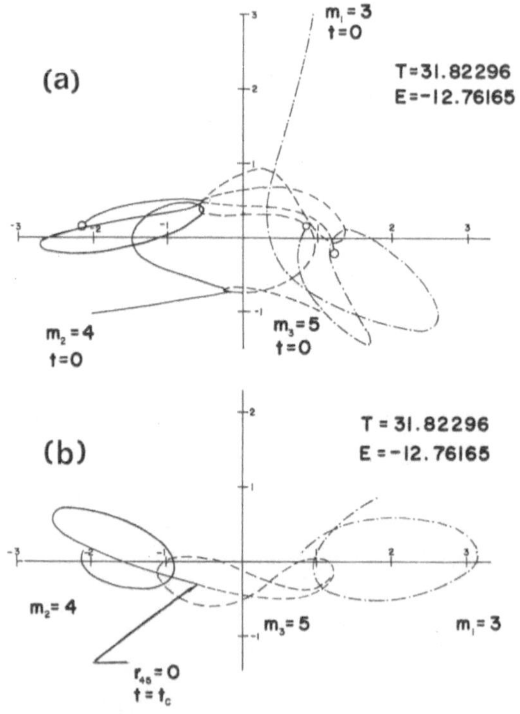

Fig. 2. Family 1 of periodic orbits; $m_1 = 3.00$.

to Figure 2 which shows the periodic orbit S-P found previously. The masses are $m_1=3$, $m_2=4$, $m_3=5$. The family is generated by changing the mass of m_1. In Figure 1, $m_1=3.47$ and in Figures 3 and 4 this value decreases to 1.5 and 1.39. In Figure 2 the orbit S-P is shown in one single plot while in Figures 1, 3, and 4 in order to increase

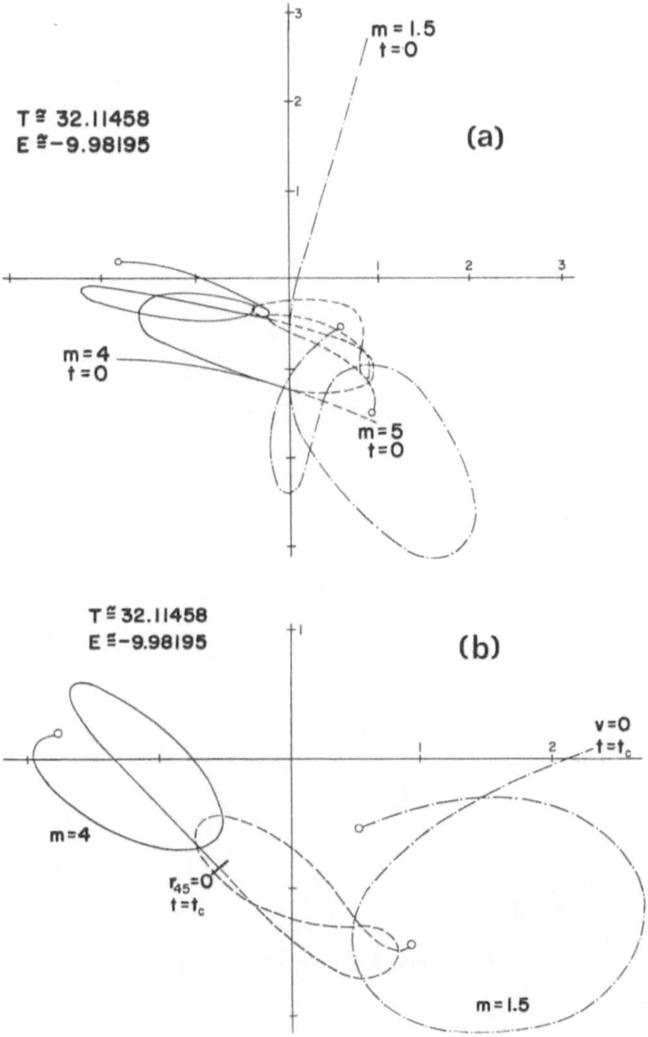

Fig. 3. Family 1 of periodic orbits; $m_1 = 1.50$.

the clarity of the presentation, the orbits are shown from $t=0$ to $t \cong 7$ and from $t \cong 7$ to $t \cong T/2$ on separate plots. Only half of the period ($T/2$) is shown since the paths of the bodies are retraced after this. At $T/2$ two of the bodies (the ones with masses of $m_2=4$ and $m_3=5$) collide and at the same time m_1 comes to rest. This condition, together with the initial condition of zero velocities assures periodicity. Note that $m_2=4$ and $m_3=5$ display several close approaches prior to their collision at $T/2$. Considering orbit S-P in Figure 2 for instance, we see that these close approaches occur approximately at $t_1=1.9$, $t_2=3.8$, $t_3=8.7$, and at $t_4=11.6$. The fifth close

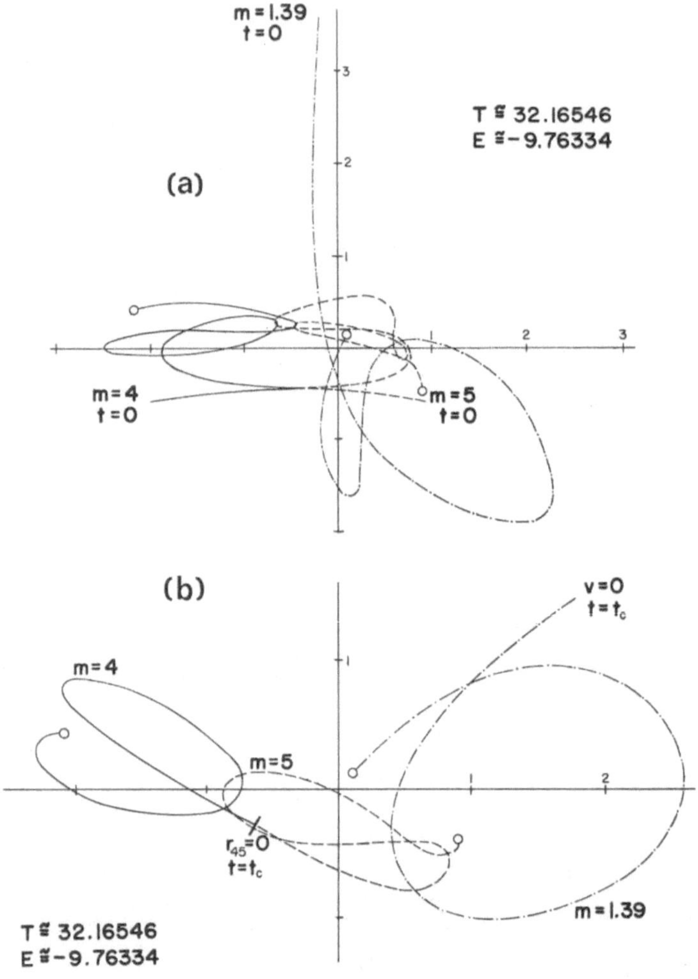

Fig. 4. Family 1 of periodic orbits; $m_1 = 1.39$.

approach is actually a collision at $t_5=t_c=T/2\cong 15.9$. This process appears on all members of Family 1.

In an attempt to classify the periodic orbits presented in this paper, one may consider the number of close approaches observed by those two bodies which actually collide. This number (including the collision) for half of a period is 5 for members of Family 1, consequently, we may call the orbits belonging to Family 1, Class 5 orbits. Note that 'Family' is a subclass of 'Class'.

Table I gives the periods and total energies of members of this Family. The total energy is defined by

$$E = \sum_{1 \leq i \leq 3} \tfrac{1}{2} m_i v_i^2 - \sum_{1 \leq i < j \leq 3} \frac{m_i m_j}{r_{ij}}$$

and consequently $E < 0$ since the kinetic energy is zero initially.

TABLE I
Family 1 of periodic orbits

Figure	m_1	T	E
1	3.47	31.773 52	−13.580 53
2	3	31.822 96	−12.761 65
3	1.5	32.114 58	−9.981 95
4	1.39	32.165 46	−9.763 34

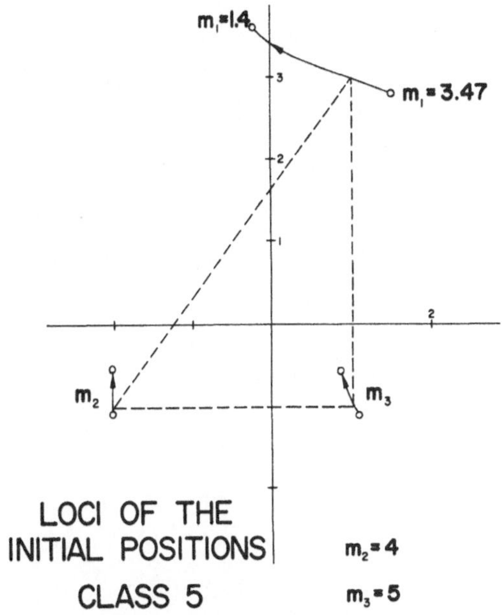

Fig. 5. Initial positions for Family 1 of periodic orbits; $1.4 < m_1 < 3.47$.

Figure 5 shows the original Pythagorean triangle as well as the initial positions of the bodies which give periodic motions as m_1 varies from $m_1 = 1.4$ *to* $m_1 = 3.47$.

The orbits shown in Figures 1 to 4 are representative samples which were selected from a large set of periodic orbits computed in order to establish Family 1. The step size of changing the value of m_1 was selected so that the general trend of the numbers of the family was established without doubt. When $m_1 < 1.4$, the convergence of the iteration process which furnishes the proper initial conditions for periodic orbits becomes slow. Continuation of the family to values of $m_1 > 3.5$ presents no difficulty.

At this time these investigations continue and it is expected that the family will be completed in the near future. Continuation of the family to $m_1 = 0$, or rather to say, analytic continuation from $m_1 = 0$, corresponding to the problem of two bodies for $m_1 \neq 0$, shall be discussed in the last section of this paper.

3. Description of Family 2 of Periodic Orbits

According to the method of classification mentioned in the previous chapter, this family belongs to Class 2. Consider Figure 6 in which a periodic orbit with $m_1 = 3$,

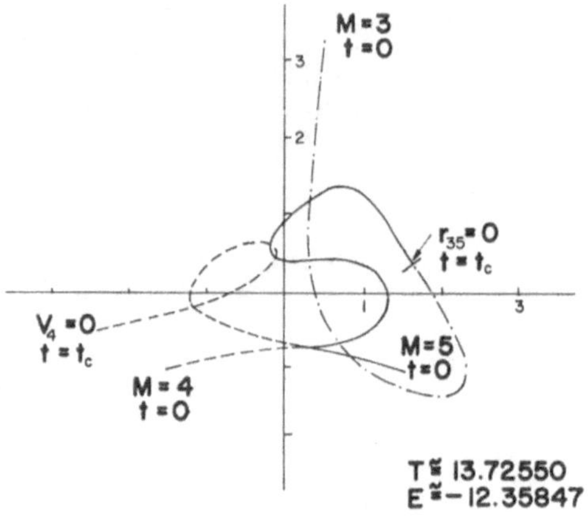

Fig. 6. Family 2 of periodic orbits; $m_2 = 4.0$.

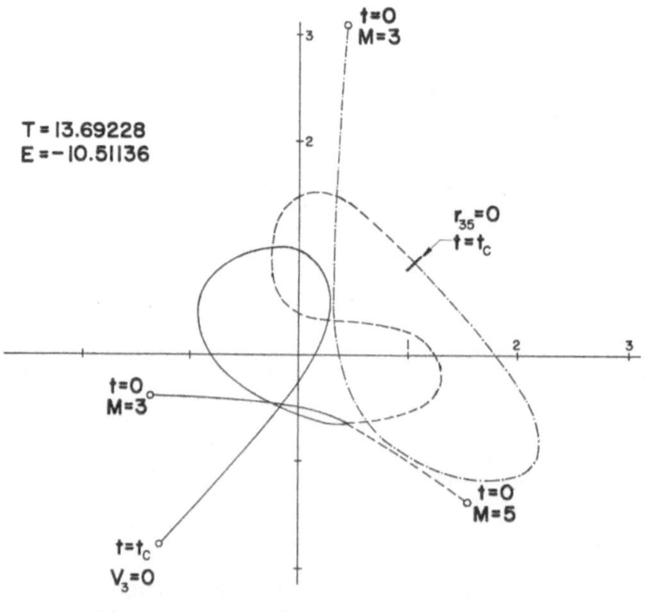

Fig. 7. Family 2 of periodic orbits; $m_2 = 3.0$.

$m_2=4$, $m_3=5$ is shown. The initial positions are slightly displaced from those shown on orbit S-P in Figure 2. Soon after the beginning of the motion, there is a close approach between $m_2=4$ and $m_3=5$ and another one appears between the same bodies before collision. Bodies $m_2=4$ and $m_3=5$ *do not*, however, collide, consequently, the two close approaches between these bodies is irrelevant regarding classification. Considering the colliding bodies ($m_1=3$ and $m_3=5$), one may observe *one* close approach and then a collision. Consequently, this family may be considered belonging to Class 2. The approach between $m_1=3$ and $m_3=5$ occurs between the two previously mentioned close approaches of $m_2=4$ and $m_3=5$.

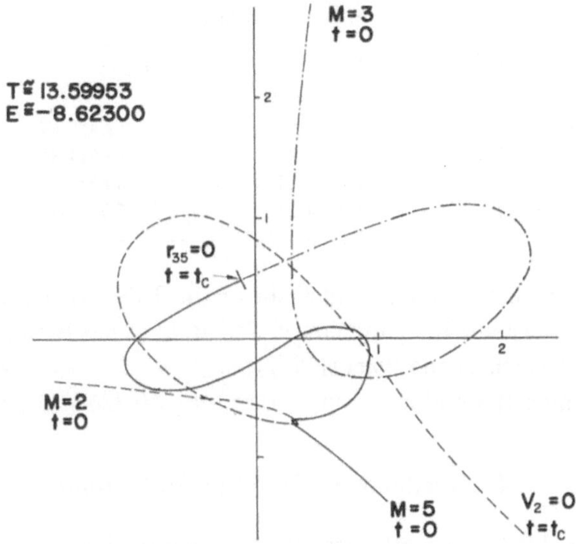

Fig. 8. Family 2 of periodic orbits; $m_2 = 2.0$.

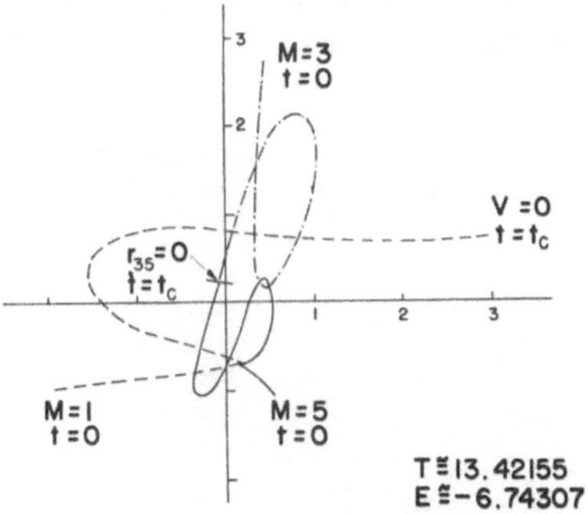

Fig. 9. Family 2 of periodic orbits; $m_2 = 1.0$.

Figures 6 to 9 show the development of Family 2. The collision always occurs between $m_1=3$ and $m_3=5$ and the members of the family are generated by changing the value of m_2 from 4 to 1. Table II shows the pertinent characteristics. Normalization regarding the size of the orbit and the total mass of the bodies participating in the system is important if the actual values shown in Table II are studied instead of their trend only. A normalizing parameter is T^2GM/L^3 as has been pointed out before (see, for instance, Szebehely and Peters (1967)).

TABLE II
Family 2 of periodic orbits

Figure	m_2	T	E
6	4	13.725 50	−12.358 47
7	3	13.692 28	−10.511 36
8	2	13.599 57	−8.623 00
9	1	13.421 55	−6.743 07

Results concerning the regions of $m_2>4$ and $m_2<0.726$ will be given in a forthcoming paper.

4. Description of Two Generating Orbits

Two periodic orbits are shortly described in this section. These may be used to build families by changing the values of masses in the same way as was described in discussing Families 1 and 2. The orbits are members of two different families, of course, but, nevertheless, they refer to masses $m_1=3$, $m_2=4$, and $m_3=5$, i.e. the Pythagorean values of the masses are used.

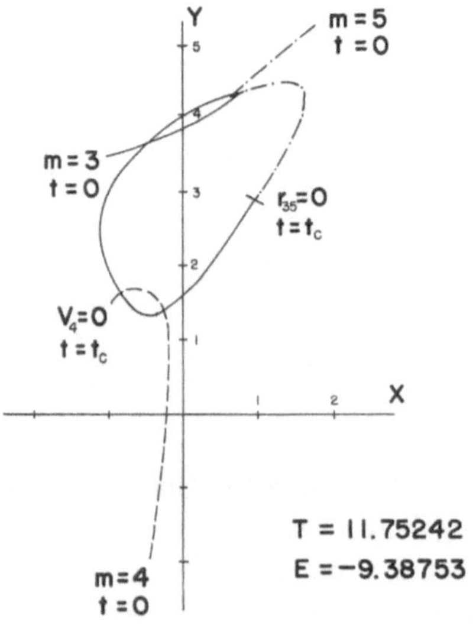

Fig. 10. Periodic orbit of Class 2; $m_1=3$, $m_2=4$, $m_3=5$.

Figure 10 shows a Class 2 orbit. Observe that collision between $m_1=3$ and $m_3=5$ occurs after only one close approach between these bodies. The orbit of $m_2=4$ is strikingly simple. Note that if $y\to -y$ and $m_1\to m_2=4$, $m_2\to m_3=5$, $m_3\to m_1=3$ the approximate beginning of orbit S-P in Figure 2 is obtained.

Figure 11 shows a Class 3 orbit. Collision and close approaches occur between $m_2=4$

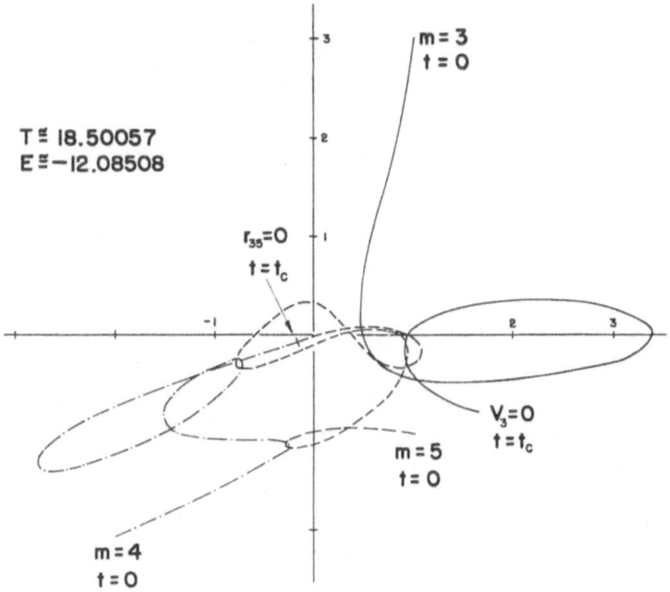

Fig. 11. Periodic orbit of Class 3; $m_1=3$, $m_2=4$, $m_3=5$.

and $m_3=5$. In comparing this figure with orbit S-P on Figure 2, it may be seen that Figure 11 corresponds to a modification of the beginning of the orbit S-P.

Orbit S-P in Figure 2 displays three close approaches between $m_2=4$ and $m_3=5$, the third occurring at approximately $t_3=8.7$. By changing the initial positions of the bodies in Figure 11 a collision occurs between $m_2=4$ and $m_3=5$ instead of the close approach of orbit S-P.

5. Conjecture Regarding the Proof of Existence

Consider first one of the few known solutions of the restricted problem of three bodies in three dimensions in which the mass-less particle moves on the (fixed) z-axis while the primaries with equal masses revolve in the (x, y) plane. That the general solution of this problem may be represented by an elliptic integral of the third kind is known since 1907, if not earlier. (Furthermore, the problem has been treated in its non-restricted form also.)

A possible modification of this problem is when the primaries move on elliptic orbits. It is of special interest when the eccentricities of these ellipses become one. In this case the dynamical system consists of the mass-less particle moving on the (fixed) z-axis and the primaries of equal masses performing oscillatory collisions move along the (fixed) x-axis. Figure 12 shows the general arrangement.

The motion of the primaries is well known and it is described by the differential equation

$$\ddot{x} = -\frac{G(m_1 + m_2)}{x^2}, \tag{1}$$

where dots denote derivatives with respect to the time, G is the constant of gravitation, m_1 and m_2 are the masses of the primaries, and $x = x_1 - x_2$ is the distance

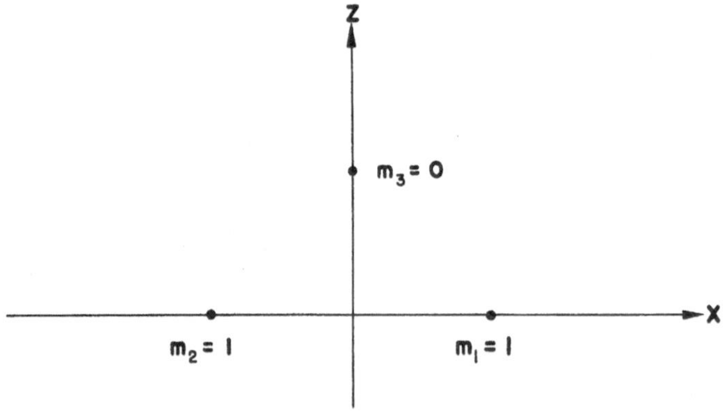

Fig. 12. Coordinate system for generating orbits.

between the primaries located at $P_1(x_1, 0, 0)$ and $P_2(x_2, 0, 0)$. Note that because of symmetry, $m_1 = m_2$ and $x = 2x_1$. The origin of the coordinates is the (fixed) center of mass of the system. The period of the motion of the primaries may be obtained after a simple regularizing transformation is applied.

Let

$$d\tau = \frac{dt}{x} \tag{2}$$

the equation defining a new independent variable τ. Then Eq. (1) becomes

$$x'' = 2Gm_1 - Cx, \tag{3}$$

where primes denote derivatives with respect to τ and C is the constant of energy. In fact, from Equation (1) we obtain

$$\dot{x}^2 = \frac{2G(m_1 + m_2)}{x} - \frac{C}{2}, \tag{4}$$

which equation defines C. Note that $C > 0$ corresponds to elliptic motion (with unit eccentricity in our case).

The energy integral in the regularized system becomes

$$(x')^2 = 4Gm_1x - Cx^2. \tag{5}$$

The period is

$$T = 2\pi/\sqrt{C} \tag{6}$$

as may be seen from Equation (3). Here the value of C may be obtained from the initial conditions using Equation (4). At $t=\tau=0$ let $x_1=-x_2=1$ and $\dot{x}_1=-\dot{x}_2=0$. Then $C=2m_1G/x_1=2$, if $G=1$ and $m_1=m_2=1$. Consequently, the period becomes $T=\pi\sqrt{2}$.

We wish to select the initial conditions of $m_3=0$ so that m_3 reaches the origin of the coordinate system before m_1 and m_2 collide. In other words $z_3(0)>0$ at $t=0$ with $\dot{z}_3(0)=0$ and at collision ($t=t_c=T/2$) the third body is on the negative z-axis, $z_3<0$ with $\dot{z}_3=0$. After collision of m_1 with m_2, the third body moves with $\dot{z}_3>0$ and when the primaries reach their original position (i.e. at $t=2t_c=T$), the third body has also reached $z_3(0)>0$ with $\dot{z}_3(0)=0$.

The motion of the third body is governed by the differential equation

$$\ddot{z} = -2m_1Gz/r^3, \tag{7}$$

where $r^2=z^2+x_1^2$ and $z=z_3$. Furthermore,

$$x_1 = (1 + \cos\sqrt{C}\tau)/C,$$

or in the special case mentioned before

$$x_1 = \tfrac{1}{2}(1 + \cos\sqrt{2}\tau) \tag{8}$$

Note that when x_1, as given by Equation (8), is substituted in the expression for r and this in turn is substituted into Equation (7); this latter equation will contain two independent variables, t and τ. In order to avoid this, one may utilize Equation (3) and obtain a relation between t and τ (which is essentially Kepler's equation):

$$t = \tau + \frac{1}{\sqrt{2}}\sin\sqrt{2}\tau. \tag{9}$$

There are several forms of Equation (7), one being

$$2x_1(1-x_1)\frac{d^2z}{dx_1^2} - \frac{dz}{dx_1} = -\frac{8zx_1^2}{(z^2+x_1^2)^{3/2}}. \tag{10}$$

In Equation (10) the quantity x_1 is the independent variable. Consequently, substituting its expression given by Equation (8) is not required.

The restricted problem of three bodies described above seems to be germane to the periodic orbits presented in this paper. As seen from Figure 13 the solution of Equation (10) is a Class 1 periodic orbit, since m_1 and m_2 have no close approaches, only a collision. By changing the geometry of the initial conditions, one may expect two collisions between m_1 and m_2 before m_3 returns to its original position – with zero velocity. In this way, Class 2 periodic orbits are generated.

When $m_3\neq0$, the straight-line motion of the primaries must be changed into perturbed collision orbits. Class 1 periodic orbits require collision between m_1 and m_2 at the first close approach. Higher class periodic orbits will have several actual close approaches before collision takes place. The essential properties of the derived motion are as follows:

(1) at $t=0$, all velocities are zero;

(2) at $t=t_c=T/2$, the primaries collide and simultaneously the third body comes to rest;

(3) at $t=T$, all velocities are again zero and all bodies occupy their original positions. If the symmetry is disturbed by using different masses for the primaries, i.e. $m_1 \neq m_2$,

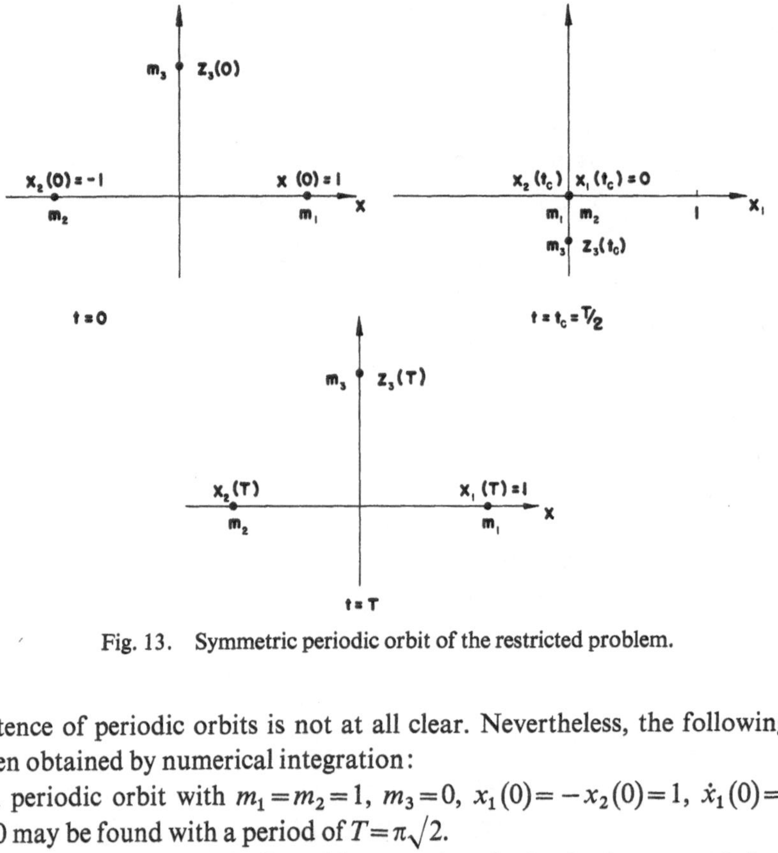

Fig. 13. Symmetric periodic orbit of the restricted problem.

the existence of periodic orbits is not at all clear. Nevertheless, the following results have been obtained by numerical integration:

(1) A periodic orbit with $m_1=m_2=1$, $m_3=0$, $x_1(0)=-x_2(0)=1$, $\dot{x}_1(0)=\dot{x}_2(0)=\dot{z}_3(0)=0$ may be found with a period of $T=\pi\sqrt{2}$.

(2) A family of such periodic orbits may be obtained when $m_3 \neq 0$ but all the symmetry conditions are the same as in (1). The triple close approach occurring in this family prevents at the present time to go higher with the value of m_3 than approximately 0.3775.

(3) Asymmetric periodic orbits with $m_3=0$, $m_1 \neq m_2$ may be numerically established.

The conjecture is therefore put forward according to which the periodic orbits discussed in this paper may be generated by analytic continuation from the simple model of the restricted problem of three bodies mentioned in this section. All these periodic orbits contain binary collisions and all appear to be isolated periodic orbits. Families may be generated by changing the values of the masses corresponding to changing the value of the mass-parameter μ in the restricted problem of three bodies.

6. Location of Regions for the Existence of Periodic Orbits

The research work described several tools which were used to locate periodic orbits, i.e. to find approximate initial conditions. One method was already mentioned which furnishes periodic orbits somewhat accidentally since the purpose of the investigation is not to find periodic orbits but to establish initial conditions which lead to the formation of binary systems with an escaping third body.

There is a more direct method which, in conjunction with the integration of any system of three bodies, has the potential to focus attention to those regions of the phase-space which may contain the initial conditions for periodic orbits.

Consider the vector representing the dynamical system in phase space. The dimensionality of this vector is $2 \times n \times m$ where n is the number of bodies with $m=2$ in the planar and $m=3$ in the three-dimensional motion. (For the periodic orbits treated in the paper we have a 12-dimensional phase space.) Note that the components of this vector are not homogeneous since six of the components are the 3×2 positions and six are velocities. Nevertheless, we might introduce a time-scale which (arbitrarily) will allow manipulations in a dimensionally meaningful manner. Compute now a state vector $s(x_1, y_1, x_2, y_2, x_3, y_3, \dot{x}_1, \dot{y}_1, \dot{x}_2, \dot{y}_2, \dot{x}_3, \dot{y}_3)$ as a function of time and consider the quantity

$$|\Delta \bar{s}| = \Delta s = |\bar{s}(t) - \bar{s}(0)|, \tag{11}$$

which is a measure of the deviation of the system in the phase space from its original state. If at a time $t=T$, $\Delta s=0$, the system returned to its original position at $t=T$, in other words, the motion repeated itself. A simple (scalar) plot of Δs versus time will furnish by inspection the times when the state is close to a set of initial conditions giving a periodic orbit. Since $\Delta s(0)=0$ the close approach of the $\Delta s=\Delta s(t)$ curve to the t-axis will direct attention to those conditions which might result in periodic orbits. There may be several such close approaches and these all may be treated in relation to $\Delta s(0)$, or in relation to each other. Some of the periodic orbits presented in this paper were discovered by the above described method.

Acknowledgements

This paper was presented at the invitation of the International Symposium of Astronomy in Brazil and the author wishes to acknowledge his deepest gratitude for the partial support received from the Symposium.

The preparation of the paper was made possible by the partial sponsorships of the Office of Naval Research and the Theoretical Division of the Goddard Space Flight Center, NASA.

Several of the author's graduate students contributed to the computation of the orbits presented in this paper. Before detailed and complete accounts will be given of these investigations in other publications, it is the author's pleasure to acknowledge the interest and help received from Messrs. T. Feagan and D. Jones, graduate students of the University of Texas.

References

Agekyan, T. A. and Anosova, Zh. P.: 1967, *Astron. J. U.S.S.R.* **44**, 1261.
Agekyan, T. A. and Anosova, Zh. P.: 1968, *Akad. Nauk. Arm. U.S.S.R.*, Astrophysics, **4**, 31.
Burrau, C.: 1913, *Astron. Nachr.* **195**, 113.
Deprit, A.: 1953, Thesis, University of Louvain.
MacMillan, W. D.: 1911, *Astron. J.* **27**, 11.
Pavanini, G.: 1907, *Ann. Mat.* [3], **13**, 184.
Sitnikov, K.: 1960, *Dokl. Akad. Nauk. U.S.S.R.* **133**, 303.
Szebehely, V.: 1967, *Proc. Natl. Acad. Sci. U.S.* **58**, 60.
Szebehely, V. and Peters, F.: 1967a, *Astron. J.* **72**, 876.
Szebehely, V. and Peters, F.: 1967b, *Astron. J.* **72**, 1187.

Discussion

M. Hénon: Is it not possible, at least in principle, to have a continuous transition from, say, Class 3 to Class 4 along a family of periodic orbits?

V. Szebehely: While I am inclined to agree with Dr. Hénon that it is possible to have continuous transitions between the classes of periodic orbits mentioned in the paper, at present we do not have enough information to either support or refute this conjecture.

A. Deprit: (question 1) Would it simplify the figures if the orbits were transferred from Cartesian coordinates to isoperimetric coordinates as defined by Lemaître's conformal mapping? This transformation converts the binary collisions into fixed singularities.

(question 2) Has not Schubart computed some periodic orbits in the general problem of three bodies?

V. Szebehely: (to Dr. Deprit – 1st question) I appreciate Dr. Deprit's suggestion regarding representation of the results in Lemaître's coordinates. We shall, in the future, look into this.

(to Dr. Deprit – 2nd question) In answer to Dr. Deprit's second question, it is my understanding that Dr. Schubart's periodic orbits were all collinear.

STABILITY AND RESONANCES IN THE RESTRICTED PROBLEM

WILLIAM H. JEFFERYS

University of Texas at Austin, Austin, Tex., U.S.A.

Abstract. Results to date of an investigation of the Restricted Problem of Three Bodies by the method of Surface-of-Section are presented. Six mass ratios have been investigated over a wide range of the Jacobi Constant. A representative sample of the resulting figures is presented showing the evolution of one resonance with the mass ratio. Another example showing the evolution of stability and resonance with Jacobi Constant for the case of a Lunar Orbiter is shown.

The Restricted Problem of Three Bodies (planar, circular case) is a classical example of a nonintegrable dynamical system. Because of its importance to astronomy, in that it incorporates many of the features of perturbed motion, it has been extensively studied and many results are known. One of the most fruitful methods has been to study the periodic solutions of the system, a major topic of this conference. Another, recent technique has been to apply the method of Surface-of-Section originally invented by Poincaré and Birkhoff, using a high-speed computer (Hénon and Heiles, 1964; Hénon, 1966; Jefferys, 1966). This is the technique of the current work, and a description follows.

Given a dynamical system of two degrees of freedom (such as the Restricted Problem) which possesses an integral of the motion (the Jacobi integral, in the present case) let us fix our attention on a given value of the integral. Then the solutions will be restricted to a three-dimensional submanifold of the four-dimensional phase space. Thus by using the known integral we could eliminate one of the variables, getting a reduced phase space of three dimensions, and the integral itself would become a parameter.

If a second, independent integral existed*, then the motion would be further restricted to at most a two-dimensional submanifold of the reduced phase space. In such a case, if we introduce any arbitrary two-dimensional surface S into the reduced phase space, it will intersect the submanifold in a set (possibly empty) of simple closed curves.

The method employed here is to detect the possible existence of these simple closed curves, and hence the integral, by numerically integrating the equations of motion and plotting the successive intersections of the orbit with the surface S. After a suitable time one often finds that the points of intersection are apparently arranged (like beads on a string) on a number of simple closed curves. Furthermore, the arrangement of these curves may show the presence of powerful low-order resonances.

On the other hand, it often happens that the points are completely scattered. This is an indication of the breakdown of integrability and the onset of instability.

Normally one integrates a number of different orbits for a given value of the known

* It should be remarked that in general, as shown by Moser, no such isolating integral exists. However, there may be a 'quasi-integral', i.e., a smooth function conserved on all but a set of small measure.

G. E. O. Giacaglia (ed.), Periodic Orbits, Stability and Resonances, 397–409.

integral and plots all the results in the same figure. Therefore the method can give, at a glance, information on many orbits, and it is a powerful tool for the study of these systems.

In this study of the Restricted Problem it has been the aim to make the resulting figures as intuitively appealing as possible by a careful choice of the surface S. This has led to the following conventions: We choose units of mass, length and time such that the sum of the masses of the two primaries is unity and they circle about each other one unit of distance apart with mean motion unity. We choose a coordinate system rotating counterclockwise with the mass μ located at $(-1, 0)$ and the mass $(1-\mu)$ at the origin. In this coordinate system we define the surface S by plotting the position of the massless particle whenever it passes through one of its (osculating) apsides, relative to the primary at the origin. This leads to the easily checked condition $\mathrm{d}r/\mathrm{d}t=0$ for plotting a point, where r is the distance from the origin.

This choice leads to an ambiguity in that direct and retrograde orbits (in the rotating coordinate system) are not distinguished (and hence their corresponding invariant curves will cross each other). This ambiguity can be easily resolved, however, because there is a symmetry in the equations of motion which leads to a symmetry of the plots over the x-axis. Therefore, only half of a plot need be shown. For this reason we distinguish between direct and retrograde apsides by plotting the former only in the upper half-plane (reflecting in the x-axis where necessary) and the latter in the lower half-plane. Thus there is a discontinuity in the figures along the x-axis.

Another peculiarity is that since on a zero velocity curve (the solid curves enclosing some figures) the particle is moving neither direct nor retrograde, but has a cusp, the upper and lower zero velocity curves should be topologically identified at symmetric points.

It is felt that despite these peculiarities, the simple physical interpretation of the resulting curves as the loci of the apsides far outweighs the disadvantages.

In the case $\mu=0$, we can readily predict what the results are. Since all orbits of the test particle are Keplerian ellipses, the apsides are always at the same point in an inertial coordinate system. In the rotating coordinate system, however, the apsides appear to regress, generally at a rate incommensurable with the rotation of the coordinate system. Thus for each orbit of nonzero eccentricity we get two circles (actually semicircles because of the above conventions), one for the pericenters and one for the apocenters. (Note that it is possible for the same orbit to have its pericenters in direct motion and its apocenters in retrograde motion due to the uniform rotation of the coordinate system; for example, an orbit of mean motion one.)

When $\mu \neq 0$ things can get more complicated. Many of the curves for $\mu=0$ persist, though distorted (Arnol'd, 1963; Moser, 1962). We may also get 'chains of islands' (Hénon and Heiles, 1964) associated with stable periodic orbits. Some curves may disintegrate (this is often due to the effects of unstable periodic orbits; see Figure 2(B)*). As the mass ratio μ gets larger, the distortions of the curves can be quite

* In Figures 1–6 and 8, one division equals 0.1 of the dimensionless unit of length. In all others, one division equals 0.01 units.

dramatic (Figures 1 to 7 where corresponding curves are shown for different mass ratios). Nevertheless, there is a remarkable amount of similarity between the curves, even for quite different values of μ.

The evolution of the curves with the Jacobi constant C follows a general pattern, regardless of μ. This is illustrated in Figures 7 to 15 for the case of a Lunar Orbiter. (The definition of the Jacobi constant used is the same as that given by Szebehely, 1967.) For low values of C (high energies) there are no curves. Then at some particular value curves appear for a few retrograde orbits, invariably in resonance, and with low eccentricity (Figure 8(A)). As C is increased, orbits with greater eccentricity are found, some of which are not resonant (Figure 10(B)). Other resonances may appear, and the orbits become more stable (Figures 11 and 12). Eventually, the zero velocity curve closes in on the region of interest. It is generally speaking not until this happens that any curves for direct orbits appear; the latter orbits seem to be less stable than the retrograde ones. About the same time as the zero velocity curve closes around the orbits, most direct orbits become stable (i.e., curves appear) although quite complicated resonances also are present (Figures 13, 7, 14). Generally by this time the retrograde orbits are not strongly perturbed (i.e., the curves corresponding to them are nearly circular). As C is increased further, there will be more evolution of the resonances as they appear and then dissolve, and eventually the direct orbits also become only slightly perturbed (Figure 15).

The relatively greater stability of the retrograde orbits versus the direct ones can be understood easily. For a given value of C, the velocity of a particle in the rotating coordinate system depends only on its position, through the Jacobi integral. Thus particles in instantaneously direct motion have the same velocity as those in retrograde motion. But in the *fixed* coordinate system it is quite different, because of the rotation of the adopted coordinate frame. Then, the orbits which were direct in the rotating system have higher velocities than the corresponding retrograde ones. Ignoring the other body, for the moment, they will have higher energy and be less strongly bound. Essentially, the direct orbits are not bound until the 'neck' of the zero-velocity curve closes. This is the value of C for which the primary around which the particle moves actually becomes the dominant mass.

The results described here are a sample chosen from nearly 200 such figures for six mass ratios that have been tried ($\mu = 0.012\,150\,67 = \frac{1}{82.3}$, the value for the Earth-Moon system; $\frac{1}{11}$, G. Darwin's value; $\frac{1}{5}$; $\frac{1}{3}$; $\frac{1}{2}$, the Copenhagen value; and $0.987\,849\,33 = 1 - \frac{1}{82.3}$, the value for a satellite of the Moon). All integrations have been done using the Levi-Civita regularization (Szebehely, 1967).

Three other mass ratios are currently under investigation, viz: Asteroid perturbed by Jupiter, Satellite of Jupiter perturbed by the Sun, and $\frac{4}{5}$ (the last bridging the gap between $\frac{1}{2}$ and 0.99). Preliminary results indicate that these values should give a good representation of the whole range of masses in the Restricted Problem.

The method used to compute these results is rather interesting and will be described briefly. It soon became apparent, after trial, that simply supplying the computer with initial conditions and retrieving the results later would not be satisfactory. Disad-

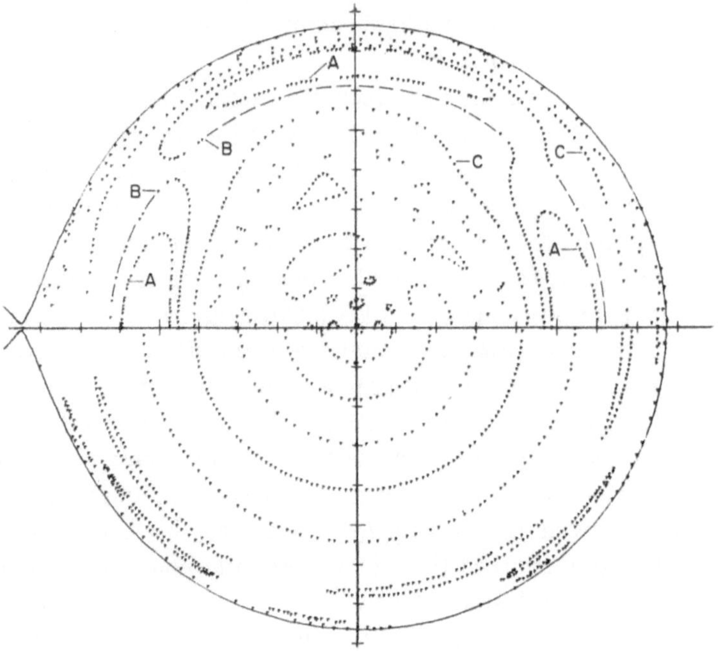

MASS RATIO = .01215067 JACOBI CONSTANT = 3.20000000

Fig. 1.

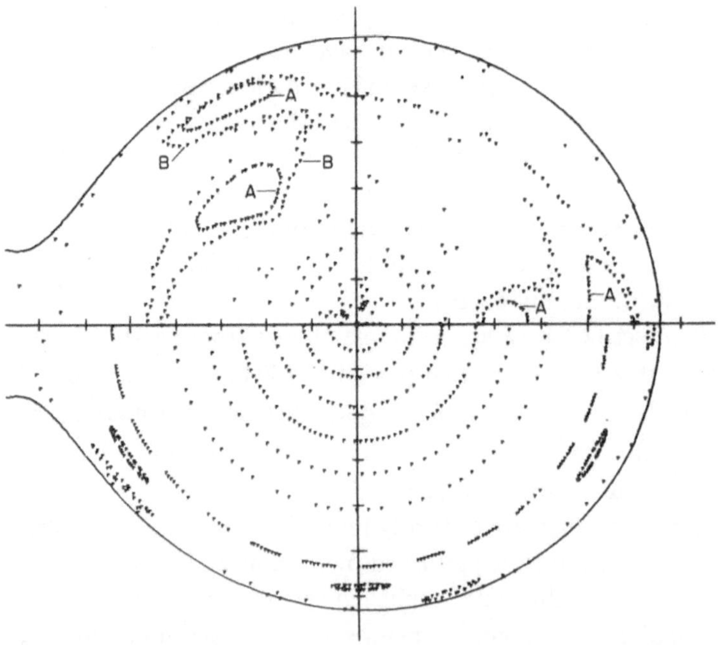

MASS RATIO = .09090909 JACOBI CONSTANT = 3.52500000

Fig. 2.

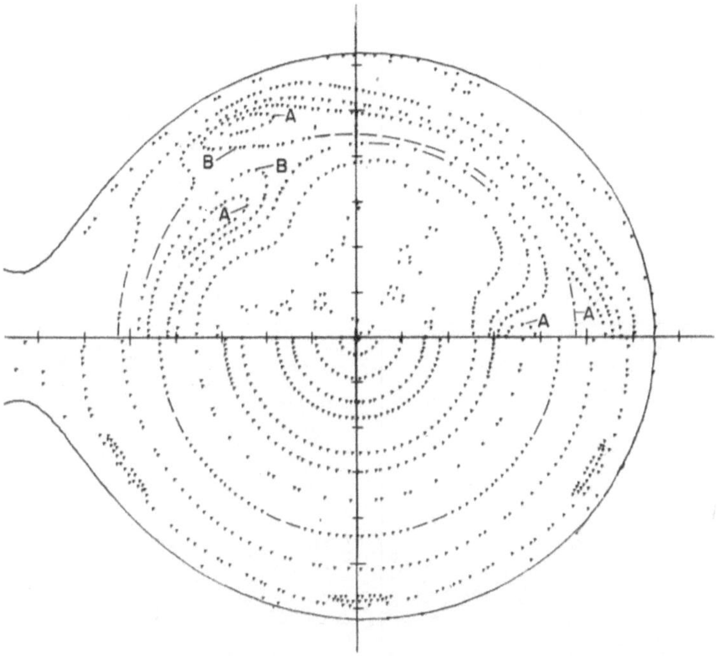

MASS RATIO = .09090909 JACOBI CONSTANT = 3.55000000

Fig. 3.

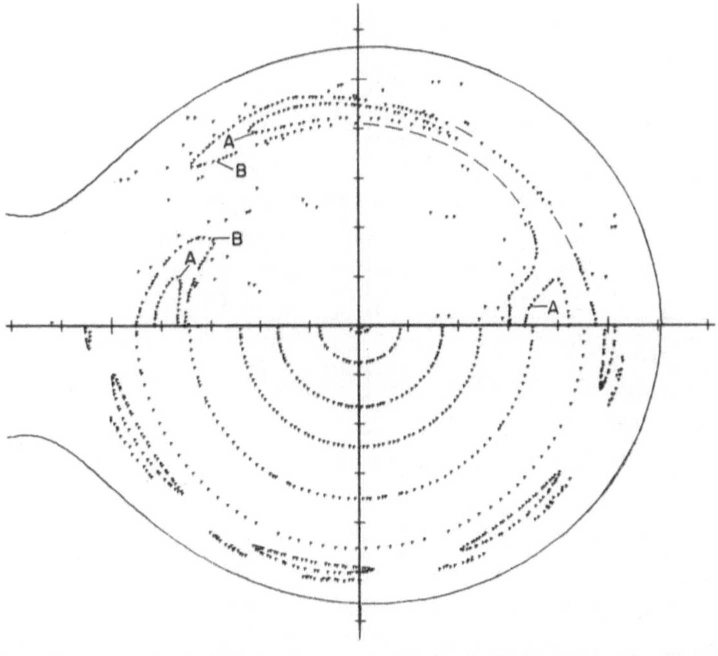

MASS RATIO = .20000000 JACOBI CONSTANT = 3.70000000

Fig. 4.

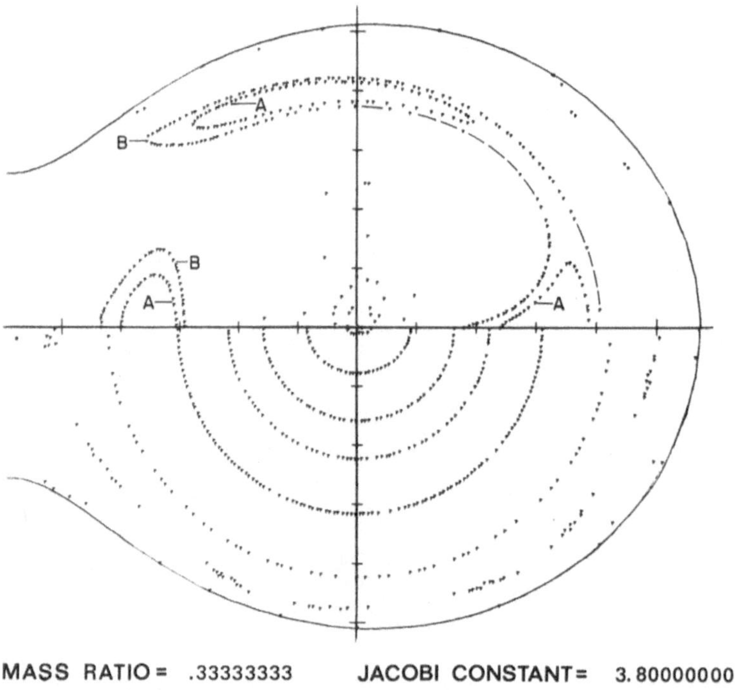
A
B
B
A
A
MASS RATIO = .33333333 JACOBI CONSTANT = 3.80000000

Fig. 5.

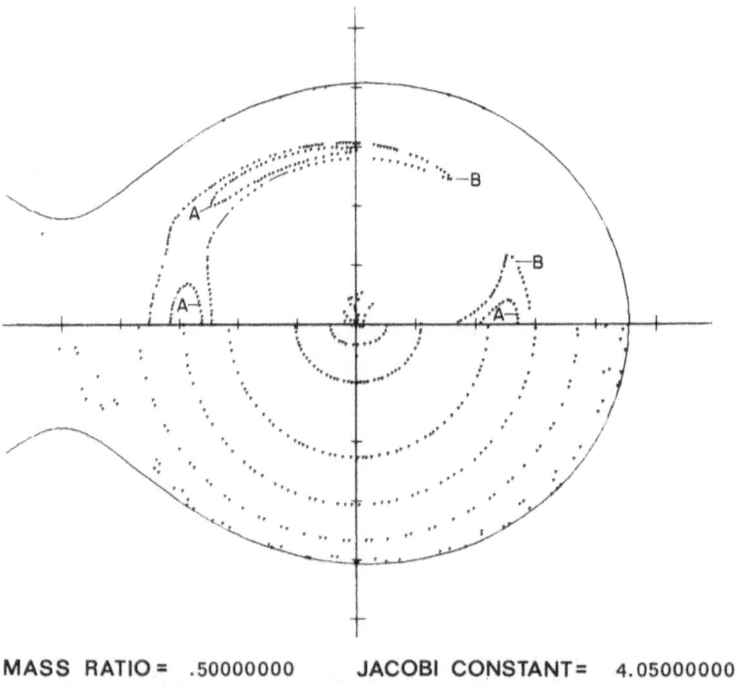
B
A
B
A
A
MASS RATIO = .50000000 JACOBI CONSTANT = 4.05000000

Fig. 6.

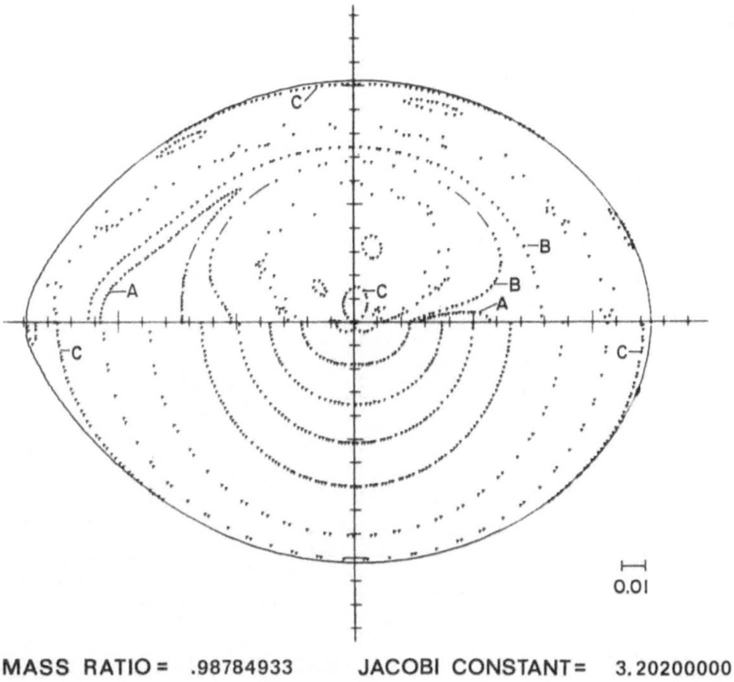
C
A
B
B
C
A
C
C
0.01
MASS RATIO = .98784933 JACOBI CONSTANT = 3.20200000

Fig. 7.

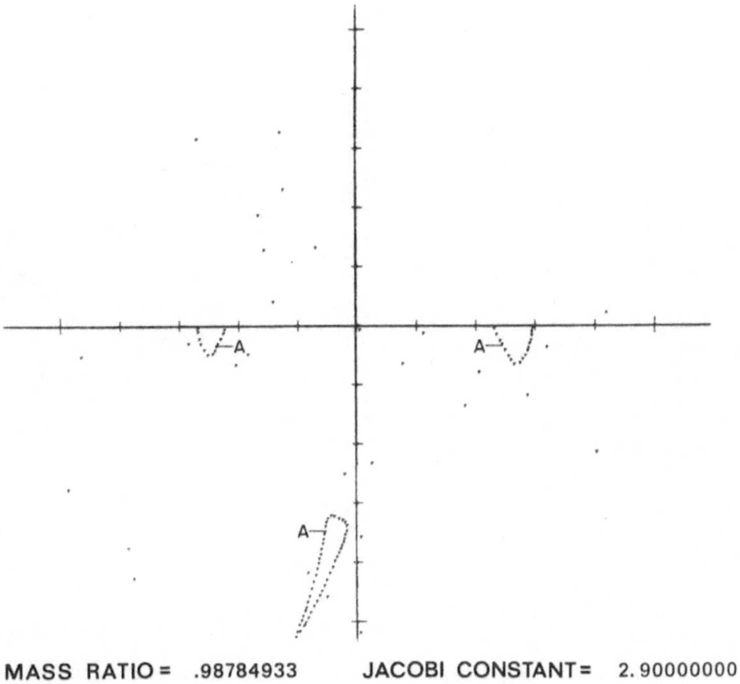
A
A
A
MASS RATIO = .98784933 JACOBI CONSTANT = 2.90000000

Fig. 8.

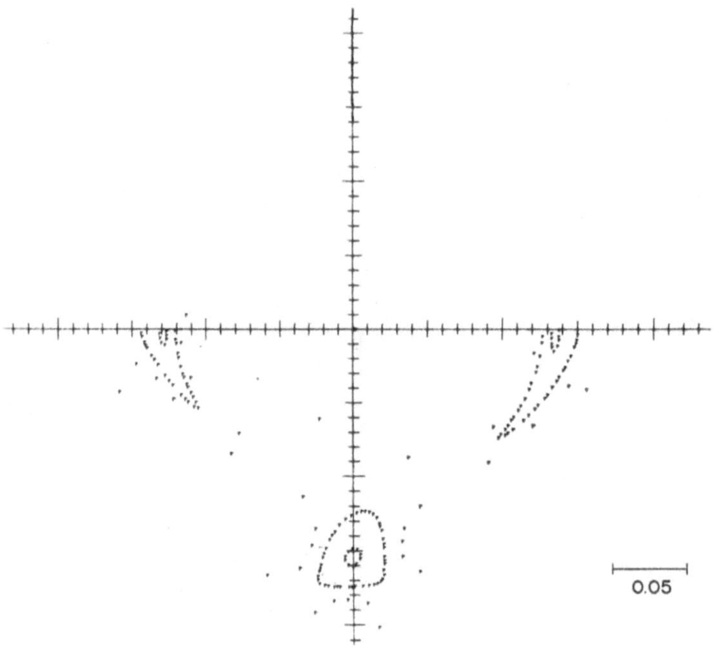

MASS RATIO = .98784933 JACOBI CONSTANT = 2.98000000

Fig. 9.

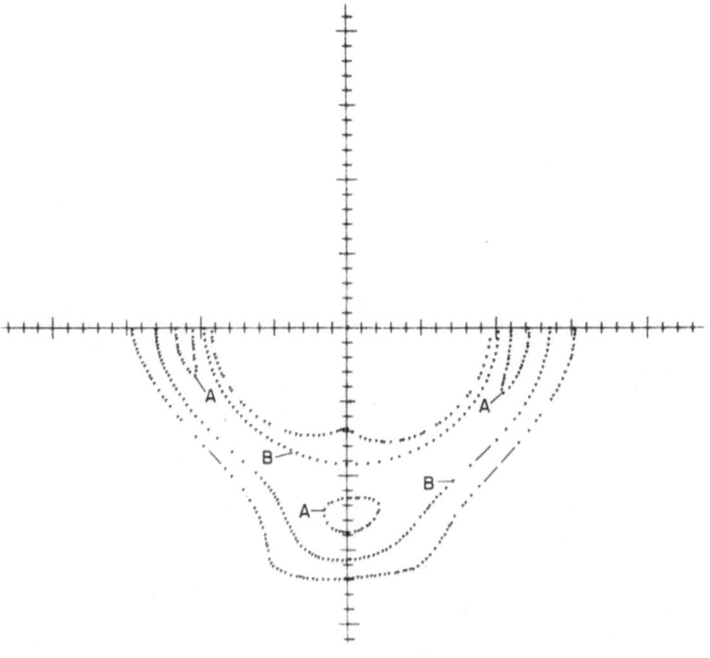

MASS RATIO = .98784933 JACOBI CONSTANT = 3.00000000

Fig. 10.

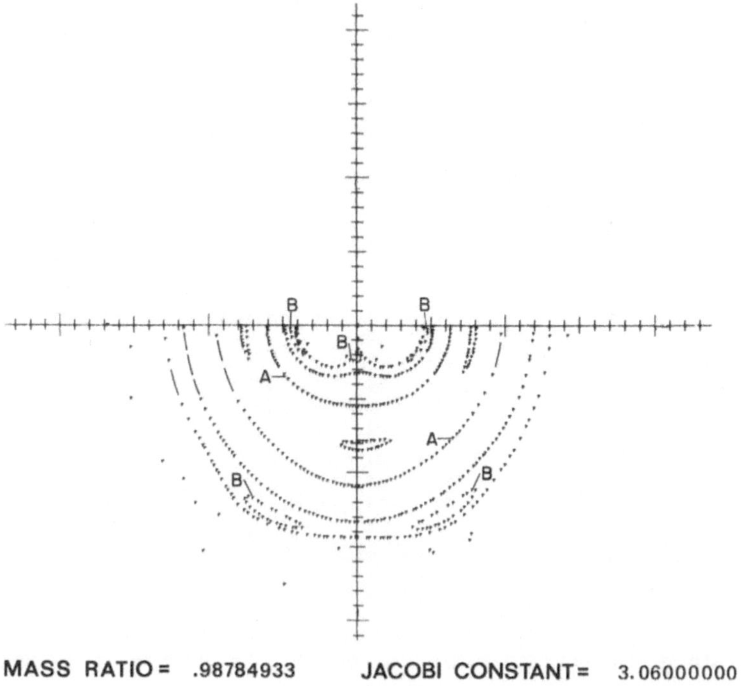
B
B
B
A
A
B
B
MASS RATIO = .98784933
JACOBI CONSTANT = 3.06000000

Fig. 11.

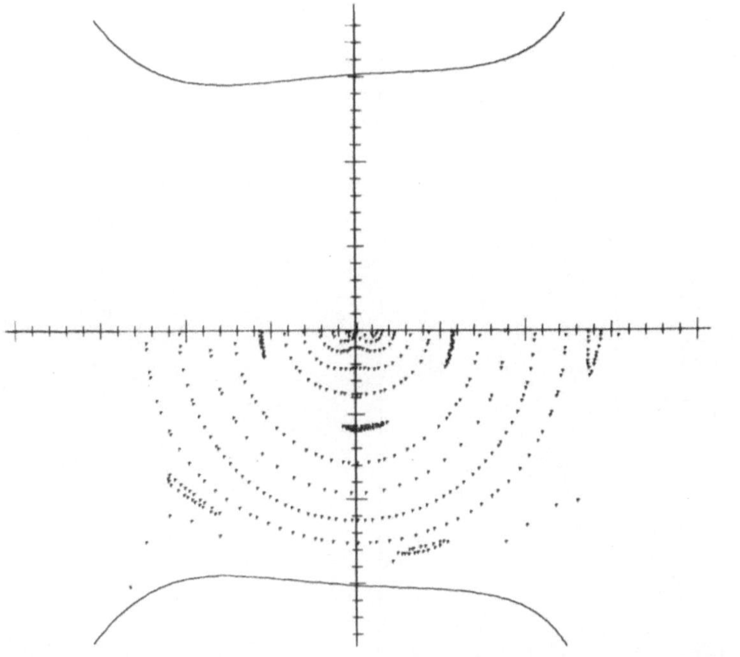
MASS RATIO = .98784933
JACOBI CONSTANT = 3.12500000

Fig. 12.

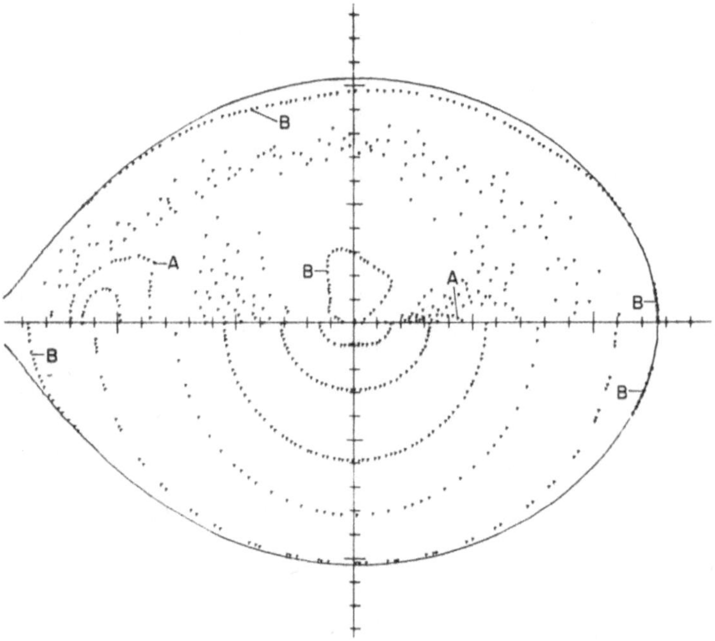

MASS RATIO = .98784933 JACOBI CONSTANT = 3.20000000

Fig. 13.

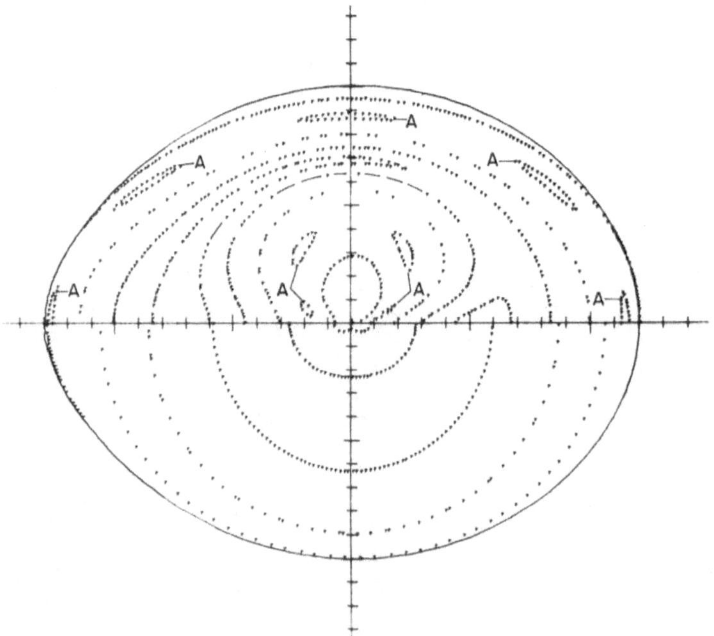

MASS RATIO = .98784933 JACOBI CONSTANT = 3.20600000

Fig. 14.

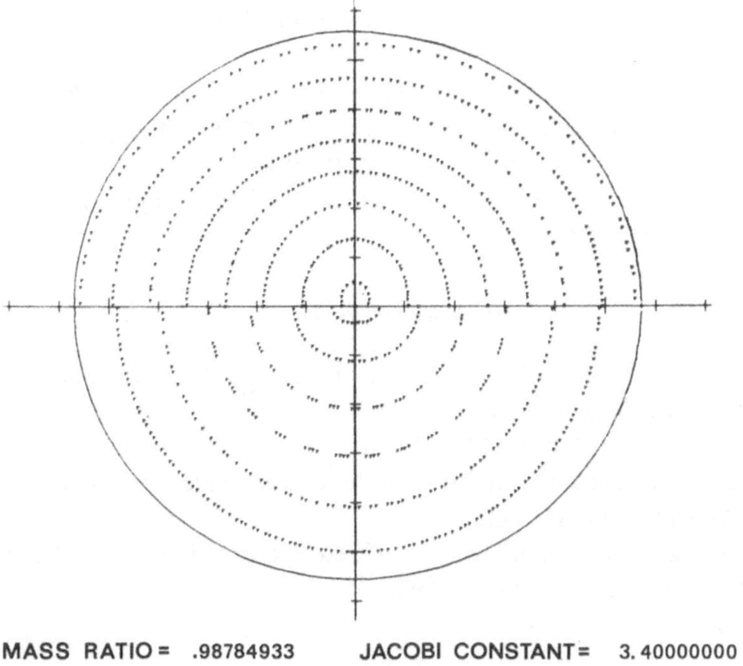

Fig. 15.

vantages were that it was difficult to prescribe *a priori* how many steps would be required; the optimum scale of the plot; and even initial conditions most likely to produce 'interesting' results. Therefore the program was adapted to an interactive mode so that the investigator could see the results as they were computed and make modifications in real time.

The current version of the program displays the points as they are calculated, on a cathode ray tube attached to the computer. After integrating for a while, it stops and awaits further instructions. The investigator may decide that more steps should be integrated for that orbit and may so order through a keyboard; or he may decide that the orbit is not going to be interesting and may delete it or any previously computed orbit. At any time he may change the scale of the picture. New initial conditions may be indicated by positioning a 'tracking cross' on the screen with a light-sensitive pen (to indicate an apsis of an orbit). When he is satisfied, a command can be given to record the data on microfilm and punch cards. (These punch cards may later be read in for revision in the same interactive manner, if needed.) In this manner, guided by clues in the current figure (such as bulges in curves indicating possible resonances) or in previously computed ones, a representative selection of orbits can be chosen as the figure is built up. This interactive mode of operation has proven to be very effective in finding interesting orbits and in producing a figure which represents the nature of the possible orbits well.

Upon completion of the remaining three mass ratios all the figures will be published together with an explanation. It is hoped that it will be useful as a reference work, where one can 'look up' certain types of orbits much as one uses a dictionary. (This concept was first suggested by Professor Danby.)

The existence of curves, even in highly perturbed orbits, may also be used as the starting point for a perturbation theory (Jefferys, 1968). Essentially one uses an approximation to the orbit derived from a study of the invariant curves. Programs are under development at the University of Texas for the manipulation of the series expansions needed to apply this method.

Acknowledgements

It is a pleasure to thank Professor Danby, whose suggestion to the author of the usefulness of a 'Dictionary of the Restricted Problem' was the motivation for this work. Thanks also go to Professor Moser for his continued interest in the calculation of invariant curves by machine, and to Dr. Hénon, who made some valuable suggestions concerning the format of the figures.

This work has been supported by an Alfred P. Sloan Foundation Fellowship, and their support is gratefully acknowledged.

References

Arnol'd, V. I.: 1963, *Usp. Math. Nauk U.S.S.R.* **18**, 13.
Hénon, M.: 1966, *Bull. Astron. Ser.* 3, **1**, 49.
Hénon, M. and Heiles, C.: 1964, *Astron. J.* **69**, 73.
Jefferys, W. H.: 1966, *Astron. J.* **71**, 306.
Jefferys, W. H.: 1968, *Astron. J.* **73**, 522.
Moser, J.: 1962, *Nachr. Akad. Wiss. Göttingen, Math. Phys. Kl. I.*
Szebehely, V.: 1967, *Theory of Orbits*, Academic Press, New York.

Discussion

J. M. A. Danby: May I first congratulate you for a job which I would not have had the courage to attempt.

J. M. A. Danby: (1) Could you in some of your sections indicate by numbers the actual mapping?

(2) Do you have any ideas from your experience about the continuation of quasi-periodic orbits?

(3) You spoke of resonances 'disappearing'. Are you sure that this is the case, or do the macroscopic archipelagos become too small to compute?

W. H. Jefferys: (1) Yes, this could be done, but there are certain technical difficulties due to the fact that only half of the figure is drawn. Actually, I depend on the existence of works on periodic orbits which also give this information. One could identify each orbit by its rotation number, which I have not done.

(2) Quasi-periodic orbits can be continued to large mass ratios, provided the Jacobi constant is sufficiently large and there are no powerful resonances nearby.

(3) They either become too small to compute or else they may become unstable. The resolution is only one part in 1000.

A. Deprit: One piece of identification from one Jacobi constant to the next for a given mass ratio would be the periodic orbit sitting in the middle of the islands of quasi-periodic orbits that you compute.

W. H. Jefferys: Yes, this is true. It is for this reason that I stress that this work is not intended to be complete in itself, but to complement the work being done on periodic orbits. 'The whole is greater than the sum of its parts' is a good maxim here. I have not specified the order in which the islands are described and assume this information could come from other sources.

A. Deprit: The present qualitative investigations in the Restricted Problem seem to grow vigorously in two contrasting directions. Together with Dr. Hénon, you represent what I would like to call the 'Global wing'; Lloyd Carpenter, J. Message and I are rather at the other end of spectrum, working as handcraftsmen to chisel the fine details for specific mass ratios. I wish we could establish closer collaboration to help one another in correlating and complementing our information.

W. H. Jefferys: I am very much interested in such collaboration and will undertake to get my results into the hands of others before publication, if possible, if it is requested.

M. Hénon: Does your computer operate on a time-sharing basis? This would seem to be a necessary requirement in order not to waste computer time with your interactive mode of operation.

W. H. Jefferys: That is correct. One must have a system with time sharing in order not to waste valuable computer time.

APPLICATION OF HILL'S LUNAR METHOD IN GENERAL PLANETARY THEORY*

V. A. BRUMBERG

Institute of Theoretical Astronomy, Leningrad, U.S.S.R.

Abstract. This paper suggests an algorithm for the formal solution of the N planet problem in the rectangular heliocentric coordinates. The algorithm is based on the use of the intermediate quasi-periodic solution generalizing the variation curve of Hill. This particular solution is expressed by power series in terms of the planetary masses with quasi-periodic coefficients. It contains all the inequalities that do not depend on the orbital eccentricities and inclinations. The system of the nonlinear differential equations for the deviations of the true values of the coordinates from those corresponding to the intermediate solution has been further derived. The solution of this system is presented by the series in powers of the variables slowly changing with the time. The coefficients of these series are quasi-periodic functions dependent on the mean longitudes of the planets and developable in powers of the planetary masses. The behaviour of the slowly changing variables is described by the autonomous system of the nonlinear differential equations. This final system yields the secular perturbations in the planetary motion.

1. Introduction

In this paper an algorithm for the formal solution of the N planet problem has been elaborated. The algorithm is based on the combination of Hill's Lunar method (the use of the non-Keplerian intermediate orbit) and the main principle of von Zeipel's method (separation of the short-period and long-period terms). The application of Hill's method in the planetary problem was first suggested (although in an inadequate form) by Roure (1949). The possibility of the use of von Zeipel's method to construct the general planetary theory in the elements of orbits was recently shown by Meffroy (1966). The calculations of the general planetary perturbations in the elements being performed now in the Institute of Theoretical Astronomy (Leningrad) by G. A. Krasinsky are also based on the separation of the short-period and long-period terms.

The algorithm suggested below follows the same idea of the separation of the fast and slowly changing variables but instead of elements the rectangular coordinates are used. The general expressions for the right-hand members of the equations of motion can be found, therefore, comparatively easily, which is of particular importance in performing the whole method on the electronic computers. The hamiltonian form of the equations is not used here because in spite of all theoretical advantages it involves some practical difficulties.

The paper consists of seven sections. To clarify the exposition the general features of the algorithm are outlined in the first section. The other sections are devoted to the details of it.

* Presented by Dr. S. F. Mello.

G. E. O. Giacaglia (ed.), Periodic Orbits, Stability and Resonances, 410–450.

2. Form of the Solution

Starting with the equations of the N planet problem in the rectangular heliocentric coordinates

$$\ddot{x}_i = \frac{\partial U_i}{\partial x_i}, \quad \ddot{y}_i = \frac{\partial U_i}{\partial y_i}, \quad \ddot{z}_i = \frac{\partial U_i}{\partial z_i}, \quad (i = 1, 2, \ldots, N) \tag{1}$$

let p_i, q_i, w_i be new variables representing the small deviations from the plane circular motion of the planets. Then

$$\begin{aligned} x_i + \sqrt{-1}\, y_i &= a_i(1 - p_i) \exp \sqrt{-1}\, \lambda_i, \\ x_i - \sqrt{-1}\, y_i &= a_i (1 - q_i) \exp (- \sqrt{-1}\, \lambda_i), \\ z_i &= a_i w_i, \end{aligned} \tag{2}$$

where λ_i is the mean longitude

$$\lambda_i = n_i t + \varepsilon_i, \tag{3}$$

the mean motion n_i and semi-major axis a_i being related by Kepler's third law

$$n_i^2 a_i^3 = f (m_0 + m_i). \tag{4}$$

Equations (1) expressed in the new variables take the form

$$\begin{aligned} \ddot{p}_i + 2\sqrt{-1}\, n_i \dot{p}_i + n_i^2 (1 - p_i) &= \frac{2}{a_i^2} \frac{\partial U_i}{\partial q_i}, \\ \ddot{q}_i - 2\sqrt{-1}\, n_i \dot{q}_i + n_i^2 (1 - q_i) &= \frac{2}{a_i^2} \frac{\partial U_i}{\partial p_i}, \\ \ddot{w}_i &= \frac{1}{a_i^2} \frac{\partial U_i}{\partial w_i}. \end{aligned} \tag{5}$$

Since p_i and q_i are the conjugate complex variables and w_i is real it is sufficient to consider the first and third of these equations. Expansion of the right-hand members in powers of p_i, q_i, w_i gives

$$\ddot{p}_i + 2\sqrt{-1}\, n_i \dot{p}_i - \tfrac{3}{2} n_i^2 (p_i + q_i) = n_i^2 P_i, \tag{6}$$

$$\ddot{w}_i + n_i^2 w_i = n_i^2 W_i \tag{7}$$

in which

$$\begin{aligned} P_i = {} & \sum_{k=0}^{\infty} \sum_{l=0}^{\infty} \sum_{m=0}^{\infty} \varphi_{klm} p_i^k q_i^l w_i^{2m} \\ & + \mu \sum_{j=1}^{N}{}^{(i)} \sum_{k=0}^{\infty} \sum_{l=0}^{\infty} \sum_{r=0}^{\infty} \sum_{s=0}^{\infty} \sum_{m=0}^{\infty} \sum_{t=-m}^{m} \kappa_{ij} \psi_{klrsmt}^{(ij)} p_i^k q_i^l p_j^r q_j^s w_i^{m-t} w_j^{m+t}, \end{aligned} \tag{8}$$

$$\begin{aligned} W_i = {} & \sum_{k=0}^{\infty} \sum_{l=0}^{\infty} \sum_{m=0}^{\infty} \varrho_{klm} p_i^k q_i^l w_i^{2m+1} \\ & + \mu \sum_{j=1}^{N}{}^{(i)} \sum_{k=0}^{\infty} \sum_{l=0}^{\infty} \sum_{r=0}^{\infty} \sum_{s=0}^{\infty} \sum_{m=0}^{\infty} \sum_{t=-m-1}^{m} \kappa_{ij} \theta_{klrsmt}^{(ij)} p_i^k q_i^l p_j^r q_j^s w_i^{m-t} w_j^{m+t+1} \end{aligned} \tag{9}$$

with the constant coefficients φ_{klm}, $\varrho_{klm}(\varphi_{000}=\varphi_{100}=\varphi_{010}=\varrho_{000}=0)$ and functions $\psi^{(ij)}_{klrsmt}$, $\theta^{(ij)}_{klrsmt}$ dependent on the time only by means of

$$\zeta_{ij} = \exp\sqrt{-1}(\lambda_i - \lambda_j). \tag{10}$$

The constant μ is a small parameter characterized by the ratio of the planetary masses m_j to the mass m_0 of the sun while κ_{ij} are numerical coefficients such that

$$\mu\kappa_{ij} = \frac{m_j}{m_0 + m_i}. \tag{11}$$

To generalize Hill's Lunar method, neglect in the first approximation all terms dependent on the eccentricities and inclinations and take for the intermediary of the N planet problem the particular plane quasi-periodic solution of Equations (5). For every planet this solution depends on $N-1$ angular arguments (the differences of the mean longitudes) and contains $2N$ arbitrary constants (the semi-major axes a_i and mean longitudes at the epoch ε_i). It should be noted that recently Jefferys and Moser (1966) and independently of them Krasinsky (1969) had rigorously demonstrated the existence of the quasi-periodic solutions similar to the formal solution used here. All the coordinates w_i vanish for this solutions while p_i and q_i may be represented in the form

$$p_i = \sum_{s=1}^{\infty} p_s^{(i)}\mu^s, \quad q_i = \sum_{s=1}^{\infty} q_s^{(i)}\mu^s, \tag{12}$$

where the quasi-periodic functions $p_s^{(i)}$ and $q_s^{(i)}$ depend on $\zeta_{ij}(j=1, 2, \ldots, N)$ only. When $\mu=0$ this solution leads to the circular motion $p_i=q_i=0$.

Denoting the intermediate solution (12) by $p_i^{(0)}$, $q_i^{(0)}$ put

$$p_i = p_i^{(0)} + \delta p_i, \quad q_i = q_i^{(0)} + \delta q_i \tag{13}$$

and substitute these expressions into Equations (6) and (7). The right-hand members of these equations may be developed in powers of δp_i, δq_i as follows

$$\begin{aligned} P_i = {} & \sum_{\alpha=0}^{\infty}\sum_{\beta=0}^{\infty}\sum_{m=0}^{\infty} P^{(i)}_{\alpha\beta m}(\delta p_i)^\alpha(\delta q_i)^\beta w_i^{2m} \\ & + \mu \sum_{j=1}^{N}{}^{(i)} \sum_{\alpha=0}^{\infty}\sum_{\beta=0}^{\infty}\sum_{\sigma=0}^{\infty}\sum_{\nu=0}^{\infty}\sum_{m=0}^{\infty}\sum_{t=-m}^{m} P^{(ij)}_{\alpha\beta\sigma\nu mt}(\delta p_i)^\alpha(\delta q_i)^\beta \\ & \times (\delta p_j)^\sigma(\delta q_j)^\nu w_i^{m-t} w_j^{m+t}, \end{aligned} \tag{14}$$

$$\begin{aligned} W_i = {} & \sum_{\alpha=0}^{\infty}\sum_{\beta=0}^{\infty}\sum_{m=0}^{\infty} W^{(i)}_{\alpha\beta m}(\delta p_i)^\alpha(\delta q_i)^\beta w_i^{2m+1} \\ & + \mu \sum_{j=1}^{N}{}^{(i)} \sum_{\alpha=0}^{\infty}\sum_{\beta=0}^{\infty}\sum_{\sigma=0}^{\infty}\sum_{\nu=0}^{\infty}\sum_{m=0}^{\infty}\sum_{t=-m-1}^{m} W^{(ij)}_{\alpha\beta\sigma\nu mt}(\delta p_i)^\alpha(\delta q_i)^\beta \\ & \times (\delta p_j)^\sigma(\delta q_j)^\nu w_i^{m-t} w_j^{m+t+1}, \end{aligned} \tag{15}$$

the coefficients being functions of the coordinates of the intermediate solution (12). For $\mu=0$, $P^{(i)}_{\alpha\beta m}$ and $W^{(i)}_{\alpha\beta m}$ reduce to the numerical coefficients $\varphi_{\alpha\beta m}$ and $\varrho_{\alpha\beta m}$ respectively. Using the equations for the intermediary and collecting the linear terms on δp_i, δq_i, w_i one obtains

$$\delta\ddot{p}_i + 2\sqrt{-1}\,n_i\,\delta\dot{p}_i + n_i^2 \sum_{j=1}^{N} (K_{ij}\,\delta p_j + L_{ij}\,\delta q_j) = n_i^2 P_i^*\,, \tag{16}$$

$$\ddot{w}_i + n_i^2 \sum_{j=1}^{N} M_{ij} w_j = n_i^2 W_i^*\,, \tag{17}$$

where P_i^*, W_i^* are defined by the same formulas (14), (15), the terms of zero and first powers of δp_i, δq_i, w_i being excluded. Thus, in computing P_i^*, W_i^* by these formulas

$$\begin{aligned} &P_{000}^{(i)} = P_{100}^{(i)} = P_{010}^{(i)} = W_{000}^{(i)} = 0\,,\\ &P_{000000}^{(ij)} = P_{100000}^{(ij)} = P_{010000}^{(ij)} = P_{001000}^{(ij)} = P_{000100}^{(ij)}\\ &\quad = W_{000000}^{(ij)} = W_{00000-1}^{(ij)} = 0\,. \end{aligned}$$

K_{ij}, L_{ij}, M_{ij} are quasi-periodic functions of the time dependent on the coordinates of the intermediate solution (12). For $\mu=0$ these coefficients take constant values

$$K_{ij} = L_{ij} = -\tfrac{3}{2}\,\delta_{ij}\,, \quad M_{ij} = \delta_{ij}\,,$$

Note, that assuming $P_i^* = W_i^* = 0$, i.e., omitting the terms of the second and higher powers of δp_i, δq_i, w_i (16) and (17) yield two independent systems of linear differential equations with quasi-periodic coefficients determining the principal parts of the motion of the perihelia and nodes as well as the inequalities of the first power of the eccentricities and inclinations of the planetary orbits.

Introducing the auxiliary unknowns $\delta\dot{p}_i$, $\delta\dot{q}_i$, $\dot{w}_i$ rewrite (16), (17) as the system of the first order equations and let for $\mu=0$ the linear part of this system have the Jordan form. For this purpose perform a linear transformation with constant coefficients

$$\begin{pmatrix} \delta p_i \\ \delta q_i \\ \delta\dot{p}_i \\ \delta\dot{q}_i \end{pmatrix} = C_i' \begin{pmatrix} \xi_i \\ \eta_i \\ u_i \\ \bar{u}_i \end{pmatrix},$$

$$C_i' = \begin{pmatrix} \sqrt{-1}\,n_i & -\frac{2}{3} & -\frac{1}{2} & \frac{3}{2} \\ -\sqrt{-1}\,n_i & -\frac{2}{3} & \frac{3}{2} & -\frac{1}{2} \\ 0 & \sqrt{-1}\,n_i & -\frac{1}{2}\sqrt{-1}\,n_i & -\frac{3}{2}\sqrt{-1}\,n_i \\ 0 & -\sqrt{-1}\,n_i & \frac{3}{2}\sqrt{-1}\,n_i & \frac{1}{2}\sqrt{-1}\,n_i \end{pmatrix}, \tag{18}$$

$$\begin{pmatrix} w_i \\ \dot{w}_i \end{pmatrix} = C_i'' \begin{pmatrix} v_i \\ \bar{v}_i \end{pmatrix}, \quad C_i'' = \begin{pmatrix} 1 & 1 \\ \sqrt{-1}\,n_i & -\sqrt{-1}\,n_i \end{pmatrix}, \tag{19}$$

ξ_i, η_i being the real variables while u_i, $\bar{u}_i$ and v_i, $\bar{v}_i$ are pairs of the conjugate complex variables (here and everywhere below, the line over the various symbols is to indicate the conjugate complex quantities).

After this transformation the system of (16) and (17) is changed into

$$\dot{X} = (P + Q)\,X + R(X, t)\,. \tag{20}$$

Here X is a block column-matrix

$$X = \begin{pmatrix} \xi \\ \eta \\ u \\ \bar{u} \\ v \\ \bar{v} \end{pmatrix}, \tag{21}$$

$\xi, \eta, u, \bar{u}, v, \bar{v}$ denoting $N \times 1$ column-matrices of the variables $\xi_i, \eta_i, u_i, \bar{u}_i, v_i, \bar{v}_i$ $(i=1, 2, \ldots, N)$ respectively. $R(X, t)$ is a column-matrix obtained from the right-hand sides of the preceding system by means of the transformation inverse to (18), (19). Indeed, if

$$R = \begin{pmatrix} R_1 \\ R_2 \\ R_3 \\ R_4 \\ R_5 \\ R_6 \end{pmatrix}, \tag{22}$$

each block $R_\kappa (\kappa=1, 2, \ldots, 6)$ being $N \times 1$ column-matrix, then

$$\begin{pmatrix} R_{1i} \\ R_{2i} \\ R_{3i} \\ R_{4i} \end{pmatrix} = (C_i')^{-1} \begin{pmatrix} 0 \\ 0 \\ n_i^2 P_i^* \\ n_i^2 \bar{P}_i^* \end{pmatrix},$$

$$(C_i')^{-1} = \begin{pmatrix} -\frac{\sqrt{-1}}{2n_i} & \frac{\sqrt{-1}}{2n_i} & -\frac{1}{n_i^2} & -\frac{1}{n_i^2} \\ -3 & -3 & \frac{3\sqrt{-1}}{2n_i} & -\frac{3\sqrt{-1}}{2n_i} \\ -\frac{3}{2} & -\frac{3}{2} & \frac{\sqrt{-1}}{2n_i} & -\frac{3\sqrt{-1}}{2n_i} \\ -\frac{3}{2} & -\frac{3}{2} & \frac{3\sqrt{-1}}{2n_i} & -\frac{\sqrt{-1}}{2n_i} \end{pmatrix}, \tag{23}$$

$$\begin{pmatrix} R_{5i} \\ R_{6i} \end{pmatrix} = (C_i'')^{-1} \begin{pmatrix} 0 \\ n_i^2 W_i^* \end{pmatrix}, \qquad (C_i'')^{-1} = \begin{pmatrix} \frac{1}{2} & -\frac{\sqrt{-1}}{2n_i} \\ \frac{1}{2} & \frac{\sqrt{-1}}{2n_i} \end{pmatrix}. \tag{24}$$

The elements of $R(X, t)$ are expanded in powers of X beginning with the terms of the second power. The coefficients of these expansions are quasi-periodic functions of the time through the coordinates of the intermediate solution. For $\mu=0$ these coefficients become constant. P is a constant block matrix in the Jordan form

$$P = \begin{pmatrix} 0 & E & 0 & 0 & 0 & 0 \\ 0 & 0 & 0 & 0 & 0 & 0 \\ 0 & 0 & \sqrt{-1}\,\mathcal{N} & 0 & 0 & 0 \\ 0 & 0 & 0 & -\sqrt{-1}\,\mathcal{N} & 0 & 0 \\ 0 & 0 & 0 & 0 & \sqrt{-1}\,\mathcal{N} & 0 \\ 0 & 0 & 0 & 0 & 0 & -\sqrt{-1}\,\mathcal{N} \end{pmatrix}, \tag{25}$$

each block being a square $N \times N$ matrix. $\mathcal{N}$ is a diagonal matrix of the mean motions

$$\mathcal{N} = \begin{pmatrix} n_1 & 0 & \dots & 0 \\ 0 & n_2 & \dots & 0 \\ \dots & \dots & \dots & \dots \\ 0 & 0 & \dots & n_N \end{pmatrix} \tag{26}$$

and E is a unit matrix (everywhere below E denotes unit matrix not necessary of $N \times N$ dimensions). Finally, Q represents a quasi-periodic matrix dependent on the time by means of the coordinates of the intermediate solution. The elements of Q vanish when $\mu = 0$.

Hence, for $\mu = 0$ the characteristic equation of the linear part of (20) admits N double zero roots (each having a nontrivial elementary divisor) and $2N$ pairs of purely imaginary conjugate roots.

Using the theory of Birkhoff's (1927) series and the recent results by Krasinsky (1968), perform the nonlinear substitution

$$X = (E + S)\,Y + \Gamma(Y, t), \tag{27}$$

where Y is a column-matrix of new variables, $S = S(t)$ is an unknown square $6N \times 6N$ matrix and $\Gamma(Y, t)$ is a column-matrix of unknown nonlinear functions developable in powers of Y beginning with the second power. The substitution (27) brings the system (20) into

$$\dot{Y} = HY + F(Y, t) \tag{28}$$

in which

$$H = (E + S)^{-1}\,[(P + Q)(E + S) - \dot{S}] \tag{29}$$

and $F(Y, t)$ is a column-matrix of new right-hand members being holomorphic functions of Y without constant and linear terms. S and $\Gamma(Y, t)$ are to be chosen to simplify (28) as much as possible. Of course, S and $\Gamma(Y, t)$ must be quasiperiodic with respect to the time. Moreover, the frequency bases of S and Q must be identical. Therefore, S may depend on the time only by means of ζ_{ij} $(i, j = 1, 2, \dots, N)$. S and $\Gamma(Y, t)$ contain thus all the non-resonance terms of the system (20) while the remaining resonance terms give rise to the system (28). The elements of S are expanded in powers of μ and the elements of $\Gamma(Y, t)$ may be developed in powers of Y and μ. Instead of expansion of S and $\Gamma(Y, t)$ in powers of μ it is possible to use the successive iterations of this parameter. For $\mu=0$ S reduces to the zero matrix and $\Gamma(Y, t)$ becomes independent of the time explicitly.

Having obtained S (up to some order of accuracy of μ) and $\Gamma(Y, t)$ (up to some order of accuracy of Y and μ) perform the linear substitution

$$Y = TZ \tag{30}$$

with the matrix T combined of square $N \times N$ blocks

$$T = \begin{pmatrix} E & 0 & 0 & 0 & 0 & 0 \\ 0 & E & 0 & 0 & 0 & 0 \\ 0 & 0 & \exp\sqrt{-1}\,\mathcal{N}t & 0 & 0 & 0 \\ 0 & 0 & 0 & \exp(-\sqrt{-1}\,\mathcal{N}t) & 0 & 0 \\ 0 & 0 & 0 & 0 & \exp\sqrt{-1}\,\mathcal{N}t & 0 \\ 0 & 0 & 0 & 0 & 0 & \exp(-\sqrt{-1}\,\mathcal{N}t) \end{pmatrix}, \tag{31}$$

in which $exp\sqrt{-1}\mathcal{N}t$ denotes a diagonal matrix

$$\exp\sqrt{-1}\,\mathcal{N}t = \begin{pmatrix} \exp\sqrt{-1}\,\lambda_1 & 0 & \dots & 0 \\ 0 & \exp\sqrt{-1}\,\lambda_2 & \dots & 0 \\ \dots & \dots & \dots & \dots \\ 0 & 0 & \dots & \exp\sqrt{-1}\,\lambda_N \end{pmatrix}. \tag{32}$$

Then (28) is transformed into an autonomous system

$$\dot{Z} = \Omega Z + \mathcal{M}(Z), \tag{33}$$

in which Ω is a constant square $6N \times 6N$ matrix and the elements of the column-matrix $\mathcal{M}(Z)$ are power series of Z beginning with the second power, the coefficients being constant.

If $\theta_i, \varrho_i, a_i, \bar{a}_i, b_i, \bar{b}_i$ are variables described by Y and $\sigma_i, v_i, \alpha_i, \bar{\alpha}_i, \beta_i, \bar{\beta}_i$ are those described by Z, then

$$Y = \begin{pmatrix} \theta \\ \varrho \\ a \\ \bar{a} \\ b \\ \bar{b} \end{pmatrix}, \qquad Z = \begin{pmatrix} \sigma \\ v \\ \alpha \\ \bar{\alpha} \\ \beta \\ \bar{\beta} \end{pmatrix}, \tag{34}$$

where each block represents $N \times 1$ column-matrix of the appropriate variables. In explicit form there results

$$\theta_i = \sigma_i, \ \varrho_i = v_i, \quad a_i = \alpha_i \exp\sqrt{-1}\,\lambda_i, \quad b_i = \beta_i \exp\sqrt{-1}\,\lambda_i. \tag{35}$$

It turns out that Equations (33) admit as a solution $\sigma_i = v_i = 0$ ($i = 1, 2, \dots, N$). The remaining equations for determination of α and β become

$$\begin{aligned} \dot{\alpha} &= A\alpha + \Upsilon(\alpha, \bar{\alpha}, \beta, \bar{\beta}) \\ \dot{\beta} &= B\beta + \Psi(\alpha, \bar{\alpha}, \beta, \bar{\beta}), \end{aligned} \tag{36}$$

where A and B are constant purely imaginary $N \times N$ matrices while Υ and Ψ are column-matrices with the elements being expanded in powers of α_i, $\bar{\alpha}_i$, β_i, $\bar{\beta}_i$. These expansions start with the terms of the third power and contain the forms of odd powers only. When $\mu = 0$ all the matrices A, B, Υ and Ψ vanish. If Υ, Ψ are neglected and only the terms of the first order of μ are retained in A, B, equations (36) coincide completely with those of the classic trigonometric Laplace-Lagrange's theory of the secular perturbations.

Eigenvalues of A and B determine the principal parts of the motion of the perihelia and nodes of the planetary orbits (one eigenvalue of B is equal to zero). Nonlinear terms Υ and Ψ give corrections to these frequencies due to the influence of the eccentricities and inclinations. To the first order of μ the trigonometric theory of the secular perturbations taking into account the terms of the third power of the eccentricities and inclinations has been elaborated in the paper by Anolik *et al.* (1969) by extension of the classic method based on the use of the Lagrange's equations for osculating elements and appropriate development of the disturbing function.

Integration of (36) yields $4N$ arbitrary constants. In addition with $2N$ arbitrary constants of the intermediary this gives the necessary set of $6N$ constants of the general solution of the N planet problem. Eventually, this general solution in virtue of (13), (18), (19), (27) and (30) is represented in form of series for p_i, q_i, w_i in powers of α_j, $\bar{\alpha}_j$, β_j, $\bar{\beta}_j$ $(j = 1, 2, \ldots N)$, the coefficients being developed in powers of μ and expressed by the trigonometric series in multiples of the mean longitudes. In forming the series (27) presenting X in terms of Y it is possible from the very start to use particular values $\theta_i = \rho_i = 0$ leading to the substantial simplification of the whole algorithm. Since the variables α_i, $\bar{\alpha}_i$, β_i, $\bar{\beta}_i$ are changed extremely slowly with the time it is no good to substitute into the general solution their analytical expressions which may be obtained by integration of (36) with the aid of series. This would result in the appearance of a tremendous number of terms with practically undistinguished periods. It is far more convenient to substitute into the general solution the constant values of α_i, $\bar{\alpha}_i$, β_i, $\bar{\beta}_i$ (corresponding to the present epoch, for example) and to collect the similar trigonometric terms. Then the general solution will be represented by trigonometric series in multiples of the mean longitudes with constant coefficients preserving their values for many centuries. In computation of the planetary coordinates for a longer interval of time, the changing of the values of the coefficients may be easily taken into account by performing the approximate numerical or analytical integration of the system (36).

It is hardly possible to make a definite judgment concerning the effectiveness of the algorithm suggested without the detailed actual testing of it. It is plain, for example, that the terms of the second and third orders of μ in the matrices A and B may be determined by this method much more easily than in the classic theory. However, the way to find here the purely elliptic terms in the coordinates of the planets may hardly be regarded as the effective one for the solution of the two body problem.

3. Right-Hand Sides of the Equations of Motion

To become familiar with the details of the method suggested begin with the derivation of the formulas defining the right-hand sides of the equations of motion. The force function U_i relating to (1) is

$$U_i = \frac{f(m_0 + m_i)}{r_i} + \sum_{j=1}^{N}{}^{(i)} f m_j \left(\frac{1}{\Delta_{ij}} - \frac{x_i x_j + y_i y_j + z_i z_j}{r_j^3} \right)$$

or after the substitution (2)

$$U_i = n_i^2 a_i^2 \left\{ \frac{a_i}{r_i} + \mu \sum_{j=1}^{N}{}^{(i)} \kappa_{ij} \left[\frac{a_i}{\Delta_{ij}} - a_i^2 a_j \frac{(1-p_i)(1-q_j)\zeta_{ij} + (1-q_i)(1-p_j)\zeta_{ij}^{-1} + 2w_i w_j}{2r_j^3} \right] \right\}$$

in which

$$r_i = a_i[(1-p_i)(1-q_i) + w_i^2]^{1/2},$$
$$\Delta_{ij} = \{[a_i(1-p_i) - a_j(1-p_j)\zeta_{ij}^{-1}][a_i(1-q_i) - a_j(1-q_j)\zeta_{ij}] + (a_i w_i - a_j w_j)^2\}^{1/2}.$$

Substituting this expression of U_i into

$$P_i = -1 - \tfrac{1}{2}p_i - \tfrac{3}{2}q_i + \frac{2}{n_i^2 a_i^2}\frac{\partial U_i}{\partial q_i}, \quad W_i = w_i + \frac{1}{n_i^2 a_i^2}\frac{\partial U_i}{\partial w_i}$$

and expanding in powers of p_i, q_i, w_i one obtains the developments (8), (9) for the right-hand members of Equations (6), (7). Numerical coefficients of these developments are

$$\varphi_{klm} = (-1)^m \frac{(\frac{3}{2})_m (m+\frac{1}{2})_k (m+\frac{3}{2})_l}{(1)_m (1)_k (1)_l}, \quad (\varphi_{000} = \varphi_{100} = \varphi_{010} = 0) \tag{37}$$

$$\varrho_{klm} = (-1)^{m+1} \frac{(\frac{3}{2})_m (m+\frac{3}{2})_k (m+\frac{3}{2})_l}{(1)_m (1)_k (1)_l}, \quad (\varrho_{000} = 0) \tag{38}$$

where $(\alpha)_s$ denotes the generalized factorial

$$(\alpha)_0 = 1, \quad (\alpha)_s = \alpha(\alpha+1)\dots(\alpha+s-1), \quad s = 1, 2, \dots$$

with useful relations

$$(\alpha)_s = (-1)^s (1-\alpha-s)_s, \quad (\alpha)_{r+s} = (\alpha)_r (\alpha+r)_s.$$

To write $\psi_{klrsmt}^{(ij)}$ and $\theta_{klrsmt}^{(ij)}$ in compact form introduce an auxiliary function of ζ_{ij}

$$\gamma^{(ij)}(n, x, y, v, \alpha) = \alpha^n (1 - \alpha\zeta_{ij}^{-1})^x (1 - \alpha\zeta_{ij})^y (-1)^v \zeta_{ij}^v \tag{39}$$

with real parameters n, x, y, $\alpha(|\alpha|<1)$ and integer ν. Then

$$\psi_{klrsmt}^{(ij)} = (-1)^t \frac{(\frac{3}{2})_m (m+\frac{1}{2})_{k+r} (m+\frac{3}{2})_{l+s}}{(1)_m (1)_{k+r} (1)_{l+s}}$$

$$\times \left[\left(\frac{a_i}{a_j}\right)^2 \zeta_{ij}^{-1} \delta_{k,0} \delta_{l,0} \delta_{t,m} + \frac{(1+r)_k (1+s)_l (1+m-t)_{m+t}}{(1)_k (1)_l (1)_{m+t}} \right.$$

$$\left. \times \gamma^{(ij)} \left\{ \begin{matrix} r+s+m+t, -m-k-r-\frac{1}{2}, -m-l-s-\frac{3}{2}, -r+s, \dfrac{a_j}{a_i} \\ k+l+m-t+2, -m-l-s-\frac{3}{2}, -m-k-r-\frac{1}{2}, k-l-1, \dfrac{a_i}{a_j} \end{matrix} \right\} \right], \tag{40}$$

$$\theta_{klrsmt}^{(ij)} = (-1)^t \frac{(\frac{3}{2})_m (m+\frac{3}{2})_{k+r} (m+\frac{3}{2})_{l+s}}{(1)_m (1)_{k+r} (1)_{l+s}}$$

$$\times \left[-\left(\frac{a_i}{a_j}\right)^2 \delta_{k,0} \delta_{l,0} \delta_{t,m} + \frac{(1+r)_k (1+s)_l (1+m-t)_{m+t+1}}{(1)_k (1)_l (1)_{m+t+1}} \right.$$

$$\left. \times \gamma^{(ij)} \left\{ \begin{matrix} r+s+m+t+1, -m-k-r-\frac{3}{2}, -m-l-s-\frac{3}{2}, -r+s, \dfrac{a_j}{a_i} \\ k+l+m-t+2, -m-l-s-\frac{3}{2}, -m-k-r-\frac{3}{2}, k-l, \dfrac{a_i}{a_j} \end{matrix} \right\} \right] \tag{41}$$

where $\gamma^{(ij)}$ should be evaluated for the upper row of parameters if $a_j<a_i$ and for the lower row if $a_i<a_j$. Expansion of (39) in the exponential series yields

$$\gamma^{(ij)}(n, x, y, \nu, \alpha) = \sum_{\sigma=-\infty}^{\infty} \gamma_\sigma^{(ij)}(n, x, y, \nu, \alpha)\, \zeta_{ij}^\sigma \tag{42}$$

with the coefficients determined by

$$\gamma_\sigma^{(ij)}(n, x, y, \nu, \alpha) = (-1)^\nu \frac{(-x) \max\{0, -\sigma+\nu\} (-y) \max\{0, \sigma-\nu\}}{(1)_{|\sigma-\nu|}}$$
$$\times \alpha^{|\sigma-\nu|+n} \cdot F(-x+\max\{0, -\sigma+\nu\},$$
$$-y+\max\{0, \sigma-\nu\}, 1+|\sigma-\nu|; \alpha^2) \tag{43}$$

in which $F(\alpha, \beta, \gamma; x)$ represents the hypergeometric function

$$F(\alpha, \beta, \gamma; x) = \sum_{s=0}^{\infty} \frac{(\alpha)_s (\beta)_s}{(\gamma)_s (1)_s} x^s.$$

Substitution of (42) into (40) and (41) gives

$$\psi_{klrsmt}^{(ij)} = \sum_{\sigma=-\infty}^{\infty} \psi_{\sigma}^{(ij;\, klrsmt)} \zeta_{ij}^{\sigma}, \tag{44}$$

$$\theta_{klrsmt}^{(ij)} = \sum_{\sigma=-\infty}^{\infty} \theta_{\sigma}^{(ij;\, klrsmt)} \zeta_{ij}^{\sigma} \tag{45}$$

in which

$$\psi_{\sigma}^{(ij;\, klrsmt)} = (-1)^t \frac{(\frac{3}{2})_m (m+\frac{1}{2})_{k+r} (m+\frac{3}{2})_{l+s}}{(1)_m (1)_{k+r} (1)_{l+s}}$$

$$\times \left[\left(\frac{a_i}{a_j}\right)^2 \delta_{k,0}\, \delta_{l,0}\, \delta_{t,m}\, \delta_{\sigma,-1} + \frac{(1+r)_k (1+s)_l (1+m-t)_{m+t}}{(1)_k (1)_l (1)_{m+t}} \right.$$

$$\left. \times\, \gamma_{\sigma}^{(ij)} \left\{ \begin{matrix} r+s+m+t,\ -m-k-r-\frac{1}{2},\ -m-l-s-\frac{3}{2},\ -r+s,\ \dfrac{a_j}{a_i} \\ k+l+m-t+2,\ -m-l-s-\frac{3}{2},\ -m-k-r-\frac{1}{2},\ k-l-1,\ \dfrac{a_i}{a_j} \end{matrix} \right\} \right]. \tag{46}$$

$$\theta_{\sigma}^{(ij;\, klrsmt)} = (-1)^t \frac{(\frac{3}{2})_m (m+\frac{3}{2})_{k+r} (m+\frac{3}{2})_{l+s}}{(1)_m (1)_{k+r} (1)_{l+s}}$$

$$\times \left[-\left(\frac{a_i}{a_j}\right)^2 \delta_{k,0}\, \delta_{l,0}\, \delta_{t,m}\, \delta_{\sigma,0} + \frac{(1+r)_k (1+s)_l (1+m-t)_{m+t+1}}{(1)_k (1)_l (1)_{m+t+1}} \right.$$

$$\left. \times\, \gamma_{\sigma}^{(ij)} \left\{ \begin{matrix} r+s+m+t+1,\ -m-k-r-\frac{3}{2},\ -m-l-s-\frac{3}{2},\ -r+s,\ \dfrac{a_j}{a_i} \\ k+l+m-t+2,\ -m-l-s-\frac{3}{2},\ -m-k-r-\frac{3}{2},\ k-l,\ \dfrac{a_i}{a_j} \end{matrix} \right\} \right]. \tag{47}$$

Now the general terms of the expansions (8), (9) are completely known.

To deal further with the determination of the coefficients of the expansions (14) and (15) substitute (13) into (8) and (9). There results

$$P_{\alpha\beta m}^{(i)} = \sum_{k=0}^{\infty} \sum_{l=0}^{\infty} \varphi_{\alpha+k,\, \beta+l,\, m} \frac{(1+k)_{\alpha} (1+l)_{\beta}}{(1)_{\alpha} (1)_{\beta}}\, p_i^k q_i^l, \tag{48}$$

$$W_{\alpha\beta m}^{(i)} = \sum_{k=0}^{\infty} \sum_{l=0}^{\infty} \varrho_{\alpha+k,\, \beta+l,\, m} \frac{(1+k)_{\alpha} (1+l)_{\beta}}{(1)_{\alpha} (1)_{\beta}}\, p_i^k q_i^l, \tag{49}$$

$$P_{\alpha\beta\sigma\nu mt}^{(ij)} = \kappa_{ij} \sum_{k=0}^{\infty} \sum_{l=0}^{\infty} \sum_{r=0}^{\infty} \sum_{s=0}^{\infty} \psi_{\alpha+k,\, \beta+l,\, \sigma+r,\, \nu+s,\, m,\, t}^{(ij)}$$

$$\times \frac{(1+k)_{\alpha} (1+l)_{\beta} (1+r)_{\sigma} (1+s)_{\nu}}{(1)_{\alpha} (1)_{\beta} (1)_{\sigma} (1)_{\nu}}\, p_i^k q_i^l p_j^r q_j^s, \tag{50}$$

$$W_{\alpha\beta\sigma\nu mt}^{(ij)} = \kappa_{ij} \sum_{k=0}^{\infty} \sum_{l=0}^{\infty} \sum_{r=0}^{\infty} \sum_{s=0}^{\infty} \theta_{\alpha+k,\, \beta+l,\, \sigma+r,\, \nu+s,\, m,\, t}^{(ij)}$$

$$\times \frac{(1+k)_\alpha(1+l)_\beta(1+r)_\sigma(1+s)_\nu}{(1)_\alpha(1)_\beta(1)_\sigma(1)_\nu} p_i^k q_i^l p_j^r q_j^s. \tag{51}$$

The zero superscript in $p_i^{(0)}$, $q_i^{(0)}$, indicating that the coordinates of the intermediary should be used everywhere in (48)–(51), is omitted here for brevity. The expressions for these coefficients can be also derived in closed form as follows

$$\begin{aligned} P_{\alpha\beta m}^{(i)} = &-\left[(1+\tfrac{1}{2}p_i+\tfrac{3}{2}q_i)\,\delta_{\alpha,0}\,\delta_{\beta,0}+\tfrac{1}{2}\,\delta_{\alpha,1}\,\delta_{\beta,0}+\tfrac{3}{2}\,\delta_{\alpha,0}\,\delta_{\beta,1}\right]\delta_{m,0} \\ &+(-1)^m\frac{(\frac{3}{2})_m(m+\frac{1}{2})_\alpha(m+\frac{3}{2})_\beta}{(1)_m(1)_\alpha(1)_\beta}(1-p_i)^{-1/2-m-\alpha}(1-q_i)^{-3/2-m-\beta} \end{aligned} \tag{52}$$

$$\begin{aligned} W_{\alpha\beta m}^{(i)} = {}&\delta_{\alpha,0}\,\delta_{\beta,0}\,\delta_{m,0}+(-1)^{m+1}\frac{(\frac{3}{2})_m(m+\frac{3}{2})_\alpha(m+\frac{3}{2})_\beta}{(1)_m(1)_\alpha(1)_\beta} \\ &\times(1-p_i)^{-3/2-m-\alpha}(1-q_i)^{-3/2-m-\beta}, \end{aligned} \tag{53}$$

$$\begin{aligned} P_{\alpha\beta\sigma\nu mt}^{(ij)} = \kappa_{ij}\Bigg[&\delta_{\alpha,0}\,\delta_{\beta,0}\,\delta_{m,t}\left(\frac{a_i}{a_j}\right)^2\zeta_{ij}^{-1}(-1)^t \\ &\times\frac{(\frac{3}{2})_m(\frac{1}{2}+m)_\sigma(\frac{3}{2}+m)_\nu}{(1)_m(1)_\sigma(1)_\nu}(1-p_j)^{-1/2-m-\sigma}(1-q_j)^{-3/2-m-\nu} \\ &+(-1)^t\frac{(\frac{3}{2})_m(\frac{1}{2}+m)_{\alpha+\sigma}(1+\alpha)_\sigma(\frac{3}{2}+m)_{\beta+\nu}(1+\beta)_\nu(1+m-t)_{m+t}}{(1)_m(1)_{\alpha+\sigma}(1)_\sigma(1)_{\beta+\nu}(1)_\nu(1)_{m+t}} \\ &\times\Gamma^{(ij)}\left\{\begin{matrix} m+t+\sigma+\nu,\ -\frac{1}{2}-m-\alpha-\sigma,\ -\frac{3}{2}-m-\beta-\nu,\ -\sigma+\nu, & \dfrac{a_j}{a_i};\ p_i,\ p_j \\ \alpha+\beta+m-t+2,\ -\frac{3}{2}-m-\beta-\nu,\ -\frac{1}{2}-m-\alpha-\sigma,\ \alpha-\beta-1, & \dfrac{a_i}{a_j};\ q_j,\ q_i \end{matrix}\right\}\Bigg], \end{aligned} \tag{54}$$

$$\begin{aligned} W_{\alpha\beta\sigma\nu mt}^{(ij)} = \kappa_{ij}\Bigg[&\delta_{\alpha,0}\,\delta_{\beta,0}\,\delta_{m,t}\left(\frac{a_i}{a_j}\right)^2(-1)^{t+1} \\ &\times\frac{(\frac{3}{2})_m(\frac{3}{2}+m)_\sigma(\frac{3}{2}+m)_\nu}{(1)_m(1)_\sigma(1)_\nu}(1-p_j)^{-3/2-m-\sigma}(1-q_j)^{-3/2-m-\nu} \\ &+(-1)^t\frac{(\frac{3}{2})_m(\frac{3}{2}+m)_{\alpha+\sigma}(1+\alpha)_\sigma(\frac{3}{2}+m)_{\beta+\nu}(1+\beta)_\nu(1+m-t)_{m+t+1}}{(1)_m(1)_{\alpha+\sigma}(1)_\sigma(1)_{\beta+\nu}(1)_\nu(1)_{m+t+1}} \\ &\times\Gamma^{(ij)}\left\{\begin{matrix} m+t+\sigma+\nu+1,\ -\frac{3}{2}-m-\alpha-\sigma,\ -\frac{3}{2}-m-\beta-\nu,\ -\sigma+\nu, & \dfrac{a_j}{a_i};\ p_i,\ p_j \\ \alpha+\beta+m-t+2,\ -\frac{3}{2}-m-\beta-\nu,\ -\frac{3}{2}-m-\alpha-\sigma,\ \alpha-\beta, & \dfrac{a_i}{a_j};\ q_j,\ q_i \end{matrix}\right\}\Bigg]. \end{aligned} \tag{55}$$

In (54), (55) $\Gamma^{(ij)}$ denotes the function of ζ_{ij} determined by

$$\Gamma^{(ij)}(n, x, y, \nu, \alpha; u, v) = \alpha^n \left[1 - u - \alpha(1 - v)\, \zeta_{ij}^{-1}\right]^x \times \left[1 - \bar{u} - \alpha(1 - \bar{v})\, \zeta_{ij}\right]^y (-1)^\nu \zeta_{ij}^\nu . \qquad (56)$$

For $u=v=0$ this function turns into (39). In (54), (55) $\Gamma^{(ij)}$ should be evaluated for the upper row of parameters in case $a_j < a_i$ and for the lower row in case $a_i < a_j$.

The formulas (48)–(51) and (52)–(55) derived here permit us to find the general terms of the right-hand sides of (16) and (17).

4. Intermediate Solution

Substitution (12) into (6) gives the equations determining $\underset{s}{p}^{(i)}$

$$\ddot{\underset{s}{p}}^{(i)} + 2\sqrt{-1}\, n_i \dot{\underset{s}{p}}^{(i)} - \tfrac{3}{2} n_i^2 \left(\underset{s}{p}^{(i)} + \underset{s}{q}^{(i)}\right) = n_i^2 \underset{s}{P}^{(i)} \quad (s = 1, 2, \ldots). \qquad (57)$$

At every stage of the approximation with respect to the parameter μ the functions $\underset{s}{P}^{(i)}$ are known trigonometric series of the form

$$S^{(i)}_{j_1 \ldots j_l} = \sum_{\sigma_1, \ldots, \sigma_l = -\infty}^{\infty} \alpha^{(ij_1 \ldots j_l)}_{\sigma_1 \ldots \sigma_l} \zeta_{ij_1}^{\sigma_1} \ldots \zeta_{ij_l}^{\sigma_l}, \qquad (58)$$

all the indices $i, j_1, \ldots, j_l$ being different.

Then $\underset{s}{p}^{(i)}$ can be represented by series of the same form

$$T^{(i)}_{j_1 \ldots j_l} = \sum_{\sigma_1, \ldots, \sigma_l = -\infty}^{\infty} p^{(ij_1 \ldots j_l)}_{\sigma_1 \ldots \sigma_l} \zeta_{ij_1}^{\sigma_1} \ldots \zeta_{ij_l}^{\sigma_l}. \qquad (59)$$

At the given stage s the maximal value of l is equal to $\min\{s, N-1\}$.

Transformation $S \to T$ from (58) to (59) is defined by

$$p^{(ij_1 \ldots j_l)}_{\underbrace{0 \ldots 0}_{l}} = -\tfrac{1}{3} \alpha^{(ij_1 \ldots j_l)}_{\underbrace{0 \ldots 0}_{l}},$$

$$p^{(ij_1 \ldots j_l)}_{\sigma_1 \ldots \sigma_l} = n_i^2 \left[\sum_{s=1}^{l} (n_i - n_{j_s})\, \sigma_s\right]^{-2} \left[n_i^2 - \left(\sum_{s=1}^{l} (n_i - n_{j_s})\, \sigma_s\right)^2\right]^{-1} \times \left\{\left[\left(\sum_{s=1}^{l} (n_i - n_{j_s})\, \sigma_s\right)^2 - 2n_i \sum_{s=1}^{l} (n_i - n_{j_s})\, \sigma_s + \tfrac{3}{2} n_i^2\right] \times \alpha^{(ij_1 \ldots j_l)}_{\sigma_1 \ldots \sigma_l} - \tfrac{3}{2} n_i^2 \alpha^{(ij_1 \ldots j_l)}_{-\sigma_1 \ldots -\sigma_l}\right\}. \qquad (60)$$

Thus, each pair of the coefficients of the conjugate terms of S yields a pair of the corresponding coefficients of T.

Indicate now a set of formulas permitting us to find the intermediary up to the terms of the third order of μ. For brevity put $\Upsilon^{(ij)}_{klrsmt} = \kappa_{ij} \psi^{(ij)}_{klrsmt}$ and, since the expansion (8) for the intermediary contains the terms with $m=0$, $t=0$ only, omit the last zero subscript in φ_{klm} and two last zero subscripts in $\Upsilon^{(ij)}_{klrsmt}$. The expansions (12) are substituted into (8) to give formulas for the computation of the right-hand members $P^{(i)}_s$.

Then application of the transformation $S \to T$ yields the appropriate series of the intermediate solution.

Formulas of the First Order

$$\underset{1}{P}^{(i)} = \sum_{j=1}^{N}{}^{(i)} \underset{1}{S}_j^{(i)}, \tag{61}$$

from which $\underset{1}{p}^{(i)}$ is represented by the sum of $N-1$ one-argument series

$$\underset{1}{p}^{(i)} = \sum_{j=1}^{N}{}^{(i)} \underset{1}{T}_j^{(i)}. \tag{62}$$

Also,

$$\underset{1}{S}_j^{(i)} = \Upsilon_{0000}^{(ij)}. \tag{63}$$

Formulas of the Second Order

Since

$$\underset{2}{P}^{(i)} = \varphi_{20}\, \underset{1}{p}^{(i)2} + \varphi_{11}\, \underset{1}{p}^{(i)}\, \underset{1}{q}^{(i)} + \varphi_{02}\, \underset{1}{q}^{(i)2} + \sum_{j=1}^{N}{}^{(i)} \times \left(\Upsilon_{1000}^{(ij)}\, \underset{1}{p}^{(i)} + \Upsilon_{0100}^{(ij)}\, \underset{1}{q}^{(i)} + \Upsilon_{0010}^{(ij)}\, \underset{1}{p}^{(j)} + \Upsilon_{0001}^{(ij)}\, \underset{1}{q}^{(j)}\right),$$

$\underset{2}{P}^{(i)}$ is represented by the sum of $N-1$ one- and $\frac{1}{2}(N-1)(N-2)$ two-argument series

$$\underset{2}{P}^{(i)} = \sum_{j=1}^{N}{}^{(i)} \underset{2}{S}_j^{(i)} + \tfrac{1}{2} \sum_{j=1}^{N}{}^{(i)} \sum_{k=1}^{N}{}^{(i,\,j)} \underset{2}{S}_{jk}^{(i)}. \tag{64}$$

Then $\underset{2}{p}^{(i)}$ is of the same form

$$\underset{2}{p}^{(i)} = \sum_{j=1}^{N}{}^{(i)} \underset{2}{T}_j^{(i)} + \tfrac{1}{2} \sum_{j=1}^{N}{}^{(i)} \sum_{k=1}^{N}{}^{(i,\,j)} \underset{2}{T}_{jk}^{(i)}. \tag{65}$$

The one-argument series $\underset{2}{S}_j^{(i)}$ can be found by

$$\begin{aligned} \underset{2}{S}_j^{(i)} = {} & \varphi_{20}\, \underset{1}{T}_j^{(i)2} + \varphi_{11}\, \underset{1}{T}_j^{(i)}\, \underset{1}{\bar{T}}_j^{(i)} + \varphi_{02}\, \underset{1}{\bar{T}}_j^{(i)2} + \Upsilon_{1000}^{(ij)}\, \underset{1}{T}_j^{(i)} \\ & + \Upsilon_{0100}^{(ij)}\, \underset{1}{\bar{T}}_j^{(i)} + \Upsilon_{0010}^{(ij)}\, \underset{1}{T}_i^{(j)} + \Upsilon_{0001}^{(ij)}\, \underset{1}{\bar{T}}_i^{(j)} \end{aligned} \tag{66}$$

while the double series $\underset{2}{S}_{jk}^{(i)}$ symmetric in lower indices are determined by

$$\begin{aligned} \underset{2}{S}_{jk}^{(i)} = {} & 2\varphi_{20}\, \underset{1}{T}_j^{(i)}\, \underset{1}{T}_k^{(i)} + \varphi_{11}\left(\underset{1}{T}_j^{(i)}\, \underset{1}{\bar{T}}_k^{(i)} + \underset{1}{\bar{T}}_j^{(i)}\, \underset{1}{T}_k^{(i)}\right) + 2\varphi_{02}\, \underset{1}{\bar{T}}_j^{(i)}\, \underset{1}{\bar{T}}_k^{(i)} \\ & + \Upsilon_{1000}^{(ij)}\, \underset{1}{T}_k^{(i)} + \Upsilon_{1000}^{(ik)}\, \underset{1}{T}_j^{(i)} + \Upsilon_{0100}^{(ij)}\, \underset{1}{\bar{T}}_k^{(i)} + \Upsilon_{0100}^{(ik)}\, \underset{1}{\bar{T}}_j^{(i)} \\ & + \Upsilon_{0010}^{(ij)}\, \underset{1}{T}_k^{(j)} + \Upsilon_{0010}^{(ik)}\, \underset{1}{T}_j^{(k)} + \Upsilon_{0001}^{(ij)}\, \underset{1}{\bar{T}}_k^{(j)} + \Upsilon_{0001}^{(ik)}\, \underset{1}{\bar{T}}_j^{(k)}. \end{aligned} \tag{67}$$

Formulas of the Third Order

The terms of the third order of μ in the expansion (8) being

$$\underset{3}{P}^{(i)} = 2\varphi_{20}\, \underset{1}{p}^{(i)}\, \underset{2}{p}^{(i)} + \varphi_{11}\left(\underset{1}{p}^{(i)}\, \underset{2}{q}^{(i)} + \underset{1}{q}^{(i)}\, \underset{2}{p}^{(i)}\right) + 2\varphi_{02}\, \underset{1}{q}^{(i)}\, \underset{2}{q}^{(i)}$$

$$+ \varphi_{30}\, \underset{1}{p}^{(i)^3} + \varphi_{21}\, \underset{1}{p}^{(i)^2} \underset{1}{q}^{(i)} + \varphi_{12}\, \underset{1}{p}^{(i)} \underset{1}{q}^{(i)^2} + \varphi_{03}\, \underset{1}{q}^{(i)^3}$$

$$+ \sum_{j=1}^{N}{}^{(i)} \big(\Upsilon^{(ij)}_{1000}\, \underset{2}{p}^{(i)} + \Upsilon^{(ij)}_{0100}\, \underset{2}{q}^{(i)} + \Upsilon^{(ij)}_{0010}\, \underset{2}{p}^{(j)} + \Upsilon^{(ij)}_{0001}\, \underset{2}{q}^{(j)}$$

$$+ \Upsilon^{(ij)}_{2000}\, \underset{1}{p}^{(i)^2} + \Upsilon^{(ij)}_{1100}\, \underset{1}{p}^{(i)} \underset{1}{q}^{(i)} + \Upsilon^{(ij)}_{0200}\, \underset{1}{q}^{(i)^2} + \Upsilon^{(ij)}_{1010}\, \underset{1}{p}^{(i)} \underset{1}{p}^{(j)}$$

$$+ \Upsilon^{(ij)}_{1001}\, \underset{1}{p}^{(i)} \underset{1}{q}^{(j)} + \Upsilon^{(ij)}_{0110}\, \underset{1}{q}^{(i)} \underset{1}{p}^{(j)} + \Upsilon^{(ij)}_{0101}\, \underset{1}{q}^{(i)} \underset{1}{q}^{(j)} + \Upsilon^{(ij)}_{0020}\, \underset{1}{p}^{(j)^2}$$

$$+ \Upsilon^{(ij)}_{0011}\, \underset{1}{p}^{(j)} \underset{1}{q}^{(j)} + \Upsilon^{(ij)}_{0002}\, \underset{1}{q}^{(j)^2}\big)$$

$\underset{3}{P}^{(i)}$ consists of $N-1$ one-, $\frac{1}{2}(N-1)(N-2)$ two- and $\frac{1}{6}(N-1)(N-2)(N-3)$ three-argument series

$$\underset{3}{P}^{(i)} = \sum_{j=1}^{N}{}^{(i)} \underset{3}{S}^{(i)}_{j} + \tfrac{1}{2} \sum_{j=1}^{N}{}^{(i)} \sum_{k=1}^{N}{}^{(i,\,j)} \underset{3}{S}^{(i)}_{jk} + \tfrac{1}{6} \sum_{j=1}^{N}{}^{(i)} \sum_{k=1}^{N}{}^{(i,\,j)} \sum_{l=1}^{N}{}^{(i,\,j,\,k)} \underset{3}{S}^{(i)}_{jkl}\,. \tag{68}$$

from which follows

$$\underset{3}{p}^{(i)} = \sum_{j=1}^{N}{}^{(i)} \underset{3}{T}^{(i)}_{j} + \tfrac{1}{2} \sum_{j=1}^{N}{}^{(i)} \sum_{k=1}^{N}{}^{(i,\,j)} \underset{3}{T}^{(i)}_{jk} + \tfrac{1}{6} \sum_{j=1}^{N}{}^{(i)} \sum_{k=1}^{N}{}^{(i,\,j)} \sum_{l=1}^{N}{}^{(i,\,j,\,k)} \underset{3}{T}^{(i)}_{jkl}\,. \tag{69}$$

One-argument series are computed from

$$\underset{3}{S}^{(i)}_{j} = 2\varphi_{20}\, \underset{1}{T}^{(i)}_{j} \underset{2}{T}^{(i)}_{j} + \varphi_{11} (\underset{1}{T}^{(i)}_{j} \underset{2}{\bar{T}}^{(i)}_{j} + \underset{1}{\bar{T}}^{(i)}_{j} \underset{2}{T}^{(i)}_{j}) + 2\varphi_{02}\, \underset{1}{\bar{T}}^{(i)}_{j} \underset{2}{\bar{T}}^{(i)}_{j}$$

$$+ \varphi_{30}\, \underset{1}{T}^{(i)^3}_{j} + \varphi_{21}\, \underset{1}{T}^{(i)^2}_{j} \underset{1}{\bar{T}}^{(i)}_{j} + \varphi_{12}\, \underset{1}{T}^{(i)}_{j} \underset{1}{\bar{T}}^{(i)^2}_{j} + \varphi_{03}\, \underset{1}{\bar{T}}^{(i)^3}_{j}$$

$$+ \Upsilon^{(ij)}_{1000}\, \underset{2}{T}^{(i)}_{j} + \Upsilon^{(ij)}_{0100}\, \underset{2}{\bar{T}}^{(i)}_{j} + \Upsilon^{(ij)}_{0010}\, \underset{2}{T}^{(j)}_{i} + \Upsilon^{(ij)}_{0001}\, \underset{2}{\bar{T}}^{(j)}_{i}$$

$$+ \Upsilon^{(ij)}_{2000}\, \underset{1}{T}^{(i)^2}_{j} + \Upsilon^{(ij)}_{1100}\, \underset{1}{T}^{(i)}_{j} \underset{1}{\bar{T}}^{(i)}_{j} + \Upsilon^{(ij)}_{0200}\, \underset{1}{\bar{T}}^{(i)^2}_{j} + \Upsilon^{(ij)}_{1010}\, \underset{1}{T}^{(i)}_{j} \underset{1}{T}^{(j)}_{i}$$

$$+ \Upsilon^{(ij)}_{1001}\, \underset{1}{T}^{(i)}_{j} \underset{1}{\bar{T}}^{(j)}_{i} + \Upsilon^{(ij)}_{0110}\, \underset{1}{\bar{T}}^{(i)}_{j} \underset{1}{T}^{(j)}_{i} + \Upsilon^{(ij)}_{0101}\, \underset{1}{\bar{T}}^{(i)}_{j} \underset{1}{\bar{T}}^{(j)}_{i} + \Upsilon^{(ij)}_{0020}\, \underset{1}{T}^{(j)^2}_{i}$$

$$+ \Upsilon^{(ij)}_{0011}\, \underset{1}{T}^{(j)}_{i} \underset{1}{\bar{T}}^{(j)}_{i} + \Upsilon^{(ij)}_{0002}\, \underset{1}{\bar{T}}^{(j)^2}_{i}\,. \tag{70}$$

For the two-argument series symmetric in lower indices there results

$$\underset{3}{S}^{(i)}_{jk} = \mathrm{Circ}\,(j, k)\, \{2\varphi_{20}\, \underset{1}{T}^{(i)}_{j} (\underset{2}{T}^{(i)}_{k} + \underset{2}{T}^{(i)}_{jk}) + \varphi_{11} [\underset{1}{T}^{(i)}_{j} (\underset{2}{\bar{T}}^{(i)}_{k} + \underset{2}{\bar{T}}^{(i)}_{jk})$$

$$+ \underset{1}{\bar{T}}^{(i)}_{j} (\underset{2}{T}^{(i)}_{k} + \underset{2}{T}^{(i)}_{jk})] + 2\varphi_{02}\, \underset{1}{\bar{T}}^{(i)}_{j} (\underset{2}{\bar{T}}^{(i)}_{k} + \underset{2}{\bar{T}}^{(i)}_{jk}) + 3\varphi_{30}\, \underset{1}{T}^{(i)^2}_{j} \underset{1}{T}^{(i)}_{k}$$

$$+ \varphi_{21}\, \underset{1}{T}^{(i)}_{j} \underset{1}{\bar{T}}^{(i)}_{k} (\underset{1}{T}^{(i)}_{j} + 2\, \underset{1}{T}^{(i)}_{k}) + \varphi_{12}\, \underset{1}{\bar{T}}^{(i)}_{j} \underset{1}{T}^{(i)}_{k} (\underset{1}{\bar{T}}^{(i)}_{j} + 2\, \underset{1}{\bar{T}}^{(i)}_{k})$$

$$+ 3\varphi_{03}\, \underset{1}{\bar{T}}^{(i)^2}_{j} \underset{1}{\bar{T}}^{(i)}_{k} + \Upsilon^{(ij)}_{1000} (\underset{2}{T}^{(i)}_{k} + \underset{2}{T}^{(i)}_{jk}) + \Upsilon^{(ij)}_{0100} (\underset{2}{\bar{T}}^{(i)}_{k} + \underset{2}{\bar{T}}^{(i)}_{jk})$$

$$+ \Upsilon^{(ij)}_{0010} (\underset{2}{T}^{(j)}_{k} + \underset{2}{T}^{(j)}_{ik}) + \Upsilon^{(ij)}_{0001} (\underset{2}{\bar{T}}^{(j)}_{k} + \underset{2}{\bar{T}}^{(j)}_{ik}) + \Upsilon^{(ij)}_{2000}\, \underset{1}{T}^{(i)}_{k} (\underset{1}{T}^{(i)}_{k} + 2\, \underset{1}{T}^{(i)}_{j})$$

$$+ \Upsilon^{(ij)}_{1100} (\underset{1}{T}^{(i)}_{k} \underset{1}{\bar{T}}^{(i)}_{k} + \underset{1}{T}^{(i)}_{j} \underset{1}{\bar{T}}^{(i)}_{k} + \underset{1}{T}^{(i)}_{k} \underset{1}{\bar{T}}^{(i)}_{j}) + \Upsilon^{(ij)}_{0200}\, \underset{1}{\bar{T}}^{(i)}_{k} (\underset{1}{\bar{T}}^{(i)}_{k} + 2\, \underset{1}{\bar{T}}^{(i)}_{j})$$

$$+ \Upsilon^{(ij)}_{1010} (\underset{1}{T}^{(i)}_{j} \underset{1}{T}^{(j)}_{k} + \underset{1}{T}^{(i)}_{k} \underset{1}{T}^{(j)}_{i} + \underset{1}{T}^{(i)}_{k} \underset{1}{T}^{(j)}_{k}) + \Upsilon^{(ij)}_{1001} (\underset{1}{T}^{(i)}_{j} \underset{1}{\bar{T}}^{(j)}_{k} + \underset{1}{T}^{(i)}_{k} \underset{1}{\bar{T}}^{(j)}_{i}$$

$$
\begin{aligned}
&+ \underset{1}{T_k^{(i)}}\, \underset{1}{\bar{T}_k^{(j)}}) + \Upsilon^{(ij)}_{0110} (\underset{1}{\bar{T}_j^{(i)}}\, \underset{1}{T_k^{(j)}} + \underset{1}{\bar{T}_k^{(i)}}\, \underset{1}{T_i^{(j)}} + \underset{1}{\bar{T}_k^{(i)}}\, \underset{1}{T_k^{(j)}}) \\
&+ \Upsilon^{(ij)}_{0101} (\underset{1}{\bar{T}_j^{(i)}}\, \underset{1}{\bar{T}_k^{(j)}} + \underset{1}{\bar{T}_k^{(i)}}\, \underset{1}{\bar{T}_i^{(j)}} + \underset{1}{\bar{T}_k^{(i)}}\, \underset{1}{\bar{T}_k^{(j)}}) + \Upsilon^{(ij)}_{0020}\, \underset{1}{T_k^{(j)}} (\underset{1}{T_k^{(j)}} + 2\, \underset{1}{T_i^{(j)}}) \\
&+ \Upsilon^{(ij)}_{0011} (\underset{1}{T_i^{(j)}}\, \underset{1}{\bar{T}_k^{(j)}} + \underset{1}{T_k^{(j)}}\, \underset{1}{\bar{T}_i^{(j)}} + \underset{1}{T_k^{(j)}}\, \underset{1}{\bar{T}_k^{(j)}}) + \Upsilon^{(ij)}_{0002}\, \underset{1}{\bar{T}_k^{(j)}} \\
&\times (\underset{1}{\bar{T}_k^{(j)}} + 2\, \underset{1}{\bar{T}_i^{(j)}})\} .
\end{aligned}
\tag{71}
$$

Finally, the three-argument series symmetric in all three lower indices are expressed by

$$
\begin{aligned}
\underset{3}{S_{jkl}^{(i)}} = {} & \operatorname{Circ}(j, k, l)\, \{2\varphi_{20}\, \underset{1}{T_j^{(i)}}\, \underset{2}{T_{kl}^{(i)}} + \varphi_{11} (\underset{1}{T_j^{(i)}}\, \underset{2}{\bar{T}_{kl}^{(i)}} + \underset{1}{\bar{T}_j^{(i)}}\, \underset{2}{T_{kl}^{(i)}}) \\
&+ 2\varphi_{02}\, \underset{1}{\bar{T}_j^{(i)}}\, \underset{2}{\bar{T}_{kl}^{(i)}} + 2\varphi_{30}\, \underset{1}{T_j^{(i)}}\, \underset{1}{T_k^{(i)}}\, \underset{1}{T_l^{(i)}} + 2\varphi_{21}\, \underset{1}{T_j^{(i)}}\, \underset{1}{T_k^{(i)}}\, \underset{1}{\bar{T}_l^{(i)}} \\
&+ 2\varphi_{12}\, \underset{1}{T_j^{(i)}}\, \underset{1}{\bar{T}_k^{(i)}}\, \underset{1}{\bar{T}_l^{(i)}} + 2\varphi_{03}\, \underset{1}{\bar{T}_j^{(i)}}\, \underset{1}{\bar{T}_k^{(i)}}\, \underset{1}{\bar{T}_l^{(i)}} + \Upsilon^{(ij)}_{1000}\, \underset{2}{T_{kl}^{(i)}} \\
&+ \Upsilon^{(ij)}_{0100}\, \underset{2}{\bar{T}_{kl}^{(i)}} + \Upsilon^{(ij)}_{0010}\, \underset{2}{T_{kl}^{(j)}} + \Upsilon^{(ij)}_{0001}\, \underset{2}{\bar{T}_{kl}^{(j)}} + 2\Upsilon^{(ij)}_{2000}\, \underset{1}{T_k^{(i)}}\, \underset{1}{T_l^{(i)}} \\
&+ \Upsilon^{(ij)}_{1100} (\underset{1}{T_k^{(i)}}\, \underset{1}{\bar{T}_l^{(i)}} + \underset{1}{T_l^{(i)}}\, \underset{1}{\bar{T}_k^{(i)}}) + 2\Upsilon^{(ij)}_{0200}\, \underset{1}{\bar{T}_k^{(i)}}\, \underset{1}{\bar{T}_l^{(i)}} \\
&+ \Upsilon^{(ij)}_{1010} (\underset{1}{T_k^{(i)}}\, \underset{1}{T_l^{(j)}} + \underset{1}{T_l^{(i)}}\, \underset{1}{T_k^{(j)}}) + \Upsilon^{(ij)}_{1001} (\underset{1}{T_k^{(i)}}\, \underset{1}{\bar{T}_l^{(j)}} + \underset{1}{T_l^{(i)}}\, \underset{1}{\bar{T}_k^{(j)}}) \\
&+ \Upsilon^{(ij)}_{0110} (\underset{1}{\bar{T}_k^{(i)}}\, \underset{1}{T_l^{(j)}} + \underset{1}{\bar{T}_l^{(i)}}\, \underset{1}{T_k^{(j)}}) + \Upsilon^{(ij)}_{0101} (\underset{1}{\bar{T}_k^{(i)}}\, \underset{1}{\bar{T}_l^{(j)}} + \underset{1}{\bar{T}_l^{(i)}}\, \underset{1}{\bar{T}_k^{(j)}}) \\
&+ 2\Upsilon^{(ij)}_{0020}\, \underset{1}{T_k^{(j)}}\, \underset{1}{T_l^{(j)}} + \Upsilon^{(ij)}_{0011} (\underset{1}{T_k^{(j)}}\, \underset{1}{\bar{T}_l^{(j)}} + \underset{1}{T_l^{(j)}}\, \underset{1}{\bar{T}_k^{(j)}}) + 2\Upsilon^{(ij)}_{0002}\, \underset{1}{\bar{T}_k^{(j)}}\, \underset{1}{\bar{T}_l^{(j)}}\} .
\end{aligned}
\tag{72}
$$

In (71), (72) the symbol 'Circ' denotes the circular permutation of the corresponding indices. Therefore, each term in (71) gives rise to one additive similar term and each term in (72) produces two similar terms.

Hence, at every stage s the determination of the series S requires the computation of the series (44) and multiplication of the trigonometric series obtained at the preceding steps. For the computation of $\underset{2}{S_j^{(i)}}$ and $\underset{3}{S_j^{(i)}}$ multiplication of one-argument series is demanded. For the determination of $\underset{2}{S_{jk}^{(i)}}$ one has to multiply one-argument series with different arguments. To obtain $\underset{3}{S_{jk}^{(i)}}$ multiplication of one-argument series by two-argument series is needed. Finally, finding of $\underset{3}{S_{jkl}^{(i)}}$ involves multiplication of one argument series by two-argument series, all three arguments being different.

Up to the terms of μ^3 the intermediate solution is expressed by (12) with (62), (65) and (69). The pertinent series are of the form

$$
T_j^{(i)} = \sum_{\sigma=-\infty}^{\infty} p_\sigma^{(ij)} \exp \sqrt{-1}\, \sigma (\lambda_i - \lambda_j), \tag{73}
$$

$$
T_{jk}^{(i)} = \sum_{\sigma,\, \varrho=-\infty}^{\infty} p_{\sigma\varrho}^{(ijk)} \exp \sqrt{-1}\, [\sigma (\lambda_i - \lambda_j) + \varrho (\lambda_i - \lambda_k)], \tag{74}
$$

$$\underset{3}{T_{jkl}^{(i)}} = \sum_{\sigma,\varrho,\tau=-\infty}^{\infty} \underset{3}{p_{\sigma\varrho\tau}^{(ijkl)}} \exp\sqrt{-1}\left[\sigma(\lambda_i-\lambda_j)+\varrho(\lambda_i-\lambda_k)+\tau(\lambda_i-\lambda_l)\right]. \tag{75}$$

The terms of the series $\mu \underset{1}{T_j^{(i)}}$ have the same order as m_j (accepting for a moment $m_0=1$). The terms of the one-argument series $\mu^2 \underset{2}{T_j^{(i)}}$ have orders of m_j^2 and $m_i m_j$ while the order of the terms of the two-argument series $\mu^2 \underset{2}{T_{jk}^{(i)}}$ is $m_j m_k$. The latter series are responsible for the perturbations of the second order in the motion of the planet i due to the simultaneous attraction of the planets j and k. The terms of the series $\mu^3 \underset{3}{T_j^{(i)}}$ have orders of m_j^3, $m_i m_j^2$ and $m_i^2 m_j$. The terms of the series $\mu^3 \underset{3}{T_{jk}^{(i)}}$ have orders of $m_i m_j m_k$, $m_j^2 m_k$, $m_j m_k^2$ and yield the inequalities of the third order caused by action of the planet j and k on the planet i. Finally, the terms of the series $\mu^3 \underset{3}{T_{jkl}^{(i)}}$ with the order of $m_j m_k m_l$ give the perturbations in the motion of the planet i from the disturbing planets j, k and l simultaneously. At present time it seems hardly to be necessary to take into account $\underset{3}{T_{jk}^{(i)}}$ and $\underset{3}{T_{jkl}^{(i)}}$. Therefore, there is no need to perform vast calculations related with the finding of (71) and (72).

The intermediate solution is represented here in form of the series (12) in powers of μ, the coefficients being trigonometric series in multiples of the differences of the mean longitudes. By the rearrangement of the terms, p_i, q_i may be expressed in form of the $(N-1)$-argument trigonometric series, the coefficients being power series in μ. Moreover, it is possible to combine the constant terms of these trigonometric series with the scale factors a_i. Similarly to the Lunar theory of Hill and Brown one may regard

$$\begin{aligned}\tilde{a}_i = a_i \Bigg[1 &- \mu \sum_{j=1}^{N}{}^{(i)} \underset{1}{p_0^{(ij)}} - \mu^2 \left(\sum_{j=1}^{N}{}^{(i)} \underset{2}{p_0^{(ij)}} + \tfrac{1}{2} \sum_{j=1}^{N}{}^{(i)} \sum_{k=1}^{N}{}^{(i,j)} \underset{2}{p_{00}^{(ijk)}}\right) \\ &- \mu^3 \left(\sum_{j=1}^{N}{}^{(i)} \underset{3}{p_0^{(ij)}} + \tfrac{1}{2} \sum_{j=1}^{N}{}^{(i)} \sum_{k=1}^{N}{}^{(i,j)} \underset{3}{p_{00}^{(ijk)}}\right. \\ &\left. + \tfrac{1}{6} \sum_{j=1}^{N}{}^{(i)} \sum_{k=1}^{N}{}^{(i,j)} \sum_{l=1}^{N}{}^{(i,j,k)} \underset{3}{p_{000}^{(ijkl)}}\right) - \ldots\Bigg]\end{aligned}$$

as the mean distance of the planet i from the sun. The rectangular heliocentric coordinates are expressed then as

$$\begin{aligned} x_i + \sqrt{-1}\, y_i &= \tilde{a}_i (1 - \tilde{p}_i) \exp\sqrt{-1}\, \lambda_i, \\ x_i - \sqrt{-1}\, y_i &= \tilde{a}_i (1 - \tilde{q}_i) \exp(-\sqrt{-1}\, \lambda_i), \end{aligned}$$

where $\tilde{p}_i$, $\tilde{q}_i$ represent trigonometric series without constant terms, the coefficients being obtained from the respective coefficients of p_i, q_i-series by multiplication by $a_i/\tilde{a}_i$. However, for the actual calculations it is more convenient to preserve the intermediate solution in form of the series (12) without any modifications.

In conclusion, the explicit expressions for the first order coefficients may be deduced.

The formulas (60) become in this case

$$p_0^{(ij)} = -\tfrac{1}{3}\alpha_0^{(ij)},$$
$$p_\sigma^{(ij)} = n_{ij}^2 \frac{(\sigma^2 - 2n_{ij}\sigma + \tfrac{3}{2}n_{ij}^2)\,\alpha_\sigma^{(ij)} - \tfrac{3}{2}n_{ij}^2\alpha_{-\sigma}^{(ij)}}{\sigma^2(n_{ij}^2 - \sigma^2)} \tag{76}$$
$$(\sigma = \pm 1, \pm 2, \ldots)$$

with

$$n_{ij} = \frac{n_i}{n_i - n_j}$$

and

$$\alpha_\sigma^{(ij)} = \kappa_{ij}\psi_\sigma^{(ij;\,0000)}.$$

From (46) it follows

$$\psi_\sigma^{(ij;\,0000)} = \left(\frac{a_i}{a_j}\right)^2 \delta_{\sigma,-1} + \frac{(\frac{1}{2})_{\max\{0,-\sigma\}}(\frac{3}{2})_{\max\{0,\sigma\}}}{(1)_{|\sigma|}} \left(\frac{a_j}{a_i}\right)^{|\sigma|}$$
$$\times F\left(\frac{1}{2} + \max\{0, -\sigma\}, \frac{3}{2} + \max\{0, \sigma\}, 1 + |\sigma|; \frac{a_j^2}{a_i^2}\right), \; a_j < a_i \tag{77}$$

$$\psi_\sigma^{(ij;\,0000)} = \left(\frac{a_i}{a_j}\right)^2 \delta_{\sigma,-1}$$
$$- \frac{(\frac{1}{2})_{\max\{0,\sigma+1\}}(\frac{3}{2})_{\max\{0,-\sigma-1\}}}{(1)_{|\sigma+1|}} \left(\frac{a_i}{a_j}\right)^{2+|\sigma+1|}$$
$$\times F\left(\frac{1}{2} + \max\{0, \sigma+1\}, \frac{3}{2} + \max\{0, -\sigma-1\},\right.$$
$$\left. 1 + |\sigma+1|; \frac{a_i^2}{a_j^2}\right), \; a_i < a_j. \tag{78}$$

These functions can be expressed by the frequently used tabulated coefficients $c_n^{(\sigma)}(a_i, a_j)$ (although it is no good for machine calculations). These coefficients symmetric in their arguments are Fourier coefficients of the expansion

$$(a^2 - 2aa'\cos H + a'^2)^{-n/2} = \tfrac{1}{2}\sum_{\sigma=-\infty}^{\infty} c_n^{(\sigma)}(a, a')\cos\sigma H.$$

They may be reduced to the Laplace coefficients

$$c_n^{(\sigma)}(a, a') = a'^{-n} b_n^{(\sigma)}\left(\frac{a}{a'}\right) \tag{79}$$

determined by the hypergeometric series

$$\frac{1}{2}b_n^{(k)}(x) = \frac{\left(\frac{n}{2}\right)_k}{(1)_k} x^k F\left(\frac{n}{2}, \frac{n}{2} + k, 1 + k; x^2\right), \quad k \geqslant 0. \tag{80}$$

The coefficients satisfy the relations $c_n^{(\sigma)}=c_n^{(-\sigma)}$, $b_n^{(k)}=b_n^{(-k)}$. In (79) a is meant to be the minimal of the arguments a_i, a_j. It is easy to show that with the aid of $c_n^{(\sigma)}(a_i, a_j)$ (77) and (78) may be combined into the unique expression

$$\psi_\sigma^{(ij;\,0000)} = \tfrac{1}{2} a_i^3 c_3^{(\sigma)}(a_i, a_j) - \tfrac{1}{2} a_i^2 a_j c_3^{(\sigma+1)}(a_i, a_j) + \left(\frac{a_i}{a_j}\right)^2 \delta_{\sigma,\,-1}\,. \tag{81}$$

5. Linear Differential Equations with Quasi-Periodic Coefficients

To deal with the systems of linear differential equations with quasi-periodic coefficients occurring in two next sections in investigating inequalities of the first power with respect to the eccentricities and inclinations of the planetary orbits apply a method based on the paper by Krasinsky (1968) mentioned above.

Consider a homogeneous linear matrix differential equation with a small parameter μ

$$\dot{X} = (P + Q)\, X \tag{82}$$

where X is a column-matrix of unknowns, P is a constant matrix in the Jordan form, Q is a quasi-periodic matrix reducing to the zero matrix when $\mu=0$.

Performing the linear substitution

$$X = (E + S)\, Y, \tag{83}$$

where a quasi-periodic matrix S has the same frequency basis as Q and vanishes with $\mu=0$, (82) becomes

$$Y = HY \tag{84}$$

in which H is determined by (29). Expansion of $(E+S)^{-1}$ in the matrix series gives

$$H = P + (Q + PS - SP - \dot{S}) + (QS - SQ - SPS + S^2P + S\dot{S}) + (S^2Q - SQS + S^2PS - S^3P - S^2\dot{S}) + \ldots,$$

where the terms of the first, second and third order of μ are collected in the corresponding parentheses. Let Q be expanded in powers of μ. Then the matrix S and an auxiliary matrix I may be looked for in the same form

$$Q = \sum_{k=1}^{\infty} \underset{k}{Q}\mu^k, \quad S = \sum_{k=1}^{\infty} \underset{k}{S}\mu^k, \quad I = \sum_{k=1}^{\infty} \underset{k}{I}\mu^k. \tag{85}$$

Thus

$$H = P + \sum_{k=1}^{\infty} (\underset{k}{I} + P\underset{k}{S} - \underset{k}{S}P - \underset{k}{\dot{S}})\, \mu^k \tag{86}$$

with

$$\begin{aligned}
\underset{1}{I} &= \underset{1}{Q}, \\
\underset{2}{I} &= \underset{2}{Q} + \underset{1}{Q}\underset{1}{S} - \underset{1}{S}\underset{1}{Q} - \underset{1}{S}P\underset{1}{S} + \underset{1}{S}^2P + \underset{1}{S}\underset{1}{\dot{S}}, \\
\underset{3}{I} &= \underset{3}{Q} + \underset{2}{Q}\underset{1}{S} + \underset{1}{Q}\underset{2}{S} - \underset{1}{S}\underset{2}{Q} - \underset{2}{S}\underset{1}{Q} - \underset{1}{S}P\underset{2}{S} - \underset{2}{S}P\underset{1}{S} - \underset{1}{S}\underset{2}{S}P + \underset{2}{S}\underset{1}{S}P \\
&\quad + \underset{1}{S}\underset{2}{\dot{S}} + \underset{2}{S}\underset{1}{\dot{S}} - \underset{1}{S}\underset{1}{Q}\underset{1}{S} + \underset{1}{S}^2\underset{1}{Q} + \underset{1}{S}^2P\underset{1}{S} - \underset{1}{S}^3P - \underset{1}{S}^2\underset{1}{\dot{S}}.
\end{aligned} \tag{87}$$

At every stage of approximation with respect to μ $\underset{k}{I}$ represents a known quasi-periodic matrix. Separate $\underset{k}{I}$ into two parts

$$\underset{k}{I} = \{\underset{k}{I}\} + \underset{k}{\tilde{I}} \tag{88}$$

so that the equation

$$\underset{k}{\dot{S}} + \underset{k}{S}P - P\underset{k}{S} = \underset{k}{\tilde{I}} \tag{89}$$

would always have a quasi-periodic solution. The resonance terms $\{\underset{k}{I}\}$ remain in H

$$H = P + \{I\}, \tag{90}$$

where $\{I\}$ denotes the series of $\{\underset{k}{I}\}$ as (85). In the cases considered below the separation (88) is produced by forming an averaged matrix

$$\underset{k}{\Lambda} = [T^{-1}\underset{k}{I}T] \tag{91}$$

and determining $\{\underset{k}{I}\}$ as

$$\{\underset{k}{I}\} = T\underset{k}{\Lambda}T^{-1}, \tag{92}$$

in which T is a matrix to be chosen from the condition of resolution of (89) in quasi-periodic form. The square brackets in (91) denote an averaging through the explicitly presented time. Then, the substitution

$$Y = TZ \tag{93}$$

transforms the system (84) with (90) into the system

$$\dot{Z} = \Omega Z \tag{94}$$

with

$$\Omega = \Lambda + T^{-1}(PT - \dot{T}). \tag{95}$$

Here again Λ is the series of $\underset{k}{\Lambda}$ as (85). In both cases considered below Ω is a constant matrix.

6. Principal Parts of the Motion of the Nodes

In this section consider the equations

$$\ddot{w}_i + n_i^2 \sum_{j=1}^{N} M_{ij} w_j = 0 \tag{96}$$

arising from (17) if $W_i^* = 0$, i.e., when the terms of the second and higher powers in the eccentricities and inclinations are neglected. Equations (96) yield the inequalities of the first power with respect to the inclinations as well as the principal parts of the motion of the nodes. The coefficients M_{ij} represent real quasi-periodic functions of the time to be evaluated by means of the coordinates of the intermediary. From (53), (55) and (49), (51) there results

$$M_{ii} = \left(\frac{a_i}{r_i}\right)^3 + \mu \sum_{j=1}^{N}{}^{(i)} \kappa_{ij} \left(\frac{a_i}{\Delta_{ij}}\right)^3 = 1 - \sum_{k=0}^{\infty} \sum_{l=0}^{\infty} \varrho_{klo} p_i^k q_i^l$$

$$- \mu \sum_{j=1}^{N}{}^{(i)} \kappa_{ij} \sum_{k=0}^{\infty} \sum_{l=0}^{\infty} \sum_{r=0}^{\infty} \sum_{s=0}^{\infty} \theta_{klrso-1}^{(ij)} p_i^k q_i^l p_j^r q_j^s, \tag{97}$$

$$M_{ij} = \mu \kappa_{ij} a_i^2 a_j \left(\frac{1}{r_j^3} - \frac{1}{\Delta_{ij}^3}\right)$$

$$= - \mu \kappa_{ij} \sum_{k=0}^{\infty} \sum_{l=0}^{\infty} \sum_{r=0}^{\infty} \sum_{s=0}^{\infty} \theta_{klrs00}^{(ij)} p_i^k q_i^l p_j^r q_j^s. \quad (i \neq j) \tag{98}$$

Taking into account the terms of the first order of μ only these coefficients become

$$M_{ii} = 1 + 2A_i + \mu \sum_{j=1}^{N}{}^{(i)} \sum_{\sigma=-\infty}^{\infty}{}^{(0)} \left[\tfrac{3}{2}\left(\underset{1}{p_\sigma^{(ij)}} + \underset{1}{p_{-\sigma}^{(ij)}}\right) - \kappa_{ij} \theta_\sigma^{(ij;\,00000-1)}\right] \zeta_{ij}^\sigma, \tag{99}$$

$$M_{ij} = - \mu \kappa_{ij} \sum_{\sigma=-\infty}^{\infty} \theta_\sigma^{(ij;\,000000)} \zeta_{ij}^\sigma \quad (i \neq j) \tag{100}$$

with

$$A_i = - \tfrac{1}{2} \mu \sum_{j=1}^{N}{}^{(i)} \kappa_{ij} \left(\psi_0^{(ij;\,000000)} + \theta_0^{(ij;\,00000-1)}\right).$$

From (47) it follows

$$\theta_\sigma^{(ij;\,000000)} = \tfrac{1}{2} a_i^2 a_j c_3^{(\sigma)}(a_i, a_j) - \left(\frac{a_i}{a_j}\right)^2 \delta_{\sigma,0}, \tag{101}$$

$$\theta_\sigma^{(ij;\,00000-1)} = - \tfrac{1}{2} a_i^3 c_3^{(\sigma)}(a_i, a_j). \tag{102}$$

Then, in virtue of (81)

$$A_i = \tfrac{1}{4} \mu \sum_{j=1}^{N}{}^{(i)} \kappa_{ij} a_i^2 a_j c_3^{(1)}(a_i, a_j). \tag{103}$$

Returning to the system (96) and introducing (19) this system becomes as (82) with X, P, Q being respectively

$$X = \begin{pmatrix} v \\ \bar{v} \end{pmatrix}, \tag{104}$$

$$P = \sqrt{-1} \begin{pmatrix} \mathcal{N} & 0 \\ 0 & -\mathcal{N} \end{pmatrix}, \tag{105}$$

$$Q = \tfrac{1}{2} \sqrt{-1} \begin{pmatrix} \mathcal{N}M & \mathcal{N}M \\ -\mathcal{N}M & -\mathcal{N}M \end{pmatrix} \tag{106}$$

in which $\mathcal{N}$ is defined by (26) and M is of the form

$$M = \begin{pmatrix} M_{11} - 1 & M_{12} & \dots & M_{1N} \\ M_{21} & M_{22} - 1 & \dots & M_{2N} \\ \dots & \dots & \dots & \dots \\ M_{N1} & M_{N2} & \dots & M_{NN} - 1 \end{pmatrix} \tag{107}$$

vanishing when $\mu = 0$.

Thus X is a block matrix combined of two $N \times 1$ column-matrices and P and Q are square four block matrices, every block being square $N \times N$ matrix. Four block matrices S and I are of the same form

$$S = \begin{pmatrix} S_{55} & S_{56} \\ S_{65} & S_{66} \end{pmatrix}, \qquad I = \begin{pmatrix} I_{55} & I_{56} \\ I_{65} & I_{66} \end{pmatrix}. \tag{108}$$

At every stage of approximation with respect to μ the lefthand member of Equation (89) is then

$$\dot{S} + SP - PS$$
$$= \begin{pmatrix} \dot{S}_{55} + \sqrt{-1}\, S_{55}\mathcal{N} - \sqrt{-1}\,\mathcal{N} S_{55} & \dot{S}_{56} - \sqrt{-1}\, S_{56}\mathcal{N} - \sqrt{-1}\,\mathcal{N} S_{56} \\ \dot{S}_{65} + \sqrt{-1}\, S_{65}\mathcal{N} + \sqrt{-1}\,\mathcal{N} S_{65} & \dot{S}_{66} - \sqrt{-1}\, S_{66}\mathcal{N} + \sqrt{-1}\,\mathcal{N} S_{66} \end{pmatrix} \tag{109}$$

Hence, Equation (89) will admit quasi-periodic solution if the right-hand member of this equation is of the form

$$\tilde{I} = \begin{pmatrix} I_{55}^{+} & I_{56} \\ I_{65} & I_{66}^{-} \end{pmatrix} \tag{110}$$

in which I_{55}^{+} is obtained from $I_{55}$by eliminating constant terms in the trigonometric developments of the diagonal elements and terms with ζ_{ij}^{1} in the non-diagonal elements $(I_{55})_{ij}$. I_{66}^{-} is obtained in a similar way from I_{66} by omitting constant terms in the developments of the diagonal elements and terms with ζ_{ij}^{-1} in the non-diagonal elements $(I_{66})_{ij}$. This form of $\tilde{I}$ results if T is chosen as

$$T = \begin{pmatrix} \exp \sqrt{-1}\,\mathcal{N} t & 0 \\ 0 & \exp\left(-\sqrt{-1}\,\mathcal{N} t\right) \end{pmatrix} \tag{111}$$

with the expression (32). Really, with this choice of T

$$T^{-1} I T$$
$$= \begin{pmatrix} \exp(-\sqrt{-1}\,\mathcal{N}t)\cdot I_{55}\cdot \exp\sqrt{-1}\,\mathcal{N}t & \exp(-\sqrt{-1}\,\mathcal{N}t)\cdot I_{56}\cdot \exp(-\sqrt{-1}\,\mathcal{N}t) \\ \exp\sqrt{-1}\,\mathcal{N}t\cdot I_{65}\cdot \exp\sqrt{-1}\,\mathcal{N}t & \exp\sqrt{-1}\,\mathcal{N}t\cdot I_{66}\cdot \exp(-\sqrt{-1}\,\mathcal{N}t) \end{pmatrix} \tag{112}$$

and putting, for example,

$$I_{55} = \begin{pmatrix} g_{11} & g_{12} & \cdots & g_{1N} \\ g_{21} & g_{22} & \cdots & g_{2N} \\ \cdots & \cdots & \cdots & \cdots \\ g_{N1} & g_{N2} & \cdots & g_{NN} \end{pmatrix} \tag{113}$$

one has

$$\exp\sqrt{-1}\,\mathcal{N}^*t\cdot I_{55}\cdot\exp\sqrt{-1}\,\mathcal{N}t$$
$$=\begin{pmatrix} g_{11}\exp\sqrt{-1}(\lambda_1+\lambda_1^*) & g_{12}\exp\sqrt{-1}(\lambda_2+\lambda_1^*) & \dots & g_{1N}\exp\sqrt{-1}(\lambda_N+\lambda_1^*) \\ g_{21}\exp\sqrt{-1}(\lambda_1+\lambda_2^*) & g_{22}\exp\sqrt{-1}(\lambda_2+\lambda_2^*) & \dots & g_{2N}\exp\sqrt{-1}(\lambda_N+\lambda_2^*) \\ \dots & \dots & \dots & \dots \\ g_{N1}\exp\sqrt{-1}(\lambda_1+\lambda_N^*) & g_{N2}\exp\sqrt{-1}(\lambda_2+\lambda_N^*) & \dots & g_{NN}\exp\sqrt{-1}(\lambda_N+\lambda_N^*) \end{pmatrix}. \tag{114}$$

Changing the signs of $\mathcal{N}$ and $\mathcal{N}^*$ and putting then $\mathcal{N}^*=\mathcal{N}$ all the blocks of (112) may be found at once. Averaging this matrix with respect to the time there results

$$\Lambda=\begin{pmatrix} B & 0 \\ 0 & \bar{B} \end{pmatrix} \tag{115}$$

in which

$$B=\begin{pmatrix} (g_{11})_0 & (g_{12})_1 & \dots & (g_{1N})_1 \\ (g_{21})_1 & (g_{22})_0 & \dots & (g_{2N})_1 \\ \dots & \dots & \dots & \dots \\ (g_{N1})_1 & (g_{N2})_1 & \dots & (g_{NN})_0 \end{pmatrix} \tag{116}$$

where $(g_{ii})_0$ represent the constant terms in the expansions of the diagonal elements g_{ii} of the matrix (113) and $(g_{ij})_1$ are the coefficients of ζ_{ij}^1 in the expansions of the non-diagonal elements g_{ij} of this matrix. Then from (92)

$$\{I\}=\begin{pmatrix} \exp\sqrt{-1}\,\mathcal{N}t\cdot B\cdot\exp(-\sqrt{-1}\,\mathcal{N}t) & 0 \\ 0 & \exp(-\sqrt{-1}\,\mathcal{N}t)\cdot\bar{B}\cdot\exp\sqrt{-1}\,\mathcal{N}t \end{pmatrix} \tag{117}$$

where by (114)

$$\exp\sqrt{-1}\,\mathcal{N}t\cdot B\cdot\exp(-\sqrt{-1}\,\mathcal{N}t)$$
$$=\begin{pmatrix} (g_{11})_0 & (g_{12})_1\,\zeta_{12} & \dots & (g_{1N})_1\,\zeta_{1N} \\ (g_{21})_1\,\zeta_{21} & (g_{22})_0 & \dots & (g_{2N})_1\,\zeta_{2N} \\ \dots & \dots & \dots & \dots \\ (g_{N1})_1\,\zeta_{N1} & (g_{N2})_1\,\zeta_{N2} & \dots & (g_{NN})_0 \end{pmatrix}$$

from which, in virtue of (88), follows (110).

All these formulas are valid at every stage of approximation with respect to μ. In all orders of μ the following relations hold good

$$I_{65}=\bar{I}_{56},\quad I_{66}=\bar{I}_{55},$$
$$S_{65}=\bar{S}_{56},\quad S_{66}=\bar{S}_{55}.$$

Hence, both matrices (108) have only two substantial blocks. Both matrices depend only on differences of the mean longitudes. In addition to this, I_{55} and I_{56} are represented by series in powers of ζ_{ij} with purely imaginary coefficients while the series for S_{55} and S_{56} have real coefficients. B is a constant purely imaginary matrix.

The basic matrix equation (89) at every stage of approximation breaks up into scalar equations

$$\left.\begin{aligned}(\dot{S}_{55})_{ij} + \sqrt{-1}(n_j - n_i)(S_{55})_{ij} &= (I_{55}^{+})_{ij}, \\ (\dot{S}_{56})_{ij} - \sqrt{-1}(n_j + n_i)(S_{56})_{ij} &= (I_{56})_{ij}\end{aligned}\right\} \quad (i, j = 1, 2, \ldots, N) \tag{118}$$

These equations may be solved without any complications since each element is determined independently of other. There are no constant terms in the expansions of $(S_{55})_{ii}$ and no terms containing ζ_{ij}^1 in the elements $(S_{55})_{ij}$ $(i \neq j)$.

With the given choice of T the second term in (95) vanishes and therefore $\Omega = \Lambda$. Putting

$$Z = \begin{pmatrix} \beta \\ \bar{\beta} \end{pmatrix} \tag{119}$$

the equations for the secular perturbations become

$$\dot{\beta} = B\beta. \tag{120}$$

It is easy to show that to the first order of μ these equations are the same as those for the inclinations and longitudes of the nodes in the classic theory of Laplace and Lagrange. In the first order of μ $I = Q$, therefore

$$g_{ij} = \tfrac{1}{2}\sqrt{-1}\, n_i (M_{ij} - \delta_{ij}).$$

Taking into account (99), (100) as well as (101)–(103) one obtains

$$B = \sqrt{-1}\,\mathcal{N}\begin{pmatrix} A_1 & -(1, 2) & \ldots & -(1, N) \\ -(2, 1) & A_2 & \ldots & -(2, N) \\ \ldots & \ldots & \ldots & \ldots \\ -(N, 1) & -(N, 2) & \ldots & A_N \end{pmatrix} \tag{121}$$

in which

$$(i, j) = \tfrac{1}{4}\mu \kappa_{ij} a_i^2 a_j c_3^{(1)}(a_i, a_j) \tag{122}$$

and hence

$$A_i = \sum_{j=1}^{N}{}^{(i)} (i, j). \tag{123}$$

Particular solutions of Equations (120) are available in the form

$$\beta_i = \frac{1}{2\sqrt{-1}} k_i \gamma^{(i)} \exp \sqrt{-1}\, g t,$$

where multipliers k_i are to be chosen from the requirement of symmetry of the secular determinant, g and $\gamma^{(i)}$ being eigenvalues and eigenvectors of the linear algebraic system

$$(n_i A_i - g)\gamma^{(i)} - \sum_{j=1}^{N}{}^{(i)} B_{ij}\gamma^{(j)} = 0. \tag{124}$$

Since

$$B_{ij} = \frac{k_j}{k_i} n_i \cdot (i, j)$$

it becomes

$$B_{ij} = B_{ji} = \frac{1}{4}\frac{f\sqrt{m_i m_j}}{\sqrt{n_i n_j}}\, c_3^{(1)}(a_i, a_j)$$

provided

$$k_i = \frac{\sqrt{m_0}}{\sqrt{m_i n_i a_i}}.$$

The general solution will be then

$$\beta_i = \frac{k_i}{2\sqrt{-1}} \sum_{j=1}^{N} \gamma_j \gamma_j^{(i)} \exp \sqrt{-1}(g_j t + \tau_j) \tag{125}$$

in which γ_j and τ_j are real arbitrary constants, $\gamma_j^{(i)}$ being real eigenvectors satisfying (124) for $g = g_j$. One of g_j vanishes in virtue of (123) and the other roots are real. The expression (125) completely coincides with the solution of the classic theory.

7. Principal Parts of the Motion of the Perihelia

Consider now the equations

$$\delta \ddot{p}_i + 2\sqrt{-1}\, n_i\, \delta \dot{p}_i + n_i^2 \sum_{j=1}^{N} (K_{ij}\, \delta p_j + L_{ij}\, \delta q_j) = 0 \tag{126}$$

resulting from (16) if the right-hand members P_i^* are neglected. These equations determine the principal parts of the motion of the perihelia and inequalities of the first power in the eccentricities of the planetary orbits. Quasi-periodic coefficients K_{ij}, L_{ij} are calculated from the intermediary by means of the closed expressions (52), (54) or expansions (48), (50). Hence

$$K_{ii} = -1 - \frac{1}{2}\left(\frac{a_i}{r_i}\right)^3 - \frac{1}{2}\mu \sum_{j=1}^{N}{}^{(i)} \kappa_{ij} \left(\frac{a_i}{\Delta_{ij}}\right)^3 = -\frac{3}{2} - \sum_{k=0}^{\infty}\sum_{l=0}^{\infty} (k+1)$$
$$\times\, \varphi_{k+1,l}\, p_i^k q_i^l - \mu \sum_{j=1}^{N}{}^{(i)} \kappa_{ij} \sum_{k=0}^{\infty}\sum_{l=0}^{\infty}\sum_{r=0}^{\infty}\sum_{s=0}^{\infty} (k+1)\, \psi_{k+1,l,r,s}^{(ij)}\, p_i^k q_i^l p_j^r q_j^s, \tag{127}$$

$$L_{ii} = -\frac{3}{2}\left(\frac{a_i}{r_i}\right)^5 (1-p_i)^2 - \frac{3}{2}\mu \sum_{j=1}^{N}{}^{(i)} \kappa_{ij} \left(\frac{a_i}{\Delta_{ij}}\right)^5 \left[1 - p_i - \frac{a_j}{a_i}(1-p_j)\zeta_{ij}^{-1}\right]^2$$
$$= -\frac{3}{2} - \sum_{k=0}^{\infty}\sum_{l=0}^{\infty} (l+1)\, \varphi_{k,l+1}\, p_i^k q_i^l$$
$$-\, \mu \sum_{j=1}^{N}{}^{(i)} \kappa_{ij} \sum_{k=0}^{\infty}\sum_{l=0}^{\infty}\sum_{r=0}^{\infty}\sum_{s=0}^{\infty} (l+1)\, \psi_{k,l+1,r,s}^{(ij)}\, p_i^k q_i^l p_j^r q_j^s, \tag{128}$$

$$K_{ij} = \frac{1}{2}\mu \kappa_{ij} a_i^2 a_j \left(\frac{1}{\Delta_{ij}^3} - \frac{1}{r_j^3}\right) \zeta_{ij}^{-1}$$
$$= -\mu \kappa_{ij} \sum_{k=0}^{\infty}\sum_{l=0}^{\infty}\sum_{r=0}^{\infty}\sum_{s=0}^{\infty} (r+1)\, \psi_{k,l,r+1,s}^{(ij)}\, p_i^k q_i^l p_j^r q_j^s,\ (i \neq j) \tag{129}$$

$$L_{ij} = \frac{3}{2}\mu\kappa_{ij}a_i^2a_j^3\left\{\frac{1}{\Delta_{ij}^5}\left[1 - p_j - \frac{a_i}{a_j}(1-p_i)\zeta_{ij}\right]^2 - \frac{1}{r_j^5}(1-p_j)^2\right\}\zeta_{ij}^{-1}$$

$$= -\mu\kappa_{ij}\sum_{k=0}^{\infty}\sum_{l=0}^{\infty}\sum_{r=0}^{\infty}\sum_{s=0}^{\infty}(s+1)\,\psi_{k,l,r,s+1}^{(ij)}\,p_i^kq_i^lp_j^rq_j^s,\ (i \neq j). \tag{130}$$

As in Section 3 the last zero subscripts in φ_{klm} and $\psi_{klrsmt}^{(ij)}$ are omitted.

To the first order of μ

$$K_{ii} = -\tfrac{3}{2} - A_i - \mu\sum_{j=1}^{N}{}^{(i)}\sum_{\sigma=-\infty}^{\infty}{}^{(0)}\left[\tfrac{3}{4}\left(\underset{1}{p_\sigma^{(ij)}} + \underset{1}{p_{-\sigma}^{(ij)}}\right) + \kappa_{ij}\psi_\sigma^{(ij;\,1000)}\right]\zeta_{ij}^\sigma, \tag{131}$$

$$L_{ii} = -\tfrac{3}{2} - A_i - \mu\sum_{j=1}^{N}{}^{(i)}\sum_{\sigma=-\infty}^{\infty}{}^{(0)}\left[\tfrac{3}{4}\left(\underset{1}{p_\sigma^{(ij)}} + 5\underset{1}{p_{-\sigma}^{(ij)}}\right) + \kappa_{ij}\psi_\sigma^{(ij;\,0100)}\right]\zeta_{ij}^\sigma, \tag{132}$$

$$K_{ij} = -\mu\kappa_{ij}\sum_{\sigma=-\infty}^{\infty}\psi_\sigma^{(ij;\,0010)}\zeta_{ij}^\sigma,\ (i\neq j) \tag{133}$$

$$L_{ij} = -\mu\kappa_{ij}\sum_{\sigma=-\infty}^{\infty}\psi_\sigma^{(ij;\,0001)}\zeta_{ij}^\sigma,\ (i\neq j) \tag{134}$$

where A_i is determined by (103) as before. This is evident since by (46)

$$\psi_\sigma^{(ij;\,1000)} = \tfrac{1}{4}a_i^3c_3^{(\sigma)}(a_i, a_j), \tag{135}$$

$$\psi_0^{(ij;\,0100)} = \begin{Bmatrix} \dfrac{3}{2}F\left(\dfrac{1}{2}, \dfrac{5}{2}, 1; \dfrac{a_j^2}{a_i^2}\right) \\ \dfrac{9}{16}\left(\dfrac{a_i}{a_j}\right)^5 F\left(\dfrac{5}{2}, \dfrac{5}{2}, 3; \dfrac{a_i^2}{a_j^2}\right) \end{Bmatrix}$$

$$= \frac{3}{4}a_i^3c_3^{(0)}(a_i, a_j) - \frac{1}{2}a_i^2a_jc_3^{(1)}(a_i, a_j) \tag{136}$$

and hence, taking (81) into account

$$\psi_0^{(ij;\,1000)} - \tfrac{1}{2}\psi_0^{(ij;\,0000)} = \psi_0^{(ij;\,0100)} - \tfrac{3}{2}\psi_0^{(ij;\,0000)} = \tfrac{1}{4}a_i^2a_jc_3^{(1)}(a_i, a_j).$$

The following expressions will be of use

$$\psi_\sigma^{(ij;\,0010)} = \frac{1}{2}\left(\frac{a_i}{a_j}\right)^2\delta_{\sigma,-1} - \frac{1}{4}a_i^2a_jc_3^{(\sigma+1)}(a_i, a_j), \tag{137}$$

$$\psi_0^{(ij;\,0001)} = -\tfrac{1}{4}a_i^2a_jc_3^{(1)}(a_i, a_j), \tag{138}$$

$$\psi_1^{(ij;\,0001)} = -\frac{a_j}{a_i}\psi_0^{(ij;\,0100)} = \tfrac{1}{2}a_ia_j^2c_3^{(1)}(a_i, a_j) - \tfrac{3}{4}a_i^2a_jc_3^{(0)}(a_i, a_j), \tag{139}$$

$$\psi_{-1}^{(ij;\,0001)} = \frac{3}{2}\left(\frac{a_i}{a_j}\right)^2 - \begin{Bmatrix} \dfrac{9}{16}\left(\dfrac{a_j}{a_i}\right)^3 F\left(\dfrac{5}{2}, \dfrac{5}{2}, 3; \dfrac{a_j^2}{a_i^2}\right) \\ \dfrac{3}{2}\left(\dfrac{a_i}{a_j}\right)^2 F\left(\dfrac{1}{2}, \dfrac{5}{2}, 1; \dfrac{a_i^2}{a_j^2}\right) \end{Bmatrix}$$

$$= \frac{3}{2}\left(\frac{a_i}{a_j}\right)^2 + \frac{1}{4}a_i^2a_jc_3^{(2)}(a_i, a_j) - \frac{1}{2}a_ia_j^2c_3^{(1)}(a_i, a_j). \tag{140}$$

Hence

$$3\psi_1^{(ij;\,0001)} + 3\psi_{-1}^{(ij;\,0001)} - \psi_1^{(ij;\,0010)} - 9\psi_{-1}^{(ij;\,0010)} = a_i^2 a_j c_3^{(2)}(a_i, a_j). \tag{141}$$

It is plain that in (136) and (140) upper rows in the braces are related to the case $a_j < a_i$ and lower rows are valid for $a_i < a_j$.

For treating the system (126) apply the method described in Section 4. The substitution (18) brings this system to the form (82) with

$$X = \begin{pmatrix} \xi \\ \eta \\ u \\ \bar{u} \end{pmatrix}, \tag{142}$$

$$P = \begin{pmatrix} 0 & E & 0 & 0 \\ 0 & 0 & 0 & 0 \\ 0 & 0 & \sqrt{-1}\,\mathcal{N} & 0 \\ 0 & 0 & 0 & -\sqrt{-1}\,\mathcal{N} \end{pmatrix}, \tag{143}$$

$$Q = \begin{pmatrix} \sqrt{-1}(K - L - \bar{K} + \bar{L})\mathcal{N} & -\tfrac{2}{3}(K + L + \bar{K} + \bar{L}) & \tfrac{1}{2}(-K + 3L + 3\bar{K} - \bar{L}) & \tfrac{1}{2}(3K - L - \bar{K} + 3\bar{L}) \\ \tfrac{3}{2}\mathcal{N}(K - L + \bar{K} - \bar{L})\mathcal{N} & \sqrt{-1}\,\mathcal{N}(K + L - \bar{K} - \bar{L}) & \tfrac{3}{4}\sqrt{-1}\,\mathcal{N}(K - 3L + 3\bar{K} - \bar{L}) & \tfrac{3}{4}\sqrt{-1}\,\mathcal{N}(-3K + L - \bar{K} + 3\bar{L}) \\ \tfrac{1}{2}\mathcal{N}(K - L + 3\bar{K} - 3\bar{L})\mathcal{N} & \tfrac{1}{3}\sqrt{-1}\,\mathcal{N}(K + L - 3\bar{K} - 3\bar{L}) & \tfrac{1}{4}\sqrt{-1}\,\mathcal{N}(K - 3L + 9\bar{K} - 3\bar{L}) & \tfrac{1}{4}\sqrt{-1}\,\mathcal{N}(-3K + L - 3\bar{K} + 9\bar{L}) \\ \tfrac{1}{2}\mathcal{N}(3K - 3L + \bar{K} - \bar{L})\mathcal{N} & \tfrac{1}{3}\sqrt{-1}\,\mathcal{N}(3K + 3L - \bar{K} - \bar{L}) & \tfrac{1}{4}\sqrt{-1}\,\mathcal{N}(3K - 9L + 3\bar{K} - \bar{L}) & \tfrac{1}{4}\sqrt{-1}\,\mathcal{N}(-9K + 3L - \bar{K} + 3\bar{L}) \end{pmatrix} \tag{144}$$

where K and L are square matrices

$$K = \begin{pmatrix} K_{11} + \tfrac{3}{2} & K_{12} & \cdots & K_{1N} \\ K_{21} & K_{22} + \tfrac{3}{2} & \cdots & K_{2N} \\ \cdots & \cdots & \cdots & \cdots \\ K_{N1} & K_{N2} & \cdots & K_{NN} + \tfrac{3}{2} \end{pmatrix},$$

$$L = \begin{pmatrix} L_{11} + \tfrac{3}{2} & L_{12} & \cdots & L_{1N} \\ L_{21} & L_{22} + \tfrac{3}{2} & \cdots & L_{2N} \\ \cdots & \cdots & \cdots & \cdots \\ L_{N1} & L_{N2} & \cdots & L_{NN} + \tfrac{3}{2} \end{pmatrix} \tag{145}$$

vanishing when $\mu = 0$.

Thus X is a block matrix combined of four column-matrices and P and Q are square block matrices consisting of 16 blocks. Denoting the blocks of Q by Q_{ij} $(i, j = 1, 2, 3, 4)$ one has

$$Q_{3j} + Q_{4j} = \tfrac{4}{3} Q_{2j}, \quad Q_{3j} - Q_{4j} = \sqrt{-1}\,\mathcal{N} Q_{1j} \quad (j = 1, 2, 3, 4)$$

$$Q_{14} = \bar{Q}_{13}, \quad Q_{41} = \bar{Q}_{31}, \quad Q_{43} = \bar{Q}_{34},$$

$$Q_{24} = \bar{Q}_{23}, \quad Q_{42} = \bar{Q}_{32}, \quad Q_{44} = \bar{Q}_{33}$$

from which follows that there are only four substantial blocks in Q.

Each matrix S and I consists of 16 $N\times N$ blocks

$$S=\begin{pmatrix} S_{11} & S_{12} & S_{13} & S_{14} \\ S_{21} & S_{22} & S_{23} & S_{24} \\ S_{31} & S_{32} & S_{33} & S_{34} \\ S_{41} & S_{42} & S_{43} & S_{44} \end{pmatrix}, \qquad I=\begin{pmatrix} I_{11} & I_{12} & I_{13} & I_{14} \\ I_{21} & I_{22} & I_{23} & I_{24} \\ I_{31} & I_{32} & I_{33} & I_{34} \\ I_{41} & I_{42} & I_{43} & I_{44} \end{pmatrix}. \tag{146}$$

From the equations derived below it is evident that the following relations take place

$$I_{14}=\bar{I}_{13},\quad I_{24}=\bar{I}_{23},\quad I_{41}=\bar{I}_{31},\quad I_{42}=\bar{I}_{32},\quad I_{43}=\bar{I}_{34},\quad I_{44}=\bar{I}_{33},$$
$$S_{14}=\bar{S}_{13},\quad S_{24}=\bar{S}_{23},\quad S_{41}=\bar{S}_{31},\quad S_{42}=\bar{S}_{32},\quad S_{43}=\bar{S}_{34},\quad S_{44}=\bar{S}_{33}$$

thus S and I each having 10 substantial blocks.

At every stage of approximation with respect to μ the left-hand side of Equation (89) is of the form

$$\dot{S}+SP-PS=$$
$$=\begin{pmatrix} \dot{S}_{11}-S_{21} & \dot{S}_{12}+S_{11}-S_{22} & \dot{S}_{13}+\sqrt{-1}\,S_{13}\mathcal{N}-S_{23} & \dot{S}_{14}-\sqrt{-1}\,S_{14}\mathcal{N}-S_{24} \\ \dot{S}_{21} & \dot{S}_{22}+S_{21} & \dot{S}_{23}+\sqrt{-1}\,S_{23}\mathcal{N} & \dot{S}_{24}-\sqrt{-1}\,S_{24}\mathcal{N} \\ \dot{S}_{31}-\sqrt{-1}\,\mathcal{N}S_{31} & \dot{S}_{32}-\sqrt{-1}\,\mathcal{N}S_{32}+S_{31} & \dot{S}_{33}+\sqrt{-1}\,S_{33}\mathcal{N}-\sqrt{-1}\,\mathcal{N}S_{33} & \dot{S}_{34}-\sqrt{-1}\,S_{34}\mathcal{N}-\sqrt{-1}\,\mathcal{N}S_{34} \\ \dot{S}_{41}+\sqrt{-1}\,\mathcal{N}S_{41} & \dot{S}_{42}+\sqrt{-1}\,\mathcal{N}S_{42}+S_{41} & \dot{S}_{43}+\sqrt{-1}\,S_{43}\mathcal{N}+\sqrt{-1}\,\mathcal{N}S_{43} & \dot{S}_{44}-\sqrt{-1}\,S_{44}\mathcal{N}+\sqrt{-1}\,\mathcal{N}S_{44} \end{pmatrix}. \tag{147}$$

Hence, Equation (89) admits a quasi-periodic solution provided its right-hand side is of the form

$$\tilde{I}=\begin{pmatrix} \tilde{I}_{11} & \tilde{I}_{12} & I_{13} & I_{14} \\ \tilde{I}_{21} & \tilde{I}_{22} & I_{23} & I_{24} \\ I_{31} & I_{32} & I_{33}^{+} & I_{34} \\ I_{41} & I_{42} & I_{43} & I_{44}^{-} \end{pmatrix}, \tag{148}$$

in which matrices $\tilde{I}_{ij}$ $(i,j=1, 2)$ are obtained from I_{ij} by omitting constant terms in the expansions of all elements. I_{33}^{+} follows from I_{33} if constant terms of the diagonal elements and terms with ζ_{ij}^{1} of the non-diagonal elements $(I_{33})_{ij}$ are ignored. Finally, I_{44}^{-} is obtained from I_{44} by omitting constant terms of the diagonal elements and the terms with ζ_{ij}^{-1}in the non-diagonal elements $(I_{44})_{ij}$. Such a matrix $\tilde{I}$ is formed with the aid of

$$T=\begin{pmatrix} E & 0 & 0 & 0 \\ 0 & E & 0 & 0 \\ 0 & 0 & \exp\sqrt{-1}\,\mathcal{N}t & 0 \\ 0 & 0 & 0 & \exp\left(-\sqrt{-1}\,\mathcal{N}t\right) \end{pmatrix}. \tag{149}$$

$$T^{-1}IT = \begin{pmatrix} I_{11} & I_{12} & I_{13}\cdot\exp\sqrt{-1}\,\mathcal{N}t & I_{14}\cdot\exp(-\sqrt{-1}\,\mathcal{N}t) \\ I_{21} & I_{22} & I_{23}\cdot\exp\sqrt{-1}\,\mathcal{N}t & I_{24}\cdot\exp(-\sqrt{-1}\,\mathcal{N}t) \\ \exp(-\sqrt{-1}\,\mathcal{N}t)\cdot I_{31} & \exp(-\sqrt{-1}\,\mathcal{N}t)\cdot I_{32} & \exp(-\sqrt{-1}\,\mathcal{N}t)\cdot I_{33}\cdot\exp\sqrt{-1}\,\mathcal{N}t & \exp(-\sqrt{-1}\,\mathcal{N}t)\cdot I_{34}\cdot\exp(-\sqrt{-1}\,\mathcal{N}t) \\ \exp\sqrt{-1}\,\mathcal{N}t\cdot I_{41} & \exp\sqrt{-1}\,\mathcal{N}t\cdot I_{42} & \exp\sqrt{-1}\,\mathcal{N}t\cdot I_{43}\cdot\exp\sqrt{-1}\,\mathcal{N}t & \exp\sqrt{-1}\,\mathcal{N}t\cdot I_{44}\cdot\exp(-\sqrt{-1}\,\mathcal{N}t) \end{pmatrix} \tag{150}$$

$$\{I\} = \begin{pmatrix} [I_{11}] & [I_{12}] & 0 & 0 \\ [I_{21}] & [I_{22}] & 0 & 0 \\ 0 & 0 & \exp\sqrt{-1}\,\mathcal{N}t\cdot A\cdot\exp(-\sqrt{-1}\,\mathcal{N}t) & 0 \\ 0 & 0 & 0 & \exp(-\sqrt{-1}\,\mathcal{N}t)\cdot\bar{A}\cdot\exp\sqrt{-1}\,\mathcal{N}t \end{pmatrix} \tag{154}$$

Then (see Eq. (150) on page 438).

Now the averaging similar to that of the preceding section results in

$$\Lambda = \begin{pmatrix} [I_{11}] & [I_{12}] & 0 & 0 \\ [I_{21}] & [I_{22}] & 0 & 0 \\ 0 & 0 & A & 0 \\ 0 & 0 & 0 & \bar{A} \end{pmatrix} \tag{151}$$

where $[I_{ij}]$ $(i, j = 1, 2)$ is a constant part of the matrix I_{ij} and

$$A = \begin{pmatrix} (f_{11})_0 & (f_{12})_1 & \cdots & (f_{1N})_1 \\ (f_{21})_1 & (f_{22})_0 & \cdots & (f_{2N})_1 \\ \cdots & \cdots & \cdots & \cdots \\ (f_{N1})_1 & (f_{N2})_1 & \cdots & (f_{NN})_0 \end{pmatrix} \tag{152}$$

provided that

$$I_{33} = \begin{pmatrix} f_{11} & f_{12} & \cdots & f_{1N} \\ f_{21} & f_{22} & \cdots & f_{2N} \\ \cdots & \cdots & \cdots & \cdots \\ f_{N1} & f_{N2} & \cdots & f_{NN} \end{pmatrix}. \tag{153}$$

Here $(f_{ii})_0$ represent the constant terms in the expansions of f_{ii} and $(f_{ij})_1$ are the coefficients of ζ_{ij}^1 in the expansions of $f_{ij} (i \neq j)$. By (92) (see page 429), from which in virtue of (88) one obtains the expression (148) for $\tilde{I}$.

Thus, at every stage of approximation with respect to μ one has to deal with the scalar equations

$$\left.\begin{aligned} &(\dot{S}_{21})_{ij} = (\tilde{I}_{21})_{ij}, \\ &(\dot{S}_{11})_{ij} = (S_{21})_{ij} + (\tilde{I}_{11})_{ij}, \\ &(\dot{S}_{22})_{ij} = -(S_{21})_{ij} + (\tilde{I}_{22})_{ij}, \\ &(\dot{S}_{12})_{ij} = (S_{22})_{ij} - (S_{11})_{ij} + (\tilde{I}_{12})_{ij}, \\ &(\dot{S}_{23})_{ij} + \sqrt{-1}\, n_j (S_{23})_{ij} = (I_{23})_{ij}, \\ &(\dot{S}_{13})_{ij} + \sqrt{-1}\, n_j (S_{13})_{ij} = (S_{23})_{ij} + (I_{13})_{ij}, \\ &(\dot{S}_{31})_{ij} - \sqrt{-1}\, n_i (S_{31})_{ij} = (I_{31})_{ij}, \\ &(\dot{S}_{32})_{ij} - \sqrt{-1}\, n_i (S_{32})_{ij} = -(S_{31})_{ij} + (I_{32})_{ij}, \\ &(\dot{S}_{33})_{ij} + \sqrt{-1}(n_j - n_i)(S_{33})_{ij} = (\overset{+}{I}_{33})_{ij}, \\ &(\dot{S}_{34})_{ij} - \sqrt{-1}(n_j + n_i)(S_{34})_{ij} = (I_{34})_{ij}. \\ &\quad (i, j = 1, 2, \ldots, N) \end{aligned}\right\} \tag{155}$$

All these equations can be solved without any difficulties one after another. There are no constant terms in all elements of S_{ij} $(i, j = 1, 2)$ and in the diagonal elements $(S_{33})_{ii}$, while the non-diagonal elements $(S_{33})_{ij}$ do not contain the terms with ζ_{ij}^1.

In virtue of (143) and (149)

$$T^{-1}(PT - \dot{T}) = \begin{pmatrix} 0 & E & 0 & 0 \\ 0 & 0 & 0 & 0 \\ 0 & 0 & 0 & 0 \\ 0 & 0 & 0 & 0 \end{pmatrix} \tag{156}$$

therefore, this matrix is to be added to Λ defined by (151) to obtain the matrix Ω.

To the first order of μ $I=Q$ and it is easy to find that $[I_{11}]=[I_{22}]=[I_{21}]=0$ and

$$[I_{12}]=\tfrac{8}{3}\begin{pmatrix} A_1 & -(1,2) & \dots & -(1,N) \\ -(2,1) & A_2 & \dots & -(2,N) \\ \dots & \dots & \dots & \dots \\ -(N,1) & -(N,2) & \dots & A_N \end{pmatrix},$$

where (i,j) is determined by (122) as before. The form of the first-order approximation and Equations (155) lead one to conclude that in all orders of μ $I_{12}, I_{21}, I_{13}, I_{31}, S_{11}, S_{22}, S_{23}, S_{32}, S_{33}, S_{34}$ are represented by series in terms of ζ_{ij} with real coefficients while the coefficients of the series for $I_{11}, I_{22}, I_{23}, I_{32}, I_{33}, I_{34}, S_{12}, S_{21}, S_{13}, S_{31}$ are purely imaginary. From this follows, first of all, that A is a purely imaginary matrix. Moreover, since I_{11}, I_{22} are real matrices the relations $[I_{11}]=[I_{22}]=0$ hold true for all orders. The relation $[I_{21}]=0$ is also valid in all orders since Equations (126) have to admit two sets of particular solutions representing the derivatives of the intermediate solution with respect to the parameters. Therefore putting

$$Z=\begin{pmatrix} \sigma \\ \nu \\ \alpha \\ \bar{\alpha} \end{pmatrix} \tag{157}$$

the equations for the secular perturbations must be of the form

$$\left.\begin{aligned} \dot{\sigma} &= (E+[I_{12}])\,\nu, \\ \dot{\nu} &= 0 \end{aligned}\right\} \tag{158}$$

and

$$\dot{\alpha} = A\alpha. \tag{159}$$

As a solution of (158) one admits naturally $\sigma=\nu=0$.

To the first order of μ Equations (159) are in keeping with those for the eccentricities and longitudes of the perihelia in the trigonometric theory of the secular perturbations due to Laplace and Lagrange. Since in the first order of μ

$$f_{ij}=\tfrac{1}{4}\sqrt{-1}\,n_i(K_{ij}-3L_{ij}+9\bar{K}_{ij}-3\bar{L}_{ij}+6\,\delta_{ij})$$

then with the aid of (141)

$$A=\sqrt{-1}\,\mathcal{N}\begin{pmatrix} -A_1 & [1,2] & \dots & [1,N] \\ [2,1] & -A_2 & \dots & [2,N] \\ \dots & \dots & \dots & \dots \\ [N,1] & [N,2] & \dots & -A_N \end{pmatrix}, \tag{160}$$

where

$$[i,j]=\tfrac{1}{4}\,\mu\kappa_{ij}a_i^2 a_j c_3^{(2)}(a_i, a_j). \tag{161}$$

Particular solutions of (159) may be found in the form

$$\alpha_i = k_i h^{(i)} \exp\sqrt{-1}\,ct,$$

where k_i are the multipliers of the preceding section and c and $h^{(i)}$ are eigenvalues and eigenvectors of the linear algebraic system

$$(-n_i A_i - c)\, h^{(i)} + \sum_{j=1}^{N}{}^{(i)} A_{ij} h^{(j)} = 0\,, \tag{162}$$

in which

$$A_{ij} = \frac{k_j}{k_i} n_i [i, j]$$

or, in virtue of the choice of k_i,

$$A_{ij} = A_{ji} = \frac{1}{4} \frac{f\sqrt{m_i m_j}}{\sqrt{n_i n_j}} c_3^{(2)}(a_i, a_j)\,.$$

The general solution will be then

$$\alpha_i = k_i \sum_{j=1}^{N} h_j h_j^{(i)} \exp \sqrt{-1}(c_j t + \chi_j)\,, \tag{163}$$

in which h_j and χ_j are real arbitrary constants, $h_j^{(i)}$ being real eigenvectors satisfying Equations (162) for $c=c_j$. The expression (163) coincides with the solution of the classic trigonometric theory of the secular perturbations.

8. Inequalities of Higher Powers

In this paper the application of Hill's Lunar method is limited by using for the intermediary the quasi-periodic orbits representing an extension of the variation curve. Of course, it is possible to construct a planetary theory by applying further stages of Hill-Brown's Lunar theory as well. For this purpose it is sufficient to perform the linear substitution (83) and (93) in general Equations (20), to determine furthermore the principal parts of the motion of the perihelia and nodes together with the inequalities of the first power in the eccentricities and inclinations and then to find step by step from the equations obtained the inequalities of higher powers and corrections to the roots c_j, g_j dependent on the eccentricities and inclinations. But this form of planetary theory implying all $3N-1$ frequencies n_j, c_j, g_j to be regarded equally is evidently not suitable from the practical point of view. In the planetary problem it is far more advantageous to isolate the terms related to the extremely slow motion of the perihelia and nodes. This aim is reached by application of the nonlinear transformation (27).

Denote matrices X, Y, Z, P, Q, S, T, I, Λ and Ω occurring in section 7 and relevant to the system (126) by the same letters with one prime. Let analogous matrices of Section 6 relating to the system (96) have two primes. Then in the system (20)

$$X = \begin{pmatrix} X' \\ X'' \end{pmatrix}, \quad P = \begin{pmatrix} P' & 0 \\ 0 & P'' \end{pmatrix}, \quad Q = \begin{pmatrix} Q' & 0 \\ 0 & Q'' \end{pmatrix},$$

the remaining matrices being of the similar form. Right upper block of any matrix like P is zero $4N \times 2N$ matrix and left lower block is zero $2N \times 4N$ matrix.

Substitution (27) leads from the system (20) to (28). The determination of the matrices S and H has been discussed in two preceding sections. Assuming these matrices to be known, one is now in position to find the matrices Γ and F. Differentiating of (27) gives with the aid of (20)

$$(E + S + \Gamma_Y)\,\dot{Y} = [(P + Q)(E + S) - \dot{S}]\,Y + (P + Q)\,\Gamma - \Gamma_t + R,$$

in which Γ_y is the Jacobian matrix

$$\Gamma_Y = \left\| \frac{\partial \Gamma_i}{\partial Y_j} \right\| \quad (i, j = 1, 2, \ldots, N)$$

and Γ_t denotes the partial derivative of Γ with respect to the time. Since

$$(E + S + \Gamma_Y)^{-1} = [E + (E + S)^{-1}\Gamma_Y]^{-1}(E + S)^{-1}$$

and

$$\begin{aligned}[E + (E + S)^{-1}\Gamma_Y]^{-1} &= \sum_{k=0}^{\infty} (-1)^K (E + S)^{-K}\Gamma_Y^K \\ &= E - (E + S)^{-1}\Gamma_Y[E + (E + S)^{-1}\Gamma_Y]^{-1} \\ &= E - (E + S)^{-1}\Gamma_Y + (E + S)^{-2}\Gamma_Y^2[E + (E + S)^{-1}\Gamma_Y]^{-1}\end{aligned}$$

F can be written as follows

$$\begin{aligned}F = {} & (E + S)^{-1}[(P + Q)\Gamma - \Gamma_t + R^* - \Gamma_Y H Y] \\ & - (E + S)^{-1}\Gamma_Y[E + (E + S)^{-1}\Gamma_Y]^{-1}(E + S)^{-1} \\ & \times [(P + Q)\Gamma - \Gamma_t + R^*] + (E + S)^{-2}\Gamma_Y^2[E + (E + S)^{-1}\Gamma_Y]^{-1} \\ & \times HY + [E + (E + S)^{-1}\Gamma_Y]^{-1}(E + S)^{-1}(R - R^*),\end{aligned} \tag{164}$$

in which

$$R^* = R(Y^*; t), \quad Y^* = (E + S)\,Y.$$

The difference $R - R^*$ starting with the third power of Y may be obtained from the symbolic expansion

$$R(X, t) = R^* + \left[\exp\left(\Gamma_1 \frac{\partial}{\partial Y_1^*} + \ldots + \Gamma_N \frac{\partial}{\partial Y_N^*}\right) - 1\right] R^*. \tag{165}$$

Now F is represented by the series

$$F = \sum_{m=2}^{\infty} F^{(m)}, \tag{166}$$

$F^{(m)}$ being the homogenous form of power m with respect to Y. It is at once apparent that

$$F^{(m)} = (E + S)^{-1}[(P + Q)\,\Gamma^{(m)} - \Gamma_t^{(m)} - \Gamma_Y^{(m)} H Y] + V^{(m)}, \tag{167}$$

where $\Gamma^{(m)}$ contains the unknown terms of power m in Γ, the remaining terms of power m in the Expression (164) are retained in $V^{(m)}$. Since the terms of Γ are determined successively with increasing m, $V^{(m)}$ may be regarded as known. Coefficients of $F^{(m)}$ are quasi-periodic functions of the time developable in powers of μ. Therefore,

$$F^{(m)} = \sum_{k=0}^{\infty} \underset{k}{F^{(m)}} \mu^k, \tag{168}$$

where

$$\underset{k}{F^{(m)}} = P\underset{k}{\Gamma^{(m)}} - \underset{k}{\Gamma_t^{(m)}} - \underset{k}{\Gamma_Y^{(m)}} PY + \underset{k}{U^{(m)}}, \tag{169}$$

in which $\underset{k}{\Gamma^{(m)}}$ denotes the terms of μ^κ in $\Gamma^{(m)}$ and $\underset{k}{U^{(m)}}$ consists of the terms of μ^κ from the remaining parts of the form (167). Terms of $\underset{k}{\Gamma^{(m)}}$ are determined successively with increasing k and at every stage of approximation with respect to μ $\underset{k}{U^{(m)}}$ is known. Thus, the problem is to find the forms $\underset{k}{\Gamma^{(m)}}$ leading to the most simple forms $\underset{k}{F^{(m)}}$ provided that the condition of quasi-periodicity holds good.

Let $\underset{k}{U^{(m)}}$ be divided into two parts

$$\underset{k}{U^{(m)}} = \{\underset{k}{U^{(m)}}\} + \underset{k}{\tilde{U}^{(m)}} \tag{170}$$

such that the equation

$$\underset{k}{\Gamma_t^{(m)}} + \underset{k}{\Gamma_Y^{(m)}} PY - P\underset{k}{\Gamma^{(m)}} = \underset{k}{\tilde{U}^{(m)}} \tag{171}$$

admits a quasi-periodic solution (representing as a form of power m in Y with quasi-periodic coefficients). The resonance terms remain in $\underset{k}{F^{(m)}}$ and thus

$$\underset{k}{F^{(m)}} = \{\underset{k}{U^{(m)}}\}. \tag{172}$$

Examine the equation (171) in some detail. Considering the expressions (34) for Y and (25) for P, subdividing Γ and U in blocks as

$$\Gamma = \begin{pmatrix} \Gamma_1 \\ \Gamma_2 \\ \Gamma_3 \\ \Gamma_4 \\ \Gamma_5 \\ \Gamma_6 \end{pmatrix}, \qquad U = \begin{pmatrix} U_1 \\ U_2 \\ U_3 \\ U_4 \\ U_5 \\ U_6 \end{pmatrix} \tag{173}$$

and substituting

$$PY = \begin{pmatrix} \varrho \\ 0 \\ \sqrt{-1}\,\mathcal{N} a \\ -\sqrt{-1}\,\mathcal{N} \bar{a} \\ \sqrt{-1}\,\mathcal{N} b \\ -\sqrt{-1}\,\mathcal{N} \bar{b} \end{pmatrix}, \qquad P\Gamma = \begin{pmatrix} \Gamma_2 \\ 0 \\ \sqrt{-1}\,\mathcal{N} \Gamma_3 \\ -\sqrt{-1}\,\mathcal{N} \Gamma_4 \\ \sqrt{-1}\,\mathcal{N} \Gamma_5 \\ -\sqrt{-1}\,\mathcal{N} \Gamma_6 \end{pmatrix} \tag{174}$$

it is easy to find that Equation (171) may be written in the form of six matrix equations for the blocks Γ_κ (for brevity indices m and k being omitted)

$$\frac{\partial \Gamma_\kappa}{\partial t} + \frac{\partial \Gamma_\kappa}{\partial \theta} \varrho + \sqrt{-1}\left(\frac{\partial \Gamma_\kappa}{\partial a} \mathcal{N} a - \frac{\partial \Gamma_\kappa}{\partial \bar{a}} \mathcal{N} \bar{a} + \frac{\partial \Gamma_\kappa}{\partial b} \mathcal{N} b - \frac{\partial \Gamma_\kappa}{\partial \bar{b}} \mathcal{N} \bar{b}\right) - (P\Gamma)_\kappa = \tilde{U}_\kappa, \quad (\kappa = 1, 2, \ldots, 6) \tag{175}$$

where $(P\Gamma)_\kappa$ denotes the corresponding block of $P\Gamma$. The equations for $\kappa=4$ and $\kappa=6$ may not be considered since they are conjugate complex with those for $\kappa=3$ and $\kappa=5$ (for all m and k $U_4=\bar{U}_3$, $U_6=\bar{U}_5$ and hence $\Gamma_4=\bar{\Gamma}_3$, $\Gamma_6=\bar{\Gamma}_5$). In the scalar form

$$\frac{\partial \Gamma_{1i}}{\partial t} + \sum_{j=1}^{N}\left[\varrho_j \frac{\partial}{\partial \theta_j} + \sqrt{-1}\, n_j \left(a_j \frac{\partial}{\partial a_j} - \bar{a}_j \frac{\partial}{\partial \bar{a}_j} + b_j \frac{\partial}{\partial b_j} - \bar{b}_j \frac{\partial}{\partial \bar{b}_j}\right)\right] \times \Gamma_{1i} - \Gamma_{2i} = \tilde{U}_{1i}, \tag{176}$$

$$\frac{\partial \Gamma_{2i}}{\partial t} + \sum_{j=1}^{N}\left[\varrho_j \frac{\partial}{\partial \theta_j} + \sqrt{-1}\, n_j \left(a_j \frac{\partial}{\partial a_j} - \bar{a}_j \frac{\partial}{\partial \bar{a}_j} + b_j \frac{\partial}{\partial b_j} - \bar{b}_j \frac{\partial}{\partial \bar{b}_j}\right)\right] \times \Gamma_{2i} = \tilde{U}_{2i}, \tag{177}$$

$$\frac{\partial \Gamma_{3i}}{\partial t} + \sum_{j=1}^{N}\left[\varrho_j \frac{\partial}{\partial \theta_j} + \sqrt{-1}\, n_j \left(a_j \frac{\partial}{\partial a_j} - \bar{a}_j \frac{\partial}{\partial \bar{a}_j} + b_j \frac{\partial}{\partial b_j} - \bar{b}_j \frac{\partial}{\partial \bar{b}_j}\right)\right] \times \Gamma_{3i} - \sqrt{-1}\, n_i \Gamma_{3i} = \tilde{U}_{3i}, \tag{178}$$

$$\frac{\partial \Gamma_{5i}}{\partial t} + \sum_{j=1}^{N}\left[\varrho_j \frac{\partial}{\partial \theta_j} + \sqrt{-1}\, n_j \left(a_j \frac{\partial}{\partial a_j} - \bar{a}_j \frac{\partial}{\partial \bar{a}_j} + b_j \frac{\partial}{\partial b_j} - \bar{b}_j \frac{\partial}{\partial \bar{b}_j}\right)\right] \times \Gamma_{5i} - \sqrt{-1}\, n_i \Gamma_{5i} = \tilde{U}_{5i}. \tag{179}$$

$$(i = 1, 2, \ldots, N)$$

For every m and k the functions $U_{\kappa i}(\kappa=1, 2, \ldots, 6;\ i=1, 2, \ldots, N)$ are represented by series of the form

$$U_{\kappa i} = \sum A^{(\kappa i)}_{klpqrs} \prod_{j=1}^{N} \theta_j^{k_j} \varrho_j^{l_j} a_j^{p_j} \bar{a}_j^{q_j} b_j^{r_j} \bar{b}_j^{s_j}, \tag{180}$$

the summation is to be extended over all integer non-negative values of k_j, l_j, p_j, q_j, r_j, s_j, the sum of which is equal to m. The coefficients of this homogeneous form are quasi-periodic functions of the time dependent on the mean longitudes only and developable in series of the form

$$A^{(\kappa i)}_{klpqrs} = \sum A^{(\kappa i;\, klpqrs)}_{\gamma} \exp\left(\sqrt{-1} \sum_{j=1}^{N} \gamma_j \lambda_j\right), \tag{181}$$

where the summation goes over all integer values of γ_j. In designations of the coefficients each index k, l, p, q, r, s, γ represents a set of N corresponding scalar indices. In the same form the functions $\Gamma_{\kappa i}$ will be found to be

$$\Gamma_{\kappa i} = \sum a^{(\kappa i)}_{klpqrs} \prod_{j=1}^{N} \theta_j^{k_j} \varrho_j^{l_j} a_j^{p_j} \bar{a}_j^{q_j} b_j^{r_j} \bar{b}_j^{s_j}, \tag{182}$$

$$a_{klpqrs}^{(\kappa i)} = \sum a_{\gamma}^{(\kappa i;\, klpqrs)} \exp\left(\sqrt{-1} \sum_{j=1}^{N} \gamma_j \lambda_j\right). \tag{183}$$

Now, assuming the coefficients of the series (181) to be known the problem is to find from Equations (176)–(179) the coefficients of the series (183).

Consider some typical terms of the right-hand members of these equations. First of all, select the terms which do not contain any θ_j or ϱ_j. Let one of these terms be

$$U_{\kappa i} = A^{(\kappa i)} \prod_{j=1}^{N} a_j^{p_j} \bar{a}_j^{q_j} b_j^{r_j} \bar{b}_j^{s_j} \tag{184}$$

and the appropriate term in $\Gamma_{\kappa i}$

$$\Gamma_{\kappa i} = a^{(\kappa i)} \prod_{j=1}^{N} a_j^{p_j} \bar{a}_j^{q_j} b_j^{r_j} \bar{b}_j^{s_j} \tag{185}$$

is to be found. From Equations (176)–(179) there results

$$\left.\begin{aligned}
&\dot{a}^{(1i)} + a^{(1i)} \sqrt{-1} \sum_{j=1}^{N} (p_j - q_j + r_j - s_j)\, n_j - a^{(2i)} = A^{(1i)},\\
&\dot{a}^{(2i)} + a^{(2i)} \sqrt{-1} \sum_{j=1}^{N} (p_j - q_j + r_j - s_j)\, n_j = A^{(2i)},\\
&\dot{a}^{(3i)} + a^{(3i)} \sqrt{-1} \sum_{j=1}^{N} (p_j - q_j + r_j - s_j - \delta_{ij})\, n_j = A^{(3i)},\\
&\ldots\ldots
\end{aligned}\right\} \tag{186}$$

The equation for $\kappa = 5$ is of the same form as that for $\kappa = 3$ and thus need not be written here.

From these equations it follows that, for obtaining a quasi-periodic solution, one has, after substitution of (181) into (180), to include in $\{U_{3i}\}$ the resonance term whose indices satisfy the requirements

$$\gamma_j + p_j - q_j + r_j - s_j - \delta_{ij} = 0 \quad (j = 1, 2, \ldots, N) \tag{187}$$

while the term with the relations

$$\gamma_j + p_j - q_j + r_j - s_j = 0 \quad (j = 1, 2, \ldots, N) \tag{188}$$

is resonant for the second equation and is to be included in $\{U_{2i}\}$. It is at once apparent that $a^{(3i)}$ may then be determined uniquely without any term with resonance argument (187). The coefficient $a^{(2i)}$ is also determined in a unique way but a coefficient of the term with resonance argument (188) must be included as a constant of integration. If the coefficient of this resonance term is chosen from the condition

$$a_{\gamma}^{(2i)} = - A_{\gamma}^{(1i)} \tag{189}$$

the equation for $a^{(1i)}$ will always have the quasi-periodic solution and hence $\{U_{1i}\} = 0$.

By induction one may verify that owing to the properties of the right-hand sides of the starting equations the terms of the type considered give no contribution into $\{U_{2i}\}$. Therefore, $\{U_{2i}\}$ cannot contain the terms without any θ_j or $\varrho_j (j = 1, 2, \ldots, N)$.

Consider further the terms containing θ_j or ϱ_j in the first power. Such terms are to be treated by pairs in the form

$$U_{\kappa i} = (A^{(\kappa i)}\theta_j + B^{(\kappa i)}\varrho_j)\prod_{j=1}^{N} a_j^{p_j}\bar{a}_j^{q_j}b_j^{r_j}\bar{b}_j^{s_j}. \tag{190}$$

The appropriate terms of $\Gamma_{\kappa i}$ are

$$\Gamma_{\kappa i} = (a^{(\kappa i)}\theta_j + b^{(\kappa i)}\varrho_j)\prod_{j=1}^{N} a_j^{p_j}\bar{a}_j^{q_j}b_j^{r_j}\bar{b}_j^{s_j}. \tag{191}$$

For determining $a^{(\kappa i)}$ the same Equations (186) are still valid and $b^{(\kappa i)}$ are furnished from

$$\left.\begin{aligned}
&\dot{b}^{(1i)} + b^{(1i)}\sqrt{-1}\sum_{j=1}^{N}(p_j - q_j + r_j - s_j)\,n_j - b^{(2i)} + a^{(1i)} = B^{(1i)},\\
&\dot{b}^{(2i)} + b^{(2i)}\sqrt{-1}\sum_{j=1}^{N}(p_j - q_j + r_j - s_j)\,n_j + a^{(2i)} = B^{(2i)},\\
&\dot{b}^{(3i)} + b^{(3i)}\sqrt{-1}\sum_{j=1}^{N}(p_j - q_j + r_j - s_j - \delta_{ij})\,n_j + a^{(3i)} = B^{(3i)},\\
&\ldots\ldots
\end{aligned}\right\} \tag{192}$$

Substitute the trigonometric expansions similar to (181) for $A^{(\kappa i)}$ and $B^{(\kappa i)}$ into (190). To find $a^{(3i)}$ and $b^{(3i)}$ in the quasi-periodic form it is necessary to eliminate from $\tilde{U}_{3i}$ the resonance term (187) and include it in $\{U_{3i}\}$. $\{U_{2i}\}$ has to contain the resonance term (188) of the expansion of the function

$$[A^{(2i)}\theta_j + (B^{(2i)} + A^{(1i)})\,\varrho_j]\prod_{j=1}^{N} a_j^{p_j}\bar{a}_j^{q_j}b_j^{r_j}\bar{b}_j^{s_j}.$$

Then the equations for $a^{(1i)}$ and $b^{(1i)}$ may be solved in the quasi-periodic form provided that the coefficients of the terms of Γ_{2i} with the resonance arguments are chosen so as to satisfy

$$a_\gamma^{(2i)} = -A_\gamma^{(1i)},\quad b_\gamma^{(2i)} = -B_\gamma^{(1i)}. \tag{193}$$

Thus, in this case also $\{U_{1i}\}=0$.

Considering in a similar manner the terms of the second and higher powers in θ and ϱ and returning to the general series (180) and (181) one finds

$$\{U_{\kappa i}\} = \sum{}' A_{klpqrs}^{(\kappa i)}\prod_{j=1}^{N}\theta_j^{k_j}\varrho_j^{l_j}a_j^{p_j}\bar{a}_j^{q_j}b_j^{r_j}\bar{b}_j^{s_j}\quad(\kappa = 3, 4, 5, 6) \tag{194}$$

with a prime indicating that the expansion (181) should retain only the terms satisfying (187) for $\kappa=3$ and $\kappa=5$ and the terms satisfying the conjugate relations

$$\gamma_j + p_j - q_j + r_j - s_j + \delta_{ij} = 0 \tag{195}$$

for $\kappa=4$ and $\kappa=6$. Furthermore,

$$\{U_{2i}\} = \sum{}''(A_{klpqrs}^{(2i)} + A_{klpqrs}^{(1i)}\delta)\prod_{j=1}^{N}\theta_j^{k_j}\varrho_j^{l_j}a_j^{p_j}\bar{a}_j^{q_j}b_j^{r_j}\bar{b}_j^{s_j}, \tag{196}$$

where δ denotes a differential operator

$$\delta = \varrho_1 \frac{\partial}{\partial \theta_1} + \ldots + \varrho_N \frac{\partial}{\partial \theta_N} \tag{197}$$

and double prime shows that only the terms satisfying (188) should be retained in the expansion (181). Finally,

$$\{U_{1i}\} = 0. \tag{198}$$

The terms with the resonance arguments are present only in Γ_{2i}, the corresponding coefficients of $a^{(2i)}_{klpqrs}$ being chosen so as to avoid the appearance of the secular terms in $a^{(1i)}_{klpqrs}$.

The described method to isolate the resonance terms can be represented in form of the following process of averaging. Performing the substitution (30), (31) and treating the matrices

$$\underset{k}{U}^{(m)} = \underset{k}{U}^{(m)}(Y, t)$$

as functions of Z and t take the mean with respect to t of

$$\underset{k}{\mathcal{M}}^{(m)} = [T^* \underset{k}{U}^{(m)}(TZ, t)], \tag{199}$$

in which

$$T^* = \begin{pmatrix} 0 & 0 & 0 & 0 & 0 & 0 \\ \delta & E & 0 & 0 & 0 & 0 \\ 0 & 0 & \exp(-\sqrt{-1}\mathcal{N}t) & 0 & 0 & 0 \\ 0 & 0 & 0 & \exp\sqrt{-1}\mathcal{N}t & 0 & 0 \\ 0 & 0 & 0 & 0 & \exp(-\sqrt{-1}\mathcal{N}t) & 0 \\ 0 & 0 & 0 & 0 & 0 & \exp\sqrt{-1}\mathcal{N}t \end{pmatrix}.$$

Expressions (199) yield $\underset{k}{\mathcal{M}}^{(m)}$ as functions of Z, i.e. of $T^{-1} Y$. Then the resonance terms of $\underset{k}{U}^{(m)}$ are defined by

$$\{\underset{k}{U}^{(m)}\} = T \underset{k}{\mathcal{M}}^{(m)}. \tag{201}$$

Taking into account

$$T^* U = \begin{pmatrix} 0 \\ U_2 + \delta U_1 \\ \exp(-\sqrt{-1}\mathcal{N}t) \cdot U_3 \\ \exp\sqrt{-1}\mathcal{N}t \cdot U_4 \\ \exp(-\sqrt{-1}\mathcal{N}t) \cdot U_5 \\ \exp\sqrt{-1}\mathcal{N}t \cdot U_6 \end{pmatrix} \tag{202}$$

and the explicit relation (35) between Y and Z it is easy to state that (201) gives the expressions of $\{U_{\kappa i}\}$ found above. For the appropriate components $\mathcal{M}_{\kappa i}(\kappa = 1, 2, \ldots, 6;$

$i=1, 2, \ldots, N$) there results

$$\mathcal{M}_{1i} = 0, \tag{203}$$

$$\mathcal{M}_{2i} = \sum \left(A_\gamma^{(2i;\, klpqrs)} + A_\gamma^{(1i;\, klpqrs)}\,\delta\right) \prod_{j=1}^{N} \sigma_j^{k_j} v_j^{l_j} \alpha_j^{p_j} \bar{\alpha}_j^{q_j} \beta_j^{r_j} \bar{\beta}_j^{s_j}, \tag{204}$$

$$\mathcal{M}_{\kappa i} = \sum A_\gamma^{(\kappa i;\, klpqrs)} \prod_{j=1}^{N} \sigma_j^{k_j} v_j^{l_j} \alpha_j^{p_j} \bar{\alpha}_j^{q_j} \beta_j^{r_j} \bar{\beta}_j^{s_j}. \quad (\kappa = 3, 4, 5, 6) \tag{205}$$

In (204) indices are connected by (188) and in the definition (197) for δ ϱ_i and θ_i should be replaced by v_i and σ_i respectively. In (205) indices satisfy (187) for $\kappa=3$, $\kappa=5$ and (195) for $\kappa=4$, $\kappa=6$.

When the functions $\mathcal{M}$ become known (up to some power m with respect to Z and some order k with respect to μ) the substitution (30) transforms the system (28) into Equations (33) for the secular perturbations. The matrix Ω of the linear part of these equations is determined by (95). The equations for $\kappa=1$ and $\kappa=2$ are satisfied by $\sigma_i = v_i = 0$. Therefore, the final equations for the secular perturbations take the form (36) where $\mathcal{M}_3$ and $\mathcal{M}_5$ are designated by Υ and Ψ respectively. From the investigations of Poincaré (1893) concerning purely trigonometric form of planetary theory it follows that Υ and Ψ contain the terms of odd powers only in α_i, $\bar{\alpha}_i$, β_i, $\bar{\beta}_i$ and vanish when $\mu=0$.

In determining the functions Γ one may use from the very start the solution $\sigma_i = v_i = 0$ or, which is the same, $\theta_i = \varrho_i = 0$. Then $\delta=0$ in (196), (200), (202) and (204) which considerably facilitates the elaborating of the whole algorithm. In particular, $U_{\kappa i}$ will consist of the terms of the form (184) only and the problem to determine the functions Γ will reduce to the solving of Equations (186). The formulas (27) and (30) should be regarded now not as the change of variables but rather as a particular solution permitting us to express X through the functions of the time α, $\bar{\alpha}$, β, $\bar{\beta}$ determined implicitly by Equations (36).

Concluding this section, try to handle the two body problem basing on the algorithm suggested. The right-hand members R of Equations (20) do not depend explicitly on the time when $\mu=0$. Therefore, the time is not present explicitly in the functions Γ. The matrix H reduces to its first term P while F and $\mathcal{M}$ vanish so that α_i and β_i become constant. The first terms of the right-hand members R will be

$$\begin{aligned}
R_{1i} &= 3n_i^2\xi_i^2 - \tfrac{8}{3}\eta_i^2 - \tfrac{9}{2}(u_i^2 + \bar{u}_i^2) + 6\sqrt{-1}\, n_i\xi_i(u_i - \bar{u}_i) \\
&\quad + 4\eta_i(u_i + \bar{u}_i) + 3u_i\bar{u}_i + 3(v_i + \bar{v}_i)^2 + \ldots, \\
R_{2i} &= \tfrac{9}{2}\sqrt{-1}\, n_i(u_i^2 - \bar{u}_i^2) - 6n_i^2\xi_i\eta_i + \tfrac{9}{2}n_i^2\xi_i(u_i + \bar{u}_i) \\
&\quad - 6\sqrt{-1}\, n_i\eta_i(u_i - \bar{u}_i) + \ldots, \\
R_{3i} &= \tfrac{3}{2}\sqrt{-1}\, n_i^3\xi_i^2 - \tfrac{4}{3}\sqrt{-1}\, n_i\eta_i^2 + \tfrac{3}{4}\sqrt{-1}\, n_i(u_i^2 - 7\bar{u}_i^2) \\
&\quad - 4n_i^2\xi_i\eta_i + 6n_i^2\xi_i\bar{u}_i - 2\sqrt{-1}\, n_i\eta_i(u_i - 3\bar{u}_i) + \tfrac{3}{2}\sqrt{-1}\, n_iu_i\bar{u}_i \\
&\quad + \tfrac{3}{2}\sqrt{-1}\, n_i(v_i + \bar{v}_i)^2 + \ldots, \\
R_{5i} &= \sqrt{-1}\, n_i[-\eta_i + \tfrac{3}{4}(u_i + \bar{u}_i)]\,(v_i + \bar{v}_i) + \ldots.
\end{aligned}$$

If powers beyond the second are ignored, then

$$\Gamma_{1i} = \frac{\sqrt{-1}}{n_i}\left[\tfrac{9}{8}(a_i^2 - \bar{a}_i^2) - \tfrac{3}{2}(b_i^2 - \bar{b}_i^2)\right] + \dots,$$

$$\Gamma_{2i} = \tfrac{9}{4}(a_i^2 + \bar{a}_i^2) - 3a_i\bar{a}_i - 6b_i\bar{b}_i + \dots,$$

$$\Gamma_{3i} = \tfrac{3}{4}a_i^2 + \tfrac{7}{4}\bar{a}_i^2 - \tfrac{3}{2}a_i\bar{a}_i + \tfrac{3}{2}b_i^2 - \tfrac{1}{2}\bar{b}_i^2 - 3b_i\bar{b}_i + \dots,$$

$$\Gamma_{5i} = \tfrac{3}{4}a_ib_i - \tfrac{3}{4}a_i\bar{b}_i - \tfrac{3}{4}\bar{a}_ib_i - \tfrac{1}{4}\bar{a}_i\bar{b}_i + \dots.$$

Thus

$$\xi_i = \Gamma_{1i}, \quad \eta_i = \Gamma_{2i}, \quad u_i = a_i + \Gamma_{3i}, \quad v_i = b_i + \Gamma_{5i}$$

and hence

$$\delta p_i = -\tfrac{1}{2}a_i + \tfrac{3}{2}\bar{a}_i - \tfrac{3}{8}a_i^2 - \tfrac{1}{8}\bar{a}_i^2 + \tfrac{1}{2}a_i\bar{a}_i + \bar{b}_i^2 + b_i\bar{b}_i + \dots, \tag{206}$$

$$w_i = b_i + \bar{b}_i + \tfrac{1}{2}(a_ib_i + \bar{a}_i\bar{b}_i) - \tfrac{3}{2}(a_i\bar{b}_i + \bar{a}_ib_i) + \dots. \tag{207}$$

Comparison of these expressions with the corresponding terms of the expansions

$$1 - p = \sum_{k=-\infty}^{\infty} X_k^{1,1}(e)\left\{\cos^2\frac{I}{2}\exp\sqrt{-1}\,[(k-1)(\lambda-\pi)] + \sin^2\frac{I}{2}\exp\sqrt{-1}\,[(-k+1)(\lambda-\pi) - 2(\lambda-\Omega)]\right\}, \tag{208}$$

$$w = \frac{1}{2\sqrt{-1}}\sin I\cdot\sum_{k=-\infty}^{\infty} X_k^{1,1}(e)\left\{\exp\sqrt{-1}\,[(k-1)(\lambda-\pi) + \lambda - \Omega] - \exp\sqrt{-1}\,[(-k+1)(\lambda-\pi) - (\lambda-\Omega)]\right\} \tag{209}$$

arising from the classic solution of the two body problem (e, I, π, Ω being respectively the eccentricity, inclination and longitudes of the perihelion and node, $X_k^{n,m}(e)$ representing the Hansen coefficients) establishes at once that they are equivalent if the constants α_i and β_i are connected with the elements of the orbits by the formulas

$$\alpha_i = e_i\exp(-\sqrt{-1}\,\pi_i), \quad \beta_i = \frac{1}{2\sqrt{-1}}\sin I_i\exp(-\sqrt{-1}\,\Omega_i). \tag{210}$$

9. Conclusion

Summarizing, three main stages may be emphasized in elaborating the planetary theory according to the algorithm suggested.

The first stage is to compute the intermediate quasi-periodic solution in form of the power series in μ (Section 4).

The second stage is to determine the quasi-periodic matrices S and H also in form of the series of μ (Section 5). Since at this stage the equations of motion break up into two independent homogenous systems (96) and (126), the determination of S and H reduces to finding the appropriate submatrices (Sections 6, 7).

The third stage is to find the matrices Γ and F in form of the power series in a, $\bar{a}$, b, $\bar{b}$, the coefficients being the quasi-periodic functions developable in powers of μ (Section 8).

At the present moment almost all calculations of the first stage are completed. The corresponding results are not presented here for the lack of space and may be published separately. The computations of the second and third stages may be performed in the future.

Acknowledgments

This work could not have been completed without the help of A. V. Egorova, who made the program (in Algol-60 language) and performed the calculations of the first stage, and G. A. Krasinsky who gave a great deal of assistance to the author in elaborating this algorithm. The author is most grateful to them for their considerable help.

References

Anolik, M. V., Krasinsky, G. A., Pius, L. Yu.: 1969, 'Trigonometric Theory of the Secular Perturbations of the Principal Planets', *Trans. Inst. Theor. Astron. (Leningrad)* **14**, 3 (in Russian).

Birkhoff, G. D.: 1927, *Dynamical Systems*, New York.

Jefferys, W. H., Moser, J.: 1966, 'Quasi-Periodic Solutions for the Three Body Problem', *Astron. J.* **71**, 568.

Krasinsky, G. A.: 1968, 'Parametric Resonance in Canonical Systems of Linear Differential Equations with Quasi-Periodic Coefficients', *Dokl. Akad. Nauk (U.S.S.R.)* **180**, 526 (in Russian).

Krasinsky, G. A.: 1969, 'Quasi-Periodic Solutions of the First Kind in the Planar *N* Body Problem', *Trans. Inst. Theor. Astron. (Leningrad)* **13**, 105 (in Russian).

Meffroy, J.: 1966, 'On von Zeipel's Method in General Planetary Theory', SAO Spec. Report No. 229.

Poincaré, H.: 1893, *Les méthodes nouvelles de la mécanique céleste*, Vol. 2, Paris.

Roure, H.: 1949, 'Théorie nouvelle des grosses planètes du système solaire. *Bull. Astron.* **15**, 1.

STATIONARY AND PERIODIC SOLUTIONS FOR THE RESTRICTED PROBLEM OF THREE BODIES IN THREE-DIMENSIONAL SPACE

Y. KOZAI

Tokyo Astronomical Observatory, Tokyo, Japan

1. Introduction

In this paper the restricted problem of three bodies in the three-dimensional space, that is, the motion of an asteroid in the gravitational field of the sun and Jupiter moving on circular orbits is treated. The equations of motion for the asteroid can be expressed by the following canonical variables;

$$\begin{aligned} L &= k\sqrt{a}, & &l\text{: mean anomaly,} \\ G &= L\sqrt{1-e^2}, & &g\text{: argument of perihelion,} \\ H &= G\cos i, & &h = \Omega - \lambda', \end{aligned} \tag{1}$$

where a, e, i, Ω, and λ' are, respectively, the semi-major axis, the eccentricity, the inclination to Jupiter's orbital plane, the longitude of the ascending node for the asteroid, and Jupiter's longitude, and k is the gravitational constant of Gauss.

Short-periodic terms which depend on l and/or h can be eliminated from the Hamiltonian by von Zeipel's transformation, for example, if there is no commensurable relation between the mean motions of the asteroid and Jupiter, and, therefore, the equations of motion can be reduced to those of one degree of freedom as two action variables, L and H, are constant after the transformation. Then values of G can be expressed as a function of g, since the energy integral is a function of two variables, G and g, only. Thus the secular perturbations for the asteroid are completely solved. If both the eccentricity and the inclination are small, the argument of perihelion increases secularly, and G is expressed as a Fourier series with $2g$ as argument and with very small amplitudes.

However, if the inclination is sufficiently high, stationary solutions, for which G and g are constant, are found for $2g = 180°$ as well as solutions of libration case, for which the argument of perihelion cannot make one revolution. The stationary solutions correspond to singular points, which are centers, of equi-energy curves in (G, g)-plane (Kozai, 1962). Such solutions do not exist if the inclination is less than 39°.2 for a small value of the semi-major axis, and as the value of the semi-major axis gets larger they can exist for smaller values of the inclination. There is little chance for Jupiter and the asteroid on the stationary orbit to approach very closely to each other since the perihelion and aphelion of the asteroid are at apices of the highly inclined orbital plane.

If the mean motion, n, of the asteroid is nearly commensurable with Jupiter's mean

G. E. O. Giacaglia (ed.), Periodic Orbits, Stability and Resonances, 451–468.

motion which is assumed to be unity, the degree of freedom of the equations of motion cannot be reduced to one but to two as l and h cannot be eliminated together as they are not independent for this case. For a commensurable case, in which the mean motion is expressed as,

$$n = (p + q)/p, \tag{2}$$

with two positive integers, p and q, the equations of motion are conveniently expressed by the following canonical variables:

$$\begin{aligned} X_1 &= [(p+q) L - pH]/q, & Y_1 &= \lambda - \lambda', \\ X_2 &= (L - H)/q, & Y_2 &= (p+q)\lambda' - p\lambda - q\varpi, \\ X_3 &= G - H, & Y_3 &= g, \end{aligned} \tag{3}$$

where λ and $\tilde{\omega}$ are, respectively, the mean longitude and the longitude of the perihelion for the asteroid. The equations of motion are,

$$\mathrm{d}X_i/\mathrm{d}t = \partial F/\partial Y_i, \qquad \mathrm{d}Y_i/\mathrm{d}t = -\partial F/\partial X_i \quad (i = 1, 2, 3) \tag{4}$$

with the Hamiltonian

$$F = k^4/2(X_1 - pX_2)^{-2} + X_1 - (p+q) X_2 + m'k^2 R, \tag{5}$$

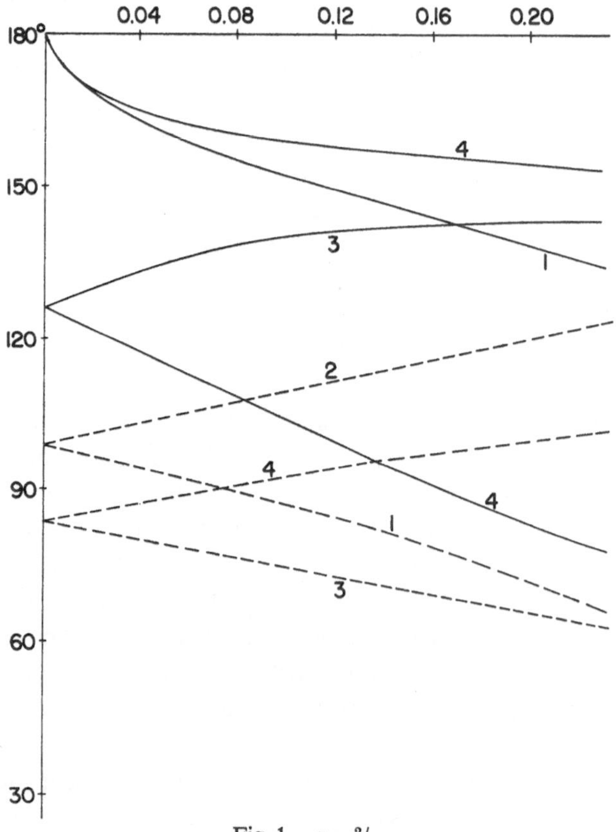

Fig. 1. $n = {}^2/_1$.

where m' is the mass of Jupiter in the unit of the mass of the sun and R is the disturbing function by Jupiter.

Short-periodic terms depending on Y_1 in the disturbing function can be eliminated from the Hamiltonian by von Zeipel's transformation which reduces the system of equations to that of two degrees of freedom. Although the equations of two degrees of freedom with one integral cannot be generally solved, particular solutions, in which X_i and Y_i ($i=2, 3$) are constant, can be derived. In this paper such particular solutions, which are stationary solutions, as well as periodic solutions generated from them are found by expanding the disturbing function by use of a high-speed computer.

2. Expansion of Disturbing Function

The disturbing function R due to Jupiter is written as

$$R = 1/\Delta - r \cos S, \tag{6}$$

where Δ and S are, respectively, the linear and the heliocentric angular distances between the asteroid and Jupiter while r is the heliocentric distance of the asteroid,

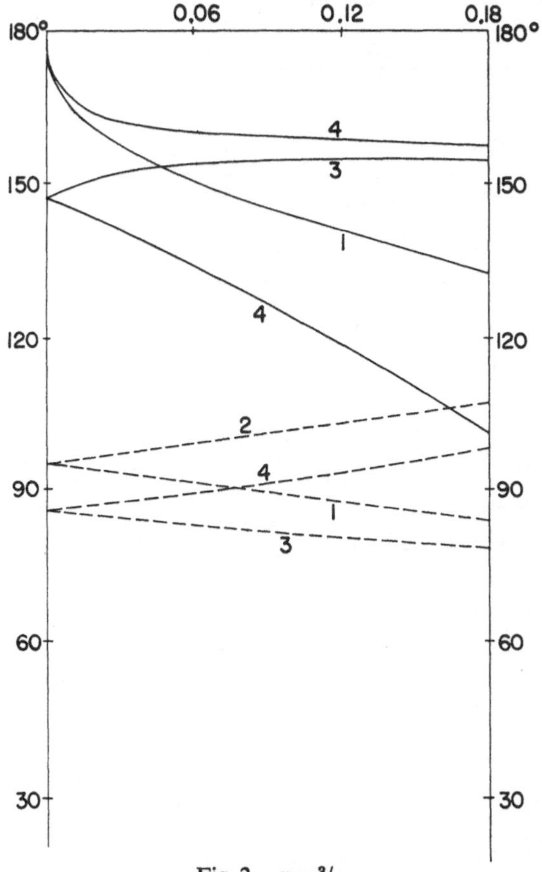

Fig. 2. $n = {}^3/_2$.

and the heliocentric distance of Jupiter has been assumed to be unity. The expressions of Δ and S are as follows:

$$\begin{aligned} \Delta^2 &= 1 + r^2 - 2r\cos S, \\ \cos S &= \cos^2(i/2)\cos(f + \varpi - \lambda') + \sin^2(i/2)\cos(f + \varpi + \lambda' - 2\Omega), \end{aligned} \tag{7}$$

where f is the true anomaly of the asteroid.

When the eccentricity of the asteroid is assumed to be zero, Δ^{-1} is expanded into a Fourier series with argument S_0, which is derived by replacing the true anomaly by the mean anomaly, as

$$1/\Delta = b_0 + 2\sum_{n=1}^{\infty} b_n \cos nS_0, \tag{8}$$

where b_n is the Laplace coefficient which is a function of a. Laplace coefficients and their Newcomb derivatives, $D^j b_n$, where D is the differential operator,

$$D = a\frac{\mathrm{d}}{\mathrm{d}a}, \tag{9}$$

can be computed by the method by Izsak and Benima (1963) when a is given.

The trigonometric terms, $\cos nS_0$, are represented by a finite sum of trigonometric

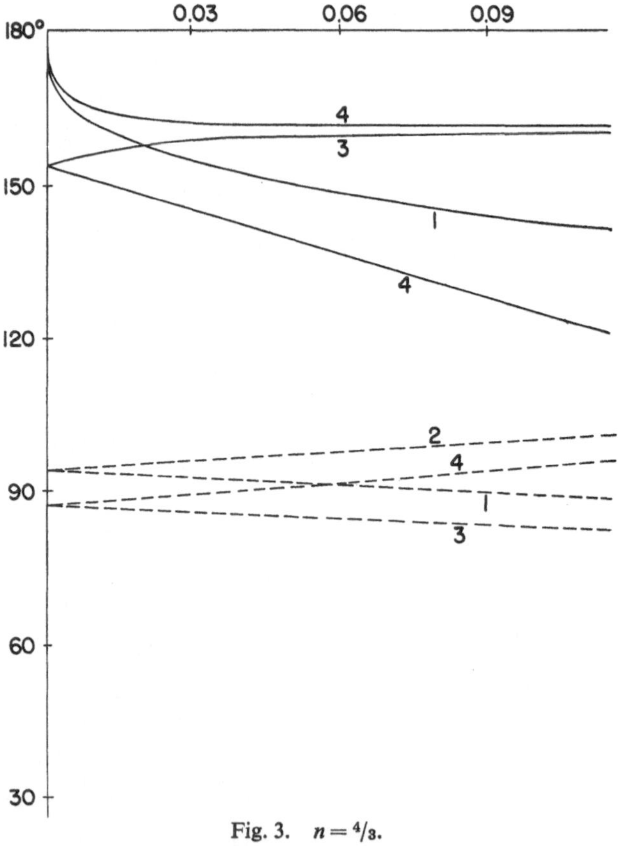

Fig. 3. $n = 4/3$.

terms as,

$$\cos nS_0 = Q_{00}^n + 2\sum_{\nu=1}^{n} Q_{0\nu}^n \cos \nu\psi + 2\sum_{\mu=1}^{n}\sum_{\nu=-n}^{n} Q_{\mu\nu}^n \cos(\mu\varphi - \nu\psi), \quad (10)$$

where

$$\varphi = \lambda - \lambda', \qquad \psi = \lambda + \lambda' - 2\Omega, \qquad \lambda = 1 + \varpi. \quad (11)$$

The coefficients $Q_{\mu\nu}^n$, which are called Tisserand's polynomials (Tisserand, 1888), are expressed as a difference of two polynomials which are squares of hypergeometric functions with negative integers or zero as the first argument and $\sin^2(i/2)$ as the fourth argument. Values of Tisserand's polynomials can be evaluated for any value of the inclination.

By combining the expressions (8) and (10) the disturbing function, R_0, for $e=0$ takes the following form:

$$R_0 = b_{00} + 2\sum_{\nu=1}^{\infty} b_{0\nu} \cos \nu\psi + 2\sum_{\mu=1}^{\infty}\sum_{\nu=-\infty}^{\infty} b_{\mu\nu} \cos(\mu\varphi - \nu\psi), \quad (12)$$

where the coefficients, $b_{\mu\nu}$, are expressed as,

$$\begin{aligned} b_{00} &= b_0 + 2\sum_{j=1}^{\infty} Q_{00}^{2j} b_{2j}, \\ b_{\mu\nu} &= 2\sum_{j=0}^{\infty} Q_{\mu\nu}^{\mu+\nu+2j} b_{\mu+\nu+2j}, \quad \text{for} \quad \mu \neq 0, \nu \neq 0. \end{aligned} \quad (13)$$

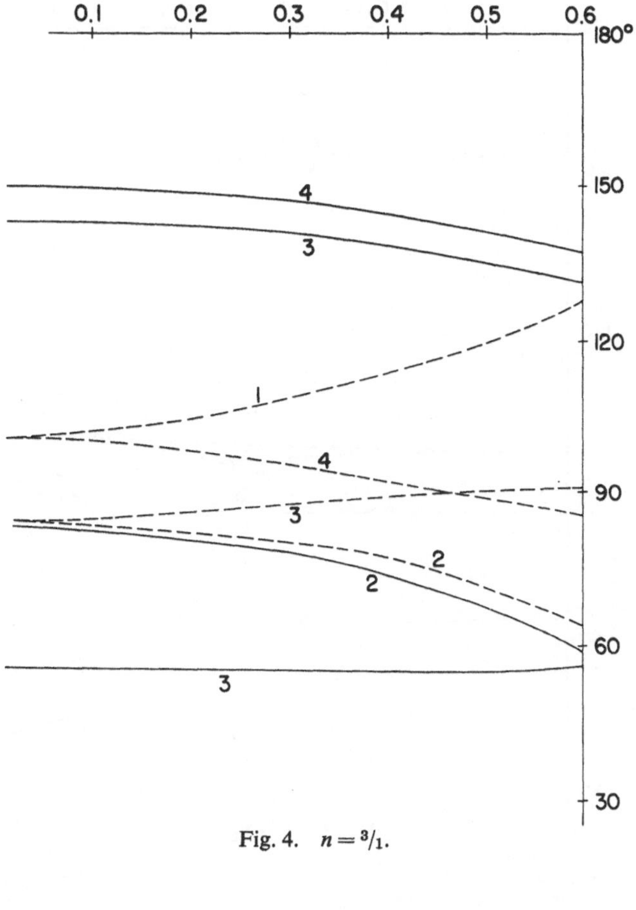

Fig. 4. $n = {}^3/_1$.

The coefficients, b_{10} and b_{01}, take the different forms since contributions from $r\cos S$ appear in these two coefficients. However, they are not used in the following computations.

The eccentricity is introduced up to the 10th power by use of Newcomb operators which are binomials of the differential operator, D, and the coefficient of λ, $\mu-\nu$, in the expression of R_0 in (12). Expressions of Newcomb operators are given in a paper by Izsak *et al.* (1964). After introducing the eccentricity by use of Newcomb operators the disturbing function is expanded in the form,

$$R = \sum_{j=-10}^{10} \sum_{\mu} \sum_{\nu} C_{j\mu\nu} \cos\left[-jl + (\mu-\nu)\lambda - (\mu+\nu)\lambda' + 2\nu\Omega\right], \tag{14}$$

where $C_{j\mu\nu}$ is a function of a, e, and i and contains $e^{|j|}\sin^{2|\nu|}(i/2)$ as a factor.

Since the short-periodic terms can be eliminated from the disturbing function, it is not necessary to compute all the terms in (14). The terms to be computed are those satisfying the following condition for μ, ν, and j;

$$(p+q)(\mu-\nu-j) = p(\mu+\nu), \quad \text{or} \quad q\mu = (2p+q)\nu + (p+q)j. \tag{15}$$

Then the disturbing function takes the following form after the short-periodic terms

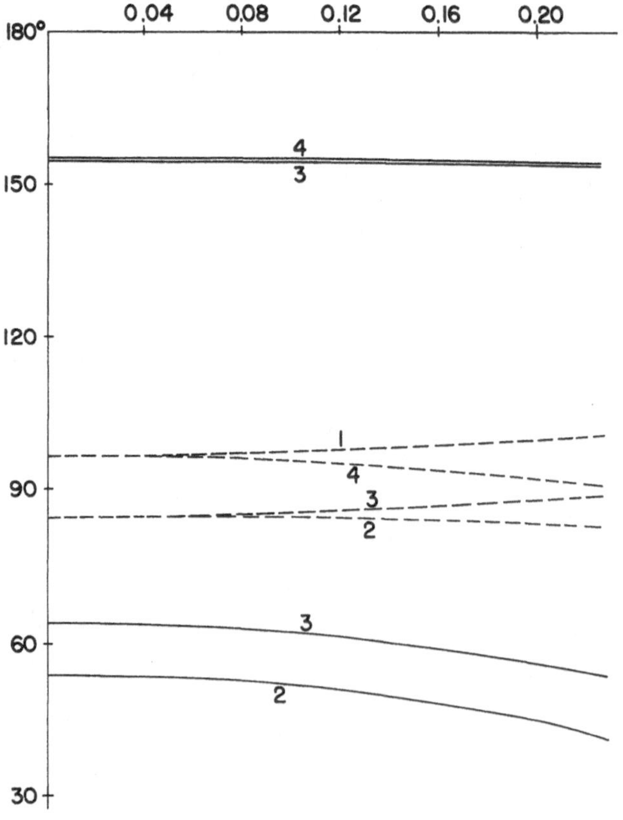

Fig. 5. $n = {}^5/_3$.

depending on Y_1 are eliminated;

$$R = \sum_{j=-\infty}^{\infty} \sum_{\nu} B_{j\nu} \cos\left(\frac{2\nu + j}{q} Y_2 + 2\nu Y_3\right), \tag{16}$$

where Y_2 is the critical argument and Y_3 is the argument of perihelion.

The computations are made for 11 commensurable cases listed in Table I which shows values of p and q and the semi-major axis. As the terms of higher than 10th degrees of the eccentricity are neglected, the expression (16) cannot be valid beyond a certain limit, e_u, of the eccentricity. Moreover, as the aphelion distance is larger than Jupiter's radius if the eccentricity is larger than e_L, the expansion (16) cannot be valid even if terms of much higher degrees are included. In Table I estimated values of e_u and e_L are also given as well as the largest value, N, of $\mu+\nu+2j$ used to compute the coefficients, $b_{\mu\nu}$, in the expression of (13). Laplace coefficients are computed up to the Nth order.

TABLE I

Mean motions and associated data

n	p	q	a	e_u	e_L	N
2	1	1	0.630	0.30	0.58	42
$^3/_2$	2	1	0.763	0.18	0.31	62
$^4/_3$	3	1	0.825	0.12	0.21	63
$^3/_1$	1	2	0.481	0.60	-	30
$^5/_3$	2	2	0.712	0.20	0.40	54
$^4/_1$	1	3	0.397	0.62	-	30
$^5/_2$	2	3	0.543	0.48	0.84	31
$^7/_4$	3	3	0.688	0.24	0.45	48
$^8/_5$	4	3	0.731	0.20	0.37	60
$^5/_1$	1	4	0.342	0.62	-	22
$^9/_5$	5	4	0.676	0.24	0.48	44

3. Stationary Solutions

Stationary solutions are derived by solving the equations,

$$\mathrm{d}X_i/\mathrm{d}t = \partial F/\partial Y_i = 0, \qquad \mathrm{d}Y_i/\mathrm{d}t = -\partial F/\partial X_i = 0. \quad (i = 2, 3) \tag{17}$$

The derivatives of X_i vanish when,

$$Y_2 = 0° \quad \text{or} \quad 180°, \qquad 2Y_3 = 0° \quad \text{or} \quad 180°. \tag{18}$$

The expression of the equation for $\dot{Y}_2$ is written as,

$$\begin{aligned}\frac{\mathrm{d}Y_2}{\mathrm{d}t} &= (p+q) - pk^4 (X_1 - pX_2)^{-3} - m'k^2 \frac{\partial R}{\partial X_2} \\ &= (p+q) - pn + m'k^2 \left[p \frac{\partial R}{\partial L} + (p+q)\left(\frac{\partial R}{\partial G} + \frac{\partial R}{\partial H}\right)\right] = 0,\end{aligned} \tag{19}$$

where the mean motion, n, is computed by $na^{\frac{3}{2}}=k$. When the values of e and i are given, this equation is satisfied by adjusting the value of a or n properly.

The last equation in (17) is written as,

$$\frac{\mathrm{d}Y_3}{\mathrm{d}t} = -m'k^2\frac{\partial R}{\partial X_3} = -m'k^2\frac{\partial R}{\partial G}$$
$$= m'k\frac{1}{a(1-e^2)}\left(\frac{1-e^2}{e}\frac{\partial R}{\partial e} - \cot i\frac{\partial R}{\partial i}\right) = 0. \tag{20}$$

The values of $\dot{Y}_3$ can be computed when a, e, and i are given for the four cases of the angular configurations of Y_2 and $2Y_3$ listed in Table II. After computing values of $\dot{Y}_3$

TABLE II
Angular configurations for stationary solutions

Cases	Y_2	$2Y_3$
1	0°	0°
2	180	0
3	0	180
4	180	180

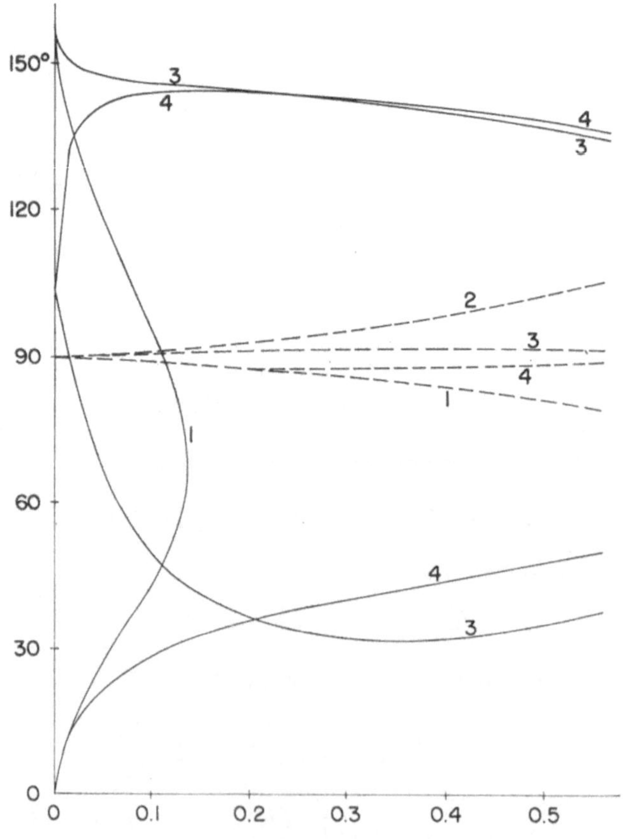

Fig. 6. $n = {}^4/_1$.

for various sets of e and i the solutions of Equation (20), $\dot{Y}_3 = 0$, can be derived by interpolation. Similar solutions have been already obtained by Jefferys and Standish (1966) by integrating the function $\partial R/\partial G$ numerically for some commensurable cases. The solutions expressed as curves in (e, i)-plane for $0° \leqq i \leqq 180°$ and $0 \leqq e \leqq e_u$ are shown in Figures 1–11 for 11 commensurable cases listed in Table I. The scales for the eccentricity are different from one figure to the other as the upper limit of the eccentricity, e_u, is different for each case. For some cases the curves near $e = e_u$ might not be reliable.

Although the solutions for larger values of the eccentricity are not given in this paper, these figures can fill gaps of small eccentricities and inclinations, for which Jefferys and Standish did not derive solutions. However, there are some discrepancies between their and my solutions for the case of $n = \frac{3}{2}$.

To a set of values of e and i a value of a which satisfies the Equation (19), $\dot{Y}_2 = 0$, exists, and these values together with those of Y_2 and Y_3 constitute mean orbital elements of the stationary orbits for a small value of m'.

In Figures 1–11 the solutions for the equation,

$$\frac{d\Omega}{dt} = -m'k^2 \frac{\partial R}{\partial H} = 0, \tag{21}$$

are also shown as broken lines in (e, i)-plane for each case of the angular configurations.

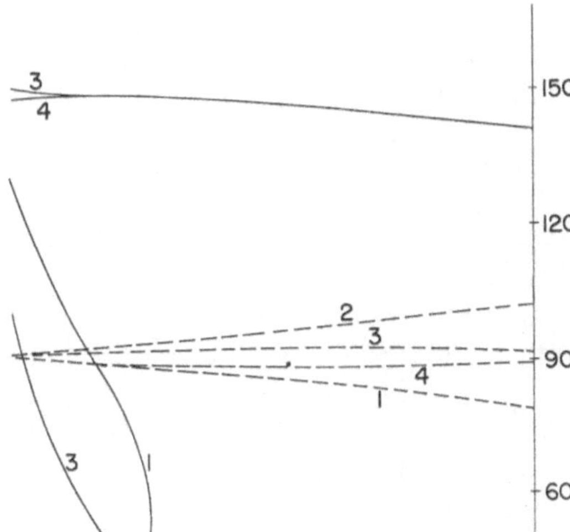

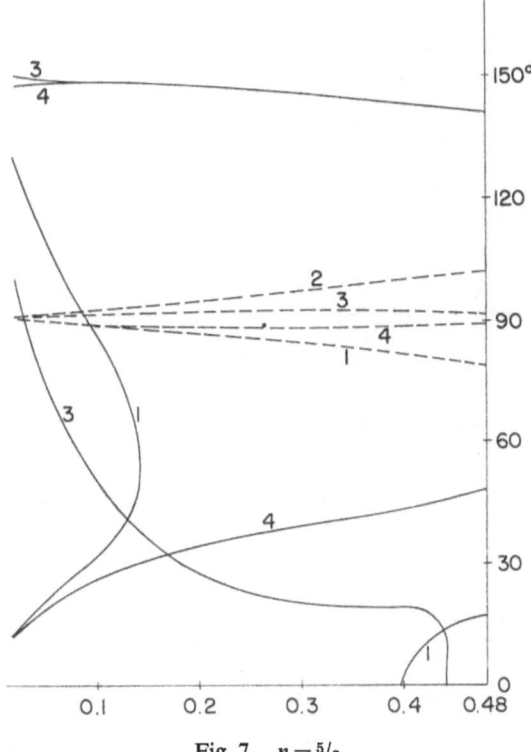

Fig. 7. $n = {}^5/_2$.

For Cases 1 and 3 of the angular configurations the oppositions take place when and only when

$$l = 0°(\text{mod. } 360°)/q\,, \tag{22}$$

as $Y_2=0°$, and for Cases 2 and 4 they take place when

$$l = 180°(\text{mod. } 360°)/q\,. \tag{23}$$

If the value of q is odd the asteroid and Jupiter can approach very closely at the opposition when the asteroid is moving on one of the stationary orbits of Cases 2 and 4 as the asteroid is then at the aphelion. They can particularly approach closely for Case 2 as both the perihelion and the aphelion are on the orbital plane of Jupiter. It must be noted here that stationary solutions of Case 2 do not exist except for $q=2$ according to my computations. When q is 2, oppositions take place only when $l=90°$ and 270° for Cases 2 and 4 and when $l=0°$ and 180° for Cases 1 and 3. Therefore, for Case 2 the oppositions take place only when the asteroid is at apices of the orbital plane, and, therefore, far from Jupiter's orbital plane. On the other hand for Case 1 the asteroid and Jupiter can approach very closely for $q=2$ as the opposition can take place at the aphelion which is on Jupiter's orbital plane, although stationary solutions of Case 1 do not exist for $n=\frac{3}{1}$ and $\frac{5}{3}$.

For odd values of q there are terms with a factor e such as $eA\cos jY_2$ in the disturbing

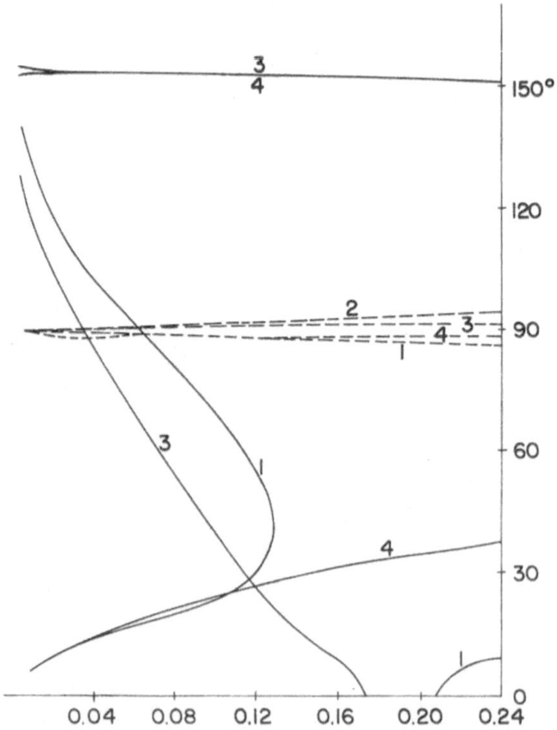

Fig. 8. $n = 7/4$.

function where the coefficient, A, is a function of a, e^2, and $\sin^2(i/2)$ with $\sin^{q-1}(i/2)$ as a factor. Therefore, terms with $1/e$ as a factor appear in the expression of $\dot{Y}_3$, and these are dominant terms if the eccentricity is very small and decrease as the eccentricity is increased. Therefore, the solutions of $\dot{Y}_3=0$ exist for Cases 1 and 3 where $Y_2=0°$ for any value of the inclination except for Case 3 ($q=1$). The values of the eccentricity for these stationary solutions are not so large for $q \geqq 3$, for which two points ($e=0$ and $i=0°$) and ($e=0$ and $i=180°$) are always solutions.

As the value of q increases the effects of terms with critical argument are diminished except for nearly polar and for very eccentric orbits as the critical terms have factors $e^j \sin^{q-j}(i/2)$ where j is an integer smaller than or equal to q. When q takes a very large value the features of the solutions are not quite different from those for non-commensurable cases, for which the Equation (20) is written as,

$$3-5\cos^2 i = 3e^2 \quad (2g = 180°), \tag{24}$$

if the value of the semi-major axis is very small. The solutions of the Equation (24) are shown in Figure 12, whereas the solutions corresponding to Cases 1 and 2 do not exist. The curves corresponding to Cases 3 and 4 in Figure 12 starts at a point ($e=0$ and $i=39°.2$), passes through a point ($e=1.0$ and $i=90°$), and ends at a point ($e=0$ and $i=180°-39°.2$). As the semi-major axis increases, the curve is distorted as the value corresponding to 39°.2 decreases. The secular motion of the node vanishes

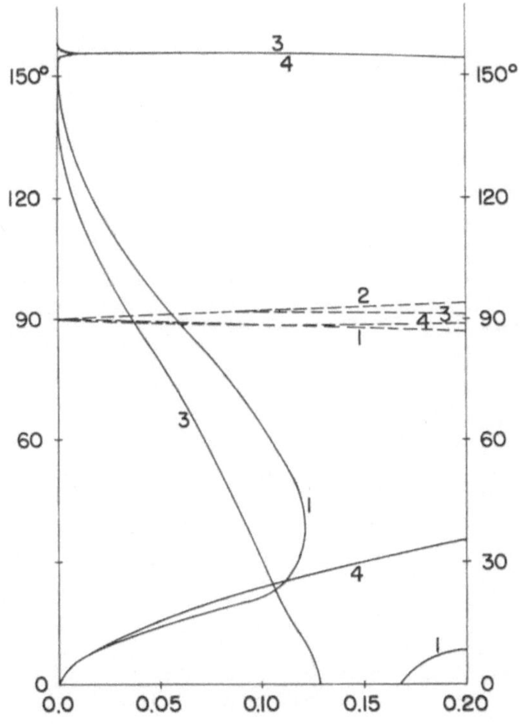

Fig. 9. $n = {}^8/_5$.

when and only when $\cos i=0$, that is, when the inclination is 90° as is shown as a broken line in Figure 12. When q is large, broken lines are very close to the straight line expressing $\cos i=0$ for non-commensurable case.

When the inclination is larger than 90°, the effects of the critical terms are also diminished as the critical argument for the retrograde orbit can be expressed as,

$$Y_2 = (p+q)\lambda' + p\lambda - (2p+q)\varpi, \tag{25}$$

if the mean motion is taken as negative and the inclination is measured in the opposite direction so that $i \leqq 90°$ always. The index expressing the importance of the critical term is not q but $(2p+q)$ for the retrograde cases. In fact curves in Figures 4–11 for $i>90°$ are not so different from that in Figure 12. When q takes an odd value, the curves with small eccentricity also appear in the figures. Therefore, we can have some ideas on features of curves for larger value of q from those for $i>90°$ for smaller value of q.

It is concluded that the asteroid on one of the stationary orbits has few chances to approach very closely to Jupiter both for commensurable and non-commensurable cases.

4. Characteristic Exponents

In order to discuss the stability of the stationary solutions, characteristic exponents

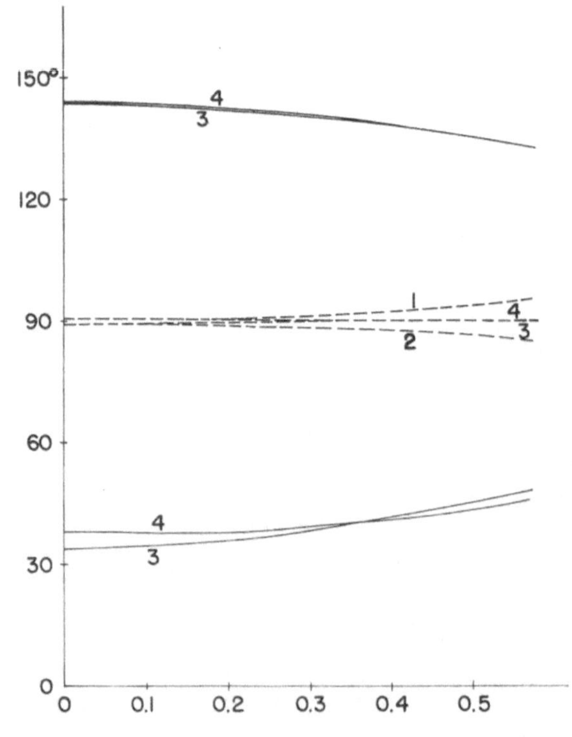

Fig. 10. $n = {}^5/_1$.

are computed as the proper values of the following linear differential equations:

$$\begin{aligned} &\frac{d\delta Y_j}{dt} + \sum_{j=2}^{3}\left(\frac{\partial^2 F}{\partial X_i \partial X_j}\,\delta X_j + \frac{\partial^2 F}{\partial X_i \partial Y_j}\,\delta Y_j\right) = 0\,, \\ &\frac{d\delta X_j}{dt} - \sum_{j=2}^{3}\left(\frac{\partial^2 F}{\partial Y_i \partial X_j}\,\delta X_j + \frac{\partial^2 F}{\partial Y_i \partial Y_j}\,\delta Y_j\right) = 0 \quad (i = 2, 3) \end{aligned} \tag{26}$$

where the variables are δX_j and δY_j and in the derivatives the constant values corresponding to the stationary solutions are put in the places of X_j and Y_j.

By putting

$$\frac{\partial^2 F}{\partial X_i \partial X_j} = -A_{ij}, \qquad \frac{\partial^2 F}{\partial Y_i \partial Y_j} = B_{ij} \qquad (i, j = 2, 3) \tag{27}$$

the equation to determine the characteristic exponent, α, is written as,

$$\begin{aligned} &\alpha^4 - \alpha^2 (A_{22}B_{22} + A_{33}B_{33} + 2A_{23}B_{23}) \\ &\quad + (A_{22}A_{33} - A_{23}^2)(B_{22}B_{33} - B_{23}^2) = 0\,, \end{aligned} \tag{28}$$

and

$$\frac{\partial^2 F}{\partial X_i \partial Y_j} = 0 \quad (i, j = 2, 3) \tag{29}$$

since each term in these equations contains $\sin(iY_2 + 2jY_3)$ as a factor and Y_2 and $2Y_3$

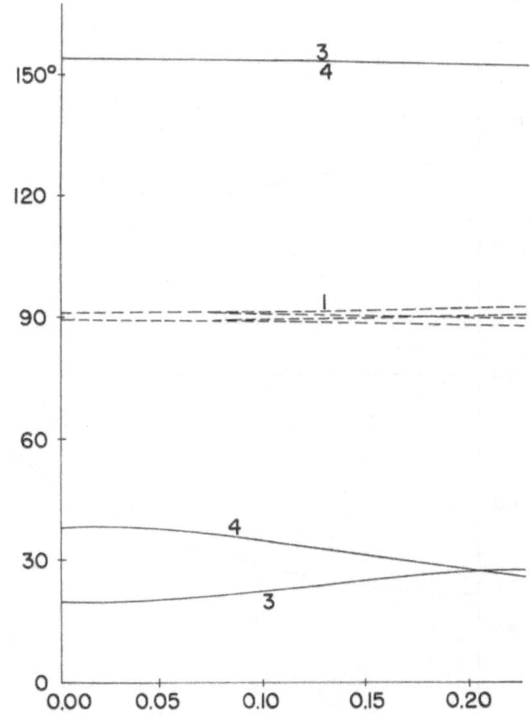

Fig. 11. $n = 9/5$.

take values of 0° or 180° for the stationary solutions. The derivatives, A_{ij} and B_{ij}, are of the order of m' except for A_{22} which is of order of 1.

Approximate solutions of the Equation (28) are derived as,

$$\begin{aligned} \alpha_1^2 &= 2A_{22}B_{22}, \\ \alpha_2^2 &= 2A_{33}(B_{22}B_{33} - B_{23}^2)/B_{22}, \end{aligned} \tag{30}$$

where α_1^2 (of order of m') corresponds to the motion of the critical argument and α_2^2 (of order of m'^2) corresponds to that of the argument of perihelion.

For various stationary solutions values of the characteristic exponents are computed by using the expression (16) of the disturbing function. One of the main conclusions is that for Case 1, where $Y_2=0°$ and $2Y_3=0°$, α_1^2 is always negative and α_2^2 is positive. In other words, the critical argument will make libration around 0° whereas the argument of perihelion will depart from 0° secularly if small perturbations are applied. Therefore, if the inclination can be neglected, it may be concluded that the motion is stable for Case 1 although it is instable in the three-dimensional space because of positive values of α_2^2. Of course, the asteroid itself cannot move out far from Jupiter's orbital plane. As the argument of perihelion varies, values of the eccentricity and the inclination can be changed considerably, namely by order of 1, particularly, for large values of the eccentricity and the inclination. On the other hand, as the critical argument is varied, the orbital elements are changed by order of $m'^{\frac{1}{2}}$.

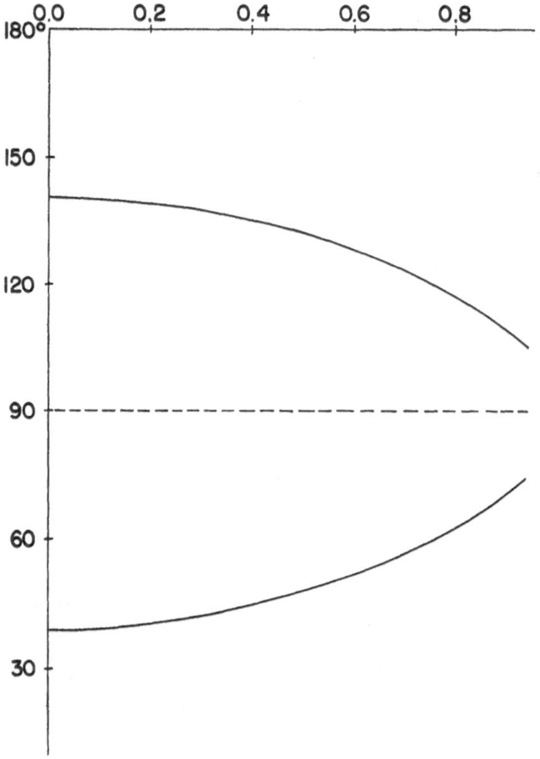

Fig. 12.

For direct orbits of $n=\frac{3}{1}$, α_1^2 and α_2^2 are, respectively, negative and positive except for a branch where $i>55°.5$ and $e>0.55$ for Case 3. For direct orbits of $n=\frac{5}{3}$, the solutions are mostly stable except for a branch of Case 2 for $i<46°$ and $e>0.19$ and that of Case 3 for $i>62°.8$ and $e<0.07$. For these exceptional cases α_2^1 is negative and α_2^2 is positive.

On the branches of Cases 3 and 4 which lie nearly horizontally in the upper parts of Figures 1–11, that is, for retrograde cases, α_2^2 is usually negative. The argument of perihelion makes libration if the orbit is very near to one of the stationary solutions both for commensurable and non-commensurable cases. On the other hand, α_1^2 is positive for Case 4 and negative for Case 3.

The solutions of Case 3 whose curves run almost vertically in the figures of odd values of q are also stable if the inclination is at least higher than 130°, although they are unstable, that is, both α_1^2 and α_2^2 are positive for direct orbits. Solutions on the other branches which appear for $q=1$ and 3 are mostly stable.

It must be emphasized that even if the orbit is found to be unstable here the values of the semi-major axis and the mean motion will be limited in small ranges of order of $m'^{\frac{1}{2}}$ although the values of the eccentricity and the inclination can be changed very widely.

5. Periodic Solutions

Even for the stationary solutions X_i and Y_i $(i=2, 3)$ as well as X_1 and $Y_1-(n-1)t$ are no more constant but oscillate around the stationary values if the short-periodic terms in the disturbing function are taken into account, and the longitude of the ascending node moves secularly even if the short-periodic perturbations are omitted. Therefore, the stationary solutions found here are not usually periodic both for commensurable and for non-commensurable cases.

However, if the mean value of $\dot{Y}_1$ is exactly equal to q/p, that is, if Y_1, which is the fundamental argument of the short-periodic terms, takes the same value as the initial one after $(p+q)$ revolutions of the asteroid, the short-periodic perturbations in all the orbital elements also take the same values as the initial ones for the stationary solutions.

As the condition $\partial R/\partial G=0$ has been already satisfied for the stationary solutions, the equation, $\dot{Y}_1=q/p$, can be written as,

$$\frac{\mathrm{d}Y_1}{\mathrm{d}t} = -\frac{\partial F}{\partial X_1} = n - 1 - m'k^2\left(\frac{\partial R}{\partial L} + \frac{\partial R}{\partial H}\right) = q/p\,, \tag{31}$$

and Equation (19) is expressed as,

$$\frac{\mathrm{d}Y_2}{\mathrm{d}t} = (p+q) - pn + m'k^2\left[p\frac{\partial R}{\partial L} + (p+q)\frac{\partial R}{\partial H}\right] = 0\,. \tag{32}$$

These equations are satisfied when and only when,

$$\partial R/\partial H = -\,\mathrm{d}\Omega/\mathrm{d}t = 0\,, \tag{33}$$

$$n = \frac{p+q}{p} + m'k^2\frac{\partial R}{\partial L}, \tag{34}$$

where n is the mean motion of the mean anomaly defined as,

$$l = \int n \, \mathrm{d}t + \varepsilon. \tag{35}$$

Here ε is the mean anomaly at the epoch and has the secular motion equal to $-m'k^2 \partial R/\partial L$. Therefore, the real mean motion of the mean longitude is exactly $(p+q)/p$ for periodic solutions.

The condition (33) shows that the secular motion of the longitude of the ascending node must vanish, and, therefore, the node is returned to the initial position after $(p+q)$ revolutions of the asteroid even if the short-periodic perturbations are included. Therefore, the stationary solutions satisfying the conditions (33) and (34) simultaneously are purely periodic in not only dynamical but also geometrical sense. The condition (33) is satisfied at points on the broken lines in Figures 1–12.

If the solid and broken curves belonging to the same case of the angular configurations in Figures 1–11 intersect at a point, a periodic solution fixed in space of the third sort by Poincaré (1892) is generated there for a small value of m'. Values of the eccentricity and the inclination of the periodic solution are derived by solving Equations (20) and (33), and then the mean motion is computed by Equation (34). By this way ten periodic solutions are found in this paper and four more solutions with large eccentricities are derived by using figures by Jefferys and Standish (1966) and numerical integrations of $\partial R/\partial H$.

Existence of such periodic solutions is verified by integrating the equations of motion numerically for a case of $m'=0.001$. The osculating orbital elements at the initial epoch when the symmetric opposition takes place are adjusted so that the orbital elements take the same values as the initial ones after $(p+q)/2$ or $(p+q)$ revolutions of the asteroid. The period of the periodic solution is $(p+q)$ times as long as the revolution period of the asteroid. After one period the asteroid and Jupiter are returned to the initial positions in the fixed coordinate system.

Table III shows the osculating orbital elements thus derived at the initial epochs and the mean elements for the fourteen periodic solutions for $m'=0.001$. The mean motion, n, given there is that defined in (35), and the actual mean motion of the mean longitude and the mean anomaly in usual sense takes exact commensurable values for the periodic solutions. The signs of the characteristic exponents are also given in Table III if they are computed.

The inclinations of the periodic solutions except for the solution No. 2 in Table III take values near 90°, although the values of the eccentricities are widely spread. The periodic solution No. 2 for $n=\frac{3}{2}$ has a large value for the eccentricity and the aphelion distance is as large as 1.09, that is, the aphelion is outside of Jupiter's orbit. However, as the oppositions take place always at the perihelion, the asteroid cannot approach to Jupiter very closely even for this case. The eccentricity of the solution No. 5 for $n=\frac{4}{3}$ is particularly large, and the aphelion is outside Jupiter's orbit.

The solution No. 6 for $n=\frac{5}{3}$ has eccentricity as large as 0.79 and the aphelion distance is 1.28. As the oppositions take place at both the perihelion and the aphelion

TABLE III

Osculating elements at the epoch on upper lines and mean elements on lower lines for periodic orbits

	Y_2	$2Y_3$	i	e	n	α^2_1	α^2_2
1	180°	180°	95°.235 3	0.171 823	1.999 301	−	−
			95°.252 1	0.172 006	1.999 723		
2	0	0	69°.617 8	0.426 509	1.498 001		
			69°.604 7	0.426 633	1.499 310		
3	180	180	98°.104 1	0.189 771	1.497 947	−	−
			98°.152 7	0.188 272	1.499 914		
4	180	180	98°.794 9	0.167 733	1.330 173	−	−
			98°.868 9	0.164 211	1.333 412		
5	0	0	85°.571 3	0.946 675	1.338 095		
			85°.160 6	0.946 898	1.331 693		
6	0	0	85°.285 8	0.793 546	1.663 531		
			85°.310 7	0.794 557	1.664 862		
7	0	0	88°.674 3	0.113 958	3.999 005	−	+
			88°.671 5	0.113 791	3.999 864		
8	0	180	90°.126 3	0.014 189	4.000 135	−	−
			90°.126 1	0.014 032	3.999 888		
9	0	0	88°.614 3	0.092 033	2.499 205	−	+
			88°.584 7	0.091 286	2.499 803		
10	0	180	90°.220 7	0.027 172	2.499 229	+	+
			90°.238 8	0.026 261	2.499 842		
11	0	0	89°.422 4	0.061 430	1.745 771	−	+
			89°.412 6	0.060 780	1.749 764		
12	0	180	90°.669 1	0.032 881	1.749 553	+	+
			90°.682 0	0.034 274	1.749 811		
13	0	0	89°.635 6	0.054 067	1.597 440	−	+
			89°.571 3	0.051 647	1.599 751		
14	0	180	90°.618 3	0.035 203	1.597 468	+	+
			90°.671 7	0.034 520	1.599 806		

for this case, the asteroid near the aphelion which is on Jupiter's orbital plane approaches very closely to Jupiter although they never collide with each other.

These three solutions with large eccentricities as well as the solution No. 4 for $n=\frac{4}{3}$ are found by numerical integrations of $\partial R/\partial H$ and by use of the figures by Jefferys and Standish (1966). The solution No. 3 for $n=\frac{3}{2}$ are also not in the figures of this paper, however, it can be found by extrapolation.

For $q=3$ two periodic solutions are found for Cases 1 and 3 for each value of the mean motion. The orbits are almost polar and the eccentricities are smaller than those for $q=1$. For odd values of q larger than three two periodic solutions of Case 1 and 3 are also found for each case. As the value of q is increased the value of the eccentricity of the periodic solution is decreased and the value of the inclination converges to 90°. In fact for $q=5$ the eccentricities of the periodic solutions of Case 1 and 3 are, respectively, 0.005 and 0.000 5 for $n=\frac{6}{1}$, and 0.008 and 0.005 for $n=\frac{11}{6}$.

For even values of q periodic solutions do not exist usually, except for one, the solution No. 6 of $n=\frac{5}{3}$. In fact features of the curves expressing $\partial R/\partial G=\partial R/\partial H=0$ for these cases, especially for large values of q, are not quite different from those in Figure 12 for non-commensurable case, where the two curves intersect only at a point ($e=0.1$ and $i=90°$).

6. Discussion

In this paper computations are made for the mean motion larger than 1, that is, for the case in which the semi-major axis of the asteroid is smaller than that of Jupiter. Similar computations for the semi-major axis larger than that of Jupiter can be made and periodic solutions will be also derived. Periodic solutions found here for $m'=0.001$ can be extended to cases with larger values of m'.

Such computations will be made in future if much more machine time will be available to me.

Acknowledgements

Computations for the expansion of the disturbing functions as well as for drawing Figures 1–11 were made by HITAC 5020E, at the Computer Center of the University of Tokyo, and the computations for the numerical integrations were made by OKITAC 5090D, at the Tokyo Astronomical Observatory. I am grateful to the staff of the computer center of the Tokyo Astronomical Observatory for their operating the OKITAC for many hours for me.

References

Izsak, I. G. and Benima, B.: 1963, Smithsonian Astrophys. Obs. Special Report No. 129.
Izsak, I. G., Gerard, J. M., Efimba, R., and Barnett, M. P.: 1964, Smithsonian Astrophys. Obs. Special Report No. 140.
Jefferys, W. H.: 1966, *Astron. J.* **71**, 99.
Jefferys, W. H. and Standish, E. M.: 1966, *Astron. J.* **71**, 982.
Kozai, Y.: 1962, *Astron. J.* **67**, 591.
Poincaré, H.: 1892, *Les méthodes nouvelles de la mécanique céleste*, tome 1, Gauthier-Villars, Paris.
Tisserand, F.: 1888, *Traité de mécanique céleste*, tome 1, Gauthier-Villars, Paris.

Discussion

G. Hori: Are the *e-i* relation and also the periodic orbits correct within an accuracy of $0(m')$, because of the averaging process?

Y. Kozai: The *e-i* relation is correct within an accuracy of $0(m')$. However, orbital elements given here for periodic solutions are derived by correcting the first approximation values by computing orbits by numerical integration methods.

W. H. Jefferys: It is very good that you have determined the stability of these orbits. Have you done this for the case when q is very large?

Y. Kozai: Yes, I did. Stability features are almost the same even if q is large, and depend on parity of q.

MOTION OF A SPACE PROBE NEAR AN OBLATE PLANET

C. FREDERICK PETERS
Naval Weapons Laboratory, Dahlgren, Va., U.S.A.

Abstract. For orbits with semi-latus rectum of the order of $\frac{1}{6}$ the planet's equatorial radius, conventional first order theories of satellite motion about an oblate planet produce errors of the order of 1300 J_2^2 outside the planet's radius, J_2 being the oblateness coefficient. It is shown that a first order theory using Kustaanheimo-Stiefel two body regularizing variables produces results with errors of the order of J_2^2 outside the planet's radius.

1. Introduction

The application of artificial satellite theories by Brouwer, Vinti, and Geyling to orbits with eccentricities around 0.8 or 0.9 and semi-major axes 0.6 the earth's radius yield results with rather large errors for ephemeris purposes. This, however, is entirely consistent with the assumptions upon which these theories were derived and constitute applications outside their range of validity.

All numerical comparisons in this paper are made with respect to orbits computed by numerical integration of the equations of motion derived from the potential

$$V = -\frac{\mu}{r}\left[1 - J_2 \frac{R^2}{r^2}\left(-\frac{1}{2} + \frac{3}{2}\sin^2 \varphi\right)\right],$$

where μ is the gravitational constant, R the equatorial radius, φ the geocentric latitude, and J_2 the dimensionless oblateness coefficient.

Under the conditions specified the motion is of interest for part of one revolution near apogee. An examination of Brouwer's results show the small parameter to be

$$\gamma_2' = \frac{1}{2} J_2 \frac{R^2}{p^2},$$

where p is the semi-latus rectum. Brouwer's theory accounts for the short period and long period perturbations to the first order in γ_2'. For 'normal' satellite orbits with $p \cong R$ numerical results show that the neglected second order terms produce periodic position errors in the orbit of about 0.01 km. For the orbits considered here $e \cong 0.85$, $a \cong 0.6\ R$, and $p = 0.17\ R$. The ratio of the neglected second order terms for the two cases is then $1/(0.17)^4$ or about 1300. This means periodic errors in position of around 13 km. Comparisons of the theories by Brouwer, Vinti, and Geyling with numerically integrated orbits confirm these results. In a few particular cases a Keplerian orbit served as a better approximation.

2. Equations of Motion Using KS Variables

The usual formulas of elliptic motion break down as the eccentricity approaches unity.

G. E. O. Giacaglia (ed.), Periodic Orbits, Stability and Resonances, 469–473.

This, of course, is due to the choice of variables. Assuming the semi-major axis remains finite, this corresponds to a rectilinear or collision orbit. A set of nonsingular variables for these orbits has been developed by Kustaanheimo and Stiefel for the regularization of the three dimensional two body problem. The KS transformation does not, however, regularize the oblateness problem. Since we are only concerned with the motion near apogee, regularization is not really the objective. It is only desirable that the unperturbed orbit be nonsingular.

The KS transformation is given by the equations

$$\begin{aligned} x &= u_1^2 - u_2^2 - u_3^2 + u_4^2, \\ y &= 2(u_1 u_2 - u_3 u_4), \\ z &= 2(u_1 u_3 + u_2 u_4), \end{aligned}$$

where x, y, z are the usual rectangular coordinates and u_i are the KS variables. The time, t, is given by

$$t = \int_{\tau_0}^{\tau} r\, d\tau,$$

where τ is the new independent variable. It can be seen that τ is equivalent to the eccentric anomaly from

$$2\omega(\tau - \tau_0) = E - E_0,$$

where

$$\omega^2 = \frac{\mu}{2r_0} - \frac{v_0^2}{4}.$$

Let prime denote differentiation with respect to τ. The KS variables must satisfy a constraint equation of the form

$$u_4 u_1' - u_3 u_2' + u_2 u_3' - u_1 u_4' = 0.$$

In the above equations r is the radius vector given by

$$r = \sum u_i^2$$

and r_0 its initial value. Also v_0 is the initial magnitude of the velocity. The transformed equations of motion are

$$u_i'' + \omega^2 u_j = \frac{\partial F}{\partial u_j},$$

where

$$\begin{aligned} F &= \frac{1}{4} r(G_0 - G), \\ G &= \mu \frac{J_2 R^2}{r^3}\left(-\frac{1}{2} + \frac{3}{2}\frac{z^2}{r^2}\right), \\ G_0 &= G(t_0), \end{aligned}$$

and G is implicitly expressed in terms of the u_i by the transformation equations. A first order solution may be obtained by the method of variation of arbitrary constants using the harmonic oscillator as the undisturbed solution. This is the solution of the two body problem in KS variables. The unperturbed solution is then

$$u_j = \alpha_j \cos \omega(\tau - \tau_0) + \beta_j \sin \omega(\tau - \tau_0)$$
$$u_j' = \omega[-\alpha_j \sin \omega(\tau - \tau_0) + \beta_j \cos \omega(\tau - \tau_0)],$$

where

$$\alpha_j = u_j(\tau_0),$$
$$\beta_j = \frac{1}{\omega} u_j'(\tau_0).$$

The constraint equation imposes the following condition on the α_j, β_j;

$$\alpha_4\beta_1 - \alpha_3\beta_2 + \alpha_2\beta_3 - \alpha_1\beta_4 = 0,$$

which must be satisfied in the perturbed case also. For the perturbed problem the differential equations for the variation of α_j, β_j are

$$\alpha_j' = -\frac{1}{\omega} \sin \omega(\tau - \tau_0) \frac{\partial F}{\partial u_j},$$
$$\beta_j' = \frac{1}{\omega} \cos \omega(\tau - \tau_0) \frac{\partial F}{\partial u_j}.$$

3. Accuracy of a First Order Theory in KS Variables

Before developing an analytic theory it would be desirable to determine whether a first order theory will produce the desired accuracy. A first order solution may be simulated by numerical integration of the above equations with the expression on the right hand side evaluated from a two body theory. From a numerical standpoint this involves more computation than solving the original problem directly, but it enables one to evaluate the accuracy of such a theory without performing the algebraic manipulations explicitely.

Computer tests indicate the position error for 'normal' orbits ($p \cong R$) is about 0.0003 km whereas for those orbits with $p = 0.17\,R$ the error is about 0.001 km. As a test on the first order simulation, the procedure was repeated with J_2 equal to $\frac{1}{10}$ of its nominal value. The errors are $\frac{1}{100}$ of those for the case with nominal J_2 confirming the computations neglected terms of order J_2^2.

The development of an analytic solution requires the evaluation of the integral

$$t = \int_{\tau_0}^{\tau} r d\tau.$$

In a first order theory the long period terms are expanded and appear with the secular terms. As a result r contains terms in the true anomaly f. This entails evaluating integrals of elliptic motion of the form

$$\int f\,dE,$$

where E is the eccentric anomaly. This cannot be evaluated in closed form and one must resort to expansions in the eccentricity. Ordinarily this would be a rather grim prospect considering the size of the eccentricity. Fortunately, the situation is not so bad. The expansion

$$f = E + 2\sum_{j=1}^{\infty} \frac{\beta_j}{j} \sin jE,$$

where $\beta = e/1 + \surd(1-e^2)$, is absolutely convergent for $e<1$ (see for example Brouwer and Clemence, p. 63). We note that for $\beta(e)$; $\beta(0)=0$, $\beta(1)=1$. The integral in question is then

$$\int f\,dE = \frac{1}{2}E^2 - 2\sum_{j=1}^{\infty} \frac{\beta^j}{j^2} \cos jE,$$

which converges rapidly for all values of e in question. Since the integral is multiplied by J_2, about 10 terms are sufficient in practice.

4. Conclusions

These studies have shown that conventional first order theories produce inaccurate results for the combination of small semi-major axis and large eccentricity. The KS variables, while not removing the singularity (through regularization), increase the stability of the solution in the vicinity of the singularity. The difference in the two theories is such that the latter can be used for the computation of accurate ephemerides of space probes, while the former may not be used for such purposes.

Acknowledgment

Professor Giorgio Giacaglia of the University of São Paulo, São Paulo, Brazil, suggested the use of the KS transformation and initiated work on this approach while he was visiting in the U.S.A. in the summer of 1968.

References

Brouwer, Dirk and Clemence, G. M.: 1961, *Methods of Celestial Mechanics*, Academic Press.

Kustaanheimo, P. and Stiefel, E.: 1965, 'Perturbation Theory of Kepler Motion Based on Spinor Regularization', *Z. Rein Angew. Math.* **218**.

Giacaglia, G. E. O.: 1968, Private communication.

Discussion

G. Hori: The big error committed by using Brouwer's theory comes from the fact that the perturbations are actually very big (for $a=\frac{1}{2}R$, $e=0.9$) so that the first order theory is not good enough. How can we get rid of this physical situation by the use of another set of variables?

F. Peters: In my opinion the trouble is caused more by the singularity than by the large perturbations. The KS approach removes the singularity from the undisturbed problem and improves the situation enough to obtain adequate accuracy.

A. Deprit: The K-S transformation consists of a conformal mapping of the space coordinates and of a time conversion. Professors Szebehely and Tapley have been successful in control problems with time conversions other than the one proposed by Stiefel and Kustaanheimo. In view of this success, I ask if you experimented with other differential forms for the time conversion.

C. F. Peters: Dr. Giacaglia and I considered some other time transformations as well as other coordinate transformations. The K-S approach appeared to be the most convenient and the others were simply left unexplored.

P. E. Nacozy: Convergence would be faster if you determined the perturbations over only the part of the trajectory above the earth. One indication of this is that only one power-series representation was necessary and included only 14 coefficients. Also, it appears that when one is interested in a trajectory that does not include pericenter, the true anomaly produces better convergence than the eccentric anomaly for large eccentricities, but less than unity.

C. F. Peters: I do not believe the unperturbed solution would be free of singularities for $e=1$ with the true anomaly as the independent variable.

Note added in proof. Also presented in the talk were the results of a numerical test using recurrent power series in KS variables. A single expansion using 14 terms converged over the entire region around apogee (for $r>R$). The accuracy was around 0.01 km in position.

ON THE IDEAL RESONANCE PROBLEM

BORIS GARFINKEL

Yale University Observatory, New Haven, Conn., U.S.A.

Abstract. The Ideal Resonance Problem is soluble in power series in $\varepsilon^{\frac{1}{2}}$, where ε is a small parameter. The zeroth order approximation, in the Hori perturbation method, is given by the motion of a simple pendulum. The perturbed solution to any order in ε is expressible in elliptic functions, and is free of singularities and mixed secular terms. This solution would provide a theoretical framework for an attack on resonance problems in celestial mechanics when the latter are reducible to the ideal form. The condition of reducibility is that the Hamiltonian can be put in the form

$$F = A_0(y_1) + A_1(y_1, y_2)\cos 2x_1 + \phi(x_1, x_2, y_1, y_2),$$

where ϕ is periodic in x_1 and x_2 with a period 2π, and the relations

$$A_0 = 0(1), \quad A_1 = 0(\varepsilon), \quad \phi = o(\varepsilon)$$

are satisfied. The rate of the convergence of the solution depends on the smallness of ϕ.

1. Introduction

Resonance in mechanics arises from the near-commensurability of two of the natural frequencies of the motion. It manifests itself in the amplification of certain oscillatory terms that appear in the solution.

The Ideal Resonance Problem has been formulated (Garfinkel, 1966) so as to admit a solution in series in powers of $\varepsilon^{\frac{1}{2}}$, where ε is a small parameter, and to provide a theoretical framework for an attack on various problems of resonance that occur in celestial mechanics if the latter are reducible to the ideal form. In this paper we formulate the problem and its extended form, discuss briefly the method and the character of the solution, and describe the process of reducing a general problem to the ideal form.

2. Formulation of the Ideal Problem

The Ideal Resonance Problem is defined by the Hamiltonian

$$F = A_0(G) + A_1(G)\cos 2g\,, \tag{1}$$

with

$$A_0 = 0(1), \quad A_1 = 0(\varepsilon), \tag{2}$$

where ε is a small parameter, and G and g a pair of canonically conjugate variables. If the equation $A_0' = 0$, where the prime denotes the differentiation with respect to G, has a real root, then there exists a libration center at some point (G^*, g^*). Let us define new variables x and y by

$$\begin{aligned} x &= g + \tfrac{1}{4}\pi(1 + \operatorname{sgn} A_1 A_0'), \\ y &= G - G(0). \end{aligned} \tag{3}$$

Here $G(0)$ is the value assumed by G on the solution curve at $x = 0$, and the argument

G. E. O. Giacaglia (ed.), Periodic Orbits, Stability and Resonances, 474–481. *All Rights Reserved.*

of the A's is $y=0$. Furthermore, let us define the constants A and B by

$$B = |A_1| \operatorname{sgn} A_0', \qquad A = A_0 - A. \tag{4}$$

Then (1) assumes the *normal form*

$$F = A(y) + 2B(y)\sin^2 x, \qquad A = 0(1), \quad B = 0(\varepsilon). \tag{5}$$

The *extended form* of the Ideal Problem is defined by

$$F = A(y_1) + 2B(y_1, y_2)\sin^2 x_1 + \phi(x_1, x_2, y_1, y_2), \tag{6}$$

where ϕ is periodic in x_1 and x_2 with a period 2π and the relations

$$A = 0(1), \quad B = 0(\varepsilon), \quad \phi = o(\varepsilon) \tag{7}$$

are satisfied. Let the constant s be defined by

$$\phi = 0(\varepsilon^{1+s}), \quad s > 0. \tag{8}$$

Then the problem is soluble by a perturbation technique based on an expansion in series in powers of parameter ε' defined by

$$\varepsilon' = \varepsilon^m, \quad m = \min\left(\tfrac{1}{2}, s\right). \tag{9}$$

For the normal form, $\phi=0$. Then $s=\infty$, and (9) leads to the usual Bohlin-type expansion, with $m=\frac{1}{2}$, $\varepsilon'=\varepsilon^{\frac{1}{2}}$. Generally, for $s>\frac{1}{2}$, we write

$$s = \left(\tfrac{1}{2}\right)^k, \quad \varepsilon' = \varepsilon^{(\frac{1}{2})^k}, \tag{10}$$

where k is an integer. Then the powers of ε' include all the powers of $\varepsilon^{\frac{1}{2}}$, and the resulting algorithm can be readily fitted into that of the *normal* form. The case $k=1$, $m=\frac{1}{2}$ for a somewhat specialized Hamiltonian of one degree of freedom was noted by Jupp (1969); if $k>1$ the convergence will be slower but it may be quite sufficient for practical purposes.

3. On the Solution of the Ideal Problem

The normal form of the problem has been treated by Garfinkel (1966) and Jupp (1969). Both authors used the von Zeipel method to solve the problem to $0(\varepsilon^{\frac{1}{2}})$. An extension of the solution to $0(\varepsilon)$ has been undertaken in two joint papers now in progress, (Garfinkel and Williams) and (Hori and Garfinkel), using the von Zeipel and the Hori method respectively.

Since the latter method (Hori, 1966) is ideally suited to the Ideal Problem we shall briefly comment on some features of this application. A preliminary transformation of the Hamiltonian F in (5) is based on the Taylor series expansion (cf. Garfinkel,

1960) of the coefficients $A(y)$ and $B(y)$ about $y=0$:

$$A(y) = A + Ay' + \tfrac{1}{2}A''y^2 + \tfrac{1}{6}A'''y^3 + \dots.$$
$$B(y) = B + B'_y + \dots. \tag{11}$$

Absolute constants Q, ω_0, α, a_i, and b_i are then defined in terms of A, B, and their derivatives by

$$\begin{aligned} Q &= |4B/A''|^{1/2} \\ \omega_0 &= |4A''B|^{1/2} \\ \alpha_0 &= -A'/\omega_0 \\ a_i &= A^{(i)}/A'' \\ b_i &= B^{(i)}/B. \end{aligned} \tag{12}$$

New canonical variables ξ, η, with the new time t_* and new Hamiltonian F_* are then introduced by

$$\begin{aligned} x &= \xi \\ y &= Q(\alpha_0 - \eta) \\ \omega_0 t &= t_* \\ F &= 4BF_* + \text{const.} \end{aligned} \tag{13}$$

If we agree to drop the asterisks on t and F, (5) takes the form

$$F = F_0 + F_{1/2} + F_1 + \dots. \tag{14}$$

with

$$\begin{aligned} F_0 &= \tfrac{1}{2}(\sin^2\xi + \eta^2), \\ F_{1/2} &= Q\left[\tfrac{1}{6}a_3(\alpha_0 - \eta)^3 + \tfrac{1}{2}b_1(\alpha_0 - \eta)\sin^2\xi\right]. \end{aligned} \tag{15}$$

The unperturbed Hamiltonian F_0 corresponds to the problem of the simple pendulum, the solution of which is shown in Table I.

TABLE I

The simple-pendulum analogy

Libration	Circulation	Asymptotic motion
$\sin\xi = \lvert\alpha\rvert \operatorname{sn} t$	$\sin\xi = \operatorname{sn}\alpha t$	$\sin\xi = \tanh t$
$\eta = \alpha \operatorname{cn} t$	$\eta = \alpha \operatorname{dn}\alpha t$	$\eta = \operatorname{sech} t$
$\lvert\alpha\rvert < 1$	$\lvert\alpha\rvert > 1$	$\lvert\alpha\rvert = 1$

An integral of the motion is

$$\sin^2\xi + \eta^2 = \alpha^2. \tag{16}$$

In the perturbation theory, the parameter $\alpha = \alpha_0$ is to be treated as a constant of integration, in contrast to α_0 entering explicitly in (15). The modulus of the elliptic functions is defined by

$$k = \min(|\alpha|, |\alpha|^{-1}). \tag{17}$$

The *resonance parameter* α characterizes the type of motion and also distinguishes the *shallow* and the *deep resonance*, as shown in Table II.

TABLE II

Types of motion, $\alpha = 0(\varepsilon^m)$

		$\lvert\alpha\rvert$	m	
shallow resonance	classical case		$-\frac{1}{2}$	circulation
	demarcation		$-\frac{1}{4}$	
deep resonance	asymptotic motion	1	0	libration
	libration center	0		

For shallow resonance, $|\alpha| > \varepsilon^{-\frac{1}{4}}$, the classical solution loses its accuracy must be replaced by the more general *resonance solution*, which is free of singularity. It has been shown (Garfinkel, 1966) that the latter solution asymptotically includes the classical solution as $\alpha \to \infty$.

The perturbations of the simple-pendulum motion, arising from $F_{\frac{1}{2}}$, F_1, etc. can be obtained by the Hori method, based on Lie series, and can be expressed in terms of elliptic functions. The extended form (6) of the problem can be treated in a similar way.

4. On the General Problem of Resonance

Let us consider the problem defined by the Hamiltonian

$$F = \sum_j A_j(y) \cos(j \cdot x), \tag{18}$$

where x, y, and j are n-vectors, with j composed of integers, and the dot denotes the scalar product. Furthermore, let the relations

$$A_0 = 0(1), \quad \sum_{j \neq 0} |A_j| = 0(\varepsilon), \quad \varepsilon \ll 1, \tag{19}$$

be satisfied, and let A_0 be a function of y_1 and y_2 only. Then x_1 and x_2 are *fast* angle variables, and the remaining x's are *slow*. The corresponding angular frequencies can be written

$$\begin{aligned} \bar{\dot{x}}_1 &= n_1 = 0(1), \\ \bar{\dot{x}}_2 &= n_2 = 0(1), \\ \bar{\dot{x}}_i &= n_i = 0(n_1\varepsilon), \end{aligned} \tag{20}$$

where the bar denotes the time-average. If the trigonometric argument is denoted by θ_j,

$$\theta_j \equiv j \cdot x = j_1 x_1 + j_2 x_2 + \dots j_n x_n, \tag{21}$$

then the expression

$$\bar{\dot{\theta}}_j = j_1 n_1 + j_2 n_2 + \dots j_n x_n \tag{22}$$

plays the role of a critical divisor.

Let now the ratio n_1/n_2 be expanded in a sequence

$$n_1/n_2 \sim \{p_i/q_i\}, \quad i = 1, 2, \ldots \tag{23}$$

of the *convergents* of the corresponding continued fraction. For a particular i, we define the class of *resonant* terms of (18) by the condition

$$j_1 = \gamma q, \quad j_2 = -\gamma p, \tag{24}$$

where γ is an integer. For such terms (21) and (22) become

$$\begin{aligned} \theta_j &= \gamma(qx_1 - px_2) + j_3 x_3 + \ldots \\ \dot{\theta}_j &= \gamma q n_1 D + 0(\varepsilon n), \end{aligned} \tag{25}$$

where the so-called *commensurability index D* is defined by

$$D \equiv 1 - pn_2/qn_1 . \tag{26}$$

It vanishes for exact commensurability, and has the properties

$$0 \leq |D| < 1, \quad \lim_{i \to \infty} D_i = 0 . \tag{27}$$

5. Reduction of a General Problem to the Ideal Form

If a reduction to the ideal form is possible, it will be accomplished as described below. For a given choice of order i in (23), the following sequence of operations is performed upon the Hamiltonian (18):

(1) Elimination of all non-resonant periodic terms by a canonical transformation $(x, y) \rightarrow (x', y')$. The resulting *resonant Hamiltonian* will involve only the arguments θ_j of the form (24).

(2) The calculation of the *dominant* resonant term, defined by

$$A_* = \max_j |A_j| . \tag{28}$$

The process involves the maximization of the expression for the D'Alembert characteristic of A_j in the domain of integers subject to certain inequality constraints. An example of such maximization can be found in Allan (1965), who treated the problem of the tesseral harmonics resonance in the artificial satellite theory. The result of the calculation yields A_* and the *critical* argument θ_*, with

$$\gamma = \gamma_*, \quad j_3 = j_3^*, \text{ etc.} \tag{29}$$

(3) A canonical transformation $(x', y') \rightarrow (X, Y)$ defined by

$$\begin{aligned} \theta_* = \gamma(qx_1 - px_2) + j_3 x_3 + \ldots &= 2X_1 , \\ x_2 &= X_2 , \\ x_3 &= X_3 \\ \vdots \quad & \quad \vdots \end{aligned} \tag{30}$$

$$
\begin{aligned}
y_1 &= \tfrac{1}{2}\gamma q Y_1\,, \\
y_2 &= -\tfrac{1}{2}\gamma q Y_1 + Y_2 \\
y_3 &= \tfrac{1}{2}j_3 Y_1 + Y_3\,. \\
&\vdots \qquad \vdots
\end{aligned}
$$

This transformation converts the critical argument into an angular variable of the new Hamiltonian. Since X_2 is ignorable, Y_2 is a constant of the motion, and the problem is thereby reduced to $n-1$ degrees of freedom.

(4) The calculation of the resonance parameter α corresponding to the dominant term.

Here formula (12.3) is used, with the secular term A_0 playing the role of A, and A_* that of B. The derivatives A' and A'' are taken with respect to the momentum Y_1 conjugate to X_1.

(5) Search for deep resonance. Steps 1–4 are repeated for $i=1, 2, 3, \ldots$, yielding the function $\alpha(i)$, whose general behavior is shown in Figure 1.

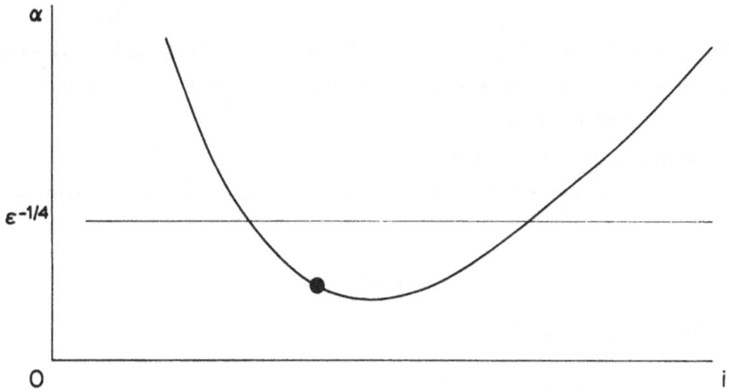

Fig. 1. The function $\alpha(i)$

Deep resonance corresponds to points lying below the straight line $\alpha=\varepsilon^{-\frac{1}{4}}$. Let N be the number of such points. Three cases are distinguished:

$$
\begin{aligned}
&\text{(a)}\ \ N = 0 \\
&\text{(b)}\ \ N > 1 \qquad\qquad (31)\\
&\text{(c)}\ \ N = 1
\end{aligned}
$$

Case (a), representing shallow resonance, is amenable to the classical treatment, and is of no interest to us here. Case (b), with two or more independent deep resonances, lies beyond the scope of the present theory. We shall therefore concern ourselves with case (c), involving a single deep resonance.

Assuming that we are in case (c), we proceed to the next step.

(6) Elimination of all shallow resonant terms by a canonical transformation $(X, Y)\rightarrow(X', Y')$.

In the special case $n=3$, the resulting deep-resonant Hamiltonian will contain periodic arguments of the form

$$\theta = j_1 X_1' + j_3 X_3'. \tag{32}$$

The dominant term corresponds to $j_1 = 1$ and $j_3 = 0$. Let A_{**} be the coefficient of greatest absolute value among the other periodic terms. Then the constant s of Equation (8) is given by

$$|A_{**}/A_*| = \varepsilon^s. \tag{33}$$

Rapid convergence is assured if s is sufficiently large; i.e. if the dominant term strongly dominates the other periodic terms.

(7) Testing the Dominance Condition,

$$|A_{**}/A_*| \ll 1. \tag{34}$$

6. Summary

If $n=2$, the absence of the variable x_3 does not affect the reasoning, or the general conclusion.

In summary, a deep-resonance problem for a conservative periodic Hamiltonian F can be reduced to the Ideal form if the following conditions are satisfied:

(1) there is only one deep resonance;

(2) F is dominated by its constant term;

(3) the deep-resonant part of F is strongly dominated by the dominant resonant term.

References

Allan, R. R.: 1966, *Planet. Space Sci.* **15**, 53.
Garfinkel, B.: 1960, *Astron. J.* **65**, 612.
Garfinkel, B.: 1966, *Astron. J.* **71**, 657.
Hori, G.: 1966, *J. Astron. Soc. Japan* **18**, 287.
Jupp, A.: 1969, *Astron. J.* **74**, 35.

Discussion

J. Kevorkian: I would like to point out that in two recent papers dealing with the resonances due to the tesseral harmonics of the earth potential, Shi and Eckstein solved a problem involving more than one small divisor using modern techniques. The terminology of 'shallow' and 'deep' resonance as used by the speaker is misleading. A deep resonance should be one that exists as a consequence of the existence of a shallow resonance. According to the work of Vagners, Shi and Eckstein the resonance argument of the shallow resonance in turn resonates with some other slow variable.

J. Vinti: In reply to Prof. Kevorkian, may I say that I have been hearing for some years that the problem of a double resonance in celestial mechanics is its outstanding unsolved problem. If I understand his remark about Shi and Eckstein, he is now asserting that this problem has now been completely solved.

J. Kevorkian: Yes, the problem has been solved for the case of an artificial earth satellite for a realistic potential. This is a case having more than one commensurability for which Dr. Garfinkel admits his method fails.

J. Vagners: (1st question) Did I understand correctly that you termed the case $s=0$ as 'non-realistic'? We have ample evidence in this Symposium of problems – lunar orbiter motion, resonant motion of the asteroids – where $s=0$ and these are certainly 'realistic' problems.

(2nd question) I do not believe that the procedure proposed by Hori will yield analytical expressions for the general motion of a lunar orbiter. Even a simplified model of the problem will lead to a Hamiltonian wherein the 'ideal resonance Hamiltonian' part in turn leads to a Hamilton-Jacobi equation which does not admit solutions in terms of known functions except under restrictive assumptions on the orbit geometry i.e. inclination and /or eccentricity. Therefore, the analytical extension to inclusion of ϕ does not appear probable. Expansion of F_0 on G is an expansion on eccentricity and is valid near critical points of the center type in the phase plane $e-g$ or for circulatory motion of g.

B. Garfinkel: I did not say that the case $s=0$ is not realistic, and I am in perfect agreement with you that it is of frequent practical occurrence.

Hori's procedure has, indeed, yielded analytical solutions for the general motion of the lunar orbiter. Perhaps, Dr. Hori himself would clarify the mystery.

J. Kevorkian: The problem of two commensurabilities has been solved by Eckstein *et al.* in a paper in the *Astronomical Journal.*

B. Garfinkel: How did they do it?

J. Kevorkian: They matched the two asymptotic expansions.

PARAMETRIC RESONANCE IN CERTAIN NONLINEAR SYSTEMS

PETER HAGEDORN

COPPE, Universidade Federal do Rio de Janeiro, Brasil

Abstract. In a number of mechanical problems, systems of differential equations with periodic coefficients have to be considered, which, in general, possess not only linear but also nonlinear terms. In this paper, the first order instability region of the Mathieu equation is examined when additional nonlinear terms are present. These terms can be of the damping or restoring type.

A case of combination resonance, in the presence of nonlinear terms, is discussed for a system with two degrees of freedom. The influence of the nonlinear terms on the instability region of the first order and of first type is considered. In both problems Bogoliubov and Mitropolsky's asymptotic method is used.

1. Introduction

A number of mechanical problems can be reduced to systems of ordinary differential equations with periodic coefficients, corresponding in a certain way to a generalization of the Mathieu equation. In many practical problems, linear damping and nonlinear terms also are to be considered in the differential equations, which influence the periodic solutions and the instability regions. The nonlinear terms may represent forces of the damping or of the restoring type.

In this paper two different cases will be considered. First, the equation

$$\ddot{x} + \omega_1^2 x(1 - \varepsilon \cos \omega t) + \sum_{i=0}^{n} e_i |x^i| \dot{x} + \sum_{j=1}^{m} a_j |x^j| x = 0 \tag{1}$$

will be discussed, where e_i and a_j are the coefficients of the nonlinear damping and restoring terms, respectively, and ε is a small parameter. As is well known, the trivial solution of Equation (1) becomes unstable for certain values of the parameters. For the equation with linear terms only, the stability chart (Figure 1) shows that for small values of ε instability occurs only if ω lies in the neighborhood of one of the critical frequencies $\omega_0 = 2\omega_1/p$, where p is a natural number. The corresponding region of instability is then said to be of the order p. In this paper, the influence of the nonlinearities on the instability region of the first order will be examined.

The second case to be examined corresponds to the equations

$$\begin{aligned} \ddot{x}_1 + \omega_1^2 x_1 + \varepsilon b_1 x_2 \cos \omega t + e_1 \dot{x}_1 + c_1 x_1^2 \dot{x}_1 + a_1 x_1^3 &= 0, \\ \ddot{x}_2 + \omega_2^2 x_2 + \varepsilon b_2 x_1 \cos \omega t + e_2 \dot{x}_2 + c_2 x_2^2 \dot{x}_2 + a_2 x_2^3 &= 0. \end{aligned} \tag{2}$$

In systems of this kind, instability may occur if the exciting frequency ω lies in the neighborhoods of one of the critical frequencies $\omega_0 = (\omega_1 \pm \omega_2)/p$. If in (2) the coefficients b_1 and b_2 are of the same sign, as assumed for the purpose of this discussion, then the critical frequencies are given by $\omega_0 = (\omega_1 + \omega_2)/p$. As two frequencies are involved, the corresponding instability region is said to be of the second type, and of the order p. The expression 'combination resonance' is used for the characterization of the related resonance phenomena. A systematic discussion of such combination

G. E. O. Giacaglia (ed.), Periodic Orbits, Stability and Resonances, 482–492.

resonances in linear systems was first given by Mettler [1], and has recently been expanded by Schmidt [2]. In (2), e_i and c_i are the coefficients of linear and cubic damping, and the a_i terms represent the nonlinear restoring forces.

Equations similar to (1) and (2) have already been discussed in [3, 4, 5, 6], but only in the case of exclusively positive damping terms; in this paper, negative 'damping' will also be discussed. Cases of negative damping occur not only in mechanics but also in active network theory, where this consideration is more important.

The principal resonance of (1) and the combination resonance of (2) can be adequately examined by Bogoliubov and Mitropolsky's asymptotic method; its first approximation proved to be sufficiently accurate, for practical purposes, at least for most problems occurring in elastodynamics [7, 8].

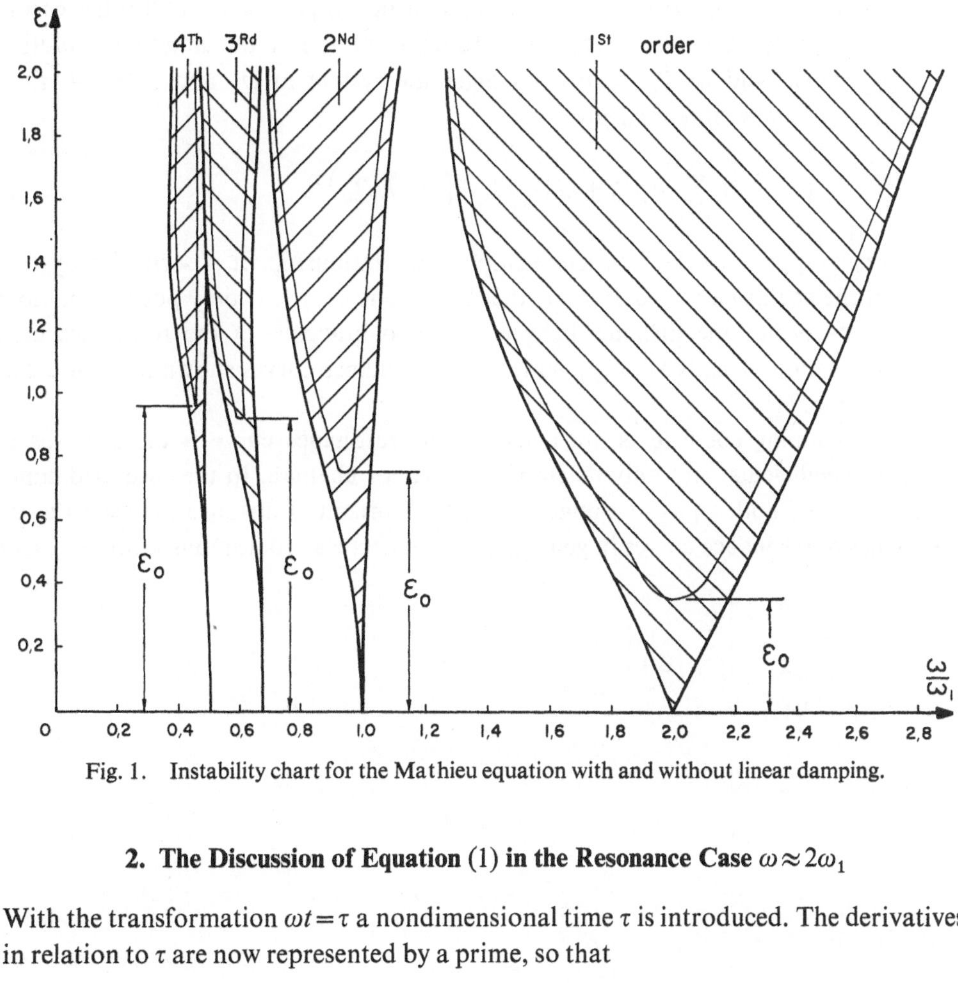

Fig. 1. Instability chart for the Mathieu equation with and without linear damping.

2. The Discussion of Equation (1) in the Resonance Case $\omega \approx 2\omega_1$

With the transformation $\omega t = \tau$ a nondimensional time τ is introduced. The derivatives in relation to τ are now represented by a prime, so that

$$\dot{x} = \omega x' . \tag{3}$$

The solutions of Equation (1) will be studied in the neighborhood of the critical frequency $\omega_0 = 2\omega_1$, so that $\omega = \omega_0(1 - \lambda)$, where λ is a small factor of the same order of magnitude as ε. Substituting

$$x = Q(\tau)\sin(\tau/2 + \vartheta(\tau)), \quad x' = Q(\tau)/2\cdot\cos(\tau/2 + \vartheta(\tau)), \tag{4}$$

the equations corresponding to the first approximation can be obtained from (1) in the usual manner, giving

$$Q' = \varepsilon/8\cdot Q\sin 2\vartheta - \varepsilon/8\cdot\sum_{i=0}^{n} E_i Q^{i+1}, \tag{5}$$

$$\vartheta' = \varepsilon/8\cdot\cos 2\vartheta + \lambda/2 + \varepsilon\sum_{j=1}^{m} A_j Q^j, \tag{6}$$

where E_i and A_j are proportional to e_i and a_j (see [6]). Above it was assumed that e_i and a_j are also small and of the magnitude of ε. Of course, the system (5), (6) has the trivial solution $Q=0$, $\vartheta=\vartheta^*(\tau)$, $\vartheta^*(\tau)$ being constant if $|4\lambda/\varepsilon|<1$, in all other cases it will be an easily obtainable monotonic function of τ. For the stationary solutions $Q=Q_0=$const. and $\vartheta=\vartheta_0=$const. an algebraic system in Q_0 and ϑ_0 is obtained, which, after elimination of ϑ_0, gives

$$\lambda = -2\varepsilon\sum_{j=1}^{m} A_j Q_0^j \pm \varepsilon/4\cdot\sqrt{1-\left(\sum_{i=0}^{n} E_i Q_0^i\right)^2}, \tag{7}$$

which corresponds to the frequency-amplitude relationship. It is seen directly from (7) that the coefficients A_j determine the 'back bone curve'. If only positive damping coefficients occur, the distance between the two branches of the resonance curve decreases monotonically with Q_0, so that the resonance curves shown in Figure 2 are obtained (see [4]).

If all damping coefficients are negative, the resonance curve is the same as for positive coefficients, and only its stability behavior changes. In the case of damping coefficients of both signs in Equation (1), the horizontal distance between the two branches of resonance curves in general does not decrease continuously with Q_0. Con-

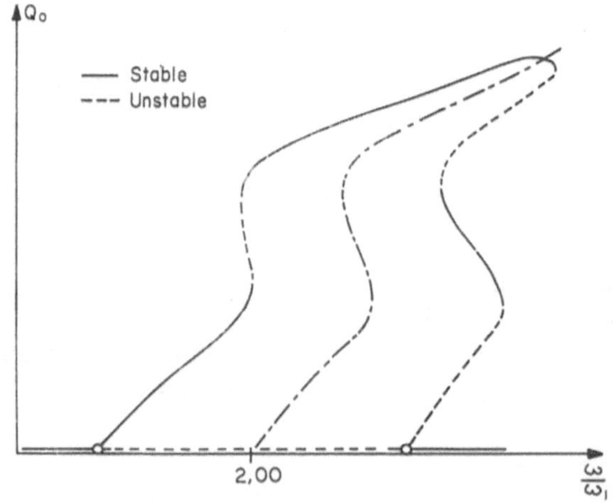

Fig. 2. Resonance curve for Equation (1) – positive damping coefficients only.

strictions can then appear in the resonance curve, which may become so strong as to divide it in several independent curves. The maximum number of these curves in each case is equal to the highest power n appearing in the damping terms of (1).

In Figure 3, resonance curves are shown for the case $e_0>0$, $e_1<0$, $e_2>0$. From

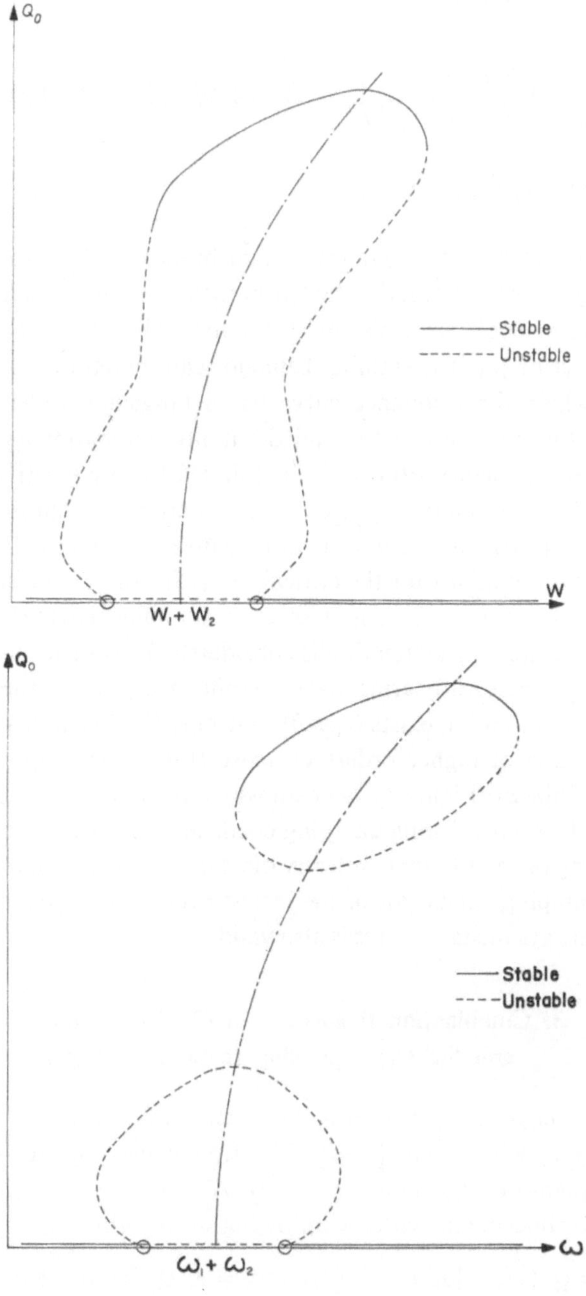

Fig. 3. Resonance curves for Equation (2) – positive and negative damping coefficients.

Equation (7) it can easily be seen, that the location of the branching points on the ω-axis depends only on $|e_0|$.

To study the stability behavior of the nontrivial stationary solutions of Equation (1), the substitution $Q=Q_0+\bar{Q}$, $\vartheta=\vartheta_0+\bar{\vartheta}$ is made in (5), (6), so that $\bar{Q}$, $\bar{\vartheta}$ represent the deviations from the stationary solutions. The differential equations in $\bar{Q}$, $\bar{\vartheta}$ obtained in this way from (5), (6) are linearized in $\bar{Q}$, $\bar{\vartheta}$. The Hurwitz criterion then gives the inequalities

$$\left(\sum_{i=0}^{n} E_i Q_0^i\right)\left(\sum_{i=0}^{n} iE_i Q_0^i\right) \pm 8 \sum_{j=1}^{m} jA_j Q_0^j \sqrt{1-\left(\sum_{i=0}^{n} E_i Q_0^i\right)^2} > 0\,, \tag{8}$$

and

$$\sum_{i=0}^{n} (i+2)\, E_i Q_0^i > 0\,. \tag{9}$$

It can be seen from (7) and (8) that the right branch of the resonance curve can be stable only if $\mathrm{d}Q_0/\mathrm{d}\omega<0$; similarly, the left branch can only be stable for $\mathrm{d}Q_0/\mathrm{d}\omega>0$. If negative damping coefficients also exist, the inequality (9) is not automatically fulfilled for all values of Q_0. The stability behavior can therefore change at other points than those at which the resonance curve has a tangent parallel to the Q_0-axis. If all damping coefficients are negative, all the nontrivial stationary solutions become unstable, because inequality (9) cannot be fulfilled by any positive values of Q_0. In Figures 2 and 3 the stability behavior of the stationary solutions is also shown.

The stability of the trivial solution can also be investigated as in [4], where Liapunov functions have been obtained for the critical cases. It can then be concluded that only the nonvanishing damping coefficient of lowest order influences the first order instability regions, if only the first approximation is considered. The coefficients A_j, corresponding to the nonlinear restoring terms, have no influence at all on the instability region.

If positive linear damping exists ($e_0>0$), not only the instability region of the first order but also these of higher orders decrease (Figure 1). Limit values $\varepsilon=\varepsilon_0$ then occur in case of finite e_0 (Figure 1), beneath which the trivial solution is always stable. If $e_0=0$, and if the nonvanishing damping coefficient of lowest order is positive, then the instability region of the first order is the same as in the undamped case. If the nonvanishing damping coefficient of the lowest order is negative, the trivial solution turns out to be always unstable. This is also valid for $e_0<0$.

3. Combination Resonance in (2) for $\omega\approx\omega_1+\omega_2$ and the Corresponding Instability Region

The critical frequency is now $\omega_0=\omega_1+\omega_2$, and the deviations from it are again characterized by λ, with $\omega=\omega_0(1-\lambda)$. The frequencies ω_i are represented by the dimensionless quantities $K_i=\omega_i/\omega_0$ $(i=1, 2)$. Also in this case the nondimensional time $\tau=\omega t$ is introduced, and with the following substitution

$$\begin{aligned} x_1 &= Q_1(\tau)\sin\left[K_1\tau+\vartheta_1(\tau)\right], & x_1' &= K_1 Q_1(\tau)\cos\left[K_1\tau+\vartheta_1(\tau)\right],\\ x_2 &= Q_2(\tau)\sin\left[K_2\tau+\vartheta_2(\tau)\right], & x_2' &= K_2 Q_2(\tau)\cos\left[K_2\tau+\vartheta_2(\tau)\right], \end{aligned} \tag{10}$$

in (2), the equations corresponding to the first approximation of the asymptotic method are obtained as in [4]:

$$\begin{aligned} Q_1' &= \varepsilon B_1/4 \cdot Q_2 \sin(\vartheta_1 + \vartheta_2) - \varepsilon E_1/2 \cdot Q_1 - 3\varepsilon/8 \cdot C_1 Q_1^3, \\ Q_2' &= \varepsilon B_2/4 \cdot Q_1 \sin(\vartheta_1 + \vartheta_2) - \varepsilon E_2/2 \cdot Q_2 - 3\varepsilon/8 \cdot C_2 Q_2^3, \\ Q_1\vartheta_1' &= \lambda K_1 Q_1 + \varepsilon B_1/4 \cdot Q_2 \cos(\vartheta_1 + \vartheta_2) + 3\varepsilon/8 \cdot A_1 Q_1^3, \\ Q_2\vartheta_2' &= \lambda K_2 Q_2 + \varepsilon B_2/4 \cdot Q_1 \cos(\vartheta_1 + \vartheta_2) + 3\varepsilon/8 \cdot A_2 Q_2^3, \end{aligned} \tag{11}$$

where

$$\varepsilon A_i = a_i/(K_i\omega_0^2), \quad B_i = b_i/(K_i\omega_0^2), \quad \varepsilon C_i = c_i K_i^2 \omega_0, \quad \varepsilon E_i = e_i/\omega_0,$$

with $i = 1, 2$. If $Q_1(\tau)$ and $Q_2(\tau)$ are different from zero, the two last equations of (11) can be combined, giving

$$\vartheta' = \lambda + \varepsilon/4 \cdot (B_1 Q_2/Q_1 + B_2 Q_1/Q_2)\cos\vartheta + 3\varepsilon/8 \cdot (A_1 Q_1^2 + A_2 Q_2^2), \tag{12}$$

with $\vartheta = \vartheta_1 + \vartheta_2$. The nontrivial stationary solutions $Q_1 = Q_{10} = \text{const.}$, $Q_2 = Q_{20} = \text{const.}$, $\vartheta = \vartheta_0 = \text{const.}$ can now be examined by means of (12) and the two first equations of (11). This was done in [4], for the case of positive damping coefficients, and the resonance curves shown in Figure 4 were obtained. If negative damping coefficients are also permitted, the resonance curves again assume relatively complicated forms, and can degenerate in several distinct curves. In case of only negative damping coefficients, all the resonance curves are unstable. These results are obtained in the same way as in the case of Equation (1).

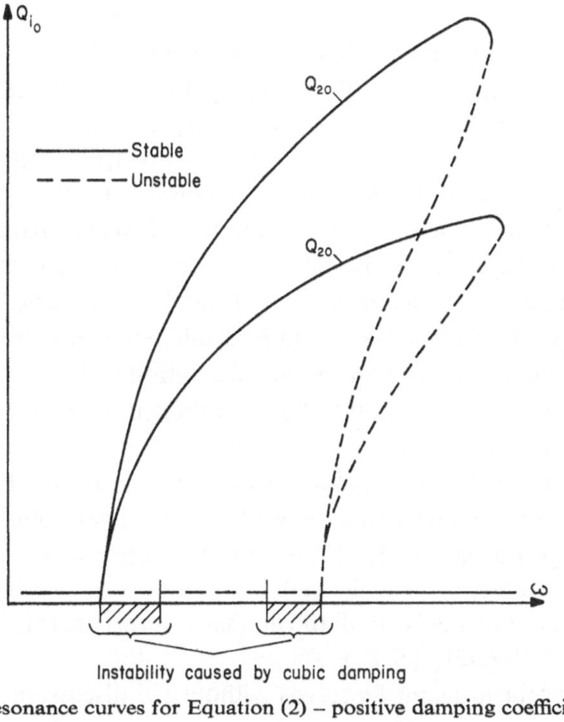

Fig. 4. Resonance curves for Equation (2) – positive damping coefficients only.

For the discussion of the stability of the trivial solution in (11) the substitution

$$y_i = Q_i \cos \vartheta_i, \quad z_i = Q_i \sin \vartheta_i, \quad i = 1, 2 \tag{13}$$

is used. In this way, the system

$$\begin{aligned}
y_1' &= -\varepsilon E_1/2 \cdot y_1 - \lambda K_1 z_1 + \varepsilon B_1/4 \cdot z_2 \\
&\qquad - 3\varepsilon/8 \cdot C_1 y_1 (y_1^2 + z_1^2) - 3\varepsilon/8 \cdot A_1 z_1 (y_1^2 + z_1^2), \\
y_2' &= -\varepsilon E_2/2 \cdot y_2 + \varepsilon B_2/4 \cdot z_1 - \lambda K_2 z_2 \\
&\qquad - 3\varepsilon/8 \cdot C_2 y_2 (y_2^2 + z_2^2) - 3\varepsilon/8 \cdot A_2 z_2 (y_1^2 + z_2^2), \\
z_1' &= \lambda K_1 y_1 + \varepsilon B_1/4 \cdot y_2 - \varepsilon E_1/2 \cdot z_1 \\
&\qquad - 3\varepsilon/8 \cdot C_1 z_1 (y_1^2 + z_1^2) + 3\varepsilon/8 \cdot A_1 y_1 (y_1^2 + z_1^2), \\
z_2' &= \varepsilon B_2/4 \cdot y_1 + \lambda K_2 y_2 - \varepsilon E_2/2 \cdot z_2 \\
&\qquad - 3\varepsilon/8 \cdot C_2 z_2 (y_2^2 + z_2^2) + 3\varepsilon/8 \cdot A_2 y_2 (y_2^2 + z_2^2),
\end{aligned} \tag{14}$$

is obtained (see [4]).

If the Hurwitz criterion is applied to the linear part of (14), it can be seen, that for E_1, $E_2 > 0$ the trivial solution is asymptotically stable for

$$|\lambda| > \frac{\varepsilon}{4} \sqrt{B_1 B_2} \left(\sqrt{\frac{E_1}{E_2}} + \sqrt{\frac{E_2}{E_1}} \right) \sqrt{1 - \frac{4E_1 E_2}{B_1 B_2}}, \tag{15}$$

and is unstable for

$$|\lambda| < \frac{\varepsilon}{4} \sqrt{B_1 B_2} \left(\sqrt{\frac{E_1}{E_2}} + \sqrt{\frac{E_2}{E_1}} \right) \sqrt{1 - \frac{4E_1 E_2}{B_1 B_2}}. \tag{16}$$

From (15), (16) it can be shown, that the trivial solution is unstable only in the region between the branching points on the ω-axis of Figure 4. The instability region disappears for $4E_1E_2 \geqslant B_1B_2$. It is known from [1] that for the linear undamped system the limits of the stability region are given by $\lambda = \pm(\varepsilon/4) \cdot \sqrt{(B_1 B_2)}$. From (15), (16) it can therefore be concluded that linear damping may widen the instability region. The phenomenon was first discovered by Schmidt and Weidenhammer [9]. For finite linear damping coefficients, the instability region assumes the form of Figure 5, where a limit value $\varepsilon = \varepsilon_0$ occurs, beneath which no instability is possible.

If at least one of the coefficients E_1 and E_2 is different from zero, and if they are not both positive, it can be shown, that the trivial solution is unstable for any value of λ. Liapunov's theory shows that all the above mentioned results remain valid also if the nonlinear terms are considered.

The stability problem becomes more involved if $E_1 = E_2 = 0$. From the linear part of (14) it can then be concluded only that the trivial solution is unstable for $|\lambda| < (\varepsilon/2)\sqrt{(B_1 B_2)}$. For $|\lambda| > (\varepsilon/2)\sqrt{(B_1 B_2)}$ a critical case with two pairs of conjugate imaginary roots occurs, so that the stability question can no longer be decided by means of the linear terms only. If all the nonlinear terms vanish, the trivial solution is weakly stable for $|\lambda| > (\varepsilon/2)\sqrt{(B_1 B_2)}$.

The stability problem might be solved without the discussion of a critical case, if not only the first but also the second or even some higher approximation is used in the asymptotic method. This, however, proved to be very laborious [7]. Therefore, it seems to be easier to solve the critical case by means of Liapunov's direct method.

The critical case with two parts of conjugated imaginary roots is discussed by Malkin [10] and by Salvadori [11], and the procedure given by the latter seems to be

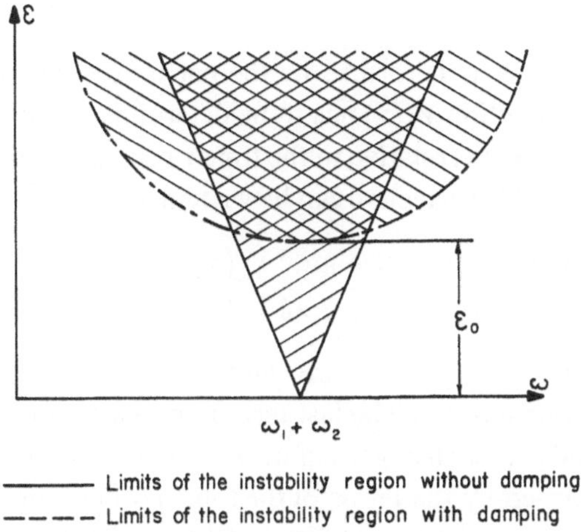

Fig. 5. Widening of the instability region of (2) by finite linear damping.

more practical. For the application of this method, Equation (14) may be written in matrix form as

$$\mathbf{x}' = \mathbf{A}\mathbf{x} + \mathbf{f}_3(\mathbf{x}), \tag{17}$$

where $\mathbf{x}$ is the column matrix of the unknown functions y_i and z_i $(i=1, 2)$, the matrix $\mathbf{A}$ corresponds to the coefficients of the linear terms, and the column matrix $\mathbf{f}_3(\mathbf{x})$ represents the nonlinear terms in (14). In the critical case under consideration, always a linear transformation can be found so that the transformed matrix assumes the form

$$\mathbf{A} = \begin{pmatrix} 0 & 0 & \nu_1 & 0 \\ 0 & 0 & 0 & \nu_2 \\ -\nu_1 & 0 & 0 & 0 \\ 0 & -\nu_2 & 0 & 0 \end{pmatrix} \tag{18}$$

where $\pm i\nu_1$ and $\pm i\nu_2$ are the roots of the characteristic equation.

Assuming that (17) is already represented in this form, the Liapunov function is investigated as

$$V(\mathbf{x}) = V_2(\mathbf{x}) + V_4(\mathbf{x}), \tag{19}$$

where $V_j(\mathbf{x})$ with $j=2, 4$ is a homogeneous polynomial of order j in the four variables y_i, z_i $(i=1, 2)$.

The function

$$V_2(\mathbf{x}) = \alpha_1(y_1^2 + z_1^2) + \alpha_2(y_2^2 + z_2^2) \tag{20}$$

is the Liapunov function of the linearized system (17), where α_1 and α_2 are arbitrary constants with the same sign, in order to make the function definite. The derivative of $V_2(\mathbf{x})$ relative to the time τ turns out to be zero on the trajectories of the solutions of the linearized system, because it is

$$V_2'(\mathbf{x}) = \mathbf{grad}\ V_2(\mathbf{x})\cdot\mathbf{Ax} = \alpha_1 v_1(y_1 z_1 - y_1 z_1) + \alpha_2 v_2(y_2 z_2 - y_2 z_2) \equiv 0 \tag{21}$$

In (21) the gradient matrix is formed by the partial derivatives of $V_2(\mathbf{x})$ relative to y_1, y_2, z_1 and z_2. The function $V_4(\mathbf{x})$ has to be determined in such a way that the derivative of $V(\mathbf{x}) = V_2(\mathbf{x}) + V_4(\mathbf{x})$ relative to the time τ becomes definite on the trajectories of the solution of (17). This derivative can be written as

$$\begin{aligned} V'(\mathbf{x}) &= \mathbf{grad}\ V_2(\mathbf{x})\cdot\mathbf{Ax} + \mathbf{grad}\ V_2(\mathbf{x})\cdot\mathbf{f}_3(\mathbf{x}) \\ &\quad + \mathbf{grad}\ V_4(\mathbf{x})\cdot\mathbf{Ax} + \mathbf{grad}\ V_4(\mathbf{x})\cdot\mathbf{f}_3(\mathbf{x}). \end{aligned} \tag{22}$$

The first term in (22) is equal to zero, and the next two terms are homogeneous polynomials of the fourth order. The last term represents a homogeneous polynomial of the sixth order, and can be disregarded in the examination of definiteness of $V'(\mathbf{x})$, if the polynomial formed by the terms of the fourth order is definite. In [11] Salvadori formulates theorems on the solutions of partial differential equations of the type

$$v_1\left(\frac{\partial V_4}{\partial y_1} z_1 - \frac{\partial V_4}{\partial z_1} y_1\right) + v_2\left(\frac{\partial V_4}{\partial y_2} z_2 - \frac{\partial V_4}{\partial z_2} y_2\right) = R_4, \tag{23}$$

where R_4 is a given polynomial of fourth order. These theorems can be used for the construction of the Liapunov function $V(\mathbf{x}) = V_2(\mathbf{x}) + V_4(\mathbf{x})$ provided v_1 and v_2 are incommensurable as will be initially supposed. This means that the case of 'internal resonance' is excluded. If the computations are carried out as in [4], it can finally be concluded that the trivial solution is unstable for

$$\frac{\varepsilon}{2}\sqrt{B_1 B_2} < |\lambda| < \frac{\varepsilon}{4}\left(B_1 \sqrt[4]{\frac{C_1 B_2}{C_2 B_1}} + B_2 \sqrt[4]{\frac{C_2 B_1}{C_1 B_2}}\right), \tag{24}$$

and asymptotically stable for

$$|\lambda| > \frac{\varepsilon}{4}\left(B_1 \sqrt[4]{\frac{C_1 B_2}{C_2 B_1}} + B_2 \sqrt[4]{\frac{C_2 B_1}{C_1 B_2}}\right), \quad \text{for} \quad C_1, C_2 > 0. \tag{25}$$

In this case the trivial solution is only unstable between the branching points (Figure 4). From (24) and (25) it is seen that positive cubic damping as in (2) always widens the instability region, except for the case in which $C_1 B_1 = C_2 B_2$, when the instability region is the same as for $C_1 = C_2 = 0$. While with linear damping a limit value ε_0 appeared, this is no longer true for cubic damping (Figure 6). Instability now occurs at arbitrarily small perturbing forces.

If at least one of the coefficients C_1 and C_2 is different from zero, and if not both are positive, the trivial solution will always be unstable, as can be seen from [4].

Up to this point in the discussion, the above mentioned results are only applicable for such values of λ which correspond to incommensurable roots ν_1 and ν_2.

In the present case, it can be shown that the described stability criterion is valid also for commensurable ν_1 and ν_2, by means of the theorems on continuity of the solution of ordinary differential equations relative to parameters. In the demonstration, it is essential that the set $S(\lambda)$ of the values of λ is everywhere dense. This set is formed by the values of λ which satisfy the relations (24), (25) and which correspond to incommensurable roots ν_1 and ν_2.

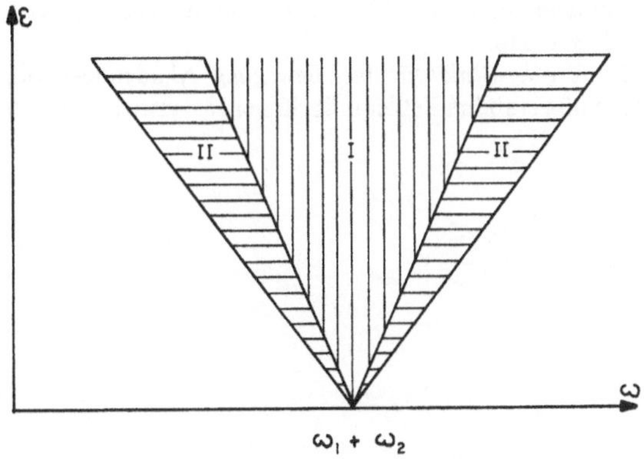

Fig. 6. Widening of the instability region of (2) by cubic damping.

4. Final Remarks

The resonance cases $\omega \approx 2\omega_1$ and $\omega \approx \omega_1 + \omega_2$ have been discussed for the nonlinear Equations (1) and (2) respectively. It has been shown that in both cases only the nonvanishing damping terms of the lowest order can influence the instability regions. While the instability region of Equation (1) can only be narrowed by positive damping, in the case of Equation (2) the instability region can be widened not only by linear but also by nonlinear damping. If the nonvanishing damping terms of the lowest order are negative, the trivial solution becomes always unstable.

It is interesting to note that nonlinear restoring terms as those appearing in (1) and (2) have no influence on the instability regions, and that the resonance curves corresponding to the nontrivial stationary solutions may degenerate into several distinct curves, if negative damping terms are present.

References

[1] Mettler, E.: 1949, 'Allgemeine Theorie der Stabilität erzwungener Schwingungen elastischer Körper', *Ing-Arch.* **17**, 418–419.

[2] Schmidt, G.: 1967, 'Instabilitätsbereiche bei rheolinearen Schwingungen', *Monatsber. Deut. Akad. Wiss.* **9**, 405–411.

[3] Hagedorn, P.: 1968, 'Zum Instabilitätsbereich erster Ordnung der Mathieugleichung mit quadratischer Dämpfung', *ZAMM* **48**, T256–T259.

[4] Hagedorn, P.: 1969, 'Kombinationsresonanz und Instabilitätsbereiche zweiter Art bei parameter-erregten Schwingungen mit nichtlinearer Dämpfung', *Ing-Arch.* **38**, 80–96.

[5] Hagedorn, P.: Über Kombinationsresonanz bei parametererregten Systemen mit Coulombscher Dämpfung, paper read at GAMM meeting 1969 in Aachen (to be published in *ZAMM*).

[6] Hagedorn, P.: 'Die Mathieugleichung mit nichtlinearen Dämpfungs- und Rückstellgliedern' (to be published).

[7] Leiss, F.: 1966, Zur Berechnung von Resonanzschwingungen quasilinearer mechanischer Systeme mit der asymptotischen Methode, Doctoral thesis, TH Karlsruhe.

[8] Benz, G.: 1962, Schwingungen nichtlinearer gedämpfter Systeme mit pulsierenden Speicherkennwerten, Doctoral thesis, TH Karlsruhe.

[9] Schmidt, G. and Weidenhammer, F.: 1961, 'Instabilitäten gedämpfter rheolinearer Schwingungen, *Math. Nachr.* **23**, 301–318.

[10] Malkin, G.: 1959, *Theorie der Stabilität einer Bewegung*, Akademie-Verlag, Berlin.

[11] Salvadori, L.: 1965, Sulla stabilità dell'equilibrio nei casi critici, *Ann. Mat. Pura Appl.*, *Bologna*, 1–32.

RESONANCES IN DUFFING'S PROBLEM

GEN-ICHIRO HORI

Dept. of Astronomy, University of Tokyo, Tokyo, Japan

1. Introduction

The forced oscillation of a pendulum is given by the solution of the differential equation,

$$\ddot{x} + \alpha \sin x = \beta \cos \omega t \quad (\alpha, \beta, \omega > 0) \tag{1}$$

and is called the Duffing's problem. The general solution of (1) is not periodic in general, but (1) admits the periodic solutions of the type

$$x = \sum_{j=0} A_{2j+1} \cos\left[(2j+1)\frac{\omega}{n} t\right], \quad n = 1, 3, 5, \ldots. \tag{2}$$

When $n=1$, the solution is called the harmonic solution, while the other cases $n=3, 5, \ldots$, are called the subharmonic solutions of the orders $\frac{1}{3}, \frac{1}{5}, \ldots$ respectively. The conditions for the harmonic or the subharmonic solutions are referred as the harmonic or the subharmonic responses. For example, the harmonic response is expressed as

$$\omega^2 A_1 - 2\alpha J_1(A_1) + \beta = 0. \tag{3}$$

(See Stoker, 1950.)

In the present article we assume that β is small in (1), and apply the theory of general perturbations (Hori, 1966) in order to see the long-range behavior of the solution when ω is close to $n\sqrt{\alpha}$ ($n=1, 3, \ldots$). We may obtain the periodic solutions (2) as the equilibrium solutions of the new equations of motion after non-critical terms are eliminated by a suitable canonical transformation.

Equation (1) is written in the canonical form,

$$\dot{x}_j = \partial F / \partial y_j, \quad \dot{y}_j = - \partial F / \partial x_j \quad (j = 1, 2) \tag{4}$$

with the Hamiltonian

$$\begin{aligned} F &= F_0 + F_1 \\ F_0 &= -\tfrac{1}{2}x_1^2 - \omega x_2 + \alpha \cos y_1, \quad F_1 = \beta y_1 \cos y_2 \end{aligned} \tag{5}$$

if

$$y_1 = x, \quad x_1 = \dot{x}, \quad y_2 = \omega t. \tag{6}$$

2. Elimination of non-critical terms

We consider the canonical transformation $(x_j, y_j) \to (\xi_j, \eta_j)$ with a determining function $S(\xi, \eta) = S_1(\xi, \eta) + \ldots$, the subscript 1 denoting $O(\beta)$. The transformation is

G. E. O. Giacaglia (ed.), Periodic Orbits, Stability and Resonances, 493–500.

given by

$$x_j = \xi_j + S_{1\eta_j} + O(\beta^2)$$
$$y_j = \eta_j - S_{1\xi_j} + O(\beta^2) \tag{7}$$

and, if $f(x, y)$ is any function of x_j and y_j, we have

$$f(x, y) = f(\xi, \eta) + \{f, S_1\} + O(\beta^2). \tag{8}$$

We first solve the auxiliary equations

$$\frac{d\xi_j}{d\tau} = \frac{\partial F_0^*}{\partial \eta_j}, \quad \frac{d\eta_j}{d\tau} = -\frac{\partial F_0^*}{\partial \xi_j} \quad (j = 1, 2) \tag{9}$$

where

$$F_0^* = F_0(\xi, \eta) = -\tfrac{1}{2}\xi_1^2 - \omega\xi_2 + \alpha \cos \eta_1. \tag{10}$$

Their solution is

$$\xi_1 = 2\sqrt{\alpha}\, k \,\mathrm{cn}\, \sqrt{\alpha}\,\tau, \quad \sin(\eta_1/2) = k \,\mathrm{sn}\, \sqrt{\alpha}\,\tau$$
$$\xi_2 = c_2, \qquad \eta_2 = \omega\tau + c_1 \tag{11}$$

where the modulus of the elliptic function is given by

$$k^2 = (2\alpha - c_3)/4\alpha = \xi_1^2/4\alpha + \sin^2(\eta_1/2). \tag{12}$$

By the theory of the elliptic function, we have

$$\xi_1 = 8\nu\, \Sigma\, Q_j \cos[(2j + 1)\nu\tau] \tag{13}$$

with

$$\nu = \frac{\pi\sqrt{\alpha}}{2K}, \tag{14}$$

then

$$\eta_1 = \int_0^\tau \xi_1 \, d\tau = 8\, \Sigma \frac{Q_j}{2j + 1} \sin(2j + 1)\nu\tau, \tag{15}$$

and

$$F_1(\xi, \eta) = \beta\eta_1 \cos \eta_2 = 4\beta\, \Sigma_\pm \frac{Q_j}{2j + 1} \sin[(2j + 1)\nu\tau \pm \eta_2]. \tag{16}$$

In (13), (15), and (16), Q_j is given by

$$Q_j = q^{j+1/2}/(1 + q^{2j+1}), \tag{17}$$

q being the nome.

We see that the critical argument in the right-hand member of (16) is $(2m+1)\nu\tau - \eta_2$ when we assume the resonance of the type $\omega \approx (2m+1)\sqrt{\alpha}$. So, according to the algorithm given by the general perturbation theory, we have the new Hamiltonian of the first order by

$$F_1^* = \frac{4Q_m}{2m + 1} \sin[(2m + 1)\nu\tau - \eta_2] \tag{18}$$

and the determining function by

$$S_1 = \int (F_1 - F_1^*)\,\mathrm{d}\tau = -4\beta\,\Sigma'_{\pm}\,\frac{Q_j \cos\left[(2j+1)\,\nu\tau \pm \eta_2\right]}{(2j+1)\left[(2j+1)\,\nu \pm \omega\right]}, \tag{19}$$

where the prime in $\Sigma_\pm$ denotes that the term with the critical argument is removed.

Equations (7) then give the transformation if we can evaluate the partial derivatives of S_1 with respect to ξ_1 and η_1. The evaluation is in fact possible with the use of (11) and (12). Here we only note that no mixed secular term occurs in $S_{1\xi_1}$ or $S_{1\eta_1}$ because of the equations,

$$\begin{aligned} \frac{\partial}{\partial\xi_1}\frac{\tau}{K} &= \frac{1}{2\alpha k k'^2 K}\left[\mathrm{cn}\,u\cdot Z(u) - \mathrm{sn}\,u\cdot\mathrm{dn}\,u\right], \\ \frac{\partial}{\partial\eta_1}\frac{\tau}{K} &= \frac{1}{2\sqrt{\alpha k k'^2}K}\left[\mathrm{sn}\,u\cdot\mathrm{dn}\,u\cdot Z(u) + \mathrm{cn}\,u\,(k'^2 - k^2\,\mathrm{sn}^2\,u)\right], \end{aligned} \tag{20}$$

where $Z(u)$ is Jacobi's zeta function.

3. Equations for the Long-Range Motion

The new equations of motion are

$$\dot{\xi}_j = \frac{\partial F^*}{\partial\eta_j}, \quad \dot{\eta}_j = -\frac{\partial F^*}{\partial\xi_j} \quad (j = 1, 2) \tag{21}$$

with the Hamiltonian

$$F^* = F_0^* + F_1^* + \mathrm{O}(\beta^2)$$

$$F_0^* = -\tfrac{1}{2}\xi_1^2 - \omega\xi_2 + \alpha\cos\eta_1 = -2\alpha k^2 - \omega\xi_2 + \alpha \tag{22}$$

$$F_1^* = \frac{4\beta Q_m}{2m+1}\sin\left[(2m+1)\,\nu\tau - \eta_2\right], \tag{23}$$

where k and τ in F_1^* are functions of ξ_1 and η_1 by the relations (11) and (12).

In order to emphasize the fact that two arguments $\nu\tau$ and η_2 enter in F^* only in the combination of $(2m+1)\nu\tau - \eta_2$, we adopt this critical argument as one of our variables and find its canonical conjugate. First consider the canonical transformation

$$\xi_1, \xi_2, \eta_1, \eta_2 \to \xi_1' = k^2, \quad \xi_2' = \xi_2, \eta_1', \eta_2' = \eta_2 \tag{24}$$

by choosing a suitable form for η_1'. If $W(\xi_1', \eta_1)$ is the determining function, the transformation is given by

$$\xi_1 = \frac{\partial W}{\partial\eta_1}, \quad \eta_1' = \frac{\partial W}{\partial\xi_1'}.$$

But by (22)

$$\xi_1 = \left[4\alpha\left(\xi_1' - \sin^2\frac{\eta_1}{2}\right)\right]^{1/2}$$

then

$$W = \int \left[4\alpha \left(\xi_1' - \sin^2 \frac{\eta_1}{2} \right) \right]^{1/2} \mathrm{d}\eta_1$$

and we find

$$\eta_1' = \frac{\partial W}{\partial \xi_1'} = \sqrt{\alpha} \int \left(\xi_1' - \sin^2 \frac{\eta_1}{2} \right)^{-1/2} \mathrm{d}\eta_1 = 2\alpha\tau . \tag{25}$$

Next we consider another canonical transformation

$$\xi_1', \xi_2', \eta_1', \eta_2' \to \xi_1'', \xi_2'' = \xi_2', \eta_1'' = \frac{\pi \eta_1'}{4\sqrt{\alpha} K}, \eta_2'' = \eta_2' \tag{26}$$

by choosing a suitable form for ξ_1''. If $W(\xi_1', \eta_1'')$ is the determining function, we have

$$\xi_1'' = -\frac{\partial W}{\partial \eta_1''}, \quad \eta_1' = -\frac{\partial W}{\partial \xi_1'},$$

therefore

$$W = -\frac{4\sqrt{\alpha}}{\pi} \eta_1'' \int K \,\mathrm{d}k^2 = -\frac{8\sqrt{\alpha}}{\pi} \eta_1'' (E - k'^2 K)$$

and

$$\xi_1'' = \frac{8\sqrt{\alpha}}{\pi} (E - k'^2 K) . \tag{27}$$

Finally we consider the last canonical transformation

$$\xi_1'', \xi_2'', \eta_1'', \eta_2'' \to p_1, p_2, q_1 = (2m+1)\,\eta_1'' - \eta_2'', q_2 = \eta_2'' \tag{28}$$

by choosing p_1 and p_2 properly. From

$$\xi_1'' \,\mathrm{d}\eta_1'' + \xi_2'' \,\mathrm{d}\eta_2'' = p_1 \,\mathrm{d}q_1 + p_2 \,\mathrm{d}q_2$$

we find

$$\xi_1'' = (2m+1)\, p_1, \quad \xi_2'' = -p_1 + p_2$$

or

$$p_1 = \frac{\xi_1''}{2m+1}, \quad p_2 = \frac{\xi_1''}{2m+1} + \xi_2'' . \tag{29}$$

The equations of the long-range motion are written in the form

$$\dot{p}_j = \frac{\partial F}{\partial q_j}, \quad \dot{q}_j = -\frac{\partial F}{\partial p_j} \quad (j = 1, 2) \tag{30}$$

$$F = -2\alpha k^2 + \omega (p_1 - p_2) + \frac{4\beta Q_m}{2m+1} \sin q_1 , \tag{31}$$

where k^2 is the function of p_1 by the relation

$$p_1 = \frac{8\sqrt{\alpha}}{(2m+1)\,\pi} (E - k'^2 K) . \tag{32}$$

We see immediately that $p_2 = \text{const.}$ and $q_2 = \omega t$, the latter of which is equivalent to

$$\eta_2 = \omega t . \tag{33}$$

4. Equilibrium Solutions

We examine the equilibrium solutions of (30), that is, the solutions $p_1 = \text{const.}$, $q_1 = \text{const.}$. We find that

$$q_1 = \pm (2m + 1) \frac{\pi}{2} \tag{34}$$

and that p_1 or k^2 should satisfy the relation

$$\omega = (2m + 1) \sqrt{\alpha} \cdot \frac{\pi}{2K} \mp \frac{(2m + 1)}{\sqrt{\alpha}\, k^2 k'^2} \left(\frac{\pi}{2K}\right)^3 \frac{1 - q^{2m+1}}{1 + q^{2m+1}} Q_m ,$$

$$(+ \text{ sign if } \sin q_1 = -1) . \tag{35}$$

The equilibrium solutions are stable if $\sin q_1 = -1$ and unstable if $\sin q_1 = +1$, since we see in (31) that

$$\frac{\mathrm{d}^2}{\mathrm{d}p_1^2} (- 2\alpha k^2 + \omega p_1) = \frac{(2m + 1)^2 \pi^2 (E - k'^2 K)}{16 k^2 k'^2 K^3} > 0 .$$

We first consider the stable equilibrium solutions, in which

$$q_1 = (-1)^{m+1} (2m + 1) \frac{\pi}{2}$$

or

$$\nu\tau = \frac{\eta_2}{2m + 1} + (-1)^{m+1} \frac{\pi}{2} . \tag{36}$$

Equation (15) then takes the form

$$\eta_1 = (-1)^{m+1}\, 8 \Sigma (-1)^j \frac{Q_j}{2j + 1} \cos \left(\frac{2j + 1}{2m + 1} \eta_2\right), \quad (\eta_2 = \omega t) . \tag{37}$$

On the other hand, the arguments of cosine series expression of $S_{1\xi_1}$ are found to be $2j\nu\tau \pm \eta_2$. Then, in view of (36), we have

$$\cos (2j\nu\tau \pm \eta_2) = (-1)^j \cos \left(\frac{2j}{2m + 1} \pm 1\right) \eta_2 , \quad (\eta_2 = \omega t) . \tag{38}$$

Therefore we see that $x = y_1 = \eta_1 - S_{1\xi_1}$ is periodic in t with the period $(2m+1)\, 2\pi/\omega$, and called the subharmonic solution of the order $1/(2m+1)$. Equation (35) with $+$ sign is the corresponding response, and is reduced to (3) if $m = 0$ (harmonic case). In fact when k^2 is small, the response is written in approximation as

$$\omega = \sqrt{\alpha} (1 - \tfrac{1}{4} k^2) + \frac{\beta}{4 \sqrt{\alpha}\, k} \tag{39}$$

but, in view of (2), (7), and (37), we have

$$A_1 = -8Q_0 + O(\beta) = -2k + O(\beta) \tag{40}$$

so that

$$\frac{\omega}{\sqrt{\alpha}} = 1 - \frac{A_1^2}{16} + \frac{\beta}{2\alpha |A_1|}$$

or

$$\frac{\omega^2}{\alpha} = 1 - \frac{A_1^2}{8} + \frac{\beta}{\alpha |A_1|} \tag{41}$$

which is equivalent to (3) if $J_1(A_1)$ is approximated by

$$\frac{A_1}{2}(1 - \tfrac{1}{8}A_1^2).$$

We note that when the response relation is satisfied only approximately, the deviation from the periodic oscillation remains small since our equilibrium solution is stable. We further note that the leading harmonic oscillation, $A_1 \cos \omega t$ is out of phase with respect to the external driving oscillation $\beta \cos \omega t$.

We next examine the unstable solutions. In this case we have

$$q_1 = (-1)^m (2m + 1)\frac{\pi}{2}$$

and

$$\eta_1 = (-1)^m \, 8 \, \Sigma \, (-1)^j \frac{Q_j}{2j + 1} \cos\left(\frac{2j + 1}{2m + 1} \eta_2\right),$$

then

$$A_1 = 2k + O(\beta) \quad \text{when} \quad m = 0.$$

So, the harmonic response may be written as

$$\frac{\omega^2}{\alpha} = 1 - \frac{A_1^2}{8} - \frac{\beta}{\alpha A_1}. \tag{42}$$

This unstable periodic oscillation is in phase with respect to the driving oscillation.

5. Long-Range Behavior

When the equilibrium condition is not satisfied exactly, p_1 (or k_2) and q_1 $(= (2m+1)\nu\tau - \eta_2)$ are not constant but change gradually. In fact the changes occur slowly as we see, from (30), that

$$\dot{p}_1 = O(\beta) \quad \text{and} \quad \dot{q}_1 = (2m + 1)\sqrt{\alpha} \cdot \frac{\pi}{2K} - \omega + O(\beta), \tag{43}$$

and that we are treating the case $\omega \approx (2m+1)\sqrt{\alpha}$, as far as $\pi/2K$ does not differ much from 1. Such long-range changes of the variables can be discussed with the help of the

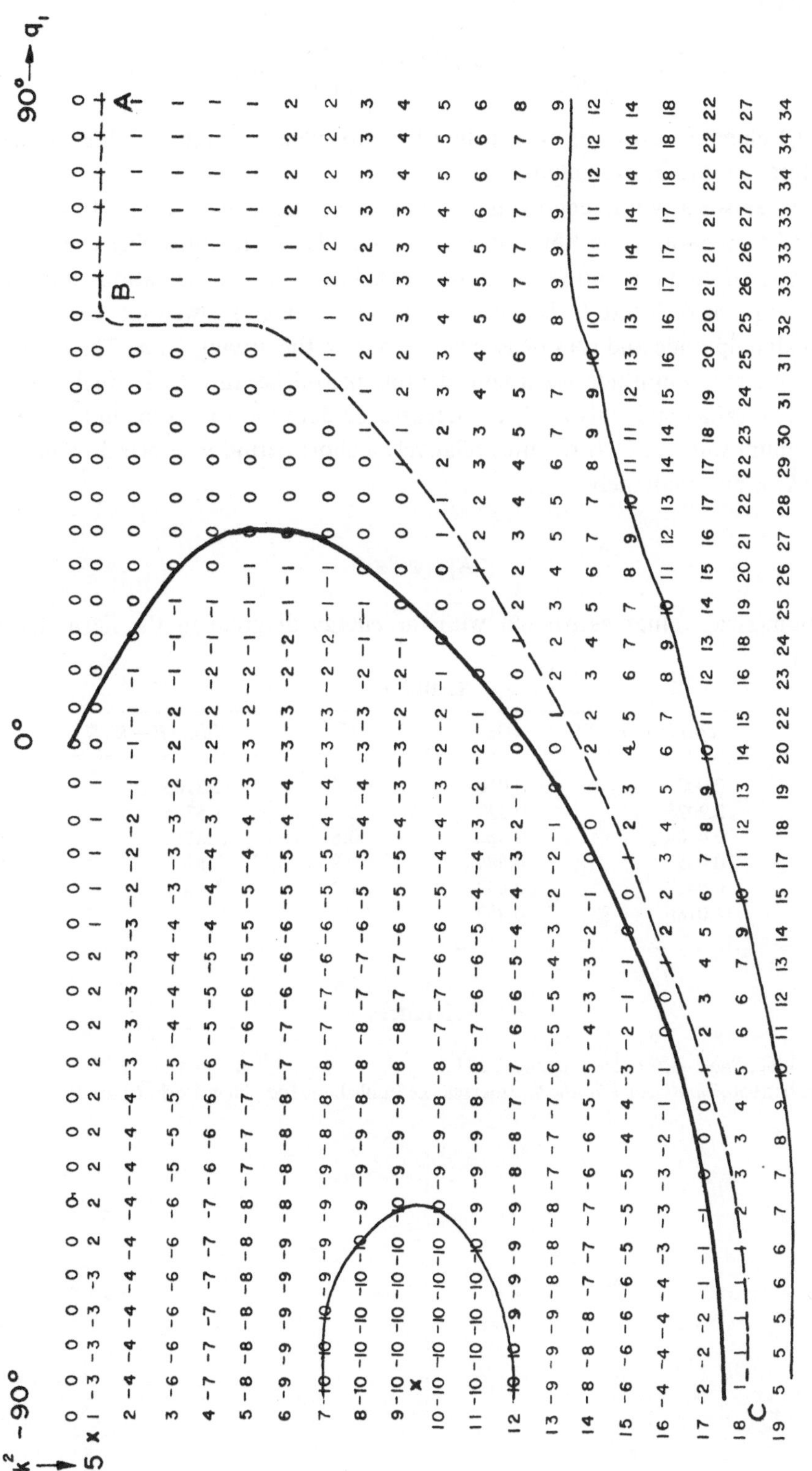

Fig. 1. $\omega = 0.9, \beta = 0.1.$

energy diagram

$$- 2\alpha k^2 + \frac{8\sqrt{\alpha\omega}}{(2m+1)\pi}(E - k'^2K) + \frac{4\beta Q_m}{2m+1}\sin q_1 = \text{const.} \tag{44}$$

in (k^2, q_1) plane. In order to obtain the time dependence of the variables, however, quadratures have to be evaluated numerically.

Figure 1 shows a typical energy diagram in the harmonic case, $m=0$. In the figure, the bold line separates the libration and the circulation regions. The energy curve drawn by the dotted line shows an interesting behavior of the oscillation: suppose we start at the point A. There k^2 is small and q_1 is 90°, which shows that the oscillation has a small amplitude and is in phase with respect to the driving force. The oscillation remains in small amplitude for a considerable period because k^2 is small and $\dot{q}_1$ is small. Then, rather suddenly at B, k^2 increases, so does the amplitude until it reaches the maximum value at C. It requires relatively a short period to proceed from B to C because k^2 is not small there.

Appendix

Some numerical values associated with the energy diagram in the harmonic case (Table I).

TABLE I

k^2	$(8/\pi)(E-k'^2K)$	$4Q_0$	k^2	$(8/\pi)(E-k'^2K)$	$4Q_0$
0.0	0.0000	0.000	0.6	1.3205	0.904
0.1	0.2028	0.322	0.7	1.5765	1.017
0.2	0.4108	0.466	0.8	1.8514	1.146
0.3	0.6256	0.584	0.9	2.1568	1.313
0.4	0.8477	0.692	1.0	2.5465	2.000
0.5	1.0786	0.797			

References

Hori, G.: 1966, *Publ. Astron. Soc. Japan* **18**, 287.
Stoker, J. J.: 1950, *Nonlinear Vibrations*, Interscience Publishers, Inc., New York, Chap. IV.

PASSAGE THROUGH RESONANCE*

W. T. KYNER

Dept. of Mathematics, University of Southern California, Los Angeles, Calif., U.S.A.

Abstract. The effect of both gravitational and tidal torques on the spin rate of the planet Mercury is used as a model for a passage through resonance problem. Both the resonance and nonresonance aspects of the spin rate are studied by means of the method of averaging. On the basis of this analysis, it is concluded that the present 3:2 resonant state is due to nonresonant capture which evolved into resonant capture as the orbital eccentricity increased.

1. Introduction

One of the most striking results of recent radar studies of the planets is the discovery that the rate at which Mercury rotates about its axis is very different from the rate which had been inferred from earlier optical observations. It is now known that the rotation rate is approximately three halves of its mean motion instead of being approximately equal to its mean motion. Stated differently, instead of rotating so that the same side always faces the sun, Mercury completes three rotations every two revolutions.

It is not surprising that as soon as the new information about Mercury's rotation rate became available articles began to appear in *Science* and *Nature* (two journals which quickly publish short technical articles) explaining the phenomenon. These early explanations are quite intuitive but, nevertheless, contain the basis of the subsequent analyses.

The radar observations of Mercury were made in April 1965 by Pettingill and Dyce at the Arecibo Ionospheric Observatory in Puerto Rico. They announced their results in the 19 June 1965 issue of *Nature* [1]. In the same issue, in fact on the same page, Peale and Gold [2] stated that the observed rotation rate is due to the torque exerted by the sun on the (assumed) tidal bulge it raises on Mercury. They claimed that in the absence of a tidal torque only synchronous motion is possible. This last statement was contradicted by Goldreich (*Nature*, 23 Oct. 1965 [3]) who pointed out that if Mercury is asymmetric about its polar axis (as is the Moon), then the resulting torque would be larger than the tidal torque – therefore, both effects must be considered. Next, Colombo (*Nature*, 6 Nov. 1965 [4]) and Liu and O'Keefe (*Science*, 24 Dec. 1965 [5]) wrote that the radar observations were consistent with a 3:2 resonant state and that this resonance could be explained by Mercury being asymmetric. None of the brief notes contained analytical arguments to substantiate the author's conclusions.

Subsequently, Colombo together with Shapiro (Oct. – Nov. 1965 [6]) and Bellomo (March 1966 [7]) developed a mathematical model which included both torques,

* This research was supported by the National Aeronautics and Space Administration, NASA Research Grant NGR-05-018-079.

G. E. O. Giacaglia (ed.), Periodic Orbits, Stability and Resonances, 501–514.

constructed approximate solutions, and attempted to compute the probability of capture into the 3:2 state. Goldreich and Peale (April 1966 [8]) published parallel results. Both groups also studied the closely related problem of explaining the rotation rate of Venus whose motion seems to be coupled with that of the earth. We should also mention the papers of Laslett and Sessler (Jan. 1966 [9]), Jefferys (April 1966 [10]), and Blitzer (May 1967 [11]) who studied the asymmetric problem (no tidal torque). More recently, Counselman (Feb. 1969 [12]) made a careful study of the Mercury problem in which he computed the effects of long term oscillations in the orbital eccentricity on the capture probability and also discussed the effects of different tidal models.

The justification of the use of the concept of capture probability in a problem in celestial mechanics rests on our ignorance of initial conditions and parameter values which influence the final state of the system. Although Counselman presents a persuasive defense of this point of view, we shall take a different approach in this paper and introduce the concept of *capture measure*, i.e., the measure of the set of spin rates which are the preimages of asymptotically stable equilibrium states. This concept is essentially equivalent to capture probability but seems more appropriate to a deterministic problem.

The differential equations governing the rotation rate of Mercury can be written as a Hamiltonian system with a dissipative term, the tidal torque. We shall use them as an example of a more general problem which will be discussed briefly. The Mercury problem will be discussed in detail with particular emphasis of the phenomenon of *passage through resonance*.

2. The Mercury Problem

We shall assume that Mercury is revolving around the sun in a fixed plane whose normal is the rotation axis. The principal moments of inertia are denoted by A, B, C, with C being the moment about the rotation axis. Using MacCullagh's formula (see Chapter 14 of Danby [13]), we write Mercury's potential energy as

$$\begin{aligned} V &= -m\left[\frac{\mu}{r} + \frac{\mu(A+B+C-3I)}{2mr^3}\right] \\ &= -m\left[\frac{\mu}{r} + \frac{\mu\xi}{2r^3}\left(1 - 3\sin^2(v-\theta)\right)\right], \end{aligned} \tag{2.1}$$

where m = the mass of Mercury, μ = the gravitational constant times the mass of the sun, r = the distance from Mercury to the sun, the origin of our inertial coordinate system, $I = A\cos^2(v-\theta) + B\sin^2(v-\theta)$, v = the angle between the Mercury-sun line and a fixed reference line in the orbital plane, θ = the angle between Mercury's long axis and the reference line, and $\xi = (C/m)\lambda$.

We assume that

$$A < B < C, \quad 0 < \lambda = \frac{B-A}{C} \ll 1 .$$

In the absence of tidal torque and other perturbations, the Lagrangian of the sun-Mercury system is

$$\mathbf{L} = \frac{m}{2}\left[\left(\frac{\mathrm{d}r}{\mathrm{d}t}\right)^2 + \left(r\frac{\mathrm{d}v}{\mathrm{d}t}\right)^2\right] + \frac{C}{2}\left(\frac{\mathrm{d}\theta}{\mathrm{d}t}\right)^2 - V.$$

Hence, the equations of motion are

$$\begin{aligned} &m\frac{\mathrm{d}^2 r}{\mathrm{d}t^2} - \frac{\partial}{\partial r}\mathbf{L} = 0, \\ &m\frac{\mathrm{d}}{\mathrm{d}t}\left(r^2\frac{\mathrm{d}v}{\mathrm{d}t}\right) - \frac{\partial}{\partial v}\mathbf{L} = 0, \\ &C\frac{\mathrm{d}^2\theta}{\mathrm{d}t^2} - \frac{\partial}{\partial\theta}\mathbf{L} = 0. \end{aligned} \tag{2.2}$$

We take Mercury's orbital semimajor axis as the unit of length and its year as the unit of time. Then (see p. 71 and p. 330 of [13]), $C/m \doteq 2\times 10^{-5}$. The constant λ has not yet been measured, but if Mercury's ellipticity is roughly that of the moon, the $\lambda \doteq 2\times 10^{-4}$. Hence $\xi \doteq 4\times 10^{-9}$. Clearly, the asymmetric term in the potential can have little effect on the orbit. However, as we shall see, it is sufficiently large to determine the stable resonance states. This seeming paradox is resolved by the fact that λ, rather than ξ, is the relevant parameter for the resonance study.

We set aside the first two equations of (2.2) and write the third as

$$\frac{\mathrm{d}^2\theta}{\mathrm{d}t^2} + \frac{3}{2}\lambda\frac{\mu}{r^3}\sin 2(\theta - v) = 0, \tag{2.3}$$

where, because of the size of ξ, we take the inplane angle v as the true anomaly and

$$\frac{1}{r} = \frac{1 + e\cos v}{1 - e^2} \tag{2.4}$$

(recall that $a=1$, the unit of length). In other words, we have assumed that Mercury moves in a fixed ellipse in accordance with Kepler's laws. Therefore, we can write v as a known periodic function of M, the true anomaly,

$$v = M + \sum_{k=1}^{\infty} C_k(e)\sin kM, \tag{2.5}$$

where $(C_k(e))$ is a transcendental function of e of order e^k (p. 207 of [14]). Since the Mercury year is our unit of time, $M = 2\pi t + M_0$.

If M is taken as the independent variable, then (using the notation of [7]) the differential Equation (2.3) can be written

$$\frac{\mathrm{d}^2\theta}{\mathrm{d}M^2} = \beta\mathbf{P}(e, \theta, M), \tag{2.6}$$

where $\beta=\frac{3}{2}\lambda$, $\mu=(2\pi)^2$, and

$$\mathbf{P}(e,\theta,M)=-\left(\frac{1+e\cos v}{1-e^2}\right)^3\sin 2(\theta-v). \tag{2.7}$$

It is easy to verify that $\mathbf{P}$ can be represented by a Fourier series,

$$\mathbf{P}(e,\theta,M)=\sum_{j=0}^{\infty}\mathbf{P}_j^-(e)\sin(jM-2\theta)+\mathbf{P}_j^+(e)\sin(jM+2\theta). \tag{2.8}$$

Therefore, the only resonant states of interest are those defined by

$$2\frac{\mathrm{d}\theta}{\mathrm{d}M}=k,\quad k=0,\pm 1,\pm 2,\ldots \tag{2.9}$$

Several models for the tidal torque have been studied (see [12]). Here, we shall consider three models,

$$T_a\left(e,\frac{\mathrm{d}\theta}{\mathrm{d}M},M\right)=-\frac{1}{r^6}\operatorname{sgn}\left(\frac{\mathrm{d}\theta}{\mathrm{d}M}-\frac{\mathrm{d}v}{\mathrm{d}M}\right)$$

(constant tidal lag angle), (2.10a)

$$T_b\left(e,\frac{\mathrm{d}\theta}{\mathrm{d}M},M\right)=-\frac{1}{r^9}\operatorname{sgn}\left(\frac{\mathrm{d}\theta}{\mathrm{d}M}-\frac{\mathrm{d}v}{\mathrm{d}M}\right)$$

(tidal lag angle proportional to the strain amplitude), (2.10b)

$$T_c\left(e,\frac{\mathrm{d}\theta}{\mathrm{d}M},M\right)=-\frac{1}{r^6}\left(\frac{\mathrm{d}\theta}{\mathrm{d}M}-\frac{\mathrm{d}v}{\mathrm{d}M}\right)$$

(tidal lag angle proportional to the rate of tidal strain). (2.10c)

So that all three models have the same qualitative properties when the rotation rate is large, we shall arbitrarily change the third model and require that the dependence on strain rate be piecewise linear, i.e.,

$$T_c\left(e,\frac{\mathrm{d}\theta}{\mathrm{d}M},M\right)=-\frac{1}{r^6}\operatorname{sgn}\left(\frac{\mathrm{d}\theta}{\mathrm{d}M}-\frac{\mathrm{d}v}{\mathrm{d}M}\right)\quad\text{if}\quad\frac{\mathrm{d}\theta}{\mathrm{d}M}\geq\max\frac{\mathrm{d}v}{\mathrm{d}M}$$

$$\text{or}\quad\frac{\mathrm{d}\theta}{\mathrm{d}M}\leq\min\frac{\mathrm{d}v}{\mathrm{d}M}. \tag{2.10c'}$$

Once the torque due to tidal friction is included in the physical model, the equations governing the rotation of Mercury become

$$\frac{\mathrm{d}^2\theta}{\mathrm{d}M^2}=\beta\mathbf{P}(e,\theta,M)+\alpha T\left(e,\frac{\mathrm{d}\theta}{\mathrm{d}M},M\right), \tag{2.11}$$

where the positive parameter α depends on the dissipative mechanism. We shall assume that $\alpha=\mathrm{O}(\beta^2)$. Again, Counselman [12] can be consulted for details.

Finally, we note that (2.11) can be written as a Hamiltonian system with a dissipative term. Let

$$x = \frac{d\theta}{dM}$$

$$H(x, \theta, M, e) = \frac{x^2}{2} + \beta F(e, \theta, M), \tag{2.12}$$

$$F(e, \theta, M) = -\frac{1}{2}\left(\frac{1 + e\cos v}{1 - e^2}\right)^3 \cos 2(\theta - v).$$

Then,

$$\begin{aligned} \frac{d\theta}{dM} &= \frac{\partial H}{\partial x}(x, \theta, M, e), \\ \frac{dx}{dM} &= -\frac{\partial H}{\partial \theta}(x, \theta, M, e) + \alpha T(e, x, M). \end{aligned} \tag{2.13}$$

It follows from these equations that along trajectories,

$$\frac{dH}{dM} = \alpha x T(e, x, M) + \beta \frac{\partial F}{\partial M}(e, \theta, M). \tag{2.14}$$

As we shall see, the size of the orbital eccentricity plays an important role in the Mercury problem. It is for this reason that we write the perturbation terms in the preceding equations as function of e as well as of θ, x, and M.

3. The Abstract Problem

In this section we shall consider a Hamiltonian system with a dissipative term,

$$\frac{dx}{dt} = -\frac{\partial H}{\partial y}(x, y, t), \quad \frac{dy}{dt} = \frac{\partial H}{\partial x}(x, y, t) + \alpha g(x, t), \tag{3.1}$$

where α and β are perturbation parameters,

$$H(x, y, t) = \frac{x^2}{2} + \alpha F(x, y, t), \tag{3.2}$$

F has period 2π in y and t and is analytic in all its arguments. The function 9 has period 2π in t, is continuous, and g has the property that

$$g(x, t) = \begin{cases} h(t) & \text{if} \quad x \le 1, \\ -h(t) & \text{if} \quad x \ge 2, \end{cases} \tag{3.3}$$

where h is strictly positive. In addition, $\bar{g}(x)$, the time average of $g(x, t)$,

$$\bar{g}(x) = \frac{1}{2\pi} \int_0^{2\pi} g(x, t)\, dt,$$

is strictly monotonic decreasing on the interval $1 < x < 2$. This last condition is consistent with the tidal models.

We also consider another Hamiltonian,

$$H^*(x, y, t) = H(x, y, t) - \alpha g(x, t)\, y, \tag{3.4}$$

and the differential equations,

$$\frac{dx}{dt} = -\frac{\partial H^*}{\partial y}(x, y, t), \quad \frac{dy}{dt} = \frac{\partial H^*}{\partial x}(x, y, t). \tag{3.5}$$

Clearly, (3.1) and (3.5) are equivalent if $x < 1$ or $x > 2$.

By definition, resonance is a rational relationship between the fundamental frequencies of the unperturbed problem ($\alpha = \beta = 0$), i.e., $\bar{x}$ is a resonant frequency if $\bar{x} = p/q$, p, q integers. A *basic* (or first order) resonant frequency satisfies the condition

$$\bar{x} = p/q; \quad F_{p,q}(\bar{x}) \neq 0, \tag{3.6}$$

where

$$F(x, y, t) = \sum_{-\infty}^{\infty} F_{p,q}(x) \exp i(pt + qy). \tag{3.7}$$

We shall assume that the basic frequencies are *uniformly sparse*, i.e., the absolute value of the difference between any two basic frequencies is greater than a positive number. For the Mercury problem, the basic resonant frequencies are integral multiples of one half.

If $\alpha = 0$, the dissipative term is absent and the solutions to the Hamiltonian equations,

$$\frac{dx}{dt} = -\frac{\partial H}{\partial y}(x, y, t), \quad \frac{dy}{dt} = \frac{\partial H}{\partial x}(x, y, t). \tag{3.8}$$

have many interesting geometric properties. For example, by the Poincaré-Birkhoff theorem (a good reference is [15]), each basic frequency, $\bar{x} = p/q$, in the interval (1, 2) generates $2q$ periodic solutions with period $2\pi q$. As usual, we must assume that the perturbation parameter is sufficiently small.

Furthermore, if x and y are taken as polar coordinates, then these periodic solutions are represented by closed curves in the solid torus, $0 \leqslant x \leqslant 2$, y, t (mod 2π). By the Kolmogorov-Arnold-Moser theory [15], these closed curves lie between invariant tori which are characterized by the condition that the radius of the generating cross section, $x = \text{const.}$, be 'sufficiently irrational'. The cross sections of the tori are invariant curves under the mapping $M_{\alpha,0}$ induced by the differential Equations (3.8).

We argue (but do not prove) that the qualitative effect of the dissipative term in (3.1) is to destroy the invariant tori and to change some of the stable periodic solutions into attractors, in this case asymptotically stable periodic solutions. Another attractor, and for the Mercury problem perhaps the most important, is the zero of $\bar{g}(x)$, the time average of the dissipative term. The concept of capture can then be formulated in terms of the preimages of the attractors.

The solutions of the full system (3.1) (with $\alpha \neq 0$, $\beta \neq 0$) are strongly influenced by the properties of $\bar{g}(x)$. By assumption, $\bar{g}(x)$ is strictly monotonically decreasing on the interval (1, 2) and must change sign at an interior point, $x = a$ (say). Therefore, the map $M_{\alpha,\beta}$ induced by (3.1) will tend to expand those circles with radii less than a and contract those with radii greater than a. Since the oscillatory term $F(x, y, t)$ can oppose this effect, we do not assert that a circle and its image cannot intersect, but rather that a sufficiently high iterate of the mapping has this property. Furthermore, the time interval needed before the dissipative effect dominates the oscillatory effect is strongly influenced by the basic resonant frequencies.

In order to clarify these somewhat vague assertions, let us return to the Mercury problem and study the qualitative effects of the tidal torque on the rotation rate.

4. The Mercury Problem - Nonresonant Frequencies

In the analysis of the differential Equation (2.11) which describes the rotation of Mercury, we must distinguish between the resonant and nonresonant frequencies. We shall first discuss the nonresonant problem.

We write (2.1) as a system,

$$\begin{aligned} \frac{d\theta}{dM} &= x, \\ \frac{dx}{dM} &= \beta \mathbf{P}(e, \theta, M) + \alpha T(e, x, M), \end{aligned} \tag{4.1}$$

with initial conditions $x(0) = x_0$, $(x_0 \neq k/2, k$ an integer$)$, $\theta(0) = \theta_0$. We shall assume that $\alpha = O(\beta^2)$ and employ the second order method of averaging to obtain approximate solutions to (4.1).

A change of variables from θ, x to *averaged* or *mean* variables φ, w is given by

$$\begin{aligned} \theta &= \varphi + \beta \Phi^{(1)}(e, \varphi, w, M) + \beta^2 \Phi^{(2)}(e, \varphi, w, M), \\ x &= w + \beta W^{(1)}(e, \varphi, w, M) + \beta^2 W^{(2)}(e, \varphi, w, M), \end{aligned} \tag{4.2}$$

where the functions $\Phi^{(j)}$, $W^{(j)}$, $j = 1, 2$, have period 2π in φ and M and are constructed so that φ and w satisfy the averaged differential equations

$$\begin{aligned} \frac{d\theta}{dM} &= w + O(\beta^3), \\ \frac{dw}{dM} &= \alpha T_0(e, w) + O(\beta^3). \end{aligned} \tag{4.3}$$

The function $T_0(e, w)$, the averaged tidal torque, is defined by

$$T_0(e, w) = \frac{1}{2\pi} \int_0^{2\pi} T(e, w, M)\, dM. \tag{4.4}$$

The simple form of (4.3) is due to the fact that the gravitational torque $\boldsymbol{P}(e, \theta, M)$ does not depend on the rotation rate and has zero mean with respect to M and θ, i.e.,

$$\overline{\overline{\mathbf{P}}}(e) \underset{=}{\Delta} \frac{1}{(2\pi)^2} \int_0^{2\pi}\int_0^{2\pi} \mathbf{P}(e, \theta, M)\, \mathrm{d}\theta\, \mathrm{d}M = 0. \tag{4.5}$$

The functions $\Phi^{(j)}$, $W^{(j)}$, $j=1, 2$, are required to be solutions of the following partial differential equations:

$$\mathbf{P}(e, \varphi, M) = w \frac{\partial W^{(1)}}{\partial \varphi} + \frac{\partial W^{(1)}}{\partial M}, \tag{4.6a}$$

$$W^{(1)}(e, \varphi, w, M) = w \frac{\partial \boldsymbol{\Phi}^{(1)}}{\partial \varphi} + \frac{\partial \boldsymbol{\Phi}^{(1)}}{\partial M}, \tag{4.6b}$$

$$\frac{\partial \mathbf{P}}{\partial \varphi}(e, \varphi, M)\, \boldsymbol{\Phi}^{(1)}(e, \varphi, w, M) + \frac{\alpha}{\beta^2}\left[T(e, w, M) - T_0(e, w)\right] = w \frac{\partial W^{(2)}}{\partial \varphi} + \frac{\partial W^{(2)}}{\partial M}, \tag{4.6c}$$

$$W^{(2)}(e, \varphi, w, M) = w \frac{\partial \boldsymbol{\Phi}^{(2)}}{\partial \varphi} + \frac{\partial \boldsymbol{\Phi}^{(2)}}{\partial M}. \tag{4.6d}$$

The derivation of these equations is straightforward and will not be given here.

These equations are of the type (suppressing the nonperiodic variables)

$$m(\varphi, M) = w \frac{\partial f}{\partial \varphi} + \frac{\partial f}{\partial M}, \tag{4.7}$$

where, because of (2.8), we can write

$$m(\varphi, M) = \sum_{-\infty}^{\infty} m_j \exp i(2\varphi + jM). \tag{4.8}$$

Therefore,

$$f(\varphi, M) = \sum_{-\infty}^{\infty} m_j \frac{\exp i(2\varphi + jM)}{i(2w + j)} \tag{4.9}$$

is a solution to (4.7) if $\overline{\overline{m}}=0$ and $2w+j\neq 0, j=0, \pm 1, \pm 2, \ldots$, i.e., if $m(\varphi, M)$ has zero mean with respect to φ and M, and w is a nonresonant frequency.

It is convenient to introduce a linear operator Q, the inverse of the differential operator $w(\partial/\partial\varphi)+(\partial/\partial M)$ and write (4.9) as

$$f(\varphi, M) = Qm(\varphi, M), \tag{4.10}$$

where, by construction, $\overline{\overline{f}} = 0$.

Using Q, we can denote the solutions to (4.6) as

$$W^{(1)}(e, \varphi, w, M) = Q\mathbf{P}(e, \varphi, M), \tag{4.11a}$$

$$\Phi^{(1)}(e, \varphi, w, M) = Q^2\mathbf{P}(e, \varphi, M), \tag{4.11b}$$

$$W^{(2)}(e, \varphi, w, M) = Q\left\{\frac{\partial}{\partial\varphi}\mathbf{P}(e, \varphi, M)\, Q^2\mathbf{P}(e, \varphi, M) + \frac{\alpha}{\beta^2}\left[T(e, w, M) - T_0(e, w)\right]\right\}, \tag{4.11c}$$

$$\Phi^{(2)}(e, \varphi, w, M) = Q^2\left\{\frac{\partial}{\partial\varphi}\mathbf{P}(e, \varphi, M)\, Q^2\mathbf{P}(e, \varphi, M) + \frac{\alpha}{\beta^2}\left[T(e, w, M) - T_0(e, w)\right]\right\}. \tag{4.11d}$$

It is easy to verify that the left terms of (4.6) have zero mean with respect to φ and M, and are therefore in the domain of the operator Q. Note that each application of Q raises the power of the denominator term $(2w+j)^{-1}$.

Approximate solutions to (4.1) are now constructed by solving the averaged equations,

$$\frac{d\varphi}{dM} = w, \tag{4.12a}$$

$$\frac{dw}{dM} = \alpha T_0(e, w), \tag{4.12b}$$

and substituting the solutions into (4.2).

It is easy to establish error bounds for the resulting approximate solutions which are valid on a time interval of length $O(1/\alpha)$. In these bounds, however, the expressions $(2w+j)^{-s}$, $s=1, 2, 3, 4, j=0, \pm 1, \pm 2, \ldots$, occur. Therefore, a separate analysis is required near a basic resonant frequency such as $w=\frac{3}{2}$. This will be discussed in the next section.

It can be verified (see [12]) that $T_0(e, w)$ is a monotonic decreasing function with a simple zero $w^*(e)$ such that

$$\min \frac{dv}{dM} = \frac{(1-e)^2}{(1-e^2)^{3/2}} < w^*(e) < \frac{(1+e)^2}{(1-e^2)^{3/2}} = \max \frac{dv}{dM}. \tag{4.13}$$

For each of the tidal torque models considered here, $w^*(e)$ is a monotonic increasing function of e, the orbital eccentricity (see p. 425 of [8]). Furthermore, $w=w^*(e)$ is an asymptotically stable equilibrium point of the differential Equation (4.12b). Therefore, any solution which has not been captured by a resonant state approaches $w=w^*(e)$, $\varphi=\varphi_0+w^*(e)M$, as time increases. The analysis of the resonance states is more complicated and is our next topic.

5. The Mercury Problem - Resonant Frequencies

In this section, we shall discuss the qualitative properties of the solutions to the differential Equation (4.1) for the most important case, the 3:2 resonant state. The analysis of the other basic resonances is similar.

As is standard in resonant problems, we introduce new variables by the transformation,

$$\begin{aligned} x &= \tfrac{3}{2} + \varepsilon z, \\ \theta &= \tfrac{3}{2} M + \eta, \\ \varepsilon &= \sqrt{\beta}. \end{aligned} \tag{5.1}$$

The differential Equations (4.1) become

$$\begin{aligned} \frac{d\eta}{dM} &= \varepsilon z, \\ \frac{dz}{dM} &= \varepsilon \mathbf{P}(e, \tfrac{3}{2} M + \eta, M) + \frac{\alpha}{\varepsilon} T(e, \tfrac{3}{2} + \varepsilon z, M). \end{aligned} \tag{5.2}$$

Since $\alpha = O(\beta^2)$, $\alpha/\varepsilon = O(\varepsilon^3)$.

From (2.8), we have

$$\begin{aligned} \mathbf{P}(e, \tfrac{3}{2} M + \eta, M) = \sum_{j=0}^{\infty} & \mathbf{P}_j^-(e) \sin[(j-3) M - 2\eta] \\ & + \mathbf{P}_j^+(e) \sin[(j+3) M + 2\eta], \end{aligned} \tag{5.3}$$

or

$$\mathbf{P}(e, \tfrac{3}{2} M + \eta, M) = -A(e) \sin 2\eta + B(e, \eta, M), \tag{5.4}$$

where $A(e) = P_3^-(e)$, and B has zero mean with respect to M, i.e.,

$$\bar{B}(e, \eta) \underset{=}{\Delta} \frac{1}{2\pi} \int_0^{2\pi} B(e, \eta, M) \, dM = 0. \tag{5.5}$$

It is convenient to write

$$T(e, \tfrac{3}{2} + \varepsilon z, M) = T(e, \tfrac{3}{2}, M) + \varepsilon T'(e, \tfrac{3}{2}, M) z + O(\varepsilon^2), \tag{5.6}$$

where the prime denotes differentiation with respect to z. Hence,

$$\begin{aligned} \frac{d\eta}{dM} &= \varepsilon z, \\ \frac{dz}{dM} &= -\varepsilon A(e) \sin 2\eta + \varepsilon B(e, \eta, M) \\ &\quad + \frac{\alpha}{\varepsilon} T(e, \tfrac{3}{2}, M) + \alpha T'(e, \tfrac{3}{2}, M) z \\ &\quad + O(\varepsilon^5). \end{aligned} \tag{5.7}$$

To be consistent with the analysis of the nonresonant problem we must carry out a fourth order expansion in ε. Let the averaged variables ζ and s be defined by

$$\begin{aligned} \eta &= \zeta + \sum_{j=1}^{4} e^j Z^{(j)}(e, \zeta, s, M), \\ z &= s + \sum_{j=1}^{4} e^j S^{(j)}(e, \zeta, s, M). \end{aligned} \tag{5.8}$$

As before, the functions $Z^{(j)}$, $S^{(j)}$, $j=1, 2, 3, 4$, have period 2π in ζ and M and are constructed so that ζ and s satisfy the averaged differential equations

$$\begin{aligned} \frac{d\zeta}{dM} &= \varepsilon s + O(\varepsilon^5), \\ \frac{ds}{dM} &= -\varepsilon A(e)\sin 2\zeta + \frac{\alpha}{\varepsilon} T_0(e) \\ &\quad + \alpha T_0'(e) s + O(\varepsilon^5), \end{aligned} \tag{5.9}$$

where $T_0 = \bar{T}$, $T_0' = \bar{T}'$.

Since η and z are constant if $\varepsilon = 0$, the partial differential equations satisfied by $Z^{(j)}$, $S^{(j)}$, $j=1, 2, 3, 4$, are simpler than the corresponding equations for the nonresonant problem. They are of the type (again suppressing the nonperiodic variables),

$$m(\zeta, M) = \frac{\partial f}{\partial M}. \tag{5.10}$$

If we have $\bar{m}=0$, then f has period 2π in M. It can be verified that requiring $\bar{Z}^{(j)}=0$, $\bar{S}^{(j)}=0$, $j=1, 2, 3, 4$, will insure that the transformation (5.8) is consistent with the differential Equation (5.9) and that no small denominators are introduced. We remark that since the system (5.7) is equivalent to a single second order equation, much of the analysis is simplified, e.g., we have

$$0 = \frac{\partial Z^{(1)}}{\partial M}, \quad \text{so} \quad Z^{(1)} \equiv 0. \tag{5.11}$$

Approximate solutions to (5.2) are now constructed by solving the averaged equations,

$$\begin{aligned} \frac{d\zeta}{dM} &= \varepsilon s, \\ \frac{ds}{dM} &= -\varepsilon A(e)\sin 2\zeta + \frac{\alpha}{\varepsilon} T_0(e) + \alpha T_0'(e) s, \end{aligned} \tag{5.12}$$

and substituting the solutions into (5.8). In contrast to the nonresonant case, the initial conditions greatly influence the qualitative properties of the solution. Therefore, they must be carefully studied.

We assume that the initial spin is greater than $\frac{3}{2}$. Hence by the nonresonant analysis, the averaged spin rate is decreasing. In other words, there exists M_0 such that for all $M < M_0$, $s(M) > 0$. If $w^*(e)$, the zero of the averaged torque, is less than $\frac{3}{2}$, then

$T_0(e)<0$, and $T_0'(e)\leqslant 0$. This implies that the energy-like function,

$$K(\zeta, s) = \tfrac{1}{2}s^2 + \frac{A(e)}{2}(1 - \cos 2\zeta) - \frac{\alpha}{\varepsilon^2} T_0(e)\zeta, \tag{5.13}$$

is monotonic decreasing along solutions to the differential Equation (5.12). In fact, if $T_0'(e)<0$, and $s^2>0$, then

$$\frac{dK}{dM} = \alpha T_0'(e) s^2 < 0. \tag{5.14}$$

It follows that there exists $M_1>M_0$ such that $s(M_1)=0$, and $(ds/dM)(M_1)\leqslant 0$. We shall show that the value of $\zeta(M_1)$ (mod π) will determine if $s(M)$, $M>M_1$, approaches zero as M approaches infinity or if $s(M)$ becomes and stays negative and thereby escapes the resonant state.

From (5.12), we have that the equilibrium states are given by the transcendental equations,

$$\frac{d\zeta}{dM} = 0, \tag{5.15a}$$

$$A(e)\sin 2\zeta = \frac{\alpha}{\beta} T_0(e), \tag{5.15b}$$

(recall that $\varepsilon^2=\beta$ and $\alpha=O(\beta^2)$).

Let

$$\begin{aligned} c_j &= c + j\pi, && \text{(stable)} \\ b_j &= -c + \frac{\pi}{2} + j\pi, && \text{(unstable)} \quad j = 0, \pm 1, \pm 2, \ldots \end{aligned} \tag{5.16}$$

denote the roots of (5.15b) with $c=O(\beta)$.

For convenience, we replace M_1 by 0 and let $\zeta(M,\sigma)$, $s(M,\sigma)$ denote the solution to (5.12) such that $\zeta(0,\sigma)=\sigma$, $s(0,\sigma)=0$, $ds/dM(0,\sigma)\leqslant 0$. The values of $\zeta(M_1)$ (mod π) lie in the set

$$\begin{aligned} \mathbf{A} = \{&\sigma, c < \sigma < b_0, \text{ such that there exist } \underline{M} < 0 \text{ with the property} \\ &\text{that } \zeta(\underline{M}, \sigma) = b_{-1}, s(\underline{M}, \sigma) > 0\}. \end{aligned} \tag{5.17}$$

A subset of **A**,

$$\begin{aligned} \mathbf{B} = \{&\sigma \text{ in } \mathbf{A} \text{ such that there exists } M^* > 0 \text{ with the property that} \\ &b_{-1} < \zeta(M^*, \sigma) \le c, s(M^*, \sigma) = 0\}, \end{aligned} \tag{5.18}$$

is called the *capture set* since

$$\sigma \text{ in } \mathbf{B} \text{ implies that } \lim_{M \to +\infty} \zeta(M, \sigma) = c. \quad \text{(stable capture)} \tag{5.19}$$

Furthermore,

$$\begin{aligned}
&\sigma \text{ in the set theoretic difference } \mathbf{A}-\mathbf{B} \text{ implies that}\\
&\text{either } \lim_{M\to+\infty} \zeta(M,\sigma) = -\infty \quad \text{(escape)}\\
&\text{or } \quad \lim_{M\to+\infty} \zeta(M,\sigma) = b_{-1} \quad \text{(unstable capture)}.
\end{aligned} \tag{5.20}$$

The proof of these two assertions is elementary and will not be given here.

Next we note that $l(\mathbf{A})$, the length of $\mathbf{A}$ is equal to $b_0 - \underline{\sigma}$, where

$$\lim_{M\to-\infty} \zeta(M,\underline{\sigma}) = b_{-1}, \tag{5.21a}$$

while $l(\mathbf{B})$, the length of $\mathbf{B}$, is equal to $\sigma^* - \underline{\sigma}$, where

$$\lim_{M\to+\infty} \zeta(M,\sigma^*) = b_{-1}. \tag{5.21b}$$

From (5.14), we have

$$\begin{aligned}
K(b_{-1},0) &= K(\underline{\sigma},0) - \alpha T_0'(e) \int_{-\infty}^{0} s^2(M,\underline{\sigma})\,\mathrm{d}M,\\
K(b_{-1},0) &= K(\sigma^*,0) + \alpha T_0'(e) \int_{0}^{\infty} s^2(M,\sigma^*)\,\mathrm{d}M.
\end{aligned} \tag{5.22}$$

Equations (5.22) can be used to estimate the length of $l(\mathbf{A})$ and $l(\mathbf{B})$. Following a similar analysis in [8], we approximate the $s^2(M,\sigma)$ by the separatrix solution of the pendulum equation. The result is

$$\underline{\sigma} \cong b_0 - \left(\frac{\alpha}{\beta}\right)^{1/2} \left\{\frac{T_0(e)\pi + \beta T_0'(e)\, I_-}{-A(e)\cos 2c}\right\}^{1/2}, \tag{5.23a}$$

$$\sigma^* \cong b_0 - \left(\frac{\alpha}{\beta}\right)^{1/2} \left\{\frac{T_0(e)\pi - \beta T_0'(e)\, I_+}{-A(e)\cos 2c}\right\}^{1/2}, \tag{5.23b}$$

where

$$\beta I_- = \beta \int_{-\infty}^{0} s^2(M,\sigma)\,\mathrm{d}M \cong \sqrt{8\beta A(e)},$$

$$\beta I_+ = \beta \int_{0}^{\infty} s^2(M,\sigma^*)\,\mathrm{d}M \cong \sqrt{8\beta A(e)}.$$

If the second term in (5.23b) is pure imaginary, then $\sigma^* = b_0$, i.e., $\mathbf{B} = \mathbf{A}$.

Note that $T_0'(e)=0$ implies that $l(\mathbf{B})=0$, moreover, $T_0(e)=0$ and $T_0'(e)<0$ implies that $l(\mathbf{A}-\mathbf{B})=0$. Therefore (see (2.10)), capture computations are meaningful only if the spin rate is less than the cutoff frequency,

$$\max \frac{\mathrm{d}v}{\mathrm{d}M} = \frac{(1+e)^2}{(1-e^2)^{3/2}}. \tag{5.24}$$

Because of this, we define the capture measure $mC(\frac{3}{2}, e)$ as the normalized measure of the set of points on the circle $x = \max \mathrm{d}v/\mathrm{d}M$, $-\pi \leqslant \theta \leqslant \pi$, which are the preimages of the 3:2 stable equilibrium states. Clearly,

$$mC(\tfrac{3}{2}, e) \cong \frac{l(\mathbf{B})}{l(\mathbf{A})}. \tag{5.25}$$

The escape measure,

$$mE(\tfrac{3}{2}, e) = 1 - mC(\tfrac{3}{2}, e) \cong \frac{l(\mathbf{A}-\mathbf{B})}{l(\mathbf{A})} \tag{5.26}$$

is the square root of the escape probability defined by Goldreich and Peale [8]. They showed that for the admissible values of Mercury's orbital eccentricity the escape probability (and therefore the escape measure) was too large to give confidence in the simple physical model which was postulated. However, those solutions which are not captured by the 3:2 resonant state are, by the nonresonant analysis, captured by the zero of the averaged tidal torque if $w^*(e) \neq \frac{3}{2}$. If, after the capture by $w^*(e)$, the eccentricity changes slowly then $w^*(e)$ may be sufficiently close to $\frac{3}{2}$ so that the gravitational effects become dominant and the tidal torque capture is transferred into gravitational torque capture. Once this happens, the gravitational torque capture is permanent. Preliminary calculation of $w^*(e)$ using the changes in the eccentricity allowed by planetary theory indicate that this capture mechanism is reasonable.

We shall not return to the abstract problem of Section 3, except to note that the resonance, nonresonance, and capture measure concepts are applicable and may serve as the framework for a rigorous analysis.

References

[1] Pettengill, G. H. and Dyce, R. B.: 1965, *Nature* **206**, 1240.
[2] Peale, S. J. and Gold, T.: 1965, *Nature* **206**, 1240.
[3] Goldreich, P.: 1965, *Nature* **208**, 375.
[4] Colombo, G.: 1965, *Nature* **208**, 575.
[5] Liu, H. S. and O'Keefe, J. A.: 1965, *Science* **150**, 1717.
[6] Colombo, G. and Shapiro, I. I.: 1965, Smithsonian Astrophys. Obs. Spec. Rept. 188R, 1.
[7] Bellomo, E., Colombo, G., and Shapiro, I.: 1966, paper presented before Symposium on Mantles of the Earth and Terrestrial Planets, March, 1966.
[8] Goldreich, P. and Peale, S. J.: 1966, *Astron. J.* **71**, 425.
[9] Laslett, L. J. and Sessler, A. M.: 1966, *Science* **151**, 1384.
[10] Jefferys, W. H.: 1966, *Science* **152**, 201.
[11] Blitzer, L.: 1967, *Astron. J.* **72**, 988.
[12] Counselman, C. C.: 1969, Ph.D. Thesis, Massachusetts Institute of Technology, Cambridge.
[13] Danby, J. M. A.: 1962, *Fundamentals of Celestial Mechanics*, MacMillan, New York, p. 98.
[14] Wintner, A.: 1947, *The Analytical Foundations of Celestial Mechanics*, Princeton Univ. Press, Princeton.
[15] Moser, J.: 1968, *Mem. Am. Math. Soc., Providence* **81**.

TWO CENTERS OF LIBRATION*

G. E. O. GIACAGLIA

University of São Paulo and University of Campinas, São Paulo, Brasil

Abstract. The problem where the Hamiltonian of a one-dimensional system has two maxima and two minima, in the angular variable, is briefly discussed when certain conditions are assumed. A tentative of generalization of Garfinkel's formalism is presented. Analogy to the $^1/_1$ resonance case of the restricted problem is discussed.

1. Background

Consider a dynamical system defined by the equations

$$\begin{aligned} \dot{x} &= -H_y \\ \dot{y} &= +H_x, \end{aligned} \tag{1}$$

where $H = H(x, y)$ has the form

$$H = A_0(x) + \sum_{j=1}^{\infty} A_j(x) \cos jy, \tag{2}$$

and we assume that $A_j(x)$ for $j = 1, 2, \ldots$ are bounded by a small quantity ε, which fact we write

$$A_j(x) = 0(\varepsilon), \quad j = 1, 2, \ldots \tag{3}$$

for x in some domain Ω. We also assume that $A_0(x) = 0(1)$, and $A_0'(x) = \mathrm{d}A_0/\mathrm{d}x = 0(\varepsilon^{\frac{1}{2}})$ in that same domain Ω.

The ideal case corresponding to Equations (1) and (2) has been defined by Garfinkel (1966) as given by the first approximation (in some sense, which assumes rapid convergence of series (2)) of the Hamiltonian (2):

$$H_G = A_0(x) + A_1(x) \cos y. \tag{4}$$

Under certain conditions to be obtained in what follows, there exists a canonical transformation to variables (ξ, η) such that form (2) is reduced to form (4), that is

$$H(x(\xi, \eta)), y(\xi, \eta) = \mathscr{H}(\xi, \eta) = P(\xi) + Q(\xi) \cos \eta. \tag{5}$$

We assume initially that:

(A) For any x in Ω and $-\pi < y < \pi$ there are only two solutions $y = 0$ and $y = \pi$ of the equation $\dot{x} = -H_y = 0$.

(B) $A_0(x), A_1(x) > 0$ for x in Ω.

(C) The maximum value of H is attained, for x in Ω, at $y = 0$.

(D) We assume

$$M(x) = \sum_{j=1}^{\infty} A_j(x) > 0.$$

* This research was partially sponsored by ONR, Contract N00014-C-67-347.

G.E. O. Giacaglia (ed.), Periodic Orbits, Stability and Resonances, 515–530.

(E) The minimum value of H is attained for x in Ω, at $y=\pi$.
(F) We assume

$$m(x)=\sum_{j=1}^{\infty} A_j(x)(-1)^j=0.$$

Consider now a generating function

$$\text{(G)}\qquad S(\xi,y)=\xi y+S_{1/2}(\xi,y)+S_1(\xi,y)+\ldots,$$

where subscripts indicate order with respect to the 'small parameter' ε. Assuming H to be transformed into form (5), we must have

$$H\left(\frac{\partial S}{\partial y},y\right)=\mathcal{H}\left(\xi,\frac{\partial S}{\partial \xi}\right)$$

or

$$H(\xi+S_{1/2,y}+S_{1,y}+\ldots,y)=\mathcal{H}(\xi,y+S_{1/2,\xi}+S_{1,\xi}+\ldots).$$

Expanding in Taylor series (assumed convergent) around the point (ξ, y), we have

$$\begin{aligned}H(\xi,y)+H_x(\xi,y)\,[S_{1/2,y}+S_{1,y}+\ldots]+\frac{1}{2!}H_{xx}(\xi,y)\\ \times[S_{1/2,y}+S_{1,y}+\ldots]^2+\ldots=\mathcal{H}(\xi,y)+\mathcal{H}_\eta(\xi,y)\\ \times[S_{1/2,\xi}+S_{1,\xi}+\ldots]+\frac{1}{2!}\mathcal{H}_{\eta\eta}(\xi,y)\\ \times[S_{1/2,\xi}+S_{1,\xi}+\ldots]^2+\ldots\end{aligned}\tag{6}$$

We now assume that

$$\text{(H)}\qquad \begin{aligned}P(\xi)&=P_0(\xi)+P_{1/2}(\xi)+P_1(\xi)+\ldots\\ Q(\xi)&=Q_0(\xi)+Q_{1/2}(\xi)+Q_1(\xi)+\ldots,\end{aligned}$$

where subscripts indicate order in ε.

Equating terms of same order in development (6) we find:

Order 1: $P_0(\xi)=A_0(\xi),\ Q_0(\xi)=0;$

Order $\varepsilon^{1/2}$: $P_{1/2}(\xi)=0=Q_{1/2}(\xi);$

Order ε: $A_0'(\xi)\,S_{1/2,y}+\tfrac{1}{2}A_0''(\xi)S_{1/2,y}^2+\sum_{j=1}^{\infty}A_j(\xi)\cos jy$
$=P_1(\xi)+Q_1(\xi)\cos y;$

Order $\varepsilon^{3/2}$: $[A_0'(\xi)+A_0''(\xi)\,S_{1/2,y}]\,S_{1,y}+\tfrac{1}{2}A_0''(\xi)\,S_{1,y}^2+\tfrac{1}{6}A_0'''(\xi)\,S_{1/2,y}^3$
$+S_{1/2,y}\sum_{j=1}^{\infty}A_j'(\xi)\cos jy=P_{3/2}(\xi)+Q_{3/2}(\xi)\cos y$
$-Q_1(\xi)\,S_{1/2,\xi}\sin y;$

and so on. In general, the equation to be considered is

$$\begin{aligned}[A_0'(\xi)+A_0''(\xi)\,S_{1/2,y}]\,S_{n,y}+\tfrac{1}{2}A_0''(\xi)\,S_{n,y}^2\\ =P_{n+1/2}(\xi)+Q_{n+1/2}(\xi)\cos y+R_n(\xi,y),\end{aligned}\tag{7}$$

where $R_n(\xi, y)$ is a known function if the previous terms $S_{\frac{1}{2}}, S_1, \ldots, S_{n-\frac{1}{2}}; P_0, P_{\frac{1}{2}}, \ldots, P_n; Q_0, Q_{\frac{1}{2}}, \ldots, Q_n$ have been determined. Equation (7) is valid for $n=1, \frac{3}{2}, 2, \frac{5}{2}, \ldots$. We may also write Equation (7) including terms of higher order, in the form

$$[A_0'(\xi) + A_0''(\xi)\, S_y^{(n-1/2)}]\, S_{n,y} + \tfrac{1}{2} A_0''(\xi)\, S_{n,y}^2 = P_{n+1/2}(\xi) + Q_{n+1/2}(\xi)\cos y + R_n(\xi, y), \tag{8}$$

where

$$S^{(p)} = S_{1/2} + S_1 + S_{3/2} + \cdots + S_p. \tag{9}$$

Summing Equation (8) from $n=1$ to $n=p$ and adding the order ε equation, we obtain

$$A_0'(\xi)\, S_y^{(p)} + \tfrac{1}{2} A_0''(\xi)\, S_y^{(p)^2} = P^{(p+1/2)} + Q^{(p+1/2)}\cos y + R^{(p)}(\xi, y) \tag{10}$$

for $p=\frac{1}{2}, 1, \frac{3}{2}, \ldots$, and where

$$P^{(k)} = P_1 + P_{3/2} + \cdots + P_k$$
$$Q^{(k)} = Q_1 + Q_{3/2} + \cdots + Q_k$$
$$R^{(k)} = R_{1/2} + R_1 + \cdots + R_k.$$

Evidently, Equation (10) cannot be solved before all previous approximations are found but is useful to show that $S^{(p)}$ satisfies an equation which is formally the same for any $p=\frac{1}{2}, 1, \frac{3}{2}, \ldots$.

Consider the case $p=\frac{1}{2}$ (order ε equation). Solving for $S_{\frac{1}{2},y}$ we find

$$S_{1/2,y} = -\frac{A_0'}{A_0''} \pm \left\{\left(\frac{A_0'}{A_0''}\right)^2 + \frac{2}{A_0''} \times \left[P_1 - (A_1 - Q_1)\cos y - \sum_{j=2}^{\infty} A_j \cos jy\right]\right\}^{1/2}, \tag{11}$$

and we choose

$$P_1(\xi) = Q_1(\xi) = A_1(\xi), \tag{12}$$

so that

$$S_{1/2,y} = -\frac{A_0'}{A_0''} \pm \left\{\left(\frac{A_0'}{A_0''}\right)^2 + \frac{2}{A_0''}\left[2A_1\cos^2\frac{y}{2} - \sum_{j=1}^{\infty} A_j \cos jy\right]\right\}^{1/2}. \tag{13}$$

If $y=0$, the quantity under square root becomes

$$\left(\frac{A_0'}{A_0''}\right)^2 + \frac{2}{A_0''}[2A_1 - M(\xi)]. \tag{14}$$

If this quantity is positive $S_{\frac{1}{2},y}$ is always real and y undergoes a circulation. If this quantity is negative, y cannot reach the value $y=0$ and undergoes a libration around $y=+\pi$. If $y=\pi$, the quantity under square root becomes simply

$$\left(\frac{A_0'}{A_0''}\right)^2$$

and we choose the sign in such a way that $S_{\frac{1}{2},y}=0$ and therefore

$$S_{1/2} = f(\xi) = \text{arbitrary function},$$

and we choose $f(\xi)=0$. These conditions are satisfied if $S_{\frac{1}{2},y}$ is a sine series of some integer multiple of y but this cannot be assumed in general. Consider now the equation

$$\dot{y} = H_x = 0,$$

that is

$$A_0'(x) + \sum_{j=1}^{\infty} A_j'(x) \cos jy = 0.$$

For $y=\pi$,

$$A_0'(\bar{x}) + \sum_{j=1}^{\infty} (-1)^j A_j'(\bar{x}) = 0, \tag{15}$$

and because of the order of magnitude assumed we suppose that an approximate (good to order $\varepsilon^{\frac{1}{2}}$) solution is given by

(I) $\bar{x} \simeq \bar{x}_{\frac{1}{2}}$, where $A_0'(\bar{x}_{\frac{1}{2}})=0$. The point $(\bar{x}, y=\pi)$ is a *libration center* and the point $(\bar{x}_{\frac{1}{2}}, y=\pi)$ is a first approximation of its location.

For $y=0$,

$$A_0'(x^*) + \sum_{j=1}^{\infty} A_j'(x^*) = 0, \tag{16}$$

and again we assume that an approximate solution is given by

(J) $x^* \simeq x^*_{\frac{1}{2}}$, where $A_0'(x^*_{\frac{1}{2}})=0$. The point $(x^*, y=0)$ is a saddle point and the point $(x^*_{\frac{1}{2}}, y=0)$ is a first approximation of its location.

When $y=\pi$ and $x=\bar{x}$, the function $S_{\frac{1}{2}}$ is not operative since we assumed in this case that $S_{\frac{1}{2},y}=0$ and, moreover, $S_{\frac{1}{2},\xi}=0$.

In the next order equation (order $\varepsilon^{\frac{3}{2}}$), in order to have $S_{1,y}=0$ $(S_{1,\xi}=0)$ we shall choose again

$$P_{3/2}(\xi) = Q_{3/2}(\xi),$$

and now $Q_{\frac{3}{2}}$ is defined by the coefficients of the known terms in $\cos y$, that is,

$$Q_{3/2}(\xi) = \frac{1}{\pi} \int_0^{2\pi} \times \left\{ \frac{1}{6} A_0''' S_{1/2,y}^3 + S_{1/2,y} \sum_{j=1}^{\infty} A_j'(\xi) \cos jy + Q_1 S_{1/2,\xi} \sin y \right\} \cos y \, dy.$$

Evidently, considering Equation (10), having reached the approximation $(p-1)$, the choice for p will be

$$P^{(p+1/2)} = Q^{(p+1/2)} = -\frac{1}{\pi} \int_0^{2\pi} R^{(p)}(\xi, y) \, dy, \tag{17}$$

and it is easy to see that, in general, proceeding in this way,

$$R^{(p)}(\xi, \pi) = 0.$$

The equation for $S_y^{(p)}$ being

$$S_y^{(p)} = -\frac{A_0'}{A_0''} \pm \left\{\left(\frac{A_0'}{A_0''}\right)^2 + \frac{2}{A_0''}\left[2P^{(p+1/2)}\cos^2\frac{y}{2} + R^{(p)}(\xi, y)\right]\right\}^{1/2},$$

it follows that for $y=\pi$,

$$S_y^{(p)} = 0,$$

valid for any p. If for $y=0$ the quantity

$$\left(\frac{A_0'}{A_0''}\right)^2 + \frac{2}{A_0''}\left[2P^{(p+1/2)} + R^{(p)}(\xi, 0)\right]$$

is positive, y undergoes a circulation and if it is negative, a libration. Only when the limit $p\to\infty$ is considered, one can define the *asymptotic case* when the above quantity is zero.

We have now defined a *formal series* $S(\xi, y)$, which reduces the system to the form

$$\begin{aligned} \mathscr{H}(\xi, \eta) &= (P_0 + P_1 + P_{3/2} + \ldots) + (Q_1 + Q_{3/2} + \ldots)\cos\eta \\ &= A_0(\xi) + 2P(\xi)\cos^2\frac{\eta}{2}, \end{aligned} \tag{18}$$

where $A_0(\xi)=O(1)$, $A_0'(\xi)=O(\varepsilon^{\frac{1}{2}})$, $P(\xi)=O(\varepsilon)$ and for $\eta=\pi$, $\mathscr{H}(\xi, \eta)$ has a minimum, and for $\eta=0$, $\mathscr{H}(\xi, \eta)$ has a maximum, since $A_0(\xi)>0$, $P(\xi)>0$.

It is also true that Hamilton's principal function $W(\xi_0, \eta)$ satisfying the equation

$$A_0\left(\frac{\partial W}{\partial \eta}\right) + 2P\left(\frac{\partial W}{\partial \eta}\right)\cos^2\frac{\eta}{2} = C(\xi_0) \tag{19}$$

will be such that the variable

$$\xi = \varrho(\xi_0, \eta) = \xi_0 + \frac{\partial W}{\partial \eta}(\xi_0, \eta) \tag{20}$$

will show the general features indicated in Figure 1 (polar coordinates ϱ, η) and Figure 2 (Cartesian coordinates η, ξ), in correspondence to the assumed existence of the equilibrium solutions defined by

$$\begin{aligned} &\eta = 0, \pi \\ &\bar{\eta} = \pi, A_0'(\bar{\xi}) = 0 \\ &\eta^* = 0, A_0'(\xi^*) + 2P'(\xi^*) = 0. \end{aligned}$$

It is also seen that the libration center is now exactly defined by the condition

$$A_0'(\bar{\xi}) = 0,$$

so that $\bar{\xi}=\bar{x}_{\frac{1}{2}}$. The transformation has shifted the solution $\bar{x}$ to the value $\bar{x}_{\frac{1}{2}}$ in the (ξ, η) space (of course, if the series S converges).

In case of libration we may perform the transformation

$$X = \frac{(\xi - \bar{\xi})^2}{\varepsilon} + X_0 \sin^2 \frac{\eta - \bar{\eta}}{2},$$

where we assume $X_0 > 0$ and $X/X_0 < 1$. If we put

$$\sin^2 \frac{\eta - \bar{\eta}}{2} = k^2 \operatorname{sn}^2 (u, k), \quad k^2 = \frac{X}{X_0}$$

the variables (X, Y) are canonically conjugate if $Y = \frac{1}{2}\sqrt{\frac{\varepsilon}{X_0}}\, u$. This leads to the formalism introduced by Garfinkel (1966) and Jupp (1969), the transformation being

$$\xi = \bar{\xi} + \sqrt{\varepsilon X} \operatorname{cn} (u, k)$$
$$\sin \frac{\eta - \bar{\eta}}{2} = \sqrt{\frac{X}{X_0}} \operatorname{sn} (u, k). \tag{21}$$

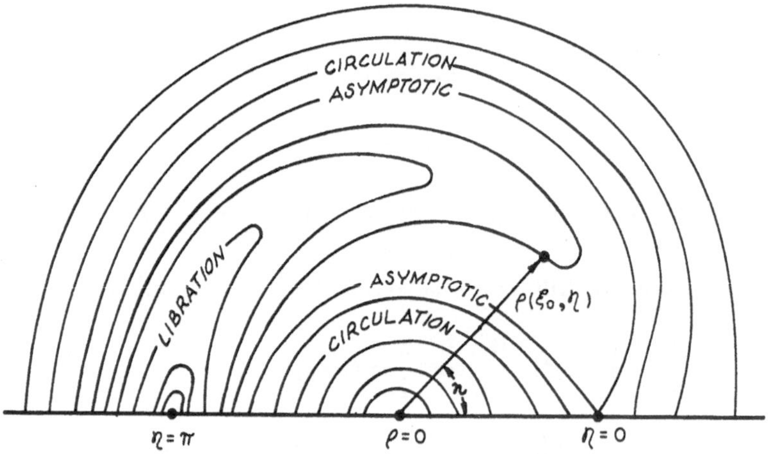

Fig. 1. Phase trajectories in the ideal resonance case.

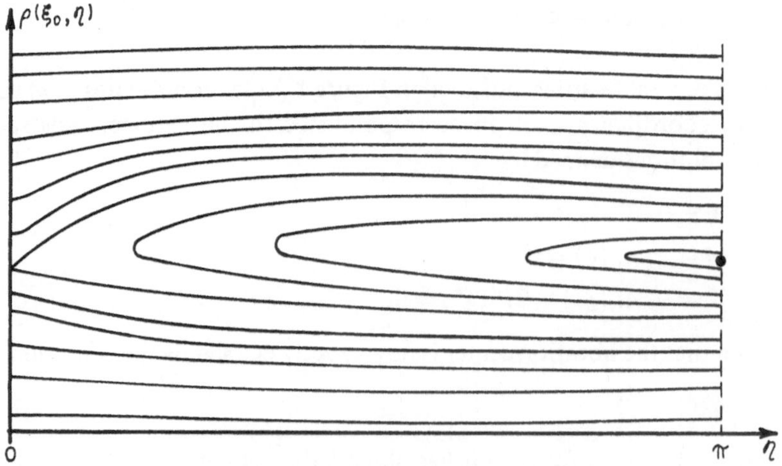

Fig. 2. Isoenergetic curves in the ideal resonance case.

In case of circulation $X > X_0 > 0$ and defining $k^2 = X_0/X$, the proper canonical transformation is

$$\xi = \bar{\xi} + \sqrt{\varepsilon X}\,\mathrm{dn}\,(u, k)$$

$$\sin \frac{\eta - \bar{\eta}}{2} = \mathrm{sn}\,(u, k) \tag{22}$$

and

$$Y = \frac{1}{2}\sqrt{\frac{\varepsilon}{X}}\,u\,.$$

2. Two Libration Centers

In this case we assume that the Hamiltonian defined by Equation (2) has two maxima and two minima in $-\pi < y \leqslant \pi$. In the present situation the appropriate definition of the 'ideal problem' is given by

$$H_G = A_0(x) + A_1(x)\cos y + A_2(x)\cos 2y\,. \tag{23}$$

We will show that, under certain conditions, there exists a canonical transformation to variables (ξ, η) such that the Hamiltonian (2) reduces to the form (23), that is,

$$H(x(\xi, \eta), y(\xi, \eta)) = \mathscr{H}(\xi, \eta) = P(\xi) + Q(\xi)\cos\eta + R(\xi)\cos 2\eta\,. \tag{24}$$

Consider the equation:

$$\dot{x} = -H_y = 0\,. \tag{25}$$

We assume that

(A′) With x in a certain domain Ω and $-\pi < y \leqslant \pi$ there are four solutions:

$$y^* = 0, \quad y^{**} = \pi, \quad \bar{y} = \alpha, \quad \bar{\bar{y}} = -\alpha$$

of Equation (25), with $0 < |\alpha| < \pi$.

(B′) $A_0(x) > 0$ for x in Ω.

(C′) The maximum values of $H(x, y)$ are attained, for x in Ω, at $y = y^*$ and $y = y^{**}$.

(D′)
$$M(x) = \sum_{j=1}^{\infty} A_j(x)\cos jy^* = \sum_{j=1}^{\infty} A_j(x) > \sum_{j=1}^{\infty} A_j(x)\cos jy^{**}$$
$$= \sum_{j=1}^{\infty} (-1)^j A_j(x) = m(x) > 0\,.$$

Note that, in general, if $M(x) = m(x)$, then $A_1 = A_3 = A_5 = \ldots = 0$.

(E′) The minimum values of $H(x, y)$ are attained, for x in Ω, at $y = \bar{y}$ and $y = \bar{\bar{y}}$.

(F′)
$$\sum_{j=1}^{\infty} A_j(x)\cos j\bar{y} = \sum_{j=1}^{\infty} A_j(x)\cos j\alpha = \sum_{j=1}^{\infty} A_j(x)\cos j\bar{\bar{y}} = 0\,.$$

(G′) The generating function of the canonical transformation $(x, y) \rightarrow (\xi, \eta)$ may be expanded in a formal series

$$S(\xi, y) = \xi y + S_{1/2}(\xi, y) + S_1(\xi, y) + \dots$$

(H′) The coefficients P, Q, R of $\mathscr{H}(\xi, \eta)$ may be expanded in the form

$$P(\xi) = P_0(\xi) + P_{1/2}(\xi) + P_1(\xi) + \dots$$
$$Q(\xi) = Q_0(\xi) + Q_{1/2}(\xi) + Q_1(\xi) + \dots$$
$$R(\xi) = R_0(\xi) + R_{1/2}(\xi) + R_1(\xi) + \dots$$

Proceeding as in the previous case we find that, for $p = \frac{1}{2}, 1, \frac{3}{2}, \dots$,

$$A_0'(\xi)\, S_y^{(p)} + \tfrac{1}{2} A_0''(\xi)\, S_y^{(p)2} = P^{(p+1/2)} + Q^{(p+1/2)} \cos y + R^{(p+1/2)} \cos 2y + T^{(p)}(\xi, y) \tag{26}$$

and the same discussion made for Equation (10) applies in the present case.

In particular we have

$$P_0(\xi) = A_0(\xi)$$
$$Q_0(\xi) = 0 = R_0(\xi)$$
$$P_{1/2}(\xi) = Q_{1/2}(\xi) = R_{1/2}(\xi) = 0,$$

and for $p = \frac{1}{2}$ we have

$$S_{1/2,y} = -\frac{A_0'}{A_0''} \pm \left\{ \left(\frac{A_0'}{A_0''}\right)^2 + \frac{2}{A_0''} \times \left[P_1 + Q_1 \cos y + R_1 \cos 2y - \sum_{j=1}^{\infty} A_j \cos jy \right] \right\}^{1/2}.$$

We choose, in analogy to the previous case,

$$Q_1(\xi) = A_1(\xi)$$
$$R_1(\xi) = A_2(\xi)$$
$$P_1(\xi) = -A_1(\xi) \cos \alpha - A_2(\xi) \cos 2\alpha,$$

so that

$$S_{1/2,y} = -\frac{A_0'}{A_0''} \pm \left\{ \left(\frac{A_0'}{A_0''}\right)^2 + \frac{2}{A_0''} \times \left[A_1(\cos y - \cos \alpha) + A_2(\cos 2y - \cos 2\alpha) - \sum_{j=1}^{\infty} A_j \cos jy \right] \right\}^{1/2},$$

and for $y = \alpha$, $S_{\frac{1}{2},y} = 0$.

When $y = 0$ we have

$$S_{1/2,y} = -\frac{A_0'}{A_0''} \pm \left\{ \left(\frac{A_0'}{A_0''}\right)^2 + \frac{2}{A_0''} \left[2A_1 \sin^2 \frac{\alpha}{2} + 2A_2 \sin^2 \alpha - M(\xi) \right] \right\}^{1/2}.$$

If

$$K \equiv \left(\frac{A_0'}{A_0''}\right)^2 + \frac{4}{A_0''} \left(A_1 \sin^2 \frac{\alpha}{2} + A_2 \sin^2 \alpha \right) \geq M(\xi). \tag{27}$$

$S_{\frac{1}{2},y}$ is real and the variable y circulates or it is in asymptotic motion to $y=0$ (equality satisfied). If that quantity is $<M(\xi)$ the angle y librates around the two libration centers $(\bar{x}, \alpha)$ and $(\bar{x}, -\alpha)$. When $y=\pi$,

$$S_{1/2,y} = -\left(\frac{A_0'}{A_0''}\right)^2 \pm \left\{\left(\frac{A_0'}{A_0''}\right)^2 + \frac{2}{A_0''} \times \left[-2A_1\cos^2\frac{\alpha}{2} - 2A_2\cos^2\alpha - m(\xi)\right]\right\}^{1/2},$$

and if

$$K > k \equiv \left(\frac{A_0'}{A_0''}\right)^2 - \frac{4}{A_0''}\left(A_1\cos^2\frac{\alpha}{2} + A_2\cos^2\alpha\right) \geq m(\xi), \tag{28}$$

the angle y librates around a single libration center or it is in doubly asymptotic motion to $y=\pi$ (equality satisfied). The possible cases are illustrated in Figure 3.

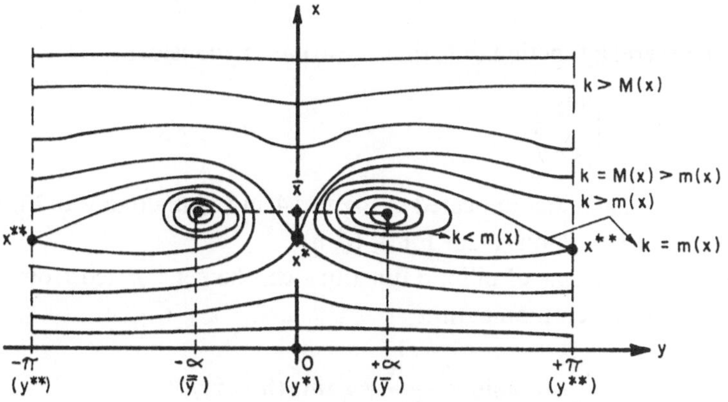

Fig. 3. Two centers of libration.

Continuing the process of approximation we will choose at any stage

$$P^{(p+1/2)}(\xi) = -Q^{(p+1/2)}(\xi)\cos\alpha - R^{(p+1/2)}(\xi)\cos 2\alpha, \tag{29}$$

and Q and R are again defined by

$$Q^{(p+1/2)}(\xi) = -\frac{1}{\pi}\int_0^{2\pi} T^{(p)}(\xi, y)\cos y\,dy$$

$$R^{(p+1/2)}(\xi) = -\frac{1}{\pi}\int_0^{2\pi} T^{(p)}(\xi, y)\cos 2y\,dy.$$

Because of the choice of P, Q, R it will result $S_y^{(p)}=0$ at $y=\pm\alpha$.

We have therefore constructed the formal series

$$\mathscr{H}(\xi, \eta) = A_0(\xi) + [A_1(\xi) + Q_{3/2}(\xi) + \ldots](\cos\eta - \cos\alpha) + [A_2(\xi) + Q_{3/2}(\xi) + \ldots](\cos 2\eta - \cos 2\alpha),$$

and because of the choice of S, as before $\bar{\xi}=\bar{\bar{\xi}}=\bar{x}=\bar{\bar{x}}$ and $\bar{\eta}=\bar{y}=\alpha$. $\bar{\bar{\eta}}=\bar{\bar{y}}=-\alpha$. At the libration centers the transformation S is therefore the identity.

The new Hamiltonian, defined by a *formal series*, can be written in the form

$$\mathscr{H}(\xi,\eta)=A_0(\xi)+U(\xi)(\cos\eta-\cos\alpha)+V(\xi)(\cos 2\eta-\cos 2\alpha), \quad (30)$$

where $A_0'(\xi)=O(\sqrt{\varepsilon})$ and $U(\xi)$ and $V(\xi)$ are $O(\varepsilon)$, the libration centers being defined by $A_0'(\bar{\xi})=0$, and $\eta=\pm\alpha$. Also because of the choice of S,

$$\mathscr{H}(\xi,\alpha)=A_0(\xi)>0,$$

for ξ in the image of Ω in the ξ-space (linear).

In this case, Hamilton's function $W(\xi_0,\eta)$ satisfying the equation

$$A_0\left(\frac{\partial W}{\partial\eta}\right)+U\left(\frac{\partial W}{\partial\eta}\right)(\cos\eta-\cos\alpha)+V\left(\frac{\partial W}{\partial\eta}\right)$$
$$\times(\cos 2\eta-\cos 2\alpha)=C(\xi_0)=\text{const},$$

where α is in general a function of ξ, that is, we have to consider

$$\alpha(\xi)=\alpha\left(\frac{\partial W}{\partial\eta}\right),$$

will show the characteristics indicated in Figures 4 and 5, analogous to Figures 1 and 2, and based as the assumptions made for $\mathscr{H}(\xi,\eta)$.

Representation in terms of elliptic functions can be generalized from the previous section by defining a new momentum

$$X=\frac{(\xi-\bar{\xi})^2}{\varepsilon}+p\sin^2\frac{\eta-\alpha}{2}+q\sin^2(\eta-\alpha), \quad (31)$$

where p and q are conveniently chosen constants.

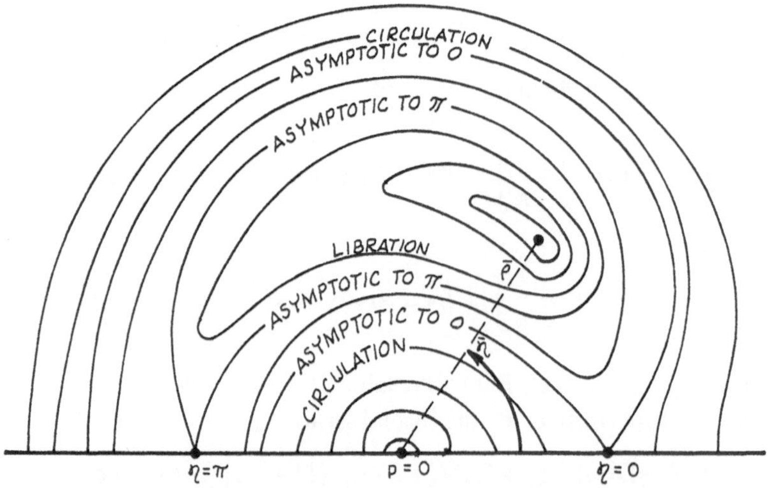

Fig. 4. Phase trajectories in case of two centers of libration.

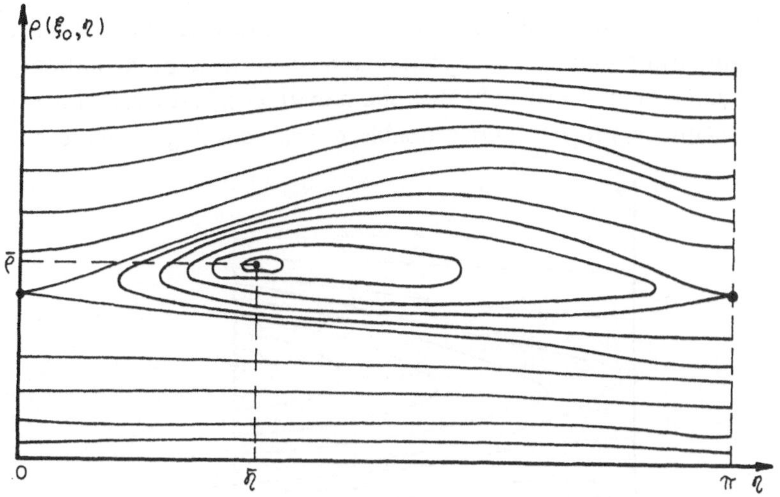

Fig. 5. Isoenergetic curves for two centers.

Let

$$\Phi(\eta, \alpha) = p \sin^2 \frac{\eta - \alpha}{2} + q \sin^2 (\eta - \alpha).$$

We require that, being zero the minimum of Φ (when $\eta = \alpha$, i.e., at the libration centers), the following conditions should be satisfied:

$$\Phi(0, \alpha) > \Phi(\pi, \alpha) > 0, \tag{32}$$

so that p and q must be chosen so as to obtain

$$p \sin^2 \frac{\alpha}{2} + q \sin^2 \alpha > p \cos^2 \frac{\alpha}{2} + q \sin^2 \alpha > 0.$$

In general the choice of p and q depends on the value of α, but if e.g. $\pi/2 < \alpha < \pi$, we see that it is sufficient to have p and q positive.

Exceptional cases are:

(a) $\phi(0, \alpha) = \phi(\pi, \alpha) > 0$, will correspond to the case when the asymptotic orbit to $\eta = 0$ coincides with the asymptotic orbit to $\eta = \pi$, as shown in Figure 6. In this case is included the situation $\mathscr{H} = \mathscr{H}(\xi, 2\eta)$.

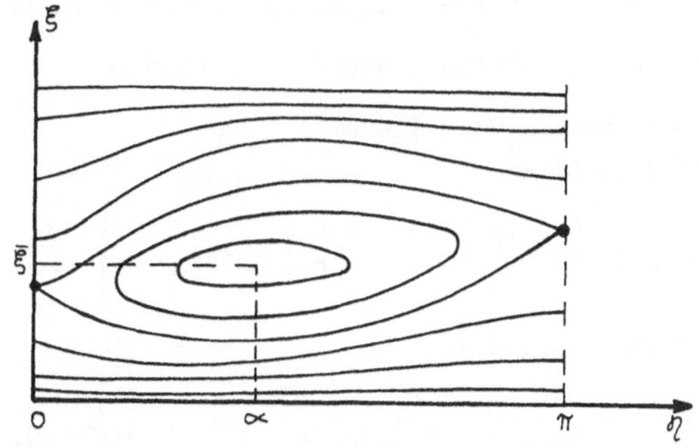

Fig. 6. Two libration centers $\varphi(0, \alpha) = \varphi(\pi, \alpha)$.

(b) $\phi(0, \alpha) > \phi(\pi, \alpha) = 0$ will correspond to the case when there is no chance of libration around one only of the libration center. In fact, by hypothesis, since we assumed a minimum in $\alpha(=0)$ and a maximum in $\pi(=0)$, between α and π the Hamiltonian is constant, or, which is more probable, $\alpha = \pi$, according to Figure 7.

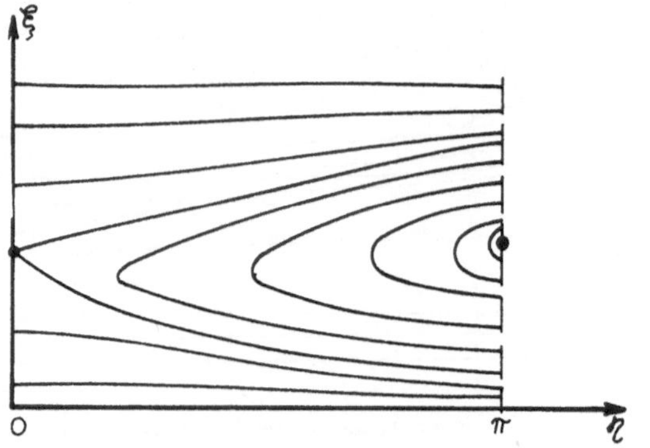

Fig. 7. Two libration centers $\varphi(\pi, \alpha) = 0$.

(c) $\phi(0, \alpha) = \phi(\pi, \alpha) = 0$ will correspond to the case of a constant everywhere Hamiltonian which has no interest.

The normal situation is therefore

$$H \equiv \Phi(0, \alpha) > h \equiv \Phi(\pi, \alpha) > 0,$$

where again should be noted that since α is in general a function of ξ, $H = H(\xi)$ and $h = h(\xi)$. There are situations in which α is independent of ξ, so that H and h are absolute constants.

All possible orbits corresponding to the normal situation are now defined as follows:

$X > H > h > 0$: circulation.

$X = H > h > 0$: asymptotic motion to $\eta = 0$.

$H > X > h > 0$: libration around both centers.

$H > X = h > 0$: asymptotic motion to $\eta = \pi$.

$H > h > X > 0$: libration around one center (any one of the two).

$H > h > X = 0$: libration center (any of the two).

They correspond to all cases shown in Figure 4.

From Equation (31) we obtain

$$\xi = \bar{\xi} + \sqrt{\varepsilon X}\left[1 + \frac{p}{X}\sin^2\frac{\eta - \alpha}{2} + \frac{q}{X}\sin^2(\eta - \alpha)\right]^{1/2}. \tag{33}$$

We also define an angular variable by

$$u = \int\left[1 + \frac{p}{X}\sin^2\frac{\eta - \alpha}{2} + \frac{q}{X}\sin^2(\eta - \alpha)\right]^{-1/2} \mathrm{d}(\eta - \alpha),$$

which is evidently reducible to an elliptic integral of the first kind, the inversion of which gives $\eta - \alpha$ as function of u. There is a factor (function of X) which reduces u to a variable Y, canonical conjugate to X.

Take for instance the case when α does not depend on ξ and (33) may be written in the form

$$\xi = \bar{\xi} + \sqrt{\varepsilon X}\left[\left(1 - n_1^2 \sin^2 \frac{\eta - \alpha}{2}\right)\left(1 - n_2^2 \sin^2 \frac{\eta - \alpha}{2}\right)\right]^{1/2},$$

where $1 > n_2^1 > n_2^2$. Considering the modulus

$$k^2 = \frac{n_1^2 - n_2^2}{1 - n_2^2} < 1,$$

we find that

$$\xi = \bar{\xi} + \sqrt{\varepsilon X}\,\frac{(1 - n_2^2)\,\mathrm{dn}\,(u, k)}{1 - n_2^2\,\mathrm{cn}^2\,(u, k)} \qquad \text{(circulation)}$$

$$\sin^2 \frac{\eta - \alpha}{2} = \frac{\mathrm{sn}^2\,(u, k)}{1 - n_2^2\,\mathrm{cn}^2\,(u, k)}$$

$$Y = \frac{1}{2}\sqrt{\frac{X}{\varepsilon(1 - n_2^2)}}\,u\,.$$

If $n_1^2 > 1 > n_2^2$, we consider the modulus

$$k^2 = \frac{1 - n_2^2}{n_1^2 - n_2^2}$$

and it follows that

$$\xi = \bar{\xi} + \sqrt{\varepsilon X}\,\frac{(1 - n_2^2)\,\mathrm{cn}\,(u, k)}{1 - n_2^2\,\mathrm{dn}^2\,(u, k)}$$

$$\sin^2 \frac{\eta - \alpha}{2} = \frac{k^2\,\mathrm{sn}^2\,(u, k)}{1 - n_2^2\,\mathrm{dn}^2\,(u, k)}$$

$$Y = \sqrt{\frac{\varepsilon}{X(n_1^2 - n_2^2)}}\,u$$

Other cases are $n_1^2 > n_2^2 > 1$ (libration, two centers), $n_1^2 < 1 = n_2^2$ (asymptotic to $\eta = \pi$), $n_1^2 = n_2^2 = 1$ (asymptotic to $\eta = 0$), $n_1^2 = n_2^2 = 0$ (libration centers).

3. Note on the Restricted Problem

Consider the case of $\frac{1}{1}$ resonance ($n \simeq n'$) and let

$$y_0 = l, \qquad L = x_0 + x$$
$$y = \lambda - \lambda', \quad G = x\,.$$

The average (over y_0) Hamiltonian may be represented in the form

$$H(x, y; x_0) = \tfrac{1}{2}(x_0 + x)^{-2} + n'x + \varepsilon B_0(x_0, x) + \varepsilon \sum_{j=1}^{\infty} B_j(x_0, x)\cos jy,$$

where the roles of momenta and angles have been interchanged so that

$$\dot{x} = H_y, \quad \dot{y} = - H_x.$$

In this problem we have 5 stationary solutions, which are circular orbits corresponding to $\lambda - \lambda' = y = 0$ (collinear points L_1 and L_2), $y = \pi$ (collinear point L_3) and $y = \pm\pi/3$ (triangular points). If we consider the largest primary (mass $1 - \varepsilon$) at the origin and the other at unit distance on the X-axis, with mass ε, the semi-major axes of these orbits are given respectively by (Szebehely, 1967)

$$a_1 = 1 + \left(\frac{\varepsilon}{3}\right)^{1/3} \left[1 + \frac{1}{3}\left(\frac{\varepsilon}{3}\right)^{1/3} - \frac{1}{9}\left(\frac{\varepsilon}{3}\right)^{2/3} + \dots\right]$$

$$a_2 = 1 - \left(\frac{\varepsilon}{3}\right)^{1/3} \left[1 - \frac{1}{3}\left(\frac{\varepsilon}{3}\right)^{1/3} - \frac{1}{9}\left(\frac{\varepsilon}{3}\right)^{2/3} + \dots\right]$$

$$a_3 = 1 - \frac{7}{12}\varepsilon\left[1 + \frac{1127}{12086}\varepsilon^2 + \dots\right]$$

$$a_4 = a_5 = 1.$$

Stationary solutions correspond also to $x_0 = \bar{x}_0 = 0$, $x = \bar{x} = \text{const.} = \sqrt{a_j}$. It is also known that $H(L_2) > H(L_1) > H(L_3) > H(L_4) = H(L_5)$. The situation is more complex than the one discussed in the previous section and a diagram which supposes $x_0 = 0$ could show very badly the overall situation in the (x, y) plane. We have also to consider the fact that H has singular points corresponding to $x_0 = 0, x = 1, y = 0$ (small primary) and to $x_0 = x = 0$ (y undefined), which is the origin (other primary) of the rotating system. These singularities correspond to $H \to \infty$. Imposing the condition $x_0 = 0$ we may represent orbits in the (y, x) plane and we obtain a configuration which is approximately given in Figure 8.

In this figure the shaded areas indicate regions where collision occurs, or, in this case, circular orbits of infinite velocity around the primary at the origin ($x \to 0$) or collision orbit to the smaller primary ($x \to 1, y = 0$). We see that excluding the rectangular region $(-\beta < y < +\beta, 1 - k < x < 1 + k)$ and the strip $(0 \leqslant x < k')$ where β, k and k' are small (positive) quantities the phase space diagram shows the similarity with the corresponding one in the previous paragraph.

In fact, it may be a reasonable approximation out of the collision regions to consider a diagram as given by Figure 5, where, of course, L_1 and L_2 have been joined together. The corresponding Hamiltonian ($x_0 = 0$)

$$H = -\frac{1}{2x^2} - n'x + \sum_{j=0}^{\infty} B_j(x)\cos jy$$

satisfies (out of those regions) the hypotheses described before, where ($\bar{x}=1, \bar{y}=\alpha=$ $=\pi/3$), ($x^{**}=\sqrt{a_3}, y^{**}=\pi$), ($x^*=\sqrt{a_1}$ or $\sqrt{a_2}, y^*=0$). This is not a completely satisfactory analogy but may serve to study the behavior of librations around L_4 or L_5 or both, giving a description of horse-shoe shaped orbits when sufficiently far from collision with Jupiter. The critical point may be the apparently existing doubly asymptotic orbits to L_3. This result should be considered with care because of the instability of such orbits when the real problem is considered and also when the short period perturbations are taken into account.

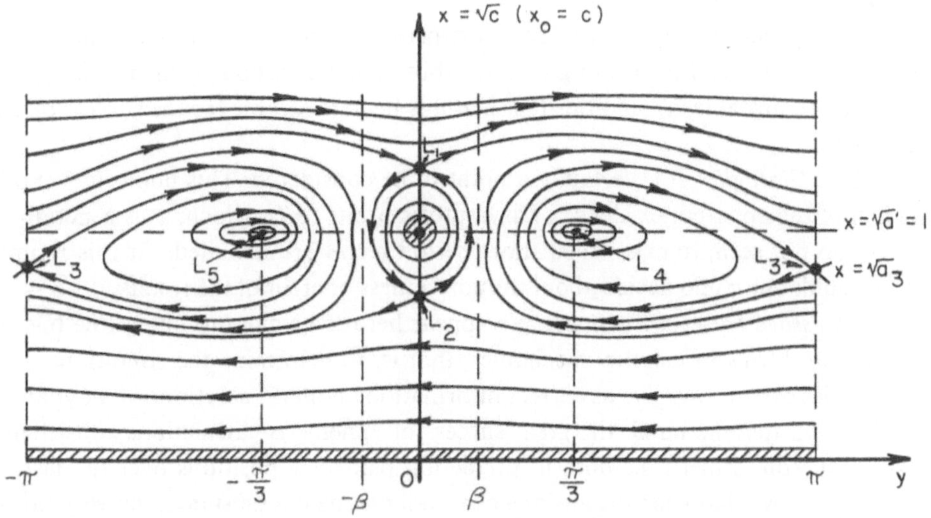

Fig. 8. Circular orbits in the $^1/_1$ resonance case of restricted problem.

References

Garfinkel, B.: 1966, *Astron. J.* **71**, 657.
Jupp, A. H.: 1969, *Astron. J.* **74**, 35.
Szebehely, V.: 1967, *Theory of Orbits*, Academic Press, New York.

Discussion

G. Hori (1) Why the new Hamiltonian is required to have the form
$F^{**}(X, Y, X')=A(X, X')+B(X, X')\cos Y+C(X, X')\cos 2y$?
(2) Why the two centered energy diagram comes out regardless of the functional form of $F^*(x, y, x')=\sum C_j(x, x')\cos jy$? Especially is a neighborhood of the origin correct because the origin $e=0$, $\lambda=\lambda'$ indicates a close approach to Jupiter?

The reduction $\sum C_j(x, x')\cos jy \to A(X, X')+B(X, X')\cos Y+C(X, X')\cos 2Y$ may be possible if $C_0(x, x')+C_1(x, x')\cos y+C_2(x, x')\cos 2y$ dominates over the remaining terms by applying the method which I presented in my talk, or by other methods.

G. E. O. Giacaglia: (to Dr. Hori – 1st question) This form is necessary and sufficient for F^{**} to present the characteristics assumed.

(To Dr. Hori – 2nd question) The diagram corresponds to particular values of the parameters of the problem and is not valid in case of close approaches.

With respect to the comment: I agree, but by 'dominates' we do not need to require an order of magnitude (with respect to ε) smaller.

J. M. A. Danby: My understanding of the 'averaged' results is that a principal application is that the resulting orbits can be used as reference orbits in perturbations theories. No claim is made that these orbits satisfy the original equations, but they are still constructive.

E. Rabe: Since you averaged the short-period terms out, I think your method would essentially lead to the long-period orbits. The horseshoe orbits of this type are known to be unstable. Would this fact perhaps be unfavorable for the success of your approach?

G. E. O. Giacaglia: It is certainly a point to be considered. This point may explain the fact of the apparent existence of asymptotic orbits to L_3 which, as is generally accepted, do not seem to exist when short period terms are included. At this moment I don't think one could decide about the correctness or error of the results.

P. J. Message (after discussion): We appear here to be close to one of the big outstanding problems of celestial mechanics, that is, to determine the conditions under which series expansions give us correct information about the solutions of a dynamical problem. On the one hand, the great success of ephemeris calculations consistent in basis with Von Zeipel's method, in predicting planetary positions over the last few centuries, shows that such expansions can under some circumstances be very powerful tools. On the other hand, the phenomena called 'wildness' by Dr. Danby, and found by many investigators, shows that there also exist circumstances under which things are much more complicated than such series would lead us to believe. We do not, I believe, yet know how to tell one of these sorts of situation from the other in advance.